Lecture Notes in Artificial Intelligence   3721

Edited by J. G. Carbonell and J. Siekmann

Subseries of Lecture Notes in Computer Science

Alípio Jorge   Luís Torgo
Pavel Brazdil   Rui Camacho
João Gama (Eds.)

Knowledge Discovery in Databases: PKDD 2005

9th European Conference on Principles and Practice
of Knowledge Discovery in Databases
Porto, Portugal, October 3-7, 2005
Proceedings

Series Editors

Jaime G. Carbonell, Carnegie Mellon University, Pittsburgh, PA, USA
Jörg Siekmann, University of Saarland, Saarbrücken, Germany

Volume Editors

Alípio Jorge
Luís Torgo
Pavel Brazdil
João Gama
LIACC/FEP, University of Porto
Rua de Ceuta, 118, 6°, 4050-190 Porto, Portugal
E-mail: {amjorge,ltorgo,pbrazdil,jgama}@liacc.up.pt

Rui Camacho
LIACC/FEUP, University of Porto
Rua de Ceuta, 118, 6°, 4050-190 Porto, Portugal
E-mail: rcamacho@fe.up.pt

Library of Congress Control Number: 2005933047

CR Subject Classification (1998): I.2, H.2, J.1, H.3, G.3, I.7, F.4.1

ISSN   0302-9743
ISBN-10   3-540-29244-6 Springer Berlin Heidelberg New York
ISBN-13   978-3-540-29244-9 Springer Berlin Heidelberg New York

This work is subject to copyright. All rights are reserved, whether the whole or part of the material is concerned, specifically the rights of translation, reprinting, re-use of illustrations, recitation, broadcasting, reproduction on microfilms or in any other way, and storage in data banks. Duplication of this publication or parts thereof is permitted only under the provisions of the German Copyright Law of September 9, 1965, in its current version, and permission for use must always be obtained from Springer. Violations are liable to prosecution under the German Copyright Law.

Springer is a part of Springer Science+Business Media

springeronline.com

© Springer-Verlag Berlin Heidelberg 2005
Printed in Germany

Typesetting: Camera-ready by author, data conversion by Scientific Publishing Services, Chennai, India
Printed on acid-free paper   SPIN: 11564126   06/3142   5 4 3 2 1 0

Preface

The European Conference on Machine Learning (ECML) and the European Conference on Principles and Practice of Knowledge Discovery in Databases (PKDD) were jointly organized this year for the fifth time in a row, after some years of mutual independence before. After Freiburg (2001), Helsinki (2002), Cavtat (2003) and Pisa (2004), Porto received the 16th edition of ECML and the 9th PKDD in October 3–7.

Having the two conferences together seems to be working well: 585 different paper submissions were received for both events, which maintains the high submission standard of last year. Of these, 335 were submitted to ECML only, 220 to PKDD only and 30 to both. Such a high volume of scientific work required a tremendous effort from Area Chairs, Program Committee members and some additional reviewers. On average, PC members had 10 papers to evaluate, and Area Chairs had 25 papers to decide upon. We managed to have 3 highly qualified independent reviews per paper (with very few exceptions) and one additional overall input from one of the Area Chairs. After the authors' responses and the online discussions for many of the papers, we arrived at the final selection of 40 regular papers for ECML and 35 for PKDD. Besides these, 32 others were accepted as short papers for ECML and 35 for PKDD. This represents a joint acceptance rate of around 13% for regular papers and 25% overall. We thank all involved for all the effort with reviewing and selection of papers.

Besides the core technical program, ECML and PKDD had 6 invited speakers, 10 workshops, 8 tutorials and a Knowledge Discovery Challenge. Our special thanks to the organizers of the individual workshops and tutorials and to the workshop and tutorial chairs Floriana Esposito and Dunja Mladenić and to the challenge organizer Petr Berka. A very special word to Richard van de Stadt for all his competence and professionalism in the management of CyberChairPRO. Our thanks also to everyone from the Organization Committee mentioned further on who helped us with the organization. Our acknowledgement also to Rodolfo Matos and Assunção Costa Lima for providing logistic support.

Our acknowledgements to all the sponsors, Fundação para a Ciência e Tecnologia (FCT), LIACC-NIAAD, Faculdade de Engenharia do Porto, Faculdade de Economia do Porto, KDubiq –Knowledge Discovery in Ubiquitous Environments— Coordinated Action of FP6, Salford Systems, Pascal Network of Excellence, PSE/SPSS, ECCAI and Comissão de Viticultura da Região dos Vinhos Verdes. We also wish to express our gratitude to all other individuals and institutions not explicitly mentioned in this text who somehow contributed to the success of these events.

Finally, our word of appreciation to all the authors who submitted papers to the main conferences and their workshops, without whom none of this would have been possible.

July 2005   Alípio Jorge, Luís Torgo, Pavel Brazdil,
                                                 Rui Camacho and João Gama

Organization

ECML/PKDD 2005 Organization

Executive Committee

General Chair

Pavel Brazdil (LIACC/FEP, Portugal)

Program Chairs

ECML
Rui Camacho (LIACC/FEUP, Portugal)
João Gama (LIACC/FEP, Portugal)

PKDD
Alípio Jorge (LIACC/FEP, Portugal)
Luís Torgo (LIACC/FEP, Portugal)

Workshop Chair

Floriana Esposito (University of Bari, Italy)

Tutorial Chair

Dunja Mladenić (Jozef Stefan Institute, Slovenia)

Challenge Chairs

Petr Berka (University of Economics, Czech Republic)
Bruno Crémilleux (Université de Caen, France)

Local Organization Committee

Pavel Brazdil, Alípio Jorge, Rui Camacho, Luís Torgo and João Gama, with the help of people from LIACC-NIAAD, University of Porto, Portugal, Rodolfo Matos, Pedro Quelhas Brito, Fabrice Colas, Carlos Soares, Pedro Campos, Rui Leite, Mário Amado Alves, Pedro Rodrigues; and from IST, Lisbon, Portugal, Cláudia Antunes.

Steering Committee

Nada Lavrač, Jozef Stefan Institute, Slovenia
Dragan Gamberger, Rudjer Boskovic Institute, Croatia
Ljupčo Todorovski, Jozef Stefan Institute, Slovenia
Hendrik Blockeel, Katholieke Universiteit Leuven, Belgium
Tapio Elomaa, Tampere University of Technology, Finland
Heikki Mannila, Helsinki Institute for Information Technology, Finland

Hannu T.T. Toivonen, University of Helsinki, Finland
Jean-François Boulicaut, INSA-Lyon, France
Floriana Esposito, University of Bari, Italy
Fosca Giannotti, ISTI-CNR, Pisa, Italy
Dino Pedreschi, University of Pisa, Italy

Area Chairs

Michael R. Berthold, Germany
Elisa Bertino, Italy
Ivan Bratko, Slovenia
Pavel Brazdil, Portugal
Carla E. Brodley, USA
Rui Camacho, Portugal
Luc Dehaspe, Belgium
Peter Flach, UK
Johannes Fürnkranz, Germany
João Gama, Portugal
Howard J. Hamilton, Canada
Thorsten Joachims, USA

Alípio Jorge, Portugal
Hillol Kargupta, USA
Pedro Larranaga, Spain
Ramon López de Mántaras, Spain
Dunja Mladenić, Slovenia
Hiroshi Motoda, Japan
José Carlos Príncipe, USA
Tobias Scheffer, Germany
Michele Sebag, France
Peter Stone, USA
Luís Torgo, Portugal
Gerhard Widmer, Austria

Program Committee

Jesus Aguilar, Spain
Paulo Azevedo, Portugal
Michael Bain, Australia
José Luís Balcázar, Spain
Elena Baralis, Italy
Bettina Berendt, Germany
Petr Berka, Czech Republic
Michael R. Berthold, Germany
Elisa Bertino, Italy
Hendrik Blockeel, Belgium
José Luís Borges, Portugal
Francesco Bonchi, Italy
Henrik Boström, Sweden
Marco Botta, Italy
Jean-François Boulicaut, France
Pavel Brazdil, Portugal
Carla E. Brodley, USA
Wray Buntine, Finland
Rui Camacho, Portugal
Amilcar Cardoso, Portugal
Barbara Catania, Italy
Jesús Cerquides, Spain

Susan Craw, Scotland
Bruno Cremilleux, France
James Cussens, UK
Luc Dehaspe, Belgium
Vasant Dhar, USA
Sašo Džeroski, Slovenia
Tapio Elomaa, Finland
Floriana Esposito, Italy
Jianping Fan, USA
Ronen Feldman, Israel
Cèsar Ferri, Spain
Peter Flach, UK
Eibe Frank, New Zealand
Alex Freitas, UK
Johannes Fürnkranz, Germany
Thomas Gabriel, Germany
João Gama, Portugal
Dragan Gamberger, Croatia
Minos N. Garofalakis, USA
Bart Goethals, Belgium
Gunter Grieser, Germany
Marko Grobelnik, Slovenia

Howard J. Hamilton, Canada
Colin de la Higuera, France
Melanie Hilario, Switzerland
Haym Hirsh, USA
Frank Höppner, Germany
Andreas Hotho, Germany
Eyke Hullermeier, German
Nitin Indurkhya, Australia
Thorsten Joachims, USA
Aleks Jakulin, Slovenia
Alípio Jorge, Portugal
Murat Kantarcioglu, USA
Hillol Kargupta, USA
Samuel Kaski, Finland
Mehmet Kaya, Turkey
Eamonn Keogh, USA
Roni Khardon, USA
Jorg-Uwe Kietz, Switzerland
Joerg Kindermann, Germany
Ross D. King, UK
Joost N. Kok, The Netherlands
Igor Kononenko, Slovenia
Stefan Kramer, Germany
Marzena Kryszkiewicz, Poland
Miroslav Kubat, USA
Nicholas Kushmerick, Ireland
Pedro Larranaga, Spain
Nada Lavrač, Slovenia
Jure Leskovec, USA
Jinyan Li, Singapore
Charles Ling, Canada
Donato Malerba, Italy
Giuseppe Manco, Italy
Ramon López de Mántaras, Spain
Stan Matwin, Canada
Michael May, Germany
Rosa Meo, Italy
Dunja Mladenić, Slovenia
Eduardo Morales, Mexico
Katharina Morik, Germany
Shinichi Morishita, Japan
Hiroshi Motoda, Japan
Steve Moyle, England
Gholamreza Nakhaeizadeh, Germany
Claire Nédellec, France

Richard Nock, France
Andreas Nürnberger, Germany
Masayuki Numao, Japan
Arlindo Oliveira, Portugal
Georgios Paliouras, Greece
Dino Pedreschi, Italy
Johann Petrak, Austria
Bernhard Pfahringer, New Zealand
Lubomir Popelinsky, Czech
José Carlos Príncipe, USA
Jan Ramon, Belgium
Zbigniew W. Ras, USA
Jan Rauch, Czech Republic
Christophe Rigotti, France
Gilbert Ritschard, Switzerland
John Roddick, Australia
Marques de Sá, Portugal
Lorenza Saitta, Italy
Daniel Sanchez, Spain
Yücel Saygin, Turkey
Tobias Scheffer, Germany
Michele Sebag, France
Marc Sebban, France
Giovanni Semeraro, Italy
Arno Siebes, The Netherlands
Andrzej Skowron, Poland
Carlos Soares, Portugal
Maarten van Someren, The Netherlands
Myra Spiliopoulou, Germany
Nicolas Spyratos, France
Ashwin Srinivasan, India
Olga Stepankova, Czech Republic
Reinhard Stolle, Germany
Peter Stone, USA
Gerd Stumme, Germany
Einoshin Suzuki, Japan
Domenico Talia, Italy
Ah-Hwee Tan, Singapore
Evimaria Terzi, Finland
Ljupčo Todorovski, Slovenia
Luís Torgo, Portugal
Shusaku Tsumoto, Japan
Peter Turney, Canada
Jaideep Vaidya, USA
Maria Amparo Vila, Spain

Dietrich Wettschereck, UK
Gerhard Widmer, Austria
Rüdiger Wirth, Germany
Mohammed J. Zaki, USA

Carlo Zaniolo, USA
Djamel A. Zighed, France

Additional Reviewers

Markus Ackermann
Anastasia Analyti
Fabrizio Angiulli
Orlando Anunciação
Giacometti Arnaud
Stella Asiimwe
Maurizio Atzori
Korinna Bade
Miriam Baglioni
Juergen Beringer
Philippe Bessières
Matjaž Bevk
Abhilasha Bhargav
Steffen Bickel
Stefano Bistarelli
Jan Blaťák
Axel Blumenstock
Janez Brank
Ulf Brefeld
Klaus Brinker
Michael Brückner
Robert D. Burbidge
Toon Calders
Michelangelo Ceci
Tania Cerquitelli
Eugenio Cesario
Vineet Chaoji
Silvia Chiusano
Fang Chu
Chris Clifton
Luís Coelho
Carmela Comito
Gianni Costa
Juan-Carlos Cubero
Agnieszka Dardzinska
Damjan Demšar
Nele Dexters

Christian Diekmann
Isabel Drost
Nicolas Durand
Mohamed Elfeky
Timm Euler
Nicola Fanizzi
Pedro Gabriel Ferreira
Francisco Ferrer
Daan Fierens
Sergio Flesca
Francesco Folino
Blaž Fortuna
Kenichi Fukui
Feng Gao
Yuli Gao
Paolo Garza
Aristides Gionis
Paulo Gomes
Gianluigi Greco
Fabian Güiza
Amaury Habrard
Hakim Hacid
Mohand-Said Hacid
Niina Haiminen
Mark Hall
Christoph Heinz
José Hernández-Orallo
Jochen Hipp
Juan F. Huete
Ali Inan
Ingo Mierswa
Stephanie Jacquemont
François Jacquenet
Aleks Jakulin
Tao Jiang
Wei Jiang
Xing Jiang

Sachindra Joshi
Pierre-Emmanuel Jouve
Michael Steinbach
George Karypis
Steffen Kempe
Arto Klami
Christian Kolbe
Stasinos Konstantopoulos
Matjaz Kukar
Minseok Kwon
Lotfi Lakhal
Carsten Lanquillon
Dominique Laurent
Yan-Nei Law
Roberto Legaspi
Aurélien Lemaire
Claire Leschi
Jure Leskovec
Francesca A. Lisi
Antonio Locane
Corrado Loglisci
Claudio Lucchese
Sara C. Madeira
Alain-Pierre Manine
M.J. Martin-Bautista
Stewart Massie
Cyrille Masson
Carlo Mastroianni
Nicolas Meger
Carlo Meghini
Taneli Mielikäinen
Mummoorthy Murugesan
Mirco Nanni
Mehmet Ercan Nergiz
Siegfried Nijssen
Janne Nikkilä
Blaž Novak
Lenka Novakova
Merja Oja
Riccardo Ortale
Ignazio Palmisano

Raffaele Perego
Sergios Petridis
Viet Phan-Luong
Frederic Piat
Dimitrios Pierrakos
Ulrich Rückert
María José Ramírez
Ganesh Ramakrishnan
Domenico Redavid
Chiara Renso
Lothar Richter
François Rioult
Céline Robardet
Pedro Rodrigues
Domingo S. Rodriguez–Baena
Luka Šajn
Saeed Salem
D. Sanchez
Eerika Savia
Karlton Sequeira
J.M. Serrano
Shengli Sheng
Javed Siddique
Georgios Sigletos
Fabrizio Silvestri
Arnaud Soulet
Hendrik Stange
Jan Struyf
Henri-Maxime Suchier
Andrea Tagarelli
Julien Thomas
Panayiotis Tsaparas
Yannis Tzitzikas
Jarkko Venna
Celine Vens
Nirmalie Wiratunga
Ghim-Eng Yap
Justin Zhan
Lizhuang Zhao
Igor Zwir

ECML/PKDD 2005 Tutorials

Ontology Learning from Text
Paul Buitelaar, Philipp Cimiano, Marko Grobelnik, Michael Sintek

Learning Automata as a Basis for Multi-agent Reinforcement Learning
Ann Nowe, Katja Verbeeck, Karl Tuyls

Web Mining for Web Personalization
Magdalini Eirinaki, Michalis Vazirgiannis

A Practical Time-Series Tutorial with MATLAB
Michalis Vlachos

Mining the Volatile Web
Myra Spiliopoulou, Yannis Theodoridis

Spectral Clustering
Chris Ding

Bioinspired Machine Learning Techniques
André Carlos Ponce de Leon Ferreira de Carvalho

Probabilistic Inductive Logic Programming
Luc De Raedt, Kristian Kersting

ECML/PKDD 2005 Workshops

Sub-symbolic Paradigms for Learning in Structured Domains
Marco Gori, Paolo Avesani

European Web Mining Forum 2005 (EWMF 2005)
Bettina Berendt, Andreas Hotho, Dunja Mladenić, Giovanni Semeraro, Myra Spiliopoulou, Gerd Stumme, Maarten van Someren

Knowledge Discovery in Inductive Databases (KDID 2005)
Francesco Bonchi, Jean-François Boulicaut

Mining Spatio-temporal Data
Gennady Andrienko, Donato Malerba, Michael May, Maguelonne Teisseire

Cooperative Multiagent Learning
Maarten van Someren, Nikos Vlassis

Data Mining for Business
Carlos Soares, Luís Moniz, Catarina Duarte

Mining Graphs, Trees and Sequences (MGTS 2005)
Siegfied Nijssen, Thorsten Meinl, George Karypis

Knowledge Discovery and Ontologies (KDO 2005)
Markus Ackermann, Bettina Berendt, Marko Grobelnik, Vojtech Svátek

Knowledge Discovery from Data Streams
Jesús Aguilar, João Gama

Reinforcement Learning in Non-stationary Environments
Ann Nowé, Timo Honkela, Ville Könönen, Katja Verbeeck

Discovery Challenge
Petr Berka, Bruno Cremilleux

Table of Contents

Invited Talks

Data Analysis in the Life Sciences — Sparking Ideas —
Michael R. Berthold 1

Machine Learning for Natural Language Processing (and Vice Versa?)
Claire Cardie 2

Statistical Relational Learning: An Inductive Logic Programming Perspective
Luc De Raedt 3

Recent Advances in Mining Time Series Data
Eamonn Keogh 6

Focus the Mining Beacon: Lessons and Challenges from the World of E-Commerce
Ron Kohavi 7

Data Streams and Data Synopses for Massive Data Sets
Yossi Matias 8

Long Papers

k-Anonymous Patterns
Maurizio Atzori, Francesco Bonchi, Fosca Giannotti, Dino Pedreschi 10

Interestingness is Not a Dichotomy: Introducing Softness in Constrained Pattern Mining
Stefano Bistarelli, Francesco Bonchi 22

Generating Dynamic Higher-Order Markov Models in Web Usage Mining
José Borges, Mark Levene 34

TREE^2 - Decision Trees for Tree Structured Data
Björn Bringmann, Albrecht Zimmermann 46

Agglomerative Hierarchical Clustering with Constraints: Theoretical and Empirical Results
    Ian Davidson, S.S. Ravi 59

Cluster Aggregate Inequality and Multi-level Hierarchical Clustering
    Chris Ding, Xiaofeng He 71

Ensembles of Balanced Nested Dichotomies for Multi-class Problems
    Lin Dong, Eibe Frank, Stefan Kramer 84

Protein Sequence Pattern Mining with Constraints
    Pedro Gabriel Ferreira, Paulo J. Azevedo 96

An Adaptive Nearest Neighbor Classification Algorithm for Data Streams
    Yan-Nei Law, Carlo Zaniolo 108

Support Vector Random Fields for Spatial Classification
    Chi-Hoon Lee, Russell Greiner, Mark Schmidt 121

Realistic, Mathematically Tractable Graph Generation and Evolution, Using Kronecker Multiplication
    Jurij Leskovec, Deepayan Chakrabarti, Jon Kleinberg, Christos Faloutsos 133

A Correspondence Between Maximal Complete Bipartite Subgraphs and Closed Patterns
    Jinyan Li, Haiquan Li, Donny Soh, Limsoon Wong 146

Improving Generalization by Data Categorization
    Ling Li, Amrit Pratap, Hsuan-Tien Lin, Yaser S. Abu-Mostafa 157

Mining Model Trees from Spatial Data
    Donato Malerba, Michelangelo Ceci, Annalisa Appice 169

Word Sense Disambiguation for Exploiting Hierarchical Thesauri in Text Classification
    Dimitrios Mavroeidis, George Tsatsaronis, Michalis Vazirgiannis, Martin Theobald, Gerhard Weikum 181

Mining Paraphrases from Self-anchored Web Sentence Fragments
    Marius Paşca 193

M^{2}SP: Mining Sequential Patterns Among Several Dimensions
    Marc Plantevit, Yeow Wei Choong, Anne Laurent, Dominique Laurent, Maguelonne Teisseire 205

A Systematic Comparison of Feature-Rich Probabilistic Classifiers for
NER Tasks
  Benjamin Rosenfeld, Moshe Fresko, Ronen Feldman 217

Knowledge Discovery from User Preferences in Conversational
Recommendation
  Maria Salamó, James Reilly, Lorraine McGinty,
  Barry Smyth 228

Unsupervised Discretization Using Tree-Based Density Estimation
  Gabi Schmidberger, Eibe Frank 240

Weighted Average Pointwise Mutual Information for Feature Selection
in Text Categorization
  Karl-Michael Schneider 252

Non-stationary Environment Compensation Using Sequential EM
Algorithm for Robust Speech Recognition
  Haifeng Shen, Jun Guo, Gang Liu, Qunxia Li 264

Hybrid Cost-Sensitive Decision Tree
  Shengli Sheng, Charles X. Ling 274

Characterization of Novel HIV Drug Resistance Mutations Using
Clustering, Multidimensional Scaling and SVM-Based Feature Ranking
  Tobias Sing, Valentina Svicher, Niko Beerenwinkel,
  Francesca Ceccherini-Silberstein, Martin Däumer, Rolf Kaiser,
  Hauke Walter, Klaus Korn, Daniel Hoffmann, Mark Oette,
  Jürgen K. Rockstroh, Gert Fätkenheuer, Carlo-Federico Perno,
  Thomas Lengauer 285

Object Identification with Attribute-Mediated Dependences
  Parag Singla, Pedro Domingos 297

Weka4WS: A WSRF-Enabled Weka Toolkit for Distributed Data
Mining on Grids
  Domenico Talia, Paolo Trunfio, Oreste Verta 309

Using Inductive Logic Programming for Predicting Protein-Protein
Interactions from Multiple Genomic Data
  Tuan Nam Tran, Kenji Satou, Tu Bao Ho 321

ISOLLE: Locally Linear Embedding with Geodesic Distance
  Claudio Varini, Andreas Degenhard, Tim Nattkemper 331

Active Sampling for Knowledge Discovery from Biomedical Data
*Sriharsha Veeramachaneni, Francesca Demichelis,
Emanuele Olivetti, Paolo Avesani* 343

A Multi-metric Index for Euclidean and Periodic Matching
*Michail Vlachos, Zografoula Vagena, Vittorio Castelli,
Philip S. Yu* 355

Fast Burst Correlation of Financial Data
Michail Vlachos, Kun-Lung Wu, Shyh-Kwei Chen, Philip S. Yu 368

A Propositional Approach to Textual Case Indexing
*Nirmalie Wiratunga, Rob Lothian, Sutanu Chakraborti,
Ivan Koychev* 380

A Quantitative Comparison of the Subgraph Miners MoFa, gSpan, FFSM, and Gaston
*Marc Wörlein, Thorsten Meinl, Ingrid Fischer,
Michael Philippsen* 392

Efficient Classification from Multiple Heterogeneous Databases
Xiaoxin Yin, Jiawei Han 404

A Probabilistic Clustering-Projection Model for Discrete Data
Shipeng Yu, Kai Yu, Volker Tresp, Hans-Peter Kriegel 417

Short Papers

Collaborative Filtering on Data Streams
Jorge Mario Barajas, Xue Li 429

The Relation of Closed Itemset Mining, Complete Pruning Strategies and Item Ordering in Apriori-Based FIM Algorithms
Ferenc Bodon, Lars Schmidt-Thieme 437

Community Mining from Multi-relational Networks
Deng Cai, Zheng Shao, Xiaofei He, Xifeng Yan, Jiawei Han 445

Evaluating the Correlation Between Objective Rule Interestingness Measures and Real Human Interest
Deborah R. Carvalho, Alex A. Freitas, Nelson Ebecken 453

A Kernel Based Method for Discovering Market Segments in Beef Meat
*Jorge Díez, Juan José del Coz, Carlos Sañudo, Pere Albertí,
Antonio Bahamonde* 462

Corpus-Based Neural Network Method for Explaining Unknown Words
by WordNet Senses
  Bálint Gábor, Viktor Gyenes, András Lőrincz 470

Segment and Combine Approach for Non-parametric Time-Series
Classification
  Pierre Geurts, Louis Wehenkel 478

Producing Accurate Interpretable Clusters from High-Dimensional
Data
  Derek Greene, Pádraig Cunningham 486

Stress-Testing Hoeffding Trees
  Geoffrey Holmes, Richard Kirkby, Bernhard Pfahringer 495

Rank Measures for Ordering
  Jin Huang, Charles X. Ling 503

Dynamic Ensemble Re-Construction for Better Ranking
  Jin Huang, Charles X. Ling 511

Frequency-Based Separation of Climate Signals
  Alexander Ilin, Harri Valpola 519

Efficient Processing of Ranked Queries with Sweeping Selection
  Wen Jin, Martin Ester, Jiawei Han 527

Feature Extraction from Mass Spectra for Classification of Pathological
States
  *Alexandros Kalousis, Julien Prados, Elton Rexhepaj,
  Melanie Hilario* 536

Numbers in Multi-relational Data Mining
  Arno J. Knobbe, Eric K.Y. Ho 544

Testing Theories in Particle Physics Using Maximum Likelihood and
Adaptive Bin Allocation
  Bruce Knuteson, Ricardo Vilalta 552

Improved Naive Bayes for Extremely Skewed Misclassification Costs
  Aleksander Kołcz, Abdur Chowdhury 561

Clustering and Prediction of Mobile User Routes from Cellular
Data
  Kari Laasonen 569

Elastic Partial Matching of Time Series
  *Longin Jan Latecki, Vasilis Megalooikonomou, Qiang Wang,
  Rolf Lakaemper, Chotirat Ann Ratanamahatana, Eamonn Keogh* 577

An Entropy-Based Approach for Generating Multi-dimensional
Sequential Patterns
  Chang-Hwan Lee 585

Visual Terrain Analysis of High-Dimensional Datasets
  Wenyuan Li, Kok-Leong Ong, Wee-Keong Ng 593

An Auto-stopped Hierarchical Clustering Algorithm for Analyzing 3D
Model Database
  *Tian-yang Lv, Yu-hui Xing, Shao-bing Huang, Zheng-xuan Wang,
  Wan-li Zuo* 601

A Comparison Between Block CEM and Two-Way CEM Algorithms to
Cluster a Contingency Table
  Mohamed Nadif, Gérard Govaert 609

An Imbalanced Data Rule Learner
  Canh Hao Nguyen, Tu Bao Ho 617

Improvements in the Data Partitioning Approach for Frequent Itemsets
Mining
  Son N. Nguyen, Maria E. Orlowska 625

On-Line Adaptive Filtering of Web Pages
  Richard Nock, Babak Esfandiari 634

A Bi-clustering Framework for Categorical Data
  Ruggero G. Pensa, Céline Robardet, Jean-François Boulicaut 643

Privacy-Preserving Collaborative Filtering on Vertically Partitioned
Data
  Huseyin Polat, Wenliang Du 651

Indexed Bit Map (IBM) for Mining Frequent Sequences
  Lionel Savary, Karine Zeitouni 659

STochFS: A Framework for Combining Feature Selection Outcomes
Through a Stochastic Process
  Jerffeson Teixeira de Souza, Nathalie Japkowicz, Stan Matwin 667

Speeding Up Logistic Model Tree Induction
  Marc Sumner, Eibe Frank, Mark Hall 675

A Random Method for Quantifying Changing Distributions in Data Streams
  Haixun Wang, Jian Pei 684

Deriving Class Association Rules Based on Levelwise Subspace Clustering
  Takashi Washio, Koutarou Nakanishi, Hiroshi Motoda 692

An Incremental Algorithm for Mining Generators Representation
  Lijun Xu, Kanglin Xie 701

Hybrid Technique for Artificial Neural Network Architecture and Weight Optimization
  Cleber Zanchettin, Teresa Bernarda Ludermir 709

Author Index 717

Data Analysis in the Life Sciences
— Sparking Ideas —

Michael R. Berthold

ALTANA-Chair for Bioinformatics and Information Mining,
Dept. of Computer and Information Science, Konstanz University, Germany
Michael.Berthold@uni-konstanz.de

Data from various areas of Life Sciences have increasingly caught the attention of data mining and machine learning researchers. Not only is the amount of data available mind-boggling but the diverse and heterogenous nature of the information is far beyond any other data analysis problem so far. In sharp contrast to classical data analysis scenarios, the life science area poses challenges of a rather different nature for mainly two reasons. Firstly, the available data stems from heterogenous information sources of varying degrees of reliability and quality and is, without the interactive, constant interpretation of a domain expert, not useful. Furthermore, predictive models are of only marginal interest to those users – instead they hope for new insights into a complex, biological system that is only partially represented within that data anyway. In this scenario, the data serves mainly to create new insights and generate new ideas that can be tested. Secondly, the notion of feature space and the accompanying measures of similarity cannot be taken for granted. Similarity measures become context dependent and it is often the case that within one analysis task several different ways of describing the objects of interest or measuring similarity between them matter.

Some more recently published work in the data analysis area has started to address some of these issues. For example, data analysis in parallel universes [1], that is, the detection of patterns of interest in various different descriptor spaces at the same time, and mining of frequent, discriminative fragments in large, molecular data bases [2]. In both cases, sheer numerical performance is not the focus; it is rather the discovery of interpretable pieces of evidence that lights up new ideas in the users mind. Future work in data analysis in the life sciences needs to keep this in mind: the goal is to trigger new ideas and stimulate interesting associations.

References

1. Berthold, M.R., Wiswedel, B., Patterson, D.E.: Interactive exploration of fuzzy clusters using neighborgrams. Fuzzy Sets and Systems 149 (2005) 21–37
2. Hofer, H., Borgelt, C., Berthold, M.R.: Large scale mining of molecular fragments with wildcards. Intelligent Data Analysis 8 (2004) 376–385

Machine Learning for Natural Language Processing (and Vice Versa?)

Claire Cardie

Department of Computer Science, Cornell University, USA
cardie@cs.cornell.edu
http://www.cs.cornell.edu/home/cardie/

Over the past 10-15 years, the influence of methods from machine learning has transformed the way that research is done in the field of natural language processing. This talk will begin by covering the history of this transformation. In particular, learning methods have proved successful in producing stand-alone text-processing components to handle a number of linguistic tasks. Moreover, these components can be combined to produce systems that exhibit shallow text-understanding capabilities: they can, for example, extract key facts from unrestricted documents in limited domains or find answers to general-purpose questions from open-domain document collections. I will briefly describe the state of the art for these practical text-processing applications, focusing on the important role that machine learning methods have played in their development.

The second part of the talk will explore the role that natural language processing might play in machine learning research. Here, I will explain the kinds of text-based features that are relatively easy to incorporate into machine learning data sets. In addition, I'll outline some problems from natural language processing that require, or could at least benefit from, new machine learning algorithms.

Statistical Relational Learning: An Inductive Logic Programming Perspective

Luc De Raedt

Institute for Computer Science, Machine Learning Lab,
Albert-Ludwigs-University, Georges-Köhler-Allee, Gebäude 079,
D-79110 Freiburg i. Brg., Germany
deraedt@informatik.uni-freiburg.de

In the past few years there has been a lot of work lying at the intersection of probability theory, logic programming and machine learning [14,18,13,9,6,1,11]. This work is known under the names of statistical relational learning [7,5], probabilistic logic learning [4], or probabilistic inductive logic programming. Whereas most of the existing works have started from a probabilistic learning perspective and extended probabilistic formalisms with relational aspects, I shall take a different perspective, in which I shall start from inductive logic programming and study how inductive logic programming formalisms, settings and techniques can be extended to deal with probabilistic issues. This tradition has already contributed a rich variety of valuable formalisms and techniques, including probabilistic Horn abduction by David Poole, PRISMs by Sato, stochastic logic programs by Muggleton [13] and Cussens [2], Bayesian logic programs [10,8] by Kersting and De Raedt, and Logical Hidden Markov Models [11].

The main contribution of this talk is the introduction of three probabilistic inductive logic programming settings which are derived from the learning from entailment, from interpretations and from proofs settings of the field of inductive logic programming [3]. Each of these settings contributes different notions of probabilistic logic representations, examples and probability distributions. The first setting, probabilistic learning from entailment, is incorporated in the well-known PRISM system [19] and Cussens's Failure Adjusted Maximisation approach to parameter estimation in stochastic logic programs [2]. A novel system that was recently developed and that fits this paradigm is the nFOIL system [12]. It combines key principles of the well-known inductive logic programming system FOIL [15] with the naïve Bayes' approach. In probabilistic learning from entailment, examples are ground facts that should be probabilistically entailed by the target logic program. The second setting, probabilistic learning from interpretations, is incorporated in Bayesian logic programs [10,8], which integrate Bayesian networks with logic programs. This setting is also adopted by [6]. Examples in this setting are Herbrand interpretations that should be a probabilistic model for the target theory. The third setting, learning from proofs [17], is novel. It is motivated by the learning of stochastic context free grammars from tree banks. In this setting, examples are proof trees that should be probabilistically provable from the unknown stochastic logic programs. The sketched settings (and their instances presented) are by no means the only possible settings for probabilistic inductive logic programming, but still – I hope – provide useful insights into the state-of-the-art of this exciting field.

For a full survey of statistical relational learning or probabilistic inductive logic programming, the author would like to refer to [4], and for more details on the probabilistic inductive logic programming settings to [16], where a longer and earlier version of this contribution can be found.

Acknowledgements

This is joint work with Kristian Kersting. The author would also like to thank Niels Landwehr and Sunna Torge for interesting collaborations on nFOIL and the learning of SLPs, respectively. This work is part of the EU IST FET project APRIL II (Application of Probabilistic Inductive Logic Programming II).

References

1. C. R. Anderson, P. Domingos, and D. S. Weld. Relational Markov Models and their Application to Adaptive Web Navigation. In D. Hand, D. Keim, O. R. Zaïne, and R. Goebel, editors, *Proceedings of the Eighth International Conference on Knowledge Discovery and Data Mining (KDD-02)*, pages 143–152, Edmonton, Canada, 2002. ACM Press.
2. J. Cussens. Loglinear models for first-order probabilistic reasoning. In K. B. Laskey and H. Prade, editors, *Proceedings of the Fifteenth Annual Conference on Uncertainty in Artificial Intelligence (UAI-99)*, pages 126–133, Stockholm, Sweden, 1999. Morgan Kaufmann.
3. L. De Raedt. Logical settings for concept-learning. *Artificial Intelligence*, 95(1):197–201, 1997.
4. L. De Raedt and K. Kersting. Probabilistic Logic Learning. *ACM-SIGKDD Explorations: Special issue on Multi-Relational Data Mining*, 5(1):31–48, 2003.
5. T. Dietterich, L. Getoor, and K. Murphy, editors. *Working Notes of the ICML-2004 Workshop on Statistical Relational Learning and its Connections to Other Fields (SRL-04)*, 2004.
6. N. Friedman, L. Getoor, D. Koller, and A. Pfeffer. Learning probabilistic relational models. In T. Dean, editor, *Proceedings of the Sixteenth International Joint Conferences on Artificial Intelligence (IJCAI-99)*, pages 1300–1309, Stockholm, Sweden, 1999. Morgan Kaufmann.
7. L. Getoor and D. Jensen, editors. *Working Notes of the IJCAI-2003 Workshop on Learning Statistical Models from Relational Data (SRL-03)*, 2003.
8. K. Kersting and L. De Raedt. Adaptive Bayesian Logic Programs. In C. Rouveirol and M. Sebag, editors, *Proceedings of the Eleventh Conference on Inductive Logic Programming (ILP-01)*, volume 2157 of *LNCS*, Strasbourg, France, 2001. Springer.
9. K. Kersting and L. De Raedt. Bayesian logic programs. Technical Report 151, University of Freiburg, Institute for Computer Science, April 2001.
10. K. Kersting and L. De Raedt. Towards Combining Inductive Logic Programming and Bayesian Networks. In C. Rouveirol and M. Sebag, editors, *Proceedings of the Eleventh Conference on Inductive Logic Programming (ILP-01)*, volume 2157 of *LNCS*, Strasbourg, France, 2001. Springer.

11. K. Kersting, T. Raiko, S. Kramer, and L. De Raedt. Towards discovering structural signatures of protein folds based on logical hidden markov models. In R. B. Altman, A. K. Dunker, L. Hunter, T. A. Jung, and T. E. Klein, editors, *Proceedings of the Pacific Symposium on Biocomputing*, pages 192 – 203, Kauai, Hawaii, USA, 2003. World Scientific.
12. N. Landwehr, K. Kersting, and L. De Raedt. nfoil: Integrating naive bayes and foil. In *Proceedings of the 20th National Conference on Artificial Intelligence*. AAAI Press, 2005.
13. S. H Muggleton. Stochastic logic programs. In L. De Raedt, editor, *Advances in Inductive Logic Programming*. IOS Press, 1996.
14. D. Poole. Probabilistic Horn abduction and Bayesian networks. *Artificial Intelligence*, 64:81–129, 1993.
15. J. R. Quinlan and R. M. Cameron-Jones. Induction of logic programs:FOIL and related systems. *New Generation Computing*, pages 287–312, 1995.
16. L. De Raedt and K. Kersting. Probabilistic inductive logic programming. In *Proceedings of the 15th International Conference on Algorithmic Learning Theory*. Springer, 2004.
17. L. De Raedt, K. Kersting, and S. Torge. Towards learning stochastic logic programs from proof-banks. In *Proceedings of the 20th National Conference on Artificial Intelligence*. AAAI Press, 2005.
18. T. Sato. A Statistical Learning Method for Logic Programs with Distribution Semantics. In L. Sterling, editor, *Proceedings of the Twelfth International Conference on Logic Programming (ICLP-1995)*, pages 715 – 729, Tokyo, Japan, 1995. MIT Press.
19. T. Sato and Y. Kameya. Parameter learning of logic programs for symbolic-statistical modeling. *Journal of Artificial Intelligence Research*, 15:391–454, 2001.

Recent Advances in Mining Time Series Data

Eamonn Keogh

Department of Computer Science & Engineering,
University of California, Riverside, USA
eamonn@cs.ucr.edu
http://www.cs.ucr.edu/~eamonn

Much of the world's supply of data is in the form of time series. Furthermore, as we shall see, many types of data can be meaningfully converted into "time series", including text, DNA, video, images etc. The last decade has seen an explosion of interest in mining time series data from the academic community. There has been significant work on algorithms to classify, cluster, segment, index, discover rules, visualize, and detect anomalies/novelties in time series.

In this talk I will summarize the latest advances in mining time series data, including:

- New representations of time series data.
- New algorithms/definitions.
- The migration from static problems to online problems.
- New areas and applications of time series data mining.

I will end the talk with a discussion of "what's left to do" in time series data mining.

References

1. E. Keogh. Exact indexing of dynamic time warping. In *Proceedings of the 8th International Conference on Very Large Data Bases*, pages 406–417, 2002.
2. E. Keogh and S. Kasetty. On the need for time series data mining benchmarks: A survey and empirical demonstration. In *Proceedings of the 8th ACM SIGKDD International Conference on Knowledge Discovery and Data Mining*, pages 102–111, 2002.
3. E. Keogh, J. Lin, and W. Truppel. Clustering of time series subsequences is meaningless: Implications for past and future research. In *Proceedings of the 3rd IEEE International Conference on Data Mining*, pages 115–122, 2003.
4. E. Keogh, S. Lonardi, and C. Ratanamahatana. Towards parameter-free data mining. In *Proceedings of the tenth ACM SIGKDD International Conference on Knowledge Discovery and Data Mining*, 2004.
5. C.A. Ratanamahatana and E. Keogh. Everything you know about dynamic time warping is wrong. In *Proceedings of the Third Workshop on Mining Temporal and Sequential Data, in conjunction with the Tenth ACM SIGKDD International Conference on Knowledge Discovery and Data Mining (KDD-2004)*, 2004.

Focus the Mining Beacon: Lessons and Challenges from the World of E-Commerce

Ron Kohavi

Microsoft Corporation, USA
ronnyk@cs.stanford.edu
http://www.kohavi.com

Electronic Commerce is now entering its second decade, with Amazon.com and eBay now in existence for ten years. With massive amounts of data, an actionable domain, and measurable ROI, multiple companies use data mining and knowledge discovery to understand their customers and improve interactions. We present important lessons and challenges using e-commerce examples across two dimensions: (i) business-level to technical, and (ii) the mining lifecycle from data collection, data warehouse construction, to discovery and deployment. Many of the lessons and challenges are applicable to domains outside e-commerce.

Data Streams and Data Synopses for Massive Data Sets
(Invited Talk)

Yossi Matias

Tel Aviv University,
HyperRoll Inc., Stanford University
matias@tau.ac.il

Abstract. With the proliferation of data intensive applications, it has become necessary to develop new techniques to handle massive data sets. Traditional algorithmic techniques and data structures are not always suitable to handle the amount of data that is required and the fact that the data often streams by and cannot be accessed again. A field of research established over the past decade is that of handling massive data sets using data synopses, and developing algorithmic techniques for data stream models. We will discuss some of the research work that has been done in the field, and provide a decades' perspective to data synopses and data streams.

1 Summary

In recent years, we have witnessed an explosion in data used in various applications. In general, the growth rate in data is known to exceed the increase rate in the size of RAM, and of the available computation power (a.k.a. Moore's Law). As a result, traditional algorithms and data structures are often no longer adequate to handle the massive data sets required by these applications.

One approach to handle massive data sets is to use *external memory algorithms*, designed to make an effective utilization of I/O. In such algorithms the data structures are often implemented in external storage devices, and the objective is in general to minimize the number of I/Os. For a survey of works on external memory algorithms see [6]. Such algorithms assume that the entire input data is available for further processing. There are, however, many applications where the data is only seen once, as it "streams by". This may be the case in, e.g., financial applications, network monitoring, security, telecommunications data management, web applications, manufacturing, and sensor networks. Even in data warehouse applications, where the data may in general be available for additional querying, there are many situations where data analysis needs to be done as the data is loaded into the data warehouse, since the cost of accessing the data in a fully loaded production system may be significantly larger than just the basic cost of I/O. Additionally, even in the largest data warehouses, consisting of hundreds of terabytes, data is only maintained for a limited time, so access to historical data may often be infeasible.

It had thus become necessary to address situations in which massive data sets are required to be handled as they "stream by", and using only limited memory. Motivated by this need, the research field of data streams and data synopses has emerged and established over the last few years. We will discuss some of the research work that has been done in the field, and provide a decades' perspective to data streams and data synopses. A longer version of this abstract will be available at [4].

The data stream model is quite simple: it is assumed that the input data set is given as a sequence of data items. Each data item is seen only once, and any computation can be done utilizing the data structures maintained in main memory. These memory resident data structures are substantially smaller than the input data. As such, they cannot fully represent the data as is the case for traditional data structures, but can only provide a synopsis of the input data; hence they are denoted as *synopsis data structures*, or *data synopses* [3].

The use of data synopses implies that data analysis that is dependent on the entire streaming data will often be approximated. Furthermore, ad hoc queries that are dependent on the entire input data could only be served by the data synopses, and as a result only approximate answers to queries will be available. A primary objective in the design of data synopses is to have the smallest data synopses that would guarantee small, and if possible bounded, error on the approximated computation.

As we have shown in [1], some essential statistical data analysis, the so-called *frequency moments,* can be approximated using synopses that are as small as polynomial or even logarithmic in the input size. Over the last few years there has been a proliferation of additional works on data streams and data synopses. See, e.g., the surveys [2] and [5]. These works include theoretical results, as well as applications in databases, network traffic analysis, security, sensor networks, and program profiling; synopses include samples, random projections, histograms, wavelets, and XML synopses, among others. There remain a plethora of interesting open problems, both theoretical as well as applied.

References

1. Alon, N., Matias, Y., Szegedy, M.: The space complexity of approximating the frequency moments. J. of Computer and System Sciences 58 (1999), 137-147. STOC'96 Special Issue
2. Babcock, B., Babu, S., Datar, M., Motwani, R. Widom, J.: Models and issues in data stream systems. In Proc. Symposium on Principles of Database Systems (2002), 1-16
3. Gibbons, P.B., Matias, Y.: Synopses data structures for massive data sets. In: External memory algorithms, DIMACS Series Discrete Math. & TCS, AMS, 50 (1999). Also SODA'99
4. Matias, Y.: Data streams and data synopses for massive data sets.
   http://www.cs.tau.ac.il/~matias/streams/
5. Muthukrishnan, S.: Data streams: Algorithms and applications.
   http://www.cs.rutgers.edu/~muthu/stream-1-1.ps
6. Vitter, J.S.: External memory algorithms and data structures. ACM Comput Surv. 33(2): 209-271 (2001)

k-Anonymous Patterns

Maurizio Atzori[1,2], Francesco Bonchi[2], Fosca Giannotti[2], and Dino Pedreschi[1]

[1] Pisa KDD Laboratory, Computer Science Department, University of Pisa, Italy
{atzori, pedre}@di.unipi.it
[2] Pisa KDD Laboratory, ISTI - CNR, Pisa, Italy
{francesco.bonchi, fosca.giannotti}@isti.cnr.it

Abstract. It is generally believed that data mining results do not violate the *anonymity* of the individuals recorded in the source database. In fact, data mining models and patterns, in order to ensure a required statistical significance, represent a large number of individuals and thus conceal individual identities: this is the case of the minimum support threshold in association rule mining. In this paper we show that this belief is ill-founded. By shifting the concept of k-*anonymity* from data to patterns, we formally characterize the notion of a threat to anonymity in the context of pattern discovery, and provide a methodology to efficiently and effectively identify all possible such threats that might arise from the disclosure of a set of extracted patterns.

1 Introduction

Privacy Preserving Data Mining, i.e., the analysis of data mining side-effects on privacy, has recently become a key research issue and is receiving a growing attention from the research community [1,3,9,16]. However, despite such efforts, a common understanding of what is meant by "privacy" is still missing. This fact has led to the proliferation of many completely different approaches to privacy preserving data mining, all sharing the same generic goal: producing a valid mining model without disclosing "private" data. As highlighted in [9], the approaches pursued so far leave a privacy question open: do the data mining results themselves violate privacy? Put in other words, do the disclosure of extracted patterns open up the risk of privacy breaches that may reveal sensitive information? During the last year, few works [7,9,11] have tried to address this problem by some different points of view, but they all require some *a priori* knowledge of what is sensitive and what is not.

In this paper we study when data mining results represent *per se* a threat to privacy, independently of any background knowledge of what is sensitive. In particular, we focus on *individual privacy*, which is concerned with the *anonymity* of individuals.

A prototypical application instance is in the medical domain, where the collected data are typically very sensitive, and the kind of privacy usually required is the anonymity of the patients in a survey. Consider a medical institution where the usual hospital activity is coupled with medical research activity. Since physicians are the data collectors and holders, and they already know everything about their patients, they have unrestricted access to the collected information. Therefore, they can perform real mining on all available information using traditional mining tools – not necessarily the privacy preserving ones. This way they maximize the outcome of the knowledge discovery process, without any concern about privacy of the patients which are recorded in the data. But the anonymity of patients becomes a key issue when the physicians want to share their discoveries (e.g., association rules holding in the data) with their scientific community.

At a first sight, it seems that data mining results do not violate the anonymity of the individuals recorded in the source database. In fact, data mining models and patterns, in order to ensure a required statistical significance, represent a large number of individuals and thus conceal individual identities: this is the case of the minimum support threshold in association rule mining. The next example shows that the above belief is ill-founded.

Example 1. Consider the following association rule:

$$a_1 \wedge a_2 \wedge a_3 \Rightarrow a_4 \quad [sup = 80, conf = 98.7\%]$$

where sup and $conf$ are the usual interestingness measures of *support* and *confidence* as defined in [2]. Since the given rule holds for a number of individuals (80), which seems large enough to protect individual privacy, one could conclude that the given rule can be safely disclosed. But, is this all the information contained in such a rule? Indeed, one can easily derive the support of the premise of the rule:

$$sup(\{a_1, a_2, a_3\}) = \frac{sup(\{a_1, a_2, a_3, a_4\})}{conf} \approx \frac{80}{0.987} = 81.05$$

Given that the pattern $a_1 \wedge a_2 \wedge a_3 \wedge a_4$ holds for 80 individuals, and that the pattern $a_1 \wedge a_2 \wedge a_3$ holds for 81 individuals, we can infer that in our database there is just one individual for which the pattern $a_1 \wedge a_2 \wedge a_3 \wedge \neg a_4$ holds.

The knowledge inferred is a clear threat to the anonymity of that individual: on one hand the pattern identifying the individual could itself contain sensitive information; on the other hand it could be used to re-identify the same individual in other databases.

It is worth noting that this problem is very general: the given rule could be, instead of an association, a classification rule, or the path from the root to the leaf in a decision tree, and the same reasoning would still hold. Moreover, it is straightforward to note that, unluckily, the more accurate is a rule, the more unsafe it may be w.r.t. anonymity. As shown later, this anonymity problem can not be simply solved by discarding the most accurate rules: in fact, more complex kinds of threats to anonymity exist which involve more than simply two itemsets.

1.1 Related Works

During the last years a novel problem has emerged in privacy-preserving data mining [7,9,11]: do the data mining results themselves violate privacy? Only little preliminary work is available. The work in [9] studies the case of a classifier trained over a mixture of different kind of data: *public* (known to every one including the adversary), *private/sensitive* (should remain unknown to the adversary), and *unknown* (neither sensitive nor known by the adversary). The authors propose a model for privacy implication of the learned classifier.

In [11] the data owner, rather than sharing the data, prefers to share the mined association rules, but requires that a set of *restricted* association rules are not disclosed. The authors propose a framework to sanitize the output of association rules mining, while blocking some inference channels for the restricted rules.

In [7] a framework for evaluating classification rules in terms of their perceived privacy and ethical sensitivity is described. The proposed framework empowers the data miner with alerts for sensitive rules that can be accepted or dismissed by the user as appropriate. Such alerts are based on an aggregate *sensitivity combination function*, which assigns to each rule a value of sensitivity by aggregating the sensitivity value (an integer between 0 and 9) of each attribute involved in the rule. The process of labelling each attribute with its sensitivity value must be accomplished by the domain expert.

The fundamental difference of these approaches with ours lies in generality: we propose a novel, objective definition of privacy compliance of patterns without any reference to a preconceived knowledge of sensitive data or patterns, on the basis of the rather intuitive and realistic constraint that the anonymity of individuals should be guaranteed.

An important method for protecting individual privacy is k-*anonymity*, introduced in [14], a notion that establishes that the cardinality of the answer to any possible query should be at least k. In this work, it is shown that protection of individual sources does not guarantee protection when sources are cross-examined: a sensitive medical record, for instance, can be uniquely linked to a *named* voter record in a publicly available voter list through some shared attributes. The objective of k-anonymity is to eliminate such opportunities of inferring private information through cross linkage. In particular, this is obtained by a "sanitization" of the source data that is transformed in such a way that, for all possible queries, at least k tuples will be returned. Such a sanitization is obtained by generalization and suppression of attributes and/or tuples [15].

Trivially, by mining a k-anonymized database no patterns threatening the anonymity can be obtained. But such mining would produce models impoverished by the information loss which is intrinsic in the generalization and suppression techniques. Since our objective is to extract valid and interesting patterns, we propose to postpone k-anonymization after the actual mining step. In other words, we do not to enforce k-anonymity onto the source data, but instead we move such a concept to the extracted patterns.

1.2 Paper Contributions

In this paper we study the privacy problem described above in the very general setting of patterns which are boolean formulas over a binary database. Our contribution is twofold:

- we define k-anonymous patterns and provide a general characterization of inference channels holding among patterns that may threat anonymity of source data;
- we develop an effective and efficient algorithm to detect such potential threats, which yields a methodology to check whether the mining results may be disclosed without any risk of violating anonymity.

We emphasize that the capability of detecting the potential threats is extremely useful for the analyst to determine a trade-off among the quality of mining result and the privacy guarantee, by means of an iterative interaction with the proposed detection algorithm. Our empirical experiments, reported in this paper, bring evidence to this observation. It should also be noted the different setting w.r.t. the other works in privacy preserving data mining: in our context no data perturbation or sanitization is performed; we allow real mining on the real data, while focussing on the *anonymity preservation properties of the extracted patterns*. We have also developed possible strategies to eliminate the threats to anonymity by introducing distortion on the dangerous patterns in a controlled way: for lack of space these results are omitted here but can be found in [5].

2 k-Anonymous Patterns and σ-Frequent Itemsets

We start by defining binary databases and patterns following the notation in [8].

Definition 1. *A binary database $\mathcal{D} = (\mathcal{I}, \mathcal{T})$ consists of a finite set of binary variables $\mathcal{I} = \{i_1, \ldots, i_p\}$, also known as* items, *and a finite multiset $\mathcal{T} = \{t_1, \ldots, t_n\}$ of p-dimensional binary vectors recording the values of the items. Such vectors are also known as* transactions. *A pattern for the variables in $\mathcal{I}$ is a logical (propositional) sentence built by AND ($\wedge$), OR ($\vee$) and NOT ($\neg$) logical connectives, on variables in $\mathcal{I}$. The domain of all possible patterns is denoted $\mathcal{P}at(\mathcal{I})$.*

According to Def. 1, $e \wedge (\neg b \vee \neg d)$, where $b, d, e \in \mathcal{I}$, is a pattern. One of the most important properties of a pattern is its frequency in the database, i.e. the number of individuals (transactions) in the database which make the given pattern true[1].

Definition 2. *Given a database $\mathcal{D}$, a transaction $t \in \mathcal{D}$ and a pattern p, we write $p(t)$ if t makes p true. The support of p in $\mathcal{D}$ is given by the number of transactions which makes p true: $sup_\mathcal{D}(p) = |\{t \in \mathcal{D} \mid p(t)\}|$.*

The most studied *pattern class* is the itemset, i.e., a conjunction of positive valued variables, or in other words, a set of items. The retrieval of itemsets which satisfy a minimum frequency property is the basic step of many data mining tasks, including (but not limited to) association rules [2,4].

Definition 3. *The set of all itemsets $2^\mathcal{I}$, is a pattern class consisting of all possible conjunctions of the form $i_1 \wedge i_2 \wedge \ldots \wedge i_m$. Given a database $\mathcal{D}$ and a minimum support threshold σ, the set of σ-frequent itemsets in $\mathcal{D}$ is denoted*

$$\mathcal{F}(\mathcal{D}, \sigma) = \{\langle X, sup_\mathcal{D}(X)\rangle \mid X \in 2^\mathcal{I} \wedge sup_\mathcal{D}(X) \geq \sigma\}$$

Itemsets are usually denoted in the form of set of the items in the conjunction, e.g. $\{i_1, \ldots, i_m\}$; or sometimes, simply $i_1 \ldots i_m$. Figure 1(b) shows the different notation used for general patterns and for itemsets. The problem addressed in this paper is given by the possibility of inferring from the output of frequent itemset mining, i.e, $\mathcal{F}(\mathcal{D}, \sigma)$, the existence of patterns with very low support (i.e., smaller than an anonymity threshold k, but not null): such patterns represent a threat for the anonymity of the individuals about which they are true.

---

[1] The notion of truth of a pattern w.r.t. a transaction t is defined in the usual way: t makes p true iff t is a model of the propositional sentence p.

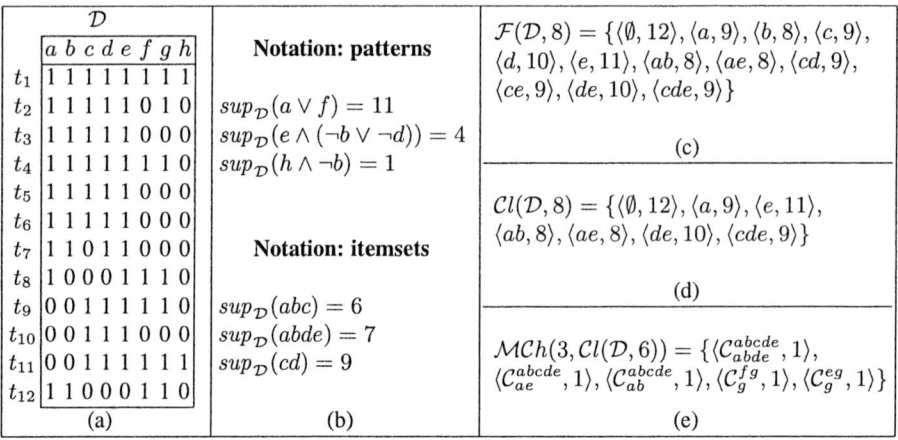

Fig. 1. Running example: (a) the binary database $\mathcal{D}$; (b) different notation used for patterns and itemsets; (c) the set of σ-frequent ($\sigma = 8$) itemsets; (d) the set of closed frequent itemsets; (e) the set of maximal inference channels for $k = 3$ and $\sigma = 6$

Definition 4. *Given a database $\mathcal{D}$ and an anonymity threshold k, a pattern p is said to be k-anonymous if $sup_{\mathcal{D}}(p) \geq k$ or $sup_{\mathcal{D}}(p) = 0$.*

2.1 Problem Definition

Before introducing our anonymity preservation problem, we need to define the inference of supports, which is the basic tool for the attacks to anonymity.

Definition 5. *A set S of pairs $\langle X, n \rangle$, where $X \in 2^{\mathcal{I}}$ and $n \in \mathbb{N}$, and a database $\mathcal{D}$ are said to be σ-compatible if $S \subseteq \mathcal{F}(\mathcal{D}, \sigma)$. Given a pattern p we say that $S \models sup(p) > x$ (respectively $S \models sup(p) < x$) if, for all databases $\mathcal{D}$ σ-compatible with S, we have that $sup_{\mathcal{D}}(p) > x$ (respectively $sup_{\mathcal{D}}(p) < x$).*

Informally, we call *inference channel* any subset of the collection of itemsets (with their respective supports), from which it is possible to infer non k-anonymous patterns. Our mining problem can be seen as frequent pattern extraction with two frequency thresholds: the usual minimum support threshold σ for itemsets (as defined in Definition 3), and an anonymity threshold k for general patterns (as defined in Definition 1).

Note that an itemset with support less than k is itself a non k-anonymous, and thus dangerous, pattern. However, since we can safely assume (as we will do in the rest of this paper) that $\sigma \gg k$, such pattern would be discarded by the usual mining algorithms.

Definition 6. *Given a collection of frequent itemsets $\mathcal{F}(\mathcal{D}, \sigma)$ and an anonymity threshold k, our problem consists in detecting all possible inference channels $\mathcal{C} \subseteq \mathcal{F}(\mathcal{D}, \sigma)$: $\exists p \in \mathcal{P}at(\mathcal{I}) : \mathcal{C} \models 0 < sup_{\mathcal{D}}(p) < k$.*

Obviously, a solution to this problem directly yields a method to formally prove that the disclosure of a given collection of frequent itemsets does not violate the anonymity constraint: it is sufficient to check that no inference channel exists for the given collection. In this case, the collection can be safely distributed even to malicious adversaries. On the contrary, if this is not the case, we can proceed in two ways:

- mine a new collection of frequent itemsets under different circumstances, e.g., higher minimum support threshold, to look for an admissible collection;
- transform (sanitize) the collection to remove the inference channels.

The second alternative opens up many interesting mining problems, which are omitted here for lack of space, and are discussed in [5].

3 Detecting Inference Channels

In this Section we study how information about non k-anonymous patterns can be possibly inferred from a collection of σ-frequent itemsets. As suggested by Example 1, a simple inference channel is given by any itemset X which has a superset $X \cup \{a\}$ such that $0 < sup_{\mathcal{D}}(X) - sup_{\mathcal{D}}(X \cup \{a\}) < k$. In this case the pair $\langle X, sup_{\mathcal{D}}(X)\rangle, \langle X \cup \{a\}, sup_{\mathcal{D}}(X \cup \{a\})\rangle$ is an inference channel for the non k-anonymous pattern $X \wedge \neg a$, whose support is directly given by $sup_{\mathcal{D}}(X) - sup_{\mathcal{D}}(X \cup \{a\})$. This is a trivial kind of inference channel. Do more complex structures of itemsets exist that can be used as inference channels? In general, the support of a pattern $p = i_1 \wedge \cdots \wedge i_m \wedge \neg a_1 \wedge \cdots \wedge \neg a_n$ can be inferred if we know the support of itemsets $I = \{i_1, \ldots, i_m\}$, $J = I \cup \{a_1, \ldots, a_n\}$, and every itemset L such that $I \subset L \subset J$.

Lemma 1. *Given a pattern $p = i_1 \wedge \cdots \wedge i_m \wedge \neg a_1 \wedge \cdots \wedge \neg a_n$ we have that:*

$$sup_{\mathcal{D}}(p) = \sum_{I \subseteq X \subseteq J} (-1)^{|X \setminus I|} sup_{\mathcal{D}}(X)$$

where $I = \{i_1, \ldots, i_m\}$ and $J = I \cup \{a_1, \ldots, a_n\}$.

Proof. (Sketch) The proof follows directly from the definition of support and the well-known *inclusion-exclusion principle* [10].

Following the notation in [6], we denote the right-hand side of the equation above as $f_I^J(\mathcal{D})$. In the database $\mathcal{D}$ in Figure 1 we have that $sup_{\mathcal{D}}(b \wedge \neg d \wedge \neg e) = f_b^{bde}(\mathcal{D}) = sup_{\mathcal{D}}(b) - sup_{\mathcal{D}}(bd) - sup_{\mathcal{D}}(be) + sup_{\mathcal{D}}(bde) = 8 - 7 - 7 + 7 = 1$.

Definition 7. *Given a database $\mathcal{D}$, and two itemsets $I, J \in 2^{\mathcal{I}}$, $I = \{i_1, \ldots, i_m\}$ and $J = I \cup \{a_1, \ldots, a_n\}$, if $0 < f_I^J(\mathcal{D}) < k$, then the set of itemsets $\{X | I \subseteq X \subseteq J\}$ constitutes an inference channel for the non k-anonymous pattern $p = i_1 \wedge \cdots \wedge i_m \wedge \neg a_1 \wedge \cdots \wedge \neg a_n$. We denote such inference channel $\mathcal{C}_I^J$ and we write $sup_{\mathcal{D}}(\mathcal{C}_I^J) = f_I^J(\mathcal{D})$.*

Example 2. Consider the database $\mathcal{D}$ of Figure 1, and suppose $k = 3$. We have that $\mathcal{C}_{ab}^{abcde}$ is an inference channel of support 1. This means that there is only one transaction $t \in \mathcal{D}$ is such that $a \wedge b \wedge \neg c \wedge \neg d \wedge \neg e$.

The next Theorem states that if there exists a non k-anonymous pattern, then there exists a pair of itemsets $I \subseteq J \in 2^{\mathcal{I}}$ such that $\mathcal{C}_I^J$ is an inference channel.

Theorem 1. $\forall p \in \mathcal{P}at(\mathcal{I}) : 0 < sup_\mathcal{D}(p) < k \,.\, \exists I \subseteq J \in 2^\mathcal{I} : \mathcal{C}_I^J$.

Proof. The case of a conjunctive pattern p is a direct consequence of Lemma 1. Let us now consider a generic pattern $p \in \mathcal{P}at(\mathcal{I})$. Without loss of generality p is in *normal disjunctive form*: $p = p_1 \vee \ldots \vee p_q$, where $p_1 \ldots p_q$ are conjunctive patterns. We have that:

$$sup_\mathcal{D}(p) \geq \max_{1 \leq i \leq q} sup_\mathcal{D}(p_i).$$

Since $sup_\mathcal{D}(p) < k$ we have for all patterns p_i that $sup_\mathcal{D}(p_i) < k$. Moreover, since $sup_\mathcal{D}(p) > 0$ is there at least a pattern p_i such that $sup_\mathcal{D}(p_i) > 0$. Therefore, there is at least a conjunctive pattern p_i such that $0 < sup_\mathcal{D}(p_i) < k$.

From Theorem 1 we conclude that all possible threats to anonymity are due to inference channels of the form $\mathcal{C}_I^J$. However we can divide such inference channels in two subgroups:

1. inference channels involving only frequent itemsets;
2. inference channels involving also infrequent itemsets.

The first problem, addressed in the rest of this paper, is the most essential. In fact, a malicious adversary can easily find inference channels made up only of elements which are present in the disclosed output. However, these inference channels are not the unique possible source of inference: further inference channels involving also infrequent itemsets could be possibly discovered, albeit in a much more complex way.

In fact, in [6] deduction rules to derive tight bounds on the support of itemsets are introduced. Given an itemset J, if for each subset $I \subset J$ the support $sup_\mathcal{D}(I)$ is known, such rules allow to compute lower and upper bounds on the support of J. Let l be the greatest lower bound we can derive, and u the smallest upper bound we can derive: if we find that $l = u$ then we can infer that $sup_\mathcal{D}(J) = l = u$ without actual counting. In this case J is said to be a *derivable itemset*. We transpose such deduction techniques in our context and observe that they can be exploited to discover information about infrequent itemsets, and from these to infer non k-anonymous patterns. For lack of space, this higher-order problem is not discussed here, and left to the extended version of this paper. However, here we can say that the techniques to detect this kind of inference channels and to block them are very similar to the techniques for the first kind of channels. This is due to the fact that both kinds of channels rely on the same concept: inferring supports of larger itemsets from smaller ones. Indeed, the key equation of our work (Lemma 1) is also the basis of the deduction rules proposed in [6].

From now on we restrict our attention to the essential form of inference channel, namely those involving frequent itemsets only.

Definition 8. *The set of all $\mathcal{C}_I^J$ holding in $\mathcal{F}(\mathcal{D}, \sigma)$, together with their supports, is denoted* $\mathcal{C}h(k, \mathcal{F}(\mathcal{D}, \sigma)) = \{\langle \mathcal{C}_I^J, f_I^J(\mathcal{D}) \rangle \mid 0 < f_I^J(\mathcal{D}) < k \wedge \langle J, sup_\mathcal{D}(J) \rangle \in \mathcal{F}(\mathcal{D}, \sigma)\}$.

Algorithm 1 detects all possible inference channels $\mathcal{C}h(k, \mathcal{F}(\mathcal{D}, \sigma))$ that hold in a collection of frequent itemsets $\mathcal{F}(\mathcal{D}, \sigma)$ by checking all possible pairs of itemsets $I, J \in \mathcal{F}(\mathcal{D}, \sigma)$ such that $I \subseteq J$. This could result in a very large number of checks. Suppose that $\mathcal{F}(\mathcal{D}, \sigma)$ is formed only by a maximal itemset Y and all its subsets (an

Algorithm 1 Naïve Inference Channel Detector

Input: $\mathcal{F}(\mathcal{D}, \sigma), k$
Output: $Ch(k, \mathcal{F}(\mathcal{D}, \sigma))$
1: $Ch(k, \mathcal{F}(\mathcal{D}, \sigma)) = \emptyset$
2: **for all** $\langle J, sup(J)\rangle \in \mathcal{F}(\mathcal{D}, \sigma)$ **do**
3:    **for all** $I \subseteq J$ **do**
4:       compute f_I^J;
5:       **if** $0 < f_I^J < k$ **then**
6:          insert$\langle C_I^J, f_I^J\rangle$ in $Ch(k, \mathcal{F}(\mathcal{D}, \sigma))$;

---

itemset is maximal if none of its proper supersets is in $\mathcal{F}(\mathcal{D}, \sigma)$). If $|Y| = n$ we get $|\mathcal{F}(\mathcal{D}, \sigma)| = 2^n$ (we also count the empty set), while the number of possible C_I^J is $\sum_{1 \leq i \leq n} \binom{n}{i}(2^i - 1)$. In the following Section we study some interesting properties that allow to dramatically reduce the number of checks needed to retrieve $Ch(k, \mathcal{F}(\mathcal{D}, \sigma))$.

4 A Condensed Representation of Inference Channels

In this section we introduce a condensed representation of $Ch(k, \mathcal{F}(\mathcal{D}, \sigma))$, i.e., a subset of $Ch(k, \mathcal{F}(\mathcal{D}, \sigma))$ which is more efficient to compute, and sufficient to reconstruct the whole $Ch(k, \mathcal{F}(\mathcal{D}, \sigma))$. The benefits of having such condensed representation go far beyond mere efficiency. In fact, removing the redundancy existing in $Ch(k, \mathcal{F}(\mathcal{D}, \sigma))$, we also implicitly avoid redundant sanitization, when we will block inference channels holding in $\mathcal{F}(\mathcal{D}, \sigma)$ (recall that, as stated before, the issue of how to block inference channels is not covered in this paper).

Consider, for instance, the two inference channels $\langle C_{ad}^{acd}, 1\rangle$ and $\langle C_{abd}^{abcd}, 1\rangle$ holding in the database in Fig. 1(a): one is more specific than the other, but they both uniquely identify transaction t_7. It is easy to see that many other families of equivalent, and thus redundant, inference channels can be found. *How can we directly identify one and only one representative inference channel in each family of equivalent ones?* The theory of *closed itemsets* can help us with this problem.

Closed itemsets were first introduced in [12] and since then they have received a great deal of attention especially by an algorithmic point of view [17,13]. They are a concise and lossless representation of all frequent itemsets, i.e., they contain the same information without redundancy. Intuitively, a closed itemset groups together all its subsets that have its same support; or in other words, it groups together itemsets which identify the same group of transactions.

Definition 9. *Given the function $f(T) = \{i \in \mathcal{I} \mid \forall t \in T, i \in t\}$, which returns all the items included in the set of transactions T, and the function $g(X) = \{t \in \mathcal{T} \mid \forall i \in X, i \in t\}$ which returns the set of transactions supporting a given itemset X, the composite function $c = f \circ g$ is the closure operator. An itemset I is closed iff and only if $c(I) = I$. Given a database $\mathcal{D}$ and a minimum support threshold σ, the set of frequent closed itemsets is denoted $Cl(\mathcal{D}, \sigma)$. An itemset $I \in Cl(\mathcal{D}, \sigma)$ is said to be maximal iff $\nexists J \supset I$ s.t. $J \in Cl(\mathcal{D}, \sigma)$.*

Analogously to what happens for the pattern class of itemsets, if we consider the pattern class of conjunctive patterns we can rely on the *anti-monotonicity property of frequency*. For instance, the number of transactions for which the pattern $a \wedge \neg c$ holds is always larger than the number of transactions for which the pattern $a \wedge b \wedge \neg c \wedge \neg d$ holds.

Definition 10. *Given two inference channels C_I^J and C_H^L we say that $C_I^J \preceq C_H^L$ when $I \subseteq H$ and $(J \setminus I) \subseteq (L \setminus H)$.*

Proposition 1. $C_I^J \preceq C_H^L \Rightarrow \forall \mathcal{D} \cdot f_I^J(\mathcal{D}) \geq f_H^L(\mathcal{D})$.

Therefore, when detecting inference channels, whenever we find a C_H^L such that $f_H^L(\mathcal{D}) \geq k$, we can avoid checking the support of all inference channels $C_I^J \preceq C_H^L$, since they will be k-anonymous.

Definition 11. *An inference channel C_I^J is said to be maximal w.r.t. $\mathcal{D}$ and σ, if $\forall H, L$ such that $I \subseteq H$ and $(J \setminus I) \subseteq (L \setminus H)$, $f_H^L = 0$. The set of maximal inference channels is denoted $\mathcal{MCh}(k, Cl(\mathcal{D}, \sigma))$.*

Proposition 2. $C_I^J \in \mathcal{MCh}(k, Cl(\mathcal{D}, \sigma)) \Rightarrow I \in Cl(\mathcal{D}, \sigma) \wedge J$ *is maximal.*

Proof. i) $I \in Cl(\mathcal{D}, \sigma)$: if I is not closed then consider its closure $c(I)$ and consider $J' = J \cup (c(I) \setminus I)$. For the definition of closure, the set of transactions containing I is the same of the set of transactions containing $c(I)$, and the set of transactions containing J' is the same of the set of transactions containing J. It follows that $C_{c(I)}^{J'} \succeq C_I^J$ and $f_{c(I)}^J = f_I^J > 0$. Then, if I is not closed, C_I^J is not maximal.

ii) J is maximal: if J is not maximal then consider its frequent superset $J' = J \cup \{a\}$ and consider $I' = I \cup a$. It is straightforward to see that $f_I^J = f_I^{J'} + f_{I'}^{J'}$ and that $C_I^{J'} \succeq C_I^J$ and $C_{I'}^{J'} \succeq C_I^J$. Therefore, since $f_I^J > 0$, at least one among $f_I^{J'}$ and $f_{I'}^{J'}$ must be not null. Then, if J is not maximal, C_I^J is not maximal as well.

The next Theorem shows how the support of any channel in $Ch(k, \mathcal{F}(\mathcal{D}, \sigma))$ can be reconstructed from $\mathcal{MCh}(k, Cl(\mathcal{D}, \sigma))$.

Theorem 2. *Given $C_I^J \in Ch(k, \mathcal{F}(\mathcal{D}, \sigma))$, let M be any maximal itemset such that $M \supseteq J$. The following equation holds:*

$$f_I^J(\mathcal{D}) = \sum_{c(X)} f_{c(X)}^M(\mathcal{D})$$

where $c(I) \subseteq c(X) \subseteq M$ and $c(X) \cap (J \setminus I) = \emptyset$.

Proof. See [5].

From Theorem 2 we conclude that all the addends needed to compute $f_I^J(\mathcal{D})$ for an inference channel are either in $\mathcal{MCh}(k, Cl(\mathcal{D}, \sigma))$ or are null. Therefore, as the set of all closed frequent itemsets $Cl(\mathcal{D}, \sigma)$ contains all the information of $\mathcal{F}(\mathcal{D}, \sigma)$ in a more compact representation, analogously the set $\mathcal{MCh}(k, Cl(\mathcal{D}, \sigma))$ represents, without redundancy, all the information in $Ch(k, \mathcal{F}(\mathcal{D}, \sigma))$.

In the database $\mathcal{D}$ of Figure 1(a), given $\sigma = 6$ and $k = 3$, $|\mathcal{C}h(3, \mathcal{F}(\mathcal{D}, 6))| = 58$ while $|\mathcal{MC}h(3, \mathcal{C}l(\mathcal{D}, 6))| = 5$ (Figure 1(e)), a reduction of one order of magnitude which is also confirmed by our experiments on real datasets, as reported in Figure 2(a). Moreover, in order to detect all inference channels holding in $\mathcal{F}(\mathcal{D}, \sigma)$, we can limit ourselves to retrieve only the inference channels in $\mathcal{MC}h(k, \mathcal{C}l(\mathcal{D}, \sigma))$, thus taking in input $\mathcal{C}l(\mathcal{D}, \sigma)$ instead of $\mathcal{F}(\mathcal{D}, \sigma)$ and thus performing a much smaller number of checks. Algorithm 2 exploits the anti-monotonicity of frequency (Prop. 1) and the property of maximal inference channels (Prop. 2) to compute $\mathcal{MC}h(k, \mathcal{C}l(\mathcal{D}, \sigma))$ from $\mathcal{C}l(\mathcal{D}, \sigma)$. Thanks to these two properties, Algorithm 2 is much faster, dramatically outperforming the naive inference channel detector (Algorithm 1), and scaling well even for very low support thresholds, as reported in Figure 2(b).

Algorithm 2 Optimized Inference Channel Detector

Input: $\mathcal{C}l(\mathcal{D}, \sigma), k$
Output: $\mathcal{MC}h(k, \mathcal{C}l(\mathcal{D}, \sigma))$
1: $M = \{I \in \mathcal{C}l(\mathcal{D}, \sigma) | I \text{ is maximal}\}$;
2: $\mathcal{MC}h(k, \mathcal{C}l(\mathcal{D}, \sigma)) = \emptyset$;
3: **for all** $J \in M$ **do**
4:    **for all** $I \in \mathcal{C}l(\mathcal{D}, \sigma)$ such that $I \subseteq J$ **do**
5:      compute f_I^J;
6:      **if** $0 < f_I^J < k$ **then**
7:        insert $\langle C_I^J, f_I^J \rangle$ in $\mathcal{MC}h(k, \mathcal{C}l(\mathcal{D}, \sigma))$;

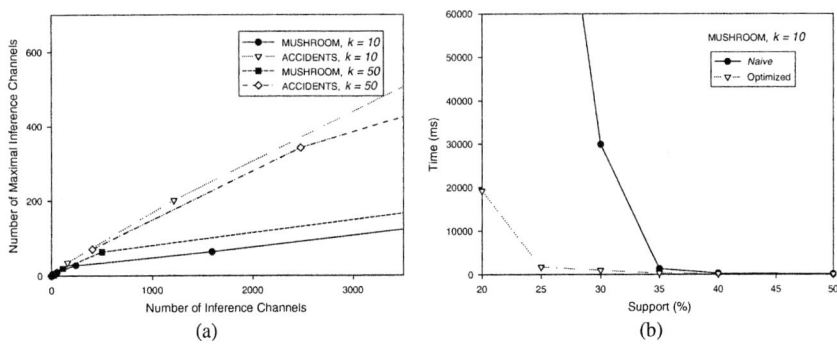

Fig. 2. Benefits of the condensed representation: size of the representations (a), and run time (b)

5 Anonymity vs. Accuracy: Empirical Observations

Algorithm 2 represents an optimized way to identify all threats to anonymity. Its performance revealed adequate in all our empirical evaluations using various datasets from the FIMI repository[2]; in all such cases the time improvement from the Naïve (Algorithm 1)

---

[2] http://fimi.cs.helsinki.fi/data/

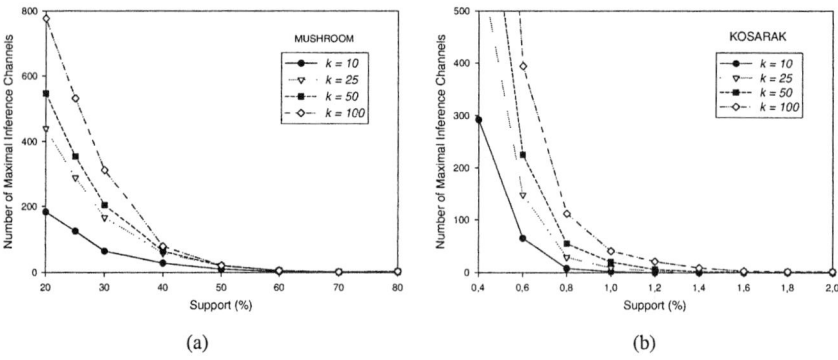

Fig. 3. Experimental results on cardinality of $\mathcal{MC}h(k, \mathcal{C}l(\mathcal{D}, \sigma))$ on two datasets

to the optimized algorithm is about one order of magnitude. This level of efficiency allows an interactive-iterative use of the algorithm by the analyst, aimed at finding the best trade-off among privacy and accuracy of the collection of patterns. To be more precise, there is a conflict among keeping the support threshold as low as possible, in order to mine all interesting patterns, and avoiding the generation of anonymity threats. The best solution to this problem is precisely to find out the minimum support threshold that generates a collection of patterns with no threats. The plots in Figure 3 illustrate this point: on the x-axis we report the minimum support threshold, on the y-axis we report the total number of threats (the cardinality of $\mathcal{MC}h(k, \mathcal{C}l(\mathcal{D}, \sigma))$), and the various curves indicate such number according to different values of the anonymity threshold k. In Figure 3(a) we report the plot for the MUSHROOM dataset (a dense one), while in Figure 3(b) we report the plot for the KOSARAK dataset which is sparse. In both cases, it is evident the value of the minimum support threshold that represents the best trade-off, for any given value of k. However, in certain cases, the best support threshold can still be too high to mine a sufficient quantity of interesting patterns. In such cases, the only option is to allow lower support thresholds and then to block the inference channels in the mining outcome. This problem, as stated before, is not covered in this paper for lack of space, and will be presented in a forthcoming paper.

6 Conclusions

We introduced in this paper the notion of k-anonymous patterns. Such notion serves as a basis for a formal account of the intuition that a collection of patterns, obtained by data mining techniques and made available to the public, should not offer any possibilities to violate the privacy of the individuals whose data are stored in the source database. To the above aim, we formalized the threats to anonymity by means of inference channel through frequent itemsets, and provided practical algorithms to detect such channels.

Other issues, emerging from our approach, are worth a deeper investigation and are left to future research. These include: (i) a thorough comparison of the various different approaches that may be used to block inference channels; (ii) a comprehensive empirical evaluation of our approach: to this purpose we are conducting a large-scale experiment with real life bio-medical data about patients to assess both applicability and scalability of the approach in a realistic, challenging domain; (iii) an investigation whether the proposed notion of privacy-preserving pattern discovery may be generalized to other forms of patterns and models.

In any case, the importance of the advocated form of privacy-preserving pattern discovery is evident: demonstrably trustworthy data mining techniques may open up tremendous opportunities for new knowledge-based applications of public utility and large societal and economic impact.

References

1. D. Agrawal and C. C. Aggarwal. On the design and quantification of privacy preserving data mining algorithms. In *Proceedings of the twentieth ACM PODS*, 2001.
2. R. Agrawal, T. Imielinski, and A. N. Swami. Mining association rules between sets of items in large databases. In *Proceedings of the 1993 ACM SIGMOD*.
3. R. Agrawal and R. Srikant. Privacy-preserving data mining. In *Proceedings of the 2000 ACM SIGMOD on Management of Data*.
4. R. Agrawal and R. Srikant. Fast Algorithms for Mining Association Rules in Large Databases. In *Proceedings of the Twentieth VLDB*, 1994.
5. M. Atzori, F. Bonchi, F. Giannotti, and D. Pedreschi. k-anonymous patterns. Technical Report 2005-TR-17, ISTI - C.N.R., 2005.
6. T. Calders and B. Goethals. Mining all non-derivable frequent itemsets. In *Proceedings of the 6th PKDD*, 2002.
7. P. Fule and J. F. Roddick. Detecting privacy and ethical sensitivity in data mining results. In *Proc. of the 27th conference on Australasian computer science*, 2004.
8. D. Hand, H. Mannila, and P. Smyh. *Principles of Data Mining*. The MIT Press, 2001.
9. M. Kantarcioglu, J. Jin, and C. Clifton. When do data mining results violate privacy? In *Proceedings of the tenth ACM SIGKDD*, 2004.
10. D. Knuth. *Fundamental Algorithms*. Addison-Wesley, Reading, Massachusetts, 1997.
11. S. R. M. Oliveira, O. R. Zaiane, and Y. Saygin. Secure association rule sharing. In *Proc.of the 8th PAKDD*, 2004.
12. N. Pasquier, Y. Bastide, R. Taouil, and L. Lakhal. Discovering frequent closed itemsets for association rules. In *Proc. ICDT '99*, 1999.
13. J. Pei, J. Han, and J. Wang. Closet+: Searching for the best strategies for mining frequent closed itemsets. In *SIGKDD '03*, 2003.
14. L. Sweeney. k-anonymity: a model for protecting privacy. *International Journal on Uncertainty Fuzziness and Knowledge-based Systems*, 10(5), 2002.
15. L. Sweeney. k-anonymity privacy protection using generalization and suppression. *International Journal on Uncertainty Fuzziness and Knowledge-based Systems*, 10(5), 2002.
16. V. S. Verykios, E. Bertino, I. N. Fovino, L. P. Provenza, Y. Saygin, and Y. Theodoridis. State-of-the-art in privacy preserving data mining. *SIGMOD Rec.*, 33(1):50–57, 2004.
17. M. J. Zaki and C.-J. Hsiao. Charm: An efficient algorithm for closed itemsets mining. In *2nd SIAM International Conference on Data Mining*, 2002.

Interestingness is Not a Dichotomy: Introducing Softness in Constrained Pattern Mining

Stefano Bistarelli[1,2] and Francesco Bonchi[3]

[1] Dipartimento di Scienze, Università degli Studi "G. D'Annunzio", Pescara, Italy
[2] Istituto di Informatica e Telematica, CNR, Pisa, Italy
[3] Pisa KDD Laboratory, ISTI - C.N.R., Pisa, Italy
bista@sci.unich.it, francesco.bonchi@isti.cnr.it

Abstract. The paradigm of pattern discovery based on constraints was introduced with the aim of providing to the user a tool to drive the discovery process towards potentially *interesting* patterns, with the positive side effect of achieving a more efficient computation. So far the research on this paradigm has mainly focussed on the latter aspect: the development of efficient algorithms for the evaluation of constraint-based mining queries. Due to the lack of research on methodological issues, the constraint-based pattern mining framework still suffers from many problems which limit its practical relevance. As a solution, in this paper we introduce the new paradigm of pattern discovery based on *Soft Constraints*. Albeit simple, the proposed paradigm overcomes all the major methodological drawbacks of the classical constraint-based paradigm, representing an important step further towards practical pattern discovery.

1 Background and Motivations

During the last decade a lot of researchers have focussed their (mainly algorithmic) investigations on the computational problem of *Frequent Pattern Discovery*, i.e. mining patterns which satisfy a user-defined constraint of minimum frequency [1].

The simplest form of a frequent pattern is the frequent itemset.

Definition 1 (Frequent Itemset Mining). *Let $\mathcal{I} = \{x_1, ..., x_n\}$ be a set of distinct items, where an item is an object with some predefined attributes (e.g., price, type, etc.). An itemset X is a non-empty subset of $\mathcal{I}$. A transaction database $\mathcal{D}$ is a bag of itemsets $t \in 2^{\mathcal{I}}$, usually called transactions. The support of an itemset X in database $\mathcal{D}$, denoted $supp_{\mathcal{D}}(X)$, is the number of transactions which are superset of X. Given a user-defined minimum support σ, an itemset X is called frequent in $\mathcal{D}$ if $supp_{\mathcal{D}}(X) \geq \sigma$. This defines the minimum frequency constraint: $\mathcal{C}_{freq[\mathcal{D},\sigma]}(X) \Leftrightarrow supp_{\mathcal{D}}(X) \geq \sigma$.*

Recently the research community has turned its attention to more complex kinds of frequent patterns extracted from more structured data: *sequences, trees*, and *graphs*. All these different kinds of pattern have different peculiarities and application fields, but they all share the same computational aspects: a usually very large input, an exponential search space, and a too large solution set. This situation – too many data yielding too many patterns – is harmful for two reasons. First, performance degrades: mining generally becomes inefficient or, often, simply unfeasible. Second, the identification of the fragments of interesting knowledge, blurred within a huge quantity of mostly useless patterns, is difficult. The paradigm of *constraint-based pattern mining* was introduced as a solution to both these problems. In such paradigm, it is the user which specifies to the system what is *interesting* for the current application: constraints are a tool to drive the mining process towards potentially interesting patterns, moreover they can be pushed deep inside the mining algorithm in order to fight the exponential search space curse, and to achieve better performance [15,20,25].

When instantiated to the pattern class of itemsets, the constraint-based pattern mining problem is defined as follows.

Definition 2 (Constrained Frequent Itemset Mining). *A constraint on itemsets is a function* $\mathcal{C} : 2^{\mathcal{I}} \to \{true, false\}$. *We say that an itemset I satisfies a constraint if and only if* $\mathcal{C}(I) = true$. *We define the theory of a constraint as the set of itemsets which satisfy the constraint:* $Th(\mathcal{C}) = \{X \in 2^{\mathcal{I}} \mid \mathcal{C}(X)\}$. *Thus with this notation, the frequent itemsets mining problem requires to compute the set of all frequent itemsets* $Th(\mathcal{C}_{freq[\mathcal{D},\sigma]})$. *In general, given a conjunction of constraints $\mathcal{C}$ the constrained frequent itemsets mining problem requires to compute* $Th(\mathcal{C}_{freq}) \cap Th(\mathcal{C})$.

Example 1. The following is an example mining query:

$$\mathcal{Q} : supp_{\mathcal{D}}(X) \geq 1500 \;\wedge\; avg(X.weight) \leq 5 \;\wedge\; sum(X.price) \geq 20$$

It requires to mine, from database $\mathcal{D}$, all patterns which are frequent (have a support larger than 1500), have average weight less than 5 and a sum of prices greater than 20.

So far constraint-based frequent pattern mining has been seen as a query optimization problem, i.e., developing efficient, sound and complete evaluation strategies for constraint-based mining queries. Or in other terms, designing efficient algorithms to mine all and only the patterns in $Th(\mathcal{C}_{freq}) \cap Th(\mathcal{C})$. To this aim, properties of constraints have been studied comprehensively, and on the basis of such properties (e.g., anti-monotonicity, succinctness [20,18], monotonicity [11,17,6], convertibility [22], loose anti-monotonicity [9]), efficient computational strategies have been defined. Despite such effort, the constraint-based pattern mining framework still suffers from many problems which limit its practical relevance.

First of all, consider the example mining query $\mathcal{Q}$ given above: *where do the three thresholds (i.e., 1500, 5 and 20) come from?* In some cases they can be precisely imposed by the application, but this is rarely the case. In most of the cases, they come from an exploratory mining process, where they are iteratively adjusted until a solution set of reasonable size is produced. This practical way of proceeding is in contrast with the basic philosophy of the constraint-based paradigm: constraints should represent what is a priori interesting, given the application background knowledge, rather than be adjusted accordingly to a preconceived output size. Another major drawback of the constraint-based pattern mining paradigm is its rigidity. Consider, for instance, the following three patterns (we use the notation $\langle v_1, v_2, v_3 \rangle$ to denote the three values corresponding to the three constraints in the conjunction in the example query $\mathcal{Q}$): $p_1 : \langle 1700, 0.8, 19 \rangle$, $p_2 : \langle 1550, 4.8, 54 \rangle$, and $p_3 : \langle 1550, 2.2, 26 \rangle$. The first pattern, p_1, largely satisfies two out of the three given constraints, while slightly violates the third one. According to the classical constraint-based pattern mining paradigm p_1 would be discarded as non interesting. Is such a pattern really *less interesting* than p_2 and p_3 which satisfy all the three constraints, but which are much less frequent than p_1? Moreover, is it reasonable, in real-world applications, that all constraints are equally important?

All these problems flow out from the same source: the fact that in the classical constraint-based mining framework, a constraint is a function which returns a boolean value $\mathcal{C} : 2^{\mathcal{I}} \rightarrow \{true, false\}$. Indeed, *interestingness is not a dichotomy*.

This consideration suggests us a simple solution to overcome all the main drawbacks of constraint-based paradigm.

Paper Contributions and Organization

In this paper, as a mean to handle interestingness [26,16,24], we introduce the *soft constraint based pattern mining* paradigm, where constraints are no longer rigid boolean functions, but are "soft" functions, i.e., functions with value in a set A, which represents the set of interest levels or costs assigned to each pattern.

- The proposed paradigm is not rigid: a potentially interesting pattern is not discarded for just a slight violation of a constraint.
- Our paradigm creates an order of patterns w.r.t. interestingness (level of constraints satisfaction): this allows to say that a pattern is *more interesting* than another, instead of strictly dividing patterns in interesting and not interesting.
- From the previous point it follows that our paradigm allows to express *top-k* queries based on constraints: the data analyst can ask for the top-10 patterns w.r.t. a given description (a conjunction of soft constraints).
- Alternatively, we can ask to the system to return all and only the patterns which exhibit an interest level larger than a given threshold λ.
- The proposed paradigm allows to assign different weights to different constraints, while in the classical constraint-based pattern discovery paradigm all constraints were equally important.
- Last but not least, our idea is very simple and thus very general: it can be instantiated to different classes of patterns such as itemsets, sequences, trees or graphs.

For the reasons listed above, we believe that the proposed paradigm represents an important step further towards practical pattern discovery.

A nice feature of our proposal is that, by adopting the soft constraint based paradigm, we do not reject all research results obtained in the classical constraint-based paradigm; on the contrary, we fully exploit such algorithmic results. In other terms, our proposal is merely methodological, and it exploits previous research results that were mainly computational.

The paper is organized as follows. In the next Section we briefly review the theory of soft constraints and we define the soft constraint based pattern mining paradigm. In Section 3 we discuss possible alternative instances of the paradigm. In Section 4 we formally define the Soft Constraint Based Pattern Discovery paradigm. We then focus on one of the many possible instances of the proposed paradigm, and we implement it in a concrete Pattern Discovery System. Such a system is built as a wrapper around a classical constraint pattern mining system.

2 Introducing Soft Constraints

Constraint Solving is an emerging software technology for declarative description and effective solving of large problems. Many real life systems, ranging from network management [14] to complex scheduling [2], are analyzed and solved using constraint related technologies. The constraint programming process consists of the generation of requirements (constraints) and solution of these requirements, by specialized constraint solvers. When the requirements of a problem are expressed as a collection of boolean predicates over variables, we obtain what is called the *crisp* (or classical) Constraint Satisfaction Problem (CSP). In this case the problem is solved by finding any assignment of the variables that satisfies all the constraints.

Sometimes, when a deeper analysis of a problem is required, *soft constraints* are used instead. Several formalizations of the concept of soft constraints are currently available. In the following, we refer to the formalization based on *c-semirings* [5]: a semiring-based constraint assigns to each instantiation of its variables an associated value from a partially ordered set. When dealing with crisp constraints, the values are the boolean *true* and *false* representing the admissible and/or non-admissible values; when dealing with soft constraints the values are interpreted as preferences/costs. The framework must also handle the combination of constraints. To do this one must take into account such additional values, and thus the formalism must provide suitable operations for combination ($\times$) and comparison ($+$) of tuples of values and constraints. This is why this formalization is based on the mathematical concept of semiring.

Definition 3 (c-semirings [5,3]). *A semiring is a tuple $\langle A, +, \times, 0, 1 \rangle$ such that: A is a set and $0, 1 \in A$; $+$ is commutative, associative and 0 is its unit element; $\times$ is associative, distributes over $+$, 1 is its unit element and 0 is its absorbing element. A c-semiring ("c" stands for "constraint-based") is a semiring $\langle A, +, \times, 0, 1 \rangle$ such that $+$ is idempotent with 1 as its absorbing element and $\times$ is commutative.*

Definition 4 (soft constraints [5,3]). *Given a c-semiring $S = \langle A, +, \times, 0, 1 \rangle$ and an ordered set of variables V over a finite domain D, a constraint is a function which, given an assignment $\eta : V \to D$ of the variables, returns a value of the c-semiring. By using this notation we define $C = \eta \to A$ as the set of all possible constraints that can be built starting from S, D and V.*

In the following we will always use the word semiring as standing for c-semiring, and we will explain this very general concept by the point of view of pattern discovery.

Example 2. Consider again the mining query $\mathcal{Q}$. In this context we have that the ordered set of variables V is $\langle supp_\mathcal{D}(X), avg(X.weight), sum(X.price) \rangle$, while the domain D is: $D(supp_\mathcal{D}(X)) = \mathbb{N}$, $D(avg(X.weight)) = \mathbb{R}^+$, and $D(sum(X.price)) = \mathbb{N}$. If we consider the classical *crisp* framework (i.e., hard constraints) we have the semiring $S_{Bool} = \langle \{true, false\}, \vee, \wedge, false, true \rangle$. A constraint C is a function $V \to D \to A$; for instance, $supp_\mathcal{D}(X) \to 1700 \to true$.

The $+$ operator is what we use to compare tuples of values (or patterns, in our context). Let us consider the relation $\leq_S$ (where S stands for the specified semiring) over

A such that $a \leq_S b$ iff $a + b = b$. It is possible to prove that: $\leq_S$ is a partial order; $+$ and $\times$ are monotone on $\leq_S$; $\mathbf{0}$ is its minimum and $\mathbf{1}$ its maximum, and $\langle A, \leq_S \rangle$ is a complete lattice with least upper bound operator $+$. In the context of pattern discovery $a \leq_S b$ means that the pattern b is *more interesting* than a, where interestingness is defined by a combination of soft constraints. When using (soft) constraints it is necessary to specify, via suitable combination operators, how the level of interest of a combination of constraints is obtained from the interest level of each constraint. The combined weight (or interest) of a combination of constraints is computed by using the operator $\otimes : \mathcal{C} \times \mathcal{C} \to \mathcal{C}$ defined as $(C_1 \otimes C_2)\eta = C_1\eta \times_S C_2\eta$.

Example 3. If we adopt the classical crisp framework, in the mining query $\mathcal{Q}$ of Example 1 we have to combine the three constraints using the $\wedge$ operator (which is the $\times$ in the boolean semiring S_{Bool}). Consider for instance the pattern $p_1 : \langle 1700, 0.8, 19 \rangle$ for the ordered set of variables $V = \langle supp_{\mathcal{D}}(X), avg(X.weight), sum(X.price) \rangle$. The first and the second constraint are satisfied leading to the semiring level *true*, while the third one is not satisfied and has associated level *false*. Combining the three values with $\wedge$ we obtain *true* $\wedge$ *true* $\wedge$ *false* = *false* and we can conclude that the pattern $\langle 1700, 0.8, 19 \rangle$ is not interesting w.r.t. our purposes. Similarly, we can instead compute level *true* for pattern $p_3 : \langle 1550, 2.2, 26 \rangle$ corresponding to an interest w.r.t. our goals. Notice that using crisp constraints, the order between values only says that we are interested to patterns with semiring level *true* and not interested to patterns with semiring level *false* (that is semiring level *false* $\leq_{S_{Bool}}$ *true*).

3 Instances of the Semiring

Dividing patterns in *interesting* and *non-interesting* is sometimes not meaningful nor useful. Most of the times we can say that each pattern is interesting with a specific level of preference. Soft constraints can deal with preferences by moving from the two values semiring S_{Bool} to other semirings able to give a finer distinction among patters (see [3] for a comprehensive guide to the semiring framework). For our scope the fuzzy and the weighted semiring are the most suitable.

Example 4 (fuzzy semiring). When using fuzzy semiring [12,23], to each pair constraint-pattern is assigned an interest level between 0 and 1, where 1 represents the best value (maximum interest) and 0 the worst one (minimum interest). Therefore the $+$ in this semiring is given by the max operator, and the order $\leq_S$ is given by the usual $\leq$ on real numbers. The value associated to a pattern is obtained by combining the constraints using the minimum operator among the semiring values. Therefore the $\times$ in this semiring is given by the min operator. Recapitulating, the fuzzy semiring is given by $S_F = \langle [0,1], max, min, 0, 1 \rangle$. The reason for such a max-min framework relies on the attempt to maximize the value of the least preferred tuple. Fuzzy soft constraints are able to model partial constraint satisfaction [13], so to get a solution even when the problem is overconstrained, and also prioritized constraints, that is, constraints with different levels of importance [10]. Figure 1 reports graphical representations of possible fuzzy instances of the constraints in $\mathcal{Q}$. Consider, for instance, the graphical representation of the frequency constraint in Figure 1(C_1). The dotted line describes the behavior

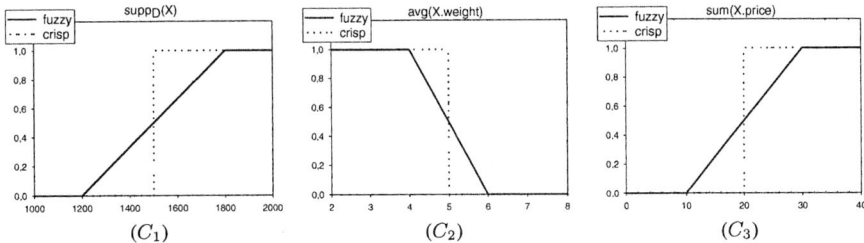

Fig. 1. Graphical representation of possible fuzzy instances of the constraints in $\mathcal{Q}$

of the *crisp* version (where $1 = true$ and $0 = false$) of the frequency constraint, while the solid line describes a possible fuzzy instance of the same constraint. In this instance domain values smaller than 1200 yield 0 (uninteresting patterns); from 1200 to 1800 the interest level grows linearly reaching the maximum value of 1. Similarly the other two constraints in Figure 1(C_2) and (C_3). In this situation for the pattern $p_1 = \langle 1700, 0.8, 19 \rangle$ we obtain that: $C_1(p_1) = 1$, $C_2(p_1) = 1$ and $C_3(p_1) = 0.45$. Since in the fuzzy semiring the combination operator $\times$ is min, we got that the interest level of p_1 is 0.45. Similarly for p_2 and p_3:

- $p_1 : C_1 \otimes C_2 \otimes C_3(1700, 0.8, 19) = min(1, 1, 0.45) = 0.45$
- $p_2 : C_1 \otimes C_2 \otimes C_3(1550, 4.8, 54) = min(1, 0.6, 1) = 0.6$
- $p_3 : C_1 \otimes C_2 \otimes C_3(1550, 2.2, 26) = min(1, 1, 0.8) = 0.8$

Therefore, with this particular instance we got that $p_1 \leq_{S_F} p_2 \leq_{S_F} p_3$, i.e., p_3 is the most interesting pattern among the three.

Example 5 (weighted semiring). While fuzzy semiring associate a level of preference with each tuple in each constraint, in the weighted semiring tuples come with an associated cost. This allows one to model optimization problems where the goal is to minimize the total cost (time, space, number of resources, ...) of the proposed solution. Therefore, in the weighted semiring the cost function is defined by summing up the costs of all constraints. According to the informal description given above, the weighted semiring is $S_W = \langle \mathbb{R}^+, min, sum, +\infty, 0 \rangle$. Consider, for instance, the graphical representation of the constraints in the query $\mathcal{Q}$ in Figure 2. In this situation we got that:

- $p_1 : C_1 \otimes C_2 \otimes C_3(1700, 0.8, 19) = sum(50, 20, 205) = 275$
- $p_2 : C_1 \otimes C_2 \otimes C_3(1550, 4.8, 54) = sum(200, 120, 30) = 350$
- $p_3 : C_1 \otimes C_2 \otimes C_3(1550, 2.2, 26) = sum(200, 55, 190) = 445$

Therefore, with this particular instance we got that $p_3 \leq_{S_W} p_2 \leq_{S_W} p_1$ (remember that the order $\leq_{S_W}$ correspond to the $\geq$ on real numbers). In other terms, p_1 is the most interesting pattern w.r.t. this constraints instance.

The weighted and the fuzzy paradigm, can be seen as two different approaches to give a meaning to the notion of optimization. The two models correspond in fact to two definitions of social welfare in utility theory [19]: *"egalitarianism"*, which maximizes the minimal individual utility, and *"utilitarianism"*, which maximizes the sum of the

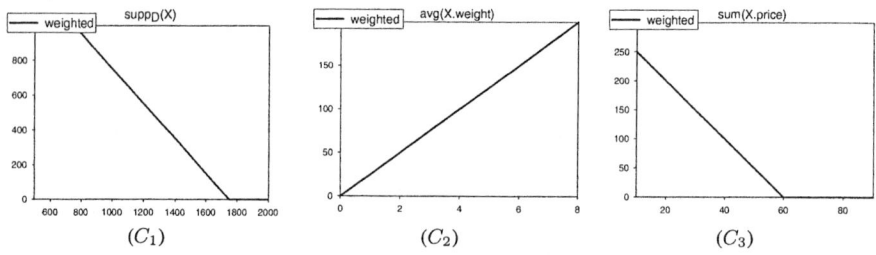

Fig. 2. Graphical representation of possible weighted instances of the constraints in $\mathcal{Q}$

individual utilities. The fuzzy paradigm has an egalitarianistic approach, aimed at maximizing the overall level of interest while balancing the levels of all constraints; while the weighted paradigm has an utilitarianistic approach, aimed at getting the minimum cost globally, even though some constraints may be neglected presenting a big cost. We believe that both approaches present advantages and drawbacks, and may preferred to the other one depending on the application domain. Beyond the fuzzy and the weighted, many other possible instances of the semiring exist, and could be useful in particular applications. Moreover, it is worth noting that the cartesian product of semirings is a semiring [5] and thus it is possible to use the framework also to deal with multicriteria pattern selection.

Finally, note that the soft constraint framework is very general, and could be instantiated not only to unary constraints (as we do in this paper) but also to binary and k-ary constraints (dealing with two or more variables). This could be useful to extend the soft constraint based paradigm to association rules with "2-var" constraints [18].

4 Soft Constraint Based Pattern Mining

In this Section we instantiate soft constraint theory to the pattern discovery framework.

Definition 5 (Soft Constraint Based Pattern Mining). *Let $\mathcal{P}$ denote the domain of possible patterns. A soft constraint on patterns is a function $\mathcal{C} : \mathcal{P} \rightarrow A$ where A is the carrier set of a semiring $S = \langle A, +, \times, \mathbf{0}, \mathbf{1} \rangle$. Given a combination of soft constraints $\otimes \mathcal{C}$, we define two different problems:*

λ**-interesting:** *given a minimum interest threshold $\lambda \in A$, it is required to mine the set of all λ-interesting patterns, i.e., $\{p \in \mathcal{P} | \otimes \mathcal{C}(p) \geq \lambda\}$.*
top-k: *given a threshold $k \in \mathbb{N}$, it is required to mine the top-k patterns $p \in \mathcal{P}$ w.r.t. the order $\leq_S$.*

Note that the Soft Constraint Based Pattern Mining paradigm just defined, has many degrees of freedom. In particular, it can be instantiated: (i) on the domain of patterns $\mathcal{P}$ in analysis (e.g., itemsets, sequences, trees or graphs), (ii) on the semiring $S = \langle A, +, \times, \mathbf{0}, \mathbf{1} \rangle$ (e.g., fuzzy, weighted or probabilistic), and (iii) on one of the two possible mining problems, i.e., λ-interesting or top-k mining.

In the rest of this paper we will focus on concretizing a simple instance of this very general paradigm: λ-interesting$_{fuzzy}$ on the pattern class of itemsets.

4.1 Mining λ-Interesting Itemsets on the Fuzzy Semiring

Definition 6. *Let $\mathcal{I} = \{x_1, ..., x_n\}$ be a set of items, where an item is an object with some predefined attributes (e.g., price, type, etc.). A soft constraint on itemsets, based on the fuzzy semiring, is a function $\mathcal{C} : 2^{\mathcal{I}} \to [0,1]$. Given a combination of such soft constraints $\otimes \mathcal{C} \equiv \mathcal{C}_1 \otimes \ldots \otimes \mathcal{C}_n$, we define the interest level of an itemset $X \in 2^{\mathcal{I}}$ as $\otimes \mathcal{C}(X) = min(\mathcal{C}_1(X), \ldots, \mathcal{C}_n(X))$. Given a minimum interest threshold $\lambda \in\,]0,1]$, the λ-interesting itemsets mining problem, requires to compute $\{X \in 2^{\mathcal{I}} | \otimes \mathcal{C}(X) \geq \lambda\}$.*

In the following we describe how to build a concrete *pattern discovery system* for λ-interesting $_{fuzzy}$ itemsets mining, as a wrapper around a classical constraint pattern mining system. The basic components which we use to build our system are the following:

A *crisp* constraints solver - i.e., a system for mining constrained frequent itemsets, where constraints are classical binary functions, and not soft constraints. Or in other terms, a system for solving the problem in Definition 2. To this purpose we adopt the system which we have developed at Pisa KDD Laboratory within the P^3D project[1]. Such a system is a general Apriori-like algorithm which, by means of *data reduction* and *search space pruning*, is able to push a wide variety of constraints (practically all possible kinds of constraints which have been studied and characterized so far [9]) into the frequent itemsets computation. Based on the algorithmic results developed in the last years by our lab (e.g., [6,7,8,9,21]), our system is very efficient and robust, and to our knowledge, is the unique existing implementation of this kind.

A language of constraints - to express, by means of queries containing conjunctions of constraints, what is interesting for the given application. The wide repertoire of constraints that we admit, comprehends the frequency constraint ($supp_{\mathcal{D}}(X) \geq \sigma$), and all constraints defined over the following aggregates[2]: $min, max, count, sum, range, avg, var, median, std, md$.

A methodology to define the interest level - that must be assigned to each pair itemset-constraint. In other terms, we need to provide the analyst with a simple methodology to define how to assign for each constraint and each itemset a value in the interval $[0,1]$, as done, for instance, by the graphical representations of constraints in Figure 1. This methodology should provide the analyst with a knob to adjust the *softness level* of each constraint in the conjunction, and a knob to set the *importance* of each constraint in the conjunction.

Let us focus on the last point. Essentially we must describe how the user can define the fuzzy behavior of a soft constraint. We restrict our system to constraints which behave as those ones in Figure 1: they return a value which grows linearly from 0 to 1 in a certain interval, while they are null before the interval and equal to 1 after the interval. To describe such a simple behavior we just need two parameters: a value associated to the center of the interval (corresponding to the 0.5 fuzzy semiring value), and a parameter to adjust the width of the interval (and consequently the gradient of the function).

---
[1] http://www-kdd.isti.cnr.it/p3d/index.html
[2] *Range* is $(max - min)$, var is for variance, std is for standard deviation, md is for mean deviation.

Definition 7. *A soft constraint $\mathcal{C}$ on itemsets, based on the fuzzy semiring, is defined by a quintuple $\langle Agg, Att, \theta, t, \alpha \rangle$, where:*

- $Agg \in \{supp, min, max, count, sum, range, avg, var, median, std, md\}$;
- Att is the name of the attribute on which the aggregate agg is computed (or the transaction database, in the case of the frequency constraint);
- $\theta \in \{\leq, \geq\}$;
- $t \in \mathbb{R}$ corresponds to the center of the interval and it is associated to the semiring value 0.5;
- $\alpha \in \mathbb{R}^+$ is the softness parameter, which defines the inclination of the preference function (and thus the width of the interval).

In particular, if $\theta = \leq$ (as in Figure 1(C_2)) then $\mathcal{C}(X)$ is 1 for $X \leq (t - \alpha t)$, is 0 for $X \geq (t + \alpha t)$, and is linearly decreasing from 1 to 0 within the interval $[t - \alpha t, t + \alpha t]$. The other way around if $\theta = \geq$ (as, for instance, in Figure 1(C_3)). Note that if the softness parameter α is 0, then we obtain the crisp (or hard) version of the constraint.

Example 6. Consider again the query $\mathcal{Q}$ given in Example 1, and its fuzzy instance graphically described by Figure 1. Such query can be expressed in our constraint language as: $\langle supp, \mathcal{D}, \geq, 1500, 0.2\rangle, \langle avg, weight, \leq, 5, 0.2\rangle, \langle sum, price, \geq, 20, 0.5\rangle$.

Since the combination operator $\times$ in min, increasing the importance of a constraint w.r.t. the others in the combination means to force the constraint to return lower values for not really satisfactory patterns. By decreasing the softness parameter α, we increase the gradient of the function making the shape of the soft constraint closer to a crisp constraint. This translates in a better value for patterns X which were already behaving well w.r.t. such constraint($\mathcal{C}(X) > 0.5$), and in a lower value for patterns which were behaving not so well ($\mathcal{C}(X) < 0.5$). Decreasing the gradient (increasing α) instead means to lower the importance of the constraint itself: satisfying or not satisfying the constraint does not result in a big fuzzy value difference. Additionally, by operating on t, we can increase the "severity" of the constraint w.r.t. those patterns which were behaving not so well. Therefore, the knob to increase or decrease the importance of a constraint is not explicitly given, because its role, in the fuzzy semiring, can be played by a combined action on the two knobs α and t.

Example 7. Consider again the query $\mathcal{Q}$ given in Example 1, and its fuzzy instance: $\langle supp, \mathcal{D}, \geq, 1500, 0.2\rangle, \langle avg, weight, \leq, 5, 0.2\rangle, \langle sum, price, \geq, 20, 0.5\rangle$. As we stated in Example 4, it holds that $p_2 \leq_{S_F} p_3$. In particular, p_2 is better than p_3 w.r.t. constraint C_3, while p_3 is better than p_2 w.r.t. constraint C_2. Suppose now that we increase the importance of C_3, e.g., $\langle sum, price, \geq, 28, 0.25\rangle$. We obtain that $p_3 \leq_{S_F} p_2$:

- $p_2 : C_1 \otimes C_2 \otimes C_3(1550, 4.8, 54) = min(1, 0.6, 1) = 0.6$
- $p_3 : C_1 \otimes C_2 \otimes C_3(1550, 2.2, 26) = min(1, 1, 0.35) = 0.35$

In [5,4] it has been proved that, when dealing with the fuzzy framework, computing all the solution better than a threshold λ can be performed by solving a crisp problem where all the constraint instances with semiring level lower than λ have been assigned level *false*, and all the instances with semiring level greater or equal to λ have been assigned level *true*. Using this theoretical result, and some simple arithmetic we can transform each soft constraint in a corresponding crisp constraint.

Definition 8. *Given a fuzzy soft constraint* $\mathcal{C} \equiv \langle Agg, Att, \theta, t, \alpha \rangle$, *and a minimum interest threshold* λ, *we define the crisp translation of* $\mathcal{C}$ *w.r.t.* λ *as:*

$$\mathcal{C}^{\lambda}_{crisp} \equiv \begin{cases} Agg(Att) \geq t - \alpha t + 2\lambda\alpha t, & \text{if } \theta = \geq \\ Agg(Att) \leq t + \alpha t - 2\lambda\alpha t, & \text{if } \theta = \leq \end{cases}$$

Example 8. The crisp translation of the soft constraint $\langle sum, price, \geq, 20, 0.5\rangle$ is $sum(X.price) \geq 26$ for $\lambda = 0.8$, while it is $sum(X.price) \geq 18$ for $\lambda = 0.4$.

Proposition 1. *Given the vocabulary of items* $\mathcal{I}$, *a combination of soft constraints* $\otimes \mathcal{C} \equiv \mathcal{C}1 \otimes \ldots \otimes \mathcal{C}n$, *and a minimum interest threshold* λ. *Let* $\mathcal{C}'$ *be the conjunction of crisp constraints obtained by conjoining the crisp translation of each constraint in* $\otimes \mathcal{C}$ *w.r.t.* λ: $\mathcal{C}' \equiv \mathcal{C}1^{\lambda}_{crisp} \wedge \ldots \wedge \mathcal{C}n^{\lambda}_{crisp}$. *It holds that:* $\{X \in 2^{\mathcal{I}} | \otimes \mathcal{C}(X) \geq \lambda\} = Th(\mathcal{C}')$.

Proof (sketch). The soundness of the mapping come from the result in [5]. We here have to only give a justification of the formula in Definition 8. This is done by means of Figure 3(b), that shows a graphical representation of the simple arithmetic problem and its solutions.

Therefore, if we adopt the fuzzy semiring, we can fully exploit a classical constraint-based pattern discovery system (and all algorithmic results behind it), by means of a simple translation from soft to crisp constraints. This is exactly what we have done, obtaining a pattern discovery system based on soft constraints built as a wrapper around a classical constraint-based mining system.

4.2 Experimental Analysis

We have conducted some experiments in order to asses the concrete effects obtained by manipulating the α, t and λ parameters. To this purpose we have compared 5 different instances (described in Figure 3(a)) of the query $\mathcal{Q}$:

$$\langle supp, \mathcal{D}, \geq, t, \alpha\rangle \langle avg, weight, \leq, t, \alpha\rangle, \langle sum, price, \geq, t, \alpha\rangle$$

where the transactional dataset $\mathcal{D}$, is the well known RETAIL dataset, donated by Tom Brijs and contains the (anonymized) retail market basket data from an anonymous Belgian retail store[3]; and the two attributes $weight$ and $price$ have been randomly generated with a gaussian distribution within the range $[0, 150000]$.

Figure 3(c) reports the number of solutions for the given five queries at different λ thresholds. Obviously as λ increases the number of solutions shrinks accordingly. This behavior is also reflected in queries evaluation times, reported in Figure 3(d): the bigger is the size of the solution set, the longer is the associated computation.

Comparing queries $\mathcal{Q}_1$, $\mathcal{Q}_2$ and $\mathcal{Q}_3$, we can gain more insight about the α parameter. In fact, the three queries differ only by the α associated with one constraint (the frequency constraint). We can observe that, if the λ threshold is not too much selective, increasing the α parameter (i.e., the size of the soft interval), the number of solutions grows. Notice however that, when λ becomes selective enough (i.e., $\lambda > 0.5$), increasing the softness parameter we obtain an opposite behavior. This is due to the fact that, if on one hand a more soft constraint is less severe with patterns not good enough, on the other hand it is less generous with good patterns, which risk to be discarded by an high λ threshold.

---
[3] http://fimi.cs.helsinki.fi/data/

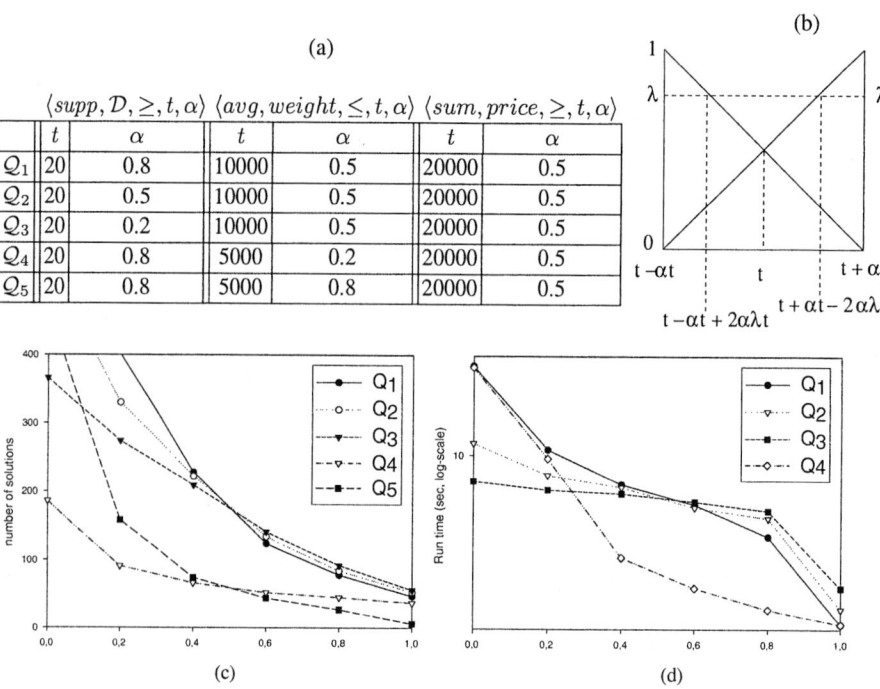

Fig. 3. (a) description of queries experimented, (b) graphical proof to Proposition 1, (c) and (d) experimental results with λ ranging in $[0,1]$

References

1. R. Agrawal and R. Srikant. Fast Algorithms for Mining Association Rules in Large Databases. In *Proceedings of the Twentieth International Conference on Very Large Databases (VLDB'94)*, 1994.
2. J. Bellone, A. Chamard, and C. Pradelles. Plane - an evolutive planning system for aircraft production. In *Proc. 1st Interantional Conference on Practical Applications of Prolog (PAP92)*, 1992.
3. S. Bistarelli. *Semirings for Soft Constraint Solving and Programming*, volume 2962 of *Lecture Notes in Computer Science*. Springer, 2004.
4. S. Bistarelli, P. Codognet, and F. Rossi. Abstracting soft constraints: Framework, properties, examples. *Artificial Intelligence*, (139):175–211, July 2002.
5. S. Bistarelli, U. Montanari, and F. Rossi. Semiring-based Constraint Solving and Optimization. *Journal of the ACM*, 44(2):201–236, Mar 1997.
6. F. Bonchi, F. Giannotti, A. Mazzanti, and D. Pedreschi. ExAMiner: Optimized level-wise frequent pattern mining with monotone constraints. In *Proceedings of the Third IEEE International Conference on Data Mining (ICDM'03)*, 2003.
7. F. Bonchi, F. Giannotti, A. Mazzanti, and D. Pedreschi. ExAnte: Anticipated data reduction in constrained pattern mining. In *Proceedings of the 7th European Conference on Principles and Practice of Knowledge Discovery in Databases (PKDD'03)*, 2003.
8. F. Bonchi and C. Lucchese. On closed constrained frequent pattern mining. In *Proceedings of the Fourth IEEE International Conference on Data Mining (ICDM'04)*, 2004.

9. F. Bonchi and C. Lucchese. Pushing tougher constraints in frequent pattern mining. In *Proceedings of the Ninth Pacific-Asia Conference on Knowledge Discovery and Data Mining (PAKDD'05)*, Hanoi, Vietnam, 2005.
10. A. Borning, M. Maher, A. Martindale, and M. Wilson. Constraint hierarchies and logic programming. In *Proc. 6th International Conference on Logic Programming*, 1989.
11. C. Bucila, J. Gehrke, D. Kifer, and W. White. DualMiner: A dual-pruning algorithm for itemsets with constraints. In *Proceedings of the 8th ACM International Conference on Knowledge Discovery and Data Mining (SIGKDD'02)*, 2002.
12. D. Dubois, H. Fargier, and H. Prade. The calculus of fuzzy restrictions as a basis for flexible constraint satisfaction. In *Proc. IEEE International Conference on Fuzzy Systems*, pages 1131–1136. IEEE, 1993.
13. E. Freuder and R. Wallace. Partial constraint satisfaction. *AI Journal*, 58, 1992.
14. T. Frühwirth and P. Brisset. Optimal planning of digital cordless telecommunication systems. In *Proc. PACT97*, London, UH, 1997.
15. J. Han, L. V. S. Lakshmanan, and R. T. Ng. Constraint-based, multidimensional data mining. *Computer*, 32(8):46–50, 1999.
16. R. Hilderman and H. Hamilton. *Knowledge Discovery and Measures of Interest*. Kluwer Academic, Boston, 2002.
17. S. Kramer, L. D. Raedt, and C. Helma. Molecular feature mining in hiv data. In *Proceedings of the 7th ACM International Conference on Knowledge Discovery and Data Mining, (SIGKDD'01)*, 2001.
18. L. V. S. Lakshmanan, R. T. Ng, J. Han, and A. Pang. Optimization of constrained frequent set queries with 2-variable constraints. In *Proceedings of the ACM International Conference on Management of Data (SIGMOD'99)*, 1999.
19. H. Moulin. *Axioms for Cooperative Decision Making*. Cambridge University Press, 1988.
20. R. T. Ng, L. V. S. Lakshmanan, J. Han, and A. Pang. Exploratory mining and pruning optimizations of constrained associations rules. In *Proceedings of the ACM International Conference on Management of Data (SIGMOD'98)*, 1998.
21. S. Orlando, P. Palmerini, R. Perego, and F. Silvestri. Adaptive and Resource-Aware Mining of Frequent Sets. In *Proc. of the 2002 IEEE Int. Conference on Data Mining (ICDM'02)*, pages 338–345, Maebashi City, Japan, Dec. 2002.
22. J. Pei and J. Han. Can we push more constraints into frequent pattern mining? In *Proceedings of the 6th ACM International Conference on Knowledge Discovery and Data Mining (SIGKDD'00)*, 2000.
23. Z. Ruttkay. Fuzzy constraint satisfaction. In *Proc. 3rd IEEE International Conference on Fuzzy Systems*, pages 1263–1268, 1994.
24. S. Sahar. Interestingness via what is not interesting. In *Proc. of the Fifth ACM SIGKDD International Conference on Knowledge Discovery and Data Mining (SIGKDD'99)*.
25. R. Srikant, Q. Vu, and R. Agrawal. Mining association rules with item constraints. In *Proceedings of the 3rd ACM International Conference on Knowledge Discovery and Data Mining, (SIGKDD'97)*, 1997.
26. P.-N. Tan, V. Kumar, and J. Srivastava. Selecting the right interestingness measure for association patterns. In *Proc. of the Eighth ACM SIGKDD International Conference on Knowledge Discovery and Data Mining (SIGKDD'2002)*.

Generating Dynamic Higher-Order Markov Models in Web Usage Mining

José Borges[1] and Mark Levene[2]

[1] School of Engineering, University of Porto,
R. Dr. Roberto Frias, 4200 - Porto, Portugal
jlborges@fe.up.pt
[2] School of Computer Science and Information Systems, Birkbeck,
University of London, Malet Street, London WC1E 7HX, UK
mlevene@dcs.bbk.ac.uk

Abstract. Markov models have been widely used for modelling users' web navigation behaviour. In previous work we have presented a dynamic clustering-based Markov model that accurately represents second-order transition probabilities given by a collection of navigation sessions. Herein, we propose a generalisation of the method that takes into account higher-order conditional probabilities. The method makes use of the state cloning concept together with a clustering technique to separate the navigation paths that reveal differences in the conditional probabilities. We report on experiments conducted with three real world data sets. The results show that some pages require a long history to understand the users choice of link, while others require only a short history. We also show that the number of additional states induced by the method can be controlled through a probability threshold parameter.

1 Introduction

Modelling user web navigation data is a challenging task that is continuing to gain importance as the size of the web and its user-base increase. Data characterising web navigation can be collected from server or client-based log files, enabling the reconstruction of user navigation sessions [15]. A session is usually defined as a sequence of pages viewed by a user within a given time window. The subarea that studies methods to extract patterns from navigation data has been called *web usage mining* and such methods have been applied in several contexts including personalisation, link prediction, e-commerce analysis, adaptive web site organisation and web page pre-fetching [10].

Several authors have proposed the use of Markov models to represent a collection of user web navigation sessions. Pitkow et al. [12] proposed a method to induce the collection of longest repeating sub-sequences, while Deshpande et al. [7] proposed a technique that builds $k^{th}-order$ Markov models and combines them to include the highest order model covering each state. On the other hand, Sarukkai [13] presented a study showing that Markov models have potential use in link prediction applications, while Zhu et al. [16] inferred a Markov model from user navigation data to measure page co-citation and coupling similarity.

An alternative method of modeling navigation sessions are tree-based models. Schechter et al. [14] use a tree-based data structure that represents the collection of paths inferred from log data to predict the next page accessed, while Dongshan and Junyi [8] proposed a hybrid-order tree-like Markov model to predict web page access. In addition, Chen and Zhang [6] use a Prediction by Partial Match tree that restricts the roots to popular nodes.

In previous work we proposed to model a collection of user web navigation sessions as a Hypertext Probabilistic Grammar (HPG) [1,2]. A HPG corresponds to a first-order Markov model, which makes use of the N-gram concept [5] to achieve increased accuracy by increasing the order of the Markov chain; for the full details on the HPG concept see [2]. In [2] an algorithm to extract the most frequent traversed paths from user data was proposed, and in [3] we have shown that the algorithm's complexity is, on average, linear time in the number of states of the grammar. In [4] we extended the HPG model with a dynamic clustering-based method that uses state cloning [9] to accurately represent second-order conditional probabilities; the method is presented in Section 2. In this work we generalise the method given in [4] to higher-order conditional probabilities.

Most current web mining systems use techniques such as clustering, association rule mining and sequential pattern mining to search for patterns in navigation records [10], and do not take into account the order in which pages were accessed. This limitation has been tackled by building a sequence of higher-order Markov models with a method that chooses the best model to use in each case [7]. However, we argue that a method to produce a single model representing the variable length history of pages has, so far, been missing.

The method we propose in Section 3 aims to fill that gap. By using the cloning operation we duplicate states corresponding to pages that require a longer history to understand the choice of link that users made. In this way the out-links from a given state reflect the n-order conditional probabilities of the in-paths to the state. In addition, the proposed model maintains the fundamental properties of the HPG model [1], while providing a suitable platform for utilising an algorithm for mining the navigation patterns that takes into account the order of page views.

In Section 2 we review the essential of the dynamic clustering method, in Section 3 we extend the method to model higher-order probabilities, and in Section 4 we present the experimental results. Finally, in Section 5 we give our concluding remarks.

2 Background

In previous work [1,2] we proposed to model user navigation data as a Hypertext Probabilistic Grammar (HPG), which corresponds to a first-order Markov model. We now review the HPG model with the aid of an example.

Consider a web site with seven web pages, $\{A_1, A_2, \ldots, A_7\}$, and the collection of navigation sessions given on the left-side of Figure 1 (NOS represents the number of occurrences of each session). A navigation session gives rise to a sequence of pages viewed by a user within a given time window. To each web page there corresponds a state in the model. In addition, the start state, S, represents the first state of every navigation session, and the a final state, F, represents the last state of every navigation session. There is a transition corresponding to each pair of pages visited in sequence, a transition from S to the first state of a session, and a transition from the last state of a session to F. The model is incrementally built by processing the complete collection of navigation sessions.

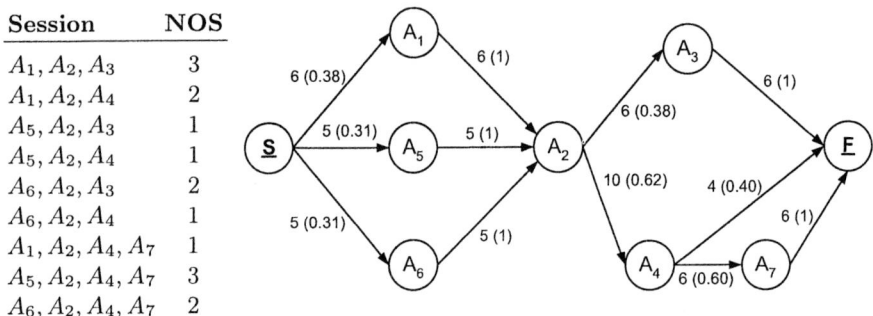

Session	NOS
A_1, A_2, A_3	3
A_1, A_2, A_4	2
A_5, A_2, A_3	1
A_5, A_2, A_4	1
A_6, A_2, A_3	2
A_6, A_2, A_4	1
A_1, A_2, A_4, A_7	1
A_5, A_2, A_4, A_7	3
A_6, A_2, A_4, A_7	2

Fig. 1. A collection of user navigation sessions and the corresponding first-order model

A transition probability is estimated by the ratio of the number of times the transition was traversed and the number of times the anchor state was visited. The right-side of Figure 1 shows a representation of the first-order model corresponding to the input sessions. Next to a link, the first number gives the number of times the link was traversed and the number in parentheses gives its estimated probability.

In [4] we proposed a method to increase the HPG precision in order to accurately represent second-order probabilities. The method makes use of a cloning operation, where a state is duplicated if first-order probabilities diverge from the corresponding second-order probabilities. In addition, the method uses a clustering algorithm to identify the best way to distribute a state's in-links between a state and its clones. We now present the essential properties of the model proposed in [4], which we extend to higher-order probabilities in Section 3.

Given a model with states $\{S, A_1, ..., A_n, F\}$, we let w_i represent the number of times the page corresponding to A_i was visited, $w_{i,j}$ be the number of times the link from A_i to A_j was traversed, and $w_{i,j,k}$ be the number of times the sequence A_i, A_j, A_k was traversed. In addition, we let $p_{i,j} = w_{i,j}/w_i$ be the first-order transition probability from A_i to A_j, and $p_{i,k\,j} = w_{i,k,j}/w_{i,k}$, be the second-order transition probability. Also, the accuracy threshold, γ, sets the highest admissible difference between a first-order and a second-order probability; a model is said to be *accurate* if there is no link that violates the constraint set by γ.

In the example given in Figure 1, the user's past navigation behaviour implies that $p_{1,23} = p_{1,24} = 0.5$. Therefore, for $\gamma = 0.1$, the state A_2 is not accurate, since $|p_{1,2\,3} - p_{2,3}| > 0.1$, and needs to be cloned. To clone state A_2, we let each in-link define a vector of second-order probabilities; each of the vector's components corresponds to an out-link from state A_2. In the example, state A_2 has three in-links and two out-links, inducing three vectors of second-order probabilities: for $i = \{3, 4\}$ we have $P_{1,2i} = \{0.5, 0.5\}$, $P_{5,2i} = \{0.2, 0.8\}$ and $P_{6,2i} = \{0.4, 0.6\}$.

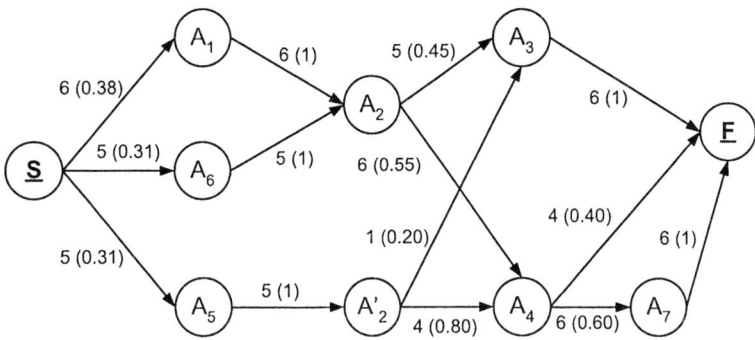

Fig. 2. The second-order HPG model obtained when applying the dynamic clustering method with $\gamma = 0.1$ to the first-order model given in Figure 1

The method applies a k-means clustering algorithm to the collection of second-order vectors, in order to identify groups of similar vectors with respect to γ. Figure 2 shows the result of applying the method to state A_2. Since $|p_{1,2i} - p_{6,2i}| < 0.1$, for $i = \{3, 4\}$, links from A_1 and A_6 are assigned to one clone, the link from A_5 is assigned to the other clone. The transition counts for the out-links are updated as follows: $w_{2,3} = w_{1,2,3} + w_{6,2,3}$, $w_{2,4} = w_{1,2,4} + w_{6,2,4}$, $w_{2',3} = w_{5,2,3}$, $w_{2',4} = w_{5,2,4}$. Note that, state A_4 is accurate, since all its in-links have an equivalent source state, and, moreover, every state having just one out-link is accurate by definition. Therefore, the model given in Figure 2 accurately represents every second-order transition probability.

3 A Dynamic Clustering Method to Model Higher-Order Probabilities

We now extend the method presented in [4] to incorporate higher-order probabilities. In a second-order HPG model, the transition probabilities from a given state are considered to be accurate, if all in-links to it induce identical second-order probabilities. Similarly, in a third-order model every two-link path to a state must induce identical third-order probabilities. In general, to accurately model n-order probabilities each $(n-1)$-length path to a state must induce identical n-order conditional probabilities. Estimates of the n-order conditional probabilities are obtained from the n-gram counts.

In the following, we let the length of a path be measured by the number of links it is composed of, and we call the length of the path from a state to the target state the *depth* of this state; $d = 0$ corresponds to the target state and $d = n - 1$ corresponds to the farthest state from the target when assessing the model for order n. We let $w_{1,...,n}$ represent the n-gram counts, and $p_{i...j,kt} = w_{i,...,j,k,t}/w_{i,...,j,k}$ represent the n-order conditional probability of going to state A_t given that the $(n-1)$-length path $A_i, \ldots, A_j, A_k$ was followed. Also, we let $\vec{l}$ represent a path and $p_{\vec{l},kt}$ the conditional probability of transition (A_k, A_t) given the path $\vec{l}$. Also, we let $\vec{l}_{[d]}$ be the state at depth d on $\vec{l}$ and $v_{\vec{l}}$ be the vector of n-order conditional probabilities given path $\vec{l}$. If a state y needs c_y clones, we let y_i, with $i = \{1, \ldots, c_y\}$, represent y and its $c_y - 1$ additional clones. Finally, we let $\vec{l}_c$ be the cluster to which path $\vec{l}$ was assigned. For a state x, the n-order conditional probabilities are assessed in three steps:

(i) Apply a breath-first search procedure to induce the $(n-1)$-length in-paths to state x, estimate the corresponding n-order conditional probabilities and, for each path, $\vec{l}$, store the conditional probabilities in a vector $v_{\vec{l}}$ (the vector's dimension is given by the number of out-links from x). If the difference between a conditional probability and the corresponding transition probability is greater than γ, label the state as needing to be cloned.

(ii) If x needs cloning, apply the k-means algorithm to the probability vectors, $v_{\vec{l}}$. The number of clusters k is incremented until in the final solution, and in every cluster, the distance between each vector and its centroid is smaller than γ.

(iii) Identify states that need to be cloned to separate the paths to x. States included in paths to x are assessed in descending depth order from $d = n-1$ to $d = 0$. For depth d, we let a prefix of a path to x, whose last state is y, be named a y path-prefix to x. Thus, to separate paths to x, state y at depth, d, needs as many clones as the number of distinct path prefixes with the same length that are assigned to different clusters. The weights of the in and out-links of y and its clones are determined by the n-gram counts. After cloning y the in-paths to x need to be updated.

We now present an example of the method and a pseudo-code description. In particular, we evaluate the third-order probabilities for the model in Figure 2. The conditional probabilities induced by the paths to A_4 are: for $i = \{7, F\}$ we have $p_{12,4i} = \{0.33, 0.67\}$, $p_{62,4i} = \{0.67, 0.33\}$ and $p_{52,4i} = \{0.75, 0.25\}$. Thus, since these probabilities are not close to the corresponding second-order probabilities, A_4 is not third-order accurate for $\gamma = 0.1$. Table 1 gives the in-paths to A_4, the third-order conditional probabilities and the resulting clustering assignment. As result, state A_2 for $d = 1$ needs one clone, and for $d = 0$ state A_4 also needs one clone. Figure 3 gives the resulting third-order model.

In Figure 3, the path S, A_1, A_2, A_4 has probability estimate of $0.38 \cdot 1.00 \cdot 0.50 = 0.19$. It can be seen that in Figure 1, from a total of 16 sessions, 3 begin with the 3-gram A_1, A_2, A_4 resulting in a probability estimate of 0.19. Also, according to the third-order model, path S, A_5, A_2, A_4, A_7 has probability $0.31 \cdot 1.00 \cdot 0.80 \cdot 0.71 = 0.18$. It can be seen that in the input data 3 sessions

Table 1. The paths to A_4, the third-order conditional probabilities and the resulting clustering assignment

$d=2$	$d=1$	$d=0$	3rd order vectors		cluster
A_1	A_2	A_4	0.33	0.67	1
A_6	A_2	A_4	0.67	0.33	2
A_5	A'_2	A_4	0.75	0.25	2

begin with A_5, A_2, A_4, A_7, resulting in a probability estimate of 0.19. In both cases the difference between the two estimates is below 0.1, which is the value specified for the accuracy probability threshold, γ.

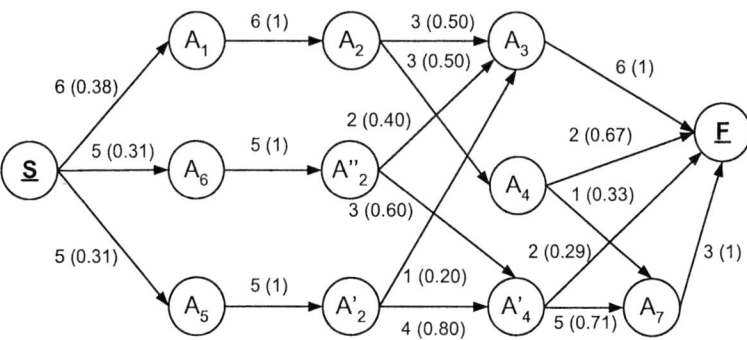

Fig. 3. The third-order model obtained when applying the dynamic clustering method with $\gamma = 0.1$ to the model given in Figure 2

Alternatively, the initial probability of a state can be estimated as $w_i / \sum_j w_j$ and every state has a link from S. In Figure 3 there is a total of 54 page views, and, for example, $p_{S,2} = 6/54 = 0.11$ and $p_{S,4'} = 7/54 = 0.13$. For the path A_2, A_4, A_7 the probability estimate is given by the sum of the probabilities of path A_2, A_4, A_7, path A'_2, A'_4, A_7 and path A''_2, A'_4, A_7, which is 0.12. In the input sessions, shown in Figure 1, we have a total of 54 3-gram counts (including 3-grams starting with S and ending with F) and the count of A_2, A_4, A_7 is 6, therefore, its estimate is $6/54 = 0.11$. For path A_1, A_2, A_3 the model gives 0.05, while the session analysis gives 0.05. Both cases are accurate with respect to $\gamma = 0.1$.

The pseudo-code description of the algorithm, which implements the method, is now given. We let n be the order with which to evaluate the model, $HPG_{(n-1)}$ be the previous order model, and $(n+1)$-grams be the n-gram counts of size $n+1$.

```
Algorithm (HPG_(n-1), n, γ, (n+1)-grams)
  begin:
    for each state x
      induce in-paths of length n − 1 to x
      for each in-path $\vec{l}$
        for each out-link i from x
          estimate $p_{\vec{l},xi}$ and store in $v_{\vec{l}}$
          if ($|p_{\vec{l},xi} − p_{x,i}| > γ$) the state needs to be cloned
        end for
      end for
      if state needs to be cloned
        apply k-means to collection of vectors $v_{\vec{l}}$
        for depth d = (n − 1) to d = 0
          for each state y at depth d
            $c_y$ = num. distinct path prefixes assigned to different clusters
            create $c_y$ − 1 clones of state y
            for each in-path $\vec{l}$ to x
              if ($\vec{l}_{[d]} = y$ and $\vec{l}_c > 1$) redirect link to corresponding clone
            end for
            for state $y_i$ with i = {1,...,$c_y$}
              for each in-link t to $y_i$
                for each out-link r from $y_i$
                  $w_{t,y_i} = w_{t,y_i} + w_{t,y_i,r}$ , $w_{y_i,r} = w_{y_i,r} + w_{t,y_i,r}$
                end for
              end for
              remove out-links from $y_i$ such that $w_{y_i,r} = 0$
            end for
            update ngram counts to take into account clones
          end for
          update in-paths with clone references
        end for
      end if
    end for
  end.
```

4 Experimental Evaluation

For the experimental evaluation we analysed three real world data sets. By using data from three different sources we aim to assess the characteristics of our model in a wide enough variety of scenarios. Our previous experience has shown that it is difficult to create random data sets that mirror the characteristics of real world data, and therefore looking at several data sets is necessary.

The first data set (CS) is from a university site, was made available by the authors of [15] and represents two weeks of usage data in 2002. The site was cookie based, page caching was prohibited and data was made available with the sessions identified. We split the data set into three subsets in order to enhance analysis in a wider variety of scenarios. The second data set (MM) was obtained from the authors of [11] and corresponds to one month of usage from the Music Machines site (machines.hyperreal.org) in 1999. The data was organised in sessions and caching was disabled during collection. We split the data set into four subsets, each corresponding to a week of usage. The third data set (LTM) represents forty days of usage from the London Transport Museum web site in 2003 (www.ltmuseum.co.uk). The data was obtained in a raw format. We filtered .gif and .jpg requests, and requests with an error status code. Sessions were defined as consecutive requests from a given IP address within a 30 minute time window and a maximum session length of 100 requests was set. We split this data set into four subsets, each corresponding to ten days of usage data.

Table 2 gives the summary characteristics for each data set; ds identifies the data set, pg gives the number of distinct pages visited, $\%1v$ and $\%\leq 2v$ indicate, respectively, the percentage of pages with just one visit and with two or less visits. Also, aOL gives the average number of out-links per state, sOL the standard deviation, aIL the average number of in-links per page and sIL the standard deviation. Finally, ses gives the number of sessions, $aSes$ the average session length, $sSes$ the standard deviation, and req the total number of requests. The variabilty on the number of states induced by the model for a given web site can be explained by the number of pages with less than one visit. Also, when the number of pages with few visits increases the average number of out-links and in-links decreases. The average session length is stable but the standard deviation shows that the MM data has a higher variability on the session length.

Table 2. Summary characteristics of the real data sets

ds	pg	$\%1v$	$\%\leq 2v$	aOL	sOL	aIL	sIL	ses	$aSes$	$sSes$	Req
LTM_1	2998	0.62	0.68	4.5	9.6	4.4	11.6	9743	7.6	13.5	74441
LTM_2	1648	0.19	0.27	8.4	13.8	8.3	16.6	11070	7.4	13.2	82256
LTM_3	1610	0.27	0.37	7.8	12.8	7.7	15.0	9116	7.7	13.1	70558
LTM_4	1586	0.24	0.34	7.8	13.3	7.7	15.9	9965	7.8	13.4	78179
MM_1	8715	0.30	0.45	4.7	12.4	4.6	14.1	14734	6.4	37.8	94989
MM_2	5356	0.32	0.44	6.0	18.9	5.9	20.7	14770	6.1	14.7	90682
MM_3	5101	0.26	0.38	6.0	15.6	5.9	17.7	10924	6.7	35.2	73378
MM_4	6740	0.35	0.49	5.1	18.5	4.9	19.8	14080	6.3	23.8	88053
CS_1	3128	0.52	0.67	3.4	10.1	3.1	10.4	7000	4.8	6.5	33854
CS_2	3946	0.59	0.74	2.8	9.3	2.6	9.9	7000	5.0	8.4	34897
CS_3	5028	0.62	0.76	2.8	9.4	2.6	11.6	6950	5.5	12.8	38236

The left-hand side of Figure 4, shows, for the three representative data sets, the variation of the model size with its order for $\gamma = 0$. (The data sets from each source reveal almost identical behaviour). For the MM_1 data set a large percentage of state cloning is performed for second and third-order probabilities which indicates that there is no significant difference between third-order probabilities and the corresponding higher-order probabilities, and that the MM site only requires a short history when deciding which link to follow. The CS data set shows a slower increase in the model's size, and the model can be seen to reach close to full accuracy with respect to fourth-order probabilities. Finally, the LTM data set shows an increase in the number of states for up to the seventh-order probabilities, meaning that the choice of which link to follow is clearly influenced by a relatively long sequence of previously visited web pages.

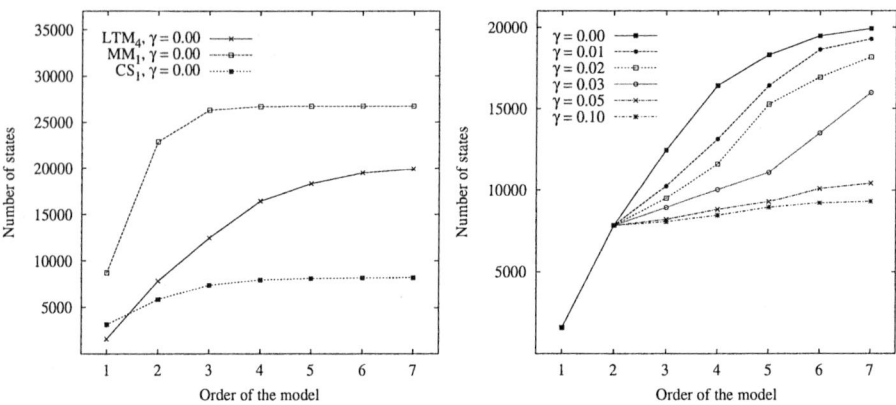

Fig. 4. The increase in model size with the model's order for $\gamma = 0$ and the increase in size for several values of the probability threshold, γ, with the LTM$_4$ data set

The right-hand side of Figure 4, shows the effect of γ, on the model size for the LTM$_4$ data set and the left-hand side of Figure 5, shows the effect for the CS$_1$ data set. In both cases it can be seen that by tuning the value of γ it is possible to control the increase on a models's number of states. For both data sets the difference in the number of states is not evident for second-order models. For third and higher-order models it is possible to reduce the number of states induced by the method by allowing some tolerance on representing the conditional probabilities (by setting $\gamma > 0$). Setting γ to a value greater than 0.1 results in almost no cloning for higher orders.

Figure 6 shows some statistics on the number of clones per state for the LTM$_4$ and CS$_1$ data sets, with $\gamma = 0.02$. The the average number of clones per state (avg) is higher for the LTM$_4$ data set than for the CS$_1$ data set, as expected by inspecting the left-side of Figure 4. The standard deviation (stdev) indicates a substantial variability in the number of clones per state, a fact that is supported by the maximum number of clones (max) and the indicated percentiles. For the LTM$_4$ data set 50% of the states were never cloned and 75% have at most six clones for the seventh order. In the CS$_1$ data set 75% of the states were never cloned and 90% of the states have at most seven clones for the seventh

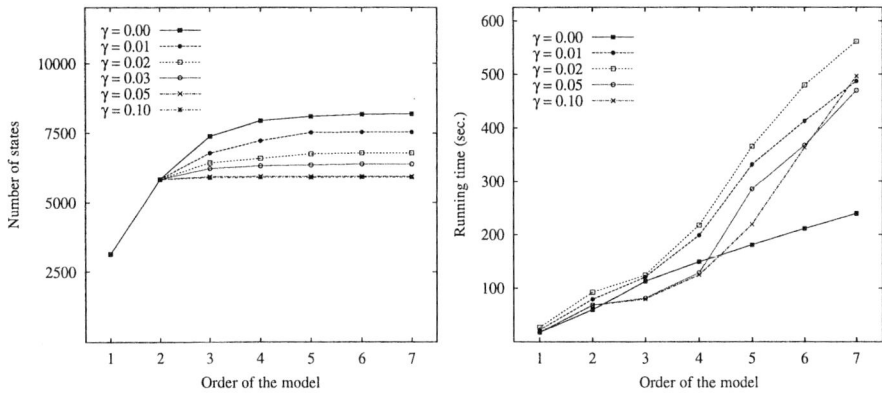

Fig. 5. The increase in model size with the model's order for several values of the probability threshold, γ, for the CS_1 data set and variation of the running time with the model's order for several values of the probability threshold for the LTM_4 data set

order. These results help to motivate our interest in the dynamic model, since while for some states the link choice of the corresponding page depends on the navigation history, for other states the link choice is completely independent of the navigation history.

	LTM_4 $\gamma = 0.02$ order							CS_1 $\gamma = 0.02$ order					
	2	3	4	5	6	7		2	3	4	5	6	7
avg	3.94	4.99	6.32	8.63	9.67	10.46	avg	0.87	1.06	1.11	1.16	1.17	1.17
stdev	10.86	15.29	24.85	40.88	47.20	50.69	stdev	5.40	7.06	7.3	7.92	7.96	7.96
max	205	307	683	989	1193	1265	max	138	175	180	208	208	208
75%	4.00	5.00	5.00	6.00	6.00	6.00	75%	0.00	0.00	0.00	0.00	0.00	0.00
85%	10.00	11.25	13.00	14.00	16.25	18.00	95%	4.00	5.00	5.00	5.00	5.00	5.00

Fig. 6. Statistics on the number of clones per state with the model's order for the LTM_4 and CS_1 data set with $\gamma = 0.02$

The right-hand side of Figure 5, and the left-hand side of Figure 7, show our analysis of the running time of the algorithm for two representative data sets. We note that, while programming the method, we did not take particular care regarding the implementation efficiency. The method is close to linear time for $\gamma = 0$, since in such case no clustering is needed. For $\gamma > 0$ the k-means method is applied and we let k increase until a solution which meets the threshold criteria is obtained. For the reported experiments, we set k to vary according to the expression $k = ceiling(1.5k)$ in order to obtain a slow increment on its value in the first stages and a larger increase of the k value in the subsequent stages. Finally, the right-hand side of Figure 7, shows, for the LTM_1 data set, the increase in number of states with the model's order for three methods used

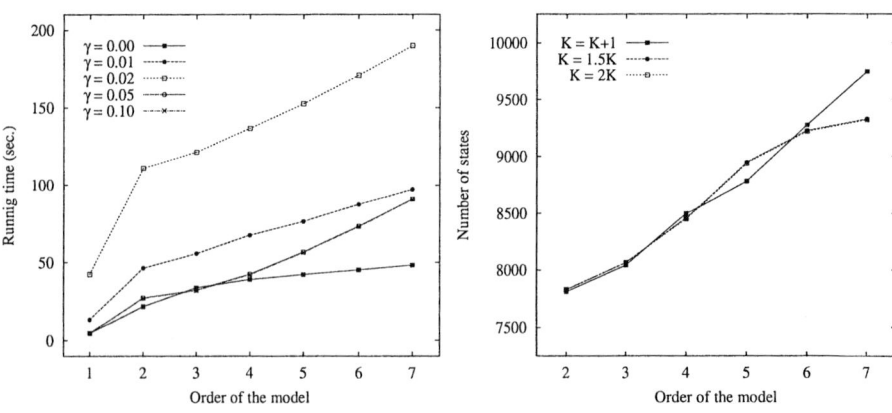

Fig. 7. The running time with the model's order for several values of γ for the CS_1 data set and the increase in model size with the model's order for three methods to set the number of clusters (k) for the LTM_1 data set

to increase k, in the k-means method. The results show that for lower orders the number of states is not very sensitive to the method, and for orders higher than five the faster methods over estimate the number of clusters but with the benefit of a faster running time (which is not shown in the plot).

5 Concluding Remarks

We have proposed a generalisation of the HPG model by using a state cloning operation that is able to accurately model higher-order conditional probabilities. The resulting dynamic high-order Markov model is such that the probabilities of the out-links from a given state reflect the n-order conditional probabilities of the paths to the state. Thus, the model is able to capture a variable length history of pages, where different history lengths are needed to accurately model user navigation. In addition, the method makes use of a probability threshold together with a clustering technique that enables us to control the number of additional states induced by the method at the cost of some accuracy. Finally, the model maintains the fundamental properties of the HPG model, [1], providing a suitable platform for an algorithm that can mine navigation patterns, taking into account the order of page views.

We reported on experiments with three distinct real world data sets. From the results we can conclude that, for some web sites users navigate with only a short history of the pages previously visited (for example, the MM site) but in other sites users hold a longer history in their memory (for example, the LTM site). The results also suggest that, in a given site, different pages require different amounts of history in order to understand the possible options users have when deciding on which link to click on. This supports our interest in the proposed dynamic method that models each state with the required history depth. The results indicate that the clustering method is interesting for large sites, where the number of states induced for high orderers, for $\gamma = 0$, becomes unmanageable.

In the short term we plan to conduct a study to analyse the semantics of the rules induced by different order probability models. We also plan to perform a statistical comparison of subsequent order probabilities aimed at determining if there is sufficient statistical evidence that the additional model complexity in moving to a higher order justifies the corresponding increase in the algorithm's complexity. It would also be interesting to be able to estimate the number of clusters necessary to achieve the required accuracy in order to speed up the method. Finally, a comparative study with tree-based models is also planned.

References

1. J. Borges. *A data mining model to capture user web navigation patterns*. PhD thesis, University College London, London University, 2000.
2. J. Borges and M. Levene. Data mining of user navigation patterns. In B. Masand and M. Spliliopoulou, editors, *Web Usage Analysis and User Profiling*, Lecture Notes in Artificial Intelligence (LNAI 1836), pages 92–111. Springer Verlag, 2000.
3. J. Borges and M. Levene. An average linear time algorithm for web usage mining. *Int. Jou. of Information Technology and Decision Making*, 3(2):307–319, June 2004.
4. J. Borges and M. Levene. A dynamic clustering-based markov model for web usage mining. cs.IR/0406032, 2004.
5. Eugene Charniak. *Statistical Language Learning*. The MIT Press, 1996.
6. X. Chen and X. Zhang. A popularity-based prediction model for web prefetching. *Computer*, 36(3):63–70, March 2003.
7. M. Deshpande and G. Karypis. Selective markov models for predicting web page accesses. *ACM Transactions on Internet Technology*, 4(2):163–184, 2004.
8. X. Dongshan and S. Junyi. A new markov model for web access prediction. *Computing in Science and Engineering*, 4(6):34–39, November/December 2002.
9. M. Levene and G. Loizou. Computing the entropy of user navigation in the web. *Int. Journal of Information Technology and Decision Making*, 2:459–476, 2003.
10. B. Mobasher. Web usage mining and personalization. In Munindar P. Singh, editor, *Practical Handbook of Internet Computing*. Chapman Hall & CRC Press, 2004.
11. Mike Perkowitz and Oren Etzioni. Towards adaptive web sites: conceptual framework and case study. *Artificial Intelligence*, 118(2000):245–275, 2000.
12. J. Pitkow and P. Pirolli. Mining longest repeating subsequences to predict world wide web surfing. In *Proc. of the 2nd Usenix Symposium on Internet Technologies and Systems*, pages 139–150, Colorado, USA, October 1999.
13. Ramesh R. Sarukkai. Link prediction and path analysis using markov chains. *Computer Networks*, 33(1-6):377–386, June 2000.
14. S. Schechter, M. Krishnan, and M. D. Smith. Using path profiles to predict http requests. *Computer Networks and ISDN Systems*, 30:457–467, 1998.
15. M. Spiliopoulou, B. Mobasher, B. Berendt, and M. Nakagawa. A framework for the evaluation of session reconstruction heuristics in web usage analysis. *IN-FORMS Journal on Computing*, (15):171–190, 2003.
16. J. Zhu, J. Hong, and J. G. Hughes. Using markov models for web site link prediction. In *Proc. of the 13th ACM Conf. on Hypertext and Hypermedia*, pages 169–170, June 2002.

TREE² - Decision Trees for Tree Structured Data

Björn Bringmann and Albrecht Zimmermann

Institute of Computer Science, Machine Learning Lab,
Albert-Ludwigs-University Freiburg, Georges-Köhler-Allee 79,
79110 Freiburg, Germany
{bbringma, azimmerm}@informatik.uni-freiburg.de

Abstract. We present TREE², a new approach to *structural classification*. This integrated approach induces decision trees that test for pattern occurrence in the inner nodes. It combines state-of-the-art tree mining with sophisticated pruning techniques to find the most discriminative pattern in each node. In contrast to existing methods, TREE² uses no heuristics and only a single, statistically well founded parameter has to be chosen by the user. The experiments show that TREE² classifiers achieve good accuracies while the induced models are smaller than those of existing approaches, facilitating better comprehensibility.

1 Introduction

Classification is one of the most important data mining tasks. Whereas traditional approaches have focused on flat representations, using feature vectors or attribute-value representations, there has recently been a lot of interest in more expressive representations, such as sequences, trees and graphs [1,2,3,4,5]. Motivations for this interest include drug design, since molecules can be represented as graphs or sequences. Classification of such data paves the way towards drug design on the screen instead of extensive experiments in the lab. Regarding documents, XML, essentially a tree-structured representation, is becoming ever more popular. Classification in this context allows for more efficient dealing with huge amounts of electronic documents.

Existing approaches to classifying structured data (such as trees and graphs) can be categorized into various categories. They differ largely in the way they derive structural features for discriminating between examples belonging to the different classes.

A first category can be described as a pure *propositionalization approach*. The propositionalization approach typically generates a very large number of features and uses an attribute-value learner to build a classifier. The resulting classifiers are often hard to understand due to the large number of features used which are possibly also combined in a non-trivial way (e.g. in a SVM).

A second class of systems can be described as the *association rule approach*, e.g. Zaki [4]. Even though the resulting rules often yield high predictive accuracy, the number of generated rules typically explodes, making the resulting classifier difficult to understand.

Both the association rule and propositionalization approaches consider feature generation and classification in two independent steps. *Integrated approaches* form a third category of systems that integrates feature construction with classification. This category includes inductive logic programming systems, such as FOIL [6] and PROGOL [7], as well as the DT-GBI approach of Motoda *et al.* [5]. For those approaches to be computationally feasible they have to perform heuristic search, possibly generating non-optimal features.

All techniques mentioned above share the need to specify a number of user-defined parameters, which is often non-trivial.

In this work we present a different approach called TREE^2. It is motivated by recent results on finding correlated patterns, allowing to find the k best features according to a convex optimization criterion such as χ^2 or *Information Gain* [8]. Rather than generating a large number of features or searching for good features in a heuristic manner, TREE^2 searches for the best features to be incorporated in a decision tree by employing a branch-and-bound search, pruning w.r.t. the best pattern seen so far. As in DT-GBI, a decision tree is induced but at each node, the *single* best feature is computed. There are several advantages: TREE^2 is an *integrated approach*, has stronger guarantees than GBI, only one parameter has to be set (the significance level), and the resulting classifiers are far smaller and easier to understand than those of the propositionalization and association rule approaches.

The paper is organized as follows: in Section 2 we describe earlier work on the topic and relate it to our approach; in Section 3, we discuss technical aspects of our method and outline our algorithm; in Section 4, the experimental evaluation is explained and its results discussed. We conclude in Section 5 and point to future work directions.

2 Related Work

Structural classification has been done with different techniques. Firstly, there are several propositionalization approaches, e.g. [2] and [3]. While details may differ, the basic mechanism in these approaches is to first mine all patterns that are unexpected according to some measure (typically frequency). Once those patterns have been found, instances are transformed into bitstrings, denoting occurrence of each pattern. Classifiers are trained using this bitstring representation. While these approaches can show excellent performance and have access to the whole spectrum of machine learning techniques there are possible problems. Obviously the decision which patterns to consider special, e.g. by fixing a minimum frequency, will have an effect on the quality of the model. The resulting feature set will probably be very large, forcing pruning of some kind. Finally, interpretation of the resulting model is not easy, especially if the classifier is non-symbolic, e.g. a SVM.

A second group of approaches is similar to the *associative classification* approach [9]. Again, outstanding patterns are mined but each of them has to associate with the class value. Zaki *et al.*'s XRULES classifier is of this variety. Each pattern is then considered as a rule predicting its class. Usually, the resulting rule set has to be post-processed and/or a conflict resolution technique employed. As in the propositionalization techniques, the choice of constraints under which to mine is not straight-forward and choosing the resolution technique can strongly influence performance, as has been shown e.g. in [10,11]. Additionally, the resulting classifier often consists of thousands of rules, making interpretation by the user again difficult.

Finally, there exist integrated techniques that do not mine *all* patterns, but construct features during building the classifier. Since structural data can be represented in predicate logic, techniques such as FOIL [6] and PROGOL [7] are capable of doing that. While ILP approaches are elegant and powerful, working on large datasets can be too computationally expensive. An approach such as DT-GBI [5], on the other hand, constructs the features it uses for the tests of the induced decision tree by doing graph-mining. What is common to these approaches is that feature induction is usually done in a heuristic way, often by greedy maximization of a correlation measure during beam search. Responsibility of deciding the parameters governing this search is placed upon the user. For instance, in FOIL decisions have to be made on the beam size and the maximum number of literals that are allowed in the rule body. Similarly, DT-GBI requires the user to specify beam size, the maximum number of specializations in each node, and possibly a minimum frequency that should not be violated. As Motoda shows in his work [5], finding the right value for the beam size and the maximum number of specializations requires essentially a meta-search in the space of possible classifiers.

In contrast, the only parameter to be specified for TREE^2 is the cut-off value for growing the decision tree. By basing this value on the p-values for the χ^2-distribution, the user has a well-founded guide-line for choosing this value.

While all the above techniques focus on directly using structural information for classification purposes, a different approach is exemplified by [12]. Instead of explicitly representing the structures used, kernels are employed that quantify similarities between entities. While the resulting classifiers are very accurate, the use of e.g. a graph kernel together with an SVM make analyzing the model difficult.

3 Methodology

In this section we explain the pattern matching notion used by the TREE^2 approach, discuss upper bound calculation, the main component of the principled search for the most discriminating pattern, and formulate the algorithm itself.

3.1 Matching Embedded Trees

Several representations for structured data such as graphs, trees and sequences exist. In this paper we will focus on tree structured data, like XML, only. Thus, we need a notion for matching tree structured data.

A rooted k-tree t is a set of k nodes V_t where each $v \in V_t$, except one called root, has a parent denoted $\pi(v) \in V_t$. We use $\lambda(v)$ to denote the label of a node and an operator $\prec$ to denote the order from left to right among the children of a node. The transitive closure of π will be denoted π^*. Let $\mathcal{L}$ be a formal language composed of all labeled, ordered, rooted trees and $\mathcal{D} \subset \mathcal{L}$ a database. To count trees $t \in \mathcal{D}$ containing a pattern p we define a function $d_t : \mathcal{L} \to \{0, 1\}$ to be 1 iff p matches the tree t and 0 otherwise.

Several notions of tree matching exist. As in Zaki et al.'s work [4] we used a notion called *tree embedding* which is defined as follows:

Definition 1. *A tree t is embedded in a tree t' iff a mapping $\varphi : V_t \to V_{t'}$ exists such that $\forall u, v \in V_t : \lambda(u) = \lambda(\varphi(u)) \wedge u \prec v \Leftrightarrow \varphi(u) \prec \varphi(v) \wedge \pi^*(u) = v \Leftrightarrow \pi^*(\varphi(u)) = \varphi(v)$.*

An example of an embedded tree is given in Figure 1.

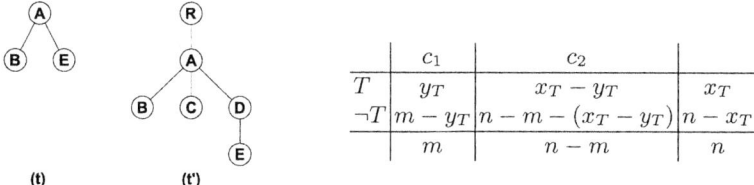

Fig. 1. The tree t is embedded in t'

Fig. 2. A Contingency Table

We use *tree embedding* to compare our approach with Zaki et al.'s technique. This notion is more flexible than simple subtrees and the mining process is still efficient. In general, other matching notions (see [1]) and even different representations could be used with our technique. This includes not only other notions of matching trees, but also graphs, sequences etc., since the general principles of our approach apply to all domains.

3.2 Correlation Measures

Popular approaches to finding relevant patterns in the data are based on the support-confidence framework, mining frequent patterns, in the hope of capturing statistically significant phenomena, with high predictive power. This framework has some problems though, namely the difficulty of choosing a "good" support and the fact that confidence tends to reward patterns occurring together with the majority class. To alleviate these problems, we use correlation measures for selecting discriminative patterns. A correlation measure compares the expected frequency of the joint occurrence of a pattern and a certain class value to the observed frequency. If the resulting value is larger than a certain threshold then the deviation from the independence assumption is considered statistically significant enough to assume a relationship between pattern and class label.

Example 1. *Consider as an example a database consisting of 50 instances, half of which are labeled with class label c_1, the other half with class label c_2. Assume furthermore a pattern T which occurs with support 10 in the database. If eight of the ten instances including T are labeled with c_1, then the χ^2 measure would give this deviation a score of 4.5. Information Gain, that quantifies only the changes in class distribution w.r.t. T, would give it a score of 0.079.*

We organize the observed frequencies of a tree pattern T in a contingency table, cf. Figure 2, with x_T denoting the total number of occurences in the dataset and y_T the occurences in the subset corresponding to the first class. Since the two variables are sufficient for calculating the value of a correlation measure on this table, we will view these measures as real-valued functions $\sigma : \mathbb{N}^2 \mapsto \mathbb{R}$ for the remainder of this paper.

While calculating the correlation value of a given pattern is relatively simple, directed search towards better solutions is somewhat more difficult since correlation measures have no desirable properties such as *anti-monotonicity*. But if they are convex it is possible to calculate an upper bound on the score that can be achieved by specializations of the current pattern T and thus to decide whether this branch in the search tree should be followed.

3.3 Convexity and Upper Bounds

It can be proved that χ^2 and *Information Gain* are convex. For the proofs of the convexity of χ^2 and *Information Gain* we refer the reader to [8].

Convex functions take their extreme values at the points forming the convex hull of their domain D. Consider the graph of $f(x)$ in Figure 3(A). Assume the function's domain is restricted to the interval $[k, l]$ which also makes those points the convex hull of D. Obviously, $f(k)$ and $f(l)$ are locally maximal, with $f(l)$ being the global maximum. Given the current value of the function at $f(c)$ and assuming that it is unknown whether c increases or decreases, evaluating f at k and l allows to check whether it is possible for any value of c to put the value of f over the threshold.

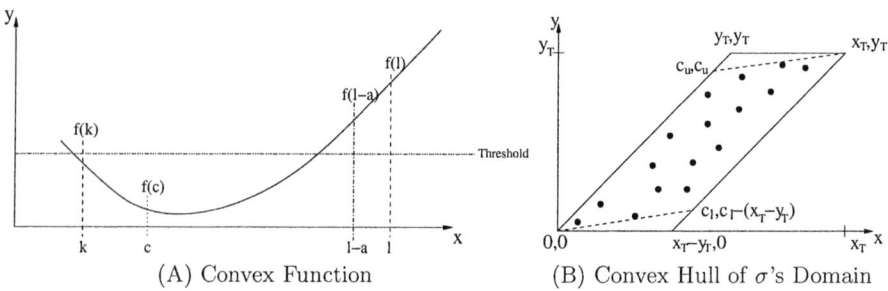

Fig. 3. Convex Function and Convex Hull of the set of possible $\langle x'_T, y'_T \rangle$

For the two-dimensional case, the extreme values are reached at the vertices of the enclosing polygon (in our case the four vertices of the parallelogram in Figure 3(B)). This parallelogram encloses all possible tuples $\langle x'_T, y'_T \rangle$ that correspond to occurence counts of specializations of the current pattern T. The tuple $\langle 0, 0 \rangle$ corresponds to a pattern that does not occur in the dataset and therefore does not have to be considered in calculating the upper bound. $\langle x_T, y_T \rangle$ represents a valid pattern, but in the context of upper bound calculation denotes a specialization of the current pattern T that is equally good in discriminative power. Since general structures have a higher expected probability of being effective on unseen data, we prefer those and thus disregard this tuple as well. Thus the upper bound on $\sigma(T')$ is $ub_\sigma(T) = \max\{\sigma(y_T, y_T), \sigma(x_T - y_T, 0)\}$. For an in-depth discussion of upper bound calculation we refer the reader to [8,11].

Example 2. *Continuing our example from 3.2, this means that for σ being χ^2, $ub_{\chi^2}(T) = \max\{9.52, 2.08\}$, given $x = 10$, $y = 8$. Since 9.52 is larger than $\chi^2(x_T, y_T) = 4.5$ there might be a specialization of T that discriminates better than T itself and therefore exploring this search path is worthwhile.*

While this upper bound calculation is correct for *Information Gain*, an additional problem w.r.t. χ^2 lies in the fact that the information provided by the score of χ^2 is not always reliable. Statistical theory says that for a contingency table with one degree of freedom, such as the one we are considering here, the expected number of occurrences has to be greater than or equal to 5 for the χ^2 score to be reliable. This means that a χ^2-value on $\langle y_T, y_T \rangle$ or $\langle x_T - y_T, 0 \rangle$ is not necessarily reliable. Thus, upper bound calculation has to be modified to achieve reliability. Based on the size of the class and of $\mathcal{D}$, upper and lower bounds c_u, c_l on x'_T for which all four cells have an expected count of 5 can be calculated and the values of the tuples adjusted accordingly. Two of the new vertices are shown as $\langle c_u, c_u \rangle$ and $\langle c_l, c_l - (x_T - y_T) \rangle$.

3.4 The TREE² Algorithm

The TREE² algorithm (shown as Algorithm 1) constructs a binary decision tree in the manner of ID3 [13]. In the root node and each inner node, the occurrence of a tree pattern is tested against the instance to be classified. A resulting tree could look like the example given in Figure 4. In each node, the subtree having the best discriminative effect on the corresponding subset is found by a systematic branch-and-bound search. The mining process is shown in the subroutine ENUMERATEBESTSUBTREE. The space of possible patterns is traversed using canonical enumeration and the value of σ calculated for each candidate pattern. If this value lies above the best score seen so far, the current pattern is the most discriminating on this subset so far and the threshold is raised to its σ-value. An upper bound on the value specializations of the current pattern can achieve is calculated and pruning of the search space using this upper bound and the threshold is performed. In this way, we separate the success of the technique from user decisions about the search strategy. The only decision a user has to

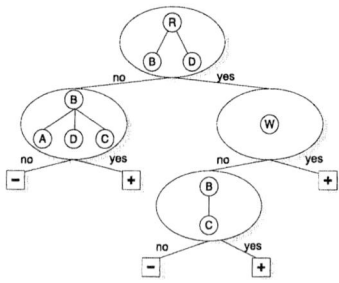

Fig. 4. A decision tree as produced by the TREE² algorithm

make is the one w.r.t. a stopping criterion for further growth of the tree. To this effect, a minimum value for the score of the correlation measure has to be specified, which can be based on statistical theory, thus giving the user a better guidance for making this decision.

Algorithm 1 The TREE² algorithm

TREE²($\mathcal{D}, \sigma, \tau_{user}$, DT)
1: p_{split} = ENUMERATEBESTSUBTREE($\top, 0, \sigma, \tau_{user}, \emptyset$)
2: **if** $p_{split} \neq \emptyset$ **then**
3:   Add node including p_{split} to the DT
4:   TREE²($\{T \in \mathcal{D}|p_{split}$ embedded in $T\}$, σ, τ_{user}, DT)
5:   TREE²($\{T \in \mathcal{D}|p_{split}$ not embedded in $T\}$, σ, τ_{user}, DT)
6: **return** DT

ENUMERATEBESTSUBTREE($t, \tau, \sigma, \tau_{user}, p$)
1: **for all** canonical expansions t' of t **do**
2:   **if** $\sigma(t') > \tau \wedge \sigma(t') \geq \tau_{user}$ **then**
3:     $p = t'$, $\tau = \sigma(t')$
4:   **if** $ub_\sigma(t') \geq \tau$ **then**
5:     p = ENUMERATEBESTSUBTREE($t', \tau, \sigma, \tau_{user}, p$)
6: **return** p

TREE² has several desirable properties. Firstly, the resulting classifier is integrated in the sense that it uses patterns directly, thus circumventing the need for the user to restrict the amount of features and making the resulting classifier more understandable. Secondly, by using correlation measures for quantifying the quality of patterns, we give the user a sounder theoretical foundation on which to base the decision about which learned tests to consider significant and include in the model. Thirdly, we avoid using heuristics that force the user to decide on the values of parameters that could have a severe impact on the resulting model's accuracy. Using principled search guarantees that TREE² finds the best discriminating pattern for each node in the decision tree w.r.t. the correlation measure used. Finally, as the experiments show, the resulting decision tree is far smaller than the rule sets produced by XRULES classifier [4], while achieving comparable accuracy, and is therefore more easily interpretable by human users.

4 Experimental Evaluation

For the experimental evaluation, we compared our approach to XRULES and a decision tree base-line approach on the XML data used in Zaki et al.'s publication [4]. Furthermore, we compared TREE^2 to a base-line approach using frequency mining for a SVM classifier and two PROGOL results on the regression-friendly subset of the Mutagenesis dataset.

XML Data. The XML data used in our experiments are log files from web-site visitors' sessions. They are separated into three weeks (CSLOG1, CSLOG2, and CSLOG3) and each session is classified as its producing visitor coming either from an .edu domain or from any other domain. Characteristics of the datasets are shown in Table 1. For the comparison we built decision trees with the

Table 1. Characteristics of Datasets (taken from [4])

DB	#Sessions	edu	other	%edu	%other
CSLOG1	8074	1962	6112	24.3	75.7
CSLOG2	7409	1687	5722	22.8	77.2
CSLOG12	13934	2969	10965	21.3	78.7
CSLOG3	7628	1798	5830	23.6	76.4

χ^2 distribution's significance value for 90%, 95% and 99% respectively. In each setting we used one set of data for training and another one for testing. Following Zaki's notation, CSLOGx-y means that we trained on set x and tested on set y. For the base-line approach we mined the 100 patterns having the highest discriminative effect on the data, transformed the data into bitstring instances according to the found patterns, and built decision trees using all 100 patterns in one run (*C4.5 - 100*) and the 50 best patterns in another run (*C4.5 - 50*) with the WEKA [14] implementation of the C4.5 [15] algorithm. We compare the accuracies of the resulting classifiers against each other as well as the complexity of the model which we measure by the number of rules used by XRULES, and by the number of leaves in the decision trees, which corresponds to the number of rules that can be derived from the trees, respectively.

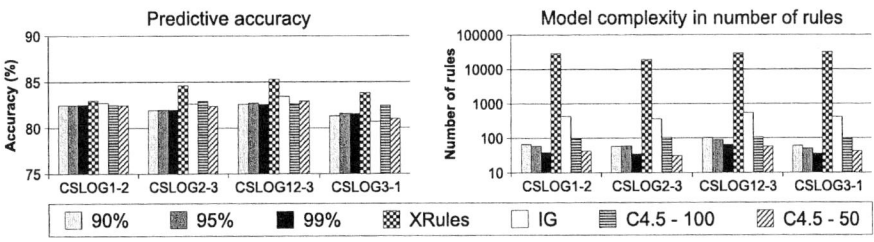

Fig. 5. Accuracies and size in rules of the different approaches

Results are summarized in Figure 5. As can be seen, the accuracies of the induced classifiers do not vary much. The only approach that significantly outperforms (by 2-3%) the other techniques on all but the CSLOG1-2 setting, is XRULES. At the same time, the size of XRULES' models is also significantly greater. While the TREE^2 trees induced with *Information Gain* have several hundred nodes and all trees induced with χ^2 (both TREE^2 and base-line) between 35 and 103 nodes, the smallest XRULES model consists of more than 19000 rules. Patterns tested against in the inner decision tree nodes consist of 3-7 nodes only. Since this is similar to the size of patterns used in XRULES' rules, complexity is really reduced and not just pushed inside the classifier. In comparing the other approaches, several things are noticeable. Raising the threshold from the 90% to the 95% significance level for χ^2-induced TREE^2 trees does not decrease accuracy (even improving it slightly in 3 cases). Raising it further to the 99% level has no clear effect. The tree size decreases, though, on average by 7.5 nodes from the 90% to the 95% setting. Raising the significance level further to 99% decreases the tree size by 18 nodes on average.

For the base-line approach we mined patterns correlating strongly with the classes and trained a classifier on them. This approach achieves competitive results w.r.t the accuracy. The clear drawback is that deciding on the number of features to use is not straightforward. Using only 50 instead of 100 features produces all kinds of behavior. In some cases the accuracy does not change. In other cases the classifier using 50 features outperforms the one using 100 or vice versa. Also, the base-line approach using 100 patterns tends to use most of these, even if TREE^2 trees of similar quality are much smaller.

Finally, using *Information Gain* as quality criterion shows mainly one thing - that it is difficult to make an informed decision on cut-off values. The accuracies and sizes shown refer to decision trees induced with a cut-off value of 0.001. For one thing, the resulting trees grow far bigger than the χ^2-trees. Additionally, the accuracies in comparison with the χ^2 approach vary, giving rise to one worse tree, one of equal quality and two better ones. None of the differences in accuracy is significant though. Inducing decision trees with a cut-off value of 0.01 lowers accuracy by 1.5 to 3 percentage points, with the induced trees still being larger than the χ^2 trees.

Mutagenicity Data. For this setting, we chose the regression-friendly subset of the well known Mutagenicity dataset used in [16]. We compare with the results of the ILP system PROGOL reported in [16,17] and the results of the base-line approach reported in [3]. Since the Mutagenicity dataset consists of molecules represented as graphs, a transformation from the SMILES representation into so-called fragment-trees is used that is explained following this paragraph.

The Smiles Encoding. The SMILES language [18] is used by computational chemists as a compact encoding of molecular structure. It is supported by many tools as OpenBabel or Daylight ([19,20]). The language contains symbols for atoms, bonds, branches, and can express cycles. Using a decomposition-algorithm by Karwath and De Raedt [21], a SMILES-String can, after some reformatting, be decomposed into a so-called *fragment tree*. Since there is no *unique* SMILES-string for a molecule, the fragment tree is not unique either. The decomposition-algorithm recursively splits the string into *cycles* $\{_xT\}_x$ and *branches* $A(B)C$. In the resulting fragment-tree the leaves contain pure cycles or linear fragments without further branches. The inner nodes of such a tree contain fragments still containing branches while the root node is the whole molecule. The edge labels denote the type of decomposition (i.e. the part of the branch or the number of the cycle). Thus, the leaves of a fragment-tree contain a lot of information decomposed into very small fragments. As in [3] we drop the edge labels and labeled all but the leaf nodes with a new, unique label. Hence, the tree-structure represents the abstract structure of the molecule with the chemical information in the leaves.

Figure 6 shows a molecule on the left-hand side which could be encoded by the SMILES-string $N - c1ccc(cc1) - O - c2ccc(cc2) - [Cl]$. This string represents the same as $N\{_0cccc(cc)_0\}O\{_1cccc(cc)_1\}[Cl]$. The corresponding fragment-tree is shown on the right-hand side of Figure 6.

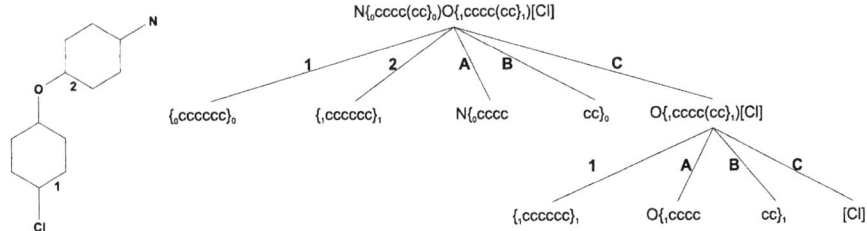

Fig. 6. A molecule with the encoding $N - c1ccc(cc1) - O - c2ccc(cc2) - [Cl]$ and the corresponding fragment-tree

Experimental Results. Predictive accuracy for each approach was estimated using ten-fold cross-validation. Reported are average accuracies and standard deviation (if known). For TREE^2, trees were induced at the 95% significance level for χ^2 and with a cut-off value of 0.01 for *Information Gain*. The results reported in [16] were achieved using PROGOL and working only on structural information, in [17], numerical values suggested by experts were used as well. This work reports only an average accuracy. The resulting accuracies and the size of the corresponding theories are shown in Table 2.

As can be seen, for both measures TREE^2 gives similar results to the purely structural PROGOL approach, with the differences being not significant. At the same time, the χ^2 induced model is far smaller than the other two. Again, the patterns tested against in the inner nodes are not overly complex (5-11 nodes). When PROGOL uses the expert-identified attributes as well, its accuracy increases. Since we do not have access to the standard deviation of these experiments, we cannot make a significance statement. Finally, the base-line approach,

Table 2. Accuracies and complexity of the models on the mutagenicity dataset

Approach	Predictive Accuracy	Average Size of the Model
TREE² χ^2	80.26±7.14	2.3 Nodes
TREE² IG	81.76±9	11.8 Nodes
PROGOL '94 [16]	80±3	9 Clauses
PROGOL '95 [17]	84	4 Clauses
FREQUENT SMILES [3]	86.70	214 Patterns

which mined all patterns frequent in one class and not exceeding a given frequency in the other class, and built a model using these features in an SVM, significantly outperforms the TREE² classifiers. On the other hand, by using a SVM, the results will hardly be interpretable for humans anymore and the amount of patterns used is larger than in the TREE² models by two orders of magnitude.

5 Conclusion and Future Work

We presented TREE², an integrated approach to structural classification. The algorithm builds a decision tree for tree structured data that tests for pattern occurrence in the inner nodes. Using an optimal branch-and-bound search, made possible by effective pruning, TREE² finds the most discriminative pattern for each subset of the data considered. This allows the user to abstract the success of the classifier from decisions about the search process, unlike in existing approaches that include heuristics. Basing the stopping criterion for growing the decision tree on statistically well founded measures rather than arbitrary thresholds whose meaning is somewhat ambiguous gives the user better guidance for selecting this parameter. It also alleviates the main problem of the support-confidence framework, namely the generation of very large rule sets that are incomprehensible to the user and possibly include uninformative rules w.r.t. classification.

As the experiments show, TREE² classifiers are effective while being less complex than existing approaches. While using χ^2 for assessing the quality of discriminative patterns, raising or lowering the significance threshold affects the induced trees in an expected manner. In contrast, using *Information Gain* is more difficult, since selecting the cut-off value has no statistical foundations. While base-line approaches, that separate feature generation and classifier construction, achieve very good results, it is not entirely clear how to justify the selected the number of features mined. Furthermore, there exists a gap in interpretability since the classifier used might combine the mined features in a way that is not easily accessible to the user.

So far, we have restricted ourselves to a single representation, *trees*, a certain type of classifier, *decision trees*, and two measures. Future work will include evaluating other correlation measures and applying our approach to different representations. Finally, the success of using effective conflict resolution strategies in the XRULES classifier suggests the expansion our approach to ensemble classifiers.

Acknowledgments. We would like to thank Mohammed J. Zaki for providing the datasets and the XRULES algorithm. Furthermore, we would like to thank Andreas Karwath, Kristian Kersting, Robert Egginton and Luc De Raedt for interesting discussions and comments to our work.

References

1. Kilpeläinen, P.: Tree Matching Problems with Applications to Structured Text Databases. PhD thesis, University of Helsinki (1992)
2. Kramer, S., Raedt, L.D., Helma, C.: Molecular feature mining in HIV data. In Provost, F., Srikant, R., eds.: Proc. KDD-01, New York, ACM Press (2001) 136–143
3. Bringmann, B., Karwath, A.: Frequent SMILES. In: Lernen, Wissensentdeckung und Adaptivität, Workshop GI Fachgruppe Maschinelles Lernen, LWA. (2004)
4. Zaki, M.J., Aggarwal, C.C.: XRules: an effective structural classifier for XML data. In Getoor, L., Senator, T.E., Domingos, P., Faloutsos, C., eds.: KDD, Washington, DC, USA, ACM (2003) 316–325
5. Geamsakul, W., Matsuda, T., Yoshida, T., Motoda, H., Washio, T.: Performance evaluation of decision tree graph-based induction. In Grieser, G., Tanaka, Y., Yamamoto, A., eds.: Discovery Science, Sapporo, Japan, Springer (2003) 128–140
6. Quinlan, J.R.: Learning logical definitions from relations. Machine Learning **5** (1990) 239–266
7. Muggleton, S.: Inverse entailment and PROGOL. New Generation Computing **13** (1995) 245–286
8. Morishita, S., Sese, J.: Traversing itemset lattices with statistical metric pruning. In: Proceedings of the Nineteenth ACM SIGACT-SIGMOD-SIGART Symposium on Principles of Database Systems, Dallas, Texas, USA, ACM (2000) 226–236
9. Liu, B., Hsu, W., Ma, Y.: Integrating classification and association rule mining. In Agrawal, R., Stolorz, P.E., Piatetsky-Shapiro, G., eds.: KDD, New York City, New York, USA, AAAI Press (1998) 80–86
10. Mutter, S., Hall, M., Frank, F.: Using classification to evaluate the output of confidence-based association rule mining. In Webb, G.I., Yu, X., eds.: Australian Conference on Artificial Intelligence, Cairns, Australia, Springer (2004) 538–549
11. Zimmermann, A., De Raedt, L.: Corclass: Correlated association rule mining for classification. [22] 60–72
12. Gärtner, T., Lloyd, J.W., Flach, P.A.: Kernels and distances for structured data. Machine Learning **57** (2004)
13. Quinlan, J.R.: Induction of decision trees. Machine Learning **1** (1986) 81–106
14. Frank, E., Hall, M., Trigg, L.E., Holmes, G., Witten, I.H.: Data mining in bioinformatics using weka. Bioinformatics **20** (2004) 2479–2481
15. Quinlan, J.R.: C4.5: Programs for Machine Learning. Morgan Kaufmann (1993)
16. Srinivasan, A., Muggleton, S., King, R., Sternberg, M.: Mutagenesis: ILP experiments in a non-determinate biological domain. In Wrobel, S., ed.: Proceedings of the 4th International Workshop on Inductive Logic Programming. Volume 237., Gesellschaft für Mathematik und Datenverarbeitung MBH (1994) 217–232

17. King, R.D., Sternberg, M.J.E., Srinivasan, A.: Relating chemical activity to structure: An examination of ILP successes. New Generation Comput. **13** (1995) 411–433
18. Weininger, D.: SMILES, a chemical language and information system 1. Introduction and encoding rules. J. Chem. Inf. Comput. Sci. **28** (1988) 31–36
19. The OpenBabel Software Community: Open Babel. http://openbabel.sourceforge.net/ (2003)
20. Daylight Chemical Information Systems, Inc. http://www.daylight.com/ (2004)
21. Karwath, A., De Raedt, L.: Predictive graph mining. [22] 1–15
22. Suzuki, E., Arikawa, S., eds.: Discovery Science, 7th International Conference, DS 2004, Padova, Italy, October 2-5, 2004, Proceedings. In Suzuki, E., Arikawa, S., eds.: DS 2004, Padova, Italy, Springer (2004)

Agglomerative Hierarchical Clustering with Constraints: Theoretical and Empirical Results

Ian Davidson and S.S. Ravi

Department of Computer Science, University at Albany - State University of New York,
Albany, NY 12222
{davidson, ravi}@cs.albany.edu

Abstract. We explore the use of instance and cluster-level constraints with agglomerative hierarchical clustering. Though previous work has illustrated the benefits of using constraints for non-hierarchical clustering, their application to hierarchical clustering is not straight-forward for two primary reasons. First, some constraint combinations make the feasibility problem (Does there exist a single feasible solution?) **NP**-complete. Second, some constraint combinations when used with traditional agglomerative algorithms can cause the dendrogram to stop prematurely in a dead-end solution even though there exist other feasible solutions with a significantly smaller number of clusters. When constraints lead to efficiently solvable feasibility problems and standard agglomerative algorithms do not give rise to dead-end solutions, we empirically illustrate the benefits of using constraints to improve cluster purity and average distortion. Furthermore, we introduce the new γ constraint and use it in conjunction with the triangle inequality to considerably improve the efficiency of agglomerative clustering.

1 Introduction and Motivation

Hierarchical clustering algorithms are run once and create a dendrogram which is a tree structure containing a k-block set partition for each value of k between 1 and n, where n is the number of data points to cluster allowing the user to choose a particular clustering granularity. Though less popular than non-hierarchical clustering there are many domains [16] where clusters naturally form a hierarchy; that is, clusters are part of other clusters. Furthermore, the popular agglomerative algorithms are easy to implement as they just begin with each point in its own cluster and progressively join the closest clusters to reduce the number of clusters by 1 until $k = 1$. The basic agglomerative hierarchical clustering algorithm we will improve upon in this paper is shown in Figure 1. However, these added benefits come at the cost of time and space efficiency since a typical implementation with symmetrical distances requires $\Theta(mn^2)$ computations, where m is the number of attributes used to represent each instance.

In this paper we shall explore the use of instance and cluster level constraints with hierarchical clustering algorithms. We believe the use of such constraints with *hierarchical* clustering is the first though there exists work that uses spatial constraints to find specific types of clusters and avoid others [14,15]. The similarly named *constrained hierarchical clustering* [16] is actually a method of combining partitional and hierarchical clustering algorithms; the method does not incorporate apriori constraints. Recent work

Agglomerative($S = \{x_1, \ldots, x_n\}$) **returns** $Dendrogram_k$ for $k = 1$ to $|S|$.

1. $C_i = \{x_i\}, \forall i$.
2. **for** $k = |S|$ **down to** 1
    $Dendrogram_k = \{C_1, \ldots, C_k\}$
    $d(i,j) = D(C_i, C_j), \forall i, j; \quad l, m = argmin_{a,b}\, d(a,b)$.
    $C_l = Join(C_l, C_m); \quad Remove(C_m)$.
   endloop

Fig. 1. Standard Agglomerative Clustering

[1,2,12] in the non-hierarchical clustering literature has explored the use of instance-level constraints. The **must-link** and **cannot-link** constraints require that two instances must both be part of or not part of the same cluster respectively. They are particularly useful in situations where a large amount of unlabeled data to cluster is available along with some labeled data from which the constraints can be obtained [12]. These constraints were shown to improve cluster purity when measured against an extrinsic class label not given to the clustering algorithm [12]. The δ **constraint** requires the distance between any pair of points in two different clusters to be at least δ. For any cluster C_i with two or more points, the ϵ-**constraint** requires that for each point $x \in C_i$, there must be another point $y \in C_i$ such that the distance between x and y is at most ϵ. Our recent work [4] explored the computational complexity (difficulty) of the *feasibility* problem: **Given** a value of k, does there exist at least one clustering solution that satisfies all the constraints and has k clusters? Though it is easy to see that there is no feasible solution for the three cannot-link constraints $CL(a,b), CL(b,c), CL(a,c)$ for $k < 3$, the general feasibility problem for cannot-link constraints is **NP**-complete by a reduction from the graph coloring problem. The complexity results of that work, shown in Table 1 (2nd column), are important for data mining because when problems are shown to be intractable in the worst-case, we should avoid them or should not expect to find an exact solution efficiently.

We begin this paper by exploring the feasibility of agglomerative **hierarchical** clustering under the above four mentioned instance and cluster-level constraints. This problem is *significantly* different from the feasibility problems considered in our previous work since the value of k for hierarchical clustering is not given. We then empirically show that constraints with a modified agglomerative hierarchical algorithm can improve the quality and performance of the resultant dendrogram. To further improve performance we introduce the γ constraint which when used with the triangle inequality can yield large computation saving that we have bounded in the best and average case. Finally, we cover the interesting result of an irreducible clustering. If we are given a feasible clustering with k_{max} clusters then for certain combination of constraints joining the two closest clusters may yield a feasible but "dead-end" solution with k clusters from which no other feasible solution with less than k clusters can be obtained, even though they are known to exist. Therefore, the created dendrograms may be incomplete.

Throughout this paper $D(x, y)$ denotes the Euclidean distance between two points and $D(X, Y)$ the Euclidean distance between the centroids of two groups of instances. We note that the feasibility and irreducibility results (Sections 2 and 5) are not neces-

Table 1. Results for Feasibility Problems for a Given k (partitional clustering) and Unspecified k (hierarchical clustering)

Constraint	Given k	Unspecified k	Unspecified k - Deadends?
Must-Link	P [9,4]	P	No
Cannot-Link	NP-complete [9,4]	P	Yes
δ-constraint	P [4]	P	No
ϵ-constraint	P [4]	P	No
Must-Link and δ	P [4]	P	No
Must-Link and ϵ	NP-complete [4]	P	No
δ and ϵ	P [4]	P	No
Must-Link, Cannot-Link, δ and ϵ	NP-complete [4]	NP-complete	Yes

sarily for Euclidean distances and are hence applicable for single and complete linkage clustering while the γ-constraint to improve performance (Section 4) is applicable to any metric space.

2 Feasibility for Hierarchical Clustering

In this section, we examine the feasibility problem for several different types of constraints, that is, the problem of determining whether the given set of points can be partitioned into clusters so that all the specified constraints are satisfied.

Definition 1. *Feasibility problem for Hierarchical Clustering* (FHC)

Instance: A set S of nodes, the (symmetric) distance $d(x, y) \geq 0$ for each pair of nodes x and y in S and a collection C of constraints.

Question: Can S be partitioned into subsets (clusters) so that all the constraints in C are satisfied?

When the answer to the feasibility question is "yes", the corresponding algorithm also produces a partition of S satisfying the constraints. We note that the FHC problem considered here is *significantly* different from the constrained non-hierarchical clustering problem considered in [4] and the proofs are different as well even though the end results are similar. For example in our earlier work we showed intractability results for some constraint types using a straightforward reduction from the graph coloring problem. The intractability proof used in this work involves more elaborate reductions. For the feasibility problems considered in [4], the number of clusters is in effect, another constraint. In the formulation of FHC, there are *no* constraints on the number of clusters, other than the trivial ones (i.e., the number of clusters must be at least 1 and at most $|S|$).

We shall in this section begin with the same constraints as those considered in [4]. They are: (a) Must-Link (ML) constraints, (b) Cannot-Link (CL) constraints, (c) δ constraint and (d) ϵ constraint. In later sections we shall introduce another cluster-level constraint to improve the efficiency of the hierarchical clustering algorithms. As observed in [4], a δ constraint can be efficiently transformed into an equivalent collection of ML-constraints. Therefore, we restrict our attention to ML, CL and ϵ constraints. We show that for any *pair* of these constraint types, the corresponding feasibility problem can be solved efficiently. The simple algorithms for these feasibility problems can be used to seed an agglomerative or divisive hierarchical clustering algorithm as is the case in our experimental results. However, when all three types of constraints are specified, we show that the feasibility problem is **NP**-complete and hence finding a clustering, let alone a good clustering, is computationally intractable.

2.1 Efficient Algorithms for Certain Constraint Combinations

When the constraint set C contains only ML and CL constraints, the FHC problem can be solved in polynomial time using the following simple algorithm.

1. Form the clusters implied by the ML constraints. (This can be done by computing the transitive closure of the ML constraints as explained in [4].) Let $C_1, C_2, \ldots, C_p$ denote the resulting clusters.
2. If there is a cluster C_i ($1 \leq i \leq p$) with nodes x and y such that x and y are also involved in a CL constraint, then there is no solution to the feasibility problem; otherwise, there is a solution.

When the above algorithm indicates that there is a feasible solution to the given FHC instance, one such solution can be obtained as follows. Use the clusters produced in Step 1 along with a singleton cluster for each node that is not involved in an ML constraint. Clearly, this algorithm runs in polynomial time. We now consider the combination of CL and ϵ constraints. Note that there is always a trivial solution consisting of $|S|$ singleton clusters to the FHC problem when the constraint set involves only CL and ϵ constraints. Obviously, this trivial solution satisfies both CL and ϵ constraints, as the latter constraint only applies to clusters containing two or more instances.

The FHC problem under the combination of ML and ϵ constraints can be solved efficiently as follows. For any node x, an ϵ-**neighbor** of x is another node y such that $D(x, y) \leq \epsilon$. Using this definition, an algorithm for solving the feasibility problem is:

1. Construct the set $S' = \{x \in S : x \text{ does not have an } \epsilon\text{-neighbor}\}$.
2. If some node in S' is involved in an ML constraint, then there is no solution to the FHC problem; otherwise, there is a solution.

When the above algorithm indicates that there is a feasible solution, one such solution is to create a singleton cluster for each node in S' and form one additional cluster containing all the nodes in $S - S'$. It is easy to see that the resulting partition of S satisfies the ML and ϵ constraints and that the feasibility testing algorithm runs in polynomial time. The following theorem summarizes the above discussion and indicates that we can extend the basic agglomerative algorithm with these combinations of constraint types to perform efficient hierarchical clustering. However, it does not mean that we can always use traditional agglomerative clustering algorithms as the closest-cluster-join operation can yield dead-end clustering solutions as discussed in Section 5.

Theorem 1. *The* FHC *problem can be solved efficiently for each of the following combinations of constraint types: (a) ML and CL (b) CL and ϵ and (c) ML and ϵ.* □

2.2 Feasibility Under ML, CL and ϵ Constraints

In this section, we show that the FHC problem is **NP**-complete when all the three constraint types are involved. This indicates that creating a dendrogram under these constraints is an intractable problem and the best we can hope for is an approximation algorithm that may **not** satisfy all constraints. The **NP**-completeness proof uses a reduction from the One-in-Three 3SAT with positive literals problem (OPL) which is known to be **NP**-complete [11]. For each instance of the OPL problem we can construct a constrained clustering problem involving ML, CL and ϵ constraints. Since complexity results are worse case, the existence of just these problems is sufficient for theorem 2.
One-in-Three 3SAT with Positive Literals (OPL)

Instance: A set $C = \{x_1, x_2, \ldots, x_n\}$ of n Boolean variables, a collection $Y = \{Y_1, Y_2, \ldots, Y_m\}$ of m clauses, where each clause $Y_j = (x_{j_1}, x_{j_2}, x_{j_3})$ has exactly three non-negated literals.

Question: Is there an assignment of truth values to the variables in C so that exactly one literal in each clause becomes true?

Theorem 2. *The* FHC *problem is **NP**-complete when the constraint set contains ML, CL and ϵ constraints.*

The proof of the above theorem is somewhat lengthy and is omitted because of space reasons. (The proof appears in an expanded technical report version of this paper [5] that is available on-line.)

3 Using Constraints for Hierarchical Clustering: Algorithm and Empirical Results

To use constraints with hierarchical clustering we change the algorithm in Figure 1 to factor in the above discussion. As an example, a constrained hierarchical clustering algorithm with must-link and cannot-link constraints is shown in Figure 2. In this section we illustrate that constraints can improve the quality of the dendrogram. We purposefully chose a small number of constraints and believe that even more constraints will improve upon these results. We will begin by investigating must-link and cannot-link constraints using six real world UCI datasets. For each data set we clustered all instances but removed the labels from 90% of the data (S_u) and used the remaining 10% (S_l) to generate constraints. We randomly selected two instances at a time from S_l and generated must-link constraints between instances with the same class label and cannot-link constraints between instances of differing class labels. We repeated this process twenty times, each time generating 250 constraints of each type. The performance measures reported are averaged over these twenty trials. All instances with missing values were

Table 2. Average Distortion per Instance and Average Percentage Cluster Purity over Entire Dendrogram

Data Set	Distortion		Purity	
	Unconstrained	Constrained	Unconstrained	Constrained
Iris	3.2	2.7	58%	66%
Breast	8.0	7.3	53%	59%
Digit (3 vs 8)	17.1	15.2	35%	45%
Pima	9.8	8.1	61%	68%
Census	26.3	22.3	56%	61%
Sick	17.0	15.6	50%	59%

ConstrainedAgglomerative(S,ML,CL) **returns** $Dendrogram_i$, $i = k_{min} \ldots k_{max}$

Notes: In Step 5 below, the term "mergeable clusters" is used to denote a pair of clusters whose merger does not violate any of the given CL constraints. The value of t at the end of the loop in Step 5 gives the value of k_{min}.

1. Construct the transitive closure of the ML constraints (see [4] for an algorithm) resulting in r connected components $M_1, M_2, \ldots, M_r$.
2. If two points $\{x, y\}$ are both a CL and ML constraint then output "No Solution" and stop.
3. Let $S_1 = S - (\bigcup_{i=1}^{r} M_i)$. Let $k_{max} = r + |S_1|$.
4. Construct an initial feasible clustering with k_{max} clusters consisting of the r clusters $M_1, \ldots, M_r$ and a singleton cluster for each point in S_1. Set $t = k_{max}$.
5. **while** (there exists a pair of mergeable clusters) **do**
   (a) Select a pair of clusters C_l and C_m according to the specified distance criterion.
   (b) Merge C_l into C_m and remove C_l. (The result is $Dendrogram_{t-1}$.)
   (c) $t = t - 1$.
   endwhile

Fig. 2. Agglomerative Clustering with ML and CL Constraints

removed as hierarchical clustering algorithms do not easily handle such instances. Furthermore, all non-continuous columns were removed as there is no standard distance measure for discrete columns.

Table 2 illustrates the quality improvement that the must-link and cannot-link constraints provide. Note that we compare the dendrograms for k values between k_{min} and k_{max}. For each corresponding level in the unconstrained and constrained dendrogram we measure the average distortion $(1/n * \sum_{i=1}^{n} D(x_i - C_{f(x_i)})$, where $f(x_i)$ returns the index of the closest cluster to x_i) and present the average over all levels. It is important to note that we are not claiming that agglomerative clustering has distortion as an objective function, rather that it is a good measure of cluster quality. We see that the distortion improvement is typically of the order of 15%. We also see that the average percentage purity of the clustering solution as measured by the class label purity improves. The cluster purity is measured against the extrinsic class labels. We believe these improvement are due to the following. When many pairs of clusters have similar short distances, the must-link constraints guide the algorithm to a better join. This type of improvement occurs at the bottom of the dendrogram. Conversely, towards the top of the dendrogram the cannot-link constraints rule out ill-advised joins. However, this preliminary explanation requires further investigation which we intend to address in the future. In particular, a study of the most informative constraints for hierarchical clustering remains an open question, though promising preliminary work for the area of non-hierarchical clustering exists [2].

We next use the cluster-level δ constraint with an arbitrary value to illustrate the great computational savings that such constraints offer. Our earlier work [4] explored ϵ and δ constraints to provide background knowledge towards the "type" of clusters we wish to find. In that paper we explored their use with the Aibo robot to find objects in images that were more than 1 foot apart as the Aibo can only navigate between such objects. For these UCI data sets no such background knowledge exists and how to set these constraint values for non-spatial data remains an active research area. Hence we test these constraints with arbitrary values. We set δ equal to 10 times the average distance between a pair of points. Such a constraint will generate hundreds even thousands of must-link constraints that can greatly influence the clustering results and algorithm efficiency as shown in Table 3. We see that the minimum improvement was 50% (for Census) and nearly 80% for Pima. This improvement is due to the constraints effectively creating a pruned dendrogram by making $k_{max} \ll n$.

Table 3. The Rounded Mean Number of Pair-wise Distance Calculations for an Unconstrained and Constrained Clustering using the δ constraint

Data Set	Unconstrained	Constrained
Iris	22,201	3,275
Breast	487,204	59,726
Digit (3 vs 8)	3,996,001	990,118
Pima	588,289	61,381
Census	2,347,305,601	563,034,601
Sick	793,881	159,801

4 Using the γ Constraint to Improve Performance

In this section we introduce a new constraint, the γ constraint and illustrate how the triangle inequality can be used to further improve the run-time performance of agglomerative hierarchical clustering. Though this improvement does not affect the worst-case analysis, we can perform a best case analysis and an expected performance improvement using the Markov inequality. Future work will investigate if tighter bounds can be found. There exists other work involving the triangle inequality but not constraints for non-hierarchical clustering [6] as well as for hierarchical clustering [10].

Definition 2. *(The γ Constraint For Hierarchical Clustering) Two clusters whose geometric centroids are separated by a distance greater than γ cannot be joined.*

IntelligentDistance (γ, $C = \{C_1, \ldots, C_k\}$)
returns $d(i,j) \; \forall i, j$.

1. **for** $i = 2$ **to** $n - 1$ $d_{1,i} = D(C_1, C_i)$ endloop
2. **for** $i = 2$ to $n - 1$
    for $j = i+1$ to $n - 1$ $\hat{d}_{i,j} = |d_{1,i} - d_{1,j}|$
    if $\hat{d}_{i,j} > \gamma$ then $d_{i,j} = \gamma + 1$; *do not join*   **else** $d_{i,j} = D(x_i, x_j)$
    endloop
  endloop
3. return $d_{i,j}, \forall i, j$.

Fig. 3. Function for Calculating Distances Using the γ Constraint and the Triangle Inequality

The γ constraint allows us to specify how geometrically well separated the clusters should be. Recall that the triangle inequality for three points a, b, c refers to the expression $|D(a,b) - D(b,c)| \leq D(a,c) \leq D(a,b) + D(c,b)$ where D is the Euclidean distance function or any other metric function. We can improve the efficiency of the hierarchical clustering algorithm by making use of the lower bound in the triangle inequality and the γ constraint. Let a, b, c now be cluster centroids and we wish to determine the closest two centroids to join. If we have already computed $D(a,b)$ and $D(b,c)$ and the value $|D(a,b) - D(b,c)|$ exceeds γ, then we need not compute the distance between a and c as the lower bound on $D(a,c)$ already exceeds γ and hence a and c cannot be joined. Formally the function to calculate distances using geometric reasoning at a particular dendrogram level is shown in Figure 3. Central to the approach is that the distance between a central point (c) (in this case the first) and every other point is calculated. Therefore, when bounding the distance between two instances (a, b) we effectively calculate a triangle with two edges with know lengths incident on c and thereby lower bound the distance between a and b. How to select the best central point and the use of multiple central points remains future important research.

If the triangle inequality bound exceeds γ, then we save making m floating point power calculations if the data points are in m dimensional space. As mentioned earlier we have no reason to believe that there will be at least one situation where the triangle inequality saves computation in *all problem instances*; hence in the worst case, there is no performance improvement. But in practice it is expected to occur and hence we can explore the best and expected case results.

4.1 Best Case Analysis for Using the γ Constraint

Consider the n points to cluster $\{x_1, \ldots, x_n\}$. The first iteration of the agglomerative hierarchical clustering algorithm using symmetrical distances is to compute the distance between each point and every other point. This involves the computation $(D(x_1, x_2)$, $D(x_1, x_3), \ldots, D(x_1, x_n)), \ldots, (D(x_i, x_{i+1}), D(x_i, x_{i+2}), \ldots, D(x_i, x_n)), \ldots, (D(x_{n-1}, x_n))$, which corresponds to an arithmetic series $n - 1 + n - 2 + \ldots + 1$ of computations. Thus for agglomerative hierarchical clustering using *symmetrical* distances the number of distance computations is $n(n-1)/2$ for the base level. At the next

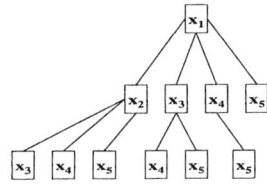

Fig. 4. A Simple Illustration for a Five Instance Problem of How the Triangular Inequality Can Save Distance Computations

level we need only recalculate the distance between the newly created cluster and the remaining $n-2$ points and so on. Therefore, the total number of distance calcluation is $n(n-1)/2+(n-1)(n-2)/2 = (n-1)^2$. We can view the base level calculation pictorially as a tree construction as shown in Figure 4. If we perform the distance calculation at the first level of the tree then we can obtain bounds using the triangle inequality for **all** branches in the second level. This is as bounding the distance between two points requires the distance between these points and a common point, which in our case is x_1. Thus in the best case there are only $n-1$ distance computations instead of $(n-1)^2$.

4.2 Average Case Analysis for Using the γ Constraint

However, it is highly unlikely that the best case situation will ever occur. We now focus on the average case analysis using the Markov inequality to determine the *expected* performance improvement which we later empirically verify. Let ρ be the average distance between any two instances in the data set to cluster. The triangle inequality provides a lower bound; if this bound exceeds γ, computational savings will result. We can bound how often this occurs if we can express γ in terms of ρ, hence let $\gamma = c\rho$.

Recall that the general form of the Markov inequality is: $P(X = x \geq a) \leq \frac{E(X)}{a}$, where x is a single value of the continuous random variable X, a is a constant and $E(X)$ is the expected value of X. In our situation since X is distance between two points chosen at random, $E(X) = \rho$ and $\gamma = a = c\rho$ as we wish to determine when the distance will exceed γ. Therefore, at the lowest level of the tree ($k = n$) then the *number of times* the triangle inequality will save us computation time is $n\frac{E(X)}{a} = n\frac{\rho}{c\rho} = n/c$, indicating a saving of a factor of $1/c$ at this lowest level. As the Markov inequality is a rather weak bound then in practice the saving may be substantially different as we shall see in our empirical section. The computation saving that are obtained at the bottom of the dendrogram are reflected at higher levels of the dendrogram. When growing the entire dendrogram we will save at least $n/c+(n-1)/c\ldots+1/c$ distance calculations. This is an arithmetic sequence with the additive constant being $1/c$ and hence the total expected computations saved is at least $n/2(2/c+(n-1)/c) = (n^2+n)/2c$. As the total computations for regular hierarchical clustering is $(n-1)^2$, the computational saving is expected to be by a approximately a factor of $1/2c$.

Consider the 150 instance IRIS data set ($n=150$) where the average distance (with attribute value ranges all being normalized to between 0 and 1) between two instances is 0.6; that is, $\rho = 0.6$. If we state that we do not wish to join clusters whose centroids are

Table 4. The Efficiency of Using the Geometric Reasoning Approach from Section 4 (Rounded Mean Number of Pair-wise Distance Calculations)

Data Set	Unconstrained	Using γ Constraint
Iris	22,201	19,830
Breast	487,204	431,321
Digit (3 vs 8)	3,996,001	3,432,021
Pima	588,289	501,323
Census	2,347,305,601	1,992,232,981
Sick	793,881	703,764

separated by a distance greater than 3.0, then $\gamma = 3.0 = 5\rho$. By not using the γ constraint and the triangle inequality the total number of computations is 22201, and the number of computations that are saved is at least $(150^2 + 150)/10 = 2265$; hence the saving is about 10%. We now show that the γ constraint can be used to improve efficiency of the basic agglomerative clustering algorithm. Table 4 illustrates the improvement that using a γ constraint equal to five times the average pairwise instance distance. We see that the average improvement is consistent with the average case bound derived above.

5 Constraints and Irreducible Clusterings

In the presence of constraints, the set partitions at each level of the dendrogram must be feasible. We have formally shown that if k_{max} is the maximum value of k for which a feasible clustering exists, then there is a way of joining clusters to reach another clustering with k_{min} clusters [5]. In this section we ask the following question: will traditional agglomerative clustering find a feasible clustering for each value of k between k_{max} and k_{min}? We formally show that in the worse case, for certain types of constraints (and combinations of constraints), if mergers are performed in an arbitrary fashion (including the traditional hierarchical clustering algorithm, see Figure 1), then the dendrogram may prematurely dead-end. A premature dead-end implies that the dendrogram reaches a stage where no pair of clusters can be merged without violating one or more constraints, even though other sequences of mergers may reach significantly higher levels of the dendrogram. We use the following definition to capture the informal notion of a "premature end" in the construction of a dendrogram. How to perform agglomerative clustering in these dead-end situations remains an important open question.

Definition 3. *A feasible clustering* $C = \{C_1, C_2, \ldots, C_k\}$ *of a set S is* **irreducible** *if no pair of clusters in C can be merged to obtain a feasible clustering with $k - 1$ clusters.*

The remainder of this section examines the question of which combinations of constraints can lead to premature stoppage of the dendrogram. We first consider each of the ML, CL and ϵ-constraints separately. It is easy to see that when only ML-constraints are used, the dendrogram can reach all the way up to a single cluster, no matter how mergers are done. The following illustrative example shows that with CL-constraints, if mergers are not done correctly, the dendrogram may stop prematurely.

Example: Consider a set S with $4k$ nodes. To describe the CL constraints, we will think of S as the union of four pairwise disjoint sets X, Y, Z and W, each with k nodes. Let $X = \{x_1, x_2, \ldots, x_k\}$, $Y = \{y_1, y_2, \ldots, y_k\}$, $Z = \{z_1, z_2, \ldots, z_k\}$ and $W = \{w_1, w_2, \ldots, w_k\}$. The CL-constraints are as follows. *(a)* There is a CL-constraint for each pair of nodes $\{x_i, x_j\}$, $i \neq j$, *(b)* There is a CL-constraint for each pair of nodes $\{w_i, w_j\}$, $i \neq j$, *(c)* There is a CL-constraint for each pair of nodes $\{y_i, z_j\}$, $1 \leq i, j \leq k$.

Assume that the distance between each pair of nodes in S is 1. Thus, nearest-neighbor mergers may lead to the following feasible clustering with $2k$ clusters: $\{x_1, y_1\}$, $\{x_2, y_2\}, \ldots, \{x_k, y_k\}$, $\{z_1, w_1\}$, $\{z_2, w_2\}, \ldots, \{z_k, w_k\}$. This collection of clusters can be seen to be irreducible in view of the given CL constraints. However, a feasible clustering with k clusters is possible: $\{x_1, w_1, y_1, y_2, \ldots, y_k\}$, $\{x_2, w_2, z_1, z_2, \ldots, z_k\}$, $\{x_3, w_3\}, \ldots, \{x_k, w_k\}$. Thus, in this example, a carefully constructed dendrogram allows k additional levels. □

When only the ϵ-constraint is considered, the following lemma points out that there is only one irreducible configuration; thus, no premature stoppages are possible. In proving this lemma, we will assume that the distance function is symmetric.

Lemma 1. *If S is a set of nodes to be clustered under an ϵ-constraint. Any irreducible and feasible collection C of clusters for S must satisfy the following two conditions.*

(a) C contains at most one cluster with two or more nodes of S.
(b) Each singleton cluster in C contains a node x with no ϵ-neighbors in S.

Proof: Suppose C has two or more clusters, say C_1 and C_2, such that each of C_1 and C_2 has two or more nodes. We claim that C_1 and C_2 can be merged without violating the ϵ-constraint. This is because each node in C_1 (C_2) has an ϵ-neighbor in C_1 (C_2) since C is feasible and distances are symmetric. Thus, merging C_1 and C_2 cannot violate the ϵ-constraint. This contradicts the assumption that C is irreducible and the result of Part (a) follows. The proof for Part (b) is similar. Suppose C has a singleton cluster $C_1 = \{x\}$ and the node x has an ϵ-neighbor in some cluster C_2. Again, C_1 and C_2 can be merged without violating the ϵ-constraint. □

Lemma 1 can be seen to hold even for the combination of ML and ϵ constraints since ML constraints cannot be violated by merging clusters. Thus, no matter how clusters are merged at the intermediate levels, the highest level of the dendrogram will always correspond to the configuration described in the above lemma when ML and ϵ constraints are used. In the presence of CL-constraints, it was pointed out through an example that the dendrogram may stop prematurely if mergers are not carried out carefully. It is easy to extend the example to show that this behavior occurs even when CL-constraints are combined with ML-constraints or an ϵ-constraint.

6 Conclusion and Future Work

Our paper made two significant theoretical results. Firstly, the feasibility problem for *unspecified* k is studied and we find that clustering under all four types (ML, CL, ϵ and δ) of constraints is **NP**-complete; hence, creating a feasible dendrogram is intractable. These results are fundamentally different from our earlier work [4] because the feasibility problem and proofs are quite different. Secondly, we proved under some constraint types (i.e. cannot-link) that traditional agglomerative clustering algorithms give rise to dead-end (irreducible) solutions. If there exists a feasible solution with k_{max} clusters then the traditional agglomerative clustering algorithm may not get all the way to a feasible solution with k_{min} clusters even though there exists feasible clusterings for each value between k_{max} and k_{min}. Therefore, the approach of joining the two "nearest" clusters may yield an incomplete dendrogram. How to perform clustering when dead-end feasible solutions exist remains an important open problem we intend to study.

Our experimental results indicate that small amounts of labeled data can improve the dendrogram quality with respect to cluster purity and "tightness" (as measured by the distortion). We find that the cluster-level δ constraint can reduce computational time between two and four fold by effectively creating a pruned dendrogram. To further improve the efficiency of agglomerative clustering we introduced the γ constraint, that allows the use of the triangle inequality to save computation time. We derived best case and expected case analysis for this situation which our experiments verified. Additional future work we will explore include constraints to create balanced dendrograms and the important asymmetric distance situation.

References

1. S. Basu, A. Banerjee, R. Mooney, Semi-supervised Clustering by Seeding, 19th ICML, 2002.
2. S. Basu, M. Bilenko and R. J. Mooney, Active Semi-Supervision for Pairwise Constrained Clustering, 4th SIAM Data Mining Conf.. 2004.
3. P. Bradley, U. Fayyad, and C. Reina, "Scaling Clustering Algorithms to Large Databases", 4th ACM KDD Conference. 1998.
4. I. Davidson and S. S. Ravi, "Clustering with Constraints: Feasibility Issues and the k-Means Algorithm", SIAM International Conference on Data Mining, 2005.
5. I. Davidson and S. S. Ravi, "Towards Efficient and Improved Hierarchical Clustering with Instance and Cluster-Level Constraints", Tech. Report, CS Department, SUNY - Albany, 2005. Available from: www.cs.albany.edu/~davidson
6. C. Elkan, Using the triangle inequality to accelerate k-means, ICML, 2003.
7. M. Garey and D. Johnson, Computers and Intractability: A Guide to the Theory of NP-completeness, Freeman and Co., 1979.
8. M. Garey, D. Johnson and H. Witsenhausen, "The complexity of the generalized Lloyd-Max problem", IEEE Trans. Information Theory, Vol. 28,2, 1982.
9. D. Klein, S. D. Kamvar and C. D. Manning, "From Instance-Level Constraints to Space-Level Constraints: Making the Most of Prior Knowledge in Data Clustering", ICML 2002.
10. M. Nanni, Speeding-up hierarchical agglomerative clustering in presence of expensive metrics, PAKDD 2005, LNAI 3518.
11. T. J. Schafer, "The Complexity of Satisfiability Problems", STOC, 1978.
12. K. Wagstaff and C. Cardie, "Clustering with Instance-Level Constraints", ICML, 2000.
13. D. B. West, Introduction to Graph Theory, Second Edition, Prentice-Hall, 2001.
14. K. Yang, R. Yang, M. Kafatos, "A Feasible Method to Find Areas with Constraints Using Hierarchical Depth-First Clustering", Scientific and Stat. Database Management Conf., 2001.
15. O. R. Zaiane, A. Foss, C. Lee, W. Wang, On Data Clustering Analysis: Scalability, Constraints and Validation, PAKDD, 2000.
16. Y. Zho & G. Karypis, "Hierarchical Clustering Algorithms for Document Datasets", Data Mining and Knowledge Discovery, Vol. 10 No. 2, March 2005, pp. 141–168.

Cluster Aggregate Inequality and Multi-level Hierarchical Clustering

Chris Ding and Xiaofeng He

Lawrence Berkeley National Laboratory,
Berkeley, California 94720, USA

Abstract. We show that (1) in hierarchical clustering, many linkage functions satisfy a cluster aggregate inequality, which allows an exact $O(N^2)$ multi-level (using mutual nearest neighbor) implementation of the standard $O(N^3)$ agglomerative hierarchical clustering algorithm. (2) a desirable close friends cohesion of clusters can be translated into kNN consistency which is guaranteed by the multi-level algorithm; (3) For similarity-based linkage functions, the multi-level algorithm is naturally implemented as graph contraction. The effectiveness of our algorithms is demonstrated on a number of real life applications.

1 Introduction

Agglomerative hierarchical clustering (AHC) is developed in 1960's and is widely used in practice. AHC produces a tree describing the hierarchical cluster structure. Such a comprehensive description of the data is quite useful for broad areas of applications. For example, in bioinformatics research, AHC is most commonly used for clustering genes in a DNA gene microarray expression data, because the resulting hierarchical cluster structure is readily recognizable by biologists. The phylogenetic tree (similar to binary clustering tree) of organisms is often built using the UPGMA (unweighted pair group method average) AHC algorithm. In social sciences, the hierarchical cluster structure often reveals gradual evolving social relationships that help explain complex social issues. Another application of AHC is in classification tasks on a large dataset using support vector machine [12]. The hierarchical cluster structure allows one to use most detailed local representation near the decision boundaries where support vectors lie; but as one moves away from the decision boundaries, the centroid representation of progressively larger clusters can be used.

Besides hierarchical clustering, many other clustering algorithms have been developed (see recent survey and text books [5,1,3]). K-means clustering is perhaps the most commonly used method and is well developed. The gaussian mixture model using EM algorithm directly improves over the K-means method by using a probabilistic model of cluster membership of each object. Both algorithms can be viewed as a global objective function optimization problem. A related set of graph clustering algorithms are developed that partition nodes into two sub-clusters based on well-motivated clustering objective functions [8]. They are typically applied in a top-down fashion (see also[7]), and thus complement the bottom-up AHC.

Standard AHC scales as $O(N^3)$. A detailed description of AHC and complete references can be found in [4,9]. A number of efficient implementation based on approximations have been proposed [10,6]. Several recent studies propose to integrate hierarchical clustering with additional information [2,11] or summary statistics[13].

In this paper, we focus on making AHC scale to large data set. Over the last 40 years, the basic AHC algorithm remains unchanged. The the basic algorithm is an iterative procedure; at each iteration, among all pairs of current clusters, the pair with largest linkage function value (smallest distance) are selected and merged.

We start with a key observation. In the AHC algorithm, at each iteration, one may merge all mutual-nearest-neighbor (1mn) pairs (defined by the linkage function) simultaneously in the same iteration. As long as the linkage function satisfies a "cluster aggregate inequality", this modified algorithm of simultaneously merge all 1mn-pairs at each iteration produces identical clustering results as the standard AHC algorithm. The cluster aggregate inequality is satisfied by most common linkage functions (see §2.2). This modified algorithm provides a natural multi-level implementation of the AHC, where at each level we merge all 1mn pairs. This *multi-level hierarchical clustering* (MLHC) algorithm provides an order-N speedup over the standard AHC.

Next we propose "close friends" cohesion as a desirable feature for clustering, which requires that for every member in a cluster, its closest friend is also in the same cluster. We show that the MLHC guarantees the close-friends cohesion, a desirable feature for clustering. We further extend this cluster membership cohesion idea to (mutual) nearest neighbor consistency, and show that MLHC improves this KNN consistency compare to other clustering algorithms. (§3)

When the linkage function is expressed in similarity (in contrast to distance or dissimilarity), the new algorithm is identical to multi-level *graph contraction*. Graph contraction provides a natural framework for hierarchical clustering (§4).

The effectiveness of our algorithms is demonstrated on a number of real life applications: DNA gene expressions for lung cancer, global climate pattern, and internet newsgroups (§5).

2 Multi-level Hierarchical Clustering (MLHC)

2.1 Algorithm

The standard agglomerative hierarchical clustering is a bottom-up process. During each step, we merge two *current* clusters C_p and C_q which are closest, or *most similar* among all pairs of current clusters:

$$\min_{<pq>} d(C_p, C_q).$$

where $d(\cdot,\cdot)$ is the dissimilarity-based (distance) linkage function between C_p and C_q. Many researches have studied the effects of different choice of *linkage* functions [4].

At each step of AHC with p current clusters, $p - 1$ new linkage functions need be computed, and we have total $O(p^2)$ pairwise linkage functions. It takes p^2 comparisons to search for the pair with the largest linkage. This is repeated $N - 1$ times. The total computation is

$$N_{\text{search}}^{\text{AHC}} = N^2 + (N - 1)^2 + \cdots + 2^2 + 1^2 = O(N^3/3).$$

The new MLHC algorithm is motivated by the following observation. In each iterative step in AHC, when all pairwise linkage functions are computed, we can form all mutual nearest neighbor pairs (1mn-pairs) of current clusters using the linkage as the distance metric. Two objects (i,j) are a 1mn-pair if j is the nearest neighbor of i and vice versa.

In this perspective, the standard AHC merges only the 1mn-pair with largest linkage value. It is then natural to ask if we may also merge all other 1mn-pairs simultaneously. Will this produces the same results? An analysis shows that if the linkage function satisfies a "cluster aggregate inequality", then the clustering results remain the same as the standard AHC.

This observation suggests a simultaneous merging algorithm which we call MLHC. At each level, all 1mn-pairs are identified and merged (not just the pair with largest linkage value). This is repeated until all objects are merged into one cluster. The total number of level is about $log_2 N$. Thus the required computation is approximately

$$N_{\text{search}}^{\text{MLHC}} = N^2 + (N/2)^2 + (N/4)^2 + 2^2 + 1^2 = O(4N^2/3).$$

The new algorithm speedups by a factor of order-N.

2.2 Cluster Aggregate Inequality

In this section, we show that the simultaneous merging of all 1mn-pairs in MLHC produces identical clustering results as the standard AHC, provided the linkage function satisfies a *Cluster Aggregate Inequality*.

Definition. *Cluster Aggregate Inequality* of the linkage function. Suppose we have three current clusters A, B, C. We try to merge B, C into a new cluster $B+C$. The cluster aggregate inequality is a property of the linkage function that the merged cluster $B + C$ is "no closer" to A than either one of its individual members B or C. More precisely, for distance (dissimilarity) based linkage $d(\cdot,\cdot)$, the cluster aggregate inequality is

$$d_{A,B+C} \geq \min(d_{A,C}, d_{A,B}) \tag{1}$$

for any triple (A, B, C).

What kind of linkage functions satisfy the cluster aggregate inequality? It is interesting to see that most commonly used linkage functions satisfy cluster aggregate inequality. Consider the four similarity-based linkage function. (i) the single linkage, defined as the closest distance among points in A, B, (ii) the complete linkage, defined as the farthest distance among points in A, B, (iii) the average linkage, defined as the average of all distances among points in A, B, (iv) the minimum variance linkage.

$$d^{\text{sgl}}_{A,B} = \min_{i \in A, j \in B} d_{ij} \qquad (2)$$

$$d^{\text{cmp}}_{A,B} = \max_{i \in A, j \in B} d_{ij} \qquad (3)$$

$$d^{\text{avg}}_{A,B} = \frac{d(A,B)}{n_A n_B} \qquad (4)$$

$$d^{\text{min-var}}_{A,B} = \frac{n_A n_B}{n_A + n_B} \|\mathbf{c}_A - \mathbf{c}_B\|^2. \qquad (5)$$

Theorem 1. The single link, the complete link and average link satisfy the strong cluster aggregate inequality.

Proof. For single link, one can easily see that

$$d^{\text{sgl}}_{A,B+C} = \min_{i \in A; j \in B+C} d_{ij} = \min(\min_{i \in A; j \in B} d_{ij}, \min_{i \in A; j \in C} d_{ij}) = \min(d^{\text{sgl}}_{A,B}, d^{\text{sgl}}_{A,C})$$

Thus the equality in Eq.(1) hold for single link. With same reasoning, one can see the inequality holds for complete linkage.

For average link, we have

$$d^{\text{avg}}_{A,B+C} = \sum_{i \in A; j \in B+C} \frac{d_{ij}}{|A||B+C|}$$

$$= \sum_{i \in A; j \in B} \frac{d_{ij}}{|A||B+C|} + \sum_{i \in A; j \in C} \frac{d_{ij}}{|A||B+C|}$$

$$= \frac{|B|}{|B+C|} d^{\text{avg}}_{A,B} + \frac{|C|}{|B+C|} d^{\text{avg}}_{A,C}$$

$$\geq \frac{|B|}{|B+C|} \min(d^{\text{avg}}_{A,B}, d^{\text{avg}}_{A,C}) + \frac{|C|}{|B+C|} \min(d^{\text{avg}}_{A,B}, d^{\text{avg}}_{A,C})$$

$$= \min(d^{\text{avg}}_{A,B}, d^{\text{avg}}_{A,C}) \qquad \square$$

2.3 Equivalence of MLHC and AHC

Cluster aggregate inequality plays a fundamental role in hierarchical clustering. It is similar to the triangle inequality in metric space: for any three vectors $\mathbf{x}_i, \mathbf{x}_j, \mathbf{x}_k$ in the Hilbert space with the distance metric $d(\cdot, \cdot)$, the metric must satisfies the triangle inequality

$$d_{ik} \leq d_{ij} + d_{jk}.$$

Triangle inequality plays a fundamental role in deriving properties of the metric Space. Now we prove the main results of this paper.

Theorem 2. If the linkage function satisfies the cluster aggregate inequality, the clustering trees produced by MLHC is identical to that produced by standard AHC.

Proof. We view the linkage function as a distance metric and build all 1mn-pairs based on the linkage function. We show that the AHC is iteratively merging 1mn-pairs, which is same as MLHC. The details is broken into two features of AHC below. □

We first note a simple feature of AHC:

Feature 1. The closest pair must be a 1mn-pair.

Proof. Suppose this is not true, i.e., the closest pair is (a, b), but a is not the closest neighbor of b. There must exist another point c which is the closest neighbor of a. Then the pair (a, c) must be the closest pair, but this contradicts the fact that (a, b) is the closest pair. Thus a must be the closest neighbor of b. Similarly, b must be the closest neighbor of a. □

Next, we prove a key feature of AHC. This shows the essence of Theorem 2.

Feature 2. If the linkage function satisfies the cluster aggregate inequality, then any 1mn-pair must be preserved and will merge eventually in standard AHC.

Proof. Suppose at certain iteration, the current clusters are listed as (C_{j_1}, C_{j_2}), $(C_{j_3}, C_{j_4}), C_{j_5}, (C_{j_6}, C_{j_7}), \cdots$, where 1mn-pair is indicated by the parenthesis. In AHC, the 1mn-pair with largest linkage value, say (C_{j_6}, C_{j_7}), is merged. Due to the cluster aggregate inequality, the newly merged cluster $C_{(j_6, j_7)}$ will be "no closer" to any other current clusters. Thus the 1mn of $C_{(j_6, j_7)}$ can not be any member of the current remaining 1mn-pairs, say C_{j_1}.

This can seen as follows. By construction, neither C_{j_6} nor C_{j_7} is closer to C_{j_1} than C_{j_2} does. Due to the cluster aggregate inequality, the newly merged cluster $C_{(j_6, j_7)}$ will be "no closer" to C_{j_1} than either C_{j_6} or C_{j_7} does. Thus 1mn of $C_{(j_6, j_7)}$ can not be C_{j_1}.

This guarantees that the 1mn-pair (C_{j_1}, C_{j_2}) will never be broken by a merged cluster. Thus in the next iteration, either a current 1mn-pair, say (C_{j_3}, C_{j_4}), is merged, or the newly-merged $C_{(j_6, j_7)}$ is merged with a singleton cluster, say C_{j_5}. Therefore, the 1mn-pair (C_{j_1}, C_{j_2}) will preserve and eventually merge at some later iteration. □

We give a simple example to illustrating some of the concepts. In Figure 1(b), we have 5 objects. They form two 1mn-pairs (a, b), (d, e) and one isolated object c. We do the standard AHC. Suppose (a, b) has the largest linkage value. So a,b are first merged into $(a+b)$. We assert that the 1mn-pair (d, e) must be preserved and will merge in later stages in AHC. This is done in two stages. First, we show that d cannot be the nearest neighbor of $(a + b)$, i.e.,

$$d(d, a + b) > d(d, e). \qquad (6)$$

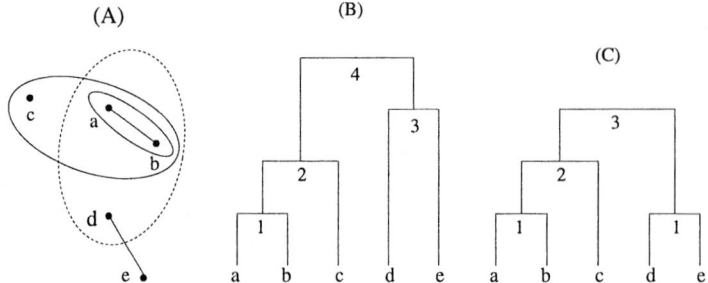

Fig. 1. (A) Dataset of 5 objects. 1mn-pairs are linked by a line. A merge is indicated by an elongated solid circle. (B) Dendrogram using AHC. Numbers indicate order of the merge. (C) Dendrogram using MLHC.

The fact that (d, e) is a 1mn-pair implies

$$d(d, a) > d(d, e), \quad (7)$$
$$d(d, b) > d(d, e), \quad (8)$$
$$d(d, c) > d(d, e). \quad (9)$$

From the cluster aggregate inequality, $d(d, a + b) \geq \min[d(d, a), d(d, b)]$. From this, and makeing use of Eqs.(7, 8), we obtain Eq.(6).

In the next round of AHC, either (i) the pair $(a+b, c)$ has the largest linkage value, or (ii) the pair (d, e) has the largest linkage value. If (ii) holds, our assertion is proved. Suppose (i) holds. Then $(a + b, c)$ are merged into $(a + b + c)$. Now we show that d cannot be the nearest neighbor of $(a + b + c)$, i.e.,

$$d(d, a + b + c) > d(d, e). \quad (10)$$

From the cluster aggregate inequality, $d(d, a + b + c) \geq \min[d(d, a + b), d(d, c)]$ From this, together with Eqs.(6, 9), we obtain Eq.(10). Therefore, (d, e) is the pair with the largest linkage value. Thus (d, e) are merged into $(d + e)$.

This example shows how 1mn pairs are preserved in AHC. Thus any cluster merge in MLHC (those 1mn-pairs) will also occur in AHC. There are total $N - 1$ cluster merges in both MLHC and AHC. So any cluster merge in AHC also occur in MLHC.

Both AHC and MLHC algorithms lead to the same binary cluster tree. The difference is the sequential order they are merged. This is illustrated in Figure 1.(B,C). If we represent the tree hight by the standard linkage function value for each merge, the dendrograms of the two algorithm remains identical.

We emphasize that the equivalence of MLHC and AHC only requires 1mn-pair preservation during AHC, as shown in Feature 2 in the above. Therefore, cluster aggregate inequality is a *sufficient* condition for 1mn-pair preservation.

For a given dataset, it is possible that a particular linkage function maybe not satisfy the generic cluster aggregate inequality for all possible triples (i, j, k), but the 1mn-pair preservation holds during AHC and thus MLHC is equivalent to AHC.

In summary, these analysis not only shows that MLHC is equivalent to AHC, thus providing a $O(n)$ speedup; but also brought out a new insight for AHC, i.e., 1mn-pair preservation during AHC. This leads to close-friends cohesion.

3 "Close Friends" Cohesion

One of the fundamental concept of data clustering is that members of the same cluster have high association with each other. One way to characterize the within cluster association is the cohesion of the cluster members via the preservation of "close friends". Suppose we divide 100 people into 10 teams. The cohesion of each team is greatly enhanced if for any member in a team, his/her close friends are also in the same team.

It is interesting to note that this close-friends cohesion is guaranteed by MLHC, if we interpret 1mn-pair relationship as close friends; by construction, this cohesion is guaranteed at all levels. We say the clustering results satisfy 1mn-consistency if for each object in a cluster, its 1mn is also in the same cluster.

By Theorem 2, clusters produced by the standard AHC also enjoy the same 1mn-consistency as in MLHC. We summarize this important result as

Theorem 3. In MLHC, 1mn-consistency is fully guaranteed. In agglomerative hierarchical clustering, if the linkage function satisfy the cluster aggregate inequality, 1mn-consistency is fully guaranteed.

3.1 Cluster Membership kNN Consistency

1mn describe the "most close" friend. Beyond the closest friend, it is desirable that other less close friends are also in the same cluster. This will increase cohesiveness of the clustering.

We thus further extend the "close friends" into k-nearest-neighbor, the cohesiveness of a cluster becomes the following knn consistency:

Cluster Membership kNN Consistency: For any data object in a cluster, its k-nearest neighbors are also in the same cluster.

Note that the relationship of "close friend" is not symmetric. Although a's closest friend is b, b's closest friend could be c, not a. Thus the "mutual closest friend" implies the tightest friendship. Thus k-Mutual-Nearest-Neighbor Consistency is more desirable.

3.2 Enforcing kNN Consistency

In general, clustering algorithms perform global optimizations, such as K-means algorithm, will gives final clusters with a smaller degree of cluster knn and kmn consistency than the bottom hierarchical clustering. This is because that in global optimizations, nearest-neighbor relations are not considered explicitly.

In HC, a pair of clusters are merged if their linkage value is high. Thus clusters being merged are typically very similar to each other. Therefore the nearest-neighbor local information are utilized to some extent, leading to higher degree of knn consistency.

What about other knn/kmn consistency? First we note that cluster knn consistency defines a "transitive relationship". If x_1 is a member of cluster C_1, and x_2 is the 1nn of x_1, then by cluster 1nn consistency, x_2 should also be a member of C_1. This transitive relation implies that 100% 1nn consistency can be achieved only if entire connected component of the 1nn graph are in C_1. To generate clusters that guarantee 100% knn consistency, at each level of the MLHC, we must first generate knn-graph, identify all connected components, and for each CC, merge all current clusters into one cluster.

Because for any object, its 2nn set always include its 1nn set. Thus 2nn consistency guarantees 1nn consistency. Similarly, because any knn set include kmn set, knn consistency implies (k-1)-nn consistency.

4 Similarity-Based Hierarchical Clustering: Multi-level Graph Contraction

The above discussion uses the distance-based linkage. All results there can be easily translate into similarity-based linkage function.

For similarity-based linkage we select the pair with the largest linkage to merge: $\max_{<pq>} s(C_p, C_q)$, where $s(C_p, C_q)$ is the aggregate similarity between clusters C_p, C_q. Let the initial pairwise similarity are $W = (w_{ij})$. The aggregate similarity has a simple form, $s(C_p, C_q) = \sum_{i \in C_p} \sum_{j \in C_q} w_{ij}$.

Cluster aggregate inequality using similarity-based linkage can be written as

$$s(A, B+C) \leq \max[s(A,B), s(A,C)] \tag{11}$$

Consider the following similarity-based linkage functions. (i) the single linkage, defined as the largest similarity among points in A, B, (ii) the complete linkage, defined as the smallest similarity among points in A, B, (iii) the average linkage, defined as the average of all similarities among points in A, B,

$$s_{\text{single}}(A,B) = \max_{\substack{i \in A \\ j \in B}} s_{ij}, \quad s_{\text{complete}}(A,B) = \min_{\substack{i \in A \\ j \in B}} s_{ij}, \quad s_{\text{avg}}(A,B) = \frac{s(A,B)}{|A||B|}.$$

With similar analysis as in the case of distance-based clustering, we can proof

Theorem 4. All three above similarity-based linkage functions satisfy the cluster aggregate inequality. The similarity based linkage functions have an advantage that merging two cluster become graph node contraction. Defining the similarity between two objects as the weight on an edge between them, this forms a similarity graph. Thus the multi-level hierarchical clustering naturally become multi-level graph contraction of the similarity graph. Many well-known results in graph theory can be applied.

Merging two current clusters into a new cluster corresponds to contracting two nodes i, j into a new node k and with edge e_{ij} being eliminated. Weights of the graph are updated according to standard graph contraction procedure. Let $W^{(t)}, W^{(t+1)}$ be the weights of the similarity graph at steps $t, t+1$. The updated weights for contracting the edge e_{ij} and merging nodes i, j into k are

$$\begin{cases} w_{kk}^{(t+1)} = w_{ii}^{(t)} + w_{jj}^{(t)} + w_{ij}^{(t)} \\ w_{kp}^{(t+1)} = w_{ip}^{(t)} + w_{jp}^{(t)}, & \forall p \notin \{i,j,k\} \\ w_{pq}^{(t+1)} = w_{pq}^{(t)}, & \forall p,q \notin \{i,j,k\} \end{cases}$$

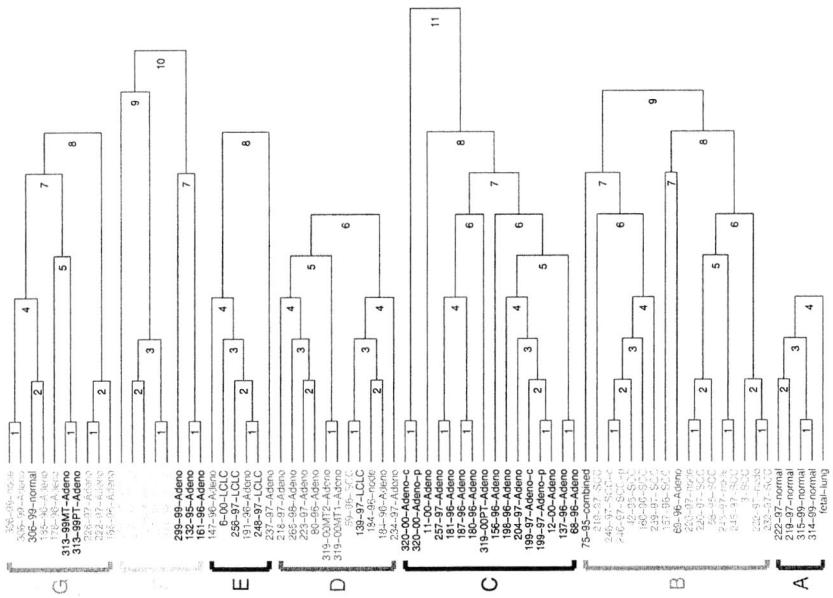

Fig. 2. MLHC clustering of lung cancer gene expressions

Using the above graph contraction operation, AHC and MLHC can be described very succinctly. AHC can be viewed as contracting one edge (with heaviest weight) at each step. The steps are repeated until all nodes merge into one. This takes exactly $N-1$ steps. MLHC can be viewed as contracting all 1mn-pairs simultaneously at each step. It is repeated about $O(\log_2 N)$ times. Now the speedup of MLHC over AHC is clear. At each step, all-pair linkage function computation is necessary. But the number of steps required in AHC is $N-1$, and it is $O(\log_2 N)$ in MLHC.

5 Experiments

5.1 DNA Gene Expression for Lung Cancer

The DNA gene expressions of lung cancer patients (available online: http://genome-www.stanford.edu/lung_cancer/adeno/) contains 73 samples of 67 lung tumors from patients whose clinical course was followed for up to 5 years. The samples comprise of 916 DNA clones representing 835 unique genes. The samples are classified into 5 groups by visual examination (41 Adenocarcinomas (ACs), 16 squamous cell carcinomas (SCCs), 5 large cell lung cancers(LCLCs), 5 small cell lung cancer (SCLCs) and 5 normal tissue with one fetal lung tissue.). The largest group ACs is further classified into three smaller groups. The purpose is to see if we can recover this 7 groups using unsupervised learning method, i.e., the hierarchical clustering method. The Pearson correlations c_{ij} among tissue samples are computed first, and the similarity metric is defined as $s_{ij} = \exp(5c_{ij})$. We use MLHC and obtain the cluster structure as shown in Fig.2.

As the Figure shows, at 1st level, after an all-pair computation, 18 1mn-pairs are formed and merged. At 2nd level, 11 1mn-pairs are formed and merged. Total 11 levels of merges are required to obtain 7 clusters. In contrast, for standard AHC, We need 66 levels of merge steps to obtain 7 clusters. The clustering result is give in the confusion matrix $T = \begin{bmatrix} 5 & . & . & . & . & . & . \\ . & 15 & . & 1 & 1 & . & . \\ . & . & 16 & . & . & . & . \\ . & 1 & . & 9 & 1 & . & . \\ . & . & . & 3 & 3 & . & . \\ . & . & 3 & . & . & 5 & . \\ 1 & . & 2 & . & . & . & 7 \end{bmatrix}$ where $T = (t_{ij})$, t_{ij} is the number of data points which are observed to be in cluster i, but was computed via the clustering method to belong to cluster j. The accuracy is defined as $Q = \sum_k t_{kk}/N = 82\%$. indicating the effectiveness of the clustering algorithm.

5.2 Climate Pattern

We tested MLHC on global precipitation data as shown in Fig.3. Regularly-spaced data points cover the surface of earth. Each data point is a 402-dimensional vector containing seasonal means over 100 years and geometric information: longitude and latitude. Similarity between two points are based two factors: (1) precipitation pattern similarity computed as Euclidean distance and (2) geometric closeness based on simple physical distance. The obtained stable regions (shown in different color and symbols) correlate well with continents, say, in Australia, south Americas.

5.3 Internet Newsgroups

We apply MLHC on Internet newsgroup articles. A 20-newsgroup dataset is from www.cs.cmu.edu /afs/cs/project/theo-11/www/naive-bayes.html. 1000 words are selected according to the mutual information between words and documents in unsupervised manner. Word - document matrix is first constructed using standard tf.idf term weighting. Cosine similarity between documents is used. We focus on two sets of 5-newsgroup combinations listed below:

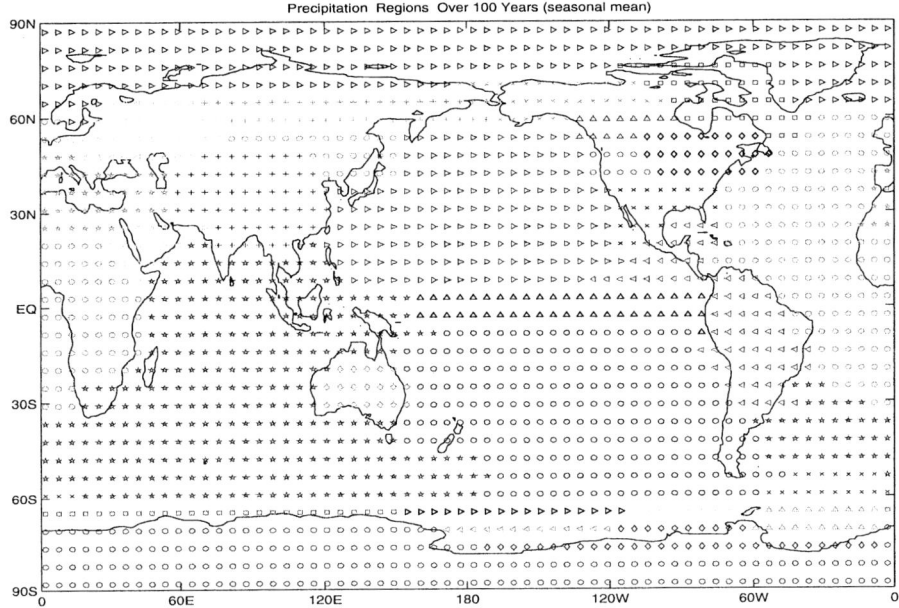

Fig. 3. Global precipitation pattern based on seasonal means over 100 years

A5:	B5:
NG2: comp.graphics	NG2: comp.graphics
NG9: rec.motorcycles	NG3: comp.os.ms-windows
NG10: rec.sport.baseball	NG8: rec.autos
NG15: sci.space	NG13: sci.electronics
NG18: talk.politics.mideast	NG19: talk.politics.misc

In A5, clusters overlap at medium level. In B5, clusters overlap substantially. Table 1 contains the results of MLHC. To accumulate sufficient statistics, for each newsgroup combination, we generate 10 datasets, each of which is a random sample of documents from the newsgroups (with 100 documents per newsgroup). The results in the table are the average over these 10 random sampled datasets. For comparison purpose, we also run K-means clustering. For each dataset, we run 10 K-means clustering from random seeds for cluster centroids and selecte the best result as determined by the K-means objective function value. Results are given in Table 1. Note that because percentage consistency results are close to 1, we give inconsistency = 1 - consistency.

From Table 1, the MLHC results have better clustering accuracy (last column of Table 1) compared to K-means clustering. More important is MLHC always provides clustering with better kmn cluster membership consistency. For 1mn-consistency, MLHC is perfect since this is guaranteed by MLHC. With this, it is not surprising that MLHC has substantially better 1nn-consistency than K-means method, about half as smaller. In all categories, MLHC has better knn/kmn consistency than K-means .

Table 1. Fractional knn and kmn inconsistency and clustering accuracy (last column) for newsgroup datasets A5 and B5. For dataset A5, 1nn inconsistency is 16.2% for K-means and 8.5% for MLHC.

	1nn	2nn	3nn	1mn	2mn	3mn	Accuracy
A5							
K-means	16.2	28.4	37.8	6.4	14.5	23.0	75.2%
MLHC	8.5	24.1	36.4	0	6.9	16.6	77.6%
B5							
K-means	23.1	39.4	50.6	8.5	21.6	32.8	56.3%
MLHC	10.2	28.9	45.0	0	9.3	21.3	60.7%

6 Summary

In this paper, we propose a modification of the standard AHC algorithm that allow an order-N faster implementation. The modification is based on the recognition that all 1mn-pairs in each iteration of AHC can be merged if the linkage function satisfies the cluster aggregate inequality. This leads to the multi-level hierarchical clustering algorithm. Many commonly used linkage functions satisfy this inequality and thus will benefit from this modification. We propose "close friends" cohesion as important feature of clustering and show that it is fully guarantees in the algorithm. This is further extended to cluster membership KNN consistency. Experiments on newsgroup show that kNN consistency is satisfied much better by MLHC than widely used algorithms such as K-means.

References

1. R. O. Duda, P. E. Hart, and D. G. Stork. *Pattern Classification, 2nd ed.* Wiley, 2000.
2. B. Fung, K. Wang, and M. Ester. Large hierarchical document clustering using frequent itemsets. *Proc. SIAM Data Mining Conf*, 2003.
3. T. Hastie, R. Tibshirani, and J. Friedman. *Elements of Statistical Learning.* Springer Verlag, 2001.
4. A.K. Jain and R.C. Dubes. *Algorithms for clustering data.* Prentice Hall, 1988.
5. A.K. Jain, M.N. Murty, and P.J. Flynn. Data clustering: a review. *ACM Computing Surveys*, 31:264–323, 1999.
6. S.Y. Jung and T.-S. Kim. An agglomerative hierarchical clustering using partial maximum array and incremental similarity computation method. *Proc. SIAM Conf. on Data Mining*, pages 265–272, 2001.
7. G. Karypis, E.-H. Han, and V. Kumar. Chameleon: Hierarchical clustering using dynamic modeling. *IEEE Computer*, 32:68–75, 1999.
8. J. Shi and J. Malik. Normalized cuts and image segmentation. *IEEE. Trans. on Pattern Analysis and Machine Intelligence*, 22:888–905, 2000.
9. S. Theodoridis and K. Koutroumbas. *Pattern Recognition.* Academic Press, 1999.

10. E.M. Voorhees. Implementing agglomerative hierarchic clustering algorithms for use in document retrieval. *Information Processing and Management*, 22:465–476, 1986.
11. H Xiong, M. Steinbach, P-N. Tan, and V. Kumar. Hicap:hierarchial clustering with pattern preservation. *Proc. SIAM Data Mining Conf*, pages 279–290, 2004.
12. H. Yu, J. Yang, and J. Han. Classifying large data sets using svms with hierarchical clusters. In *Proc. ACM Int'l Conf Knowledge Disc. Data Mining (KDD)*, pages 306–315, 2003.
13. T. Zhang, R. Ramakrishnan, and M. Livny. Birch: an efficient data clustering method for very large databases. *Proc. ACM Int'l Conf. Management of Data (SIGMOD)*, pages 103–114, 1996.

Ensembles of Balanced Nested Dichotomies for Multi-class Problems

Lin Dong[1], Eibe Frank[1], and Stefan Kramer[2]

[1] Department of Computer Science, University of Waikato, New Zealand
{ld21, eibe}@cs.waikato.ac.nz
[2] Department of Computer Science, Technical University of Munich, Germany
kramer@in.tum.de

Abstract. A system of nested dichotomies is a hierarchical decomposition of a multi-class problem with c classes into $c-1$ two-class problems and can be represented as a tree structure. Ensembles of randomly-generated nested dichotomies have proven to be an effective approach to multi-class learning problems [1]. However, sampling trees by giving each tree equal probability means that the depth of a tree is limited only by the number of classes, and very unbalanced trees can negatively affect runtime. In this paper we investigate two approaches to building balanced nested dichotomies—*class-balanced* nested dichotomies and *data-balanced* nested dichotomies—and evaluate them in the same ensemble setting. Using C4.5 decision trees as the base models, we show that both approaches can reduce runtime with little or no effect on accuracy, especially on problems with many classes. We also investigate the effect of caching models when building ensembles of nested dichotomies.

1 Introduction

Many real-world classification problems are multi-class problems: they involve a nominal class variable that has more than two values. There are basically two approaches for tackling this type of problem. One is to adapt the learning algorithm to deal with multi-class problems directly, and the other is to create several two-class problems and form a multi-class prediction based on the predictions obtained from the two-class problems. The latter approach is appealing because it does not involve any changes to the underlying two-class learning algorithm. Well-known examples of this type of approach are error-correcting output codes [2] and pairwise classification [3], and they often result in significant increases in accuracy.

Recently, it has been shown that ensembles of nested dichotomies are a promising alternative to pairwise classification and standard error-correcting output codes. In experiments with a decision tree learner and logistic regression, their performance was less dependent on the base learner used, and they yield probability estimates in a natural and well-founded way if the base learner can generate two-class probability estimates [1].

A drawback of ensembles of nested dichotomies, at least compared to pairwise classification, is the significant increase in runtime. Although pairwise classification requires applying the base learner $c*(c-1)/2$ times for a learning problem with c classes, each learning problem is much smaller than the original problem because only data from the relevant pair of classes is considered [3]. Assuming a learning algorithm that scales linearly in the number of instances, and assuming that every class has the same number of instances, the overall runtime for pairwise classification is linear in the number of classes.[1]

Building a single system of nested dichotomies in the same setting also requires time linear in the number of classes in the worst case, but the algorithm must be applied a fixed, user-specified number of times to build an ensemble of trees (10 to 20 ensemble members were found to be generally sufficient to achieve maximum accuracy on the UCI datasets investigated in [1]).

In this paper we are looking at approaches to reducing the time required to build an ensemble of nested dichotomies (END). More specifically, we propose *class-balanced* or *data-balanced* systems of nested dichotomies (ECBNDs or EDBNDs, respectively). Using C4.5 as the base learner, we show that they can improve runtime, especially on problems with many classes, with little or no effect on classification accuracy. We also investigate the effect of caching models: the same two-class learning problem can occur multiple times in an ensemble and it makes sense to re-use two-class base models that have already been built for previous systems of nested dichotomies.

The paper is structured as follows. In Section 2 we discuss the basic method of building ENDs, our two modified versions of the algorithm (ECBNDs and EDBNDs), and the use of caching models. Section 3 presents empirical results, obtained from 21 multi-class UCI datasets, and several artificial domains with a varying number of classes. Section 4 summarizes our findings.

2 Balanced Nested Dichotomies

A system of nested dichotomies is a statistical model that is used to decompose a multi-class problem into multiple two-class problems (e.g. [4] introduces it as a method for performing multi-class logistic regression). The decomposition can be represented as a binary tree (Figure 1). Each node of the tree stores a set of class labels, the corresponding training data and a binary classifier. At the very beginning, the root node contains the whole set of the original class labels corresponding to the multi-class classification problem. This set is then split into two subsets. These two subsets of class labels are treated as two "meta" classes and a binary classifier is learned for predicting them. The training dataset is split into two subsets corresponding to the two meta classes and one subset of training data is regarded as the positive examples while the other subset is regarded as the negative examples. The two successor nodes of the root inherit the two subsets of the original class labels with their corresponding training datasets and a tree is built by applying this process recursively. The process finally reaches a leaf node if the node contains only one class label.

---

[1] Pairwise classification is actually even more beneficial when the base learner's runtime is worse than linear in the number of instances.

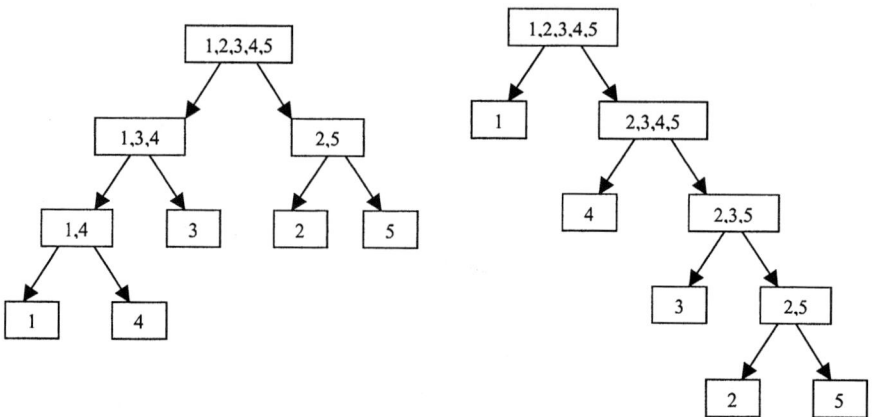

Fig. 1. Two different nested dichotomies for a classification problem with five classes

It is obvious that for any given c-class problem, the tree contains c leaf nodes (one for each class) and $c-1$ internal nodes. Each internal node contains a binary classifier. A nice feature of using a system of nested dichotomies for multi-class problems is that it yields class probability estimates in a straightforward fashion. Assuming the individual two-class classifiers built by the base learner produce class probability estimates for the corresponding two-class problems—one for each branch extending from the corresponding internal node—we can obtain a class probability estimate for a particular leaf node (i.e. the class label in the multi-class problem that is associated with this leaf node) by simply multiplying together the probability estimates obtained from the binary classifiers along the particular path from the root node to that leaf [1].

However, there is a problem with the application of nested dichotomies to standard multi-class problems: there are many possible tree structures for a given set of classes, and in the absence of prior knowledge about whether a particular decomposition is more appropriate, it is not clear which one to use. Figure 1 shows two different systems of nested dichotomies for a five-class problem. The problem is that two different trees will often lead to different predictions because the binary classifiers for each node are dealing with different two-class problems. The selection of the tree structure will influence the classification results. Given this observation and the success of ensemble learning in yielding accurate predictions, it makes sense to use all possible nested dichotomies for a given problem and average their probability estimates. Unfortunately this is infeasible because the number of possible systems of nested dichotomies for a c-class problem is $(2c-3)!!$ [1]. Hence it is necessary to take a subset.

Frank and Kramer [1] sample randomly from the space of all possible trees, giving each tree equal probability. However, it is not clear whether this is the best approach. In the absence of prior knowledge, any sampling scheme that does not give preferential treatment to a particular class can be considered a suitable candidate. The problem with random sampling based on a uniform distribution

Table 1. Comparison of the number of possible trees

Number of classes	Number of nested dichotomies	Number of class-balanced nested dichotomies
2	1	1
3	3	3
4	15	3
5	105	30
6	945	90
7	10,395	315
8	135,135	315
9	2,027,025	11,340
10	34,459,425	113,400
11	654,729,075	1,247,400
12	13,749,310,575	3,742,200

over trees is that the tree depth is only limited by the number of classes, and deep trees can take a long time to build. Consider the case where the tree is a list, as in the second tree shown in Figure 1, and assume the two largest classes are separated out last (classes 2 and 5 in the example). Then all binary learning problems will involved the two largest classes, incurring a high computational cost for the process of building the binary classifiers in the tree.

2.1 Class-Balanced Nested Dichotomies

In light of these observations we consider two different sampling strategies in this paper. The first method is based on balancing the number of classes at each node. Instead of sampling from the space of all possible trees, we sample from the space of all *balanced* trees, and build an ensemble of balanced trees. The advantage of this method is that the depth of the tree is guaranteed to be logarithmic in the number of classes. We call this an ensemble of class-balanced nested dichotomies (ECBND).

The number of possible class-balanced nested dichotomies is obviously smaller than the total number of nested dichotomies. The following recurrence relation defines the number of possible class-balanced trees:

$$T(c) = \begin{cases} \frac{1}{2}\binom{c}{c/2}T(\frac{c}{2})T(\frac{c}{2}) & : \text{if } c \text{ is even} \\ \binom{c}{(c+1)/2}T(\frac{c+1}{2})T(\frac{c-1}{2}) & : \text{if } c \text{ is odd} \end{cases}$$

where $T(1) = 1$ and $T(2) = 1$.

Table 1 shows the number of possible systems of nested dichotomies for up to 12 classes for the class-balanced (CBND) and the unconstrained case (ND). It shows that a non-trivial number of CBNDs can be generated for classification problems with five or more classes.

Figure 2 shows the algorithm for building a system of class-balanced nested dichotomies. At each node the set of classes is split into equal size subsets (of course, if the number of classes is odd, the size will not be exactly equal), and the

```
method buildClassBalancedNestedDichotomies(Dataset D, Set of classes C)
    if |C| = 1 then return
    P = subset of C, randomly chosen from all subsets of size ⌊|C|/2⌋
    N = C \ P
    D_p = all instances in D apart from those pertaining to classes in P
    buildClassBalancedNestedDichotomies(D_p, P)
    D_n = all instances in D apart from those pertaining to classes in N
    buildClassBalancedNestedDichotomies(D_n, N)
    D' = a two-class version of D created based on N and P
    classifierForNode = buildClassifier(D')
```

Fig. 2. Algorithm for generating class-balanced nested dichotomies

base learning algorithm is applied to the data corresponding to these two subsets. The algorithm then recurses until only one class is left. It is applied repeatedly with different random number seeds to generate a committee of trees.

2.2 Data-Balanced Nested Dichotomies

There is a potential problem with the class-balanced approach: some multi-class problems are very unbalanced and some classes are much more populous than others. In that case a class-balanced tree does not imply that it is also *data-balanced* (i.e. that it exhibits about the same number of instances in the two successor nodes of an internal node). This can negatively affect runtime if the base learning algorithm has time complexity worse than linear in the number of instances. Hence we also consider a simple algorithm for building data-balanced nested dichotomies in this paper. Note that this method violates the condition that the sampling scheme should not be biased towards a particular class: based on this scheme, leaf nodes for larger classes will be located higher up in the tree structure. Despite this potential drawback we decided to investigate this scheme empirically because it is difficult to say how important this condition is in practice.

Figure 3 shows our algorithm for building a system of data-balanced nested dichotomies. It randomly assigns classes to two subsets until the size of the training data in one of the subsets exceeds half the total amount of training data at the node. One motivation for using this simple algorithm was that it is important to maintain a degree of randomness in the assignment of classes to subsets in order to preserve diversity in the committee of randomly generated systems of nested dichotomies. Given that we are aiming for an ensemble of nested dichotomies it would not be advisable, even if it were computationally feasible, to aim for an optimum balance because this would severely restrict the number of trees that can be generated. Even with our simple algorithm diversity suffers when the class distribution is very unbalanced. However, it is difficult to derive a general expression for the number of trees that can potentially generated by this method because this number depends on the class distribution in the dataset.

```
method buildDataBalancedNestedDichotomies(Dataset D, List of classes C)
    if |C| = 1 then return
    C = random permutation of C
    D_p = ∅,  D_n = ∅
    do
        if (|C| > 1) then
            add all instances from D pertaining to first class in C to D_p
            add all instances from D pertaining to last class in C to D_n
            remove first and last class from C
        else
            add all instances from D pertaining to remaining class in C to D_p
            remove remaining class from C
    while (|D_p| < ⌊|D|/2⌋) and (|D_n| < ⌊|D|/2⌋)
    if ((|D_p| ≥ ⌊|D|/2⌋) then
        add instances from D pertaining to remaining classes in C to D_n
    else
        add instances from D pertaining to remaining classes in C to D_p
    P = all classes present in D_p,  N = all classes present in D_n
    buildDataBalancedNestedDichotomies(D_p, P)
    buildDataBalancedNestedDichotomies(D_n, N)
    D' = a two-class version of D created based on N and P
    classifierForNode = classifier learned by base learner from D'
```

Fig. 3. Algorithm for generating data-balanced nested dichotomies

2.3 Computational Complexity

The motivation for using balanced nested dichotomies is that this reduces runtime. In the following we analyze the computational complexity of completely random and balanced nested dichotomies. Let c be the number of classes in the dataset, and n be the number of training instances. For simplicity, assume that all classes have an approximately equal number of instances in them (i.e. that the number of instances in each class is approximately n/c). We assume that the time complexity of the base learning algorithm is linear in the number of training instances, and that we can ignore the effect of the number of attributes.

In the worst case, a completely random system of nested dichotomies can degenerate into a list, and the total runtime for building a multi-class classifier based on this kind of structure and the above assumptions is

$$\sum_{i=0}^{c-2} \frac{c-i}{c} n = \frac{n}{c} \sum_{i=0}^{c-2} c - i = \frac{n}{c} \left((c-1)c - \sum_{i=0}^{c-2} i \right) = \frac{n}{c} \left((c-1)c - \frac{(c-2)(c-1)}{2} \right)$$

$$> \frac{n}{c} \left((c-1)c - \frac{c(c-1)}{2} \right) = \frac{(c-1)}{2} n.$$

Hence the worst-case time complexity is linear in the number of instances and classes.

Let us now consider the balanced case. Assuming c is even, we have $\log c$ layers of internal nodes. In each layer, all the training data needs to be processed (because the union of all subsets in each layer is the original dataset). Given that we have assumed that the base learner scales linearly in the number of instances, the overall runtime becomes $n \log c$, i.e. it is logarithmic in the number of classes and linear in the number of instances.

Assuming a base learning algorithm whose time complexity is worse than linear, the advantage of the balanced scheme becomes even more pronounced (because the size of the subsets of data considered at each node decreases more quickly in this scheme). Note also that the assumption of evenly distributed classes is not strictly necessary. This can be seen by considering the worst case, where one class has almost all the instances. The worst-case time complexity for the unbalanced case remains linear in the number of classes in this situation, and the one for the balanced case logarithmic in the number of classes.

However, in the case of a skewed class distribution it is possible to improve on the class-balanced scheme when the base learning algorithm's runtime is worse than linear. In that case it makes sense to attempt to divide the number of instances as evenly as possible at each node, so as to reduce the maximum amount of data considered at a node as quickly as possible. This is why we have investigated the data-balanced approach discussed above.

2.4 Caching Models

There is a further opportunity to improve the training time for ensembles of nested dichotomies. It arises from the fact that ensemble members may share some two-class problems. Consider Figure 1. In both trees, a classifier has to be learned that separates classes 2 and 5. These classifiers will be identical because they are based on exactly the same data. It is not sensible to build them twice. Consequently we can cache models that have been built in a hash table and reduce computational complexity further.

As explained by Frank and Kramer [1], there are $(3^c - (2^{c+1} - 1))/2$ possible two-class problems for a c-class problem, i.e. growth is exponential in the number of classes. Hence we can expect that caching only makes a difference for relatively small numbers of classes. If we consider balanced dichotomies, the number of possible two-class problems is reduced. Consequently caching will be more beneficial in the balanced case.

3 Experiments

In the following we empirically investigate the effect of our proposed modifications on runtime and accuracy. We used 21 multi-class UCI datasets [5]. The number of classes varies from 3 to 26. We also performed some experiments with artificial data that exhibits a larger number of classes. For each scheme, we used 10 ensemble members (i.e. 10 systems of nested dichotomies are generated and their probability estimates are averaged to form a prediction). J48, the implementation of the C4.5 decision tree learner [6] from the Weka workbench [7] was used as the base learner for the experiments.

Table 2. Effect of model caching on ENDs for UCI datasets

Dataset	Number of classes	Number of instances	Training time for ENDs w/o caching	with caching	
iris	3	150	0.03 ± 0.02	0.01 ± 0.00	
balance-scale	3	625	0.28 ± 0.06	0.09 ± 0.04	•
splice	3	3190	4.56 ± 0.32	1.37 ± 0.14	•
waveform	3	5000	38.78 ± 0.65	11.42 ± 0.96	•
lymphography	4	148	0.06 ± 0.02	0.04 ± 0.02	•
vehicle	4	846	1.87 ± 0.11	1.08 ± 0.16	•
hypothyroid	4	3772	3.13 ± 0.47	1.75 ± 0.29	•
anneal	6	898	0.99 ± 0.08	0.82 ± 0.13	•
zoo	7	101	0.07 ± 0.02	0.06 ± 0.02	
autos	7	205	0.40 ± 0.05	0.36 ± 0.05	
glass	7	214	0.30 ± 0.03	0.27 ± 0.03	
segment	7	2310	6.61 ± 0.27	5.87 ± 0.37	•
ecoli	8	336	0.25 ± 0.04	0.23 ± 0.04	
optdigits	10	5620	72.53 ± 3.30	68.70 ± 3.00	•
pendigits	10	10992	49.30 ± 2.00	47.07 ± 2.12	•
vowel	11	990	4.21 ± 0.11	4.04 ± 0.16	•
arrhythmia	16	452	21.14 ± 1.01	20.76 ± 1.09	
soybean	19	683	1.02 ± 0.07	0.99 ± 0.06	
primary-tumor	22	339	0.63 ± 0.06	0.63 ± 0.06	
audiology	24	226	0.74 ± 0.06	0.74 ± 0.05	
letter	26	20000	317.53 ± 11.44	315.74 ± 11.53	

All experimental results are averages from 10 runs of stratified 5-fold cross-validation (UCI datasets) or 3-fold cross-validation (artificial data). We also report standard deviations for the 50 (UCI data) or 30 (artificial data) individual estimates. Runtime was measured on a machine with a Pentium 4 3 GHz processor running the Java HotSpot Client VM (build 1.4.2_03) on Linux, and is reported in seconds. We tested for significant differences using the corrected resampled t-test [8].

3.1 Applying Caching to ENDs

In this section we discuss the effect of caching individual classifiers in an ensemble of nested dichotomies. Table 2 has the average training time for ENDs with and without caching based on a hash table. Significant improvements in training time obtained by caching are marked with a •.

The results show that the runtime decreases significantly for 14 of the 21 UCI datasets (of course, accuracy remains identical). The improvement is especially obvious on datasets with a small number of classes and a large number of instances. With a small number of classes one is more likely to encounter the same binary classifier in different systems of nested dichotomies in the ensemble. With a large number of instances, more time is saved by avoiding rebuilding the same binary classifier. For instance, the training time on the waveform dataset, which

Table 3. Comparison of training time on UCI datasets

Dataset	Number of classes	Training time ENDs	ECBNDs	EDBNDs
iris	3	0.01 ± 0.00	0.03 ± 0.02	0.02 ± 0.00
balance-scale	3	0.09 ± 0.04	0.09 ± 0.04	0.09 ± 0.04
splice	3	1.37 ± 0.14	1.45 ± 0.12	1.33 ± 0.20
waveform	3	11.42 ± 0.96	11.31 ± 0.56	11.24 ± 1.10
lymphography	4	0.04 ± 0.02	0.03 ± 0.02	0.03 ± 0.00
vehicle	4	1.08 ± 0.16	0.51 ± 0.06 •	0.52 ± 0.05 •
hypothyroid	4	1.75 ± 0.29	0.86 ± 0.12 •	0.92 ± 0.22 •
anneal	6	0.82 ± 0.13	0.63 ± 0.09 •	0.50 ± 0.13 •
zoo	7	0.06 ± 0.02	0.06 ± 0.03	0.06 ± 0.02
autos	7	0.36 ± 0.05	0.26 ± 0.04 •	0.25 ± 0.05 •
glass	7	0.27 ± 0.03	0.21 ± 0.04 •	0.20 ± 0.04 •
segment	7	5.87 ± 0.37	4.88 ± 0.34 •	4.98 ± 0.41 •
ecoli	8	0.23 ± 0.04	0.20 ± 0.03	0.21 ± 0.04
optdigits	10	68.70 ± 3.00	55.17 ± 1.91 •	55.03 ± 2.16 •
pendigits	10	47.07 ± 2.12	37.95 ± 1.53 •	38.40 ± 1.52 •
vowel	11	4.04 ± 0.16	3.62 ± 0.11 •	3.70 ± 0.12 •
arrhythmia	16	20.76 ± 1.09	19.20 ± 1.08 •	17.39 ± 1.56 •
soybean	19	0.99 ± 0.06	0.87 ± 0.06 •	0.85 ± 0.07 •
primary-tumor	22	0.63 ± 0.06	0.54 ± 0.06 •	0.54 ± 0.06 •
audiology	24	0.74 ± 0.05	0.64 ± 0.08 •	0.63 ± 0.09 •
letter	26	315.74 ± 11.53	273.45 ± 16.75 •	274.07 ± 16.84 •

has 5000 instances and only 3 classes, decreases dramatically from 38.78 seconds to 11.42 seconds by using hash tables. On the other hand, for the arrhythmia dataset, which has 16 classes and only 452 instances, the training time decreases only slightly, from 21.14 seconds to 20.76 seconds. From Table 2, we also see that there is no significant improvement for the training time when the number of classes exceeds 11. The chance to encounter the same binary classifier in those situations becomes limited as there are so many possible two-class problems. Moreover, the number of instances in these datasets (excluding the letter data) is small so that the time saved by using hash tables is not noticeable. We also performed experiments with artificial data with even more classes and there was essentially no difference in runtime on that data.

3.2 Comparing ENDs, ECBNDs, and EDBNDs

As we have seen, caching does not help when there are many classes. In the following we will see that using balanced nested dichotomies helps in those cases. We will first look at training time and then the effect on accuracy.

Training time. Table 3 shows the training times for ENDs, class-balanced ENDs (ECBNDs), and data-balanced ENDs (EDBNDs), on the UCI datasets. Model caching was applied in all three versions of ENDs. A • indicates a significant reduction in runtime compared to ENDs.

Table 4. Comparison of training time on artificial datasets

Number of classes	Number of instances	Training time ENDs	ECBNDs	EDBNDs
10	820	0.60 ± 0.09	0.58 ± 0.07	0.58 ± 0.07
20	1390	1.50 ± 0.12	1.42 ± 0.08	1.44 ± 0.09
30	1950	2.72 ± 0.11	2.31 ± 0.12 •	2.33 ± 0.12 •
40	2410	3.87 ± 0.16	3.18 ± 0.14 •	3.24 ± 0.13 •
50	3090	5.55 ± 0.23	4.54 ± 0.17 •	4.57 ± 0.20 •
60	3660	7.48 ± 0.29	5.86 ± 0.17 •	5.90 ± 0.25 •
70	4560	10.41 ± 0.36	8.23 ± 0.28 •	8.35 ± 0.30 •
80	5010	12.32 ± 0.43	9.56 ± 0.33 •	9.67 ± 0.31 •
90	5840	15.75 ± 0.53	12.62 ± 0.44 •	12.78 ± 0.34 •
100	6230	18.25 ± 0.61	13.61 ± 0.38 •	13.98 ± 0.44 •
150	9590	40.65 ± 1.90	27.63 ± 0.77 •	28.19 ± 0.65 •
200	12320	66.41 ± 2.95	42.37 ± 1.30 •	42.70 ± 1.30 •

The results show that using class-balanced nested dichotomies results in significantly reduced training times on 14 of the 21 datasets. Using the data-balanced scheme also helps: EDBNDs are significantly more efficient than ENDs on 14 datasets, just like ECBNDs. Compared to class-balanced trees, data-balanced trees are significantly more efficient on one dataset (arrhythmia). (This information is not included in Table 3.) This dataset has an extremely unbalanced class distribution and this is why the data-balanced approach helps.

The advantage of the balanced schemes is restricted to datasets with more than 3 classes. On three-class datasets, all nested dichotomies are class-balanced, so we would not expect any significant difference between ENDs and ECBNDs. The experimental results bear this out.

Table 4 shows the training times for our 12 artificial datasets. To generate these datasets we used a cluster generator and varied the number of clusters from 10 to 200. Instances in the same cluster were assigned the same class label. Each instance in these datasets consists of one boolean attribute and two numeric attributes. The attribute value ranges were set to be different but could overlap. Attribute values were generated randomly within each cluster. The number of instances in each cluster (i.e. class) was also randomly generated and varied between 20 and 110.

The results on the artificial datasets show that the balanced schemes exhibit a significant advantage in terms of running time when 30 or more classes are present in the data. There was no significant difference in running time for the two balanced schemes (class-balanced vs. data-balanced) on any of the datasets. This indicates that the class distribution in our artificial datasets is not skewed enough for the data-balanced approach to help.

Accuracy. Improvements in runtime are less useful if they affect accuracy in a significant fashion. Hence it is important to evaluate the effect of our proposed modifications on accuracy. Table 5 shows the estimated accuracy for ENDs, ECBNDs, and EDBNDs on the UCI datasets. We can see that there is no dataset with a significant difference in accuracy for ENDs and ECBNDs. This is the

Table 5. Comparison of accuracy on UCI datasets

Dataset	Number of classes	Percent correct ENDs	ECBNDs	EDBNDs
iris	3	94.13 ± 3.84	94.13 ± 3.72	94.27 ± 3.81
balance-scale	3	79.92 ± 2.37	79.49 ± 2.41	79.78 ± 2.31
splice	3	94.75 ± 1.01	94.55 ± 0.98	93.07 ± 1.33 •
waveform	3	77.89 ± 1.88	77.53 ± 1.91	77.85 ± 2.06
lymphography	4	77.73 ± 7.47	76.63 ± 6.35	76.90 ± 6.93
vehicle	4	73.20 ± 2.92	72.36 ± 2.30	72.36 ± 2.30
hypothyroid	4	99.54 ± 0.26	99.51 ± 0.27	99.54 ± 0.28
anneal	6	98.63 ± 0.80	98.44 ± 0.75	98.53 ± 0.62
zoo	7	93.66 ± 5.67	93.87 ± 4.61	93.88 ± 4.50
autos	7	76.20 ± 6.11	74.83 ± 6.62	75.32 ± 7.10
glass	7	72.82 ± 7.42	73.51 ± 6.17	72.25 ± 6.84
segment	7	97.45 ± 0.83	97.35 ± 0.80	97.39 ± 0.87
ecoli	8	85.60 ± 4.11	85.36 ± 4.06	84.88 ± 4.13
optdigits	10	96.99 ± 0.49	97.14 ± 0.45	97.18 ± 0.50
pendigits	10	98.59 ± 0.27	98.76 ± 0.25	98.76 ± 0.26
vowel	11	88.31 ± 2.66	89.98 ± 2.47	89.24 ± 2.79
arrhythmia	16	72.59 ± 3.24	72.82 ± 4.11	71.51 ± 3.55
soybean	19	93.90 ± 1.63	94.49 ± 1.69	94.36 ± 1.78
primary-tumor	22	44.72 ± 5.04	46.28 ± 4.61	45.96 ± 4.62
audiology	24	78.46 ± 5.44	79.66 ± 5.12	79.48 ± 5.23
letter	26	94.33 ± 0.37	94.50 ± 0.36	94.51 ± 0.35

Table 6. Comparison of accuracy on artificial datasets

Number of classes	Number of instances	Percent correct ENDs	ECBNDs	EDBNDs
10	820	78.08 ± 1.94	78.34 ± 2.35	78.32 ± 2.32
20	1390	77.79 ± 1.87	77.21 ± 1.44	77.47 ± 1.66
30	1950	77.09 ± 1.61	76.93 ± 1.53	76.85 ± 1.46
40	2410	76.64 ± 1.24	76.56 ± 1.39	76.46 ± 1.24
50	3090	76.26 ± 1.09	76.17 ± 1.26	76.25 ± 1.19
60	3660	76.43 ± 1.08	76.33 ± 1.04	76.37 ± 0.95
70	4560	73.58 ± 1.12	73.27 ± 0.97	73.50 ± 0.90
80	5010	75.85 ± 1.06	75.61 ± 0.94	75.71 ± 0.87
90	5840	76.41 ± 0.84	76.40 ± 0.91	76.41 ± 0.87
100	6230	76.59 ± 0.77	76.54 ± 0.73	76.50 ± 0.85
150	9590	75.92 ± 0.66	75.89 ± 0.72	75.86 ± 0.62
200	12320	75.89 ± 0.51	75.67 ± 0.51	75.73 ± 0.49

desired outcome. For EDBNDs, there is one three-class dataset (splice) where the accuracy is significantly reduced compared to ENDs. The splice data has a skewed class distribution, where one class has about half the instances and the rest is evenly distributed among the remaining two classes. We measured the diversity of the three types of ensembles on this dataset using the kappa statistic. This statistic can be used to measure agreement between pairs of ensemble members [9]. For EDBNDs, the mean kappa value over all pairs, measured on the training data, was 0.96, which was indeed higher than the mean kappa values for ENDs and ECBNDs (0.94 and 0.93 respectively). This indicates that reduction in diversity is the reason for the drop in performance.

Table 6 shows the same information for the artificial datasets. In this case there is not a single dataset where there is a significant difference in accuracy between any of the schemes.

4 Conclusions

Ensembles of nested dichotomies have recently been shown to be a very promising meta learning scheme for multi-class problems. They produce accurate classifications and yield class probabilities estimates in a natural way. In this paper we have shown that it is possible to improve the runtime of this meta learning scheme without affecting accuracy. A simple way to improve runtime for problems with a small number of classes is to cache two-class models and re-use them in different members of an ensemble of nested dichotomies. On problems with many classes we have shown that using class-balanced nested dichotomies significantly improves runtime, with no significant change in accuracy. We have also presented a data-balanced scheme that can help to improve runtime further when there are many classes and the class distribution is highly skewed.

References

1. Frank, E., Kramer, S.: Ensembles of nested dichotomies for multi-class problems. In: Proc Int Conf on Machine Learning, ACM Press (2004) 305–312
2. Dietterich, T., Bakiri, G.: Solving multiclass learning problems via error-correcting output codes. Journal of Artificial Intelligence Research **2** (1995) 263–286
3. Fürnkranz, J.: Round robin classification. Journal of Machine Learning Research **2** (2002) 721–747
4. Fox, J.: Applied Regression Analysis, Linear Models, and Related Methods. Sage (1997)
5. Blake, C., Merz, C.: UCI repository of machine learning databases. University of California, Irvine, Dept. of Inf. and Computer Science (1998) [www.ics.uci.edu/~mlearn/MLRepository.html].
6. Quinlan, J.: C4.5: Programs for Machine Learning. Morgan Kaufmann, Los Altos, CA (1992)
7. Witten, I.H., Frank, E.: Data Mining: Practical Machine Learning Tools and Techniques with Java Implementations. Morgan Kaufmann (2000)
8. Nadeau, C., Bengio, Y.: Inference for the generalization error. Machine Learning **52** (2003) 239–281
9. Dietterich, T.G.: An experimental comparison of three methods for constructing ensembles of decision trees: Bagging, boosting, and randomization. Machine Learning **40** (1998) 139–157

Protein Sequence Pattern Mining with Constraints

Pedro Gabriel Ferreira* and Paulo J. Azevedo**

University of Minho, Department of Informatics,
Campus of Gualtar, 4710-057 Braga, Portugal
{pedrogabriel, pja}@di.uminho.pt

Abstract. Considering the characteristics of biological sequence databases, which typically have a small alphabet, a very long length and a relative small size (several hundreds of sequences), we propose a new sequence mining algorithm (gIL). gIL was developed for linear sequence pattern mining and results from the combination of some of the most efficient techniques used in sequence and itemset mining. The algorithm exhibits a high adaptability, yielding a smooth and direct introduction of various types of features into the mining process, namely the extraction of rigid and arbitrary gap patterns. Both breadth or a depth first traversal are possible. The experimental evaluation, in synthetic and real life protein databases, has shown that our algorithm has superior performance to state-of-the art algorithms. The use of constraints has also proved to be a very useful tool to specify user interesting patterns.

1 Introduction

In the development of sequence pattern mining algorithms, two communities can be considered: the *Data Mining* and the *Bioinformatics* community. The algorithms from the Data Mining community inherited some characteristics from the association rule mining algorithms. They are best suited for data with many (from hundred of thousands to millions) sequences with a relative small length (from 10 to 20), and an alphabet of thousands of events, e.g. [9,7,11,1]. In the bioinformatics community, algorithms are developed in order to be very efficient when mining a small number of sequences (in the order of hundreds) with large lengths (few hundreds). The alphabet size is typically very small (ex: 4 for DNA and 20 for protein sequences). We emphasize the algorithm Teiresias [6] as a standard.

The major problem with Sequence pattern mining is that it usually generates too many patterns. When databases attain considerable size or when the average

---

* Supported by a PhD Scholarship (SFRH/BD/13462/2003) from Fundação Ciência e Tecnologia.
** Supported by Fundação Ciência e Tecnologia - Programa de Financiamento Plurianual de Unidades de I & D, Centro de Ciências e Tecnologias da Computação - Universidade do Minho.

length of the sequences is very long, the mining process becomes computationally expensive or simply infeasible. This is often the case when we are mining biological data like proteins or DNA. Additionally, the user interpretation of the results turns out to be a very hard task since the interesting patterns are blurred into the huge amount of outputted patterns. The solution to this problem can be achieved through the definition of alternative interesting measures besides support, or with user imposed restrictions to the search space. When properly integrated in the mining process these restrictions reduce the computation demands in terms of time and memory, allowing to deal with datasets that are otherwise potentially untractable. These restrictions are expressed through what is typically called as *Constraints*. The use of Constraints enhances the database queries. The runtime reduction grants the user with the opportunity to interactively refine the query specification. This can be done until an expected answer is found.

2 Preliminaries

We consider the special case of linear sequences databases. A database D is as a collection of linear sequences. A *linear sequence* is a sequence composed by successive atomic elements, generically called events. Examples of this type of databases are protein or DNA sequences or website navigation paths. The term *linear* is used to make the distinction from the transactional sequences, that consist in sequences of *EventSets*(usually called as *ItemSets*). Given a sequence S, S' is subsequence of S if S' can be obtained by deleting some of the events in S. A sequence pattern is called a *frequent sequence pattern* if it is found to be subsequence of a number of sequences in the dataset greater or equal to a specified threshold value. This value is called *minimum support*, σ, and is defined as an user parameter. The *cover* represents the list of sequence identifiers where the pattern occurs. The cardinality of this list corresponds to the support of that pattern.

Considering patterns in the form $A_1 - x(p_1, q_1) - A_2 - x(p_2, q_2) - ... A_n$, a sequence pattern is an *arbitrary gap sequence pattern* when a variable (zero or more) number of gaps exist between adjacent events in the pattern, i.e. $p_i \leq q_i, \forall i$. Typically a variable gap with n minimum and m maximum number of gaps is described as $-x(n,m)-$. In the sequences $< 1\ 5\ 3\ 4\ 5 >$ and $< 1\ 2\ 2\ 3 >$ exists an arbitrary gap pattern $1-x(1,2)-3$. A *rigid gap pattern* is a pattern where gaps contain a fixed size for all the database occurrences of the sequence pattern, i.e. $p_i = q_i, \forall i$. To denote a rigid gap the $-r(n)-$ notation is used, where n is the size of the gap. The $1-r(2)-3$ is a pattern of length 4, in the sequences $< 1\ 2\ 5\ 3\ 4\ 5 >$ and $< 1\ 1\ 6\ 3 >$. Each gap position is denoted by the "." (wildcard) symbol, meaning that it matches any symbol of the alphabet. A pattern belongs to one of three classes: *maximal, closed* or *all*. A sequence pattern is *maximal* if it is not contained in any other pattern, and *closed* when all its extensions have an inferior support than itself. The *all* refers to when all the patterns are enumerated. When extending a sequence pattern $S =< s_1\ s_2 \ldots s_n >$, with a new event s_{n+1}, then

S is called a *base sequence* and $S' = <s_1 \ s_2 \ \ldots s_n \ s_{n+1}>$ the *extended sequence*. If an event b occurs after a in a certain sequence, we denoted it as: $a \rightarrow b$, and a is called the *predecessor*, $pred(a \rightarrow b) = a$, and b the *successor*, $succ(a \rightarrow b) = b$. The pair is frequent if it occurs in at least σ sequences of the database.

Constraints represent an efficient way to prune the search space [9,10]. Considering the user's point of view, it also enables to focus the search on more interesting sequence patterns. The most common and generic types of constraints are:

- *Item Constraints*: restricts the set of the events (*excludedEventsSet*) that may appear in the sequence patterns,
- *Gap Constraints*: defines the (*minGap*) minimum distance or the maximum distance (*maxGap*) that may occur between two adjacent events in the sequence patterns,
- *Duration or Window Constraints*: defines the maximum distance (*window*) between the first and the last event of the sequence patterns.
- *Start Events Constraints*: determines that the extracted patterns start with the specified events (*startEvents*).

Another useful feature in sequence mining, in particular to protein pattern mining, is the use of *Equivalent/Substitution Sets*. When used during the mining process an event can be substituted by another event belonging to the same set. A "*is-a*" hierarchy of relations can be represented through substitution sets.

Depending on the target application of the frequent sequence patterns other measures of interest and scoring can be applied as posterior step of the mining process. Since the closed and the maximal patterns are not necessarily the most interesting we designed our algorithm in order to find all the frequent patterns. From the biological point of view, rigid patterns allow to find more well conserved regions, while arbitrary patterns permit the cover of a large number of sequences in the database.

The problem we address in this paper can be formulated as follow: given a database D of linear sequences, a minimum support, σ, and the optional parameters *minGap*, *maxGap*, *window*, *excludedEventsSet*, *startEventsSets* and *substitutionSets*, find *all* the *arbitrary* or *rigid gap* frequent sequence patterns that respect the defined constraints.

3 Algorithm

The proposed algorithm uses a Bottom-Up search space enumeration and a combination of frequent pairs of events to extend and find all the frequent sequence patterns. The algorithm is divided in two phases: *scanning phase* and *sequence extension phase*. Since the frequent sequences are obtained from the set of frequent pairs, the first phase of the algorithm consists in traversing all the sequences in the database and building two auxiliary data structures. The first structure contains the set of all pairs of events found in the database. Each pair representation points to the sequences where they appear (through a sequence identifier bitmap). The second data structure consists of a vertical representation of the database. It contains the positions or offsets of the events in the sequences where they occur. This information is required to ensure that the order of the events along the data sequence is respected. Both data structures are thought for quick information retrieval. At the end of the scanning phase we obtain a map of all the pairs of events present in the database and a vertical format representation of the original database. In the second phase the pairs of events are successively combined to find all the sequence patterns. These operations are fundamentally based on two properties:

Property 1 (Anti-Monotonic). *All supersequences of an infrequent sequence are infrequent.*

Property 2 (Sequence Transitive Extension). *Let $S = <s_1 \ldots s_n>$, C_S is its cover list and O_S the list of the offset values of S for all the sequences in C_S. Let $P = (s_j \rightarrow s_m)$, C_P is it cover list and O_P the offset list of $succ(P)$ for all sequences in C_P. If $succ(S) = pred(P)$, i.e., $s_n = s_j$, then the extended sequence $E = <s_1 \ldots s_n s_m>$ will occur in C_E, where $C_E = \{X : \forall X \text{ in } C_S \cap C_P, O_P(X) > O_S(X)\}$.*

Hence, the basic idea is to successively extend a frequent pair of events with another frequent pair, as long as the predecessor of one pair is equal to the successor of the other. This joining step is sound provided that the above mentioned properties (1 and 2) are respected. The joining of pairs combined with a breadth first or a depth first traversal yields all the frequent sequences patterns in the database.

3.1 Scanning Phase

The first phase of the algorithm consists in the following procedure: For each sequence in D, obtain all ordered pairs of events, without repetitions. Consider the sequence 5 in the example database of table 1(a). The obtained pairs are: $1 \rightarrow 2$, $1 \rightarrow 3$, $1 \rightarrow 4$, $2 \rightarrow 2$, $2 \rightarrow 3$, $2 \rightarrow 4$ and $3 \rightarrow 4$. During the determination of the pairs of events the first auxiliary data structure, that consists of an N-bidimensional matrix, is built and updated. N corresponds to the size of the alphabet. The N^2 cells in the matrix correspond to the N^2 possible combinations of pairs. We call this structure the *Bitmap Matrix*. Each $Cell(i, j)$ contains the information relative to the pair $i \rightarrow j$. This information consists of a bitmap that indicates the presence (1) or the absence (0) in the respective sequence

Table 1. (a) Parameters used in the synthetic data generator; (b) Properties of the proteins datasets

Symbol	Meaning
S	Number of Sequences (x 10^3)
L	Avg. Length of the sequences
R	Alphabet Size
P	Distribution Skewness

DataSet	NumSeq	AlphabetSize	AvgLen	MinLen	MaxLen
Yeast	393	21	256	15	1859
PSSP	396	22	158	21	577
nonOM	60	20	349	53	1161
mushroom	8124	120	23	23	23

(i-th bit corresponds to the sequence i in D) and an integer that contains the support count. This last value allows a fast support checking. For each pair $i \rightarrow j$ we update the respective $Cell(i,j)$ in the Bitmap Matrix, by activating the bit corresponding to the sequence where the pair occurs and incrementing the support counter. As an example, for the pair $1 \rightarrow 3$, the $Cell(1,3)$ is represented in figure 1(b):

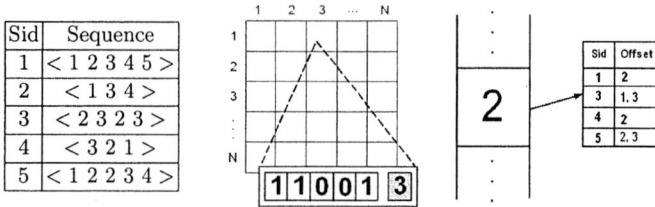

Fig. 1. (a) An example database; (b) Content of the Cell(1,3) in the Bitmap Matrix; (c) Representation of event 2 in the Offset Matrix

This means that the pair occurs in the database sequence 1, 2 and 5 and has a support of 3. Simultaneously, as each event in the database is being scanned, a second data structure called *Offset Matrix* is also being built. Conceptually, this data structure consists of an adjacency matrix that will contain all the offset (positions) of all the events in the entire database. Each event is a key that points to a list of pairs <$Sid, OffsetList$>, where *OffsetList* is a list of all the positions of the event in the sequence Sid. Thus, the Offset Matrix is a vertical representation of the database. Figure 1(c) shows the information stored in the Offset Matrix for the event 2.

3.2 Sequence Extension Phase

We start by presenting how arbitrary gap patterns are extracted. In section 3.4 we will show how easily our algorithm can be adapted to extract rigid gap patterns. For implementing the extension phase we present two tests (algorithms) that conjunctively are necessary and sufficient conditions to consider as frequent a new extended sequence.

This is a quick test that implements property 1. The *bitmap* function gets the correspondent bitmaps of S and P. The intersection operation is also very fast and simple and the support function retrieves the support of the intersection bitmap. This test allows the verification of a *necessary but not sufficient* condition for the extended sequence to be frequent. A second test is necessary

input : $S(BaseSequence); P(ExtensionPair); \sigma(Min.Support)$
$C_S = bitmap(S)$ and $C_P = bitmap(P)$
$C_{S'} = C_S \cap C_P$
if $support(C_{S'}) \geq \sigma$, then return OK.

Algorithm 1: Support Test

```
input   : C_{S'}(Bitmap); S(Base Seq); P(Ext. Pair); σ(Sup.)
1  seqLst = getSeqIdLst(C_{S'});
2  E_v = succ(P);
3  cnt = 0;
4  foreach Sid in seqLst do
5      O_v = offsetLst(Sid, E_v);
6      Y = offsetLastEvent(Sid, S);
7      W = offsetStartEvent(Sid, S);
8      if ∃X ∈ O_v, X > Y then
9          cnt = cnt + 1;
10         if (X − Y) < n then n = (X − Y)
11         if (X − Y) > m then m = (X − Y)
12     end
13     diffTest(cnt, σ);
14 end
15 if cnt ≥ σ then
16     return OK;
17 end
```

Algorithm 2: Order Test

to guarantee that the order of the events is kept along the sequences that $C_{S'}$ bitmap points to.

Algorithm 2 assumes that, for each frequent sequence, additional information besides the sequence event list is kept during the extension phase. Namely, the corresponding bitmap that for the case exposed in algorithm 1 will be $C_{S'}$ if S' is determined to be frequent. Also two offset lists in the form $<Sid, offset>$ are kept. One will contain the offset of the last event of the sequence, *offsetLastEvent*, and will be used for the "Order Test". The second, *offsetStartEvent*, contains the offset of the first event of the sequence pattern in all the Sid where it appears. This will be used when the verification of the window constraint is performed. In the Order Test, given a bitmap resulted from the support test, the *getSeqIdLst* function returns the list of the sequence identifiers for the bitmap. The function *offsetLst* returns a list of offset values of the event in the respective Sid. For each sequence identifier it is tested whether the extension pair has an offset greater than the offset value of the extended sequence. This implements the computation of C_E and the offsetList of $succ(E)$ as in property 2. At line 13 the $diffTest$ function performs a simple test to check whether the minimum support is still reachable. At the end of the procedure (lines 15 to 17) it is tested whether the order of the extended sequence pattern is respected in a sufficient number of database sequences. In the positive case the extended sequence is considered frequent. Given algorithm 1 and 2, property 3 guarantees the necessary and sufficient conditions to safely extend a base sequence into a frequent one.

Property 3 (Frequent Extended Sequence). *Given a minimum support σ, a frequent base sequence $S = <E_1 \ldots E_n>$, where $|S| \geq 2$ and a pair $P = E_k \rightarrow E_w$. If $E_n = E_k$, then $S' = <E_1 \ldots E_n \, g_{n,k} E_k>$, where $g_{n,k} = -x(n,m)-$ if in arbitrary gap mode or $-r(n)-$ if in rigid gap mode, is frequent if algorithm 1 and 2 return OK.*

3.3 Space Search Traversal

Guided by the Bitmap Matrix the search space can be traversed using two possible approaches: *breadth first* or *depth first*. For both cases the set of the frequent sequences starts as the set of frequent pairs. In the depth first mode it starts with a sequence of size 2 that is successively expand until it can not be further extended. Then we backtrack and start extending another sequence. The advantage of this type of traversal is that we don't need to keep all the intermediary frequent sequence information, in contrast with the breadth first traversal where all the information of the sequences size k need to be kept before the sequences of size k+1 are generated. This yields is some cases, a significant memory reduction.

3.4 Rigid Gap Patterns

The algorithm described in 2 is designed to mine arbitrary gap patterns. Using gIL to mine rigid gap patterns requires only minor changes in the Order Test algorithm. Lines 4 to 11 in algorithm 2 are rewritten in algorithm 3. In this algorithm, first it is collected (in gapLst) the size of all the gaps for a certain sequence extension. Next, for each gap size it is tested whether the extended sequence is frequent. One should note that for rigid gap patterns, two sequence patterns with the same events are considered different if the gaps between the events have different size, e.g., $< 1 \cdot\cdot\, 2 >$ is different from $< 1 \cdot\cdot\cdot\, 2 >$.

```
1  foreach Sid in seqLst do
2      O_v = offsetLst(Sid, E_v);
3      Y = offsetLastEvent(Sid, S);
4      W = offsetStartEvent(Sid, S);
5      if ∃X ∈ O_v, X > Y then
6          gap = X − Y; gapLst.add(gap);
7      end
8  end
9  foreach R in gapLst do
10     foreach Sid in seqLst do
11         Repeat Step 2 to 4;
12         if ∃X ∈ O_v, (X − Y) = R then
13             cnt = cnt + 1;
14         end
15     end
16     if cnt ≥ σ then
17         return OK;
18     end
19 end
```
Algorithm 3: Algorithm changes to mine rigid gap patterns

4 Constraints

The introduction of constraints in the *gIL* algorithm like *min/max gap, window size, items exclusion* is a straightforward process and translates into a considerable performance gain. These efficiency improvements are naturally expected since (depending on the values of the constraints) the search space can be greatly reduced. The introduction of *substitution sets* is also very easy to achieve. Implementing events exclusion constraint and substitution sets turns out to be a natural operation. Simple changes in the Bitmap Matrix (that guides the sequence extension) and in the Offset Matrix (discriminates the positions of the events in every sequence where they occur) enable this implementations. The new features are introduced between the scanning phase and the sequence extension phase. The min/max gap and window constraints constitute an additional to be applied when the sequence is extended.

4.1 Events Exclusion, Start Events and Substitution Sets

The event exclusion constraint is applied by traversing the rows and columns of the Bitmap Matrix where the excluded events occurs. At that positions the support[1] count variable in the respective cells is set to zero. Start events constraints are also straightforwardly implemented by allowing extensions only to the events in *StartEventSets*.

When substitution sets are activated, one or more sets of equivalent events are available. For each set of equivalent events one has to form the union of the rows (horizontal union) and columns (vertical union) in the Bitmap Matrix, where those events occur. The vertical union is similar to the horizontal union. Moreover, for all the equivalent events, one needs to pairwisely intersect the sequences where they occur and then perform the union of the offsetLists for the intersected sequences. This results in the new offsetLists of the equivalent events.

4.2 Min / Max Gap and Window Size

These constraints are trivially introduced in the "Order Test". In algorithm 2, the test in line 8 is extended with three additional tests: $(X - Y) < maxGap$ AND $(X - Y) > minGap$ AND $(X - W) < windowSize$.

5 Experimental Evaluation

We evaluated our algorithm along different variables using two collections of synthetic and real datasets. To generate the synthetic datasets we developed a sequence generator based on the Zipfian distribution. This generator receives the following parameters (see table 1(a)): number of sequences, average length of the sequences, alphabet size and a parameter p that expresses the skewness of the distribution. This generator has allowed us to generate sequences with a relative small alphabet. The evaluated variables for this datasets were: *support*, *dataset size (number of sequences)*, and *sequence size*. Additionally, we tested the mushroom dataset used at the FIMI workshop [4]. To represent real life data, we used several datasets of proteins. The Yeast (*saccharomyces cerevisiae*) dataset is available at [5] and PSSP used for protein secondary structure prediction [3]. We also used a subset of the non Outer Membrane proteins obtained from [8]. The properties for this datasets are summarized in table 1(b). It is interesting to notice that, for all datasets, gIL's scanning phase time is residual (less than 0.4 seconds).

Since gIL finds two types of patterns we performed evaluation against two different algorithms. Both are in memory algorithms, assuming that the database completely fits into main memory. For the arbitrary gap patterns from the all patterns class we compared gIL with the SPAM [1] algorithm. SPAM has shown to outperform SPADE [11] and PrefixSpan [7] and is a state-of-the-art algorithm

---

[1] Future interactions on this dataset still have the Bitmap Matrix intact since the bitmaps remain unchanged.

in transactional sequence pattern mining. The datasets suffer a conversion into the transactional dataset format, in order to be processed by SPAM. In this conversion each customer is considered as a sequence and each *itemset* contains a unique item (event).

For the rigid gap patterns we compared gIL with Teiresias [6], a well known algorithm from the bioinformatics community. It can be obtained from [2]. It is, as far as we know, the most complete and efficient algorithm for mining closed (called "most specific" in their paper) frequent rigid gap patterns. Closed patterns are a subset of all frequent sequence patterns. In this sense, gIL (which derives all patterns) tackles a more general problem and consequently considers a much larger search space than Teiresias. Besides minimum support, Teiresias uses two additional parameters. L and W are respectively the number of non-wild cards events in a pattern and the maximum spanning between two consecutive events. Since gIL starts by enumerating patterns with size 2, we will set L=2 and W to the maxGap value. All the experiments[2] were performed using exact discovery, i.e. without the use of substitution sets, and on a 1.5GHz Intel Centrino machine with 512MB of main memory, running windows XP Professional. The applications were written in C/C++ language.

5.1 Arbitrary Gap Patterns Evaluation

We start by comparing the efficiency of SPAM with the gIL algorithm without constraints. In figure 2(a) and 2(b) we tested different values of support for two datasets of $1K$ and $2K$ respectively. The sequences have an average length of 60 and an alphabet of 20 events. It was clear in these two experiments that for relative smaller dataset sizes and lower support values gIL becomes more efficient than SPAM. Figure 2(c) shows the scalability of the algorithms in respect to the dataset size for a support of 30%. This graphic shows that gIL scales well in relation to the dataset size.

In order to test a dataset with different characteristics, namely larger alphabet size, small length and greater dataset size, we used the Mushroom dataset, see figure 3(a). In figure 3(b) we have runtimes of gIL for datasets with one thousand sequences and different values of average sequence length. It was imposed a maxGap constraint of 15. As we observed during all the experiments, there is a critical point in the support variation, typically between 10% and 20%, that translates into an explosion of number of frequent patterns. This leads to an exponential behaviour in the algorithm's runtime. Even so, we can see that gIL shows similar behaviour for the different values of sequence length. Figure 3(c) measures the relative performance time, i.e. the ratio between the mining time with constraints and without constraints. These values were obtained for a support of 70%. Runtime without constraints was 305 seconds. It describes the behaviour of the algorithm when decreasing the maxGap and the Window values.

---

[2] Further details and results can be obtained from an extended version of this paper.

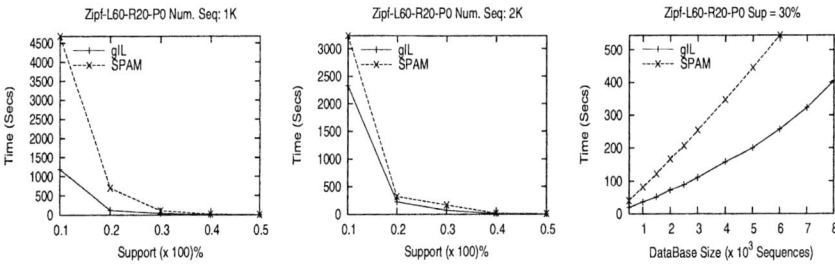

Fig. 2. (a) Support variation with Zipf database size=1K; (b) Support variation with Zipf database size=2K; (c) Scalability of gIL w.r.t. database size with a support of 30%

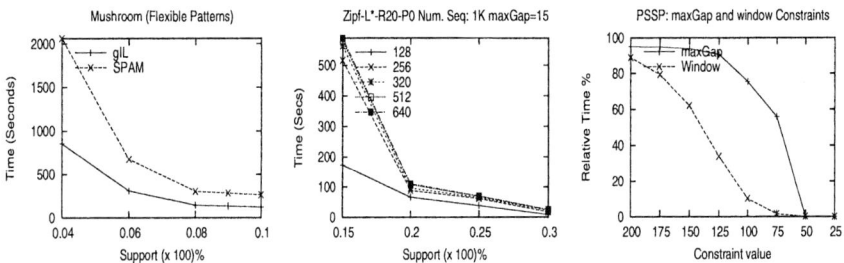

Fig. 3. (a) Support variation for the Mushroom dataset; (b) Scalability of gIL w.r.t sequence size for different support values (c) Performance evaluation using maxgap and windowgap constraints

In respect to memory usage both algorithms showed a low memory demand for all the datasets. For the Mushroom dataset which was the most demanding in terms of memory, SPAM used a maximum of 9 MB for a support of 4% and gIL a constant memory usage of 26 MB for all the support values. gIL shows a constant and support independent memory usage since once the data structures are built for a given dataset they remain unchanged.

5.2 Rigid Gap Patterns Evaluation

In order to assess the performance of gIL in the mining of rigid gap patterns we compared it with Teiresias [6], for different proteins datasets. In figure 4(a) and 4(b) the Yeast dataset was evaluated for two values of maxGap(W), 10 and 15. The results showed that gIL outperforms Teiresias by an order of magnitude. When comparing the performance of the algorithms in relation to the PSSP (figure 4(c)) and the nonOM (figure 5(a)) datasets, for a maxGap of 15, gIL outperforms Teiresias by a factor of 2 in the first case. This difference becomes more significant in the second case. The nonOM dataset has a greater average sequence length, but a small dataset size. This last characteristic results into a smaller bitmap length yielding a significant performance improvement. As

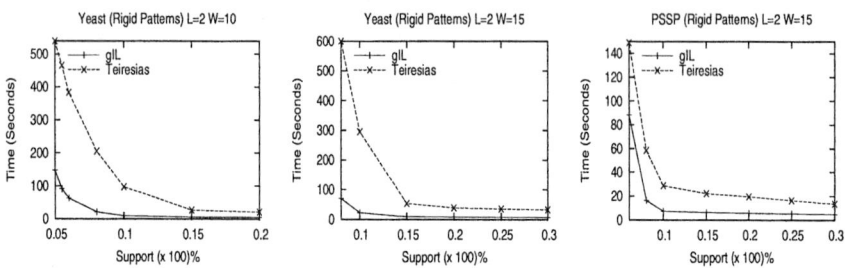

Fig. 4. (a) Support variation for the Yeast dataset, with L=2 and W(maxGap) = 10; (b) Support variation for the Yeast dataset, with L=2 and W(maxGap) = 15; (c) Support variation for the PSSP dataset, with L=2 and W(maxGap) = 15

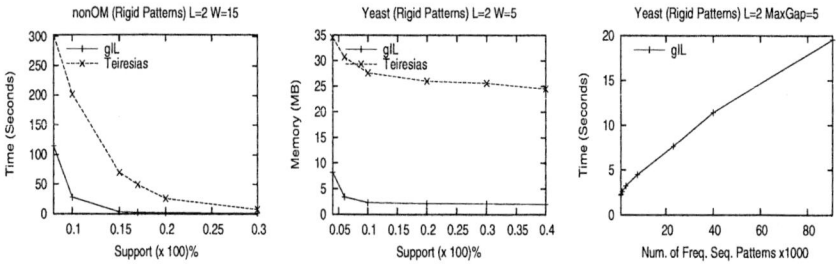

Fig. 5. (a) Support variation for the nonOM dataset, with L=2 and W(maxGap) = 15; (b) Memory usage for the Yeast dataset, with L=2 and W(maxGap) = 5; (c) Scalability of gIL w.r.t number of sequences for the Yeast dataset

we already verified in the arbitrary gap experiments, gIL memory usage maintains nearly constant for all the tested support values (figure 5(b)). Figure 5(c) shows the linear scalability of gIL in relation to the number of frequent sequence patterns.

6 Conclusions

We presented an algorithm called *gIL*, suitable to work with databases of linear sequences with a long average length and a relative small alphabet size. Our experiments showed that for the particular case of the proteins datasets, gIL exhibits superior performance to state-of-the-art algorithms. The algorithm has a high adaptability, and thus it was easily changed to extract two different types of patterns: arbitrary and rigid gap patterns. Furthermore, the data organization allows a straightforward implementation of constraints and substitution sets. These features are pushed directly into the mining process, which in some cases enables the mining in useful time of otherwise untractable problems. In this sense gIL is an interesting and powerful algorithm to be applied in a broader range of domains and in particular suitable for biological data. Thus, even when performing extensions an event at a time (using a smart combination of some of the most efficient techniques that have been used in the task of itemset and sequence mining) one can obtain an algorithm that efficiently handles the explosive nature of pattern search, inherent to the biological sequence datasets.

References

1. J. Ayres, J. Flannick, J. Gehrke, and T. Yiu. Sequential pattern mining using a bitmap representation. In *Proceedings of the 8th SIGKDD International Conference on KDD and Data Mining, 2002*.
2. IBM Bioinformatics. Teiresias. http://www.research.ibm.com/bioinformatics/.
3. James Cuff and Geoffrey J. Barton. Evaluation and improvement of multiple sequence methods for protein secondary structure prediction. In *PROTEINS: Structure, Function, and Genetics*, number 34. WILEY-LISS, INC, 1999.
4. Fimi. Fimi workshop 2003 (mushroom dataset). http://fimi.cs.helsinki.fi/fimi03.
5. GenBank. yeast (saccharomyces cerevisiae). www.maths.uq.edu.au.
6. A.Floratos I. Rigoutsos. Combinatorial pattern discovery in biological sequences: the teiresias algorithm. *Bioinformatics*, 1(14), January 1998.
7. J. Pei, J. Han, B. Mortazavi-Asl, H. Pinto, Q. Chen, U. Dayal, and M.-C. Hsu. PrefixSpan: Mining sequential patterns efficiently by prefix projected pattern growth. In *Proceedings of the International Conference on Data Engineering, ICDE 2001*.
8. Psort. Psort database. http://www.psort.org/.
9. Ramakrishnan Srikant and Rakesh Agrawal. Mining sequential patterns: Generalizations and performance improvements. In *Proceedings 5th International Conference on Extending DataBase Technology*, 1996.
10. Mohammed J. Zaki. Sequence mining in categorical domains: Incorporating constraints. In *In Proceedings of 9th International Conference on Information and Knowledge Management, CIKM 2000*.
11. Mohammed J. Zaki. Spade: An efficient algorithm for mining frequent sequences. *Machine Learning*, 42(1-2):31–60, 2001.

An Adaptive Nearest Neighbor Classification Algorithm for Data Streams

Yan-Nei Law and Carlo Zaniolo

Computer Science Dept., UCLA, Los Angeles, CA 90095, USA
{ynlaw, zaniolo}@cs.ucla.edu

Abstract. In this paper, we propose an incremental classification algorithm which uses a multi-resolution data representation to find adaptive nearest neighbors of a test point. The algorithm achieves excellent performance by using small classifier ensembles where approximation error bounds are guaranteed for each ensemble size. The very low update cost of our incremental classifier makes it highly suitable for data stream applications. Tests performed on both synthetic and real-life data indicate that our new classifier outperforms existing algorithms for data streams in terms of accuracy and computational costs.

1 Introduction

A significant amount of recent research has focused on mining data streams for applications such as financial data analysis, network monitoring, security, sensor networks, and many others [3,8]. Algorithms for mining data streams have to address challenges not encountered in traditional mining of stored data: at the physical level, these include fast input rates and unending data sets, while, at the logical level, there is the need to cope with concept drift [18]. Therefore, classical classification algorithms must be replaced by, or modified into, incremental algorithms that are fast and light and gracefully adapt to changes in data statistics [17,18,5].

Related Works. Because of their good performance and intuitive appeal, decision tree classifiers and nearest neighborhood classifiers have been widely used in traditional data mining tasks [9]. For data streams, several decision tree classifiers have been proposed—either as single decision trees, or as ensembles of such trees. In particular, VFDT [7] and CVFDT [10] represent well-known algorithms for building single decision tree classifiers, respectively, on stationary, and time-changing data streams. These algorithms employ a criterion based on Hoeffding bounds to decide when a further level of the current decision tree should be created. While this approach assures interesting theoretical properties, the time required for updating the decision tree can be significant, and a large amount of samples is needed to build a classifier with reasonable accuracy. When the size of the training set is small, the performance of this approach can be unsatisfactory.

Another approach to data stream classification uses ensemble methods. These construct a set of classifiers by a base learner, and then combine the predictions of these base models by voting techniques. Previous research works [17,18,5] have shown that ensembles can often outperform single classifiers and are also suitable for coping with concept drift. On the other hand, ensemble methods suffer from the drawback that they often fail to provide a simple model and understanding of the problem at hand [9].

In this paper, we focus on building nearest neighbor (NN) classifiers for data streams. This technique works well in traditional data mining applications, is supported by a strong intuitive appeal, and it rather simple to implement. However, the time spent for finding the exact NN can be expensive and, therefore, a significant amount of previous research has focused on this problem. A well-known method for accelerating the nearest neighbor lookup is to use k-d trees [4]. A k-d tree is a balanced binary tree that recursively splits a d-dimensional space into smaller subregions. However, the tree can become seriously unbalanced by massive new arrivals in the data stream, and thus lose the ability of expediting the search. Another approach to NN classifiers attempts to provide approximate answers with error bound guarantees. There are many novel algorithms [11,12,13,14] for finding approximate K-NN on stored data. However, to find the $(1 + \epsilon)$-approximate nearest neighbors, these algorithms must perform multiple scans of the data. Also, the update cost of the dynamic algorithms [11,13,14] depends on the size of the data set, since the entire data set is needed for the update process. Therefore, they are not suitable for mining data streams.

Our ANNCAD Algorithm. In this paper, we introduce an Adaptive NN Classification Algorithm for Data-streams. It is well-known that when data is non-uniform, it is difficult to predetermine K in the KNN classification [6,20]. So, instead of fixing a specific number of neighbors, as in the usual KNN algorithm, we adaptively expand the nearby area of a test point until a satisfactory classification is obtained. To save the computation time for finding adaptive NN, we first preassigning a class to every subregion (cell). To achieve this, we decompose the feature space of a training set and obtain a multi-resolution data representation. There are many decomposition techniques for multi-resolution data representations. The averaging technique used in this paper can be thought of Haar Wavelets Transformation [16]. Thus, information from different resolution levels can then be used for adaptively preassigning a class to every cell. Then we determine to which cell the test point belongs, in order to predict its class. Moreover, because of the compact support property inherited from wavelets, the time spent updating a classifier when a new tuple arrives is a small constant, and it is independent of the size of the data set. Unlike VDFT, which requires a large data set to decide whether to expand the tree by one more level, ANNCAD does not have this restriction.

In the paper, we use grid-based approach for classification. The main characteristic of this approach is the fast processing time and small memory usage, which is independent of the number of data points. It only depends on the number of cells of each dimension in the discretized space, which is easy to adjust in order to fulfill system constraints. Therefore, this approach has been widely employed in clustering problem. Some examples of novel clustering algorithms are *STING* [19], *CLIQUE* [1] and *WaveCluster* [15]. However, there is not much work using this approach for classification.

Paper Organization. In this paper, we present our algorithm ANNCAD and discuss its properties in §2. In §3, we compare ANNCAD with some existing algorithms. The results suggest that ANNCAD will outperform existing algorithms. Finally, conclusions and suggestions for future work will be given in §4.

2 ANNCAD

In this section, we introduce our proposed algorithm ANNCAD, which includes four main stages: (1) Quantization of the Feature Space; (2) Building classifiers; (3) Finding predictive label for a test point by adaptively finding its neighboring cells; (4) Updating classifiers for newly arriving tuples. This algorithm only read each data tuple at most once, and only requires a small constant time to process it. We then discuss its properties and complexity.

2.1 Notation

We are given a set of d-dimensional data D with attributes $X_1, X_2, ..., X_d$. For each $i = 1, ..., d$, the domain of X_i is bounded and totally ordered, and ranges over the interval $[L_i, H_i)$. Thus, $X = [L_1, H_1) \times ... \times [L_d, H_d)$ is the feature space containing our data set D.

Definition 1. *A discretized feature space is obtained by dividing the domain of each dimension into g open intervals of equal length. The discretized feature space so produced consists of g^d disjoint rectangular blocks, of size $\Delta x_i = (H_i - L_i)/g$ in their i^{th} dimension.*

Let $B_{i_1,...,i_d}$ denote the block:

$$[L_1 + (i_1 - 1)\Delta x_1, L_1 + i_1 \Delta x_1) \times ... \times [L_d + (i_d - 1)\Delta x_d, L_d + i_d \Delta x_d).$$

Alternatively, we denote $B_{i_1,...,i_d}$ by $B_{\mathbf{i}}$, with $\mathbf{i} = (i_1, ..., i_d)$ the unique identifier for the block. Then, two blocks $B_{\mathbf{k}}$ and $B_{\mathbf{h}}$, $\mathbf{k} \neq \mathbf{h}$, are said to be *adjacent* if $|k_i - h_i| \leq 1$, for each $i = 1, ..., d$. In this case, $B_{\mathbf{k}}$ is said to be a neighbor of $B_{\mathbf{h}}$. $Ctr_{B_{\mathbf{i}}}$ denotes the center of block $B_{\mathbf{i}}$, computed as the average of its vertices:

$$Ctr_{B_{\mathbf{i}}} = (L_1 + (i_1 - 1/2)\Delta x_1, ..., L_d + (i_d - 1/2)\Delta x_d).$$

Definition 2. *Let x be a point and $B_{\mathbf{i}}$ be a block in the same feature space. The distance between x and $B_{\mathbf{i}}$ is defined as the distance between x and $Ctr_{B_{\mathbf{i}}}$.*

Note that the distance in Def. 2 can be any kind of distance. In the following, we use Euclidean distance to be the distance between a point and a block.

2.2 Quantization of the Feature Space

The first step of ANNCAD is to partition the feature space into a discretized space with g^d blocks as in Def. 1. It is advisable to choose different sizes of grid according to system resource constraints and desirable fineness of a classifier. For each nonempty block, we count the number of training points contained in it for each class. Now we get the distribution of the data entities in each class. To decide whether we need to start with a finer resolution feature space, we then count the number of training points that do not belong to the majority class of its block as a measure of the training error. We then calculate the coarser representations of the data by averaging the 2^d corresponding blocks in the next finer level. We illustrate the above process by Example 1.

Example 1. A set of 100 two-class training points in the 2-D unit square is shown in Fig. 1(a). There are two classes for this data set, where a circle (resp. triangle) represents a training point of class I (resp. II). First we separate the training points of each class, discretize them using a 4 × 4 grid and count the number of training points for each block to get the data distribution of each class (see Fig. 1(b)). Moreover, Fig. 1(c)-(d) show the coarser representations of the data.

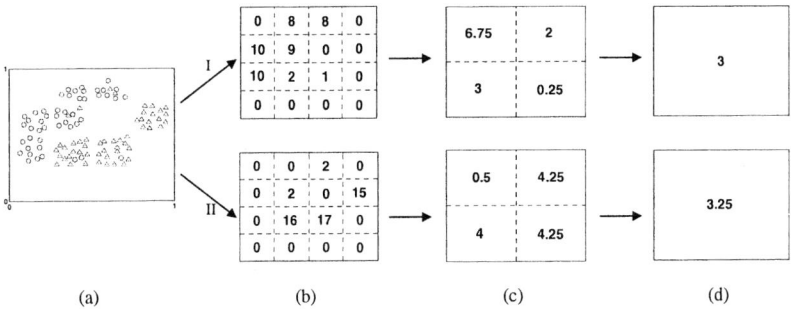

Fig. 1. Multi-resolution representation of a two-class data set

Due to the problem of the curse of dimensionality, the storage amount is exponential in the number of dimensions. To deal with this, we store the nonempty blocks in the leaf nodes of a B^+-tree using their z-values [21] as keys. Thus the required storage space is much smaller and is bounded by $O(\min(N, g^d))$ where N is the number of training samples. For instance, in Fig. 1, we only need to store information for at most 8 blocks even though there are 100 training points in the 4 × 4 blocks feature space. To reduce space usage, we may only store the data array of the finest level and calculate the coarser levels on the fly when building a classifier. On the other hand, to reduce time complexity, we may pre-calculate and store the coarser levels. In the following discussion, we assume that the system stores the data representation of each level.

2.3 Building a Classifier and Classifying Test Points

The main idea of ANNCAD is to use a multi-resolution data representation for classification. Notice that the neighborhood relation strongly depends on the quantization process. This will be addressed in next subsection by building several classifier ensembles using different grids obtained by subgrid displacements. Observe that in general, the finer level the block can be classified, the shorter distance between this block and the training set. Therefore, to build a classifier and classify a test point (see Algorithms 1 and 2), we start with the finest resolution for searching nearest neighbors and progressively consider the coarser resolutions, in order to find nearest neighbors adaptively.

We first construct a single classifier as a starting point (see Algorithm 1). We start with setting every block to have a default tag U (Non-visited). In the finest level, we classify any nonempty block with its majority class label. We then classify any nonempty block of every lower level as follows: We label the block by its majority class label if the majority class label has more points than the second majority class by a threshold percentage. If not, we use a specific tag M (Mixed) to label it.

Algorithm 1. BuildClassifier($\{\mathbf{x}, y\}|\mathbf{x}$ is a vector of attributes, y is a class label.)

Quantize the feature space containing $\{\mathbf{x}\}$
Label majority class for each nonempty block in the finest level
For each level $i = \log(g)$ downto 1
    For each nonempty block B
        If $|\text{majority } c_a| - |\text{2nd majority } c_b| > \text{threshold \%}$, label class c_a
        else label tag M
Return Classifier

Algorithm 2. TestClass(test point: **t**)

For each level $i = \log(g) + 1$ downto 1
    If label of $B^i(\mathbf{t}) <> U$ /*$B^i(\mathbf{t})$ is nonempty */
        If label of $B^i(\mathbf{t}) <> M$, class of **t** = class of $B^i(\mathbf{t})$
        else class of **t** = class of NN of $B^{i+1}(\mathbf{t})$ /*$B^{i+1}(\mathbf{t})$ contains **t** in level $i+1$*/
        Break
Return class label for **t**, $B^i(\mathbf{t})$

Example 2. We build a classifier for the data set of Example 1 and set the threshold value to be 80%. Fig. 2(a), (b) and (c) show the class label of each nonempty block in the finest, intermediate and coarsest resolution respectively.

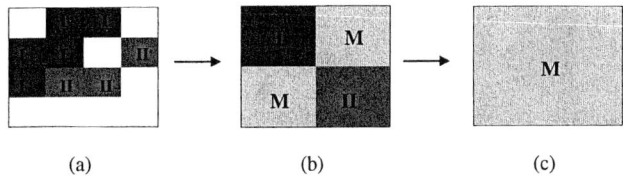

(a) (b) (c)

Fig. 2. Hierarchical structure of classifiers

For each level i, a test point **t** belongs to a unique block $B^i(\mathbf{t})$. We search from the finest to the coarsest level until reaching a nonempty block $B^i(\mathbf{t})$. If the label of $B^i(\mathbf{t})$ is one of the classes, we label the test point by this class. Otherwise, if $B^i(\mathbf{t})$ has tag M, we find the nearest neighbor block of $B^{i+1}(\mathbf{t})$ where $B^{i+1}(\mathbf{t})$ is a block containing **t** in level $i+1$. To reduce the time spent, we only consider the neighbors of $B^{i+1}(\mathbf{t})$ which belong to $B^i(\mathbf{t})$ in level i. It is very easy to access these neighbors as they are also neighbors of $B^{i+1}(\mathbf{t})$ in the B$^+$-tree with their z-values as keys. Note that $B^{i+1}(\mathbf{t})$ must be empty, otherwise we should classify it at level $i+1$. But some of the neighbors of $B^{i+1}(\mathbf{t})$ must be nonempty as $B^i(\mathbf{t})$ is nonempty. We simply calculate the distance between test point **t** and each neighbor of $B^{i+1}(\mathbf{t})$ and label **t** by the class of NN.

Example 3. We use the classifier built in Example 2 to classify a test point $\mathbf{t} = (0.6, 0.7)$. Starting with the finest level, we found that the first nonempty block containing **t** is $[0.5, 1) \times [0.5, 1)$ (see Fig. 3(b)). Since it has tag M, we calculate the distance between **t** and each nonempty neighboring block in the next finer level ($[0.75, 1) \times [0.5, 0.75), [0.5, 0.75) \times [0.75, 1)$). Finally, we get the nearest neighboring block $[0.75, 1) \times [0.5, 0.75)$ and label **t** to be class I (see Fig.

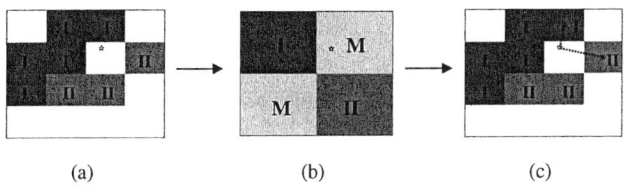

(a) (b) (c)

Fig. 3. Hierarchical classifier access

Fig. 4. The combined classifier

3(c)). When we combine the multi-resolution classifier of each level, we get a classifier for the whole feature space (see Fig. 4).

2.4 Incremental Updates of Classifiers

The main requirement of a data stream classification algorithm is that it is able to update classifiers incrementally and effectively when a new tuple arrives. Moreover, updated classifier should be adapt to concept drift behaviors. In this subsection, we present incremental update process of ANNCAD for a stationary data, without re-scanning the data and discuss an exponential forgetting technique to adapt to concept drifts.

Because of the compact support property, arrival of a new tuple only affects the blocks of the classifier in each level containing this tuple. Therefore, we only need to update the data array of these blocks and their classes if necessary. During the update process, the system may run out of memory as the number of nonempty blocks may increase. To deal with this, we may simply remove the finest data array, multiple the entries of the remaining coarser data arrays by 2^d, and update the quantity g. A detailed description of updating classifiers can be found in Algorithm 3. This solution can effectively meet the memory constraint.

Algorithm 3. UpdateClassifier(new tuple: **t**)

For each level $i = \log(g) + 1$ downto 1
    Add $\delta_t / 2^{d \times (\log(g)+1-i)}$ to data array Φ^i
    /*δ_t is a matrix with value 1 in the corr. entry of t and 0 elsewhere.*/
    If i is the finest level, label $B^i(\mathbf{t})$ with the majority class
    else if |majority c_a| − |2nd majority c_b| > threshold %, label $B^i(\mathbf{t})$ by c_a
    else label $B^i(\mathbf{t})$ by tag M
If memory runs out,
    Remove the data array of level $\log(g) + 1$
    For each level $i = \log(g)$ downto 1, $\Phi^i = 2^d \cdot \Phi^i$
    Label each nonempty block of the classifier in level $\log(g)$ by its majority class
    Set $g = g/2$
Return updated classifier

Exponential Forgetting. If the concept of the data changes over time, a very common technique called exponential forgetting may be used to assign less weight to the old data to adapt to more recent trend. To achieve this, we multiply an exponential forgetting factor λ to the data array, where $0 \leq \lambda \leq 1$. For each level i, after each time interval t, we update the data array Φ^i to be:

$$\Phi^i|_{(n+1)t} \leftarrow \lambda \Phi^i|_{n \cdot t}$$

where $\Phi^i|_{n \cdot t}$ is the data array at time $n \cdot t$. Indeed, if there is no concept change, the result of classifier will not be affected. If there is a concept drift, the classifier can adapt to the change quickly since the weight of the old data is exponentially decreased. In practice, an exponential forgetting technique is easier to implement than a sliding window because we need extra memory buffer to store the data of the most current window for implementing the sliding window.

2.5 Building Several Classifiers Using Different Grids

As mentioned above, the neighborhood relation strongly depends on the quantization process. For instance, consider the case that there is a training point u which is close to the test point v but they are located in different blocks. Then the information on u may not affect the classification of v.

To overcome the problem of initial quantization process, we build several classifier ensembles starting with different quantization space. In general, to build n^d different classifiers, each time we shift $\frac{1}{n}$ of the unit length of feature space for a set of selected dimensions. Fig. 5 shows a reference grid and its 3 different shifted grids for a feature space with 4×4 blocks. For a given test point $\mathbf{t}$, we use these n^d classifiers to get n^d class labels and selected blocks $B^i(\mathbf{t})$ of $\mathbf{t}$ in each level i, starting from the finest one. We then choose the majority class label. If there is tie, we calculate the distance between each selected block $B^i(\mathbf{t})$ with majority class label and $\mathbf{t}$ to find the closest one. Algorithm 4 shows this classifying process using n^d classifiers.

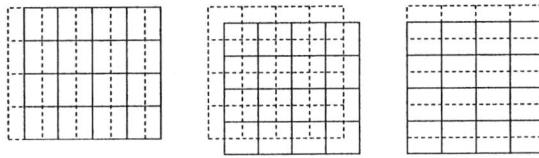

Fig. 5. An example of 4 different grids for building 4 classifiers

---

Algorithm 4. Testn^dClass(objects: **t**)

For each level $i = \log(g) + 1$ downto 1
    Get the label of **t** for each classifier
    If there is a label $<>$ U, choose the majority label
        If there is a tie, label **t** by class of $B^i(\mathbf{t})$ with closest center to **t**
    Break
Return class label for **t**

The following theorem shows that the approximation error of finding nearest neighbors decreases as the number of classifier ensembles increases.

Theorem 1. *For d attributes, let x be the test point and Y be the set of training points which are in blocks containing x of those n^d classifiers. Then, for every training point $z \notin Y$, $dist(x,y) < (1 + \frac{1}{n-1})*dist(x,z)$ for every $y \in Y$.*

Proof. For simplicity, we consider the case when $d = 1$. This proof works for any d. For $d = 1$, we build n classifiers, where each classifier i use the grid that is shifted $\frac{i}{n}$ unit length from the original grid. Let ϵ be the length of a block. Consider a test point x, x belongs to an interval I_k for classifier k. Note that $[x - \frac{n-1}{n}\epsilon, x + \frac{n-1}{n}\epsilon] \subset \bigcup I_k \subset [x - \epsilon, x + \epsilon]$. Hence, the distance between x and its nearest neighbor that we found must be less than ϵ. Meanwhile, the points that we do not consider should be at least $\frac{n-1}{n}\epsilon$ far away from x. If $z \notin Y$, $\frac{dist(x,y)}{dist(x,z)} < \frac{\epsilon}{(n-1)\epsilon/n} = (1 + \frac{1}{n-1})$ for every $y \in Y$.

The above theorem shows that the classification result using one classifier does not have any guarantee about the quality of the nearest neighbors that it found because the ratio of approximation error will tend to infinity. When n is large enough, the set of training points selected by those classifier ensembles are exactly the set of training points with distance ϵ from the test point. To achieve an approximation error bound guarantee, theoretically we need an exponential number of classifiers. However, in practice, we only use two classifiers to get a good result. Indeed, experiments in §3 show that few classifiers can obtain a significant improvement at the beginning. After this stage, the performance will become steady even though we keep increasing the number of classifiers.

2.6 Properties of ANNCAD

As ANNCAD is a combination of multi-resolution and adaptive nearest neighbors techniques, it inherits both their properties and their advantages.

- *Compact support:* The locality property allows a fast update. As a new tuple arrival only affects the class of the block containing it in each level, the incremental update process only costs a constant time (number of levels).
- *Insensitivity to noise:* We may set a threshold value for classifying decisions to remove noise.
- *Multi-resolution:* This algorithm makes it easy to build multi-resolution classifiers. Users can specify the number of levels to efficiently control the fineness of the classifier. Moreover, one may optimize the system resource constraints and easy to adjust on the fly when the system runs out of memory.
- *Low complexity:* Let g, N and d be the number of blocks of each dimension, training points and attributes respectively. The time spent on building a classifier is $O(\min(N, g^d))$ with constant factor $\log(g)$. For the time spent on classifying a test point, the worst case complexity is $O(\log_2(g) + 2^d)$ where the first part is for classifying a test point using classifiers and the second part is for finding its nearest neighbor which is optional. Also, the time spent for updating classifiers when a new tuple arrives is $\log_2(g) + 1$. Comparing with the time spent in VFDT, our method is more attractive.

3 Performance Evaluation

In this section, we first study the effects on parameters for ANNCAD by using two synthetic data sets. We then compare ANNCAD with VFDT and CVFDT on three real-life data sets. To illustrate the approximation power of ANNCAD, we include the results of *Exact ANN*, which computes ANN exactly, as controls. *Exact ANN*: For each test point t, we search the area within 0.5 block side length distance. If the area is nonempty, we classify t as the majority label of all these points in this area. Otherwise, we expand the searching area by doubling the radius until we get a class for t. Note that the time and space complexities of *Exact ANN* are very expensive making it impractical to use.

3.1 Synthetic Data Sets

The aim of this experiment is to study the effect on the initial resolution for ANNCAD. In this synthetic data set, we consider a 3-D unit cube. We randomly pick 3k training points and assign those points which are inside a sphere with center (0.5, 0.5, 0.5) and radius 0.5 to be class 0, and class 1 otherwise. This data set is effective to test the performance of a classifier as it has a curve-like decision boundary. We then randomly draw 1k test points and run ANNCAD starting with different initial resolution and 100% threshold value. In Fig. 6(a), the result shows that a finer initial resolution gets a better result. This can be explained by the fact that we can capture a curve-like decision boundary if we start with a finer resolution. On the other hand, as discussed in last section, the time spent for building a classifier increases linearly for different resolutions. In general, we should choose a resolution according to system resource constraints.

The aim of this experiment is to study the effect on number of classifier ensembles for ANNCAD. As in the previous experiment, we randomly pick 1k training examples and assign them labels. We then randomly draw 1k test points and test them based on the voting result of these classifiers. We set $16 \times 16 \times 16$ blocks for the finest level and 100% threshold value. In Fig. 6(b), the result shows that having more classifiers will get a better result in the beginning. The performance improvement becomes steady even though we keep increasing the number of classifiers. It is because there is no further information given when we increase the number of classifiers. In this experiment, we only use 2 or 3 classifiers to obtain a competitive result with the *Exact ANN* (90.4%).

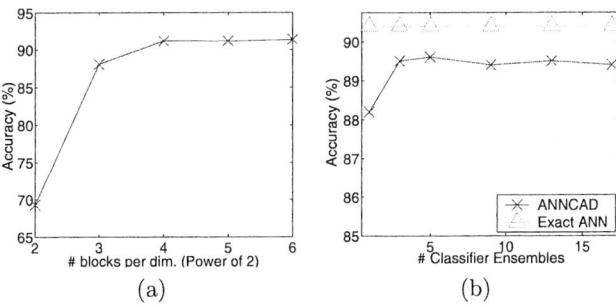

Fig. 6. Effect on initial resolutions and number of classifiers

3.2 Real Life Data Sets

The aim of this set of experiments is to compare the performance of ANNCAD with that of VFDT and CFVDT on stationary and time-changing real-life data sets respectively. We first used a letter recognition data set from the UCI machine learning repository web site [2]. The objective is to identify a black-and-white pixel displays as one of the 26 English alphabet. In this data set, each entity is a pixel display for an English alphabet and has 16 numerical attributes to describe its pixel displays. The detail description of this data set is provided in [2]. In this experiment, we use 15k tuples for training set with 5% noise added and 5k for test set. We obtain noisy data by randomly assigning a class label for 5% training examples. For ANNCAD, we set g for the initial grid to be 16 units and build two classifiers. Moreover, since VFDT needs a very large training set to get a fair result, we rescan the data sets up to 500 times for VFDT. So the data set becomes 7,500,000 tuples. In Fig. 7(a), the performance of ANNCAD dominates that of VFDT. Moreover, ANNCAD only needs one scan to achieve this result, which shows that ANNCAD even works well for a small training set.

The second real life data set we used is the Forest Cover Type data set which is another data set from [2]. The objective is to predict forest cover type (7 types). For each observation, there are 54 variables. Neural network (backpropagation) was employed to classify this data set and got 70% accuracy, which is the highest one recorded in [2]. In our experiment, we used all the 10 quantitative variables. There are 12k examples for training set and 90k examples for testing set. For ANNCAD, we scaled each attribute to the range $[0, 1)$. We set g for the initial grid to be 32 units and build two classifiers. As the above experiment, we rescan the training set up to 120 times for VFDT, until its performance becomes steady. In Fig. 7(b), the performance of ANNCAD dominates that of VFDT. These two experiments show that ANNCAD works well in different kinds of data sets.

We further tested ANNCAD in the case when there are concept drifts in data set. The data we used was extracted from the census bureau database [2]. Each observation represents a record of an adult and has 14 attributes including age, race etc. The prediction task is to determine whether a person makes over 50K a year. Concept drift is simulated by grouping records with same race (Amer-Indian-Eskimo(AIE), Asian-Pac-Islander(API), Black(B), Other(O), White(W)). The distribution of training tuples of each race is shown in Fig. 7(c). Since the models for different races of people should be different, concept drifts are introduced when $n = 311, 1350, 4474, 4746$. In this experiment, we used the 6 continuous attributes. We used 7800 examples for learning and tested the classifiers for every 300 examples. For ANNCAD, we build two classifiers and set λ to be 0.98 and g for the initial grid to be 64 units. We scaled the attribute values as mentioned in the previous experiment. The results are shown in Fig. 7(c). The curves show that ANNCAD keeps improving in each region. Also, as mentioned in §2.6, computations required for ANNCAD are much lower than CVFDT.

Moreover, notice that ANNCAD works almost as well as *Exact ANN* on these three data sets, which demonstrates its excellent approximation ability.

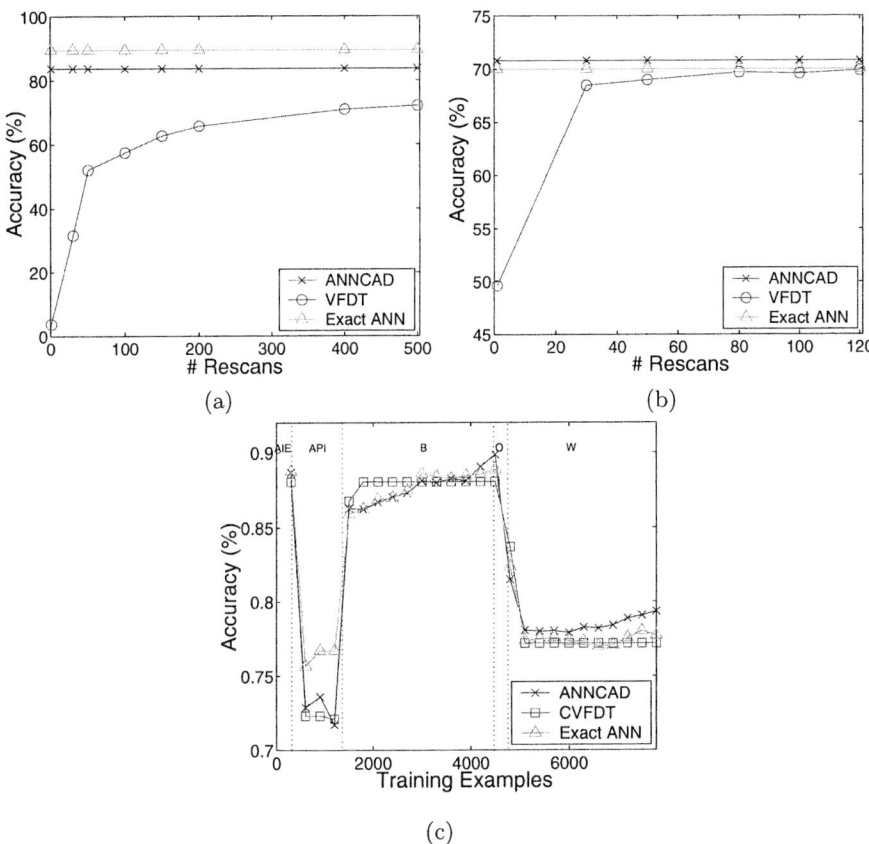

Fig. 7. Three real-life data sets:(a) Letter Recognition (b) Forest Covertype (c) Census

4 Conclusion and Future Work

In this paper, we proposed an incremental classification algorithm ANNCAD using a multi-resolution data representation to find adaptive nearest neighbors of a test point. ANNCAD is very suitable for mining data streams as its update speed is very fast. Also, the accuracy compares favorably with existing algorithms for mining data streams. ANNCAD adapts to concept drift effectively by the exponential forgetting approach. However, the very detection of sudden concept drift is of interest in many applications. The ANNCAD framework can also be extended to detect concept drift–e.g. changes in class label of blocks is a good indicator of possible concept drift. This represents a topic for our future research.

Acknowledgement. This research was supported in part by NSF Grant No. 0326214.

References

1. R. Agrawal, J. Gehrke, D. Gunopulos and P. Raghavan. Automatic Subspace Clustering of High Dimensional Data for Data Mining Applications. *SIGMOD 1998: 94–105*.
2. C. L. Blake and C. J. Merz. UCI Repository of machine learning databases. http://www.ics.uci.edu/~mlearn/MLRepository.html.
3. B. Babcock, S. Babu, R. Motawani and J. Widom. Models and issues in data stream systems. *PODS 2002: 1–16*.
4. J. Bentley. Multidimensional binary search trees used for associative searching. *Communication of the ACM 18(9): 509–517 (1975)*.
5. F. Chu and C. Zaniolo. Fast and light boosting for adaptive mining of data streams. *PAKDD 2004: 282–292*.
6. C. Domeniconi, J. Peng and D. Gunopulos, Locally adaptive metric nearest-neighbor classification, *IEEE Transactions on Pattern Analysis and Machine Intelligence 24(9): 1281–1285 (2002)*.
7. P. Domingos and G. Hulten. Mining high-speed data streams. *KDD 2000: 71–80*.
8. L. Golab and M. Özsu. Issues in data stream management. *ACM SIGMOD 32(2): 5–14 (2003)*.
9. J. Han and M. Kamber. *Data Mining – Concepts and Techniques (2000)*. Morgan Kaufmann Publishers.
10. G. Hulten, L. Spence and P. Domingos. Mining time-changing data streams. *KDD 2001: 97–106*.
11. P. Indyk, R. Motwani. Approximate nearest neighbors: towards removing the curse of dimensionality. *STOC 1998: 604–613*.
12. P. Indyk. Dimensionality reduction techniques for proximity problems. *ACM-SIAM symposium on Discrete algorithms 2000: 371–378*.
13. P. Indyk. High-dimensional computational geometry. *Dept. of Comput. Sci., Stanford Univ., 2001*.
14. E. Kushilevitz, R. Ostrovsky, Y. Rabani. Efficient Search for Approximate Nearest Neighbor in High Dimensional Spaces. *SIAM J. Comput. 30(2): 457–474 (2000)*.
15. G. Sheikholeslami, S. Chatterjee and A. Zhang. WaveCluster: A Multi-Resolution Clustering Approach for Very Large Spatial Databases. *VLDB 1998: 428–439*.
16. G. Strang and T. Nguyen. *Wavelets and Filter Banks (1996)*. Wellesley-Cambridge Press.
17. W. Street and Y. Kim. A streaming ensemble algorithm (SEA) for large-scale classification. *SIGKDD 2001: 377–382*.
18. H. Wang, W. Fan, P. Yu and J. Han. Mining concept-drifting data streams using ensemble classifiers. *SIGKDD 2003: 226–235*.
19. W. Wang, J. Yang and R. Muntz. STING: A Statistical Information Grid Approach to Spatial Data Mining *VLDB 1997: 186–195*.
20. D. Wettschereck and T. Dietterich. Locally Adaptive Nearest Neighbor Algorithms. *Advances in Neural Information Processing Systems 6: 184–191 (1994)*.
21. C. Zaniolo, S. Ceri, C. Faloutsos, R. Snodgrass, V. Subrahmanian and R. Zicari. *Advanced Database Systems (1997)*. Morgan Kaufmann Press.

Support Vector Random Fields for Spatial Classification

Chi-Hoon Lee, Russell Greiner, and Mark Schmidt

Department of Computing Science,
University of Alberta,
Edmonton AB, Canada
{chihoon, greiner, schmidt}@cs.ualberta.ca

Abstract. In this paper we propose Support Vector Random Fields (SVRFs), an extension of Support Vector Machines (SVMs) that explicitly models spatial correlations in multi-dimensional data. SVRFs are derived as Conditional Random Fields that take advantage of the generalization properties of SVMs. We also propose improvements to computing posterior probability distributions from SVMs, and present a local-consistency potential measure that encourages spatial continuity. SVRFs can be efficiently trained, converge quickly during inference, and can be trivially augmented with kernel functions. SVRFs are more robust to class imbalance than Discriminative Random Fields (DRFs), and are more accurate near edges. Our results on synthetic data and a real-world tumor detection task show the superiority of SVRFs over both SVMs and DRFs.

1 Introduction

The task of classification has traditionally focused on data that is "independent and identically distributed" (iid), in particular assuming that the class labels for different data points are conditionally independent (ie. knowing that one patient has cancer does not mean another one will). However, real-world classification problems often deal with data points whose labels are correlated, and thus the data violates the iid assumption. There is extensive literature focusing on the 1-dimensional 'sequential' case (see [1]), where correlations in the labels of data points in a linear sequence exist, such as in strings, sequences, and language. This paper focuses on the more general 'spatial' case, where these correlations exist in data with two-dimensional (or higher-dimensional) structure, such as in images, volumes, graphs, and video.

Classifiers that make the iid assumption often produce undesirable results when applied to data with spatial dependencies in the labels. For example, in the task of image labeling, a classifier could classify a pixel as 'face', even if all adjacent pixels were classified as 'non-face'. This problem motivates the use of Markov Random Fields (MRFs) and more recently Conditional Random Fields (CRFs) for spatial data. These classification techniques augment the performance of an iid classification technique (often a Mixture Model for MRFs, and Logistic Regression for CRFs) by taking into account spatial class dependencies.

Support Vector Machines (SVMs) are classifiers that have appealing theoretical properties [2], and have shown impressive empirical results in a wide variety of tasks. However, this technique makes the critical iid assumption. This paper proposed an extension to SVMs that considers spatial correlations among data instances (as in Random Field models), while still taking advantage of the powerful discriminative properties of SVMs. We refer to this technique as Support Vector Random Fields (SVRFs)

The remaining sections of this paper are organized as follows. Section 2 formalizes the task and reviews related methods for modeling dependencies in the labels of spatial data. Section 3 reviews Support Vector Machines, and presents our Support Vector Random Field extension. Experimental results on synthetic and real data sets are given in Sect. 4, while a summary of our contribution is presented in Sect. 5.

2 Related Work

The challenge of performing classification while modeling class dependencies is often divided into two perspectives: Generative and Discriminative models [1]. Generative classifiers learn a model of the joint probability, $p(x, y) = p(x|y)p(y)$, of the features x and corresponding labels y. Predictions are made using Bayes rule to compute $p(y|x)$, and finding an assignment of labels maximizing this probability. In contrast, discriminative classifiers model the posterior $p(y|x)$ directly without generating any prior distributions over the classes. Thus, discriminative models solely focus on maximizing the conditional probability of the labels, given the features. For many applications, discriminative classifiers often achieve higher accuracy than generative classifiers [1]. There has been much related work on using random field theory to model class dependencies in generative and more recently discriminative contexts [3,4]. Hence, we will first review *Markov Random Fields* (typically formulated as a generative classifier), followed by *Conditional Random Fields* (a state-of-the-art discriminative classifier built upon the foundations of Markov Random Fields).

2.1 Problem Formulation

In this work, we will focus on the task of classifying elements (pixels or regions) of a two-dimensional image, although the methods discussed also apply to higher-dimensional data. An image is represented with an M by N matrix of elements. For an instance $X = (x_{11}, x_{12}, \ldots, x_{1N}, \ldots, x_{M1}, x_{M2}, \ldots, x_{MN})$, we seek to infer the most likely joint class labels:

$$Y^* = (y_{11}^*, y_{12}^*, \ldots, y_{1N}^*, \ldots, y_{M1}^*, y_{M2}^*, \ldots, y_{MN}^*)$$

If we assume that the labels assigned to elements are independent, the following joint probability can be formulated: $P(Y) = \prod_{i=1}^{M} \prod_{j=1}^{N} P(y_{ij})$. However, conditional independency does not hold for image data, since spatially adjacent elements are likely to receive the same labels. We therefore need to explicitly consider this local dependency. This involves addressing three important issues: How should the optimal solution be defined, how are spatial dependencies considered, and how should we search the (exponential size) configuration space.

2.2 Markov Random Fields (MRFs)

Markov Random Fields (MRFs) provide a mathematical formulation for modeling local dependencies, and are defined as follows [3]:

Definition 1. *A set of random variables Y is called a Markov Random Field on S with respect to a neighborhood N, if and only if the following two conditions are satisfied, where $S - \{i\}$ denotes the set difference, $y_{S-\{i\}}$ denotes random variables in $S - \{i\}$, and N_i denotes the neighboring random variables of random variable i:*

1. $P(Y) > 0$
2. $P(y_i | y_{S-\{i\}}) = P(y_i | y_{N_i})$

Condition 2 (Markovianity) states that the conditional distribution of an element y_i is dependent only on its neighbors. Markov Random Fields have traditionally sought to maximize the joint probability $P(Y^*)$ (a generative approach). In this formulation, the posterior over the labels given the observations is formulated using Bayes' rule as:

$$P(Y|X) \propto P(X|Y)P(Y) = P(Y) \prod_i^n P(x_i|y_i) \qquad (1)$$

In (1), the equivalence between MRFs and Gibbs Distributions [5] provides an efficient way to factor the prior $P(Y)$ over cliques defined in the neighborhood Graph G. The prior $P(Y)$ is written as

$$P(Y) = \frac{\exp(\sum_{c \in C} V_c(Y))}{\sum_{Y' \in \Omega} exp(\sum_{c \in C} V_c(Y'))} \qquad (2)$$

where $V_c(Y)$ is a clique potential function of labels for clique $c \in C$, C is a set of cliques in G, and Ω is the space of all possible labelings. From (1) and (2), the target configuration Y^* is a realization of a locally dependent Markov Random Field with a specified prior distribution. Based on (1) and (2) and using Z to denote the (normalizing) "partition function", if we assume Gaussian likelihoods then the posterior distribution can be factored as:

$$P(Y|X) = \frac{1}{Z} \exp\left[\sum_{i \in S} \log(P(x_i|y_i)) + \sum_{c \in C} V_c(Y_c)\right] \qquad (3)$$

The Gaussian assumption for $P(X|Y)$ in (1) allows straightforward Maximum Likelihood parameter estimation. Although there have been many approximation algorithms designed to find the optimal Y^*, we will focus on a local method called *Iterated Conditional Modes* [5], written as:

$$y_i^* = \arg\max_{y_i \in L} P(y_i | y_{N_i}, x_i) \quad (4)$$

Assuming Gaussians for the likelihood and a pairwise neighborhood system for the prior over labels, (4) can be restated as:

$$y_i^* = \arg\max_{y_i \in L} \frac{1}{Z_i} \exp\left[\log(P(x_i|y_i)) + \sum_{j \in N_i} \beta y_i y_j\right] \quad (5)$$

where β is a constant and L is a set of class labels.

This concept has proved to be applicable in a wide variety of domains where there exists correlations among neighboring instances. However, the generative nature of the model and the assumption that the likelihood is Gaussian can be too restrictive to capture complex dependencies between neighboring elements or between observations and labels. In addition, the prior over labels is completely independent from the observations, thus the interactions between neighbors are not proportional to their similarity.

2.3 Conditional Random Fields (CRFs)

CRFs avoid the Gaussian assumption by using a model that seeks to maximize the conditional probability of the labels given the observations $P(Y^*|X)$ (a discriminative model), and are defined as follows [1]:

Definition 2. *Let $G = (S, E)$ be a graph such that Y is indexed by the vertices S of G. Then (X, Y) is said to be a CRF if, when conditioned on Y, the random variables y_i obey the Markov property with respect to the graph: $P(y_i|X, y_{s\setminus i}) = P(y_i|X, y_{N_i})$.*

This model alleviates the need to model the observations $P(X)$, allowing the use of arbitrary attributes of the observations without explicitly modeling them. CRFs assume a 1-dimensional chain-structure where only adjacent elements are neighbors. This allows the factorization of the joint probability over labels. Discriminative Random Fields (DRFs) extend 1-dimensional CRFs to 2-dimensional structures [6]. The conditional probability of the labels Y in the Discriminative Random Field framework is defined as:

$$P(Y|X) = \frac{1}{Z} \exp\left(\sum_{i \in S} A_i(y_i, X) + \sum_{i \in S} \sum_{j \in N_i} I_{ij}(y_i, y_j, X)\right) \quad (6)$$

A_i is the 'Association' potential that models dependencies between the observations and the class labels, while I_i is the 'Interaction' potential that models dependencies between the labels of neighboring elements (and the observations). Note that this is a much more powerful model than the assumed Gaussian Association potential and the indicator function used for the Interaction potential (that doesn't consider the observations) in MRFs. Parameter learning in DRFs involves maximizing the log likelihood of (6), while inference uses ICM [6].

DRFs are a powerful method for modeling dependencies in spatial data. However, several problems associated with this method include the fact that it is hard to find a good initial labeling and stopping criteria during inference, and it is sensitive to issues of class imbalance. Furthermore, for some real-world tasks the use of logistic regression as a discriminative method in DRFs often does not produce results that are as accurate as powerful classification models such as Support Vector Machines (that make the iid assumption).

3 Support Vector Random Fields (SVRFs)

This section presents Support Vector Random Fields (SVRFs), our extension of SVMs that allows the modelling of non-trivial two-dimensional (or higher) spatial dependencies using a CRF framework. This model has two major components: The *observation-matching* potential function and the *local-consistency* potential function. The *observation-matching* function captures relationships between the observations and the class labels, while the *local-consistency* function models relationships between the labels of neighboring data points and the observations at data points. Since the selection of the observation-matching potential is critical to the performance of the model, the Support Vector Random Field model employs SVMs for this potential, providing a theoretical and empirical advantage over the logistic model used in DRFs and the Gaussian model used in MRFs, that produce unsatisfactory results for many tasks. SVRFs can be formulated as follows:

$$P(Y|X) = \frac{1}{Z} \exp \left\{ \sum_{i \in S} \log(O(y_i, \varUpsilon_i(X))) + \sum_{i \in S} \sum_{j \in N_i} V(y_i, y_j, X) \right\} \quad (7)$$

In this formulation, $\varUpsilon_i(X)$ is a function that computes features from the observations X for location i, $O(y_i, \varUpsilon_i(X))$ is the observation-potential, and $V(y_i, y_j, X)$ is the local-consistency potential. The pair-wise neighborhood system is defined as a local dependency structure. In this work, interactions between pixels with a Euclidean distance of 1 were considered (ie. the radius 1 von Neumann neighborhood). We will now examine these potentials in more detail.

3.1 Observation-Matching

The observation-matching potential seeks to find a posterior probability distribution that maps from the observations to corresponding class labels. DRFs employ a Generalized Linear Models (GLM) for this potential. However, GLMs often do not estimate appropriate parameters. This is especially true in image data where feature sets may have a high number of dimensions and/or several features have a high degree of correlation. This can cause problems in parameter estimation and approximations to resolve these issues may not produce optimal parameters [7].

Fortunately, the CRF framework allows a flexible choice of the observation-matching potential function. We overcome the disadvantages of the GLM by employing a Support Vector Machine classifier, seeking to find the margin maximizing hyperplane between the classes. This classifier has appealing properties in high-dimensional spaces and is less sensitive to class imbalance. Furthermore, due to the properties of error bounds, SVMs tends to outperform GLMs, especially when the classes overlap in the feature space (often the case with image data). Parameter estimation for SVMs involves optimizing the following Quadratic Programming problem for the training data x_i (where C is a constant that bounds the misclassification error):

$$\max \sum_{i=1}^{N} \alpha_i - \frac{1}{2} \sum_{i}^{N} \sum_{j}^{N} \alpha_i \alpha_j y_i y_j x_i^T x_j$$

$$subject\ to\ 0 \leq \alpha_i \leq C\ and\ \sum_{i=1}^{N} \alpha_i y_i = 0 \quad (8)$$

Consequently, the decision function, given the parameters α_i for the l training instances and bias term b, is (for a more thorough discussion of SVMs, we refer to [2]): $f(x) = \sum_{i=1}^{l}(\alpha_i y_i x \cdot x_i) + b$

Unfortunately, the decision function $f(x)$ produced by SVMs measures distances to the decision boundary, while we require a posterior probability function. We adopted the approach of [8] to convert the decision function to a posterior probability function. This approach is efficient and minimizes the risk of overfitting during the conversion, but has some ambiguities and potential difficulties in numerical computation. We have addressed these issues in our approach, which will be briefly outlined here.

We estimate a posterior probability from the Support Vector Machine decision function using the sigmoid function:

$$O(y_i = 1, \Upsilon_i(X)) = \frac{1}{1 + \exp(Af(\Upsilon_i(X)) + B)} \quad (9)$$

The parameters A and B are estimated from training data represented as pairs $(f(\Upsilon_i(X)), t_i)$, where $f(\cdot)$ is the Support Vector Machine decision function, and t_i denotes a relaxed probability that $y_i = 1$ as in (9). We could set $t_i = 1$, if the class label at i is 1(ie. $y_i = 1$). However, this ignores the possibility that $\Upsilon_i(X)$ has the opposite class label (ie. -1). Thus, we employed the relaxed probability: $t_i = \frac{N_+ + 1}{N_+ + 2}$, if $y_i = 1$, and $t_i = \frac{1}{N_- + 2}$, if $y_i = -1$ (N_+ and N_- being the number of positive and negative class instances). By producing the new forms of training instances, we can solve the following optimization problem to estimate parameters:

$$\min - \sum_{i=1}^{l} \left[t_i \log p(\Upsilon_i(X)) + (1 - t_i) \log(1 - p(\Upsilon_i(X))) \right] \quad (10)$$

where

$$p(\Upsilon_i(X)) = \frac{1}{1 + exp(Af(\Upsilon_i(X)) + B)}$$

[8] adopted a Levenberg-Marquardt approach to solve the optimization problem, finding an approximation of the Hessian matrix. However, this may cause incorrect computations of the Hessian matrix (especially for unconstrained optimizations [7]). Hence, we employed Newton's method with backtracking line search to solve the optimization. In addition, in order to avoid overflows and underflows of *exp* and *log* functions, we reformulate Eq.10 as follows:

$$-\Big(t_i \log p(\Upsilon_i(X)) + (1-t_i)\log(1 - p(\Upsilon_i(X)))\Big)$$

$$= t_i(Af(\Upsilon_i(X)) + B) + \log(1 + \exp(-Af(\Upsilon_i(X)) - B)) \qquad (11)$$

3.2 Local-Consistency

In MRFs, local-consistency considers correlations between neighboring data points, and is considered to be observation independent. CRFs provide more powerful modelling of local-consistency by removing the assumption of observation independence. In order to use the principles of CRFs for local-consistency, an approach is needed that penalizes discontinuity between pairwise sites. For this, we use a linear function of pairwise continuity:

$$V(y_i, y_j, X) = y_i y_j \nu^T \Phi_{ij}(X) \qquad (12)$$

$\Phi_{ij}(X)$ is a function that computes features for sites i and j based on observations X. As opposed to DRFs, which penalize discontinuity by considering the absolute difference between pairwise observations [6], our approach introduces a new mapping function $\Phi(\cdot)$ that encourages continuity in addition to penalizing discontinuity (using $\max(\Upsilon(X))$ to denote the vector of max values for each feature):

$$\Phi_{ij}(X) = \frac{\max(\Upsilon(X)) - \mid \Upsilon_i(X) - \Upsilon_j(X) \mid}{\max(\Upsilon(X))} \qquad (13)$$

3.3 Learning and Inference

The proposed model needs to estimate the parameters of the observation-matching function and the local-consistency function. Although we estimate these parameters sequentially, our model outperforms the simultaneous learning approach of DRFs and significantly increases its computational efficiency.

The parameters of the Support Vector Machine decision function are first estimated by solving the Quadratic Programming problem in (8) (using SVMlight [9]). We then convert the decision function to a posterior function using (10) and the new training instances. Finally, we adopted pseudolikelihood [3] to estimate the local consistency parameters ν, due to its simplicity and fast computation. For training on l pixels from K images, pseudolikehood is formulated as:

$$\widehat{\nu} = \arg\max_{\nu} \prod_{k=1}^{K} \prod_{i=1}^{l} P(y_i^k | y_{N_i}^k, X^k, \nu) \qquad (14)$$

As in [6], to ensure that the log-likelihood is convex we assume that ν is Gaussian and compute the local-consistency parameters using its log likelihood $l(\hat{\nu})$:

$$l(\hat{\nu}) = \arg\max_{\nu} \sum_{k=1}^{K} \sum_{i=1}^{l} \left\{ O_i^n + \sum_{j \in N_i} V(y_i^k, y_j^k, X^k) - \log(z_i^k) \right\} - \frac{1}{2\tau} \nu^T \nu \quad (15)$$

In this model, z_i^k is a partition function for each site i in image k, and τ is a regularizing constant. Equation (15) is solved by gradient descent, and note that the observation matching function acts as a constant during this process. Due to the employment of SVMs, the time complexity of learning is $O(S^2)$, where S is the number of pixels to be trained, although in practice it is much faster.

The inference problem is to infer an optimal labeling Y^* given a new instance X and the estimated model parameters. We herein adopted the Iterated Conditional Modes (ICM) approach described in Section 2.2 [5], that maximizes the local conditional probability iteratively. For our proposed model and [6], ICM is expressed as,

$$y_i^* = \arg\max_{y_i \in L} P(y_i | y_{N_i}, X) \quad (16)$$

Although ICM is based on iterative principles, it often converges quickly to a high quality configuration, and each iteration has time complexity $O(S)$.

4 Experiments

We have evaluated our proposed model on synthetic and real-world binary image labeling tasks, comparing our approach to Logistic Regression, SVMs, and DRFs for these problems. Since class imbalance was present in many of the data sets, we used the Jaccard measure to quantify performance: $f = \frac{TP}{TP+FP+FN}$, where TP is the number of true positives, FP denotes the number of false positives, and FN tallies false negatives.

4.1 Experiments on Synthetic Data

We evaluated the four techniques over 5 synthetic binary image sets. These binary images were corrupted by zero mean Gaussian noise with unit standard deviation, and the task was to label the foreground objects (see the first and second columns in Fig. 1). Two of the sets contained balanced class labels (*Car* and *Objects*), while the other three contained imbalanced classes. The five 150 image sets were divided into 100 images for training and 50 for testing. Example results and aggregate scores are shown in Fig. 1. Note that the last 4 columns illustrate the outcomes from each technique– SVMs, Logistic Regression (LR), SVRFs, and DRFs.

Logistic Regression and subsequently DRFs performed poorly in all three imbalanced data sets (*Toybox*, *Size*, and *M*). In these cases, SVMs outperformed these methods and consequently our proposed SVRFs outperformed

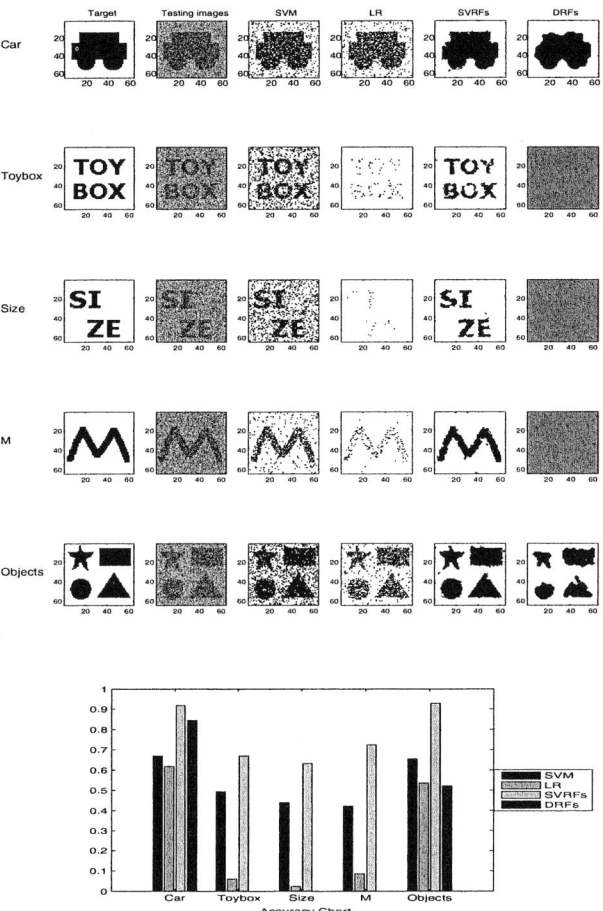

Fig. 1. Average scores on synthetic data sets

SVMs. In the first balanced data set (*Car*), DRFs and SVRFs both significantly outperformed SVMs and Logistic Regression (the iid classifiers). However, DRFs performed poorly on the second balanced data set (*Objects*). This is due to DRFs simultaneous parameter learning, that tends to overestimate the local-consistency potential. Since the observation-matching is underweighted, edges become degraded during inference (there are more edge areas in the *Objects* data). Terminating inference before convergence could reduce this, but this is not highly desirable for automatic classification. Overall, our Support Vector Random Field model demonstrated the best performance on all data sets, in particular those with imbalanced data and a greater proportion of edge areas.

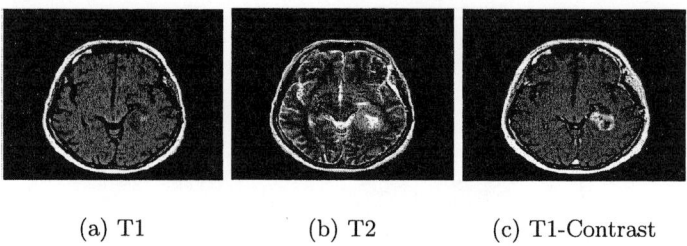

(a) T1   (b) T2   (c) T1-Contrast

Fig. 2. A multi-spectral MRI

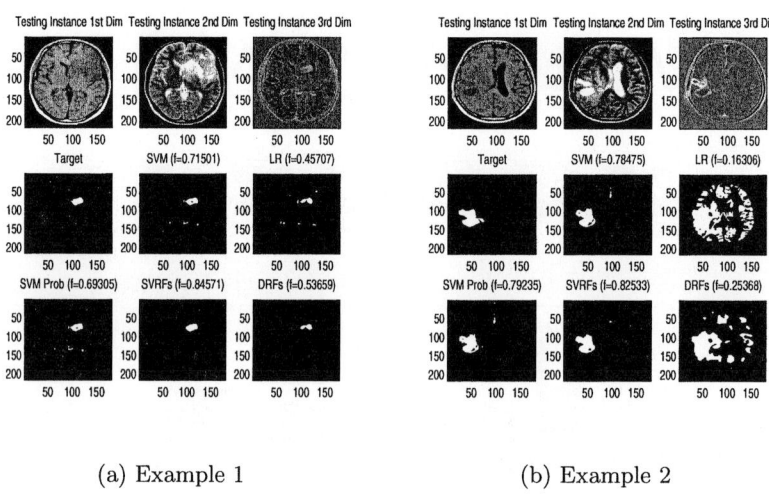

(a) Example 1   (b) Example 2

Fig. 3. An example of the classification result

4.2 Experiments on Real Data

We applied our model to the real-world problem of tumor segmentation in medical imaging. We focused on the task of brain tumor segmentation in MRI, an important task in surgical planning and radiation therapy currently being laboriously done by human medical experts. There has been significant research focusing on automating this challenging task (see [10]). Markov Random Fields have been explored previously for this task (see [10]), but recently SVMs have shown impressive performance [11,12]. This represents a scenario where our proposed Support Vector Random Field model could have a major impact. We evaluated the four classifiers from the previous section over 7 brain tumor patients. For each patient, three MRI 'modalities' were available: T1 (visualizing fat locations), T2 (visualizing water locations), and an additional T1 image with a 'contrast agent' added to enhance the visualization of metabolically active tumor areas (refer to Fig. 2).

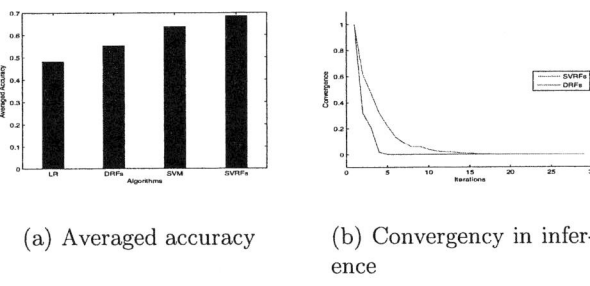

(a) Averaged accuracy (b) Convergency in inference

Fig. 4. Averaged accuracy and convergence in inference

The data was preprocessed with the Statistical Parametric Mapping software [13] to non-linearly align the images with a template in a standard coordinate system, and remove intensity inhomogeneity field effects. This non-linear template alignment approach was quantified to be highly effective in [14], and the inhomogeneity correction step computes a smooth corrective field that seeks to minimize the residual entropy after transformation of the log-intensity value's probability distribution [15]. We used 12 features that incorporate image information and domain knowledge (the raw intensities, spatial expected intensities within the coordinate system, spatial priors for the brain area and normal tissue types within the coordinate system, the template image information, and left-to-right symmetry), each measured as features at 3 scales by using 3 different sizes of Gaussian kernel filters. We used a 'patient-specific' training scenario similar to [11,12].

Results for two of the patients are shown in Fig. 3, while average scores over the 7 patients are shown in Fig. 4(a). Note that 'SVM+prob' in Fig. 3 denotes the classification results from the Support Vector Machine posterior probability estimate. The Logistic Regression model performs poorly at this task, but DRFs perform significantly better. As with the synthetic data in cases of class imbalance, SVMs outperform both Logistic Regression and the DRFs. Finally, SVRFs improve the scores obtained by the SVMs by almost 5% (a significant improvement).

We compared convergence of the DRFs and SVRFs by measuring how many label changes occured between inference iterations averaged over 21 trials (Fig. 4(a)). These results show that DRFs on average require almost 3 times as many iterations to converge, due to the overestimation of the local-consistency potential.

5 Conclusion

We have proposed a novel model for classification of data with spatial dependencies. The Support Vector Random Field combines ideas from SVMs and CRFs, and outperforms SVMs and DRFs on both synthetic data sets and an important real-world application. We also proposed an improvement to computing posterior probability distributions from SVM decision functions, and a method to encourage continuity with local-consistency potentials. Our Support Vector Random Field model is robust to class imbalance, can be efficiently trained, converges quickly during inference, and can trivially be augmented with kernel functions to further improve results.

Acknowledgment

R. Greiner is supported by the National Science and Engineering Research Council of Canada (NSERC) and the Alberta Ingenuity Centre for Machine Learning (AICML). C.H. Lee is supported by NSERC, AICML, and iCORE. Our thanks to Dale Schuurmans for helpful discussions on optimization and parameter estimation, J. Sander for hlepful discussions for the classification issues, BTGP members for help in data processing, and Albert Murtha (M.D.) for domain knowledge on the tumor data set.

References

1. Lafferty, J., Pereira, F., McCallum, A.: Conditional random fields: Probabilistic models for segmenting and labeling sequence data. ICML (2001)
2. Shawe-Taylor, Cristianini: Kernel Methods for Pattern Analysis. Cambridge University Press, Cambridge, UK (2004)
3. Li, S.Z.: Markov Random Field Modeling in Image Analysis. Springer-Verlag, Tokyo (2001)
4. Kumar, S., Hebert, M.: Discriminative random fields: A discriminative framework for contextual interaction in classification. ICCV (2003) 1150–1157
5. Besag, J.: On the statistical analysis of dirty pictures. Journal of Royal Statistical Society. Series B **48** (1986) 3:259–302
6. Kumar, S., Hebert, M.: Discriminative fields for modeling spatial dependencies in natural images. NIPS (2003)
7. R.Fletcher: Practical Methods of Optimization. John Wiley & Sons (1987)
8. Platt, J.: Probabilistic outputs for support vector machines and comparison to regularized likelihood methods. MIT Press, Cambridge, MA (2000)
9. Joachims, T.: Making large-scale svm learning practical. In Scholkopf, B., Burges, C., Smola, A., eds.: Advances in Kernel Methods - Support Vector Learning, MIT Press (1999)
10. Gering, D.: Recognizing Deviations from Normalcy for Brain Tumor Segmentation. PhD thesis, MIT (2003)
11. Zhang, J., Ma, K., Er, M., Chong, V.: Tumor segmentation from magnetic resonance imaging by learning via one-class support vector machine. Int. Workshop on Advanced Image Technology (2004) 207–211
12. Garcia, C., Moreno, J.: Kernel based method for segmentation and modeling of magnetic resonance images. LNCS **3315** (2004) 636–645
13. : Statistical parametric mapping, http://www.fil.ion.bpmf.ac.uk/spm/ (Online)
14. Hellier, P., Ashburner, J., Corouge, I., Barillot, C., Friston, K.: Inter subject registration of functional and anatomical data using spm. In: MICCAI. Volume 587-590. (2002)
15. Ashburner, J.: Another mri bias correction approach. In: 8th Int. Conf. on Functional Mapping of the Human Brain, Sendai, Japan. (2002)

Realistic, Mathematically Tractable Graph Generation and Evolution, Using Kronecker Multiplication*

Jurij Leskovec[1], Deepayan Chakrabarti[1],
Jon Kleinberg[2], and Christos Faloutsos[1]

[1] School of Computer Science, Carnegie Mellon University
{jure, deepay, christos}@cs.cmu.edu
[2] Department of Computer Science, Cornell University
kleinber@cs.cornell.edu

Abstract. How can we generate realistic graphs? In addition, how can we do so with a mathematically tractable model that makes it feasible to analyze their properties rigorously? Real graphs obey a long list of surprising properties: Heavy tails for the in- and out-degree distribution; heavy tails for the eigenvalues and eigenvectors; small diameters; and the recently discovered "Densification Power Law" (DPL). All published graph generators either fail to match several of the above properties, are very complicated to analyze mathematically, or both. Here we propose a graph generator that is mathematically tractable and matches this collection of properties. The main idea is to use a non-standard matrix operation, the *Kronecker product*, to generate graphs that we refer to as "Kronecker graphs".

We show that Kronecker graphs naturally obey all the above properties; in fact, we can rigorously *prove* that they do so. We also provide empirical evidence showing that they can mimic very well several real graphs.

1 Introduction

What do real graphs look like? How do they evolve over time? How can we generate synthetic, but realistic, time-evolving graphs? Graph mining has been attracting much interest recently, with an emphasis on finding patterns and abnormalities in social networks, computer networks, e-mail interactions, gene

---

* Work partially supported by the National Science Foundation under Grants No. IIS-0209107, SENSOR-0329549, IIS-0326322, CNS-0433540, CCF-0325453, IIS-0329064, CNS-0403340, CCR-0122581, a David and Lucile Packard Foundation Fellowship, and also by the Pennsylvania Infrastructure Technology Alliance (PITA), a partnership of Carnegie Mellon, Lehigh University and the Commonwealth of Pennsylvania's Department of Community and Economic Development (DCED). Any opinions, findings, and conclusions or recommendations expressed in this material are those of the author(s) and do not necessarily reflect the views of the National Science Foundation, or other funding parties.

regulatory networks, and many more. Most of the work focuses on static snapshots of graphs, where fascinating "laws" have been discovered, including small diameters and heavy-tailed degree distributions.

A realistic graph generator is important for at least two reasons. The first is that it can generate graphs for extrapolations, "what-if" scenarios, and simulations, when real graphs are difficult or impossible to collect. For example, how well will a given protocol run on the Internet five years from now? Accurate graph generators can produce more realistic models for the future Internet, on which simulations can be run. The second reason is more subtle: it forces us to think about the patterns that a graph generator should obey, to be realistic.

The main contributions of this paper are the following:

- We provide a generator which obeys all the main static patterns that have appeared in the literature.
- Generator also obeys the recently discovered temporal evolution patterns.
- Contrary to other generators that match this combination of properties, our generator leads to tractable analysis and rigorous proofs.

Our generator is based on a non-standard matrix operation, the *Kronecker product*. There are several theorems on Kronecker products, which actually correspond exactly to a significant portion of what we want to prove: heavy-tailed distributions for in-degree, out-degree, eigenvalues, and eigenvectors. We also demonstrate how a Kronecker Graph can match the behavior of several real graphs (patent citations, paper citations, and others). While Kronecker products have been studied by the algebraic combinatorics community (see e.g. [10]), the present work is the first to employ this operation in the design of network models to match real datasets.

The rest of the paper is organized as follows: Section 2 surveys the related literature. Section 3 gives the proposed method. We present the experimental results in Section 4, and we close with some discussion and conclusions.

2 Related Work

First, we will discuss the commonly found (static) patterns in graphs, then some recent patterns on temporal evolution, and finally, the state of the art in graph generation methods.

Static Graph Patterns: While many patterns have been discovered, two of the principal ones are heavy-tailed degree distributions and small diameters.

Degree distribution: The degree-distribution of a graph is a power law if the number of nodes c_k with degree k is given by $c_k \propto k^{-\gamma}$ $(\gamma > 0)$ where γ is called the power-law exponent. Power laws have been found in the Internet [13], the Web [15,7], citation graphs [24], online social networks [9] and many others. Deviations from the power-law pattern have been noticed [23], which can be explained by the "DGX" distribution [5]. DGX is closely related to a truncated lognormal distribution.

Small diameter: Most real-world graphs exhibit relatively small diameter (the "small- world" phenomenon): A graph has diameter d if every pair of nodes can be connected by a path of length at most d. The diameter d is susceptible to outliers. Thus, a more robust measure of the pairwise distances between nodes of a graph is the *effective diameter* [26]. This is defined as the minimum number of hops in which some fraction (or quantile q, say $q = 90\%$) of all connected pairs of nodes can reach each other. The effective diameter has been found to be small for large real-world graphs, like Internet, Web, and social networks [2,21].

Scree plot: This is a plot of the eigenvalues (or singular values) of the adjacency matrix of the graph, versus their rank, using a log-log scale. The scree plot is also often found to approximately obey a power law. The distribution of eigenvector components (indicators of "network value") has also been found to be skewed [9].

Apart from these, several other patterns have been found, including the "stress" [14,9], "resilience" [2,22], "clustering coefficient" and many more.

Temporal evolution Laws: Densification and shrinking diameter: Two very recent discoveries, both regarding time-evolving graphs, are worth mentioning [18]: (a) the "effective diameter" of graphs tends to shrink or stabilize as the graph grows with time, and (b) the number of edges $E(t)$ and nodes $N(t)$ seems to obey the *densification power law* (DPL), which states that

$$E(t) \propto N(t)^a \tag{1}$$

The *densification exponent* a is typically greater than 1, implying that the average degree of a node in the graph is *increasing* over time. This means that real graphs tend to sprout many more edges than nodes, and thus are densifying as they grow.

Graph Generators: The earliest probabilistic generative model for graphs was a random graph model, where each pair of nodes has an identical, independent probability of being joined by an edge [11]. The study of this model has led to a rich mathematical theory; however, this generator produces graphs that fail to match real-world networks in a number of respects (for example, it does not produce heavy-tailed degree distributions).

The vast majority of recent models involve some form of *preferential attachment* [1,2,28,15,16]: new nodes join the graph at each time step, and preferentially connect to existing nodes with high degree (the "rich get richer"). This simple behavior leads to power-law tails and to low diameters. The diameter in this model grows slowly with the number of nodes N, which violates the "shrinking diameter" property mentioned above.

Another family of graph-generation methods strives for small diameter, like the *small-world* generator [27] and the Waxman generator [6]. A third family of methods show that heavy tails emerge if nodes try to optimize their connectivity under resource constraints [8,12].

Summary: Most current generators focus on only one (static) pattern, and neglect the others. In addition, it is usually hard to prove properties of them. The generator we describe in the next section addresses these issues.

3 Proposed Method

The method we propose is based on a recursive construction. Defining the recursion properly is somewhat subtle, as a number of standard, related graph construction methods fail to produce graphs that densify according to the patterns observed in practice, and they also produce graphs whose diameters increase. To produce densifying graphs with constant diameter, and thereby match the qualitative behavior of real network datasets, we develop a procedure that is best described in terms of the *Kronecker product* of matrices. To help in the description of the method, the accompanying table provides a list of symbols and their definitions.

Symbol	Definition
G_1	the initiator of a Kronecker Graph
N_1	number of nodes in initiator
E_1	number of edges in initiator
$G_1^{[k]} = G_k$	the kth Kronecker power of G_1
a	densification exponent
d	diameter of a graph
$\mathcal{P}_1$	probability matrix

3.1 Main Idea

The main idea is to create self-similar graphs, recursively. We begin with an *initiator* graph G_1, with N_1 nodes and E_1 edges, and by recursion we produce successively larger graphs $G_2 \ldots G_n$ such that the kth graph G_k is on $N_k = N_1^k$ nodes. If we want these graphs to exhibit a version of the Densification Power Law, then G_k should have $E_k = E_1^k$ edges. This is a property that requires some care in order to get right, as standard recursive constructions (for example, the traditional Cartesian product or the construction of [4]) do not satisfy it.

It turns out that the *Kronecker product* of two matrices is the perfect tool for this goal. The Kronecker product is defined as follows:

Definition 1 (Kronecker product of matrices). *Given two matrices* $\mathbf{A} = [a_{i,j}]$ *and* $\mathbf{B}$ *of sizes* $n \times m$ *and* $n' \times m'$ *respectively, the Kronecker product matrix* $\mathbf{C}$ *of dimensions* $(n * n') \times (m * m')$ *is given by*

$$\mathbf{C} = \mathbf{A} \otimes \mathbf{B} \doteq \begin{pmatrix} a_{1,1}\mathbf{B} & a_{1,2}\mathbf{B} & \ldots & a_{1,m}\mathbf{B} \\ a_{2,1}\mathbf{B} & a_{2,2}\mathbf{B} & \ldots & a_{2,m}\mathbf{B} \\ \vdots & \vdots & \ddots & \vdots \\ a_{n,1}\mathbf{B} & a_{n,2}\mathbf{B} & \ldots & a_{n,m}\mathbf{B} \end{pmatrix} \quad (2)$$

We define the Kronecker product of two graphs as the Kronecker product of their adjacency matrices.

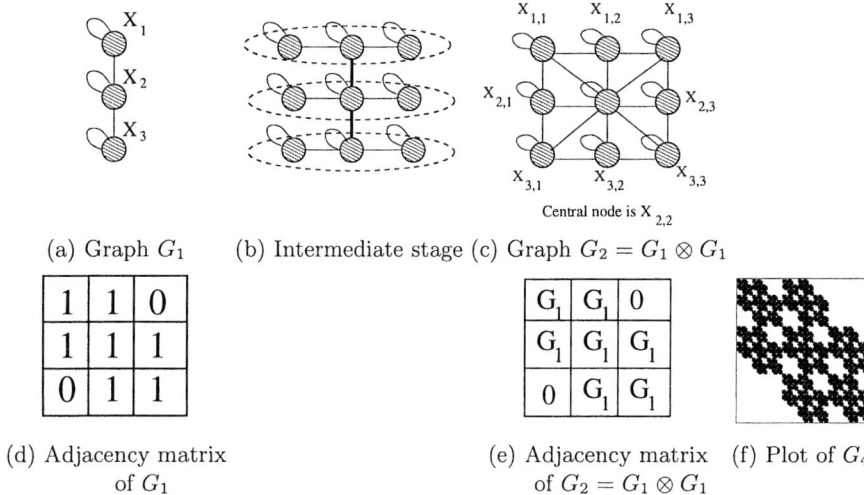

Fig. 1. *Example of Kronecker multiplication:* Top: a "3-chain" and its Kronecker product with itself; each of the X_i nodes gets expanded into 3 nodes, which are then linked using Observation 1. Bottom row: the corresponding adjacency matrices, along with matrix for the fourth Kronecker power G_4.

Observation 1 (Edges in Kronecker-multiplied graphs)

$$\text{Edge } (X_{ij}, X_{kl}) \in G \otimes H \text{ iff } (X_i, X_k) \in G \text{ and } (X_j, X_l) \in H$$

where X_{ij} and X_{kl} are nodes in $G \otimes H$, and X_i, X_j, X_k and X_l are the corresponding nodes in G and H, as in Figure 1.

The last observation is subtle, but crucial, and deserves elaboration: Figure 1(a–c) shows the recursive construction of $G \otimes H$, when $G = H$ is a 3-node path. Consider node $X_{1,2}$ in Figure 1(c): It belongs to the H graph that replaced node X_1 (see Figure 1(b)), and in fact is the X_2 node (i.e., the center) within this small H-graph.

We propose to produce a growing sequence of graphs by iterating the Kronecker product:

Definition 2 (Kronecker power). *The kth power of G_1 is defined as the matrix $G_1^{[k]}$ (abbreviated to G_k), such that:*

$$G_1^{[k]} = G_k = \underbrace{G_1 \otimes G_1 \otimes \ldots G_1}_{k \text{ times}} = G_{k-1} \otimes G_1$$

The self-similar nature of the Kronecker graph product is clear: To produce G_k from G_{k-1}, we "expand" (replace) each node of G_{k-1} by converting it into a copy of G, and we join these copies together according to the adjacencies in

G_{k-1} (see Figure 1). This process is very natural: one can imagine it as positing that communities with the graph grow recursively, with nodes in the community recursively getting expanded into miniature copies of the community. Nodes in the subcommunity then link among themselves and also to nodes from different communities.

3.2 Theorems and Proofs

We shall now discuss the properties of Kronecker graphs, specifically, their degree distributions, diameters, eigenvalues, eigenvectors, and time-evolution. Our ability to prove analytical results about all of these properties is a major advantage of Kronecker graphs over other generators. The next few theorems prove that several distributions of interest are multinomial for our Kronecker graph model. This is important, because a careful choice of the initial graph G_1 can make the resulting multinomial distribution to behave like a power-law or DGX distribution.

Theorem 1 (Multinomial degree distribution). *Kronecker graphs have multinomial degree distributions, for both in- and out-degrees.*

Proof. Let the initiator G_1 have the degree sequence $d_1, d_2, \ldots, d_{N_1}$. Kronecker multiplication of a node with degree d expands it into N_1 nodes, with the corresponding degrees being $d \times d_1, d \times d_2, \ldots, d \times d_{N_1}$. After Kronecker powering, the degree of each node in graph G_k is of the form $d_{i_1} \times d_{i_2} \times \ldots d_{i_k}$, with $i_1, i_2, \ldots i_k \in (1 \ldots N_1)$, and there is one node for each ordered combination. This gives us the multinomial distribution on the degrees of G_k. Note also that the degrees of nodes in G_k can be expressed as the kth Kronecker power of the vector $(d_1, d_2, \ldots, d_{N_1})$. □

Theorem 2 (Multinomial eigenvalue distribution). *The Kronecker graph G_k has a multinomial distribution for its eigenvalues.*

Proof. Let G_1 have the eigenvalues $\lambda_1, \lambda_2, \ldots, \lambda_{N_1}$. By properties of the Kronecker multiplication [19,17], the eigenvalues of G_k are k^{th} Kronecker power of the vector $(\lambda_1, \lambda_2, \ldots, \lambda_{N_1})$. As in Theorem 1, the eigenvalue distribution is a multinomial. □

A similar argument using properties of Kronecker matrix multiplication shows the following.

Theorem 3 (Multinomial eigenvector distribution). *The components of each eigenvector of the Kronecker graph G_k follow a multinomial distribution.*

We have just covered several of the static graph patterns. Notice that the proofs were direct consequences of the Kronecker multiplication properties.

Next we continue with the temporal patterns: the densification power law, and shrinking/stabilizing diameter.

Theorem 4 (DPL). *Kronecker graphs follow the Densification Power Law (DPL) with densification exponent $a = \log(E_1)/\log(N_1)$.*

Proof. Since the kth Kronecker power G_k has $N_k = N_1^k$ nodes and $E_k = E_1^k$ edges, it satisfies $E_k = N_k^a$, where $a = \log(E_1)/\log(N_1)$. The crucial point is that this exponent a is independent of k, and hence the sequence of Kronecker powers follows an exact version of the Densification Power Law. □

We now show how the Kronecker product also preserves the property of constant diameter, a crucial ingredient for matching the diameter properties of many real-world network datasets. In order to establish this, we will assume that the initiator graph G_1 has a self-loop on every node; otherwise, its Kronecker powers may in fact be disconnected.

Lemma 1. *If G and H each have diameter at most d, and each has a self-loop on every node, then the Kronecker product $G \otimes H$ also has diameter at most d.*

Proof. Each node in $G \otimes H$ can be represented as an ordered pair (v, w), with v a node of G and w a node of H, and with an edge joining (v, w) and (x, y) precisely when (v, x) is an edge of G and (w, y) is an edge of H. Now, for an arbitrary pair of nodes (v, w) and (v', w'), we must show that there is a path of length at most d connecting them. Since G has diameter at most d, there is a path $v = v_1, v_2, \ldots, v_r = v'$, where $r \leq d$. If $r < d$, we can convert this into a path $v = v_1, v_2, \ldots, v_d = v'$ of length exactly d, by simply repeating v' at the end for $d - r$ times By an analogous argument, we have a path $w = w_1, w_2, \ldots, w_d = w'$. Now by the definition of the Kronecker product, there is an edge joining (v_i, w_i) and (v_{i+1}, w_{i+1}) for all $1 \leq i \leq d-1$, and so $(v, w) = (v_1, w_1), (v_2, w_2), \ldots, (v_d, w_d) = (v', w')$ is a path of length d connecting (v, w) to (v', w'), as required. □

Theorem 5. *If G_1 has diameter d and a self-loop on every node, then for every k, the graph G_k also has diameter d.*

Proof. This follows directly from the previous lemma, combined with induction on k. □

We also consider the *effective diameter* d_e; we define the q-effective diameter as the minimum d_e such that, for at least a q fraction of the reachable node pairs, the path length is at most d_e. The q-effective diameter is a more robust quantity than the diameter, the latter being prone to the effects of degenerate structures in the graph (e.g. very long chains); however, the q-effective diameter and diameter tend to exhibit qualitatively similar behavior. For reporting results in subsequent sections, we will generally consider the q-effective diameter with $q = .9$, and refer to this simply as the *effective diameter*.

Theorem 6 (Effective Diameter). *If G_1 has diameter d and a self-loop on every node, then for every q, the q-effective diameter of G_k converges to d (from below) as k increases.*

Proof. To prove this, it is sufficient to show that for two randomly selected nodes of G_k, the probability that their distance is d converges to 1 as k goes to infinity.

We establish this as follows. Each node in G_k can be represented as an ordered sequence of k nodes from G_1, and we can view the random selection of a node in G_k as a sequence of k independent random node selections from G_1. Suppose that $v = (v_1, \ldots, v_k)$ and $w = (w_1, \ldots, w_k)$ are two such randomly selected nodes from G_k. Now, if x and y are two nodes in G_1 at distance d (such a pair (x, y) exists since G_1 has diameter d), then with probability $1 - (1 - 2/N_1)^k$, there is some index j for which $\{v_j, w_j\} = \{x, y\}$. If there is such an index, then the distance between v and w is d. As the expression $1 - (1 - 2/N_1)^k$ converges to 1 as k increases, it follows that the q-effective diameter is converging to d. □

3.3 Stochastic Kronecker Graphs

While the Kronecker power construction discussed thus far yields graphs with a range of desired properties, its discrete nature produces "staircase effects" in the degrees and spectral quantities, simply because individual values have large multiplicities. Here we propose a stochastic version of Kronecker graphs that eliminates this effect. counterparts.

We start with an $N_1 \times N_1$ *probability matrix* $\mathcal{P}_1$: the value p_{ij} denotes the probability that edge (i, j) is present. We compute its kth Kronecker power $\mathcal{P}_1^{[k]} = \mathcal{P}_k$; and then for each entry p_{uv} of $\mathcal{P}_k$, we include an edge between nodes u and v with probability $p_{u,v}$. The resulting binary random matrix $R = R(\mathcal{P}_k)$ will be called the *instance matrix* (or *realization matrix*).

In principle one could try choosing each of the N_1^2 parameters for the matrix $\mathcal{P}_1$ separately. However, we reduce the number of parameters to just two: α and β. Let G_1 be the initiator matrix (binary, deterministic); we create the corresponding probability matrix $\mathcal{P}_1$ by replacing each "1" and "0" of G_1 with α and β respectively ($\beta \leq \alpha$). The resulting probability matrices maintain — with some random noise — the self-similar structure of the Kronecker graphs in the previous subsection (which, for clarity, we call *deterministic Kronecker graphs*).

We find empirically that the random graphs produced by this model continue to exhibit the desired properties of real datasets, and without the staircase effect of the deterministic version. The task of setting α and β to match observed data is a very promising research direction, outside the scope of this paper. In our experiments in the upcoming sections, we use heuristics which we describe there.

4 Experiments

Now, we demonstrate the ability of Kronecker graphs to match the patterns of real-world graphs. The datasets we use are:
- *arXiv:* This is a citation graph for high-energy physics research papers, with a total of $N = 29,555$ papers and $E = 352,807$ citations. We follow its evolution from January 1993 to April 2003, with one data-point per month.

- *Patents:* This is a U.S. patent citation dataset that spans 37 years from January 1963 to December 1999. The graph contains a total of $N = 3,942,825$ patents and $E = 16,518,948$ citations. Citation graphs are normally considered as directed graphs. For the purpose of this work we think of them as undirected.
- *Autonomous systems:* We also analyze a static dataset consisting of a single snapshot of connectivity among Internet autonomous systems from January 2000, with $N = 6,474$ and $E = 26,467$.

We observe two kinds of graph patterns — "static" and "temporal." As mentioned earlier, common static patterns include the degree distribution, the scree plot (eigenvalues of graph adjacency matrix vs. rank), principal eigenvector of adjacency matrix and the distribution of connected components. Temporal patterns include the diameter over time, the size of the giant component over time, and the densification power law. For the diameter computation, we use a smoothed version of the effective diameter that is qualitatively similar to the standard effective diameter, but uses linear interpolation so as to take on non-integer values; see [18] for further details on this calculation.

Results are shown in Figures 2 and 3 for the graphs which evolve over time (*arXiv* and *Patents*). For brevity, we show the plots for only two static and two temporal patterns. We see that the deterministic Kronecker model already captures the qualitative structure of the degree and eigenvalue distributions, as well as the temporal patterns represented by the Densification Power Law and the stabilizing diameter. However, the deterministic nature of this model results in "staircase" behavior, as shown in scree plot for the deterministic Kronecker graph of Figure 2 (second row, second column). We see that the Stochastic Kronecker Graphs smooth out these distributions, further matching the qualitative structure of the real data; they also match the shrinking-before-stabilization trend of the diameters of real graphs.

For the Stochastic Kronecker Graphs we need to estimate the parameters α and β defined in the previous section. This leads to interesting questions whose full resolution lies beyond the scope of the present paper; currently, we searched by brute force over (the relatively small number of) possible initiator graphs of up to five nodes, and we then chose α and β so as to match well the edge density, the maximum degree, the spectral properties, and the DPL exponent.

Finally, Figure 4 shows plots for the static patterns in the *Autonomous systems* graph. Recall that we analyze a single, static snapshot in this case. In addition to the degree distribution and scree plot, we also show two typical plots [9]: the distribution of *network values* (principal eigenvector components, sorted, versus rank) and the *hop-plot* (the number of reachable pairs $P(h)$ within h hops or less, as a function of the number of hops h).

5 Observations and Conclusions

Here we list several of the desirable properties of the proposed Kronecker Graphs and Stochastic Kronecker Graphs.

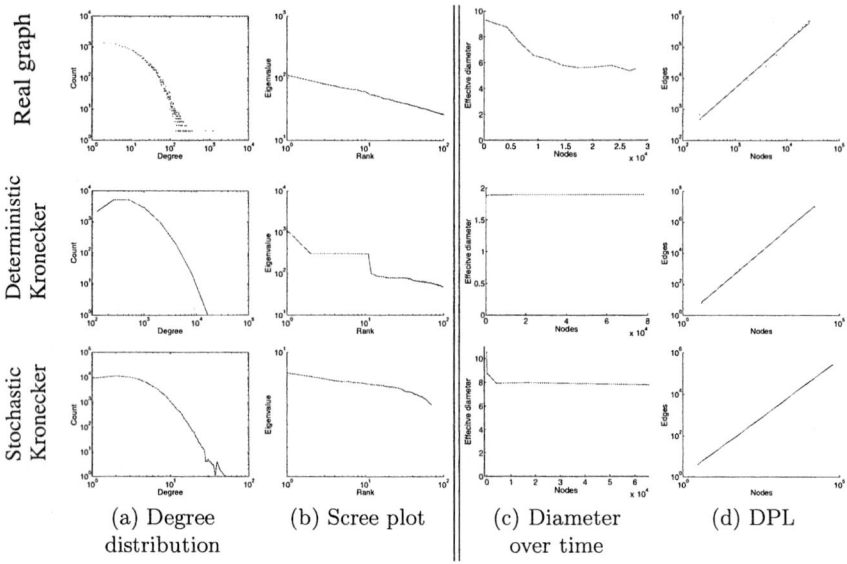

Fig. 2. *arXiv dataset:* Patterns from the real graph (top row), the deterministic Kronecker graph with G_1 being a star graph with 3 satellites (middle row), and the Stochastic Kronecker graph ($\alpha = 0.41$, $\beta = 0.11$ – bottom row). *Static* patterns: (a) is the PDF of degrees in the graph (log-log scale), and (b) the distribution of eigenvalues (log–log scale). *Temporal* patterns: (c) gives the effective diameter over time (linear-linear scale), and (d) is the number of edges versus number of nodes over time (log-log scale). Notice that the Stochastic Kronecker Graph qualitatively matches all the patterns very well.

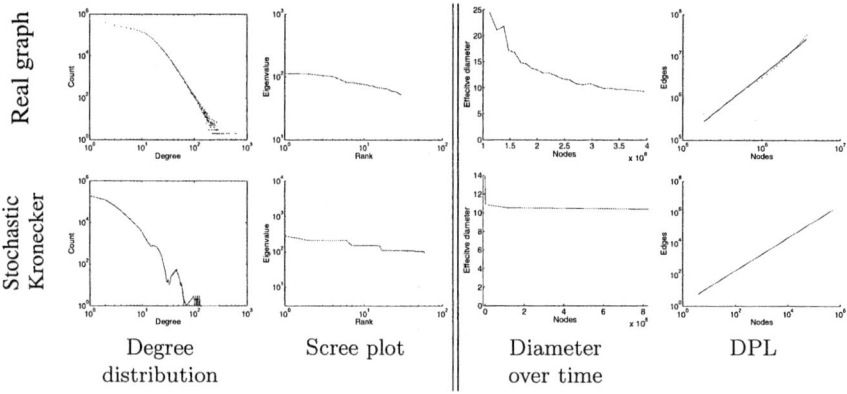

Fig. 3. *Patents:* Again, Kronecker graphs match all of these patterns. We show only the Stochastic Kronecker graph for brevity.

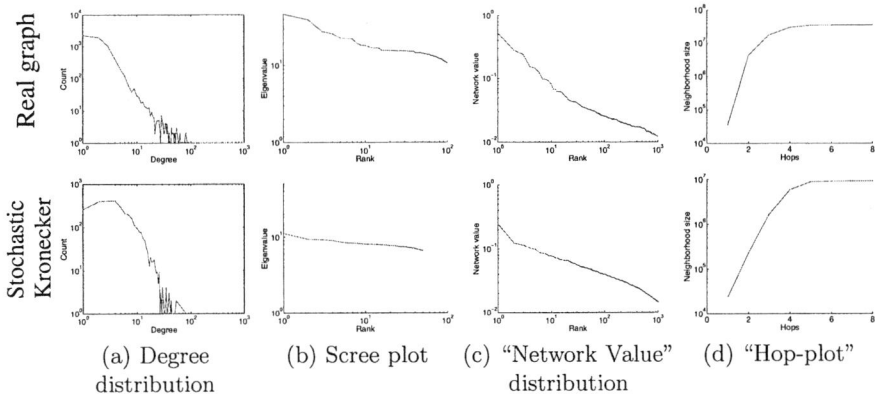

(a) Degree distribution (b) Scree plot (c) "Network Value" distribution (d) "Hop-plot"

Fig. 4. *Autonomous systems*: Real (top) versus Kronecker (bottom). Columns (a) and (b) show the degree distribution and the scree plot, as before. Columns (c) and (d) show two more static patterns (see text). Notice that, again, the Stochastic Kronecker Graph matches well the properties of the real graph.

Generality: Stochastic Kronecker Graphs include several other generators, as special cases: For $\alpha=\beta$, we obtain an Erdős-Rényi random graph; for $\alpha=1$ and $\beta=0$, we obtain a deterministic Kronecker graph; setting the G_1 matrix to a 2x2 matrix, we obtain the RMAT generator [9]. In contrast to Kronecker graphs, the RMAT cannot extrapolate into the future, since it needs to know the number of edges to insert. Thus, it is incapable of obeying the "densification law".

Phase transition phenomena: The Erdős-Rényi graphs exhibit phase transitions [11]. Several researchers argue that real systems are "at the edge of chaos" [3,25]. It turns out that Stochastic Kronecker Graphs also exhibit phase transitions. For small values of α and β, Stochastic Kronecker Graphs have many small disconnected components; for large values they have a giant component with small diameter. In between, they exhibit behavior suggestive of a phase transition: For a carefully chosen set of (α, β), the diameter is large, and a giant component just starts emerging. We omit the details, for lack of space.

Theory of random graphs: All our theorems are for the deterministic Kronecker Graphs. However, there is a lot of work on the properties of random matrices (see e.g. [20]), which one could potentially apply in order to prove properties of the Stochastic Kronecker Graphs.

In conclusion, the main contribution of this work is a family of graph generators, using a non-traditional matrix operation, the *Kronecker product*. The resulting graphs (a) have all the static properties (heavy-tailed degree distribution, small diameter), (b) all the temporal properties (densification, shrinking diameter), and in addition, (c) we can formally prove all of these properties.

Several of the proofs are extremely simple, thanks to the rich theory of Kronecker multiplication. We also provide proofs about the diameter and "effective diameter", and we show that Stochastic Kronecker Graphs can be tuned to mimic real graphs well.

References

1. R. Albert and A.-L. Barabasi. Emergence of scaling in random networks. *Science,* pages 509512, 1999.
2. R. Albert and A.-L. Barabasi. Statistical mechanics of complex networks. *Reviews of Modern Physics,* 2002.
3. P. Bak. How nature works : The science of self-organized criticality, Sept. 1996.
4. A.-L. Barabasi, E. Ravasz, and T. Vicsek. Deterministic scale-free networks. *Physica A,* 299:559-564, 2001.
5. Z. Bi, C. Faloutsos, and F. Korn. The DGX distribution for mining massive, skewed data. In KDD, pages 17-26, 2001.
6. B.M.Waxman. Routing of multipoint connections. *IEEE Journal on Selected Areas in Communications,* 6(9), December 1988.
7. A. Broder, R. Kumar, F. Maghoul1, P. Raghavan, S. Rajagopalan, R. Stata, A. Tomkins, and J. Wiener. Graph structure in the web: experiments and models. *In Proceedings of World Wide Web Conference,* 2000.
8. J. M. Carlson and J. Doyle. Highly optimized tolerance: a mechanism for power laws in designed systems. *Physics Review E,* 60(2):1412-1427, 1999.
9. D. Chakrabarti, Y. Zhan, and C. Faloutsos. R-MAT: A recursive model for graph mining. In *SIAM Data Mining,* 2004.
10. T. Chow. The Q-spectrum and spanning trees of tensor products of bipartite graphs. *Proc. Amer. Math. Soc.,* 125:3155-3161, 1997.
11. P. Erdos and A. Renyi. On the evolution of random graphs. Publication of the Mathematical Institute of the Hungarian Acadamy of Science, 5:17-67, 1960.
12. A. Fabrikant, E. Koutsoupias, and C. H. Papadimitriou. Heuristically optimized trade-offs: A new paradigm for power laws in the internet (extended abstract), 2002.
13. M. Faloutsos, P. Faloutsos, and C. Faloutsos. On power-law relationships of the internet topology. In *SIGCOMM,* pages 251-262, 1999.
14. M. Girvan and M. E. J. Newman. Community structure in social and biological networks. In *Proc. Natl. Acad. Sci.* USA, volume 99, 2002.
15. J. M. Kleinberg, S. R. Kumar, P. Raghavan, S. Rajagopalan, and A. Tomkins. The web as a graph: Measurements, models and methods. In *Proceedings of the International Conference on Combinatorics and Computing,* 1999.
16. S. R. Kumar, P. Raghavan, S. Rajagopalan, and A. Tomkins. Extracting large-scale knowledge bases from the web. VLDB, pages 639-650, 1999.
17. A. N. Langville and W. J. Stewart. The Kronecker product and stochastic automata networks. *Journal of Computation and Applied Mathematics,* 167:429-447, 2004.
18. J. Leskovec, J. Kleinberg, and C. Faloutsos. Graphs over time: densification laws, shrinking diameters and possible explanations. In *KDD'05,* Chicago, IL, USA, 2005.
19. C. F. V. Loan. The ubiquitous Kronecker product. *Journal of Computation and Applied Mathematics,* 123:85-100, 2000.
20. M. Mehta. Random Matrices. Academic Press, 2nd edition, 1991.
21. S. Milgram. The small-world problem. *Psychology Today,* 2:60-67, 1967.
22. C. R. Palmer, P. B. Gibbons, and C. Faloutsos. Anf: A fast and scalable tool for data mining in massive graphs. In *SIGKDD,* Edmonton, AB, Canada, 2002.
23. D. M. Pennock, G.W. Flake, S. Lawrence, E. J. Glover, and C. L. Giles. Winners dont take all: Characterizing the competition for links on the web. *Proceedings of the National Academy of Sciences,* 99(8):5207-5211, 2002.

24. S. Redner. How popular is your paper? an empirical study of the citation distribution. *European Physical Journal B*, 4:131-134, 1998.
25. R. Sole and B. Goodwin. *Signs of Life: How Complexity Pervades Biology*. Perseus Books Group, New York, NY, 2000.
26. S. L. Tauro, C. Palmer, G. Siganos, and M. Faloutsos. A simple conceptual model for the internet topology. *In Global Internet, San Antonio, Texas,* 2001.
27. D. J. Watts and S. H. Strogatz. Collective dynamics of 'small-world' networks. *Nature,* 393:440-442, 1998.
28. J. Winick and S. Jamin. Inet-3.0: Internet Topology Generator. Technical Report CSE-TR-456-02, University of Michigan, Ann Arbor, 2002.

A Correspondence Between Maximal Complete Bipartite Subgraphs and Closed Patterns

Jinyan Li, Haiquan Li, Donny Soh, and Limsoon Wong

Institute for Infocomm Research,
21 Heng Mui Keng Terrace, Singapore 119613
{jinyan, haiquan, studonny, limsoon}@i2r.a-star.edu.sg

Abstract. For an undirected graph G without self-loop, we prove: (i) that the number of closed patterns in the adjacency matrix of G is even; (ii) that the number of the closed patterns is precisely double the number of maximal complete bipartite subgraphs of G; (iii) that for every maximal complete bipartite subgraph, there always exists a unique pair of closed patterns that matches the two vertex sets of the subgraph. Therefore, we can enumerate all maximal complete bipartite subgraphs by using efficient algorithms for mining closed patterns which have been extensively studied in the data mining field.

1 Introduction

Interest in graphs and their applications has grown to a very broad spectrum in the past decades (see [18] and the Preface of [8]), largely due to the usefulness of graphs as models in many areas such as mathematical research, electrical engineering, computer programming, business administration, sociology, economics, marketing, biology, and networking and communications. In particular, many problems can be modeled with *maximal complete bipartite subgraphs* (see the definition below) formed by grouping two non-overlapping subsets of vertices of a certain graph that show a kind of full connectivity between them.

We consider two examples. Suppose there are p customers in a mobile communication network. Some people have a wide range of contact, while others have few. Which groups of customers (with a maximal number) have a full interaction with another group of customers? This situation can be modeled by a graph where a mobile phone customer is a node and a communication is an edge. Thus, a maximal bipartite subgraph of this graph corresponds to two groups of customers between whom there exist a full communication. This problem is similar to the one studied in web mining [4,9,12] where web communities are modeled by bipartite cores. Our second example is about proteins' interaction in a cell. There are usually thousands of proteins in a cell that interact with one another. This situation again can be modeled by a graph, where a protein is a node and an interaction between a pair of proteins forms an edge. Then, listing all maximal complete bipartite subgraphs from this graph can answer questions such as which two protein groups have a full interaction, which is a problem studied in biology [14,15].

Listing all maximal complete bipartite subgraphs has been studied theoretically in [5]. The result is that all maximal complete bipartite subgraphs of a graph can be enumerated in time $O(a^3 2^{2a} n)$, where a is the arboricity of the graph and n is the number of vertices in the graph. Even though the algorithm has a linear complexity, it is not practical for large graphs due to the large constant overhead (a can easily be around 10-20 in practice) [20]. In this paper, we study this problem from a data mining perspective: We use a heuristics data mining algorithm to efficiently enumerate all maximal complete bipartite subgraphs from a large graph. A main concept of the data mining algorithm is called *closed patterns*. There are many recent algorithms and implementations devoted to the mining of closed patterns from the so-called *transactional databases* [2,6,7,13,16,17,19]. The data structures are efficient and the mining speed is tremendously fast. Our main contribution here is the observation that the mining of closed patterns from the adjacency matrix of a graph, termed a special transactional database, is equivalent to the problem of enumerating all maximal complete bipartite subgraphs of this graph.

The rest of this paper is organized as follows: Sections 2 and 3 provide basic definitions and propositions on graphs and closed patterns. In Section 4 we prove that there is a one-to-one correspondence between closed pattern pairs and maximal complete bipartite subgraphs for any simple graph. In Section 5, we present our experimental results on a proteins' interaction graph. Section 6 discusses some other related work and then concludes this paper.

2 Maximal Complete Bipartite Subgraphs

A **graph** $G = \langle V^G, E^G \rangle$ is comprised of a set of vertices V^G and a set of edges $E^G \subseteq V^G \times V^G$. We often omit the superscripts in V^G, E^G and other places when the context is clear. Throughout this paper, we assume G is an undirected graph without any self-loops. In other words, we assume that (i) there is no edge $(u,u) \in E^G$ and (ii) for every $(u,v) \in E^G$, (u,v) can be replaced by (v,u)—that is, (u,v) is an unordered pair.

A graph H is a **subgraph** of a graph G if $V^H \subseteq V^G$ and $E^H \subseteq E^G$. A graph G is **bipartite** if V^G can be partitioned into two non-empty and non-intersecting subsets V_1 and V_2 such that $E^G \subseteq V_1 \times V_2$. This bipartite graph G is usually denoted by $G = \langle V_1 \cup V_2, E^G \rangle$. Note that there is no edge in G that joins two vertices within V_1 or V_2. G is **complete bipartite** if $V_1 \times V_2 = E^G$.

Two vertices u, v of a graph G are said to be adjacent if $(u,v) \in E^G$—that is, there is an edge in G that connects them. The **neighborhood** $\beta^G(v)$ of a vertex v of a graph G is the set of all vertices in G that are adjacent to v—that is, $\beta^G(v) = \{u \mid (u,v) \text{ or } (v,u) \in E^G\}$. The neighborhood $\beta^G(X)$ for a *non-empty* subset X of vertices of a graph G is the set of common neighborhood of the vertices in X—that is, $\beta^G(X) = \cap_{x \in X} \beta^G(x)$.

Note that for any subset X of vertices of a graph G such that X and $\beta^G(X)$ are both non-empty, it is the case that $H = \langle X \cup \beta^G(X), X \times \beta^G(X) \rangle$ is a complete bipartite subgraph of G. Note also it is possible for a vertex $v \notin X$ of G to be adjacent to every vertex of $\beta^G(X)$. In this case, the subset X can be expanded by adding the vertex v, while maintaining the same neighborhood. Where to stop the expansion? We use the following definition of maximal complete bipartite subgraphs.

Definition 1. *A graph $H = \langle V_1 \cup V_2, E \rangle$ is a **maximal complete bipartite subgraph** of G if H is a complete bipartite subgraph of G such that $\beta^G(V_1) = V_2$ and $\beta^G(V_2) = V_1$.*

Not all maximal complete bipartite subgraphs are equally interesting. Recall our earlier motivating example involving customers in a mobile communication network. We would probably not be very interested in those two groups of customers with a small size containing a single person or just a few. In contrast, we would probably be considerably more interested if one of the group is large, or both of the groups are large. Hence, we introduce the notion of density on maximal complete bipartite subgraphs.

Definition 2. *A maximal complete bipartite subgraph $H = \langle V_1 \cup V_2, E \rangle$ of a graph G is said to be (m,n)-dense if $|V_1|$ or $|V_2|$ is at least m, and the other is at least n.*

A complete bipartite subgraph $H = \langle V_1 \cup V_2, E \rangle$ of G such that $\beta^G(V_1) = V_2$ and $\beta^G(V_2) = V_1$ is maximal in the sense that there is no other complete bipartite subgraph $H' = \langle V_1' \cup V_2', E' \rangle$ of G with $V_1 \subset V_1'$ and $V_2 \subset V_2'$ such that $\beta^G(V_1') = V_2'$ and $\beta^G(V_2') = V_1'$. To appreciate this notion of maximality, we prove the proposition below.

Proposition 1. *Let $H = \langle V_1 \cup V_2, E \rangle$ and $H' = \langle V_1' \cup V_2', E' \rangle$ be two maximal complete bipartite subgraphs of G such that $V_1 \subseteq V_1'$ and $V_2 \subseteq V_2'$. Then $H = H'$.*

Proof. Suppose $H = \langle V_1 \cup V_2, E \rangle$ and $H' = \langle V_1' \cup V_2', E' \rangle$ are two maximal complete bipartite subgraphs of G such that $V_1 \subseteq V_1'$ and $V_2 \subseteq V_2'$. Since $V_1 \subseteq V_1'$ and $V_2 \subseteq V_2'$, we have $\beta^G(V_1') \subseteq \beta^G(V_1)$ and $\beta^G(V_2') \subseteq \beta^G(V_2)$. Using the definition of maximal complete bipartite subgraphs, we derive $V_2' = \beta^G(V_1') \subseteq \beta^G(V_1) = V_2$ and $V_1' = \beta^G(V_2') \subseteq \beta^G(V_2) = V_1$. Then $E = V_1 \times V_2 = V_1' \times V_2' = E'$. Thus $H = H'$ as desired.

3 Closed Patterns of an Adjacency Matrix

The adjacency matrix of a graph is important in this study. Let G be a graph with $V^G = \{v_1, v_2, \ldots, v_p\}$. The **adjacency matrix** $\mathbf{A}$ of G is the $p \times p$ matrix defined by

$$\mathbf{A}[i,j] = \begin{cases} 1 \text{ if } (v_i, v_j) \in E^G \\ 0 \text{ otherwise} \end{cases}$$

Recall that our graphs do not have self-loop and are undirected. Thus $\mathbf{A}$ is a symmetric matrix and every entry on the main diagonal is 0. Also, $\{v_j \mid \mathbf{A}[k,j] = 1, 1 \leq j \leq p\} = \beta^G(v_k) = \{v_j \mid \mathbf{A}[j,k] = 1, 1 \leq j \leq p\}$.

The adjacency matrix of a graph can be interpreted into a **transactional database** (*DB*) [1]. To define a *DB*, we first define a **transaction**. Let I be a set of **items**. Then a transaction is defined as a subset of I. For example, assume I to be all items in a supermarket, a transaction by a customer is the items that the customer bought. A *DB* is a non-empty multi-set of transactions. Each transaction T in a *DB* is assigned a unique identity $id(T)$. A **pattern** is defined as a non-empty set[1] of items of I. A pattern may be or may not be contained in a transaction. Given a *DB* and a pattern P, the number of transactions in *DB* containing P is called the **support** of P, denoted $sup^{DB}(P)$. We are often interested in patterns that occur sufficiently frequent in a *DB*. Those patterns are called **frequent** patterns—that is, patterns P satisfying $sup^{DB}(P) \geq ms$, for a threshold $ms > 0$. In this paper, unless mentioned otherwise, we consider all and only those patterns with a non-zero support, namely all those frequent pattern with the support threshold $ms = 1$. So, by a pattern of a *DB*, we mean that it is non-empty and it occurs in *DB* at least once.

Let G be a graph with $V^G = \{v_1, v_2, \ldots, v_p\}$. If each vertex in V^G is defined as an item, then the neighborhood $\beta^G(v_i)$ of v_i is a transaction. Thus,

$$\{\beta^G(v_1), \beta^G(v_2), \ldots, \beta^G(v_p)\}$$

is a *DB*. Such a special *DB* is denoted by DB_G. The identity of a transaction in DB_G is defined as the vertex itself—that is, $id(\beta^G(v_i)) = v_i$. Note that DB_G has the same number of items and transactions. Note also that $v_i \notin \beta^G(v_i)$ since we assume G to be an undirected graph without self-loop.

DB_G can be represented as a binary square matrix. This binary matrix **B** is defined by

$$\mathbf{B}[i,j] = \begin{cases} 1 \text{ if } v_j \in \beta^G(v_i) \\ 0 \text{ otherwise} \end{cases}$$

Since $v_j \in \beta^G(v_i)$ iff $(v_i, v_j) \in E^G$, it can be seen that $\mathbf{A} = \mathbf{B}$. So, "a pattern of DB_G" is equivalent to "a pattern of the adjacency matrix of G".

Closed patterns are a type of interesting patterns in a *DB*. In the last few years, the problem of efficiently mining closed patterns from a large *DB* has attracted a lot of researchers in the data mining community [2,6,7,13,16,17,19]. Let I be a set of items, and D be a transactional database defined on I. For a pattern $P \subseteq I$, let $f^D(P) = \{T \in D \mid P \subseteq T\}$—that is, $f^D(P)$ are all transactions in D containing the pattern P. For a set of transactions $D' \subseteq D$, let $g(D') = \bigcap_{T \in D'} T = \bigcap D'$—that is, the set of items which are shared by all transactions in D'. Using these two functions, we can define the notion of **closed patterns**. For a pattern P, $CL^D(P) = g(f^D(P))$ is called the **closure** of P. A pattern P is said to be **closed** with respect to a transactional database D iff $CL^D(P) = P$.

We define the **occurrence set** of a pattern P in *DB* as $occ^{DB}(P) = \{id(T) \mid T \in DB, P \subseteq T\} = \{id(T) \mid T \in f^{DB}(P)\}$. It is straightforward to see that

---

[1] The $\emptyset$ is usually defined as a valid pattern in the data mining community. However, in this paper, to be consistent to the definition of $\beta^G(X)$, it is excluded.

$id(T) \in occ^{DB}(P)$ iff $T \in f^{DB}(P)$. There is a tight connection between the notions of neighbourhood in a graph G and occurrence in the corresponding transactional database DB_G.

Proposition 2. *Given a graph G and a pattern P of DB_G. Then $occ^{DB_G}(P) = \beta^G(P)$.*

Proof. If $v \in occ(P)$, then v is adjacent to every vertex in P. Therefore, $v \in \beta(v')$ for each $v' \in P$. That is, $v \in \bigcap_{v' \in P} \beta(v') = \beta(P)$.

If $u \in \beta(P)$, then u is adjacent to every vertex in P. So, $\beta(u) \supseteq P$. Therefore, $\beta(u)$ is a transaction of DB_G containing P. So, $u \in occ(P)$.

There is also a nice connection between the notions of neighborhood in a graph and that of closure of patterns in the corresponding transactional database.

Proposition 3. *Given a graph G and a pattern P of DB_G. Then $\beta^G(\beta^G(P)) = CL^{DB_G}(P)$. Thus $\beta^G \circ \beta^G$ is a closure operation on patterns of DB_G.*

Proof. By Proposition 2, $\beta(\beta(P)) = \beta(occ(P)) = \bigcap_{id(T) \in occ(P)} T = \bigcap_{T \in f(P)} T = g(f(P)) = CL(P)$.

We discuss in the next section deeper relationships between the closed patterns of DB_G and the maximal complete bipartite subgraphs of G.

4 Results

The occurrence set of a closed pattern C in DB_G plays a key role in the maximal complete bipartite subgraphs of G. We introduce below some of its key properties.

Proposition 4. *Let G be a graph. Let C_1 and C_2 be two closed patterns of DB_G. Then $C_1 = C_2$ iff $occ^{DB_G}(C_1) = occ^{DB_G}(C_2)$.*

Proof. The left-to-right direction is trivial. To prove the right-to-left direction, let us suppose that $occ(C_1) = occ(C_2)$. It is straightforward to see that $id(T) \in occ(P)$ iff $T \in f(P)$. Then we get $f(C_1) = f(C_2)$ from $occ(C_1) = occ(C_2)$. Since C_1 and C_2 are closed patterns of DB_G, it follows that $C_1 = g(f(C_1)) = g(f(C_2)) = C_2$, and finishes the proof.

Proposition 5. *Let G be a graph and C a closed pattern of DB_G. Then C and its occurrence set has empty intersection. That is, $occ^{DB_G}(C) \cap C = \{\}$.*

Proof. Let $v \in occ(C)$. Then v is adjacent to every vertex in C. Since we assume G is a graph without self-loop, $v \notin C$. Therefore, $occ^{DB_G}(C) \cap C = \{\}$.

In fact this proposition holds for any pattern P, not necessarily a closed pattern C.

Lemma 1. *Let G be a graph. Let C be a closed pattern of DB_G. Then $f^{DB_G}(occ^{DB_G}(C)) = \{\beta^G(c) \mid c \in C\}$.*

Proof. As C is a closed pattern, by definition, then $\{c \mid c \in C\}$ are all and only items contained in every transaction of DB_G that contains C. This is equivalent to that $\{c \mid c \in C\}$ are all and only vertices of G that are adjacent to every vertex in $occ(C)$. This implies that $\{\beta(c) \mid c \in C\}$ are all and only transactions that contain $occ(C)$. In other words, $f(occ(C)) = \{\beta(c) \mid c \in C\}$.

Proposition 6. *Let G be a graph and C a closed pattern of DB_G. Then $occ^{DB_G}(C)$ is also a closed pattern of DB_G.*

Proof. By Lemma 1, $f(occ(C)) = \{\beta(c) \mid c \in C\}$. So $CL(occ(C)) = g(f(occ(C))) = \bigcap f(occ(C)) = \bigcap_{c \in C} \beta(c) = \beta(C)$. By Proposition 2, $\beta(C) = occ(C)$. Thus $occ(C)$ is a closed pattern.

The three propositions above give rise to a couple of interesting corollaries below.

Corollary 1. *Let G be a graph. Then the number of closed patterns in DB_G is even.*

Proof. Suppose there are n closed patterns (that appear at least once) in DB_G, denoted as $C_1, C_2, ..., C_n$. As per Proposition 6, $occ(C_1), occ(C_2), ..., occ(C_n)$ are all closed patterns of DB_G. As per Proposition 4, $occ(C_i)$ is different from $occ(C_j)$ iff C_i is different from C_j. So every closed pattern can be paired with a distinct closed pattern by $occ(\cdot)$ in a bijective manner. Furthermore, as per Proposition 5, no closed pattern is paired with itself. This is possible only when the number n is even.

Corollary 2. *Let G be a graph. Then the number of closed patterns C, such that both C and $occ^{DB_G}(C)$ appear at least ms times in DB_G, is even.*

Proof. As seen from the proof of Corollary 1, every closed pattern C of DB_G can be paired with $occ^{DB_G}(C)$, and the entire set of closed patterns can be partitioned into such pairs. So a pair of closed patterns C and $occ^{DB_G}(C)$ either satisfy or do not satisfy the condition that both C and $occ^{DB_G}(C)$ appear at least ms times in DB_G. Therefore, the number of closed patterns C, satisfying that both C and $occ^{DB_G}(C)$ appear at least ms times in DB_G, is even.

Note that this corollary does not imply the number of frequent closed patterns that appear at least ms times in DB_G is always even. A counter example is given below.

Example 1. Consider a DB_G given by the following matrix:

	p_1	p_2	p_3	p_4	p_5
$\beta(p_1)$	0	1	1	0	0
$\beta(p_2)$	1	0	1	1	1
$\beta(p_3)$	1	1	0	1	1
$\beta(p_4)$	0	1	1	0	0
$\beta(p_5)$	0	1	1	0	0

We list its closed patterns, their support, and their $occ(\cdot)$ counterpart patterns below:

support of X	close pattern X	$Y = occ(X)$	support of Y
3	$\{p_2, p_3\}$	$\{p_1, p_4, p_5\}$	2
4	$\{p_2\}$	$\{p_1, p_3, p_4, p_5\}$	1
4	$\{p_3\}$	$\{p_1, p_2, p_4, p_5\}$	1

Suppose we take $ms = 3$. Then there are only 3 closed patterns—an odd number—that occur at least ms times, viz. $\{p_2, p_3\}$, $\{p_2\}$, and $\{p_3\}$.

Finally, we demonstrate our main result on the relationship with closed patterns and maximal complete bipartite subgraphs. In particular, we discover that every pair of a closed pattern C and its occurrence set $occ^{DB_G}(C)$ yields a distinct maximal complete bipartite subgraph of G.

Theorem 1. *Let G be an undirected graph without self-loop. Let C be a closed pattern of DB_G. Then the graph*

$$H = \langle C \cup occ^{DB_G}(C), C \times occ^{DB_G}(C) \rangle$$

is a maximal complete bipartite subgraph of G.

Proof. By assumption, C is non-empty and C has a non-zero support in DB_G. Therefore, $occ(C)$ is non-empty. By Proposition 5, $C \cap occ^{DB_G}(C) = \{\}$. Furthermore, $\forall v \in occ(C)$, v is adjacent in G to every vertex of C. So, $C \times occ(C) \subseteq E^G$, and every edge of H connects a vertex of C and a vertex of $occ(C)$. Thus, H is a complete bipartite subgraph of G. By Proposition 2, we have $occ^{DB_G}(C) = \beta^G(C)$. By Proposition 3, $C = \beta^G(\beta^G(C))$. By Proposition 2, we derive $C = \beta^G(occ^{DB_G}(C))$. So H is maximal. This finishes the proof.

Theorem 2. *Let G be an undirected graph without self-loop. Let graph $H = \langle V_1 \cup V_2, E \rangle$ be a maximal complete bipartite subgraph of G. Then, V_1 and V_2 are both a closed pattern of DB_G, $occ^{DB_G}(V_1) = V_2$ and $occ^{DB_G}(V_2) = V_1$.*

Proof. Since H is a maximal complete bipartite subgraph of G, then $\beta(V_1) = V_2$ and $\beta(V_2) = V_1$. By Proposition 3, $CL(V_1) = \beta(\beta(V_1)) = \beta(V_2) = V_1$. So, V_1 is a closed pattern. Similarly, we can get V_2 is a closed pattern. By Proposition 2, $occ(V_1) = \beta(V_1) = V_2$ and $occ(V_2) = \beta(V_2) = V_1$, as required.

The above two theorems say that maximal complete bipartite subgraphs of G are all in the form of $H = \langle V_1 \cup V_2, E \rangle$, where V_1 and V_2 are both a closed pattern of DB_G. Also, for every closed pattern C of DB_G, the graph $H = \langle C \cup occ^{DB_G}(C), C \times occ^{DB_G}(C) \rangle$ is a maximal complete bipartite subgraph of G. So, there is a one-to-one correspondence between maximal complete bipartite subgraphs and closed pattern pairs.

We can also derive a corollary linking support threshold of DB_G to the density of maximal complete bipartite subgraphs of G.

Corollary 3. *Let G be an undirected graph without self-loop. Then*

$$H = \langle C \cup occ^{DB_G}(C), C \times occ^{DB_G}(C) \rangle$$

is a (m,n)-dense maximal complete bipartite subgraph of G iff C is a closed pattern such that C occurs at least m times in DB_G and $occ^{DB_G}(C)$ occur at least n times in DB_G.

The corollary above has the following important implication.

Theorem 3. *Let G be an undirected graph without self-loop. Then*

$$H = \langle C \cup occ^{DB_G}(C), C \times occ^{DB_G}(C) \rangle$$

is a (m,n)-dense maximal complete bipartite subgraph of G iff C is a closed pattern such that C occurs at least m times in DB_G and $|C| \geq n$.

Proof. Suppose $H = \langle C \cup occ^{DB_G}(C), C \times occ^{DB_G}(C) \rangle$ is a (m,n)-dense maximal complete bipartite subgraph of G. By Theorem 2, $C = occ(occ(C))$. By definition of $occ(\cdot)$, $sup(occ(C)) = |occ(occ(C))| = |C|$. Substitute this into Corollary 3, we get H is a (m,n)-dense maximal complete bipartite subgraph of G iff C is a closed pattern such that C occurs at least m times in DB_G and $|C| \geq n$ as desired.

Theorems 1 and 2 show that algorithms for mining closed patterns can be used to extract maximal complete bipartite subgraphs of undirected graphs without self-loop. Such data mining algorithms are usually significantly more efficient at higher support threshold ms. Thus Theorem 3 suggests an important optimization for mining (m,n)-dense maximal complete bipartite subgraphs. To wit, assuming $m > n$, it suffices to mine closed patterns at support threshold $ms = m$, and then get the answer by filtering out those patterns of length less than n.

5 Experimental Results

We use an example to demonstrate the speed of listing all maximal complete bipartite subgraphs by using an algorithm for mining closed patterns. The graph is a protein interaction network with proteins as vertices and interactions as edges. As there are many physical protein interaction networks corresponding to different species, here we take the simplest and most comprehensive yeast physical and genetic interaction network [3] as an example. This graph consists of 4904 vertices and 17440 edges (after removing 185 self loops and 1413 redundant edges from the original 19038 interactions). Therefore, the adjacency matrix is a transactional database with 4904 items and 4904 transactions. On average, the number of items in a transaction is 3.56. That is, the average size of the neighborhood of a protein is 3.56.

We use FPclose* [7], a state-of-the-art algorithm for mining closed patterns, for enumerating the maximal complete bipartite subgraphs. Our machine is a PC with a CPU clock rate 3.2GHz and 2GB of memory. The results are reported in Table 1, where the second column shows the total number of **frequent** close patterns whose support level is at least the threshold number in the column one. The third column of this table shows the number of close patterns whose cardinality and support are both at least the support threshold; all such closed patterns are termed qualified closed patterns. Only these qualified closed patterns can be used to form maximal complete bipartite subgraphs $H = \langle V_1 \cup V_2, E \rangle$ such that both of $|V_1|$ and $|V_2|$ meet the thresholds. From the table, we can see:

- The number of all closed patterns (corresponding to those with the support threshold of 1) is even. Moreover, the number of qualified close patterns with cardinality no less than any support level is also even, as expected from Corollary 2.
- The algorithm runs fast—The algorithm program can complete within 4 seconds for all situations reported here. This indicates that enumerating all maximal complete bipartite subgraphs from a large graph can be practically solved by using algorithms for mining closed patterns.
- A so-called "many-few" property [11] of protein interactions is observed again in our experiment results. The "many-few" property says that: a protein that interacts with a large number of proteins tends *not* to interact with another protein which also interacts with a large number of proteins [11]. In other words, highly connected proteins are separated by low-connected proteins. This is most clearly seen in Table 1 at the higher support thresholds. For example, at the support threshold 11, there are 12402 protein groups that have full interactions with at least 11 proteins. But there are only two groups, as seen in the third column of the table, that each contain at least 11 proteins and that have full mutual interaction.

Table 1. Close patterns in a yeast protein interaction network

support threshold	# of frequent close patterns	# of qualified close patterns	time in sec.
1	121314	121314	3.859
2	117895	114554	2.734
3	105854	95920	2.187
4	94781	80306	1.765
5	81708	60038	1.312
6	66429	36478	0.937
7	50506	15800	0.625
8	36223	3716	0.398
9	25147	406	0.281
10	17426	34	0.171
11	12402	2	0.109
12	9138	0	0.078

6 Discussion and Conclusion

There are two recent research results related to our work. The problem of enumerating all maximal complete bipartite subgraphs (called maximal bipartite cliques there) from a *bipartite graph* has been investigated by [10]. The difference is that our work is to enumerate all the subgraphs from any graphs (without self loops and undirected), but Makino and Uno's work is limited to enumerating from only bipartite graphs. So, our method is more general. Zaki [20] observed that a transactional database DB can be represented by a bipartite graph H, and also a relation that closed patterns (wrongly stated as maximal patterns in [20]) of DB one-to-one correspond to maximal complete bipartite subgraphs (called maximal bipartite clique there) of H. However, our work is to convert a graph G, including bipartite graphs, into a special transactional database DB_G, and then to discover all closed patterns from DB_G for enumerating all maximal complete bipartite subgraphs of G. Furthermore, the occurrence set of a closed pattern in Zaki's work may not be a closed pattern, but that of ours is always a closed pattern.

Finally, let's summarize the results achieved in this paper. We have studied the problem of listing all maximal complete bipartite subgraphs from a graph. We proved that this problem is equivalent to the mining of all closed patterns from the adjacency matrix of this graph. Experimental results on a large protein interactions' data show that a data mining algorithm can run very fast to find all interacted protein groups. The results will have great potential in applications such as in web mining, in communication systems, and in biological fields.

References

1. R. Agrawal, T. Imielinski, and A. Swami. Mining association rules between sets of items in large databases. In *Proceedings of the 1993 ACM-SIGMOD International Conference on Management of Data*, pages 207–216, Washington, D.C., May 1993. ACM Press.
2. Yves Bastide, Nicolas Pasquier, Rafik Taouil, Gerd Stumme, and Lotfi Lakhal. Mining minimal non-redundant association rules using frequent closed itemsets. *Computational Logic*, pages 972–986, 2000.
3. B. J. Breitkreutz, C. Stark, and M. Tyers. The grid: The general repository for interaction datasets. *Genome Biology*, 4(3):R23, 2003.
4. Andrei Z. Broder, Ravi Kumar, Farzin Maghoul, Prabhakar Raghavan, Sridhar Rajagopalan, Raymie Stata, Andrew Tomkins, and Janet L. Wiener. Graph structure in the web. *Computer Networks*, 33(1-6):309–320, 2000.
5. David Eppstein. Arboricity and bipartite subgraph listing algorithms. *Information Processing Letters*, 51:207–211, 1994.
6. Bart Goethals and Mohanned J. Zaki. FIMI'03: Workshop on frequent itemset mining implementations. In *Third IEEE International Conference on Data Mining Workshop on Frequent Itemset Mining Implementations*, pages 1–13, 2003.
7. Gosta Grahne and Jianfei Zhu. Efficiently using prefix-trees in mining frequent itemsets. In *Proceedings of FIMI'03: Workshop on Frequent Itemset Mining Implementations*, 2003.

8. Jonathan L. Gross and Jay Yellen. *Handbook of Graph Theory*. CRC Press, 2004.
9. Ravi Kumar, Prabhakar Raghavan, Sridhar Rajagopalan, and Andrew Tomkins. Trawling the web for emerging cyber-communities. *Computer Networks*, 31(11-16):1481–1493, 1999.
10. Kazuhisa Makino and Takeaki Uno. New algorithms for enumerating all maximal cliques. In *Proceedings of the 9th Scandinavian Workshop on Algorithm Theory (SWAT 2004)*, pages 260–272. Springer-Verlag, 2004.
11. Sergei Maslov and Kim Sneppen. Specificity and stability in topology of protein networks. *Science*, 296:910–913, 2002.
12. Tsuyoshi Murata. Discovery of user communities from web audience measurement data. In *Proceedings of The 2004 IEEE/WIC/ACM International Conference on Web Intelligence (WI 2004)*, pages 673–676, 2004.
13. Nicolas Pasquier, Yves Bastide, Rafik Taouil, and Lotfi Lakhal. Discovering frequent closed itemsets for association rules. In *Proceedings of the 7th International Conference on Database Theory (ICDT)*, pages 398–416, 1999.
14. D. J. Reiss and B. Schwikowski. Predicting protein-peptide interactions via a network-based motif sampler. *Bioinformatics (ISMB 2004 Proceedings)*, 20 (suppl.):i274–i282, 2004.
15. A. H. Tong, B. Drees, G Nardelli, G. D. Bader, B. Brannetti, L. Castagnoli, M. Evangelista, S. Ferracuti, B. Nelson, S. Paoluzi, M. Quondam, A. Zucconi, C. W. Hogue, S. Fields, C. Boone, and G. Cesareni. A combined experimental and computational strategy to define protein interaction networks for peptide recognition modules. *Science*, 295:321–324, 2002.
16. Takiake Uno, Masashi Kiyomi, and Hiroaki Arimura. LCM ver.2: Efficient mining algorithms for frequent/closed/maximal itemsets. In *IEEE ICDM'04 Workshop FIMI'04 (International Conference on Data Mining, Frequent Itemset Mining Implementations)*, 2004.
17. J. Wang, Jiawei Han, and Jian Pei. CLOSET+: Searching for the best strategies for mining frequent closed itemsets. In *Proceedings of the Ninth ACM SIGKDD International Conference on Knowledge Discovery and Data Mining (KDD'03), Washington, DC, USA*, pages 236–245, 2003.
18. Takashi Washio and Hiroshi Motoda. State of the art of graph-based data mining. *SIGKDD Explorations*, 5(1):59–68, 2003.
19. Mohammed Javeed Zaki and Ching-Jiu Hsiao. CHARM: An efficient algorithm for closed itemset mining. In *Proceedings of the Second SIAM International Conference on Data Mining*, 2002.
20. Mohammed Javeed Zaki and Mitsunori Ogihara. Theoretical foundations of association rules. In *Proc. 3rd SIGMOD Workshop on Research Issues in Data Mining and Knowledge Discovery*, 1998.

Improving Generalization by Data Categorization

Ling Li, Amrit Pratap, Hsuan-Tien Lin, and Yaser S. Abu-Mostafa

Learning Systems Group, California Institute of Technology, USA

Abstract. In most of the learning algorithms, examples in the training set are treated equally. Some examples, however, carry more reliable or critical information about the target than the others, and some may carry wrong information. According to their intrinsic margin, examples can be grouped into three categories: typical, critical, and noisy. We propose three methods, namely the selection cost, SVM confidence margin, and AdaBoost data weight, to automatically group training examples into these three categories. Experimental results on artificial datasets show that, although the three methods have quite different nature, they give similar and reasonable categorization. Results with real-world datasets further demonstrate that treating the three data categories differently in learning can improve generalization.

1 Introduction

Machine learning is an alternative approach to system design. Instead of the conventional way of mathematically modeling the system, the role of learning is to take a dataset of examples, such as input-output pairs from an unknown target function, and synthesize a hypothesis that best approximates the target. The dataset, acting as the information gateway between the target and the hypothesis, is thus at the heart of the learning process.

Generally, every example in the dataset is treated equally and no example is explicitly discarded. After all, each example carries its own piece of information about the target. However, if some of the examples are corrupted with noise, the information they provide would be misleading. In this case, it is better to identify and remove them, which can be performed either explicitly by an outlier detection preprocessing step [1], or implicitly in the learning algorithm via regularization [2].

Even in cases where all the examples are noiseless, there are situations in which we want to deal with examples differently. For instance, in large datasets which often contain redundant examples, a subset of informative examples that carries most of the information is usually more desirable for computational reasons [3]. In cases where none of the hypotheses can perfectly model the target, it is better to discard examples that cannot be classified correctly by any hypothesis as they may "confuse" the learning [4].

Most existing methods that treat examples differently tend to group examples into two categories based on different criteria: consistent vs. inconsistent (outliers) [1], easy vs. hard [5], typical vs. informative [3], etc. In this paper, we introduce the concept of *intrinsic margin* as a criterion for grouping data and motivate the need to have three categories instead of two. We present three methods to automate the categorizing. We show that by treating examples in these three categories differently, we can improve the generalization performance of learning on real-world problems. In addition, the categorization can be used to reduce the dataset size without affecting the learning performance.

The paper is organized as follows. The formal framework of learning is defined in Sect. 2. Then in Sect. 3, we introduce the concept of data categorization and present our methods for automatic categorization. Results on artificial and real-world datasets are presented in Sects. 4 and 5. We finally conclude in Sect. 6.

2 Learning Systems

In learning problems, *examples* are in the form of input-output pair $(\mathbf{x}, y)$. We assume that the input vectors $\mathbf{x} \in \mathcal{X}$ are generated independently from an unknown probability distribution $P_\mathcal{X}$, and the output labels $y \in \mathcal{Y}$ are computed from $y = f(\mathbf{x})$. Here the unknown function $f \colon \mathcal{X} \to \mathcal{Y}$ is called the *target function*. In this paper, we shall only focus on binary classification problems, in which $\mathcal{Y} = \{-1, 1\}$. We further assume that f comes from thresholding an *intrinsic function* $f_r \colon \mathcal{X} \to \mathbb{R}$, i.e., $f(\mathbf{x}) = \text{sign}(f_r(\mathbf{x}))$, where the magnitude of $f_r(\mathbf{x})$ corresponds to the reliability of the output $f(\mathbf{x})$. For example, if the target function $f(\mathbf{x})$ indicates whether a credit card should be issued to a person $\mathbf{x}$, the intrinsic function $f_r(\mathbf{x})$ could be the aligned credit score of the person.

For a hypothesis $g \colon \mathcal{X} \to \mathcal{Y}$ and an example $(\mathbf{x}, y)$, a commonly used error measure (loss function) is

$$e(g(\mathbf{x}), y) = [g(\mathbf{x}) \neq y],$$

where the Boolean test $[\cdot]$ is 1 if the condition is true and 0 otherwise. Then, for a target function f, we can define the *out-of-sample error* of g as

$$\pi(g) = \mathrm{E}_{\mathbf{x} \sim P_\mathcal{X}} \left[e(g(\mathbf{x}), f(\mathbf{x})) \right].$$

The goal of learning is thus to choose a hypothesis g that has a low out-of-sample error $\pi(g)$ from a set of candidate hypotheses, namely the *learning model* $\mathcal{G}$.

However, $\pi(g)$ cannot be directly computed because the distribution $P_\mathcal{X}$ and the target function f are both unknown. The only information we can access is often limited in the *training set* $\mathcal{D}$, which consists of N examples $(\mathbf{x}_i, y_i)$, $i = 1..N$. Thus, instead of looking for a hypothesis g with low $\pi(g)$ values, a learning algorithm may try to find g that minimizes an estimator of $\pi(g)$. A commonly used estimator is the *in-sample error* $\nu(g)$ on the training set $\mathcal{D}$,

$$\nu(g) = \nu(g, \mathcal{D}) = \frac{1}{N} \sum_{i=1}^{N} e(g(\mathbf{x}_i), y_i).$$

For a fixed hypothesis g, $\nu(g)$ is an unbiased estimator of $\pi(g)$, and when the size of $\mathcal{D}$ is large enough, statistical bounds guarantee that $\nu(g)$ and $\pi(g)$ would not differ by too much.

Note that the learning algorithm searches the whole learning model $\mathcal{G}$ for a suitable hypothesis rather than focusing on a fixed one. In this case, the probability that $\nu(g)$ and $\pi(g)$ differs for some $g \in \mathcal{G}$ gets magnified by the complexity of $\mathcal{G}$. Thus, the hypothesis found might fit the training set well while still having a high out-of-sample error [2]. This situation is called *overfitting*, which arises when good in-sample predictions do not relate to good out-of-sample predictions. The situation can become worse when the examples contain noise. Then, fitting the training set well means fitting the wrong information, which leads to bad out-of-sample predictions.

Learning algorithms often try to avoid overfitting through regularization [2]. Regularization usually enforces a trade-off between the complexity of $\mathcal{G}$ and the necessity to predict the training examples correctly. If we can characterize the usefulness of each training example, the learning algorithm can then be guided to focus on predicting important examples correctly, leading to a more meaningful regularization trade-off and thus a better generalization performance. This motivates our work to categorize the training examples.

3 Data Categorization

The purpose of data categorization is to group examples according to their usefulness to learning so that it is possible to treat them differently. Guyon et al. [3] grouped data into two categories, typical and informative. However, they found that the category of informative examples contained both useful examples and noisy ones. Thus, they needed human-based post-processing to eliminate the noisy examples. Similar problems are encountered in other methods that use two-group categorization. This shows that we need to have more than two categories. In this paper, we fit the need by having three categories: typical, critical and noisy.

Although all examples carry information about the target, they are not equal in the sense that some examples carry more useful information about the target than others, and some examples may misguide the learning algorithm. For instance, in classification problems, an example close to the class boundary gives more critical information than an example deep in the class territory. In addition, real-world data often contain mislabeled examples, which compromise the ability of the in-sample error to approximate the out-of-sample error and lead to bad generalization.

One way to categorize examples based on the above intuition is through the concept of *intrinsic margin*. For an example $(\mathbf{x}, y)$, its intrinsic margin is $y f_r(\mathbf{x})$, where f_r is the implicit intrinsic function defined in Sect. 2. Under some reasonable smoothness assumption, the intrinsic margin can be treated as a measure of how close the example is to the classification decision boundary. If the intrinsic margin is small positive, the example lies near the decision boundary and should be categorized as critical. If the margin is large positive, the example is far from the boundary and should be categorized as typical. Examples with negative intrinsic margin are mislabeled, and should be classified as noisy. Thus, we may use two thresholds, 0 and some small positive value, to partition the intrinsic margin and categorize the data.

In practical situations, it is impossible to calculate the intrinsic margin unless the intrinsic function is known. However, since we are only interested in thresholding the intrinsic margin, any monotonic function of the intrinsic margin can be used with appropriate thresholds. Next, we propose three different methods to estimate such functions for automatically categorizing the data.

3.1 Selection Cost

Bad generalization arises when the in-sample error is a bad indicator of the out-of-sample error. A particular example $(\mathbf{x}, y)$ may deteriorate the generalization performance if its error is a bad indicator of the out-of-sample error. Based on this intuition, Nicholson [4] suggested to use the correlation coefficient between $e(g(\mathbf{x}), y)$ and $\pi(g)$ under a prior distribution $P_{\mathcal{G}}$ for g,

$$\rho(\mathbf{x}, y) = \operatorname{corrcoef}_g \left[e(g(\mathbf{x}), y), \pi(g) \right]$$
$$= \frac{\mathrm{E}_g \left[e(g(\mathbf{x}), y) \pi(g) \right] - \mathrm{E}_g \left[e(g(\mathbf{x}), y) \right] \mathrm{E}_g \left[\pi(g) \right]}{\sqrt{\operatorname{Var}_g \left[e(g(\mathbf{x}), y) \right] \operatorname{Var}_g \left[\pi(g) \right]}},$$

to measure how well the individual error $e(g(\mathbf{x}), y)$ indicates $\pi(g)$. A positive correlation ρ indicates that if g has a low error on this example, it is likely to have a low out-of-sample error, too. This is formalized in Theorem 1.

Theorem 1. *If the learning model $\mathcal{G}$ is negation symmetric (i.e., $P_{\mathcal{G}}[g] = P_{\mathcal{G}}[-g]$ for any $g \in \mathcal{G}$),*

$$\rho(\mathbf{x}, y) \propto \mathrm{E}_g \left[\pi(g) \mid g(\mathbf{x}) \neq y \right] - \mathrm{E}_g \left[\pi(g) \mid g(\mathbf{x}) = y \right], \tag{1}$$

where the proportional constant is positive and depends only on $\mathcal{G}$.

Proof. For a given example $(\mathbf{x}, y)$ and $P_{\mathcal{G}}$, let $p_i = \Pr\left[e(g(\mathbf{x}), y) = i \right]$ and $\pi_i = \mathrm{E}_g \left[\pi(g) \mid e(g(\mathbf{x}), y) = i \right]$ for $i = 0, 1$. We have $p_0 + p_1 = 1$, and

$$\mathrm{E}_g \left[e(g(\mathbf{x}), y) \pi(g) \right] = p_1 \pi_1, \quad \mathrm{E}_g \left[\pi(g) \right] = p_0 \pi_0 + p_1 \pi_1,$$
$$\mathrm{E}_g \left[e(g(\mathbf{x}), y) \right] = p_1, \quad \operatorname{Var}_g \left[e(g(\mathbf{x}), y) \right] = p_0 p_1.$$

Hence with the definition of $\rho(\mathbf{x}, y)$,

$$\rho(\mathbf{x}, y) = \frac{p_1 \pi_1 - p_1 (p_0 \pi_0 + p_1 \pi_1)}{\sqrt{\operatorname{Var}_g \left[e(g(\mathbf{x}), y) \right] \operatorname{Var}_g \left[\pi(g) \right]}} = (\pi_1 - \pi_0) \sqrt{\frac{p_0 p_1}{\operatorname{Var}_g \left[\pi(g) \right]}}.$$

When $\mathcal{G}$ is negation symmetric, it is trivial that $p_0 = p_1 = \frac{1}{2}$ for any $(\mathbf{x}, y)$. So the proportional ratio is a constant. □

The conditional expectation π_1 (π_0) is the expected out-of-sample error of hypotheses that predict $(\mathbf{x}, y)$ wrongly (correctly). In the learning process, we can select hypotheses that agree on $(\mathbf{x}, y)$ or those that do not. The difference between the two conditional expectations is thus the relative change in the average out-of-sample error, and is called the *selection cost*. If there were only one example to learn, a positive selection cost would imply that we should choose a hypothesis that agrees with it. When a dataset is concerned, the selection cost indicates, at least qualitatively, how desirable it is to classify the example correctly. Since the correlation ρ is just a scaled version of the selection cost, we will use the name selection cost for both quantities in the following text.

In practice, the selection cost of an example $(\mathbf{x}_i, y_i)$ is inaccessible because $\pi(g)$ cannot be computed. However, we may estimate $\pi(g)$ by the leave-one-out error[1] $\nu^{(i)}(g) = \nu(g, \mathcal{D} \setminus \{(\mathbf{x}_i, y_i)\})$. The selection cost can then be estimated, by random sampling over the learning model, as the correlation coefficient between $e(g(\mathbf{x}_i), y_i)$ and $\nu^{(i)}(g)$. This works well even in the presence of noise [4].

Note that for Theorem 1 to be meaningful, the actual learning model has to be used to estimate the selection cost. Under suitable choices of model complexity, however, relaxing this requirement often does not affect the performance in experiments. In this paper, we shall use neural networks as our underlying model when computing the selection cost.

We categorize an example as typical if its selection cost is greater than a threshold t_c, noisy if the cost is less than a threshold t_n, and critical if the cost lies between the two thresholds. Since a negative selection cost implies that it is better to misclassify the example, zero is an ideal threshold to separate noisy and noiseless examples. We choose thresholds around zero, $t_c = 0.15$ and $t_n = -0.1$, to accommodate estimation errors. Better categorization may be obtained by further estimating the optimal thresholds; but for the illustration purpose of this paper, we use ad hoc thresholding for all our methods.

3.2 SVM Confidence Margin

The support vector machine (SVM) [2] is a learning algorithm that finds a large-confidence hyperplane classifier in the feature space. Such classifier is usually obtained by solving the Lagrange dual problem:

$$\min_{\alpha} \frac{1}{2} \sum_{i=1}^{N} \sum_{j=1}^{N} \alpha_i \alpha_j y_i y_j K(\mathbf{x}_i, \mathbf{x}_j) - \sum_{i=1}^{N} \alpha_i$$

s.t. $\sum_{i=1}^{N} y_i \alpha_i = 0, \quad 0 \leq \alpha_i \leq C.$

Here the kernel $K(\mathbf{x}, \mathbf{x}')$ is an inner product of the transformed $\mathbf{x}$ and $\mathbf{x}'$ in the feature space, and C is the regularization parameter. SVM predicts the label of $\mathbf{x}$ as the sign of $\tilde{g}(\mathbf{x}) = \sum_{i=1}^{N} \alpha_i y_i K(\mathbf{x}_i, \mathbf{x}) + b$. We call $y_i \tilde{g}(\mathbf{x}_i)$ the *confidence margin* of the example $(\mathbf{x}_i, y_i)$. This concept is related to the intrinsic margin, but comes specifically from the view of a learning algorithm.

---

[1] To avoid a positive bias in estimating ρ, which can be easily verified from a formula similar to (1), we do not use the in-sample error ν.

The Lagrange multiplier α_i and the confidence margin $y_i\tilde{g}(\mathbf{x}_i)$ are also closely related:

- When $\alpha_i = 0$, we have $y_i\tilde{g}(\mathbf{x}_i) \geq 1$. The example is typical because the confidence margin is large.
- When $\alpha_i > 0$, we have $y_i\tilde{g}(\mathbf{x}_i) \leq 1$. The example is a support vector and is informative for evaluating $\tilde{g}(\mathbf{x})$.

Guyon et al. [3] used the relative magnitude of α_i as a criterion for automated identification and elimination of noisy examples with $C = \infty$. Somehow they found that the criterion failed to distinguish critical examples from noisy ones cleanly, and hence they proposed human-based post-processing to manually eliminate the noisy ones. The situation becomes even more confusing when C is finite, simply because we cannot tell whether examples with $\alpha_i = C$ are critical or noisy without other means.

In this paper, we propose to use the confidence margin $y_i\tilde{g}(\mathbf{x}_i)$ as the criterion for categorization, which works well with a suitable choice of finite C. The ideal thresholds, according to the relationship between the confidence margin and α_i, would be $t_n = 0$ and $t_c = 1$. For robustness, we use slightly different ad hoc values $t_n = 0.05$ and $t_c = 0.95$. We apply the popular Gaussian kernel with grid-based parameter search [6], and train an SVM with the best parameter to compute the confidence margin.

3.3 AdaBoost Data Weight

AdaBoost [7] is an algorithm to improve the accuracy of any base learner by iteratively generating a linear ensemble of base hypotheses. During its iterations, some examples are more likely to be misclassified than others, and are thus "hard" to learn [5]. AdaBoost maintains a set of weights for the training examples and gradually focuses on hard examples by giving them higher weights. At iteration t, the ensemble $\tilde{g}_t(\mathbf{x}) = \sum_{s=1}^{t} \alpha_s h_s(\mathbf{x})$ is constructed, where h_s is a base hypothesis and α_s is the coefficient for h_s. The data weight $w_i^{(t)}$, proportional to $e^{-y_i \tilde{g}_t(\mathbf{x}_i)}$, is thus tightly related to the ensemble confidence margin $y_i \tilde{g}_t(\mathbf{x}_i)$, and shows how hard it is to get an example correct at iteration t [7]. For instance, noisy examples tend to get misclassified a lot by base hypotheses and would have very large weights for most of the iterations; Typical examples, on the contrary, are almost always classified correctly and would have small weights. Thus, the average weight over different iterations can be used for data categorization.

Note that examples with smaller average weights are usually more reliable. For consistency with the other two methods, we actually use the negative of the average weight to approximate the intrinsic margin. For a set of size N, the initial weight is $1/N$. Thus, we use $t_c = -1.05/N$ and $t_n = -2.1/N$ to categorize examples with average weights slightly above the initial weight as critical, and examples with even higher average weights as noisy. We observe that these ad hoc thresholds work well under a common AdaBoost setting: 1000 iterations with the decision stump as the base learner.

Merler et al. [5] used the concept of *misclassification ratio,* the fraction of times an example is misclassified when it is not in the training set, to detect hard examples. Since they used resampling instead of reweighting in AdaBoost, in each iteration, the training set is sampled based on the data weight, and the misclassification ratio of an example is computed only when the example is not picked for training. This method however has a drawback that hard examples tend to have large weights and so are almost always picked in the training set. Thus the misclassification ratio is computed from a very small number of cases and is not very reliable, especially for critical and noisy examples.

We compared the misclassification ratio and the average data weight, and found that the average data weight is a better indicator of the intrinsic margin, as seen from Fig. 2.

4 Experiments with Artificial Data

We first test our methods for data categorization on three artificial targets (details in Appendix A), for which the intrinsic function is known. For each target, a dataset of size 400 is randomly generated, and the outputs of 40 examples (the last 10% indices) are further flipped as injected outliers. The ability of each method to capture the intrinsic margin is examined in two steps, the two-category experiments and the three-category experiments.

4.1 Two-Category Experiments

As mentioned previously, we try to construct a measure which is monotonic in the intrinsic margin. The scatter plots of two measures used in our methods versus the intrinsic margin (Fig. 1) show the overall monotonic relationship. However, we also observe that the monotonicity is not perfectly honored locally, and the degree of inconsistency depends on the dataset and the method used.

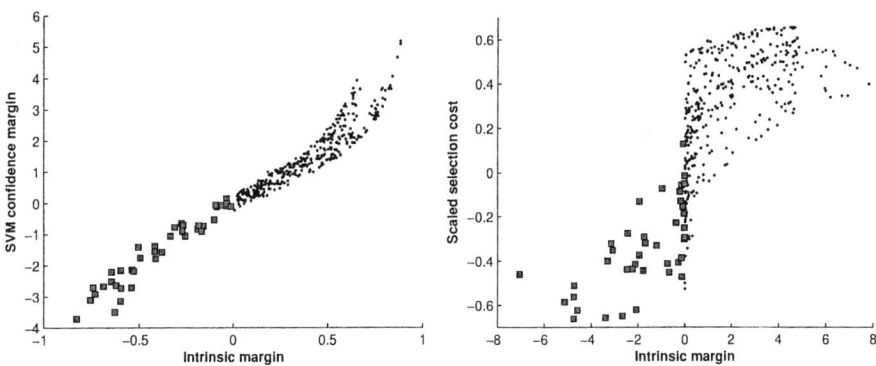

Fig. 1. Correlation between the measures and the intrinsic margin for the NNet dataset with the SVM confidence margin (**Left**) and the Sin dataset with the selection cost (**Right**). Noisy examples with negative intrinsic margins are shown as filled squares.

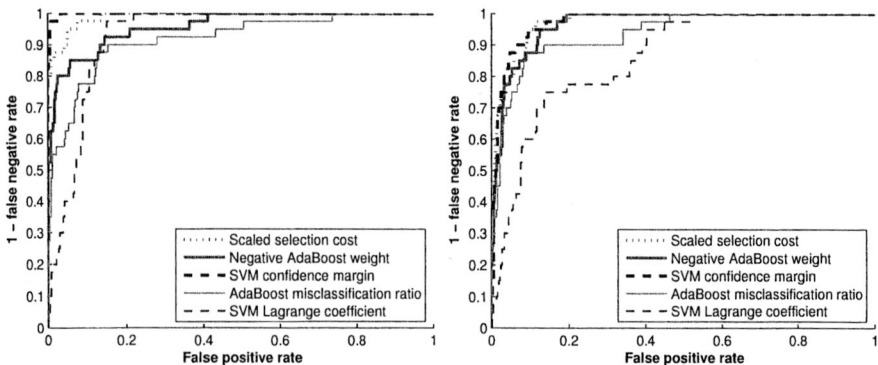

Fig. 2. ROC curves comparing the performance of all the methods on artificial datasets NNet (**Left**) and Sin (**Right**)

The scatter plots also show that our methods can reasonably group the data into noiseless (including typical and critical) and noisy categories, since the mislabeled examples are mostly cluttered in the bottom half of the plots. Figure 2 shows the receiver operating characteristic (ROC) curves for such categorization, where the false positive rate is the portion of noiseless examples being categorized as noisy. The ROC curves of two other methods, namely, the SVM Lagrange coefficient [3] and the AdaBoost misclassification ratio [5], are also plotted. These curves show that our methods, SVM confidence margin and AdaBoost data weight, surround larger area underneath and hence are much better than the related methods in literature for two-group categorization.

4.2 Three-Category Experiments

To visually study the nature of the data categorization obtained by the three methods, we design *fingerprint plots*, in which examples are positioned according to their intrinsic value $f_r(x_i)$ on the vertical axis and their index i in the dataset on the horizontal axis. Examples are also marked as typical, critical, and noisy, as assigned by the categorization method.

An ideal fingerprint plot would have the last 10% of the examples, which are mislabeled, categorized as noisy. The plot should also have a band of critical examples around the zero value. Figure 3 shows the fingerprint plot for the NNet dataset with the selection cost as the categorization method. Since the target function is in the model for estimating the selection cost, the categorization is near perfect. Figure 4 shows fingerprint plots for two other cases, where we do not have perfect categorization, and some of the critical examples are categorized as outliers and vice versa. This is partly due to the ad hoc thresholding used to categorize the examples. Similar results are obtained for the other combinations of dataset and method.

Figure 5 shows the categorization results for the two 2-D datasets visually. First, we notice that almost all the mislabeled examples are detected as noisy

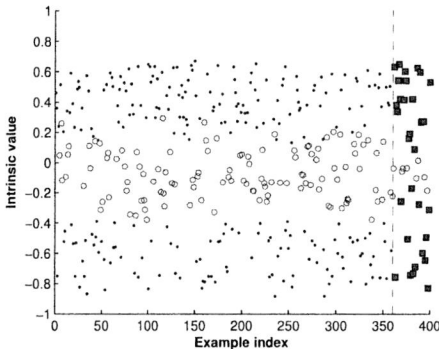

Fig. 3. Fingerprint plot of the NNet dataset with the selection cost. Critical and noisy examples are shown as empty circles and filled squares, respectively.

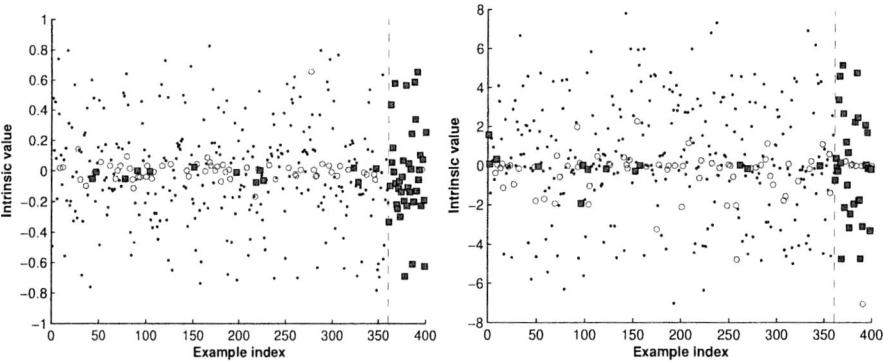

Fig. 4. Fingerprint plots of the Yin-Yang dataset with the SVM confidence margin (**Left**), and the Sin dataset with the AdaBoost data weight (**Right**)

(shown as ■), while very few of them are wrongly categorized as critical (□). Some clean examples, mostly those around the decision boundary, are also categorized as noisy (•), partly explained with the ad hoc thresholding. Secondly, we can see that most of the identified critical examples (○ or □) are around the boundary or the outliers, which is desired since examples there do provide critical information about the target.

5 Real-World Data

When the dataset has been categorized, it is possible to treat different data categories differently in learning. For example, we can remove the noisy examples and also emphasize the critical examples, and intuitively this would help

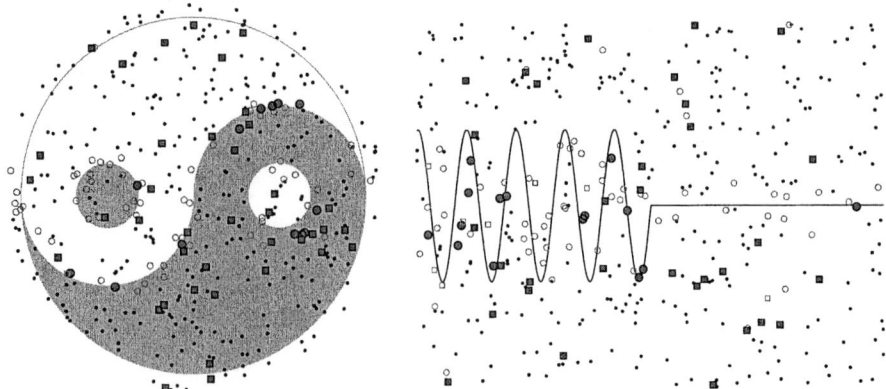

Fig. 5. 2-D categorization with the SVM confidence margin on artificial datasets Yin-Yang (**Left**) and Sin (**Right**). The 10% mislabeled examples are shown in squares and the 90% correctly labeled ones are shown in dots or circles. The three categories are shown as dots (typical), empty circles or squares (critical), and filled circles or squares (noisy).

learning.[2] To demonstrate that such a simple intuition based on our data categorization methods can be quite useful in practice, we carry out experiments on seven datasets[3] from the UCI machine learning repository [8]. Their input features are normalized to the range $[-1, 1]$. Each dataset is randomly split into training and testing parts with 60% of the data for training and rest for testing. The data categorization methods[4] are applied to the training set, and noisy examples are then removed. We further emphasize the critical examples by giving them twice the weight of the typical ones. A 500-iteration AdaBoost of decision stumps is used to learn both the full training set and the filtered one. The test error averaged over 100 runs is reported in Table 1 together with its standard error. It is observed that removing the noisy examples and emphasizing the critical ones almost always reduces the test error. The exceptions with the selection cost method should be due to a possible mismatch of complexity between the underlying model of the selection cost and the learning model used in AdaBoost.

Although details are not included here, we also observed that the test error, when we just removed the noisy examples, was not statistically different from

---

[2] We do not flip the noisy examples since the categorization may not be perfect. If a noiseless example is marked as noisy, flipping it brings a relatively high risk. So removing the noisy examples would be a safer choice.

[3] They are australian (Statlog: Australian Credit Approval), breast (Wisconsin Breast Cancer), cleveland (Heart Disease), german (Statlog: German Credit), heart (Statlog: Heart Disease), pima (Pima Indians Diabetes), and votes84 (Congressional Voting Records), with incomplete records removed.

[4] Note that the feed-forward neural networks for estimating the selection cost have one hidden layer of 15 neurons.

Table 1. Test error (%) of AdaBoost with 500 iterations

dataset	full dataset	selection cost	SVM margin	AdaBoost weight
australian	16.65 ± 0.19	15.23 ± 0.20	14.83 ± 0.18	13.92 ± 0.16
breast	4.70 ± 0.11	6.44 ± 0.13	3.40 ± 0.10	3.32 ± 0.10
cleveland	21.64 ± 0.31	18.24 ± 0.30	18.91 ± 0.29	18.56 ± 0.30
german	26.11 ± 0.20	30.12 ± 0.15	24.59 ± 0.20	24.68 ± 0.22
heart	21.93 ± 0.43	17.33 ± 0.34	17.59 ± 0.32	18.52 ± 0.37
pima	26.14 ± 0.20	35.16 ± 0.20	24.02 ± 0.19	25.15 ± 0.20
votes84	5.20 ± 0.14	6.45 ± 0.17	5.03 ± 0.13	4.91 ± 0.13

the case when we also emphasized the critical examples. However, we found that removing the critical examples almost always increased the test error, and removing as much as 50% of the typical examples did not affect the test error by much. This clearly shows the distinction between the three categories.

6 Conclusion

We proposed the concept of grouping data into typical, critical, and noisy categories according to the intrinsic margin, and presented three methods to automatically carry out the categorization. The three methods, rooted from different parts of learning theory, are quite different in the models they use and the way they approximate the intrinsic margin. However, they still gave similar categorization results on three artificial datasets, which established that the concept is independent of the methods. The categorization results can be used in conjunction with a large variety of learning algorithms for improving the generalization. The results on the UCI datasets with AdaBoost as the learning algorithm demonstrated the applicability of the methods in real-world problems. In addition, the categorization can also be used to reduce the dataset size without affecting the learning performance.

Further work needs to be done to estimate the optimal thresholds from the dataset (say, using a validation set [5]), to better utilize the categorization in learning, and to extend the framework for regression problems.

A Artificial Targets

We used three artificial target functions in the paper.

3-5-1 NNet. This is a feed-forward neural network with 3 inputs, 5 neurons in the hidden layer, and 1 output neuron. All neurons use tanh (sigmoid) as the transfer function. The weights and thresholds are randomly picked with a Gaussian distribution $\mathcal{N}(0, 0.7^2)$. The continuous output from the output neuron is used as the intrinsic value f_r.

Yin-Yang. A round plate centered at $(0,0)$ in $\mathbb{R}^2$ is partitioned into two classes (see Fig. 5). The "Yang" (white) class includes all points (x_1, x_2) that satisfy

$$(d_+ \leq r) \vee \left(r < d_- \leq \tfrac{R}{2}\right) \vee \left(x_2 > 0 \wedge d_+ > \tfrac{R}{2}\right),$$

where the radius of the plate is $R = 1$, the radius of two small circles is $r = 0.18$, $d_+ = \sqrt{(x_1 - \tfrac{R}{2})^2 + x_2^2}$, and $d_- = \sqrt{(x_1 + \tfrac{R}{2})^2 + x_2^2}$. Points out of the plate belong to the Yang class if its $x_2 > 0$. For each example, we use its Euclidean distance to the nearest boundary as its intrinsic margin.

Sin. The Sin target in [5] is also used in this paper (see Fig. 5). It partitions $[-10, 10] \times [-5, 5]$ into two class regions, and the boundary is

$$x_2 = \begin{cases} 2\sin 3x_1, & \text{if } x_1 < 0; \\ 0, & \text{if } x_1 \geq 0. \end{cases}$$

As in the Yin-Yang target, the distance to the nearest boundary is used as the intrinsic margin.

Acknowledgment

We thank Anelia Angelova, Marcelo Medeiros, Carlos Pedreira, David Solove- ichik and the anonymous reviewers for helpful discussions. This work was mainly done in 2003 and was supported by the Caltech Center for Neuromorphic Systems Engineering under the US NSF Cooperative Agreement EEC-9402726.

References

1. Hodge, V.J., Austin, J.: A survey of outlier detection methodologies. Artificial Intelligence Review **22** (2004) 85–126
2. Vapnik, V.N.: The Nature of Statistical Learning Theory. Springer-Verlag, Berlin (1995)
3. Guyon, I., Matić, N., Vapnik, V.: Discovering informative patterns and data cleaning. In Fayyad, U.M., Piatetsky-Shapiro, G., Smyth, P., Uthurusamy, R., eds.: Advances in Knowledge Discovery and Data Mining. AAAI Press / MIT Press, Cambridge, MA (1996) 181–203
4. Nicholson, A.: Generalization Error Estimates and Training Data Valuation. PhD thesis, California Institute of Technology (2002)
5. Merler, S., Caprile, B., Furlanello, C.: Bias-variance control via hard points shaving. International Journal of Pattern Recognition and Artificial Intelligence **18** (2004) 891–903
6. Hsu, C.W., Chang, C.C., Lin, C.J.: A practical guide to support vector classification. Technical report, National Taiwan University (2003)
7. Freund, Y., Schapire, R.E.: Experiments with a new boosting algorithm. In: Machine Learning: Proceedings of the Thirteenth International Conference. (1996) 148–156
8. Hettich, S., Blake, C.L., Merz, C.J.: UCI repository of machine learning databases (1998) Downloadable at http://www.ics.uci.edu/~mlearn/MLRepository.html.

Mining Model Trees from Spatial Data

Donato Malerba, Michelangelo Ceci, and Annalisa Appice

Dipartimento di Informatica, Università degli Studi di Bari,
via Orabona, 4 - 70126 Bari - Italy
{malerba, ceci, appice}@di.uniba.it

Abstract. Mining regression models from spatial data is a fundamental task in Spatial Data Mining. We propose a method, namely Mrs-SMOTI, that takes advantage from a tight-integration with spatial databases and mines regression models in form of trees in order to partition the sample space. The method is characterized by three aspects. First, it is able to capture both spatially global and local effects of explanatory attributes. Second, explanatory attributes that influence the response attribute do not necessarily come from a single layer. Third, the consideration that geometrical representation and relative positioning of spatial objects with respect to a reference system implicitly define both spatial relationships and properties. An application to real-world spatial data is reported.

1 Introduction

The rapidly expanding market for spatial databases and Geographic Information System (GIS) technologies is driven by the pressure from the public sector, environmental agencies and industries to provide innovative solutions to a wide range of data intensive applications that involve spatial data, that is, a collection of (spatial) objects organized in thematic layers (e.g., enumeration districts, roads, rivers). A thematic layer is characterized by a geometrical representation (e.g., point, line, and polygon in 2D) as well as several non-spatial attributes (e.g., number of inhabitants), called thematic attributes. A GIS provides the set of functionalities to adequately store, display, retrieve and manage both geometrical representation and thematic attributes collected within each layer and stored in a spatial database. Anyway, the range of GIS applications can be profitably extended by adding spatial data interpretation capabilities to the systems. This leads to a generation of GIS including Spatial Data Mining (SDM) facilities [11].

Spatial Data Mining investigates how interesting and useful but implicit knowledge can be extracted from spatial data [8]. Regression is a fundamental task of SDM where the goal is to mine a functional relationship between a continuous attribute Y_i (*response attribute*) and m continuous or discrete attributes $X_{j,i}$ $j = 1, ..., m$ (*explanatory attributes*). The training sample consists of spatial objects. For instance, for UK census data available at the level of Enumeration Districts (EDs), a possible goal may be estimating the response attribute "number of migrants" associated to each ED i on the basis of explanatory attributes $X_{j,i}$ (e.g., "number of inhabitants") associated to EDs.

The simplest approach to mine regression models from spatial data, is based on standard regression [18] that models a functional relationship in the form: $Y_i = \beta_0 + \beta_1 X_{1,i} + \ldots + \beta_k X_{k,i}$, where i is each ED area. The main problem with this model is that it disregards the arrangement properties due to spatial structure of data [5] (e.g., the phenomenon of migration is typically stronger in peripheral EDs). When spatially-dependent heterogeneity of the model can be anticipated by the analyst, the model can be improved by introducing a dummy variable $D_i \in \{0, 1\}$, which differentiate the behavior of the model according to a predefined partitioning of areas in two groups. In this way, the model is either $Y_i = \beta_0 + \beta_1 X_{1,i} + \ldots + \beta_k X_{k,i} + \gamma D_i$ (constant spatial variation) or $Y_i = \beta_0 + (\beta_1 + \gamma D_i)X_{1,i} + \ldots + \beta_k X_{k,i}$ (regression parameter spatial variation). However, when the areas of homogeneous dependence cannot be anticipated by the expert, a solution is represented by model trees [16] that approximate a piece-wise (linear) function by means of a tree structure, where internal nodes partition the sample space (as decision trees), while leaves are associated to (linear) functions. In this way, it is possible to automatically determine different regression model for different areas.

In this paper, we propose the model tree induction method, namely Mrs-SMOTI (Multi-relational Spatial Stepwise Model Tree Induction), that faces several degrees of complexity which characterize the regression problem from spatial data. In the next section, we discuss these problems and introduce our solution. Section 3 presents a stepwise approach to mine spatial regression models. Section 4 focuses on spatial database integration. Finally, an application is presented in Section 5 and some conclusions are drawn.

2 Spatial Regression: Background and Motivations

While model tree learning has been widely investigated in the data mining literature [16,10,17], as far as we know, no attention has been given to the problem of mining model trees from spatial data. Model tree induction from spatial data raises several distinctive problems: i) some explanatory attributes can have spatially global effect on the response attribute, while others have only a spatially local effect; ii) explanatory attributes that influence the response attribute not necessarily come from a single layer, but in most of cases they come from layers possibly spatially related with the layer that is the main subject of the analysis; iii) geometrical representation and relative positioning of spatial objects with respect to some reference system implicitly define both spatial relationships (e.g., "intersects", "distance") and spatial attributes (e.g., "area", "direction").

Concerning the first point, it would be useful to identify the global effect of some attributes (possibly) according to the space arrangement of data. Indeed, in almost all model trees induction methods, the regression model associated with a leaf is built on the basis of those training cases falling in the corresponding partition of the feature space. Therefore, models in the leaves have only a local validity and do not consider the global effects that some attributes might have in the underlying model. In model trees, global effects can be represented by

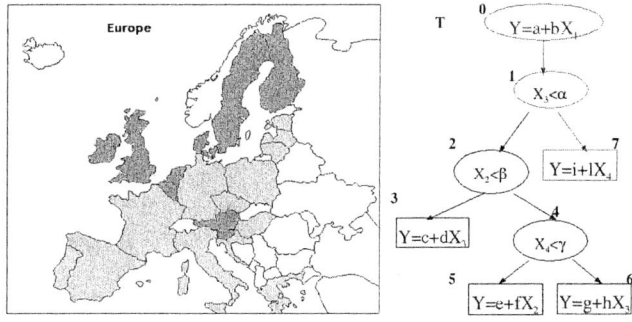

Fig. 1. An example of spatial model tree with regression and splitting nodes. Node 0 is a regression node that captures a global effect between the unemployed rate (Y) and the GDP per capita (X_1). It is associated to all countries. Node 1 splits the sample space as depicted in the map. Functions at leaves only capture local effects.

attributes that are introduced in the linear models at higher levels of the tree. This requires a different tree-structure where internal nodes can either define a partitioning of the sample space or introduce some regression attributes in the linear models to be associated to the leaves. In our previous work [10], we proposed the method SMOTI whose main characteristic is the construction of trees with two types of nodes: splitting nodes, which partition the sample space, and regression nodes, which perform only straight-line regression. The multiple model associated to a leaf is built stepwise by combining straight-line regressions along the path from the root to the leaf. In this way, internal regression nodes contribute to the definition of multiple models and capture global effects, while straight-line regressions at leaves capture only local effects. Detecting global and local effects over spatial data, allows to model phenomena, that otherwise, would be ignored. As an example we show a simplistic case: suppose we are interested in analyzing the unemployed rate in EU. In this case, it may be found that the unemployed rate of each country is proportional to its GDP (Gross Domestic Product) per capita. This behavior is independent of the specific country and represents a clear example of global effect. This global effect corresponds to a regression node in higher levels of the tree (see Fig. 1).

The second point enlightens that the value of the response attribute may go beyond the values of explanatory attributes of the spatial object to be predicted. In particular, it is possible that the response attribute depends on the attribute values of objects spatially-related to the object to be predicted and possibly belonging to a different layer. In this point of view, the response attribute is associated to the spatial objects that are the main subjects of the analysis (target objects) while each explanatory attribute refers either to the target objects to be predicted or to the spatial objects that are relevant for the task in hand and are spatially related to the target ones (non-target objects). This is coherent with the idea of exploiting intra-layer and inter-layer relationships when mining spatial data [1]. Intra-layer relationships describe a spatial interaction between two spatial objects belonging to the same layer, while inter-layer relationships describe a spatial interaction between two spatial objects belonging to different layers. According to [5], intra-layer relationships make available both spatially-lagged explanatory attributes useful when the effect of an explanatory attribute at any site is not limited to the specified site (e.g., the proportion of people suffering from respiratory diseases in an ED also depends on the high/low level of pollution of EDs where people daily move) and spatially lagged response attribute, that is, when autocorrelation affects the response values (e.g., the price for a good at a retail outlet in a city may depend on the price of the same good sold by local competitors). Differently, inter-layer relationships model the fact that the response attribute value observed from some target object may depend on explanatory attributes observed at spatially related non target objects belonging to different layers. For instance, if the "EDs" layer is the subject of the analysis and the response attribute is the mortality rate associated to an ED, mortality rate may depend on the air-pollution degree on crossing roads. Although spatial regression systems (such as the R spatial project - http://sal.uiuc.edu/csiss/Rgeo/index.html) are able to deal with user defined intra-layer spatial relationships, they ignore inter-layer relationships that can be naturally modeled by resorting to the multi-relational setting [4].

The third point is due to the fact that geometrical representation and relative positioning of spatial objects with respect to some reference system implicitly define both spatial relationships and spatial attributes . This implicit information is often responsible for the spatial variation over data and it is extremely useful in modelling [15]. Hence, spatial regression demands for the development of specific methods that, differently from traditional ones, take the spatial dimension of the data into account when exploring the spatial pattern space. In this way, thematic and spatial attribute values of target objects and spatially related non-target objects are involved in predicting the value of the response attribute.

The need of extracting and mining the information that is implicitly defined in spatial data motivates a tight-integration between spatial regression method and spatial database systems where some sophisticated treatment of real-world geometry is provided for storing, indexing and querying spatial data. This is confirmed by the fact that spatial operations (e.g., computing the topological relationships among spatial objects) are available free of charge for data analysts in several spatial database advanced facilities [6].

In this work, we present Mrs-SMOTI that extends SMOTI by taking advantage of a tight integration with a spatial database in order to mine stepwise a spatial regression model from multiple layers. The model is built taking into account all three degrees of complexity presented above.

3 Stepwise Mining of a Spatial Regression Model

Mrs-SMOTI mines a spatial regression model by partitioning training spatial data according to intra-layer and inter-layer relationships and associating different regression models to disjoint spatial areas. In particular, it mines spatial data and performs the stepwise construction of a tree-structured model with both splitting nodes and regression nodes until some stopping criterion is satisfied. In this way, it faces the spatial need of distinguishing among explanatory attributes that have some global effect on the response attribute and others that have only local effect. Both splitting and regression nodes may involve several layers and spatial relationships among them.

Spatial split. A spatial splitting test involves either a *spatial relationship condition* or a *spatial attribute condition* on some layer from S. The former partitions target objects according to some spatial relationship (either intra-layer or inter-layer). For instance, when predicting the proportion of people suffering from respiratory diseases in EDs, it may be significant to mine a different regression function according to the presence or absence of main roads crossing the territory. An extra-consequence of performing such spatial relationship condition concerns the introduction of another layer in the model. The latter is a test involving a boolean condition ("$X \leq \alpha$ vs. $X > \alpha$" in the continuous case and "$X \in \{x_1, \ldots, x_k\}$ vs. $X \notin \{x_1, \ldots, x_k\}$" in the discrete one) on a thematic attribute X of a layer already included in the model. In addition to thematic attributes, an attribute condition may involve a spatial property (e.g., the area for polygons and the extension for lines), that is implicitly defined by the geometrical structure of the corresponding layer in S. It is noteworthy that only spatial relationship conditions add new layers of S to the model. Consequently, a split on a thematic attribute or spatial property involves a layer already introduced in the model. However, due to the complexity of computing spatial relationships, we impose that a relationship between two layers can be introduced at most once in each unique path connecting the root to the leaf.

Coherently with [10], the validity of a spatial splitting test is based on an heuristic function $\sigma(t)$ that is computed on the attribute-value representation of the portion of spatial objects in S falling in t_L and t_R, that is, the left and right child of the splitting node t respectively. This attribute-value representation corresponds to the tuples of S derived according to both spatial relationship conditions and attribute conditions along the path from the root of the tree to the current node. We define $\sigma(t) = (n(t_L)/(n(t_L) + n(t_R)))R(t_L) + (n(t_R)/(n(t_L) + n(t_R)))R(t_R)$, where $n(t_L)$ $(n(t_R))$ is the number of attribute-value tuples passed down to the left (right) child. Since intra-layer and inter-layer relationships lead to a regression model that may include several layers (not necessarily separate), it may happen that $n(t) \neq n(t_L) + n(t_R)$ although the split in t satisfies the mutual exclusion requirement. This is due to the many-to many nature of intra-layer and inter-layer relationships. In fact, when several spatial objects are spatially related to the same object (e.g., a single ED may be intersected by zero, one or more roads), computing spatial relationships may return a number of attribute-value tuples greater than one. $R(t_L)$ $(R(t_R))$ is the Minimum Squared Error (MSE) computed on the left (right) child t_L (t_R) as follows:

$$R(t_L) = \sqrt{\frac{1}{n(t_L)} \sum_{i=1\ldots n(t_L)} (y_i - \widehat{y}_i)^2} \quad (R(t_R) = \sqrt{\frac{1}{n(t_R)} \sum_{i=1\ldots n(t_R)} (y_i - \widehat{y}_i)^2}),$$

such that $\hat{y}_i$ is the response value predicted according to the spatial regression model built by combining the best straight-line regression associated to t_L (t_R), with all straight-line regressions in the path from the root to t_L (t_R) [3].

Spatial regression. A spatial regression node performs a straight-line regression on either a continuous thematic attribute or a continuous spatial property not yet introduced in the model currently built. Coherently with the stepwise procedure [3], both response and explanatory attributes are replaced with their residuals. For instance, when a regression step is performed on a continuous attribute X, the response attribute is replaced with the residual $Y' = Y - \hat{Y}$, where $\hat{Y} = \hat{\alpha} + \hat{\beta}X$. The regression coefficients $\hat{\alpha}$ and $\hat{\beta}$ are estimated on the attribute-value representation of the portion of S falling in the current node.

According to the spatial structure of data, the regression attribute comes from one of the layers already involved in the model. Continuous thematic and spatial attributes of these layers, which have not yet been introduced in the model, are replaced with the corresponding residuals in order to remove the effect of the regression attribute. Whenever a new layer is added to the model (by means of a spatial relationship condition), continuous thematic and spatial attributes, introduced with it, are replaced with the corresponding residuals. Residuals are contextually computed on the attribute-value representation of the portion of S falling in the current node. In this way, the effect of regression attributes previously introduced in the model by regression steps is also removed by introduced attributes.

The evaluation of a spatial regression step $\hat{Y} = \hat{\alpha} + \hat{\beta}X$ is based on the heuristic function $\rho(t)$, that is: $\rho(t) = \min\{R(t), \sigma(t')\}$, where t' is the best spatial splitting node following the regression step in t. This look-ahead step involved in the heuristic function above depends on the fact that spatial split looks for best straight-line regression after the split condition is performed, while the regression step does not. A fairer comparison would be growing the tree at a further level to base the computation of $\rho(T)$ on the best multiple linear regressions after the regression step on X_i is performed [10].

Stopping criteria. Three different stopping criteria are implemented. The first requires that a minimal number of target objects fall in current node. The second stops the induction process when the coefficient of determination is greater than a threshold [18]. This coefficient is a scale-free one-number summary of the strength of the relation between explanatory attributes in the actual multiple model and the response attribute. Finally, the third stops the induction process when no further regression step can be performed (i.e. all continuous attributes are included in the current model) also after introducing some new layer.

4 Spatial Database Integration

Most spatial data mining systems process data in main memory. This results in high performance for computationally intensive processes when enough memory is available to store all necessary data. However, in spatial data intensive processes it is important to exploit powerful mechanisms for accessing, filtering and indexing data, such as those available in spatial DBMS (DataBase Management Systems). For instance, spatial operations (e.g., computing the topological relationships among spatial objects) supported by any spatial DBMS take advantage from spatial indexes like Quadtrees or Kd-tree [14]. This motivates a tight integration of spatial data mining systems and spatial DBMS in order to i) guarantee the applicability of spatial data mining algorithms to large spatial datasets; ii) exploit useful knowledge of spatial data model available, free of charge, in the spatial database, iii) specify directly what data stored in a database have to be mined, iv) avoid useless preprocessing leading to redundant data storage that may be unnecessary when part of space of the hypothesis may be never explored.

Some examples of integrating spatial data mining and spatial database system are presented in [11] for classification tasks and in [1] for association rules discovery tasks. In both cases, a data mining algorithm working in first-order logic is only loosely integrated with a spatial database by means of some middle layer module that extracts spatial attributes and relationships independently from the mining step and represents these features in a first-order logic formalism. Thus, data mining algorithms are practically applied to preprocessed data and this preprocessing is user-controlled. Conversely, in [6] a spatial data mining system, named SubgroupMiner, is proposed for the task of subgroup discovery in spatial databases. Subgroup discovery is here approached by taking advantage from a tight integration of the data mining algorithm with the database environment. Spatial relationships and attributes are then dynamically derived by exploiting spatial DBMS extension facilities (e.g., packages, cartridges or extenders) and used to guide the subgroup discovery.

Following the inspiration of SubgroupMiner, we assume an object-relational (OR) data representation, such that spatial patterns representing both splitting and regression nodes are expressed with spatial queries. These queries include spatial operators based on the non-atomic data type for geometry consisting in an ordered set of coordinates (X, Y) representing points, lines and polygons. Since no spatial operator is present in basic relational algebra or Datalog, we resort to an extension of the OR-DBMS Oracle Spatial Cartridge $9i$ where spatial operators to compute spatial relationships and to extract spatial attributes are made available free of charge [6]. These operators can be called in SQL queries. For example: *SELECT * FROM EDs x, Roads y   WHERE SDO_GEOM.*

RELATE(x.geometry,'ANYINTERACT',y.geometry, 0.001) = 'TRUE'
This spatial query retrieves the pairs ⟨ED, Road⟩ whose topological relationship is "not disjoint" by means or the Oracle operator "RELATE". It is noteworthy that, the use of such SQL queries, appears to be more direct and much more practical than formulating non-trivial extension of relational algebra or Datalog such that those provided in constraint database framework [9].

When running a spatial query (associated to a node of the tree), the result is a set of tuples describing both thematic attributes and spatial attributes of involved layers. The FROM clause includes layers (not necessarily different) in the model at the current node. The WHERE clause includes split conditions found along the path from the root to the current node. The negation of either a

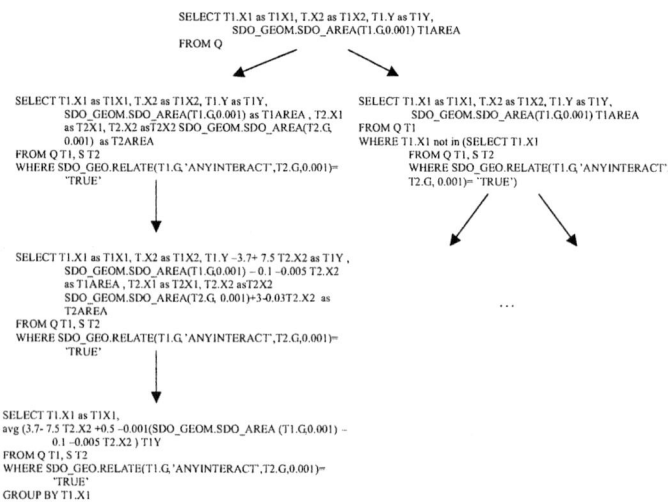

Fig. 2. An example of spatial model tree with regression, splitting and leaf nodes expressed by means of spatial queries assuming that training data are stored in spatial layers (e.g., Q and R) of a spatial database

spatial relationship condition or an attribute condition involving some attribute of a non-target layer is transformed into a negated nested spatial sub-query. This is coherent with the semantic of tests involving multiple tables of a relational database [2]. Finally, the SELECT clause includes thematic and spatial attributes (or their residuals) from the layers involved in the WHERE clause.

Leaf nodes are associated with aggregation spatial queries, that is, spatial queries where all tuples referring the same target object are grouped together. In this way, the prediction of the response variable is the average response value predicted on the set of attribute-value tuples describing the unique target object to be predicted. This means that spatial model trees can be expressed in form of a set of SQL spatial queries (see Fig. 2). Queries are stored in XML format that can be subsequently used for predicting (unknown) response attributes.

5 Spatial Regression on Stockport Census Data

In this section we present a real-world application concerning the mining of spatial regression models. We consider both 1991 census and digital map data provided in the context of the European project SPIN! (Spatial Mining for Data of Public Interest) [12]. This data concerns Stockport, one of the ten metropolitan districts in Greater Manchester (UK) which is divided into twenty-two wards for a total of 589 census EDs. Spatial analysis is enabled by the availability of vectorized boundaries for 578 Stockport EDs as well as by other Ordnance Survey digital maps of UK. Data are stored in an Oracle Spatial Cartridge $9i$ database.

The application in this study investigates the number of unemployed people in Stockport EDs according to the number of migrant people available for each ED in census data as well as geographical factors represented in topographic maps stored in form of layers. The target objects are the Stockport EDs, while other layers, such as, shopping (53 objects), housing (9 objects) and employment areas (30 objects) are the non target objects. The EDs play the role of both target objects and non target objects when considering intra-layer relationship on EDs.

Two experimental settings are defined. The first setting (BK_1) is obtained by exclusively considering the layer representing EDs. The second setting (BK_2) is obtained by considering all the layers. In both settings, intra-layer relationships on EDs make possible to model the unemployment phenomenon in Stockport EDs by taking into account the self-correlation on the spatially lagged explanatory attributes of EDs. The auto-correlation on the spatially-lagged response attribute can be similarly exploited during the mining process. In this study, we consider (intra-layer and inter-layer) spatial relationships that describe some (non disjoint) topological interaction between spatial objects. Furthermore, we consider area of polygons and extension of lines as spatial properties.

In order to prove the advantage of using intra-layer and inter-layer relationships in the mining process, we compare the spatial regression model mined by Mrs-SMOTI with the regression models mined by SMOTI and M5'[17]. Since SMOTI and M5' work under single table assumption, we transform the original object-relational representation of Stockport data in a single relational table format. Two different transformations are considered. The former (P1) creates a single table by deriving all thematic and spatial attributes from layers according to all possible intra-layer and inter-layer relationships. This transformation leads to generate multiple tuples for the same target object. The latter transformation (P2) differs from the previous one because it does not generate multiple tuples for the same target object. This is obtained by including aggregates (i.e., the average for continuous values and the mode for discrete values)[7] of the attributes describing the non target objects referring to the same target object[1].

Model trees are mined by requiring that the minimum number of spatial target objects falling in an internal node must be greater than the square root of the number of training target objects, while the coefficient of determination must be below 0.80. Comparison is performed on the basis of the average MSE, number of regression nodes and leaves obtained by means of the same five-fold cross validation of Stockport data. Results are reported in Table 1.

The non-parametric Wilcoxon two-sample paired signed rank test [13] is used for the pairwise comparison of methods. In the Wilcoxon signed rank test, the summations on both positive (W+) and negative (W-) ranks determine the winner. Results of Wilcoxon test are reported in Table 2.

---

[1] In both P1 and P2 transformations the attribute-value dataset is composed by 5 attibutes for BK_1 (6 when including the lagged response) and 11 for BK_2 (12 when including the lagged response). The number of tuples for P1 is 4033 for BK_1 and 4297 for BK_2. In the case of P2, the number of tuples is 578 in both settings.

Table 1. Average MSE, No. of leaves and regression nodes of trees induced by Mrs-SMOTI, SMOTI and M5'. L1 is "No lagged response", L2 is "Lagged response".

Setting		MSE				Leaves				RegNodes			
		BK1		BK2		BK1		BK2		BK1		BK2	
		L1	L2	L1	L2	L1	L2	L1	L2	L1	L2	L1	L2
Mrs-SMOTI		12.34	13.74	11.99	10.92	19.80	23.40	23.60	23.60	3.4	6.6	3.8	6.2
SMOTI	P1	12.91	10.23	20.11	13.0	101.6	107.6	104.0	111.8	6.2	5.0	15.0	11.4
	P2	11.89	18.17	19.71	15.80	41.00	24.80	42.40	44.20	3.4	4.0	10.2	11.6
M5'	P1	13.52	12.41	12.92	12.30	433.6	872.0	408.6	711.2	-	-	-	-
	P2	12.44	9.19	12.48	9.59	198.0	199.4	199.2	197.4	-	-	-	-

Results confirm that Mrs-SMOTI is better or at worst comparable to SMOTI in terms of predictive accuracy. This result is more impressive when we consider the regression model mined when both intra-layer and inter-layer relationships are ignored. The average MSE of model trees mined by SMOTI taking into account only the number of migrants and the area of EDs is 15.48.

Moreover, when we consider results of SMOTI on data transformed according to P1 and P2, we note that the stepwise construction takes advantage of the tight-integration of Mrs-SMOTI with the spatial DBMS that avoids the generation of useless features (relationships and attributes). The side effect of useless features may lead to models that overfit training data, but fail in predicting new data. In a deeper analysis, we note that even when SMOTI, in average, outperforms Mrs-SMOTI in terms of MSE, the Wilcoxon test does not show any statistically significant difference. Results on the two data settings show that mining the geographical distribution of shopping (housing or employment) areas over EDs (i.e., the spatial relationships between EDs and shopping areas, shopping areas and shopping areas, shopping areas and employment areas, and so on) decreases the average MSE of models mined by Mrs-SMOTI, while no significant improvement is observed in mining the same information with SMOTI. The autocorrelation on the response improves performance of Mrs-SMOTI only for BK_2 level (10.92 vs. 11.99) without significantly increasing tree size.

Table 2. Mrs-SMOTI vs SMOTI and M5': results of the Wilcoxon test on the MSE of trees. If W+ ≤ W- then results are in favour of Mrs-SMOTI. The statistically significant values ($p \leq 0.1$) are in boldface. L1 is "No lagged response", L2 is "Lagged response".

Setting		Mrs-SMOTI vs. SMOTI P1			Mrs-SMOTI vs. SMOTI P2			Mrs-SMOTI vs. M5' P1			Mrs-SMOTI vs. M5' P2		
		W+	W-	p	W+	W-	p	W+	W-	p	W+	W-	p
BK1	L1	6	9	0.81	9	6	0.81	3	12	0.310	7	8	1.000
	L2	10	5	0.63	6	9	0.81	8	7	1.000	15	0	**0.060**
BK2	L1	1	14	**0.125**	0	15	**0.06**	4	11	0.430	6	9	0.810
	L2	0	15	**0.06**	3	12	0.31	0	15	**0.060**	15	0	**0.060**

```
- split on EDs' number of migrants [≤ 47] (578 EDs)
   - regression on EDs' area (458 EDs)
      - split on EDs - Shopping areas spatial relationship (458 EDs)
         - split on Shopping areas' area (94 EDs) ...
         - split on EDs' number of migrants (364 EDs) ...
   - split on EDs' area (120 EDs)
      - leaf on EDs' area (22 EDs)
      - regression on EDs' area (98 EDs) ...
```

Fig. 3. Top-level description of a portion of the model mined by Mrs-SMOTI on the entire dataset at BK_2 level with no spatially lagged response attributes

The number of regression nodes and leaves are indicators of the complexity of the induced regression models. In this case, results show that the model induced by Mrs-SMOTI is much simpler than the model induced by SMOTI in both settings independently from data transformation. The relative simplicity of the spatial regression models mined by Mrs-SMOTI makes them easily to be interpreted. In particular, the tree structure can be easily navigated in order to distinguish among global and local effects of explanatory attributes. For instance, in Fig. 3 it is shown the top-level description of the spatial regression model mined by Mrs-SMOTI on the entire dataset at BK_2 level with no spatially lagged response attributes. Mrs-SMOTI captures the global effect of the area of EDs over Stockport covered by the 458 EDs having "number of migrants ≤ 47". The effect of this regression is shared by all nodes in the corresponding sub-tree.

Finally, the comparison of Mrs-SMOTI with M5', does not show any clear difference in terms of MSE. Anyway, M5' presents two important disadvantages with respect to Mrs-SMOTI. First, M5' cannot capture spatial global and local effects. Second, mined model trees cannot be interpreted by humans because of the complexity of the models (there is an increase of one order of magnitude in the number of leaves from Mrs-SMOTI to M5')

6 Conclusions

In this paper we have presented a spatial regression method Mrs-SMOTI that is able to capture both spatially global and local effects of explanatory attributes. The method extends the stepwise construction of model trees performed by its predecessor SMOTI in two directions. First, by taking advantage from a tight-integration with a spatial database in order to mine both spatial relationships and spatial attributes which are implicit in spatial data. Indeed, this implicit information is often responsible for the spatial variation over data and it is extremely useful in regression modelling. Second, the search strategy is modified in order to mine models that capture the implicit relational structure of spatial data. This means that spatial relationships (intra-layer and inter-layer) make possible to consider explanatory attributes that influence the response attribute but do not necessarily come from a single layer. In particular, intra-layer relationships make available spatially lagged response attributes in addition to spatially lagged explanatory attributes.

Experiments on real-world spatial data show the advantages of the proposed method with respect to SMOTI. As future work, we intend to extend the method in order to mine both geometrical (e.g., distance) and directional (e.g., north of) relationships in addition to topological relationships.

Acknowledgment

This work is partial fulfillment of the research objective of ATENEO-2005 project "Gestione dell'informazione non strutturata: modelli, metodi e architetture".

References

1. A. Appice, M. Ceci, A. Lanza, F. A. Lisi, and D.Malerba. Discovery of spatial association rules in georeferenced census data: A relational mining approach. *Intelligent Data Analysis*, 7(6):541–566, 2003.
2. A. Appice, M. Ceci, and D. Malerba. Mining model trees: A multi-relational approach. In T. Horvath and A. Yamamoto, editors, *Proceedings of ILP 2003*, volume 2835 of *LNAI*, pages 4–21. Springer-V., 2003.
3. N. R. Draper and H. Smith. *Applied regression analysis*. John Wiley & Sons, 1982.
4. S. Džeroski and N. Lavrač. *Relational Data Mining*. Springer-V., 2001.
5. R. Haining. *Spatial data analysis in the social and environmental sciences*. Cambridge University Press, 1990.
6. W. Klosgen and M. May. Spatial subgroup mining integrated in an object-relational spatial database. In T. Elomaa, H. Mannila, and H. Toivonen, editors, *Proceedings of PKDD 2002*, volume 2431 of *LNAI*, pages 275–286. Springer-V., 2002.
7. J. Knobbe, M. Haas, and A. Siebes. Propositionalisation and aggregates. In L. D. Raedt and A. Siebes, editors, *Proceedings of PKDD 2001*, volume 2168 of *LNAI*, pages 277–288. Springer-V., 2001.
8. K. Koperski. *Progressive Refinement Approach to Spatial Data Mining*. PhD thesis, Computing Science, Simon Fraser University, British Columbia, Canada, 1999.
9. G. Kuper, L. Libkin, and L. Paredaens. *Constraint databases*. Springer-V., 2001.
10. D. Malerba, F. Esposito, M. Ceci, and A. Appice. Top down induction of model trees with regression and splitting nodes. *IEEE Transactions on Pattern Analysis and Machine Intelligence*, 26(5):612–625, 2004.
11. D. Malerba, F. Esposito, A. Lanza, F. A. Lisi, and A. Appice. Empowering a gis with inductive learning capabilities: The case of ingens. *Journal of Computers, Environment and Urban Systems, Elsevier Science*, 27:265–281, 2003.
12. M. May. Spatial knowledge discovery: The spin! system. In K. Fullerton, editor, *Proceedings of the EC-GIS Workshop*, 2000.
13. M. Orkin and R. Drogin. *Vital Statistics*. McGraw Hill, New York, USA, 1990.
14. H. Samet. *Applications of spatial data structures*. Addison-Wesley longman, 1990.
15. S. Shekhar, P. R. Schrater, R. Vatsavai, W. Wu, and S. Chawla. Spatial contextual classification and prediction models for mining geospatial data. *IEEE Transactions on Multimedia*, 4(2):174–188, 2002.
16. L. Torgo. *Inductive Learning of Tree-based Regression Models*. PhD thesis, Department of Computer Science, University of Porto, Porto, Portugal, 1999.
17. Y. Wang and I. Witten. Inducing model trees for continuous classes. In M. Van Someren and G. Widmer, editors, *Proceedings of ECML 1997*, pages 128–137, 1997.
18. S. Weisberg. *Applied regression analysis*. Wiley, New York, USA, 1985.

Word Sense Disambiguation for Exploiting Hierarchical Thesauri in Text Classification

Dimitrios Mavroeidis[1], George Tsatsaronis[1], Michalis Vazirgiannis[1], Martin Theobald[2], and Gerhard Weikum[2]

[1] Department of Informatics, Athens University of Economics and Business, Greece
[2] Max-Planck Institute of Computer Science, Saarbruecken, Germany

Abstract. The introduction of hierarchical thesauri (HT) that contain significant semantic information, has led researchers to investigate their potential for improving performance of the text classification task, extending the traditional "bag of words" representation, incorporating syntactic and semantic relationships among words. In this paper we address this problem by proposing a Word Sense Disambiguation (WSD) approach based on the intuition that word proximity in the document implies proximity also in the HT graph. We argue that the high precision exhibited by our WSD algorithm in various humanly-disambiguated benchmark datasets, is appropriate for the classification task. Moreover, we define a semantic kernel, based on the general concept of GVSM kernels, that captures the semantic relations contained in the hierarchical thesaurus. Finally, we conduct experiments using various corpora achieving a systematic improvement in classification accuracy using the SVM algorithm, especially when the training set is small.

1 Introduction

It can be argued that WSD algorithms for the document classification task should differ in their design and evaluation from pure WSD algorithms. It is expected that correctly disambiguated words could improve (and certainly not degrade) the performance of a document classification task, while falsely disambiguated words would entail noise. Although the SVM algorithm [1] used in our experiments is known to be noise tolerant, it is certain that noise, above a certain level, will eventually degrade in SVM's performance. In the absence of theoretical or experimental studies on the exact level of falsely disambiguated words that can be tolerated by classification algorithms, the most appropriate performance measure for WSD algorithms designed for a classification task is precision. Choosing the WSD algorithm with the highest precision will result in the incorporation of the lowest amount of noise in the classification task.

Another important issue for the successful embedding of WSD in text classification, is the exploitation of senses' semantic relations, that are provided by the HT. These relations are essential for defining distances and kernels that reflect semantic similarities between senses. An extensive bibliography exists for measuring distances and similarities on thesauri and ontologies, which has not been taken into account by other research approaches that embed WSD in the text classification task. The need for exploiting semantic relations is illustrated in [2], where SemCor 1.7.1, a humanly-disambiguated corpus, is used in classification experiments. It is demonstrated that even with a 100% accurate disambiguation, the simple use of senses instead of keywords does not improve classification performance.

In this paper we propose an unsupervised WSD algorithm for classification, that utilizes a background HT. Our approach adopts the intuition that adjacent terms extracted from a given document are expected to be semantically close to each other and that is reflected to their pathwise distance on the HT. Thus, the objective of our WSD method is, given a set of terms, to select the senses (one for each term among many found in the HT) that overall minimize the pathwise distance and reflect the compactness of the selected sense set. The semantic compactness measure introduced is based on the concept of the Steiner Tree [3]. As opposed to other approaches that have utilized WSD for classification [4],[5],[6],[7], we have conducted extensive experiments with disambiguated corpora (Senseval 2 and 3, SemCor 1.7.1), in order to validate the appropriateness of our WSD algorithm. Experiments, using the WordNet HT, demonstrate that our WSD algorithm can be configured to exhibit very high precision, and thus can be considered appropriate for classification. In order to exploit the semantic relations inherent in the HT, we define a semantic kernel based on the general concept of GVSM kernels [8]. Finally, we have conducted experiments utilizing various sizes of training sets for the two largest Reuters-21578 categories and a corpus constructed from crawling editorial reviews of books from the Amazon website. The results demonstrate that our approach for exploiting hierarchical thesauri semantic information contributes significantly to the SVM classifier performance, especially when the training set size is small.

In the context of this paper WordNet [9] is utilized as a hierarchical thesaurus both for WSD and for classification. Although WordNet contains various semantic relations between concepts[1], our approach relies only on the hypernym/hyponym relation that orders concepts according to generality, and thus our approach can generalize to any HT that supports the hypernym/hyponym relation.

The rest of the paper is organized as follows. Section 2 discusses the preliminary notions and the related work. Section 3 presents our compactness measure for WSD that is based on the graph structure of an HT. Section 4 describes the semantic kernel that is utilized for the experiments. Section 5 discusses the experiments performed. Section 6 contains the comparison of the proposed framework to other approaches, concluding remarks and pointers to further work.

2 Preliminaries

2.1 Graph Theoretic Notions

Assuming that a document is represented by a set of senses, the semantic compactness measure that we introduce for WSD implies a similarity notion either among the senses of a sense set or between two sense sets. Its commutation is based on the notion of

---

[1] Concepts are word senses in WordNet terminology and in this paper we will use the terms word senses and concepts interchangeably.

Steiner Tree. Given a set of graph vertices, the Steiner Tree is the smallest tree that connects the set of nodes in the graph. The formal definition of the Steiner Tree is given below.

Definition 1 (Steiner Tree). *Given an undirected graph $G = (V, E)$, and a set $S \subseteq V$, then the Steiner Tree is the minimal Tree of G that contains all vertices of S.*

2.2 Semantic Kernels Based on Hierarchical Thesaurus

Since we aim at embedding WSD in the SVM classifier, we require the definition of a kernel that captures the semantic relations provided by the HT. To the extend of our knowledge the only approach that defines a semantic kernel based on a HT is [10]. The formal definition of their kernel is given below.

Definition 2 (Semantic Smoothing Kernels [10]). *The Semantic smoothing Kernel between two documents d_1, d_2 is defined as $K(d_1, d_2) = d_1 P' P d_2 = d_1 P^2 d_2$, where P is a matrix whose entries $P_{ij} = P_{ji}$, represent the semantic proximity between concepts i and j.*

The similarity matrix P is considered to be derived by a HT similarity measure. The Semantic Smoothing Kernels have similar semantics to the GVSM model defined in [8]. A kernel definition based on the GVSM model is given below.

Definition 3 (GVSM Kernel). *The GVSM kernel between two documents d_1 and d_2 is defined as $K(d_1, d_2) = d_1 D D' d_2$, where D is the term document matrix.*

The rows of matrix D, in the GVSM kernel contain the vector representation of terms, used to measure their pairwise semantic relatedness. The Semantic Smoothing Kernel has similar semantics. The Semantic Smoothing Kernel between two documents $K(d_1, d_2) = d_1 P^2 d_2$, can be regarded as a GVSM kernel, where the matrix D is derived by the decomposition of $P^2 = DD'$ (the decomposition is always possible since P^2 is guaranteed to be positive definite). The rows of D can be considered as the vector representation of concepts, used to measure their semantic proximity. Semantic Smoothing Kernels use P^2 and not P, because P is not guaranteed to be positive definite.

2.3 Related Work

WSD. The WordNet HT has been used for many supervised and unsupervised WSD algorithms. In direct comparison to our WSD approach we can find [11],[12],[13] that are unsupervised and rely on the semantic relations provided by WordNet. In the experimental section we show that our WSD algorithm can be configured to exhibit very high precision in various humanly-disambiguated benchmark corpora, and thus is more appropriate for the classification task.

Senseval (*www.senseval.org*), provides a forum, where the state of the art WSD systems are evaluated against disambiguated datasets. In the experimental sections we will compare our approach to the state of the art systems that have been submitted to the Senseval contests.

WSD and classification. In this section we shall briefly describe the relevant work done in embedding WSD in the document classification task. In [7], a WSD algorithm based on the general concept of Extended Gloss Overlaps is used and classification is performed with an SVM classifier for the two largest categories of the Reuters-25178 collection and two IMDB movie genres (*www.imdb.com*).

It is demonstrated that, when the training set is small, the use of WordNet senses together with words improves the performance of the SVM classification algorithm, however for training sets above a certain size, the approach is shown to have inferior performance to term-based classification. Moreover, the semantic relations inherent in WordNet are not exploited in the classification process. Although the WSD algorithm that is employed is not verified experimentally, its precision is estimated with a reference to [13], since the later work has a very similar theoretical basis. The experiments conducted by [13] in Senseval 2 lexical sample data, show that the algorithm exhibits low precision (around 45%) and thus may result in the introduction of much noise that can jeopardize the performance of a classification task.

In [4], the authors experiment with various settings for mapping words to senses (no disambiguation, most frequent sense as provided by WordNet and WSD based on context). Their approach is evaluated on the Reuters-25178, the OSHUMED and the FAODOC corpus, providing positive results. Their WSD algorithm has similar semantics to the WSD algorithm proposed in [12]. Although in [12] the experiments are conducted in a very restricted subset of SemCor 1.7.1, the results reported can be compared with our experiment results for the same task, as it is shown in Section 5. Moreover [4], use hypernyms for expanding the feature space.

In [5] the authors utilize the supervised WSD algorithm proposed in [14] in k-NN classification of the 20-newsgroups dataset. The WSD algorithm they employ is based on a Hidden Markov Model and is evaluated against Senseval 2, using "English all words task", reporting a maximum precision of around 60%. On the classification task of the 20-newsgroup dataset, they report a very slight improvement in the error-percentage of the classification algorithm. The semantic relations that are contained in WordNet are not exploited in the k-NN classification process.

The authors in [6] present an early attempt to incorporate semantics by means of a hierarchical thesauri in the classification process, reporting negative results on the Reuters-21578 and DigiTrad collection. While none disambiguation algorithm is employed, the use of hypernyms for extending the feature space representation is levied.

2.4 Hierarchical Thesaurus Distances – Similarities

As we have discussed in the introduction section, an important element for the successful incorporation of semantics in the classification process is the exploitation of the vast amount of semantic relations that are contained in the HT. There is an extensive bibliography that addresses the issue of defining distances and similarity measures based on the semantic relations provided by an HT [9],[15],[16],[17], which has not been related to the existing approaches for embedding WSD in classification. A common ground of most of the approaches is that the distance or similarity measure will depend on the "size" of the shortest path that connects the two concepts through a common ancestor in the hierarchy, or on the largest "depth" of a common ancestor in the hierarchy. The terms "size" and "depth" are used in an informal manner, for details one should use the references provided.

3 Compactness Based Disambiguation

In this section we present our unsupervised WSD method, as this was initially sketched in [18]. Our WSD algorithm is based on the intuition that adjacent terms extracted from a text document are expected to be semantically close to each other. Given a set of adjacent terms, our disambiguation algorithm will consider all the candidate sets of senses and output the set of senses that exhibits the highest level of semantic relatedness. Therefore, the main component of our WSD algorithm is the definition of a semantic compactness measure for sets of senses. We refer to our disambiguation approach as CoBD (Compactness Based Disambiguation). The compactness measure utilized in CoBD is defined below.

Definition 4. *Given an HT O and a set of senses $S = (s_1, ..., s_n)$, where $s_i \in O$ the compactness of S is defined as the cost of the Steiner Tree of $S \cup lca(S)$, such that there exists at least one path, using hypernym relation, from each s_i to the $lca(S)$.*

In the definition above we include one path, using the hypernym relation, for every sense to the least common ancestor $lca(S)$. The reason for imposing such a restriction is that the distance between two concepts in an HT is not defined as the shortest path that connects them in the HT, but rather as the shortest path that goes through a common ancestor. Thus, it can be argued that two concepts are connected only through a common ancestor and not through any other path in the HT. The existence of the $lca(S)$ (and of a path between every concept and the $lca(S)$ using the hypernym relation) guarantees that a path connecting all pairs of concepts (in the context discussed earlier) exists.

Although in general the problem of computing the Steiner Tree is NP-complete, the computation of the Steiner Tree (with the restriction imposed) of a set of concepts with their lca in a HT is computationally feasible and is reduced to the computation of the shortest path of the lca to every concept of the set. Another issue, potentially adding excessive computational load, is the large number of combinations of possible sets of senses, when a term set of large cardinality is considered for disambiguation. In order to address this issue, we reduce the search space by using a Simulated Annealing algorithm. The experimental setup used in this paper for the empirical evaluation of our WSD algorithm is described in detail in section 5.

4 Exploitation of Hierarchical Thesaurus Semantics in SVM Classification

We have argued in the introductory section that the exploitation of the semantics provided by an HT are important for the successful embedding of WSD in the classification task. In this section we will present the definition of the Kernel we will utilize in SVM classification. The Kernel we define is based on the general concept of GVSM kernel and depicts the semantics of the HT.

It is shown in detail in [18], that the use of hypernyms for the vector space representation of the concepts of a HT, enables the measurement of semantic distances in the vector space. More precisely, given a Tree HT, there exists a weight configuration for the hypernyms, such that standard vector space distance and similarity measures are equivalent to popular HT distances and similarities. The proofs for propositions given below can be found in [18].

Proposition 1. *Let O be a Tree HT, if we represent the concepts of the HT O, as vectors containing all their hypernyms, then there exists a configuration for the weights of the hypernyms such that the Manhattan distance (Minkowski distance with $p=1$) of any two concepts in vector space is equal to the Jiang-Conrath measure [15] in the HT.*

Proposition 2. *Let O be a Tree HT, if we represent the concepts of the HT O as vectors containing all their hypernyms, then there exists a configuration for the weights of the hypernyms such that the inner product of any two concepts in vector space is equal to the Resnik similarity measure [16] in the HT.*

The WordNet hierarchical thesaurus is composed by 9 hierarchies that contain concepts that inherit from more than one concept, and thus are not Trees. However, since only 2.28% of the concepts inherit from more than one concept [19], we can consider that the structure of WordNet hierarchies is close to the Tree structure.

From the above we conclude that, if we construct a matrix D where each row contains the vector representation of each sense containing all its hypernyms, the matrix DD' will reflect the semantic similarities that are contained in the HT. Based on D, we move on to define the kernel between two documents d_1, d_2, based on the general concept of GVSM kernels as $K_{concepts}(d_1, d_2) = d_1 D D' d_2$. In our experiments we have used various configurations for the rows of D. More precisely, we have considered the vector representation of each concept to be extended with a varying number of hypernyms. The argument for using only a limited number and not all hypernyms is that the similarity between hypernyms close to the root of the HT is considered to be very close to 0. Apart from hypernyms, in the experiments section, we have explored the potential of using hyponyms for constructing matrix the D in the GVSM kernel. The kernel that we finally utilize in our experiments is a combination of the inner product kernel for terms with the concept kernel $K(d_1, d_2) = K_{terms}(d_1, d_2) + K_{concepts}(d_1, d_2)$. This kernel was embedded into the current version of *SVMLight* [20] and replaced the standard linear kernel used for document classification with sparse training vectors.

The kernel defined implies a mapping from the original term and concept space, to a space that includes the terms, the concepts and their hypernyms. The kernel can be considered as the inner product in this feature space.

5 Experiments

5.1 Evaluation of the WSD Method

CoDB was tested in four benchmark WSD corpora; Brown 1 and Brown 2 from the SemCor 1.7.1 corpus, and the in the "English All Words" task of Senseval 2 and 3. These corpora are pre-tagged and pre-annotated. From all the parts of speech in the

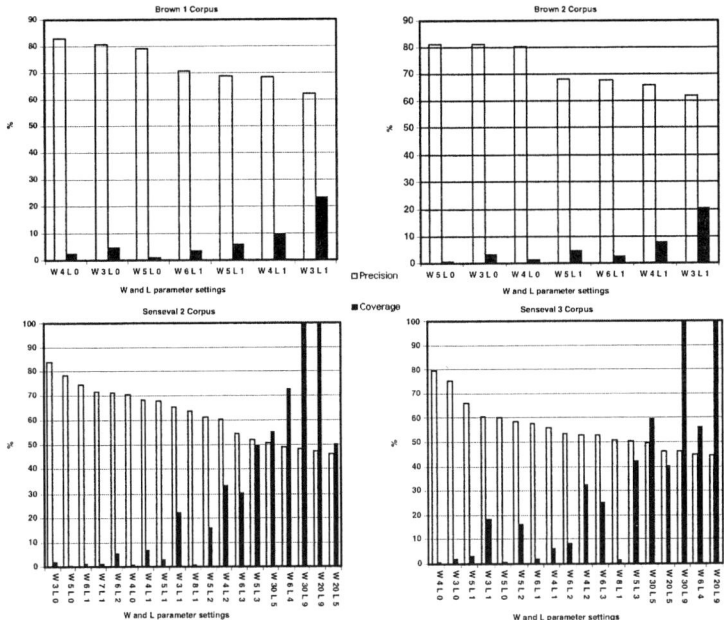

Fig. 1. WSD results on 4 benchmark datasets for different initializations of W and L

texts we only considered nouns, which are usually more informative than the rest and form a meaningful type hierarchy in WordNet. In order to implement CoBD efficiently we had to take into account that the search space of combinations to be examined for their compactness increases dramatically as the cardinality of the set of words examined increases, making exhaustive computation infeasible. Thus we adopted simulated annealing as in [21]. This approach reduced the search space and allowed us to execute the WSD using various set of words sizes in a time efficient manner. The parameters of the WSD method are:

1. Window Size (W): Set cardinality of the words to be disambiguated.
2. Allowed Lonely: Given a word set L, it is the maximum number of lonely senses [1] allowed in a WordNet noun hierarchy, for any senses combination of that window.

Figure 1 presents experiments we have conducted using various parameter settings. The results are sorted in decreasing order of precision. The precision and coverage [2] values reported do not take into account the monosemous nouns, but only the ambiguous ones. We can estimate, based on the examined corpora statistics, that the inclusion of the monosemous nouns would report an increase in precision between 3% and 4%, as well as an increase in coverage of almost 22%.

---

[1] A sense s belonging to a set of senses S is referred to as lonely if the WordNet noun hierarchy H it belongs to, does not contain any other $k \in S$.
[2] Coverage is defined as the percentage of the nouns that are disambiguated.

We observe that CoBD achieves precision greater than 80% with an associated coverage of more than 25%, if monosemous (i.e., non-ambiguous) nouns are also taken into account. Comparable experiments conducted in [12] reported a top precision result of 64,5% with an associated coverage of 86,2%. Similar experiments conducted in [11], [13] and [14] resulted as well in lower precision than CoBD. In comparing our approach to the state of the art WSD algorithms that were submitted to the "English All Words" Senseval 2 contest (*www.senseval.org*), we observe that our approach can be configured to exhibit the highest precision.

5.2 Document Collections and Preprocessing for Text Classification

Reuters. Reuters-21578 is a compilation of news articles from the Reuters newswire in 1987. We include this collection mostly for transparency reasons, since it has become the gold standard in document classification experiments. We conducted experiments on the two largest categories, namely *acquisitions* and *earnings*, in terms of using test- and training documents based on the [4] split. This split yields a total of 4,436 training and 1,779 test documents for the two categories. We extracted features from the mere article bodies, thus using whole sentences only and hiding any direct hint to the actual topic from the classifier.

Amazon. To test our methods on a collection with a richer vocabulary, we also extracted a real-life collection of natural-language text from amazon.com using Amazon's publicly available Web Service interface. From that taxonomy, we selected all the available editorial reviews for books in the three categories *Physics*, *Mathematics* and *Biological Sciences*, with a total of 6,167 documents. These reviews typically contain a brief discussion of a book's content and its rating. Since there is a high overlap among these topics' vocabulary and a higher diversity of terms within each topic than in Reuters, we expect this task to be more challenging for both the text- as well as the concept-aware classifier.

Before actually parsing the documents, we POS-annotated both the Reuters and Amazon collections, using a version of the commercial Connexor software for NLP processing. We restricted the disambiguation step to matching noun phrases in WordNet, because only noun phrases form a sufficiently meaningful HT in the ontology DAG. Since WordNet also contains the POS information for each of its concepts, POS document tagging significantly reduces the amount of choices for ambiguous terms and simplifies the disambiguation step. For example the term *run* has 52 (!) distinct senses in WordNet out of which 41 are tagged as verbs. The parser first conducts continuous noun phrase tokens in a small window of up to a size of 5 into dictionary lookups in WordNet before the disambiguation step takes place. If no matching phrase is found within the current window, the window is moved one token ahead. This sliding window technique enables us to match any composite noun phrase known in WordNet, whereupon larger phrases are typically less ambiguous. Non-ambiguous terms can be chosen directly as safe seeds for the compactness-based disambiguation step. Note that we did not perform any feature selection methods such as Mutual Information or Information Gain [22] prior to training the SVM, in order not to bias results toward a specific classification method.

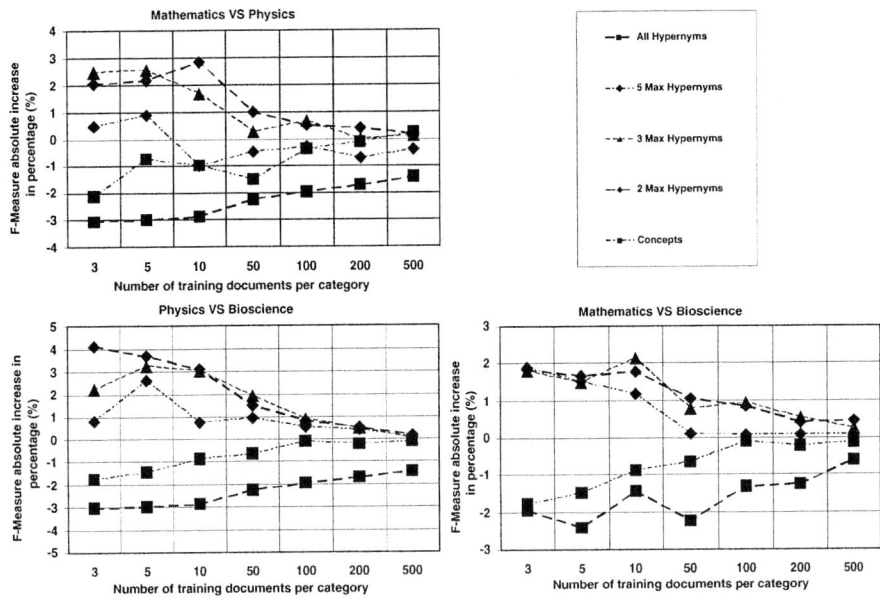

Fig. 2. Relative Improvement of F-measures scores for various Similarity Configurations in the Amazon Topics

5.3 Evaluation of Embedding CoBD in the Text Classification Task

To evaluate the embedding of CoBD in text classification, we performed binary classification tasks, only, i.e., we did not introduce any additional bias from mapping multi-class classification task onto the binary decision model used by the SVM method. The binary classification tasks were performed after forming all pairs between the three Amazon topics, and one pair between the two largest Reuters-21578 topics. The parameters' setting for CoBD was *W 3 L 0*, since it reported high percision and performed in a stable manner during the WSD evaluation experiments in the 4 benchmark corpora. Our baseline was the F-Measure [22] arising from the mere usage of term features. The baseline competed against the embedding of the term senses, whenever disambiguation was possible, and their hypernyms/hyponyms into the term feature vectors, according to the different GVSM kernel configurations shown in Figures 2,3. In our experiments, the weights of the hypernyms used in the GVSM kernel are taken to be equal to the weights of the terms they correspond to. We varied the training set sizes between 3 and 500 documents per topic. For each setup, in Figures 2,3 we report the differences of the *macro-averaged F-Measure* between the baseline and the respective configurations, using 10 iterations for each of the training set sizes of the Reuters dataset and 30 iterations for each of the training set sizes of the Amazon dataset. The variation of the differences was not too high and allowed for all the results where the absolute difference of the sample means was greater than 1% to reject the null hypothesis (that the means are equal) at a significance level of 0.05. For more than 500 documents, all our experiments

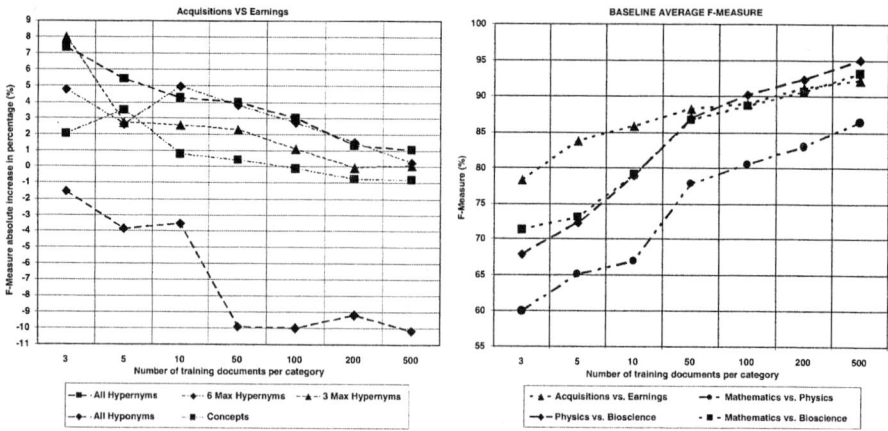

Fig. 3. Relative Improvement of F-measures scores for various Similarity Configurations in the Reuters Topics

indicate a convergence in results between the concept-aware classifier and the text classifier. The average F-measures for the baseline classifier are reported in Figure 3. For each run, the training documents were selected randomly following a uniform distribution. Since there is no split into separate documents for training and testing given in the Amazon collection, we performed cross-validation runs over the whole set, each using all the remaining documents for the test phase.

The results demonstrate that the use of CoBD and our kernel function, based on a small number of hypernyms increases consistently the classification quality especially for small training sets. In some cases, as the number of hypernyms increases we observe a performance deterioration which in some cases falls below the term-based classification.

The variance in the number of hypernyms needed for achieving better performance, can be explained by the fact that we did not employ a hypernym weighting scheme. Thus, when semantically correlated categories are considered, (such as Maths/Physics in the Amazon data), then the use of all the hypernyms with equal weights would result in many documents belonging to the Physics category to have a high similarity to documents of Maths category, degrading the performance of the classification algorithm.

6 Discussion and Conclusions

The context of the current work entails the content and structure (i.e. the senses and hierarchical relationships) of HTs and their usage for successful extension of the bag of words model for text classification. The objective is that such extensions (i.e. senses and hypenyms/hyponyms more precisely) are contributing to higher quality in the classification process.

The *contribution* of the paper is the design of a successful WSD approach to be incorporated and improve the text classification process. Our WSD approach takes into account term senses found in HTs, (in the specific case Wordnet), and for each document selects the best combination of them based on their conceptual compactness in terms of related Steiner tree costs. Apart from the senses we add to the original document feature set a controlled number of hypernyms of the senses at hand. The hypernyms are incorporated by means of the kernel utilized. The attractive features of our work are:

Appropriate WSD approach for text classification. Most of the related approaches incorporating WSD in the classification task [6],[7],[4] do not provide a sound experimental evidence on the quality of their WSD approach. On the contrary in our work, the WSD algorithm is exhaustively evaluated against various humanly disambiguated benchmark datasets and achieves very high precision (among the top found in related work) although at low coverage values (see Fig.1). This is not a problem, though since as mentioned earlier, it is essential to extend the feature space with correct features in order to prevent introduction of noise in the classification process. The experimental evaluation provides us with the assurance that our WSD algorithm can be configured to have high precision, and thus, would insert in the training set very little noise.

Similarity measure that takes into account the structure of the HT. Document classification depends on a relevant similarity measure to classify a document into the closest of the available classes. It is obvious that the similarity among sets of features (representing documents) should take into account their hierarchical relationships as they are represented in the HT. None of the previous approaches for embedding WSD in classification has taken into account the existing literature for exploiting the HT relations. Even when the use of hypernyms is used [6],[4], it is done in an ad-hoc way, based on the argument that the expansion of a concept with hypernyms would behave similar to query expansion using more general concepts. We utilize a Kernel based on the general concept of a GVSM kernel that can be used for measuring the semantic similarity between two documents. The kernel is based on the use of hypernyms for the representation of concepts - theoretically justified in the context of the related work concerning the computation of semantic distances and similarities on a HT that aligns to tree structure.

We conducted classification experiments on two real world datasets (the two largest Reuters categories and a dataset constructed by the editorial reviews of products on three categories at the *amazon.com* web site). The results demonstrate that our approach for embedding WSD in classification yields significantly better results especially when the training sets are small.

An issue that we will investigate in further work is the introduction of a weighting scheme for hypernyms favoring hypernyms that are close to the concept. A successful weighting scheme is expected to reduce the problem of the variance in the number of hypernyms needed to achieve optimal performance. We will investigate learning approaches to learn the weighting schemes for hypernyms. Moreover, we aim in conducting further experiments on other larger scale and heterogeneous data sets.

References

1. Drucker, H., Burges, C.J.C., Kaufman, L., Smola, A.J., Vapnik, V.: Support vector regression machines. In: Advances in Neural Information Processing Systems (NIPS). (1996) 155–161
2. Kehagias, A., Petridis, V., Kaburlasos, V.G., Fragkou, P.: A comparison of word- and sense-based text categorization using several classification algorithms. Journal of Intelligent Information Systems **21** (2003) 227–247

3. Hwang, R., Richards, D., Winter, P.: The steiner tree problem. Annals of Discrete Mathematics **53** (1992)
4. Bloehdorn, S., Hotho, A.: Boosting for text classification with semantic features. In: Proc. of the 10th ACM SIGKDD International Conference on Knowledge Discovery and Data Mining, Mining for and from the Semantic Web Workshop. (2004) 70–87
5. Rosso, P., Ferretti, E., Jimenez, D., Vidal, V.: Text categorization and information retrieval using wordnet senses. In: Proc. of the 2nd International WordNet Conference (GWC). (2004)
6. Scott, S., Matwin, S.: Feature engineering for text classification. In: Proc. of the 16th International Conference on Machine Learning (ICML). (1999) 379–388
7. Theobald, M., Schenkel, R., Weikum, G.: Exploiting structure, annotation, and ontological knowledge for automatic classification of xml data. In: International Workshop on Web and Databases (WebDB). (2003) 1–6
8. Wong, S.K.M., Ziarko, W., Wong, P.C.N.: Generalized vector space model in information retrieval. In: Proc. of the 8th annual international ACM SIGIR conference on Research and development in information retrieval. (1985) 18–25
9. Fellbaum, C., ed.: WordNet, An Electronic Lexical Database. The MIT Press (1998)
10. Siolas, G., d'Alche Buc, F.: Support vector machines based on semantic kernel for text categorization. In: Proc. of the International Joint Conference on Neural Networks (IJCNN). Volume 5., IEEE Press (2000) 205–209
11. Sussna, M.: Word sense disambiguation for free-text indexing using a massive semantic network. In: Proc. of the 2nd International Conference on Information and Knowledge Management (CIKM). (1993) 67–74
12. Agirre, E., Rigau, G.: A proposal for word sense disambiguation using conceptual distance. In: Proc. of Recent Advances in NLP (RANLP). (1995) 258–264
13. Banerjee, S., Pedersen, T.: Extended gloss overlaps as a measure of semantic relatedness. In: Proc. of the 18th International Joint Conference on Artificial Intelligence (IJCAI). (2003) 805–810
14. Molina, A., Pla, F., Segarra, E.: A hidden markov model approach to word sense disambiguation. In: Proc. of the 8th Iberoamerican Conference on Artificial Intelligence. (2002)
15. Jiang, J., Conrath, D.: Semantic similarity based on corpus statistics and lexical taxonomy. In: Proc. of the International Conference on Research in Computational Linguistics. (1997)
16. Resnik, P.: Using information content to evaluate semantic similarity in a taxonomy. In: Proc. of the 14th International Joint Conference on Artificial Intelligence (IJCAI). (1995)
17. Lin, D.: An information-theoretic definition of similarity. In: Proc. of the 15th International Conference on Machine Learning (ICML). (1998) 296–304
18. Mavroeidis, D., Tsatsaronis, G., Vazirgiannis, M.: Semantic distances for sets of senses and applications in word sense disambiguation. In: Proc. of the 3rd International Workshop on Text Mining and its Applications. (2004)
19. Devitt, A., Vogel, C.: The topology of wordnet: Some metrics. In: Proc. of the 2nd International WordNet Conference (GWC). (2004) 106–111
20. Klinkenberg, R., Joachims, T.: Detecting concept drift with support vector machines. In: Proc. of the 17th International Conference on Machine Learning (ICML). (2000) 487–494
21. Cowie, J., Guthrie, J., Guthrie, L.: Lexical disambiguation using simulated annealing. In: 14th International Conference on Computational Linguistics (COLING). (1992) 359–365
22. Manning, C., Schuetze, H.: Foundations of Statistical Natural Language Processing. MIT Press (2000)

Mining Paraphrases from Self-anchored Web Sentence Fragments

Marius Paşca

Google Inc. 1600 Amphitheatre Parkway,
Mountain View, California 940431 USA
mars@google.com

Abstract. Near-synonyms or paraphrases are beneficial in a variety of natural language and information retrieval applications, but so far their acquisition has been confined to clean, trustworthy collections of documents with explicit external attributes. When such attributes are available, such as similar time stamps associated to a pair of news articles, previous approaches rely on them as signals of potentially high content overlap between the articles, often embodied in sentences that are only slight, paraphrase-based variations of each other. This paper introduces a new unsupervised method for extracting paraphrases from an information source of completely different nature and scale, namely unstructured text across arbitrary Web textual documents. In this case, no useful external attributes are consistently available for all documents. Instead, the paper introduces linguistically-motivated text anchors, which are identified automatically within the documents. The anchors are instrumental in the derivation of paraphrases through lightweight pairwise alignment of Web sentence fragments. A large set of categorized names, acquired separately from Web documents, serves as a filtering mechanism for improving the quality of the paraphrases. A set of paraphrases extracted from about a billion Web documents is evaluated both manually and through its impact on a natural-language Web search application.

1 Motivation

The qualitative performance of applications relying on natural language processing may suffer, whenever the input documents contain text fragments that are lexically different and yet semantically equivalent as they are paraphrases of each other. The automatic detection of paraphrases is important in document summarization, to improve the quality of the generated summaries [1]; information extraction, to alleviate the mismatch in the trigger word or the applicable extraction pattern [2]; and question answering, to prevent a relevant document passage from being discarded due to the inability to match a question phrase deemed as very important [3].

In specialized collections such as news, the coverage of major events by distinct sources generates large numbers of documents with high overlap in their content. Thus, the task of detecting documents containing similar or equivalent information is somewhat simplified by the relative document homogeneity, use of properly-formatted text, the availability of external attributes (headlines), and knowledge of the document temporal proximity (similar article issue dates, or time stamps). When switching to unrestricted Web textual documents, all these advantages and clues are lost. Yet despite the diversity of content, the sheer size of the Web suggests that text fragments "hidden" inside quasi-anonymous documents will sometimes contain similar, or even equivalent information.

The remainder of the paper is structured as follows. After an overview of the proposed paraphrase acquisition method and a contrast to previous literature in Section 2, Section 3 provides more details and explains the need for self-anchored fragments as a source of paraphrases, as well as extensions for increasing the accuracy. Candidate paraphrases are filtered based on a large set of categorized named entities acquired separately from unstructured text. Section 4 describes evaluation results when applying the method to textual documents from a Web repository snapshot of the Google search engine. The section also evaluates the impact of the extracted paraphrases in providing search results that directly answer a standard evaluation set of natural-language questions.

2 Proposed Method for Paraphrase Acquisition

2.1 Goals

With large content providers and anonymous users contributing to the information accessible online, the Web has grown into a significant resource of implicitly-encoded human knowledge. The lightweight unsupervised method, presented in this paper, acquires useful paraphrases by mining arbitrary textual documents on the Web. The method is designed with a few goals in mind, which also represent advantages over previous methods:

1. No assumptions of any kind are made about the source, genre or structure of the input documents. In the experiments reported here, noise factors such as errors, misspellings, improperly formed sentences, or the use of HTML tags as implicit visual delimiters of sentences, are the norm rather than exceptions.
2. The method does not have access to any document-level attributes, which might otherwise hint at which pairs of documents are more likely to be sources of paraphrases. Such external attributes are simply not available for Web documents.
3. The acquisition is lightweight, robust and applicable to Web-scale collections. This rules out the use of deeper text analysis tools, e.g. syntactic [4] or semantic-role parsers [5].
4. For simplicity, the method derives paraphrases as a by-product of pairwise alignment of sentence fragments. When the extremities of the sentence fragments align, the variable parts become potential paraphrases of each other.
5. The method places an emphasis on defining the granularity (e.g., words, phrases, sentences or entire passages) and the actual mechanism for selecting the sentence fragments that are candidates for pairwise alignment. The

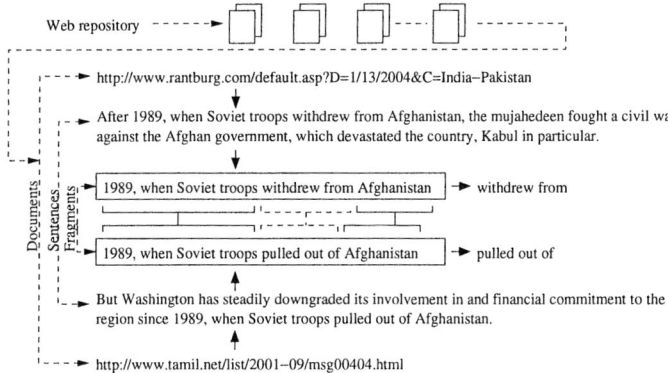

Fig. 1. Paraphrase acquisition from unstructured text across the Web

selection depends on *text anchors*, which are linguistic patterns whose role is twofold. First, they reduce the search/alignment space, which would otherwise be overwhelming (i.e., all combinations of contiguous sentence fragments). Second and more important, the anchors increase the quality of potential paraphrases, as they provide valuable linguistic context to the alignment phase, with little processing overhead.

2.2 Overview of Acquisition Method

As a pre-requisite, after filtering out HTML tags, the documents are tokenized, split into sentences and part-of-speech tagged with the TnT tagger [6]. Due to the inconsistent structure (or complete lack thereof) of Web documents, the resulting candidate sentences are quite noisy. Therefore some of the burden of identifying reliable sentences as sources of paraphrases is passed onto the acquisition mechanism.

Figure 1 illustrates the proposed method for unsupervised acquisition of paraphrases from Web documents. To achieve the goals listed above, the method mines Web documents for sentence fragments and associated text anchors. The method consists in searching for pairwise alignments of text fragments that have the same associated anchors. In the example, the anchors are identical time stamps (i.e., *1989*) of the events captured by the sentence fragments. The acquisition of paraphrases is a side-effect of the alignment.

The choice of the alignment type determines the constraints that two sentence fragments must satisfy in order to align, as well as the type of the acquired paraphrases. The example in Figure 1 is a const-var-const alignment. The two sentence fragments must have common word sequences at both extremities, as well as identical associated anchors. If that constraint holds, then the middle, variable word sequences are potential paraphrases of each other. Even if two sentences have little information content in common, their partial overlap can still produce paraphrase pairs such as ⟨*pulled out of, withdrew from*⟩ in Figure 1.

2.3 Comparison to Previous Work

Lexical resources such as WordNet [7] offer access to synonym sets, at the expense of many years of manual construction efforts. As general-purpose resources, they only cover the upper ontologies of a given language. Misspellings, idioms and other non-standard linguistic phenomena occur often in the noisy Web, but are not captured in resources like WordNet. Search engine hit counts rather than entries in lexical resources can be successfully exploited to select the best synonym of a given word, out of a small, closed-set of possible synonyms [8].

In addition to its relative simplicity when compared to more complex, sentence-level paraphrase acquisition [9], the method introduced in this paper is a departure from previous data-driven approaches in several respects. First, the paraphrases are not limited to variations of specialized, domain-specific terms as in [10], nor are they restricted to a narrow class such as verb paraphrases [11]. Second, as opposed to virtually all previous approaches, the method does not require high-quality, clean, trustworthy, properly-formatted input data. Instead, it uses inherently noisy, unreliable Web documents. The source data in [12] is also a set of Web documents. However, it is based on top search results collected from external search engines, and its quality benefits implicitly from the ranking functions of the search engines. Third, the input documents here are not restricted to a particular genre, whereas virtually all other recent approaches are designed for collections of parallel news articles, whether the articles are part of a carefully-compiled collection [13] or aggressively collected from Web news sources [14]. Fourth, the acquisition of paraphrases in this paper does not rely on external clues and attributes that two documents are parallel and must report on the same or very similar events. Comparatively, previous work has explicit access to, and relies strongly on clues such as the same or very similar timestamps being associated to two news article documents [13], or knowledge that two documents are translations by different people of the same book into the same language [15].

3 Anchored Sentence Fragments as Sources of Paraphrases

Even though long-range phrase dependencies often occur within natural-language sentences, such dependencies are not available without deeper linguistic processing. Therefore the acquisition method exploits only short-range dependencies, as captured by text fragments that are *contiguous* sequences of words. Two factors contribute substantially to the quality of the extracted paraphrases, namely the granularity of the text fragments, and the selection of their boundaries.

3.1 Fragment Granularity: Passages vs. Sentence Fragments

In principle, the granularity of the text fragments used for alignment ranges from full passages, a few sentences or a sentence, down to a sentence fragment, a phrase

Table 1. Examples of incorrect paraphrase pairs collected through the alignment of sentence fragments with arbitrary boundaries

(Wrong) Pairs	Examples of Common Sentence Fragments	
⟨city, place⟩	(to visit the _ of their birth)	(is a beautiful _ on the river),
	(live in a _ where things are)	(once the richest _ in the world)
⟨dogs, men⟩	(one of the _ took a step)	(does not allow _ to live in),
	(average age of _ at diagnosis is)	(a number of _ killed and wounded)

or a word. In practice, full text passages provide too much alignment context to be useful, as the chance of finding pairwise const-var-const alignments of any two passage pairs is very low. On the other hand, words and even phrases are useless since they are too short and do not provide any context for alignment. Sentences offer a good compromise in terms of granularity, but they are rarely confined to describing exactly one event or property as illustrated by the two sentences from Figure 1. Even though both sentences use similar word sequences to refer to a common event, i.e. the withdrawal of troops, they do not align to each other as complete sequences of words. Based on this and other similar examples, the paraphrase acquisition method relies on the alignment of contiguous chunks of sentences, or *sentence fragments*, instead of full-length sentences.

3.2 Fragment Boundaries: Arbitrary vs. Self-Anchored

It is computationally impractical to consider all possible sentence fragments as candidates for alignment. More interestingly, such an attempt would actually strongly degrade the quality of potential extractions as shown in Table 1. The pairs ⟨city, place⟩ and ⟨dogs, men⟩ are extracted from 1149 and 38 alignments found in a subset of Web documents, out of which only four alignments are shown in the table. For example, the alignment of the sentence fragments *"to visit the city of their birth"* and *"to visit the place of their birth"* results in ⟨city, place⟩ becoming a potential paraphrase pair. On the positive side, the alignments capture properties shared among the potential paraphrases, such as the fact that both *cities* and *places* can be visited, located on a river, be lived in, or be the richest among others. Similarly, both categories of *dogs* and *men* can take steps, not be allowed to live somewhere, have an average age, and be killed or wounded. Unfortunately, the sharing of a few properties is not a sufficient condition for two concepts to be good paraphrases of each other. Indeed, neither ⟨city, place⟩ not ⟨dogs, men⟩ constitute adequate paraphrase pairs.

Arbitrary boundaries are oblivious to syntactic structure, and will often span only partially over otherwise cohesive linguistic units, such as complex noun phrases, clauses, etc. Their main limitation, however, is the lack of an anchoring context, that would act as a pivot to which the information within the sentence fragments would be in strong dependency. We argue that it is both necessary and

Table 2. Types of text anchors for sentence fragment alignment

Anchor Type	Examples
Named entities for appositives	("*Scott McNealy, CEO of Sun Microsystems*", "*Scott McNealy, chief executive officer of Sun Microsystems*")
Common statements for main verbs	("*President Lincoln was killed by John Wilkes Booth*", "*President Lincoln was assassinated by John Wilkes Booth*")
Common dates for temporal clauses	("*1989, when Soviet troops withdrew from Afghanistan*", "*1989, when Soviet troops pulled out of Afghanistan*")
Common entities for adverbial relative clauses	("*Global and National Commerce Act, which took effect in October 2000*", "*Global and National Commerce Act, which came into force in October 2000*")

possible to automatically extract anchoring context from the sentences, and use it in conjunction with the sentence fragments, to decide whether the fragments are worth aligning or not. Text anchors provide additional linguistic context to the alignment phase. Generally speaking, they are linguistic units to which the sentence fragments as a whole are in a strong syntactic or semantic relation. From the types of anchors suggested in Table 2, only the temporal relative clauses and the more general adverbial relative clauses are implemented in the experiments reported in this paper.

To ensure robustness on Web document sentences, simple heuristics rather than complex tools are used to approximate the text anchors and sentence fragment boundaries. Sentence fragments are either temporal relative clauses or other types of adverbial relative clauses. They are detected with a small set of lexico-syntactic patterns, which can be summarized as:

(Temporal-Anchors): *Date* [,|-|(|nil] *when TemporalClause* [,|-|)|.]
(Adverbial-Anchors): *NamedEntity* [,|-|(|nil] *WhAdv RelativeClause* [,|-|)|.]

The patterns are based mainly on *wh*-words and punctuation. The disjunctive notation [,|-|)|.] stands for a single occurrence of a comma, a dash, a parenthesis, or a dot. *WhAdv* is one of *who*, *which* or *where*, and a *NamedEntity* is approximated by proper nouns, as indicated by part-of-speech tags. The matching clause *TemporalClause* and *RelativeClause* must satisfy a few other constraints, which aim at avoiding, rather than solving, complex linguistic phenomena. First, personal and possessive pronouns are often references to other entities. Therefore clauses containing such pronouns are discarded as ambiguous. Second, appositives and other similar pieces of information are confusing when detecting the end of the current clause. Consequently, during pattern matching, if the current clause does not contain a verb, the clause is either extended to the right, or discarded upon reaching the end of the sentence.

The time complexity for brute-force pairwise alignment is the square of the cardinality of the set of sentence fragments sharing the same anchors. A faster implementation exploits an existing parallel programming model [16] to divide the acquisition and alignment phases into three extraction stages. Each stage is distributed for higher throughput.

Table 3. Examples of siblings within the resource of categorized named entities

Phrase	Top Siblings
BMW M5	S-Type R, Audi S6, Porsche, Dodge Viper, Chevrolet Camaro, Ferrari
Joshua Tree	Tahquitz, Yosemite, Death Valley, Sequoia, Grand Canyon, Everglades
NSA	CIA, FBI, INS, DIA, Navy, NASA, DEA, Secret Service, NIST, Army
Research	Arts, Books, Chat, Fitness, Education, Finance, Health, Teaching
Porto	Lisbon, Algarve, Coimbra, Sintra, Lisboa, Funchal, Estoril, Cascais

3.3 Categorized Named Entities for Paraphrase Validation

Spurious sentences, imperfect alignments, and misleading contextual similarity of two text fragments occasionally produce incorrect paraphrases. Another contribution of the paper is the use of a novel post-filtering mechanism, which validates the candidate paraphrase pairs against a large resource of InstanceOf relations separately acquired from unstructured Web documents.

The data-driven extraction technique introduced in [17] collects large sets of categorized named entities from the Web. A categorized named entity encodes an InstanceOf relation between a named entity (e.g. *Tangerine Dream*) and a lexicalized category (e.g., *progressive rock group*) to which the entity belongs. Both the named entity and the lexicalized category are extracted from some common sentence from the Web. Even though the algorithm in [17] was developed for Web search applications, it is exploited here as one of the many possible criteria for filtering out some of the incorrect paraphrase pairs.

The key source of information derived from the categorized named entities are the siblings, i.e. named entities that belong to the same category. They are directly available in large numbers within the categorized named entities, as named entities often belong to common categories as shown in Table 3. Since siblings belong to a common class, they automatically share common properties. This results in many surrounding sentence fragments that look very similar to one another. Consequently, siblings produce a significant percentage of the incorrect paraphrase pairs. However, these errors can be detected if the phrases within a potential paraphrase pair are matched against the siblings from the categorized names. If the elements in the pair are actually found to be siblings of each other, their value as paraphrases is questionable at best, and hence the pair is discarded.

4 Evaluation

4.1 Experimental Setting

The input data is a collection of approximately one billion Web documents from a 2003 Web repository snapshot of the Google search engine. All documents are in English. The sentence fragments that are aligned to each other for paraphrase acquisition are based on two types of text anchors. In the first run,

Table 4. Top ranked paraphrases in decreasing order of their frequency of occurrence (top to bottom, then left to right)

With Temporal-Anchors		With Adverbial-Anchors	
passed, enacted	percent, per- cent	died, passed away	included, includes
percent, per cent	took, came into	percent, per cent	played, plays
figures, data	totalled, totaled	United States, US	lives, resides
passed, approved	took, came to	finished with, scored	operates, owns
statistics, figures	over, more than	over, more than	consists of, includes
statistics, data	enacted, adopted	began, started	center, centre
United States, US	information is, data are	include, includes	came, entered
figures are, data is	information is, figures are	operates, runs	takes, took
statistics are, data is	was elected, became	begins, starts	lost, won
passed, adopted	statistics are, information is	effect, force	chairs, heads

Temporal-Anchors, the sentence fragments are relative clauses that are temporally anchored through a *when* adverb to a date. In the Adverbial-Anchors sun, the sentence fragments are adverbial relative clauses anchored to named entities through other *wh*-adverbs.

For each unique date anchor (Temporal-Anchors) and named entity anchor (Adverbial-Anchors), a maximum of 100,000 associated sentence fragments are considered for pairwise alignment to one another. The extracted paraphrase pairs are combined across alignments, and ranked according to the number of unique alignments from which they are derived. Pairs that occur less than three times are discarded.

The impact of the extracted paraphrases is measured on a test set of temporal queries. The set consists of 199 *When* or *What year* queries from the TREC Question Answering track (1999 through 2002) [18]. The queries extract direct results from an existing experimental repository of 8 million factual fragments associated with dates [17]. The fragments are similar to the sentence fragments from the Temporal-Anchors run, e.g., *1953* associated to *"the first Corvette was introduced"*, and *1906* associated to *"Mount Vesuvius erupted"*. Each query receives a score equal to the reciprocal rank of the first returned result that is correct, or 0 if there is no such result [18]. Individual scores are aggregated over the entire query set.

4.2 Results

Table 4 shows the top paraphrases extracted in the two runs, after removal of pairs that contain either only stop words, or any number of non-alphabetic characters, or strings that differ only in the use of upper versus lower case. A small number of extractions occur in both sets, e.g., ⟨over, more than⟩. At

Table 5. Quality of the acquired paraphrases computed over the top, middle and bottom 100 pairs

Classification of Pairs	Temporal-Anchors			Adverbial-Anchors		
	Top	Mid	Low	Top	Mid	Low
(1) Correct; synonyms	53	37	3	33	23	6
(2) Correct; equal if case-insensitive	4	7	0	9	2	14
(3) Correct; morphological variation	0	0	0	20	15	6
(4) Correct; punctuation, symbols,spelling	22	1	10	18	11	15
(5) Correct; hyphenation	2	33	0	2	19	43
(6) Correct; both are stop words	15	0	0	1	0	0
Total correct	96	78	13	83	70	84
(7) Siblings rather than synonyms	0	10	82	5	7	7
(8) One side adds an elaboration	0	11	4	4	3	1
Total siblings	0	21	86	9	10	8
(10) Incorrect; e.g., antonyms	4	1	1	8	20	8

least one of the pairs is spurious, namely ⟨*lost, won*⟩, which are antonyms rather than synonyms. Difficulties in distinguishing between synonyms, on one side, and siblings or co-ordinate terms (e.g., *Germany* and *France*) or even antonyms, on the other, have also been reported in [11]. The occurrence of the spurious antonym pair in Table 4 suggests that temporal anchors provide better alignment context than the more general adverbial anchors, as they trade off coverage for increased accuracy.

The automatic evaluation of the acquired paraphrases is challenging despite the availability of external lexical resources and dictionaries. For example, the lexical knowledge encoded in WordNet [7] does not include the pair ⟨*abduction, kidnapping*⟩ as synonyms, or the pair ⟨*start, end*⟩ as antonyms. Therefore these and many other pairs of acquired paraphrases cannot be automatically evaluated as correct (if synonyms) or incorrect (e.g., if antonyms) based only on information from the benchmark resource. To measure the quality of the paraphrases, the top, middle and bottom 100 paraphrase pairs from each run are categorized manually into the classes shown in Table 5. Note that previous work on paraphrase acquisition including [9], [13] and [16] also relies on manual rather than automatic evaluation components. The pairs in class (1) in Table 5 are the most useful; they include ⟨*photo, picture*⟩, ⟨*passed, approved*⟩, etc. The following categories correspond to other pairs classified as correct. For instance, ⟨*Resolution, resolution*⟩ is classified in class (2); ⟨*takes, took*⟩ is classified in class (3); ⟨*world, wolrd*⟩ is classified in (4); ⟨*per-cent, percent*⟩ in (5); and ⟨*has not, hasn't*⟩ in (6). The next three classes do not contain synonyms. The pairs in (7) are siblings rather than direct synonyms, including pairs of different numbers. Class (8) contains pairs in which a portion of one of the elements is a synonym or phrasal equivalent of the other element, such as ⟨*complete data, records*⟩. Finally, the last class from Table 5 corresponds to incorrect extractions, e.g. due to antonyms like ⟨*started, ended*⟩. The results confirm that temporal anchors produce better paraphrases, at least over the first half of the ranked list of paraphrases. In comparison to the results shown in Table 5, the

Table 6. Examples of paraphrase pairs discarded by sibling-based validation

Discarded Pair	Ok?	Discarded Pair	Ok?
April, Feb.	Yes	Monday, Tuesday	Yes
season, year	Yes	country, nation	No
goods, services	Yes	north, south	Yes
Full, Twin	Yes	most, some	Yes
country, county	Yes	higher, lower	Yes
authority, power	No	Democrats, Republicans	Yes
England, Scotland	Yes	fall, spring	Yes

Table 7. Performance improvement on natural-language queries

Max. Nr. Disjunctions per Expanded Phrase	Nr. Queries with Better Scores	Nr. Queries with Lower Scores	Overall Score
1 (no paraphrases)	0	0	52.70
5 (4 paraphrases)	18	5	63.35

evaluation of a sample of 215 pairs results in an accuracy of 61.4% in [11], whereas 81.4% of a sample of 59 pairs are deemed as correct in [9].

The validation procedure, based on siblings from categorized names, identifies and discards 4.7% of the paraphrase pairs as siblings of one another. This is a very good ratio, if corroborated with the percentage of pairs classified as siblings in Table 5. Out of 200 pairs selected randomly among the discarded pairs, 28 are in fact useful synonyms, which corresponds to a projected precision of 86% for the validation procedure. Table 6 illustrates a few of the pairs discarded during validation.

The acquired paraphrases impact the accuracy of the dates retrieved from the repository of factual fragments associated with dates. All phrases from the test set of temporal queries are expanded into Boolean disjunctions with their top-ranked paraphrases. For simplicity, only individual words rather than phrases are expanded, with up to 4 paraphrases per word. For example, the inclusion of paraphrases into the query Q685: *"When did Amtrak begin operations?"* results in the expansion *"When did Amtrak (begin|start|began|continue| commence) (operations|operation|activities|business|operational)?"*. The top result retrieved for the expanded query is *1971*, which is correct according to the gold standard.

As shown in Table 7, paraphrases improve the accuracy of the returned dates, increase the number of queries for which a correct result is returned, and increase the overall score by 20%. Further experiments show that the incremental addition of more paraphrases, i.e., four versus three paraphrases per query word, results in more individual queries with a better score than for their non-expanded version, and higher overall scores for the returned dates. After reaching a peak score, the inclusion of additional paraphrases in each expansion actually degrades the overall results, as spurious paraphrases start redirecting the search towards irrelevant items.

5 Conclusion

Sophisticated methods developed to address various natural language processing tasks tend to make strong assumptions about the input data. In the case of paraphrase acquisition, many methods assume reliable sources of information, clean text, expensive tools such as syntactic parsers, and the availability of explicit document attributes. Comparatively, this paper makes no assumption of any kind about the source or structure of the input documents. The acquisition of paraphrases is a result of pairwise alignment of sentence fragments occurring within the unstructured text of Web documents. The inclusion of lightweight linguistic context into the alignment phase increases the quality of potential paraphrases, as does the filtering of candidate paraphrases based on a large set of categorized named entities also extracted from unstructured text. The experiments show that unreliable text of the Web can be distilled into paraphrase pairs of good quality, which are beneficial in returning direct results to natural-language queries.

Acknowledgments

The author would like to thank Péter Dienes for suggestions and assistance in evaluating the impact of the extracted paraphrases on natural-language queries.

References

1. Hirao, T., Fukusima, T., Okumura, M., Nobata, C., Nanba, H.: Corpus and evaluation measures for multiple document summarization with multiple sources. In: Proceedings of the 20th International Conference on Computational Linguistics (COLING-04), Geneva, Switzerland (2004) 535–541
2. Shinyama, Y., Sekine, S.: Paraphrase acquisition for information extraction. In: Proceedings of the 41st Annual Meeting of the Association of Computational Linguistics (ACL-03), 2nd Workshop on Paraphrasing: Paraphrase Acquisition and Applications, Sapporo, Japan (2003) 65–71
3. Paşca, M.: Open-Domain Question Answering from Large Text Collections. CSLI Studies in Computational Linguistics. CSLI Publications, Distributed by the University of Chicago Press, Stanford, California (2003)
4. Collins, M.: Head-Driven Statistical Models for Natural Language Parsing. PhD thesis, University of Pennsylvania, Philadelphia, Pennsylvania (1999)
5. Gildea, D., Jurafsky, D.: Automatic labeling of semantic roles. In: Proceedings of the 38th Annual Meeting of the Association of Computational Linguistics (ACL-00), Hong Kong (2000) 512–520
6. Brants, T.: TnT - a statistical part of speech tagger. In: Proceedings of the 6th Conference on Applied Natural Language Processing (ANLP-00), Seattle, Washington (2000) 224–231
7. Miller, G.: WordNet: a lexical database. Communications of the ACM **38** (1995) 39–41

8. Turney, P.: Mining the Web for synonyms: PMI-IR versus LSA on TOEFL. In: Proceedings of the 12th European Conference on Machine Learning (ECML-01), Freiburg, Germany (2001) 491–502
9. Barzilay, R., Lee, L.: Learning to paraphrase: An unsupervised approach using multiple-sequence alignment. In: Proceedings of the 2003 Human Language Technology Conference (HLT-NAACL-03), Edmonton, Canada (2003) 16–23
10. Jacquemin, C., Klavans, J., Tzoukermann, E.: Expansion of multi-word terms for indexing and retrieval using morphology and syntax. In: Proceedings of the 35th Annual Meeting of the Association of Computational Linguistics (ACL-97), Madrid, Spain (1997) 24–31
11. Glickman, O., Dagan, I.: Acquiring Lexical Paraphrases from a Single Corpus. In: Recent Advances in Natural Language Processing III. John Benjamins Publishing, Amsterdam, Netherlands (2004) 81–90
12. Duclaye, F., Yvon, F., Collin, O.: Using the Web as a linguistic resource for learning reformulations automatically. In: Proceedings of the 3rd Conference on Language Resources and Evaluation (LREC-02), Las Palmas, Spain (2002) 390–396
13. Shinyama, Y., Sekine, S., Sudo, K., Grishman, R.: Automatic paraphrase acquisition from news articles. In: Proceedings of the Human Language Technology Conference (HLT-02), San Diego, California (2002) 40–46
14. Dolan, W., Quirk, C., Brockett, C.: Unsupervised construction of large paraphrase corpora: Exploiting massively parallel news sources. In: Proceedings of the 20th International Conference on Computational Linguistics (COLING-04), Geneva, Switzerland (2004) 350–356
15. Barzilay, R., McKeown, K.: Extracting paraphrases from a parallel corpus. In: Proceedings of the 39th Annual Meeting of the Association for Computational Linguistics (ACL-01), Toulouse, France (2001) 50–57
16. Dean, J., Ghemawat, S.: MapReduce: Simplified data processing on large clusters. In: Proceedings of the 6th Symposium on Operating Systems Design and Implementation (OSID-04), San Francisco, California (2004) 137–150
17. Paşca, M.: Acquisition of categorized named entities for Web search. In: Proceedings of the 13th ACM Conference on Information and Knowledge Management (CIKM-04), Washington, D.C. (2004)
18. Voorhees, E., Tice, D.: Building a question-answering test collection. In: Proceedings of the 23rd International Conference on Research and Development in Information Retrieval (SIGIR-00), Athens, Greece (2000) 200–207

M²SP: Mining Sequential Patterns Among Several Dimensions

M. Plantevit[1], Y.W. Choong[2,3], A. Laurent[1], D. Laurent[2], and M. Teisseire[1]

[1] LIRMM, Université Montpellier 2, CNRS, 161 rue Ada, 34392 Montpellier, France
[2] LICP, Université de Cergy Pontoise, 2 av. Chauvin, 95302 Cergy-Pontoise, France
[3] HELP University College, BZ-2 Pusat Bandar Damansara, 50490 Kuala Lumpur, Malaysia

Abstract. Mining sequential patterns aims at discovering correlations between events through time. However, even if many works have dealt with sequential pattern mining, none of them considers frequent sequential patterns involving several dimensions in the general case. In this paper, we propose a novel approach, called M^2SP, to mine multidimensional sequential patterns. The main originality of our proposition is that we obtain not only intra-pattern sequences but also inter-pattern sequences. Moreover, we consider generalized multidimensional sequential patterns, called jokerized patterns, in which some of the dimension values may not be instanciated. Experiments on synthetic data are reported and show the scalability of our approach.

Keywords: Data Mining, Sequential Patterns, Multidimensional Rules.

1 Introduction

Mining sequential patterns aims at discovering correlations between events through time. For instance, rules that can be built are *A customer who bought a TV and a DVD player at the same time later bought a recorder*. Work dealing with this issue in the literature have proposed scalable methods and algorithms to mine such rules [9]. As for association rules, the efficient discovery is based on the *support* which indicates to which extend data from the database contains the patterns.

However, these methods only consider one dimension to appear in the patterns, which is usually called the *product* dimension. This dimension may also represent web pages for web usage mining, but there is normally a single dimension. Although some works from various studies claim to combine several dimensions, we argue here that they do not provide a complete framework for multidimensional sequential pattern mining [4,8,11]. The way we consider multidimensionality is indeed generalized in the sense that patterns must contain several dimensions combined over time. For instance we aim at building rules like *A customer who bought a surfboard and a bag in NY later bought a wetsuit in SF*. This rule not only combines two dimensions (*City* and *Product*) but it also combines them over time (NY appears before SF, surfboard appears before wetsuit). As far as we know, no method has been proposed to mine such rules.

In this paper, we present existing methods and their limits. Then, we define the basic concepts associated to our proposition, called M²SP, and the algorithms to build such rules. Experiments performed on synthetic data are reported and assess our proposition.

In our approach, sequential patterns are mined from a relational table, that can be seen as a fact table in a multidimensional database. This is why, contrary to the standard terminology of the relational model, the attributes over which a relational table is defined are called *dimensions*.

In order to mine such frequent sequences, we extend our approach so as to take into account partially instanciated tuples in sequences. More precisely, our algorithms are designed in order to mine frequent jokerized multidimensional sequences containing as few ∗ as possible, i.e., replacing an occurrence of ∗ with any value from the corresponding domain cannot give a frequent sequence.

The paper is organized as follows: Section 2 introduces a motivating example illustrating the goal of our work, and Section 3 reviews previous works on sequential patterns mining. Section 4 introduces our contribution, and in Section 5, we extend multidimensional patterns to *jokerized* patterns. Section 6 presents the algorithms, and experiments performed on synthetic data are reported in Section 7. Section 8 concludes the paper.

2 Motivating Example

In this section, we first briefly recall the basic ingredients of the relational model of databases used in this paper (we refer to [10] for details on this model), and we present an example to illustrate our approach. This example will be used throughout the paper as a running example.

Let $U = \{D_1, \ldots D_n\}$ be a set of attributes, which we call *dimensions* in our approach. Each dimension D_i is associated with a (possibly infinite) domain of values, denoted by $dom(D_i)$. A relational table T over universe U is a finite set of tuples $t = (d_1, \ldots, d_n)$ such that, for every $i = 1, \ldots, n$, $d_i \in dom(D_i)$. Moreover, given a table T over U, for every $i = 1, \ldots, n$, we denote by $Dom_T(D_i)$ (or simply $Dom(D_i)$ if T is clear from the context) the *active domain* of D_i in T, i.e., the set of all values of $dom(D_i)$ that occur in T.

Since we are interested in sequential patterns, we assume that U contains at least one dimension whose domain is totally ordered, corresponding to the *time dimension*.

In our running example, we consider a relational table T in which transactions issued by customers are stored. More precisely, we consider a universe U containing six dimensions (or attributes) denoted by D, CG, A, P and Q, where: D is the date of transactions (considering three dates, denoted by 1, 2 and 3), CG is the category of customers (considering two categories, denoted by $Educ$ and Ret, standing for educational and retired customers, respectively), A is the age of customers (considering three discretized values, denoted by Y (young), M (middle) and O (old)), C is the city where transactions have been issued (considering three cities, denoted by NY (New York), LA (Los Angeles) and SF (San Francisco)), P is the product of the transactions (considering four products, denoted by c, m, p and r), and Q stands for the quantity of products in the transactions (considering nine such quantities).

Fig. 1 shows the table T in which, for instance, the first tuple means that, at date 1, educational young customers bought 50 products c in New York. Let us now assume that we want to extract all multidimensional sequences that deal with the age of customers, the products they bought and the corresponding quantities, and that are frequent with respect to the groups of customers and the cities where transactions have been issued. To this end, we consider three sets of dimensions as follows: (i) the dimension D, representing the date, (ii) the three dimensions A, P and Q that we call *analysis dimensions*, (iii) the two dimensions CG and C, that we call *reference dimensions*.

Tuples over analysis dimensions are those that appear in the items that constitute the sequential patterns to be mined. The table is partitioned into blocks according to tuple values over reference dimensions and the support of a given multidimensional sequence is the ratio of the number of blocks supporting the sequence over the total number of blocks. Fig. 2 displays the corresponding blocks in our example.

In this framework, $\langle\{(Y,c,50),(M,p,2)\},\{(M,r,10)\}\rangle$ is a multidimensional sequence having support $\frac{1}{3}$, since the partition according to the reference dimensions contains 3 blocks, among which one supports the sequence. This is so because $(Y,c,50)$ and $(M,p,2)$ both appear at the same date (namely date 1), and $(M,r,10)$ appears later on (namely at date 2) in the first block shown in Figure 4.

It is important to note that, in our approach, more general patterns, called *jokerized sequences*, can be mined. The reason for this generalization is that considering partially instanciated tuples in sequences implies that more frequent sequences are mined. To see this, considering a support threshold of $\frac{2}{3}$, no sequence of the form $\langle\{(Y,c,\mu)\},\{(M,r,\mu')\}\rangle$ is frequent. On the other hand, in the first two blocks of Fig. 2, Y associated with c and M associated with r appear one after the other, according to the date of transactions. Thus, we consider that the jokerized sequence, denoted by $\langle\{(Y,c,*)\},\{(M,r,*)\}\rangle$, is frequent since its support is equal to $\frac{2}{3}$.

D	CG	C	A	P	Q
(Date)	(Customer-Group)	(City)	(Age)	(Product)	(Quantity)
1	Educ	NY	Y	c	50
1	Educ	NY	M	p	2
1	Educ	LA	Y	c	30
1	Ret.	SF	O	c	20
1	Ret.	SF	O	m	2
2	Educ	NY	M	p	3
2	Educ	NY	M	r	10
2	Educ	LA	Y	c	20
3	Educ	LA	M	r	15

Fig. 1. Table T

3 Related Work

In this section, we argue that our approach generalizes previous works on sequential patterns. In particular, the work described in [8] is said to be *intra*-pattern since sequences are mined within the framework of a single description (the so-called *pattern*). In this paper, we propose to generalize this work to *inter*-pattern multidimensional sequences.

3.1 Sequential Patterns

An early example of research in the discovering of patterns from sequences of events can be found in [5]. In this work, the idea is the discovery of rules underlying the generation of a given sequence in order to predict a plausible sequence continuation. This idea is then extended to the discovery of interesting patterns (or *rules*) embedded in a database of sequences of sets of events (items). A more formal approach in solving the problem of mining sequential patterns is the AprioriAll algorithm as presented in [6]. Given a database of sequences, where each sequence is a list of transactions ordered by transaction time, and each transaction is a set of items, the goal is to discover all sequential patterns with a user-specified minimum support, where the support of a pattern is the number of data-sequences that contain the pattern.

In [1], the authors introduce the problem of mining sequential patterns over large databases of customer transactions where each transaction consists of customer-id, transaction time, and the items bought in the transaction. Formally, given a set of sequences, where each sequence consists of a list of elements and each element consists of a set of items, and given a user-specified min support threshold, sequential pattern mining is to find all of the frequent subsequences, i.e., the subsequences whose occurrence frequency in the set of sequences is no less than min support. Sequential pattern mining discovers frequent patterns ordered by time. An example of this type of pattern is *A customer who bought a new television 3 months ago, is likely to buy a DVD player now*. Subsequently, many studies have introduced various methods in mining sequential patterns (mainly in time-related data) but almost all proposed methods are Apriori-like, i.e., based on the Apriori property which states the fact that any superpattern of a nonfrequent pattern cannot be frequent. An example using this approach is the GSP algorithm [9].

3.2 Multidimensional Sequential Patterns

As far as we know, three propositions have been studied in order to deal with several dimensions when building sequential patterns. Next, we briefly recall these propositions.

Pinto et al. [8]. This work is the first one dealing with several dimensions in the framework of sequential patterns. For instance, purchases are not only described by considering the customer ID and the products, but also by considering the age, the type of the customer (Cust-Grp) and the city where (s)he lives, as shown in Fig. 1.

Multidimensional sequential patterns are defined over the schema $A_1, ..., A_m, S$ where $A_1, ..., A_m$ are the dimensions describing the data and S is the sequence of items purchased by the customers ordered over time. A multidimensional sequential pattern is defined as $(id_1, (a_1, ..., a_m), s)$ where $a_i \in A_i \cup \{*\}$. $id_1, (a_1, ..., a_m)$ is said to be a multidimensional pattern. For instance, the authors consider the sequence $((*, NY, *), \langle bf \rangle)$ meaning that customers from NY have all bought a product b and then a product f. Sequential patterns are mined from such multidimensional databases either (i) by mining all frequent sequential patterns over the product dimension and then regrouping them into multidimensional patterns, (ii) or by mining all frequent multidimensional patterns and then mining frequent product sequences over these patterns. Note that the sequences found by this approach do not contain several dimensions since the dimension time only concerns products. Dimension product is the only dimension that can be combined over time, meaning that it is not possible to have a rule indicating that when *b* is bought in *Boston* then *c* is bought in *NY*. Therefore, our approach can seen as a generalization of the work in [8].

Yu et Chen. [11]. In this work, the authors consider sequential pattern mining in the framework of Web Usage Mining. Even if three dimensions (pages, sessions, days) are considered, these dimensions are very particular since they belong to a single hierarchized dimension. Thus, the sequences mined in this work describe correlations between objects over time by considering only one dimension, which corresponds to the web pages.

de Amo et al. [4]. This approach is based on first order temporal logic. This proposition is close to our approach, but more restricted since (i) groups used to compute the support are predefined whereas we consider the fact that the user should be able to define them (see reference dimensions below), and (ii) several attributes cannot appear in the sequences. The authors claim that they aim at considering several dimensions but they have only shown one dimension for the sake of simplicity. However, the paper does not provide hints for a complete solution with *real* multidimensional patterns, as we do in our approach.

4   M²SP: *M*ining *M*ultidimensional *S*equential *P*atterns

4.1   Dimension Partition

For each table defined on the set of dimensions D, we consider a partition of D into four sets: D_t for the temporal dimension, D_A for the *analysis* dimensions, D_R for the *reference* dimensions, and D_F for the *ignored* dimensions.

Each tuple $c = (d_1, \ldots, d_n)$ can thus be written as $c = (f, r, a, t)$ where f, r, a and t are the restrictions of c on D_F, D_R, D_A and D_t, respectively.

Given a table T, the set of all tuples in T having the same restriction r over D_R is said to be a *block*. Each such block B is denoted by the tuple r that defines it, and we denote by B_{T,D_R} the set of all blocks that can be built up from table T.

In our running example, we consider $F = \emptyset$, $D_R = \{CG, C\}$, $D_A = \{A, P, Q\}$ and $D_t = \{D\}$. Fig. 2 shows the three blocks built up from table T.

D	CG	C	A	P	Q
1	Educ	NY	Y	c	50
1	Educ	NY	M	p	2
2	Educ	NY	M	p	3
2	Educ	NY	M	r	10

a. Block (*Educ, NY*)

D	CG	C	A	P	Q
1	Educ	LA	Y	c	30
2	Educ	LA	Y	c	20
3	Educ	LA	M	r	15

b. Block (*Educ, LA*)

D	CG	C	A	P	Q
1	Ret.	SF	O	c	20
1	Ret.	SF	O	m	2

c. Block (*Ret., SF*)

Fig. 2. Blocks defined on T over dimensions CG and C

When mining multidimensional sequential patterns, the set D_R identifies the blocks of the database to be considered when computing supports. The support of a sequence is the proportion of blocks embedding it. Note that, in the case of usual sequential patterns and of sequential patterns as in [8] and [4], this set is reduced to one dimension (*cid* in [8] or *IdG* in [4]).

The set D_A describes the analysis dimensions, meaning that values over these dimensions appear in the multidimensional sequential patterns. Note that usual sequential patterns only consider one analysis dimension corresponding to the products purchased or the web pages visited. The set F describes the ignored dimensions, i.e. those that are used neither to define the date, nor the blocks, nor the patterns to be mined.

4.2 Multidimensional Item, Itemset and Sequential Pattern

Definition 1 (Multidimensional Item). *Let* $D_A = \{D_{i_1}, \ldots, D_{i_m}\}$ *be a subset of D. A multidimensional item on* D_A *is a tuple* $e = (d_{i_1}, \ldots, d_{i_m})$ *such that, for every k in* $[1, m]$, d_{i_k} *is in* $Dom(D_{i_k})$.

Definition 2 (Multidimensional Itemset). *A multidimensional itemset on* D_A *is a non empty set of items* $i = \{e_1, \ldots, e_p\}$ *where for every j in* $[1, p]$, e_j *is a multidimensional item on* D_A *and for all* j, k *in* $[1, p]$, $e_j \neq e_k$.

Definition 3 (Multidimensional Sequence). *A multidimensional sequence on* D_A *is an ordered non empty list of itemsets* $\varsigma = \langle i_1, \ldots, i_l \rangle$ *where for every j in* $[1, l]$, i_j *is a multidimensional itemset on* D_A.

In our running example, $(Y, c, 50)$, $(M, p, 2)$, $(M, r, 10)$ are three multidimensional items on $D_A = \{A, P, Q\}$. Thus, $\langle \{(Y, c, 50), (M, p, 2)\}, \{(M, r, 10)\} \rangle$ is a multidimensional sequence on D_A.

Definition 4 (Inclusion of sequence). *A multidimensional sequence* $\varsigma = \langle a_1, \ldots, a_l \rangle$ *is said to be a subsequence of a sequence* $\varsigma' = \langle b_1, \ldots, b_{l'} \rangle$ *if there exist* $1 \leq j_1 \leq j_2 \leq \ldots \leq j_l \leq l'$ *such that* $a_1 \subseteq b_{j_1}, a_2 \subseteq b_{j_2}, \ldots, a_l \subseteq b_{j_l}$.

With $\varsigma = \langle \{(Y, c, 50)\}, \{(M, r, 10)\} \rangle$ and $\varsigma' = \langle \{(Y, c, 50), (M, p, 2)\}, \{(M, r, 10)\} \rangle$, ς is a subsequence of ς'.

4.3 Support

Computing the support of a sequence amounts to count the number of blocks that *support* the sequence. Intuitively, a block supports a sequence ς if (i) for each itemset i in ς there exists a date in $Dom(D_t)$ such that all items in i appear at this date, and (ii) all itemsets in ς are successively retrieved at different and increasing dates.

Definition 5. *A table T supports a sequence* $\langle i_1, \ldots, i_l \rangle$ *if for every* $j = 1, \ldots, l$, *there exists* d_j *in* $Dom(D_t)$ *such that for every item e in* i_j, *there exists* $t = (f, r, e, d_j)$ *in T with* $d_1 < d_2 < \ldots < d_l$.

In our running example, the block $(Educ, NY)$ from Fig. 2.a supports $\varsigma = \langle \{(Y,c,50), (M,p,2)\}, \{(M,r,10)\} \rangle$ since $\{(Y,c,50),(M,p,2)\}$ appears at $date = 1$ and $\{(M,r,10)\}$ appears at $date = 2$.

The support of a sequence in a table T is the proportion of blocks of T that support it.

Definition 6 (Sequence Support). *Let D_R be the reference dimensions and T a table partitioned into the set of blocks B_{T,D_R}. The support of a sequence ς is defined by:*

$$support(\varsigma) = \frac{|\{B \in B_{T,D_R} \mid B \text{ supports } \varsigma\}|}{|B_{T,D_R}|}$$

Definition 7 (Frequent Sequence). *Let $minsup \in [0,1]$ be the minimum user-defined support value. A sequence ς is said to be frequent if $support(\varsigma) \geq minsup$. An item e is said to be frequent if so is the sequence $\langle \{e\} \rangle$.*

In our running example, let us consider $D_R = \{CG, C\}$, $D_A = \{A, P, Q\}$, $minsup = \frac{1}{5}$, $\varsigma = \langle \{(Y,c,50),(M,p,2)\},\{(M,r,10)\} \rangle$. The three blocks of the partition of T from Fig. 2 must be scanned to compute $support(\varsigma)$.

1. **Block $(Educ, NY)$** (Fig. 2.a). In this block, we have $(Y,c,50)$ and $(M,p,2)$ at date 1, and $(M,r,10)$ at date 2. Thus this block supports ς.
2. **Block $(Educ, LA)$** (Fig. 2.b). This block does not support ς since it does not contain $(M,p,2)$.
3. **Block $(Ret., SF)$** (Fig. 2.c). This block does not support ς since it contains only one date.

Thus, we have $support(\varsigma) = \frac{1}{3} \geq minsup$.

5 Jokerized Sequential Patterns

Considering the definitions above, an item can only be retrieved if there exists a frequent tuple of values from domains of D_A containing it. For instance, it can happen that neither (Y,r) nor (M,r) nor (O,r) is frequent whereas the value r is frequent. In this case, we consider $(*,r)$ which is said to be *jokerized*.

Definition 8 (Jokerized Item). *Let $e = (d_1, \ldots, d_m)$ a multidimensional item. We denote by $e_{[d_i/\delta]}$ the replacement in e of d_i by δ. e is said to be a jokerized multidimensional item if: (i) $\forall i \in [1,m], d_i \in Dom(D_i) \cup \{*\}$, and (ii) $\exists i \in [1,m]$ such that $d_i \neq *$, and (iii) $\forall d_i = *, \nexists \delta \in Dom(D_i)$ such that $e_{[d_i/\delta]}$ is frequent.*

A *jokerized* item contains at least one specified analysis dimension. It contains a $*$ only if no specific value from the domain can be set. A *jokerized* sequence is a sequence containing at least one *jokerized* item. A block is said to *support* a sequence if a set of tuples containing the itemsets satisfying the temporal constraints can be found.

Definition 9 (Support of a Jokerized Sequence). *A table T supports a jokerized sequence $\varsigma = \langle i_1, \ldots, i_l \rangle$ if: $\forall j \in [1,l], \exists \delta_j \in Dom(D_t), \forall e = (d_{i_1}, \ldots, d_{i_m}) \in i_j, \exists t = (f, r, (x_{i_1}, \ldots, x_{i_m}), \delta_j) \in T$ with $d_{i_k} = x_{i_k}$ or $d_{i_k} = *$ and $\delta_1 < \delta_2 < \ldots < \delta_l$.
The support of ς is defined by: $support(\varsigma) = \frac{|\{B \in B_{T,D_R} \text{ s.t. } B \text{ supports } \varsigma\}|}{|B_{T,D_R}|}$*

6 Algorithms

6.1 Mining Frequent Items

The computation of all frequent sequences is based on the computation of all frequent multidimensional items. When considering no joker value, a single scan of the database is enough to compute them.

On the other hand, when considering jokerized items, a levelwise algorithm is used in order to build the frequent multidimensional items having as few joker values as possible. To this end, we consider a lattice which lower bound is the multidimensional item $(*,\ldots,*)$. This lattice is partially built from $(*,\ldots,*)$ up to the frequent items containing as few $*$ as possible. At level i, i values are specified, and items at this level are combined to build a set of candidates at level $i+1$. Two frequent items are combined to build a candidate if they are $\bowtie$-compatible.

Definition 10 ($\bowtie$-compatibility). Let $e_1 = (d_1,\ldots,d_n)$ and $e_2 = (d'_1,\ldots,d'_n)$ be two distinct multidimensional items where d_i and $d'_i \in dom(D_i) \cup \{*\}$. e_1 and e_2 are said to be $\bowtie$-compatible if there exists $\Delta = \{D_{i_1},\ldots,D_{i_{n-2}}\} \subset \{D_1,\ldots,D_n\}$ such that for every $j \in [1,n-2]$, $d_{i_j} = d'_{i_j} \neq *$ with $d_{i_{n-1}} = *$ and $d'_{i_{n-1}} \neq *$ and $d_{i_n} \neq *$ and $d'_{i_n} = *$.

Definition 11 (Join). Let $e_1 = (d_1,\ldots,d_n)$ and $e_2 = (d'_1,\ldots,d'_n)$ be two $\bowtie$-compatible multidimensional items. We define $e_1 \bowtie e_2 = (v_1,\ldots,v_n)$ where $v_i = d_i$ if $d_i = d'_i$, $v_i = d_i$ if $d'_i = *$ and $v_i = d'_i$ if $d_i = *$.
Let E and E' be two sets of multidimensional items of size n, we define

$$E \bowtie E' = \{e \bowtie e' \mid (e,e') \in E \times E' \land e \text{ and } e' \text{ are } \bowtie\text{-compatible}\}$$

In our running example, $(NY,Y,*)$ and $(*,Y,r)$ are $\bowtie$-compatible. We have $(NY,Y,*) \bowtie (*,Y,r) = (NY,Y,r)$. On the contrary, $(NY,M,*)$ and $(NY,Y,*)$ are not $\bowtie$-compatible. Note that this method is close to the one used for iceberg cubes in [2,3].

Let F_1^i denote the set of 1-frequent items having i dimensions which are specified (different from $*$). F_1^1 is obtained by counting each value over each analysis dimension, i.e., $F_1^1 = \{f \in Cand_1^1, support(f) \geq minsup\}$. Candidate items of size i are obtained by joining the set of frequent items of size $i-1$ with itself: $Cand_1^i = F_1^{i-1} \bowtie F_1^{i-1}$.

Function supportcount
Data : $\varsigma, T, D_R, counting$ //counting indicates if joker values are considered or not
Result : support of ς
Integer support $\longleftarrow 0$; *Boolean seqSupported*;
$B_{T,D_R} \longleftarrow \{blocks\ of\ T\ identified\ over\ D_R\}$;
foreach $B \in B_{T,D_R}$ **do**
$\quad$ *seqSupported* $\longleftarrow$ *supportTable*$(\varsigma,B,counting)$;
$\quad$ **if** *seqSupported* **then** *support* $\longleftarrow$ *support* $+1$;
return $\left(\frac{support}{|B_{T,D_R}|}\right)$

Algorithm 1: Support of a sequence (supportcount)

```
Function supportTable
Data      : ς, T, counting
Result    : Boolean
ItemSetFound ⟵ false ; seq ⟵ ς ; itset ⟵ seq.first() ; it ⟵ itset.first()
if ς = ∅ then return (true) // End of Recursivity
while t ⟵ T.next ≠ ∅ do
    if supports(t, it, counting) then
        if (NextItem ⟵ itset.second()) = ∅ then ItemSetFound ⟵ true
        // Look for all the items from the itemset
        else
            // Anchoring on the item (date)
            T' ⟵ σ_{date=t.date}(T)
            while t' ⟵ T'.next() ≠ ∅ ∧ ItemSetFound = false do
                if supports(t', NextItem, counting) then NextItem ⟵ itset.next()
                if NextItem = ∅ then ItemSetFound ⟵ true
        if ItemSetFound = true then
            // Anchoring on the current itemset succeeded; test the other itemsets in seq
            return (supportTable(seq.tail(), σ_{date>t.date}(T), counting))
        else
            // Anchoring failure: try anchoring with the next dates
            itset ⟵ seq.first()
            T ⟵ σ_{date>t.date}(T) // Skip to next dates
return(false) // Not found
```

Algorithm 2: supportTable *(Checks if a sequence ς is supported by a table T)*

6.2 Mining Jokerized Multidimensional Sequences

The frequent items give all frequent sequences containing one itemset consisting of a single item. Then, the candidate sequences of size k ($k \geq 2$) are generated and validated against the table T. This computation is based on usual algorithms such as PSP [7] that are adapted for the treatment of joker values.

The computation of the support of a sequence ς according to the reference dimensions D_R is given by Algorithm 1. This algorithm checks whether each block of the partition supports the sequence by calling the function supportTable (Algorithm 2). *supportTable* attempts to find a tuple from the block that matches the first item of the first itemset of the sequence in order to *anchor* the sequence. This operation is repeated recursively until all itemsets from the sequence are found (return true) or until there is no way to go on further (return false). Several possible anchors may have to be tested.

7 Experiments

In this section, we report experiments performed on synthetic data. These experiments aim at showing the interest and scalability of our approach, especially in the jokerized approach. As many databases from the real world include quantitative information, we have distinguished a quantitative dimension. In order to highlight the particular role of this quantitative dimension, we consider four ways of computing frequent sequential patterns: (i) no joker (M^2SP), (ii) jokers on all dimensions but the quantitative one (M^2SP-$alpha$), (iii) jokers only on the quantitative dimension (M^2SP-mu), (iv) jokers on all dimensions (M^2SP-$alpha$-mu). Note that case (iv) corresponds to the jokerized approach presented in Section 5. Our experiments can thus be seen as being conducted in the context of a fact table of a multidimensional database, where the quantitative dimension is the *measure*. In Figures 5-12, minsup is the minimum support taken into account, nb_dim is the number of analysis dimensions being considered, DB_size is the number of tuples, and avg_card is the average number of values in the domains of the analysis dimensions.

Fig. 3 and 4 compare the behavior of the four approaches described above when the support changes. M^2SP-$alpha$ and M^2SP-$alpha$-mu have a similar behavior, the difference being due to the verification of quantities in the case of M^2SP-$alpha$. Note that these experiments are not led with the same minimum support values, since no frequent items are found for M^2SP and M^2SP-mu if the support is too high. Fig. 5 shows the scalability of our approach since runtime grows almost linearly when the database size increases (from 1,000 tuples up to 26,000 tuples).

Fig. 6 shows how runtime behaves when the average cardinality of the domains of analysis dimensions changes. When this average is very low, numerous frequent items are mined among few candidates. On the contrary, when this average is high, numerous candidates have to be considered from which few frequent items are mined. Between these two extrema, the runtime decreases. Fig. 7 and 8 show the behavior of our approach when the number of analysis dimensions changes. The number of frequent items increases as the number of analysis dimensions grows, leading to an increase of the number of frequent sequences. Fig. 9 and 10 show the differential between the number of frequent sequences mined by our approach compared to the number of frequent sequences mined by the approach described in [8], highlighting the interest of our proposition.

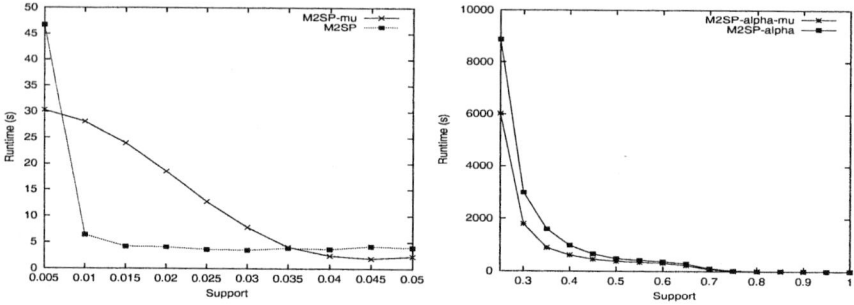

Fig. 3. Runtime over Support (DB_size=12000, nb_dim=5, avg_card=20)

Fig. 4. Runtime over Support (DB_size=12000, nb_dim=5, avg_card=20)

M²SP: Mining Sequential Patterns Among Several Dimensions   215

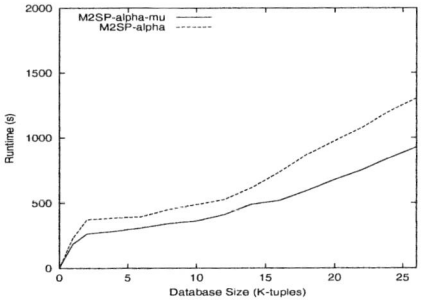 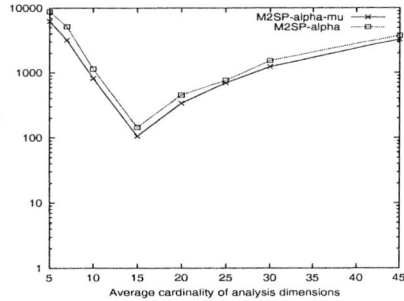

Fig. 5. Runtime over database size (minsup=0.5, nb_dim=15, avg_card = 20)

Fig. 6. Runtime over Average Cardinality of Analysis Dimensions (minsup=0.8, DB_size=12000, nb_dim=15)

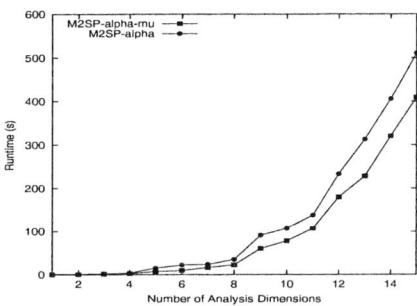 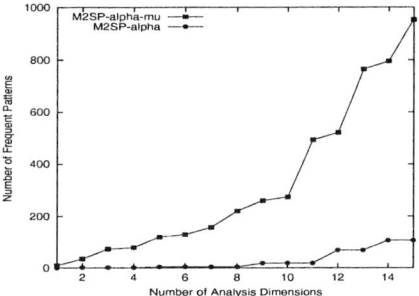

Fig. 7. Runtime over Number of Analysis Dimensions (minsup=0.5, DB_size=12000, nb_dim=15, avg_card=20)

Fig. 8. Number of Frequent patterns over number of analysis dimensions (minsup=0.5, DB_size=12000, nb_dim=15, avg_card=20)

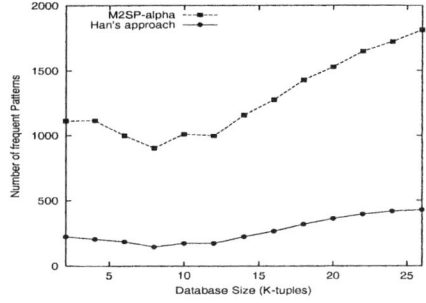 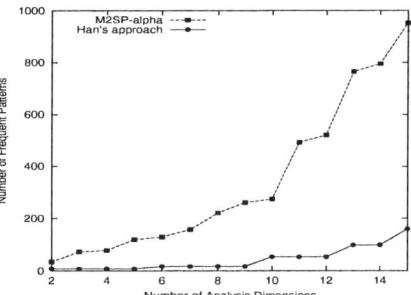

Fig. 9. Number of Frequent Sequences over Database Size (minsup=0.5, nb_dim=15, avg_card=20)

Fig. 10. Number of Frequent Sequences over Number of Analysis Dimensions (minsup=0.5, DB_size=12000, avg_card=20)

8 Conclusion

In this paper, we have proposed a novel definition for multidimensional sequential patterns. Contrary to the propositions [4,8,11], several analysis dimensions can be found in the sequence, which allows for the discovery of rules as *A customer who bought a surfboard together with a bag in NY later bought a wetsuit in LA*. We have also defined *jokerized sequential patterns* by introducing the joker value ∗ on analysis dimensions. Algorithms have been evaluated against synthetic data, showing the scalability of our approach.

This work can be extended following several directions. For example, we can take into account approximate values on quantitative dimensions. In this case, we allow the consideration of values that are not fully jokerized while remaining frequent. This proposition is important when considering data from the real world where the high number of quantitative values prevents each of them to be frequent. Rules to be built will then be like *The customer who bought a DVD player on the web is likely to buy* almost 3 *DVDs in a supermarket later*. Hierarchies can also be considered in order to mine multidimensional sequential patterns at different levels of granularity in the framework of multidimensional databases.

References

1. R. Agrawal and R. Srikant. Mining sequential patterns. In *Proc. 1995 Int. Conf. Data Engineering (ICDE'95)*, pages 3–14, 1995.
2. K. Beyer and R. Ramakrishnan. Bottom-up computation of sparse and iceberg cube. In *Proc. of ACM SIGMOD Int. Conf. on Management of Data*, pages 359–370, 1999.
3. A. Casali, R. Cicchetti, and L. Lakhal. Cube lattices: A framework for multidimensional data mining. In *Proc. of 3rd SIAM Int. Conf. on Data Mining*, 2003.
4. S. de Amo, D. A. Furtado, A. Giacometti, and D. Laurent. An apriori-based approach for first-order temporal pattern mining. In *Simpósio Brasileiro de Bancos de Dados*, 2004.
5. T.G. Dietterich and R.S. Michalski. Discovering patterns in sequences of events. *Artificial Intelligence*, 25(2):187–232, 1985.
6. H. Mannila, H. Toivonen, and A.I. Verkamo. Discovering frequent episodes in sequences. In *Proc. of Int. Conf. on Knowledge Discovery and Data Mining*, pages 210–215, 1995.
7. F. Masseglia, F. Cathala, and P. Poncelet. The PSP Approach for Mining Sequential Patterns. In *Proc. of PKDD*, volume 1510 of *LNCS*, pages 176–184, 1998.
8. H. Pinto, J. Han, J. Pei, K. Wang, Q. Chen, and U. Dayal. Multi-dimensional sequential pattern mining. In *ACM CIKM*, pages 81–88, 2001.
9. R. Srikant and R. Agrawal. Mining Sequential Patterns: Generalizations and Performance Improvements. In *Proc. of EDBT*, pages 3–17, 1996.
10. J.D. Ullman. *Principles of Database and Knowledge-Base Systems*, volume I. Computer Science Press, 1988.
11. C.-C. Yu and Y.-L. Chen. Mining sequential patterns from multidimensional sequence data. *IEEE Transactions on Knowledge and Data Engineering*, 17(1):136–140, 2005.

A Systematic Comparison of Feature-Rich Probabilistic Classifiers for NER Tasks

Benjamin Rosenfeld, Moshe Fresko, and Ronen Feldman

Bar-Ilan University, Computer Science Department, Data-Mining Lab.,
Ramat-Gan, Israel
{freskom1, feldman}@cs.biu.ac.il
http://www.cs.biu.ac.il/~freskom1/DataMiningLab

Abstract. In the CoNLL 2003 NER shared task, more than two thirds of the submitted systems used the feature-rich representation of the task. Most of them used maximum entropy to combine the features together. Others used linear classifiers, such as SVM and RRM. Among all systems presented there, one of the MEMM-based classifiers took the second place, losing only to a committee of four different classifiers, one of which was ME-based and another RRM-based. The lone RRM was fourth, and CRF came in the middle of the pack. In this paper we shall demonstrate, by running the three algorithms upon the same tasks under exactly the same conditions that this ranking is due to feature selection and other causes and not due to the inherent qualities of the algorithms, which should be ranked otherwise.

1 Introduction

Recently, feature-rich probabilistic conditional classifiers became state-of-the-art in sequence labeling tasks, such as NP chunking, PoS tagging, and Named Entity Recognition. Such classifiers build a probabilistic model of the task, which defines a conditional probability on the space of all possible labelings of a given sequence. In this, such classifiers differ from the binary classifiers, such as decision trees and rule-based systems, which directly produce classification decisions, and from the generative probabilistic classifiers, such as HMM-based Nymble [2] and SCFG-based TEG [8], which model the joint probability of sequences and their labelings. Modeling the conditional probability allows the classifiers to have all the benefits of probabilistic systems while having the ability to use any property of tokens and their contexts, if the property can be represented in the form of binary features. Since almost all local properties can be represented in such a way, this ability is very powerful.

There are several different feature-rich probabilistic classifiers developed by different researchers, and in order to compare them, one usually takes a known publicly available dataset, such as MUC-7 [23] or CoNLL shared task [12], and compares the performance of the algorithms on the dataset. However, performance of a feature-rich classifier strongly depends upon the feature sets it uses. Since systems developed by different researches are bound to use different feature sets, the differences in performance of complete systems can not reliably teach us about the qualities of the underlying algorithms.

In this work we compare the performances of three common models (all present in the CoNLL 2003 shared task) – MEMM [15], CRF [16], and RRM (regularized Winnow) [14] – within the same platform, using exactly the same set of features. We also test the effects of different training sizes, different choice of parameters, and different feature sets upon the algorithms' performance.

Our experiments indicate that CRF outperforms MEMM for all datasets and feature sets, which is not surprising, since CRF is a better model of sequence labeling. Surprisingly, though, the RRM performs at the same level or even better than CRF, despite being local model like MEMM, and being significantly simpler to build than both CRF and MEMM.

The following section of the paper we outline the three algorithms. We then present our experiments and their results.

2 Classifiers

The general sequence labeling problem can be described as follows. Given a small set Y of labels, and a sequence $x = x_1 x_2 \ldots x_{l(x)}$, the task is to find a labeling $y = y_1 y_2 \ldots y_{l(x)}$, where each label $y_i \in Y$. In the framework of feature-rich classifiers, the elements x_i of the sequence should not be thought of as simple tokens, but rather as sequence positions, or *contexts*. The contexts are characterized by a set of externally supplied binary *features*. Thus, each context x_i can be represented as a vector $x_i = (x_{i1}, x_{i2}, \ldots, x_{ik})$, where $x_{ij} = 1$ if the j-th feature is present in the i-th context, and $x_{ij} = 0$ otherwise.

The feature-rich sequence classifiers have no knowledge of the nature of features and labels. Instead, in order to make predictions, the classifiers are supplied with a training set $T = \{(x^{(t)}, y^{(t)})\}_{t=1..n}$ of sequences with their intended labelings. The classifiers use the training set to build the model of the task, which is subsequently used to label unseen sequences.

We shall describe the particular algorithms only briefly, referring to the original works to supply the details.

2.1 MEMM

A Maximum Entropy Markov Model classifier [4] builds a probabilistic conditional model of sequence labeling. Labeling each position in each sequence is considered to be a separate classification decision, possibly influenced by a small constant number of previous decisions in the same sequence. In our experiments we use a Markov model of order one, in which only the most recent previous decision is taken into account.

Maximal Entropy models are formulated in terms of *feature functions* $f(x_i, y_i, y_{i-1}) \to \{0, 1\}$, which link together the context features and the target labels. In our formulation, we have a feature function f_{jy} for each context feature j and each label y, and a feature function $f_{jyy'}$ for each context feature and each pair of labels. The functions are defined as follows:

$f_{jy}(x_i, y_i, y_{i-1}) = x_{ij} I_y(y_i)$ and $f_{jyy'}(x_i, y_i, y_{i-1}) = x_{ij} I_y(y_i) I_{y'}(y_{i-1})$, where $I_a(b)$ is one if $a = b$ and zero otherwise. The vector of all feature functions is denoted $f(x_i, y_i, y_{i-1})$.

A trained MEMM model has a real weight λ_f for each feature function f. Together, the weights form the parameter vector λ. The model has the form

$$(1) \quad P_\lambda(y_i \mid x_i, y_{i-1}) = \frac{1}{Z(x_i, y_{i-1})} \exp(\lambda \cdot f(x_i, y_i, y_{i-1})),$$

where $Z(x_i, y_{i-1}) = \sum_y \exp(\lambda \cdot f(x_i, y, y_{i-1}))$ is the factor making the probabilities for different labels sum to one.

Given a model (1), it can be used for inferring the labeling $y = y_1 y_2 \ldots y_{l(x)}$ of an unseen sequence $x = x_1 x_2 \ldots x_{l(x)}$ by calculating the most probable overall sequence of labels:

$$(2) \quad y(x) := \arg\max_{y_1 y_2 \ldots y_{l(x)}} \sum_{i=1}^{l(x)} \log P_\lambda(y_i \mid x_i, y_{i-1}).$$

This most probable sequence can be efficiently calculated using a variant of the Viterbi algorithm.

The model parameters are trained in such a way as to maximize the model's entropy while making the expected value of each feature function agree with the observed relative frequency of the feature function in the training data. Those conditions can be shown to be uniquely satisfied by the model which maximizes the log-likelihood of the training data among all models of the form (1). In order to avoid overfitting, the likelihood can be penalized with a prior $\Pr(\lambda)$. Then, the log-likelihood is

$$L_T(\lambda) = \sum_t \sum_{i=1}^{l(x^{(t)})} \log P_\lambda(y_i^{(t)} \mid x_i^{(t)}, y_{i-1}^{(t)}) - \Pr(\lambda) =$$

$$= \sum_t \sum_{i=1}^{l(x^{(t)})} \left(\lambda \cdot f(x_i^{(t)}, y_i^{(t)}, y_{i-1}^{(t)}) - \log Z(x_i^{(t)}) \right) - \Pr(\lambda)$$

and its gradient is

$$\nabla L_T(\lambda) = \sum_t \sum_{i=1}^{l(x^{(t)})} \left(f(x_i^{(t)}, y_i^{(t)}, y_{i-1}^{(t)}) - E_{P\lambda}(f(x_i^{(t)}, Y, y_{i-1}^{(t)})) \right) - \nabla \Pr(\lambda),$$

where

$$E_{P\lambda}(f(x_i^{(t)}, Y, y_{i-1}^{(t)})) = \sum_{y \in Y} P_\lambda(y \mid x_i^{(t)}, y_{i-1}^{(t)}) f(x_i^{(t)}, y, y_{i-1}^{(t)})$$

is the expectation of the feature vector under the model (1).

With a reasonably chosen prior, the function $L_T(\lambda)$ is strictly concave, and so can be maximized by any convex optimization algorithm. We use L-BFGS for this purpose.

2.2 CRF

A Conditional Random Fields (CRF) [7] classifier also builds a probabilistic model of sequence labeling. CRF uses the maximal entropy principle to model the labeling of a sequence as a whole, in contrast to MEMM, which builds a model of separate labeling decisions at different sequence positions.

The model is built upon exactly the same vector $f(x_i, y_i, y_{i-1})$ of feature functions as MEMM. The feature functions are summed along a sequence to produce a *sequence feature functions vector*

$$(3) \quad F(x,y) = \sum_{i=1}^{l(x)} f(x_i, y_i, y_{i-1}),$$

which is then used for constructing the maximal entropy model

$$P_\lambda(y \mid x) = \frac{1}{Z(x)} \exp(\lambda \cdot F(x,y)).$$

A trained model can be used for inferring the most probable labeling of an unseen sequence. The decomposition (3) allows to use the Viterbi algorithm almost identically to the MEMM case, except that in (2), instead of $\log P_\lambda(y_i \mid x_i, y_{i-1}) = \lambda \cdot f(x_i, y_i, y_{i-1}) - \log Z(x_i, y_{i-1})$, simple $\lambda \cdot f(x_i, y_i, y_{i-1})$ is used. Since $Z(x)$ does not depend on labeling, it need not be calculated at all during inference.

To train the CRF model, we need to maximize the model entropy while satisfying the expectation constrains, expressed this time in terms of the sequence feature functions. As before, this is equivalent to maximizing the log-likelihood of the training data, which can also be penalized with a prior to avoid overfitting:

$$L_T(\lambda) = \sum_t \log P_\lambda(y^{(t)} \mid x^{(t)}) - \frac{\|\lambda\|^2}{2\sigma^2} = \sum_t \left(\lambda \cdot F(x^{(t)}, y^{(t)}) - \log Z(x^{(t)}) \right) - \Pr(\lambda).$$

The gradient is

$$\nabla L_T(\lambda) = \sum_t \left(F(x^{(t)}, y^{(t)}) - E_{P_\lambda}(F(x^{(t)}, Y^{(t)})) \right) - \nabla \Pr(\lambda),$$

where $Y^{(t)}$ is the set of label sequences of length $l(x^{(t)})$, and

$$E_{P_\lambda}(F(x^{(t)}, Y)) = \sum_{y \in Y} P_\lambda(y \mid x^{(t)}) F(x^{(t)}, y)$$

is the expectation of the sequence feature functions vector under the model (3).

In order to maximize $L_T(\lambda)$, we need a way to calculate $\log Z(x)$ and $E_{P_\lambda}(F(x, Y))$ for the given sequence x. It is possible to do this efficiently, using a variant of the Forward-Backward algorithm. Details can be found in [7] and [19].

2.3 RRM

The Robust Risk Minimization classifier [14] results from regularization of the Winnow algorithm [21]. Winnow is a multiplicative-update online algorithm used for estimating the weights of a binary linear classifier, which has the following general form:

$$y = \text{sign}(w^T x),$$

where x is the input vector, w is the weight vector, and $y \in \{+1, -1\}$ is the classification decision.

It was shown in [20], that using a risk function of a special form, the regularized Winnow can produce such weights w that

$$P(y = +1 \mid \mathbf{x}) \approx (Tr_{[-1,1]}(\mathbf{w}^T\mathbf{x}) + 1)/2,$$

where $Tr_{[a,b]}(s) = \min(b, \max(a, s))$ is a truncation of s onto $[a, b]$.

Although the derivation is elaborate, the resulting algorithm is very simple. It consists of iteratively going over the training set $T = \{(\mathbf{x}^{(t)}, y^{(t)})\}_{t=1..n}$ (here, $y^{(t)} = \pm 1$), and incrementally updating

(4)
$$\alpha_t := Tr_{[0,2c]}\left(\alpha_t + \eta\left(1 - \frac{\alpha_t}{c} - \mathbf{w}^T\mathbf{x}^{(t)}y^{(t)}\right)\right)$$

$$w_j := \mu_j \exp\left(\sum_t \alpha_t x_j^{(t)} y^{(t)}\right)$$

The α_t are the *dual* weights, initialized to zero and kept between the iterations. c is the *regularization* parameter, η is the *learning rate*, and μ_j is the *prior*.

The $y^{(t)}$ in (4) are binary decisions. In order to use the RRM for sequence labeling task with more than two labels, we can build a separate classifier for each label and then combine them together within a single Viterbi search.

3 Experimental Setup

The goal of this work is to compare the three sequence labeling algorithms in several different dimensions: absolute performance, dependence upon the corpus, dependence upon the training set size and the feature set, and dependence upon the hyperparameters.

3.1 Datasets

For our experiments we used four datasets: CoNLL-E, the English CoNLL 2003 shared task dataset, CoNLL-D, the German CoNLL 2003 shared task dataset, the MUC-7 dataset [23], and the proprietary CLF dataset [8]. For the experiments with smaller training sizes, we cut training corpora into chunks of 10K, 20K, 40K, 80K, and 160K tokens. The corresponding datasets are denoted <Corpus>_<Size>, e.g. "CoNLL-E_10K".

3.2 Feature Sets

There are many properties of tokens and their contexts that could be used in a NER system. We experiment with the following properties, ordered according to the difficulty of obtaining them:

A. The exact character strings of tokens in a small window around the given position.
B. Lowercase character strings of tokens.
C. Simple properties of characters inside tokens, such as capitalization, letters vs digits, punctuation, etc.
D. Suffixes and prefixes of tokens with length 2 to 4 characters.

E. Presence of tokens in local and global dictionaries, which contain words that were classified as certain entities someplace before – either anywhere (for global dictionaries), or in the current document (for local dictionaries).
F. PoS tags of tokens.
G. Stems of tokens.
H. Presence of tokens in small manually prepared lists of semantic terms – such as months, days of the week, geographical features, company suffixes, etc.
I. Presence of tokens inside gazetteers, which are huge lists of known entities.

The PoS tags are available only for the two CoNLL datasets, and the stems are available only for the CoNLL-D dataset. Both are automatically generated and thus contain many errors.

The gazetteers and lists of semantic terms are available for all datasets except CoNLL-D.

We tested the following feature sets:
set0: checks properties A, B, C at the current and the previous token.
set1: A, B, C, B+C in a window [-2...0].
set2: A, B, C, B+C in a window [-2...+2].
set2x: Same as set2, but only properties appearing > 3 times are used.
set3: A, B, C, B+C in a window [-2...+2], D at the current token.
set4: A, B, C, B+C in a window [-2...+2], D at the current token, E.
set5: A, B, C, B+C, F, G in a window [-2...+2] , D at the current token, E.
set6: set4 or set5, H
set7: set4 or set5, H, I

3.3 Hyperparameters

The MaxEntropy-based algorithms, MEMM and CRF, have similar hyperparameters, which define the priors for training the models. We experimented with two different priors – Laplacian (double exponential) $\Pr_{LAP}(\lambda) = \alpha \Sigma_i |\lambda_i|$ and Gaussian $\Pr_{GAU}(\lambda) = (\Sigma_i \lambda_i^2) / (2\sigma^2)$. Each prior depends upon a single hyperparameter specifying the "strength" of the prior. Note, that $\nabla \Pr_{LAP}(\lambda)$ has discontinuities at zeroes of λ_i. Because of that, a special consideration must be given to the cases when λ_i approaches or is at zero. Namely,

(1) if λ_i tries to change sign, set $\lambda_i := 0$, and allow it to change sign only on the next iteration, and

(2) if $\lambda_i = 0$, and $\left|\frac{\partial}{\partial \lambda_i} L_T(\lambda)\right| < \alpha$, do not allow λ_i to change, because it will immediately be driven back toward zero.

In some of the previous works (e.g., [22]) the Laplacian prior was reported to produce much worse performance than the Gaussian prior. Our experiments show them to perform similarly. The likely reason for the difference is poor handling of the zero discontinuities.

The RRM algorithm has three hyperparameters – the prior μ, the regularization parameter c, and the learning rate η.

4 Experimental Results

It is not possible to test every possible combination of algorithm, dataset and hyperparameter. Therefore, we tried to do a meaningful series of experiments, which would together highlight the different aspects of the algorithms.

All of the results are presented as final microaveraged F1 scores.

4.1 Experiment 1

In the first series of experiments we evaluated the dependence of the performance of the classifiers upon their hyperparameters. We compared the performance of the

Table 1. RRM results on CoNLL-E dataset

	CoNLL-E_40K_set7			CoNLL-E_80K_set7			CoNLL-E_160K_set7		
$\mu=0.01$	$c=0.001$	$c=0.01$	$c=0.1$	$c=0.001$	$c=0.01$	$c=0.1$	$c=0.001$	$c=0.01$	$c=0.1$
$\eta=0.001$	78.449	78.431	78.425	81.534	81.534	81.510	84.965	84.965	84.965
$\eta=0.01$	**85.071**	**85.071**	84.922	87.766	87.774	87.721	**90.246**	90.238	90.212
$\eta=0.1$	82.918	83.025	83.733	**87.846**	87.835	88.031	89.761	89.776	89.904
$\mu=0.1$									
$\eta=0.001$	84.534	84.552	84.534	87.281	87.281	87.264	89.556	89.556	89.573
$\eta=0.01$	85.782	**85.800**	**85.800**	89.032	89.032	**89.066**	**91.175**	**91.175**	**91.150**
$\eta=0.1$	82.439	82.709	83.065	63.032	63.032	63.032	30.741	30.741	56.445
$\mu=1.0$									
$\eta=0.001$	85.973	85.973	**85.990**	**89.108**	**89.108**	89.100	**91.056**	**91.056**	**91.056**
$\eta=0.01$	83.850	83.877	83.904	88.141	88.141	88.119	90.286	90.317	90.351
$\eta=0.1$	0	0	29.937	0	0	0	0	0	0

Table 2. RRM results on other datasets

	CoNLL-D_20K_set7			MUC7_40K_set2x			CLF_80K_set2		
$\mu=0.01$	$c=0.001$	$c=0.01$	$c=0.1$	$c=0.001$	$c=0.01$	$c=0.1$	$c=0.001$	$c=0.01$	$c=0.1$
$\eta=0.001$	43.490	43.490	43.453	48.722	48.722	48.650	49.229	49.229	49.244
$\eta=0.01$	46.440	46.438	**46.472**	63.220	63.207	62.915	64.000	**64.040**	63.710
$\eta=0.1$	44.878	44.943	45.995	61.824	62.128	**63.678**	58.088	58.628	61.548
$\mu=0.1$									
$\eta=0.001$	44.674	44.674	44.671	60.262	60.249	60.221	59.943	59.943	59.943
$\eta=0.01$	44.799	44.845	**44.957**	65.529	**65.547**	65.516	**64.913**	**64.913**	64.811
$\eta=0.1$	43.453	43.520	44.192	60.415	60.958	63.120	55.040	55.677	60.161
$\mu=1.0$									
$\eta=0.001$	44.682	44.682	**44.694**	**66.231**	**66.231**	66.174	**65.408**	**65.408**	**65.408**
$\eta=0.01$	43.065	43.080	43.195	62.622	62.579	62.825	59.197	59.311	59.687
$\eta=0.1$	0	0	6.123	2.922	2.922	8.725	0	0	1.909

Table 3. CRF results on a selection of datasets

CRF	CLF			CoNLL-D			MUC7	CoNLL-E
	20K_set2	40K_set2	80K_set2	40K_set1	80K_set1	160K_set1	80K_set0	80K_set0
GAU σ = 1	**76.646**	**78.085**	80.64	29.851	35.516	**39.248**	**80.756**	69.247
GAU σ = 3	75.222	77.553	79.821	28.530	35.771	38.254	80.355	**69.693**
GAU σ = 5	75.031	77.525	79.285	29.901	35.541	38.671	79.853	69.377
GAU σ = 7	74.463	77.633	79.454	**30.975**	**36.517**	38.748	79.585	69.341
GAU σ = 10	74.352	77.05	77.705	29.269	36.091	38.833	80.625	68.974
LAP α=0.01	73.773	77.446	79.071	29.085	**35.811**	38.947	79.738	69.388
LAP α=0.03	75.023	77.242	78.810	31.082	34.097	38.454	79.044	**69.583**
LAP α=0.05	76.314	77.037	79.404	30.303	35.494	**39.248**	79.952	69.161
LAP α=0.07	74.666	76.329	**80.841**	30.675	34.530	38.882	79.724	68.806
LAP α=0.1	74.985	**77.655**	80.095	**31.161**	35.187	39.234	79.185	68.955

Table 4. MEMM results on a selection of datasets

MEMM	CLF			CoNLL-D			MUC7	CoNLL-E
	20K_set2	40K_set2	80K_set2	40K_set1	80K_set1	160K_set1	80K_set0	80K_set0
GAU σ = 1	**75.334**	**78.872**	**79.364**	**30.406**	35.013	**40.164**	**78.773**	67.537
GAU σ = 3	74.099	75.693	77.278	28.484	**35.330**	40.005	77.295	67.401
GAU σ = 5	73.959	74.685	77.316	28.526	35.043	39.799	77.489	67.870
GAU σ = 7	73.411	74.505	77.563	28.636	34.630	38.531	77.255	67.897
GAU σ = 10	73.351	74.398	77.379	28.488	33.955	37.830	77.094	**68.043**
LAP α=0.01	71.225	74.04	75.721	28.316	34.329	40.074	**78.312**	67.871
LAP α=0.03	72.603	72.967	76.540	29.086	35.159	38.621	77.385	67.401
LAP α=0.05	71.921	**75.523**	75.370	30.425	33.942	39.984	78.262	**67.908**
LAP α=0.07	72.019	74.486	**77.197**	30.118	**35.250**	39.195	76.646	67.833
LAP α=0.1	**72.695**	75.311	76.335	**30.315**	33.487	**40.861**	78.141	67.421

classifiers on a selection of datasets, with different hyperparameter values. All of the algorithms showed moderate and rather irregular dependence upon their hyperparameters. However, single overall set of values can be selected.

The RRM results are shown in the Table 1 and the Table 2. As can be seen, selecting $\mu = 0.1$, $c = 0.01$ and $\eta = 0.01$ gives reasonably close to optimal performance on all datasets. All subsequent experiments were done with those hyperparameter values.

Likewise, the ME-based algorithms have no single best set of hyperparameter values, but have close enough near-optimal values. A selection of MEMM and CRF results is shown in the Table 3 and Table 4. For subsequent experiments we use CRF with Laplacian prior with $\alpha = 0.07$ and MEMM with Gaussian prior with $\sigma = 1$.

4.2 Training Size

In this series of experiments we evaluated the performance of the algorithms using progressively bigger training datasets: 10K, 200K, 400K, 800K and 1600K tokens. The results are summarized in the Fig.1. As expected, the algorithms exhibit very similar training size vs. performance behavior.

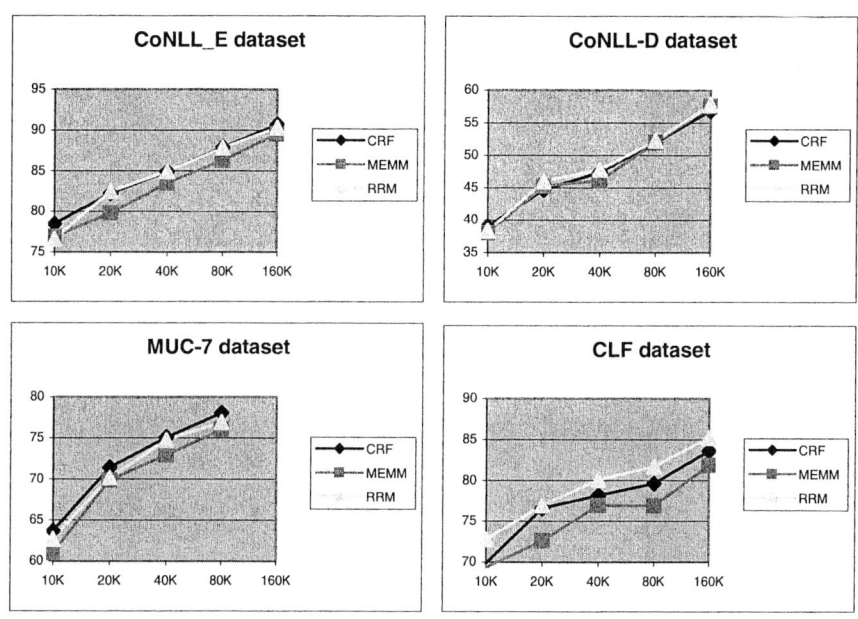

Fig. 1. Performance of the algorithms with different training sizes

Table 5. Performance of the algorithms with different feature sets

	MUC7			CoNLL-D			CoNLL-E		
	CRF	MEMM	RRM	CRF	MEMM	RRM	CRF	MEMM	RRM
set0	**75.748**	66.582	62.206	**48.988**	43.36	40.109	**87.379**	82.281	76.887
set1	**75.544**	67.075	68.405	**50.672**	49.164	48.046	**87.357**	82.516	81.788
set2	**75.288**	74.002	74.755	**52.128**	~52.01	51.537	86.891	87.089	**87.763**
set3	**76.913**	76.333	76.794	~60.172	59.526	**61.103**	88.927	88.711	**89.110**
set4	**78.336**	77.887	77.828	62.79	63.58	**65.802**	~90.037	~90.047	**90.722**
set5				~65.649	65.319	**67.813**	~90.139	~90.115	**90.559**
set6	**78.969**	78.442	78.016				~90.569	~90.492	**90.982**
set7	**81.791**	80.923	81.057				~91.414	90.88	**91.777**

4.3 Feature Sets

In this series of experiments we trained the algorithms with all available training data, but using different feature sets. The results are summarized in the Table 5. The results were tested for statistical significance using the McNemar test. All the performance differences between the successive feature sets are significant at least at the level p=0.05, except for the difference between set4 and set5 in CoNLL-E dataset for all models, and the differences between set0, set1, and set2 in CoNLL-E and MUC7 datasets for the CRF model. Those are statistically insignificant. The differences between the performance of different models that use same feature sets are also mostly significant. Exceptions are the numbers preceded by a tilda "~". Those numbers are not significantly different from the best results in their corresponding rows.

As can be seen, both CRF and RRM generally outperform MEMM. Among the two, the winner appears to depend upon the dataset. Also, it is interesting to note that CRF always wins, and by a large margin, on feature sets 0 and 1, which are distinguished from the set 2 by absense of "forward-looking" features. Indeed, using "forward-looking" features produces little or no improvement for CRF, but very big improvement for local models, probably because such features help to alleviate the *label bias problem* [7].

5 Conclusions

We have presented the experiments comparing the three common state-of-the-art feature-rich probabilistic sentence classifiers inside a single system, using completely identical feature sets. The experiments show that both CRF and RRM significantly outperform MEMM, while themselves performing roughly similarly. Thus, it shows that the comparatively poor performance of CRF in the CoNLL 2003 NER task [16] is due to suboptimal feature selection and not to any inherent flaw in the algorithm itself.

Also, we demonstrated that the Laplacian prior performs just as well and sometimes better than Gaussian prior, contrary to the results of some of the previous researches.

On the other hand, the much simpler RRM classifier performed just as well as CRF and even outperformed it on some of the datasets. The reason of such surprisingly good performance invites further investigation.

References

1. Aitken, J. S.: Learning Information Extraction Rules: An Inductive Logic Programming approach. 15th European Conference on Artificial Intelligence. IOS Press. (2002)
2. Bikel, D. M., Schwartz, R., Weischedel, R.M.: An Algorithm that Learns What's in a Name. Machine Learning. (34): (1999) 211-231.
3. Chieu, H.L., Tou Ng, H.: Named Entity Recognition: A Maximum Entropy Approach Using Global Information. Proceedings of the 17th International Conference on Computational Linguistics. (2002)
4. McCallum, A., Freitag, D., Pereira, F.: Maximum Entropy Markov Models for Information Extraction and Segmentation. Proceedings of the 17th International Conference on Machine Learning. (2000)
5. Sun A., et al.: Using Support Vector Machine for Terrorism Information Extraction. 1st NSF/NIJ Symposium on Intelligence and Security Informatics. (2003)

6. Kushmerick, N., Johnston, E., McGuinness, S.: Information extraction by text classification. IJCAI-01 Workshop on Adaptive Text Extraction and Mining. Seattle, WA. (2001)
7. Lafferty, J., McCallum, A., Pereira, F.: Conditional Random Fields: Probabilistic Models for Segmenting and Labeling Sequence Data. Proc. 18th International Conf. on Machine Learning. (2001)
8. Rosenfeld, B., Feldman, R., Fresko, M., Schler, J., Aumann, Y.: TEG - A Hybrid Approach to Information Extraction. Proc. of the 13th ACM. (2004)
9. Berger, A., della Pietra, S., della Pietra, V.: A maximum entropy approach to natural language processing. Computational Linguistics 22(1), (1996) 39-71.
10. Darroch, J.N., Ratcliff, D.: Generalized iterative scaling for log-linear models. Annals of Mathematical Statistics (1972). 43(5): 1470-1480.
11. Borthwick, A., Sterling, J., Agichtein, E., Grishman, R.: Exploiting Diverse Knowledge Sources via Maximum Entropy in Named Entity Recognition. In the proceedings of the 6th Workshop on Very Large Corpora. (1998)
12. Kim Sang, T., Erik, F., De Meulder, F.: Introduction to the CoNLL-2003 Shared Task: Language-Independent Named Entity Recognition. Edmonton, Canada (2003)
13. Florian, R., Ittycheriah, A., Jing, H., Zhang, T.: Named Entity Recognition through Classifier Combination. In: Proceedings of CoNLL-2003, Edmonton, Canada, (2003), pp. 168-171.
14. Zhang, T., Johnson, D.: A Robust Risk Minimization based Named Entity Recognition System. In: Proceedings of CoNLL-2003, Edmonton, Canada, (2003), pp. 204-207.
15. Chieu, H.L., Tou Ng, H.: Named Entity Recognition with a Maximum Entropy Approach. In: Proceedings of CoNLL-2003, Edmonton, Canada, (2003), pp. 160-163.
16. McCallum, A., Li, W.: Early results for Named Entity Recognition with Conditional Random Fields, Feature Induction and Web-Enhanced Lexicons. In: Proceedings of CoNLL-2003, Edmonton, Canada, (2003), pp. 188-191.
17. Joachims, T.: Learning to Classify Text Using Support Vector Machines. Dissertation, Kluwer, (2002)
18. Nigam, K., Lafferty, J., McCallum, A.: Using maximum entropy for text classification. In IJCAI-99 Workshop on Machine Learning for Information (1999) 61—67.
19. Sha, F., Pereira, F.: Shallow parsing with conditional random fields, Technical Report CIS TR MS-CIS-02-35, University of Pennsylvania, (2003)
20. Zhang, T., Damerau, F., Johnson, D.: Text Chunking using Regularized Winnow. Meeting of the Association for Computational Linguistics. (2001) 539-546
21. Zhang, T., Regularized Winnow Methods. NIPS, (2000) 703-709
22. Peng, F., McCallum, A.: Accurate Information Extraction from Research Papers Using Conditional Random Fields. (1997)
23. Chinchor, N. MUC-7 Named Entity Task Definition Dry Run Version, Version 3.5. Proceedings of the Seventh Message Understanding Conference. (1998)
24. Borthwick, A., Sterling, J., Agichtein, E., Grishman, R. Description of the MENE Named Entity System as Used in MUC-7. Proceedings of the Seventh Message Understanding Conference. (1998)

Knowledge Discovery from User Preferences in Conversational Recommendation

Maria Salamó, James Reilly, Lorraine McGinty, and Barry Smyth

Smart Media Institute, University College Dublin,
Belfield, Dublin 4, Ireland
{maria, james.d.reilly, lorraine.mcginty, barry.smyth}@ucd.ie

Abstract. Knowledge discovery for personalizing the product recommendation task is a major focus of research in the area of conversational recommender systems to increase efficiency and effectiveness. Conversational recommender systems guide users through a product space, alternatively making product suggestions and eliciting user feedback. Critiquing is a common and powerful form of feedback, where a user can express her feature preferences by applying a series of directional critiques over recommendations, instead of providing specific value preferences. For example, a user might ask for a *'less expensive'* vacation in a travel recommender; thus *'less expensive'* is a critique over the *price* feature. The expectation is that on each cycle, the system discovers more about the user's *soft* product preferences from minimal information input. In this paper we describe three different strategies for knowledge discovery from user preferences that improve recommendation efficiency in a conversational system using critiquing. Moreover, we will demonstrate that while the strategies work well separately, their combined effort has the potential to considerably increase recommendation efficiency even further.

1 Introduction

Recommender systems apply knowledge discovery techniques to decide which recommendations are the most suitable for each user during a live customer interaction. In this paper, we focus on conversational case-based recommenders [1] which help users navigate product spaces by combining ideas and technologies from information retrieval, artificial intelligence and user modelling. As part of their cyclic recommendation process, conversational systems aim to retrieve products that respect user preferences by requiring users to provide minimal feedback in each cycle. It is expected that over the course of a recommendation session that the recommender learns more about user preferences and therefore it can assist in the discovery of recommendation knowledge which prioritises products that best satisfy these preferences [2,3]. Advantages of the approach include: (1) users have more control over the navigation process [4]; and (2) users are guided to target products faster than standard browsing and alternative recommendation approaches [5,6].

Recommender systems can be distinguished by the type of feedback that they support; examples include *value elicitation, ratings-based feedback* and *preference-based feedback* [7]. In this paper we are especially interested in a form of user feedback called *critiquing* [8], where a user indicates a directional feature preference in relation to the current recommendation. For example, in a travel/vacation recommender, a user might indicate that she is interested in a vacation that is *longer* than the currently recommended option; in this instance, *longer* is a critique over the *duration* feature.

Within the recommender systems literature the basic idea of critiquing can be traced back to the seminal work of Burke *et al.* [4]. For example, Entrée is the quintessential recommender system that employs critiquing (also sometimes referred to as *tweaking*) in the restaurant domain. Entrée allows users to critique restaurant features such as *price, style, atmosphere* etc. Importantly, critiquing is a good example of a minimal user feedback approach where the user does not need to provide a lot of specific preference information, while at the same time it helps the recommender to narrow its search focus quite significantly [8]. As recommender systems become more commonplace, there has been renewed interest in critiquing, with the major focus of research in increasing the efficiency of recommendation dialogues [9]. Furthermore, recent research has highlighted the importance of investigating techniques for automating the discovery of implicit preference knowledge while requiring minimal information input from the user [3].

In this paper we describe three strategies for knowledge discovery from user preferences that improves the performance of a critique-based recommender. Specifically, we build upon work previously described by [10], where the idea is to consider a user's critiquing history, as well as the current critique when making new recommendations. This approach leads to significant improvements in recommendation efficiency. We continue in this paper by considering the history of critiques as a *user model* which determines the user preferences in a session. We present a case discovery strategy, a feature discovery strategy and a query discovery strategy which have the potential to focus rapidly on satisfactory product cases. Finally, we show that by combining all three strategies, we can further improve overall recommendation efficiency.

2 Background

This section describes the "incremental critiquing" [10] approach, which offers major benefits in recommendation efficiency over the basic critiquing approach as described by [11]. We consider the incremental critiquing approach as the basis for our knowledge discovery strategies because it also considers a user's critiquing history as a basic starting point.

The incremental critiquing implementation assume a conversational recommender system in the style of Entrée [11]. Each recommendation session starts with an initial user query resulting in the retrieval of a case with the highest *quality*. The user will have the opportunity to accept this case, thereby ending

```
q: query, CB: CaseBase, cq: critique, c_r : current recommendation, U : User model

1.   define Incremental_Critiquing(q, CB)          17.  define UserReview(c_r , CB)
2.     cq:= null                                    18.    cq ← user critique for some f ∈ c_r
3.     U:= null                                     19.    CB ← CB - c_r
4.     begin                                        20.    return cq
5.       do
6.         c_r ← ItemRecommend(q, CB, cq, U)        21.  define QueryRevise(q, c_r)
7.         cq ← UserReview(c_r, CB)                 22.    q ← c_r
8.         q ← QueryRevise(q, c_r)                  23.    return q
9.         U ← UpdateModel(U, cq, c_r)
10.      until UserAccepts(c_r)                     24.  define UpdateModel(U, cq, c_r)
11.    end                                          25.    U ← U - contradict(U, cq, c_r)
                                                    26.    U ← U - refine(U, cq, c_r)
12.  define ItemRecommend(q, CB, cq, U)             27.    U ← U + (<cq, c_r>)
13.    CB' ← {c ∈ CB | Satisfies(c, cq)}            28.    return U
14.    CB' ← sort cases in CB' in decreasing Quality
15.    c_r ← most quality case in CB'
16.    return c_r
```

Fig. 1. The incremental critiquing algorithm

the recommendation session, or to critique it as a means to influence the next cycle. A simplified version of the incremental critiquing algorithm is given in Figure 1.

The incremental critiquing algorithm consists of 4 key steps: (1) a new case c_r is *recommended* to the user based on the current query and the previous critiques; (2) the user *reviews* the recommendation and applies a directional feature critique, cq; (3) the query, q is *revised* for the next cycle; (4) the user model, U is updated by adding the last critique cq and pruning all the critiques that are inconsistent with it. The recommendation process terminates either when the user is presented with a suitable case, or when they give up.

Importantly, the recommendation process is influenced by the user model of previous critiques, U, that is incrementally updated on each cycle. Incremental critiquing modifies the basic critiquing algorithm. Instead of ordering the filtered cases on the basis of their similarity to the recommend case, it also computes a *compatibility* score (see Equation 1) for each candidate case. The compatibility score is essentially the percentage of critiques in the user model that this case satisfies. Then, the compatibility score and the candidate's (c') similarity to the current recommendation (c_r) are combined in order to obtain an overall *quality* score (see Equation 2, by default $\beta = 0.75$). The quality score is used to rank the filtered cases prior to the next recommendation cycle (see line 14 in Figure 1) and the case with the highest quality is then chosen as the new recommendation.

$$Compatibility(c', U) = \frac{\sum_{\forall i} satisfies(U_i, c')}{|U|} \quad (1)$$

$$Quality(c', c_r, U) = \beta \cdot Compatibility(c', U) + (1 - \beta) \cdot Similarity(c', c_r) \quad (2)$$

Algorithm 1 maintains a critique-based user model which is composed of those critiques that have been chosen by the user so far. One of the key points focused on the incremental critiquing approach is the maintenance of the user model which prevents the existence of critiques that may be inconsistent with earlier critiques. The user model is maintained using two possible actions: (1) pruning previous critiques that are inconsistent with the current critique; (2) removing all existing critiques, for which the new critique is a refinement.

3 Knowledge Discovery Strategies

This section presents three strategies for knowledge discovery from user preferences that have the potential to improve product recommendations. The first strategy discovers those cases that better satisfy the user preferences. The second strategy deals with feature dimensionality, taking into account user preferences to discover the relative importance of each feature for computing similarity. Finally, this paper presents a query discovery strategy that facilitates larger jumps through the product space based on the current user critique. All of them share a common foundation, they exploit the user's history of critiques to discover recommendation knowledge from the user preferences, in order to personalize and focus in more rapidly on satisfactory product cases.

3.1 Discovering Satisfactory Cases: Highest Compatibility Selection

A key problem with the standard incremental critiquing [10] approach is that there are no guarantees that the recommendations it returns will completely satisfy a user's preferences. This is largely due to the underlying case selection strategy which averages the compatibility (with past critiques) and similarity (with the current preference case). What we propose is a new strategy for product recommendation, *Highest Compatibility Selection* (HCS), that allows the recommender to select the most *suitable* cases; i.e., those cases that are most compatible with user preferences. This maximum compatibility strategy can be easily introduced into the incremental critiquing algorithm.

Figure 2 demonstrates that the only procedure which is affected in the incremental critiquing algorithm is the *ItemRecommend* step. As before, the list of remaining cases is filtered out using the current critique cq. In addition two new steps are added. First, the recommender computes the compatibility score, as detailed in Equation 3. It is important to note that the compatibility function has also been modified, as explained below in Equation 3. Instead of averaging the compatibility and the similarity, as is done with incremental critiquing, our second step assembles the cases with the highest compatibility from the list of remaining cases. Importantly, in our approach, only the remaining cases with highest compatibility value, CB'', influence the product recommendation. Put differently, the strategy prioritises those cases that satisfy the largest number of critiques made by the user over time.

The compatibility function. We have considered case discovery to be an optimization problem in which we are trying to recommend cases that maximally

```
q: query, CB: CaseBase, cq: critique, c_r : current recommendation, U: User Model

1. define ItemRecommend(q, CB, cq, U)
2.   CB'  ← {c ∈ CB | Satisfies(c, cq)}
3.   CB'  ← sort cases in CB' in decreasing compatibility score
4.   CB'' ← selects those cases in CB' with highest compatibility
5.   CB'' ← sort cases in CB'' in decreasing order of their sim to q
6.   c_r  ← most similar case in CB''
7. return c_r
```

Fig. 2. Adapting the incremental critiquing algorithm *ItemRecommend* procedure to improve focus on recommendation by using Highest Compatibility Selection strategy

satisfy the user preferences. For this reason, we evaluate the remaining cases as if they were a set of states in a *Reinforcement Learning Problem* (RLP) [12], which consists of maximising the sum of future rewards in a set of states. Reinforcement Learning theory is usually based on *Finite Markov Decision Processes* (FMDP).

Each case is treated as a state whose compatibility score is updated at each cycle using a Monte-Carlo value function (see Equation 3). This function evaluates the *goodness* of each state — for us the possible states are the complete set of remaining cases we want to enhance — according to the critiques the user has selected.

$$Compatibility(c', U_f) = \begin{cases} comp(c') + \alpha \times (1 - comp(c')) & \text{if } c' \text{ satisfies } U_f \\ comp(c') + \alpha \times (0 - comp(c')) & \text{if } c' \text{ dissatisfies } U_f \end{cases} \quad (3)$$

Our goal is to maximally satisfy all the user preferences. Thus, we are looking for a set of maximally compatible cases (i.e., those cases which have the highest compatibility *(comp)* value considering all the user preferences (U) or past critiques). At the beginning of each session each candidate case, c', has a default compatibility value (i.e., $comp(c') = 0.5$). This value is updated over cycles taking into account the satisfaction or not of the current critique. The α parameter in Equation 3 is the learning rate which is usually set up to 0.1 or 0.2 values; a larger value leads to a larger gap between cases in early stages. In our case, the learning rate is not important since we are looking for levels of satisfaction. In other words, we are not trying to obtain a set of states that arrive as quickly as possible to a 1.0 value, as usually is done in RLP.

It is important to note that Equation 3 updates the compatibility value stored by each case according to the last user critique (U_f) as opposed to computing all the set of critiques like the incremental approach (see Equation 1). The $Compatibility(c', U_f)$ value computed in the current cycle will be the ($comp(c')$) in the next cycle.

3.2 Discovering Important Features: Local User Preference Weighting

The previous strategy highlights the case dimensionality problem. In other words, it is focused on discovering cases that maximally satisfy user preferences.

Now, we present a strategy that concentrates on the feature dimensionality. We propose a *local user preference weighting* (LW) strategy that discovers the relative importance of each feature in each case as a weighting value for computing the similarity, taking into account user preferences.

Our LW strategy for the discovery of feature knowledge is basically motivated by the previous knowledge discovery strategy. As we have explained in Section 3.1, the discovery of case knowledge is based on maximising user preferences, which means we are looking for the most compatible cases. These cases are quite similar on their critiqued features and their differences mainly belong to those features that have not yet been critiqued. So, the aim of LW strategy is to prioritise the similarity of those features that have not yet been critiqued.

for each feature f in case c' compute:
$$weight(c'_f) = 1 - \left(\frac{\#critiques\ in\ U\ that\ satisfy\ feature_f\ in\ case\ c'}{\#total\ critiques\ feature_f\ in\ U} \times 0.5 \right) \quad (4)$$

We generate a feature weight vector for each case, as shown in Equation 4. A feature that has not been critiqued will assume a weight value of 1.0 and a decrement will be applied when a critique is satisfied by the case. As such, the feature weight will be proportional to the number of times a critique on this feature is satisfied by the case. However, as it can be seen in Equation 4 the weights never decrease to a 0 value. For example, in a travel vacation recommender with a user model that contains two critiques [price, >, 1000] and [price, >, 1500], a case with two features {duration, price} whose price is 2000 will have as price weight a 0.5 value because it satisfies both critiques whereas the duration weight will be 1.0 because there is no critique on this feature. It is important to recap that the key idea here is to prioritise the similarity of those features that have not yet been critiqued in a given session.

Our proposal is to discover the best product to recommend by exploiting the similarity of those features that best differentiate the highest compatible cases. To achieve this, a candidate's (c') similarity to the recommended case (c_r) is computed at each cycle in the *incremental* recommender system as shown by Equation 5.

$$Similarity(c', c_r) = \sum_{\forall f} weight(c'_f) \times similarity(c'_f, c_{r_f}) \quad (5)$$

The similarity between the candidate case (c') and the recommended case (c_r) for each feature f is combined with the weight for this feature. The weight is computed previously using Equation 4.

3.3 Discovering Query Knowledge: Binary Search

The incremental critiquing approach is susceptible to feature-critique repetitions that offer only a minor change in the relevant feature value from each cycle to the next. We propose that this is largely due to the linear search policy it uses to navigate through the value-space for the critiqued feature. The result is that the recommendation system takes very *short steps* through the space of possible alternatives. In this section we describe how the incremental critiquing algorithm can be easily altered to facilitate *larger jumps* through the value space for knowledge discovery of a given feature by taking a more efficient *binary search* (BS) approach.

```
q: query, CB: CaseBase, cq: critique, c_r : current recommendation
1.   define QueryRevise (q, cq, CB, c_r)
2.   begin
3.      q  ← c_r
4.      CB' ← {c ∈ CB | Satisfies(c, cq)}
5.      CB' ← eliminate cases that conflict with prior critiques
6.      f_cq ← set value in q for critiqued feature f ∈ c_r by Eq. 6
7.      return q
8.   end
```

Fig. 3. Illustrating the binary search procedure for query discovery

Figure 3 demonstrates how the incremental algorithm can be easily extended to support our proposed approach. The only procedure which is affected is the *QueryRevise* step of the incremental critiquing algorithm. The new query is updated with all of the features from the current recommendation, c_r. In addition two new steps are added. First, the recommender system gathers all of the available cases that satisfy current feature critique (see line 4 of Figure 3). The second step involves determining the value-change the critiqued feature will take on in the revised query, q, used for retrieval of the next recommendation. Importantly, in our approach all the remaining cases, CB', influence the final value. There are many approaches that could be used to compute this value. In this paper we examine the possibility of computing the median (see equation 6) for all cases in CB'.

The recommender system collects all of the alternative value possibilities for the critiqued feature from the cases covered by CB'. For instance, if the critiqued feature were [price, <, 2000] the recommender would gather all value options that were less than 2000 from the set of remaining cases (e.g., 1800, 1650, 1600, 1570, 1460, 1350, etc.). Equation 6 assigns a value for the critiqued feature $f_{cq} \in q$ by calculating the average feature value over all the relevant cases.

$$f_{cq} = \begin{cases} CB'_{n+1/2}(f \text{ of } cq) & \text{if odd } \#cases \\ \frac{CB'_{n+1/2}(f \text{ of } cq) + CB'_{(n+1/2)+1}(f \text{ of } cq)}{2} & \text{if even } \#cases \end{cases} \quad (6)$$

For Equation 6 it is assumed that the remaining case options, CB' are first sorted in ascending order. Here $CB'_i(f \text{ in } cq)$ is the feature value critiqued by cq in the ith case. The median value corresponds to a cumulative percentage of 50% (i.e., 50% of the values are below the median and 50% of the values are above the median). We place the critiqued features in ascending value order and find the middle value if the number of cases is odd or find the middle pair and compute the mean value between them if we have an even number of cases.

One important point, that also needs to be considered, is previous critiques on the same feature. For example, suppose that a user has asked in a previous cycle for a *less expensive* vacation than a €2500 recommendation and, in the current cycle, the user says that she prefers a *more expensive* than a €1000 vacation. In such situation in the current cycle, all the cases including those that exceed a €2500 vacation will satisfy the current critique *more expensive* than €1000. If we compute the median value to jump larger in the search space, we also include those cases rejected previously by the user. To avoid these situations, we use the history of critiques applied by the user in order to cut correctly off the search space. The previous critiques stored in the user model are treated as a set of *soft constraints* [13] that allow us to control the number of remaining cases that will be used to compute the median value.

So, following the earlier example, we only consider computing the median of those cases that are *more expensive* than €1000 and *less expensive* than €2500. As detailed in line 5 of Figure 3, before computing the median, we check for the existence of previously applied critiques that contest the inclusion of cases in CB', and eliminate these cases from further consideration. Put differently, we use prior critiques to decide what cases should be covered by CB', and to ultimately set the value selection bounds for f_{cq}.

The key motivation behind our *binary search* extension to incremental critiquing was to reduce critique repetition sequences, and improve recommendation efficiency by discovering satisfactory products for users more rapidly. In short, this binary search style approach enables the recommender to focus its search on those candidate cases that: (1) satisfy the current critique; (2) fulfill previously applied critiques; and (3) are similar to the current case but further away from it, and thus have the capability of navigating the search space of options quickly.

4 Evaluation

In this paper so far we have argued that the incremental form of critiquing is limited by its tendency to recommend cases that do not maximally satisfy the user preferences. We propose three strategies that aid knowledge discovery in a quest to improve retrieval accuracy and recommendation efficiency. This section describes the related evaluation methodology that we used and the results that ensued.

4.1 Setup

The evaluation was performed using the standard Travel dataset (available from *http://ww.ai-cbr.org*) which consists of 1024 vacation cases. Each case is described in terms of 9 features including *price*, *duration*, etc. The dataset was chosen because it contains numerical and nominal features and it also provides a wide search space.

We evaluate the highest compatibility selection (HCS), the local user preference weighting (LW), the binary search (BS) and also all strategies combined in our recommender (ALL) over incremental critiquing (incremental).

4.2 Methodology

We would like to have carried out an online evaluation with live-users, but unfortunately this was not possible. As an alternative we opted for an offline evaluation similar to the one described by [14]. Accordingly, each case (which are called the 'base') in the case-base is temporarily removed and used in two ways. First, it serves as a basis for a set of queries by taking random subsets of its features. We focus on subsets of 1, 3 and 5 features to allow us to distinguish between *hard*, *moderate* and *easy* queries respectively. Second, we select the case that is most similar to the original base. These cases are the recommendation targets for the experiments. Thus, the base represents the ideal query for a user, the generated query is the initial query provided by the 'user', and the target is the best available case for the user. Each generated query is a test problem for the recommender, and in each recommendation cycle the 'user' picks a critique that is compatible with the known target case; that is, a critique that when applied to the remaining cases, results in the target case being left in the filtered set of cases. Each leave-one-out pass through the case-base is repeated 10 times and the recommendation sessions terminate when the target case is returned.

Related real user studies [15] have highlighted discrepancies between the original [14] artificial user model construction and real user behaviour with respect to critiqued application and repetition. We use a modified artificial user model that is informed by our real-user studies. The new model is designed to respond to recommendations in a manner that is more consistent with the responses recorded from real-users. In particular, our artificial user model repeats critique selections during recommendation sessions until its target feature values are met. For example, suppose our artificial user is looking for a 3-week vacation and they are presented a 3-day city-break. They are likely to ask for a *longer* vacation by critiquing the *duration* feature. In this evaluation, the artificial user will continue to critique a feature until it's preferences constraint is satisfied.

4.3 Recommendation Efficiency

We analyse the recommendation efficiency — by which we mean average recommendation session length — when comparing the new strategies to incremental critiquing. Figure 4(A) presents a graph comparing the average session length of the incremental critiquing approach to the combination of all the strategies (ALL) for 3 different initial query lengths. The three strategies combined consistently reduce average session length when compared to the incremental critiquing approach, demonstrating the potential to improve recommendation efficiency. For example, for the hard queries the incremental recommender results in an average on session of 12.46 cycles while the combined recommender results in an average of 11.47 cycles.

Figure 4(B) shows the benefit of each strategy (HCS, LW, and BS) separately and the combined strategies (ALL) in our recommender when compared to the incremental critiquing. We find that all strategies separately result in a relative session length reduction of between 2.65% and under 7.5%, with some variation in the relative benefit due to the HCS, LW and BS approaches. The lowest benefit is for the highest compatibility selection (HCS) approach, which ranges between 2.65% and 3.81%, because it does the same process as the incremental critiquing approach with two little modifications that consists of using a different compatibility measure and a different strategy for discovering the set of cases available for recommendation. Similarly a 3% to 4% benefit is found using the binary search (BS) strategy. On the other hand, the local weighting approach (LW) gives the highest benefit, ranging from 4.5% to 6.73%, when applied alone. These results show that the strategy to promote uncritiqued features is able to discover and detect differences between cases that are maximally compatible to the user critiques.

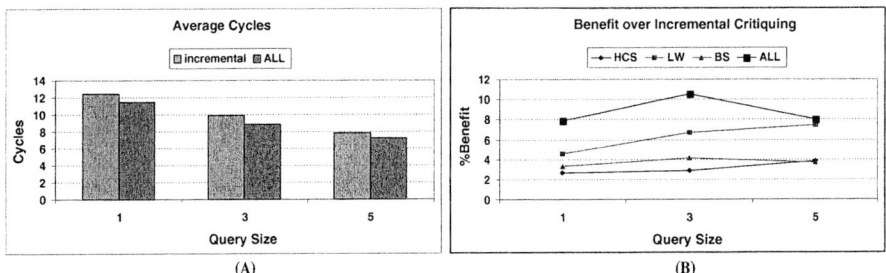

Fig. 4. Average session length and benefit over incremental critiquing

On the other hand, the combined strategies in our recommender result in a reduction in session length that ranges from nearly 8% to 10.5%. Combining all of the strategies further enhances recommendation performance, resulting in the discovery of better recommendations for all queries (hard, moderate and easy). It seems that the recommenders ability to learn user preferences is greater when combining information from these three distinct knowledge discovery resources. An important point to note is that all results show a lower benefit for easy queries. This is to be expected perhaps since the easy queries naturally result in shorter sessions and thus there are fewer opportunities to find good lower and upper critique bounds to focus the search space properly in the BS strategy, and hence fewer opportunities for the benefit to be felt.

It is worth noting the benefit of the proposed strategies over the basic critiquing algorithm, see Figure 5. We have selected incremental critiquing as a benchmark because it improves on the recommendation efficiency of the basic critiquing algorithm by over 82%. Nevertheless, our combination of approaches has the potential to deliver further reductions in session length (from 83.5% to upper 84%) even with short sessions where the BS approach does not have much of an opportunity to affect the recommendations.

To summarise, a significant efficiency benefit is enjoyed by HCS, LW and BS strategies, when compared to the incremental critiquing approach. The main contribution of this paper is that the proposed strategies assist in the discovery of useful recommendation knowledge, allowing the system to prioritise products that best satisfy the user. We have demonstrated that this approach is highly effective, even in situations where only a minimal knowledge of user preferences is available (e.g., critiquing approach). Furthermore, the results of the combined strategies show a significant increase in recommendation efficiency when compared to incremental critiquing and also to the basic critiquing approach proposed by [11].

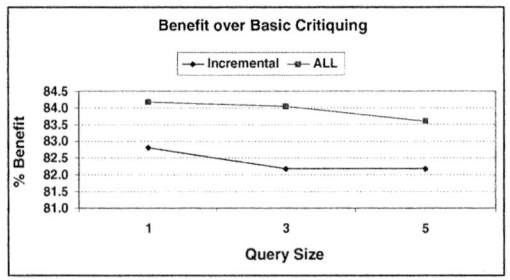

Fig. 5. Incremental and ALL benefit over basic critiquing approach

5 Conclusions

The discovery of implicit user preference knowledge is necessary to decide which product recommendations are the most suitable for each user during a live customer interaction. In this paper we have proposed three discovery strategies that aim to improve recommendation efficiency. First of all, we have presented a case prioritization strategy that maximises the user preferences over time. Secondly, we have presented a user preference weighting strategy that prioritises features locally to each case. Finally, we have presented a query strategy that has the capability of navigating the search space quickly.

Our experiments indicate that the three proposals have the potential to deliver worthwhile efficiency benefits. Reductions in the average length of recommendation sessions were noted in all of the proposals, both separately and combined, when compared to the incremental and basic critiquing setups. Importantly, the proposed strategies are sufficiently general to be applicable across a wide range of recommendation scenarios. In particular, those that assume a complex product-space where recommendation sessions are likely to be protracted, and/or domains where only minimal user feedback is likely to be available.

References

1. D.W. Aha, L.A. Breslow, and H. Muñoz-Avila. Conversational Case-Based Reasoning. *Applied Intelligence*, 14:9–32, 2000.
2. D. McSherry. Increasing Dialogue Efficiency in Case-Based Reasoning without Loss of Solution Quality. In *Proceedings of the 18th International Joint Conference on Artificial Intelligence*, pages 121–126. Morgan-Kaufmann, 2003.
3. D. McSherry and C. Stretch. Automating the Discovery of Recommendation Knowledge. In *Proceedings of the 19th International Joint Conference on Artificial Intelligence*, page forthcoming. Morgan-Kaufmann, 2005.
4. R. Burke, K. Hammond, and B.C. Young. The FindMe Approach to Assisted Browsing. *Journal of IEEE Expert*, 12(4):32–40, 1997.
5. L. McGinty and B. Smyth. Comparison-Based Recommendation. In Susan Craw, editor, *Proceedings of the 6th European Conference on Case-Based Reasoning*, pages 575–589. Springer, 2002. Aberdeen, Scotland.
6. H. Shimazu. ExpertClerk: A Conversational Case-Based Reasoning Tool for Developing Salesclerk Agents in E-Commerce Webshops. *Artificial Intelligence Review*, 18(3-4):223–244, 2002.
7. B. Smyth and L. McGinty. An Analysis of Feedback Strategies in Conversational Recommender Systems. In P. Cunningham, editor, *Proceedings of the 14th National Conference on Artificial Intelligence and Cognitive Science*, 2003. Dublin, Ireland.
8. L. McGinty and B. Smyth. Tweaking Critiquing. In *Proceedings of the Workshop on Personalization and Web Techniques at the International Joint Conference on Artificial Intelligence*. Morgan-Kaufmann, 2003.
9. R. Burke. Interactive Critiquing for Catalog Navigation in E-Commerce. *Artificial Intelligence Review*, 18(3-4):245–267, 2002.
10. J. Reilly, K. McCarthy, L. McGinty, and B. Smyth. Incremental Critiquing. In M. Bramer, F. Coenen, and T. Allen, editors, *Research and Development in Intelligent Systems XXI. Proceedings of AI-2004*, pages 101–114. Springer, 2004. Cambridge, UK.
11. R. Burke, K. Hammond, and B. Young. Knowledge-Based Navigation of Complex Information Spaces. In *Proceedings of the Thirteenth National Conference on Artificial Intelligence*, pages 462–468. AAAI Press/MIT Press, 1996. Portland, OR.
12. M.E. Harmon and S.S. Harmon. Reinforcement learning: A tutorial, 1996.
13. M. Stolze. Soft Navigation in Electronic Product Catalogs. *International Journal on Digital Libraries*, 3(1):60–66, 2000.
14. B. Smyth and L. McGinty. The Power of Suggestion. In *Proceedings of the International Joint Conference on Artificial Intelligence*. Morgan-Kaufmann, 2003.
15. K. McCarthy, L. McGinty, B. Smyth, and J. Reilly. On the Evaluation of Dynamic Critiquing: A Large-Scale User Study. In *Proceedings Twentieth National Conference on Artificial Intelligence*, pages 535–540. AAAI Press / The MIT Press, 2005.

Unsupervised Discretization Using Tree-Based Density Estimation

Gabi Schmidberger and Eibe Frank

Department of Computer Science, University of Waikato,
Hamilton, New Zealand
{gabi, eibe}@cs.waikato.ac.nz

Abstract. This paper presents an unsupervised discretization method that performs density estimation for univariate data. The subintervals that the discretization produces can be used as the bins of a histogram. Histograms are a very simple and broadly understood means for displaying data, and our method automatically adapts bin widths to the data. It uses the log-likelihood as the scoring function to select cut points and the cross-validated log-likelihood to select the number of intervals. We compare this method with equal-width discretization where we also select the number of bins using the cross-validated log-likelihood and with equal-frequency discretization.

1 Introduction

Discretization is applied whenever continuous data needs to be transformed into discrete data. We consider data that is organized into instances with a fixed number of attributes. Some or all of the attributes might be continuous. To discretize a continuous attribute the range of its values is divided into intervals. The attribute values are substituted by an identifier for each bin. Note that the new attribute is actually not categorical but ordered [1]. In this paper we consider the problem of unsupervised discretization, where the discretization is based on the distribution of attribute values alone and there are no class labels. Moreover, we consider univariate discretization: our method only considers the attribute to be discretized, not the values of other attributes.

Our algorithm treats unsupervised discretization as piece-wise constant density estimation. The intervals gained from the discretization can be used to draw a histogram or used to pre-process the data for another data mining scheme. In the former case, the height of each bin h is the density that is computed from the bin width w_i, the number of instances n_i that fall into that bin, and the total number of instances N in the dataset:

$$h = \frac{n_i}{w_i * N} \qquad (1)$$

The paper is organized as follows. In Section 2 we discuss non-parametric density estimation, of which our histogram estimator is a special case. We compare our discretization method to equal-width discretization (and variants) and equal-frequency discretization. These methods are summarized in Section 3. In Section 4 we discuss the cross-validated log-likelihood, which we use as the model selection criterion to choose an appropriate number of bins. Our method is explained in detail in Section 5. An experimental comparison is presented in Section 6. Section 7 has some concluding remarks.

2 Non-parametric Density Estimation

Density estimation, parametric or non-parametric, is about constructing an estimated density function from some given data. Parametric density estimation assumes that the data has a density function that is of a known family of distributions. For example, the distributions of the normal or Gaussian model have the parameters μ for the mean and σ^2 for the variance. The parametric method has to find the parameters for the best fit to the data. However, practical applications showed that there is often data that cannot be fit well enough with parametric methods, so non-parametric methods have been developed that work without the assumption of one of these specific distributions and fit more complex and flexible models to the data [2].

Non-parametric density estimation suffers from two major problems: the curse of dimensionality and finding a good smoothing parameter [3]. Since we work in the univariate case the curse of dimensionality does not matter in our application. What still remains is the problem of finding a good smoothing parameter. The smoothing parameter for histograms is the number of bins. If the histogram has many bins, the density curve will show many details; if the bins get fewer, the density curve will appear smoother. The question is how much is enough detail. Our method automatically finds an appropriate bin width based on cross-validation and the width is not constant for the whole range but adapts to the data (i.e. the bin width varies locally).

Histograms have been widely used because they generate an easily understandable representation of the data. They represent the density function as a piece-wise constant function. Alternative methods of non-parametric density estimation are kernel estimators, which represent the data with a smooth function [2]. Although kernel density estimators avoid the discontinuities present in histograms, they cannot be used to summarize the data in a concise form and are more difficult to explain to the non-expert.

A further disadvantage of the kernel method is its computational complexity. Assuming all kernels contribute to the density, computing the density for a test instance requires time linear in the number of training instances. In contrast a binary search on the bins is sufficient in a histogram, and the time complexity is logarithmic in the number of bins. This is particular relevant for large datasets. Kernel density estimation is a lazy method. Our method is an eager method that requires more effort at training time.

3 Existing Unsupervised Discretization Methods

We compare our method with two well-known unsupervised discretization methods: equal-width discretization and equal-frequency discretization. Equal-width discretization divides the range of the attribute into a fixed number of intervals of equal length. The user specifies the number of intervals as a parameter. A variant of this method, which we also compare against, selects the number of intervals using the cross-validated log-likelihood. This variant is implemented in Weka [4].

For equal-width histograms it is not only important to select the number of intervals but also the origin of the bins [2]. The origin is found by shifting the grid by a part of the actual bin width (e.g. one 10th of it), and selecting the best one of these shifts. We implemented this by using the cross-validated log-likelihood to select the origin and the number of bins.

Equal-frequency discretization also has a fixed number of intervals, but the intervals are chosen so that each one has the same or approximately the same number of instances in it. The number of intervals is determined by the user.

4 Cross-Validating the Log-Likelihood

Cross-validation is used in machine learning to evaluate the fit of a model to the real distribution. It is generally applied to classification or regression problems but it can also be applied to clustering [5]. The idea is to split the dataset into n equal-sized folds and repeat the training process n times using $n-1$ folds for training and the remainder for testing. This is done with every parameter value and the best value is chosen to build the final model based on the full training set. Various scoring functions are used to decide which parameter value is the best. We use the log-likelihood of the histogram, which is also used for fitting mixtures of normal distributions for clusters.

The log-likelihood is a commonly used measurement to evaluate density estimators [2]. It measures how likely the model is, given the data. Choosing the model that maximizes the likelihood on the training data results in overfitting, analogue to the classification or regression case. Cross-validation gives an (almost) unbiased estimate of performance on the true distribution and is a more suitable criterion for determining model complexity.

Leave-one-out cross-validation can be applied in the case of equal-width discretization because the log-likelihood on each test instance can be easily computed as the bins stay fixed. In our new discretization method the location of each cut point can change with one instance removed, making the leave-one-out method too expensive. Hence we use 10-fold cross-validation instead.

Let n_i be the number of training instances in bin i, n_{i-test} the number of instances of the test set that fall into this bin, w_i the bin width, and N the total number of training instances. Then the log-likelihood L on the test data is:

$$L = \sum_i n_{i-test} * log \frac{n_i}{w_i * N} \qquad (2)$$

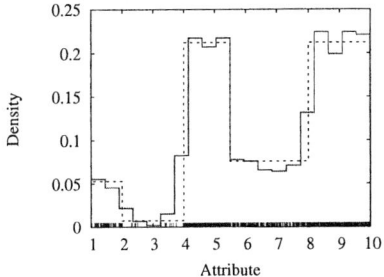
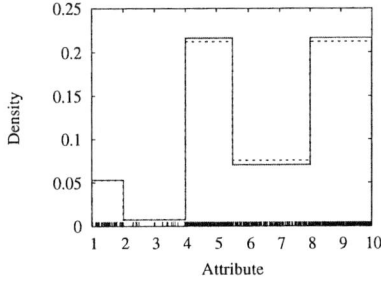

Fig. 1. Equal-width method with 20 bins

Fig. 2. Our TUBE-Method chose 5 bins of varying length

There is one problem: empty bins. If n_i equals zero the logarithm is undefined. To solve this problem we spread a single instance over the whole range of the data by adding a part of the instance to each bin that is equivalent to the relative width of the bin. Assume W denotes the total length of the range. Then the above formula becomes:

$$L = \sum_i n_{i-test} * log \frac{n_i + \frac{w_i}{W}}{w_i * (N+1)} \quad (3)$$

5 Tree-Based Unsupervised Discretization

The most common and simplest way of making histograms is the equal-width method. The range is divided into subranges or bins of equal-width. In contrast to this, our new algorithm divides the range of an attribute into intervals of varying length. The goal is to cut the range in such a way that intervals are defined that exhibit uniform density. Of course, in practical problems the true underlying density will not really be uniform in any subrange but the most significant changes in density should be picked up and result in separate intervals. Figure 1 shows an equal-width estimator for a simple artificial dataset. Figure 2 shows the density function generated with our discretization method. In both figures the "true" density (the density function that was used to generate the data) is plotted with a dotted line. The training data is shown as vertical bars at the bottom.

We call our method TUBE (Tree-based Unsupervised Bin Estimator), because it uses a tree-based algorithm to determine the cut points. More specifically, it builds a density estimation tree in a top-down fashion. Each node defines one split point. In the following we describe how a locally optimum split point can be found for a subset of data. Then we describe how the tree is built and pruned and what can be done about the problem of "small" cuts.

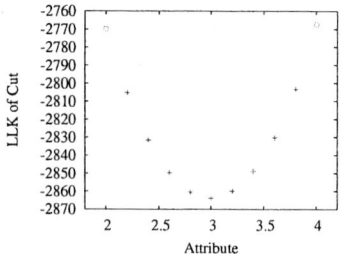

Fig. 3. The log-likelihood is minimized between instances

5.1 Where to Cut

The quality of the density estimation is measured by the log-likelihood. We choose the split point that maximizes the likelihood based on the training data and the ranges at the current node of our density estimation tree:

$$L = n_{left} * log \frac{n_{left}}{w_{left} * N} + n_{right} * log \frac{n_{right}}{w_{right} * N} \quad (4)$$

Here, n_{left} is the number of instances in the left subrange and w_{left} its width. The quantities for the right subrange are defined accordingly. In contrast to decision tree learning [6], every training instance defines two potential maximum likelihood cut points. The cut is made at the instance (and not in-between instances as in the case of classification or regression trees) and includes the instance in either (a) the left or (b) the right subset. This is because the log-likelihood of a division into two bins has a local minimum if the cut point is set in-between two points. It attains a local maximum at the points. The diagram in Figure 3 shows the log-likelihood of cut points at two values of an attribute (circles) and at nine points in-between the values (crosses). The log-likelihood is maximized at the instance values. Hence we cut at the values of the training instances and consider including a point in either the right or the left subset—i.e. we consider the interval boundaries $.., x)[x,..$ and $..x](x,..$ for an instance x.

Note that this causes problems when the data is very discontinuous and has a "spike" in its range (i.e. several identical training instances at the minimum or maximum). The estimated density would be infinity because the range would be zero. Therefore our implementation does not actually cut at the instance value itself but add, or substract, a small value (we used 10^{-4} in our experiments).

5.2 Building the Tree

The selection of k cut points can be seen as a search through a resolution space for the optimal solution. A well-known search method is the divide-and-conquer method that decision trees use. On numeric attributes this method finds the locally optimal binary split and repeats the process recursively in all subranges until a stopping criterion is met. This is a greedy search that does not find

```
maxNumBins               = [find optimal number of splits];
numSplits                = 0;
splitPriorityQueue       = empty;

firstBin                 = new Bin(bin that contains the whole attribute range);
fringe                   = [initialize with (firstBin)];

REPEAT {
  FOR (bin = all bins in fringe) {
    split = bin.[find best split in the range of this bin];
    splitPriorityQueue.[add (split)];
    fringe.[delete (bin)];
  }
  nextBestSplit = splitPriorityQueue.[give best split in queue];
  newBinLeft, newBinRight =
           nextBestSplit.[perform split on its bin and replace the bin with the
                          two new bins newBinLeft and newBinRight];
  numSplits++;
  fringe.[add the two new bins (newBinLeft, newBinRight)];
} UNTIL (numSplits == maxNumBins - 1);
```

Fig. 4. Pseudo code for the tree building algorithm

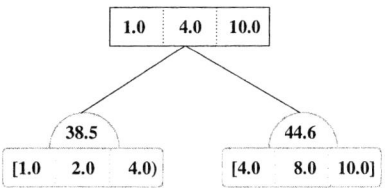

Fig. 5. Tree after the first cut

an optimal global division but a division that is a computational inexpensive estimate. We apply this method to unsupervised discretization using best-first node expansion. The pseudo code of our algorithm is shown in Figure 4.

In the following we present an example using the dataset from Figures 1 and 2. First the best cut point is found in the whole range and two new bins are formed. Within the two subranges two new locally optimal cut points are searched for. Both splits are evaluated and a log-likelihood for the division into the resulting three bins is computed for both possible splits. Figure 5 shows the discretization tree corresponding to this situation. The root node represents the first cut and the two leaf nodes represent the next two possible cuts.[1]

Each node represents a subrange, the root node the whole range. The variables written to the left and right side of the square corresponding to a node represent the minimum and maximum of the subrange. The overall minimum and maximum of this example dataset are 1.0 and 10.0. Each leaf node repre-

---
[1] All values are rounded.

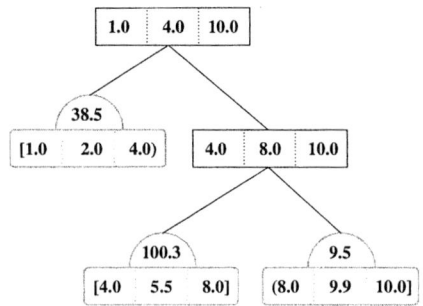

Fig. 6. Tree after the second cut

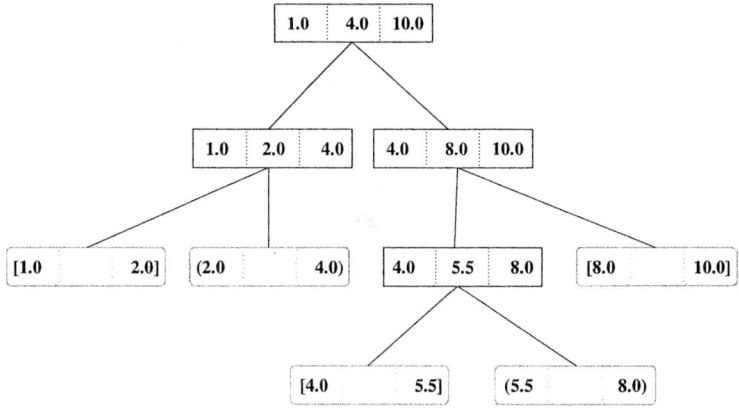

Fig. 7. Finalized tree

sents a bin and the minimum exhibits a "[" if the minimum value itself is part of the bin and a "(" if it is part of the next bin. The notation for the maximum is analogue. The variable written in the middle of the node represents the cut point.

The whole range is first cut at the value 4.0. The next possible cut point is either 2.0 or 8.0. These would split the dataset into the subranges [1.0:2.0] [2.0:4.0] [4.0:10.0] or [1.0:4.0] [4.0:8.0] [8.0:10.0] respectively. The gain in log-likelihood for each of the two possible divisions is written in the half-circle over the not-yet-exercised cuts. The cut at 2.0 results in a log-likelihood gain of 38.5 computed based on Formula 4, the cut at 8.0 has a log-likelihood gain of 44.6. Among the possible cuts the one with the largest gain in log-likelihood is selected, which in this case is the cut at 8.0. Figure 6 shows the state of the dicretization tree after two cuts.

After the cut at 8.0 is performed, two new bins are generated, and in each of them a new possible cut is searched for. These cuts are 5.5 and 9.9, with log-likelihood gains of 100.3 and 9.5 respectively. So for the third cut there is a choice between three cuts (including the cut at 2.0) and the next one chosen would be 5.5.

After four cuts our discretization tree learning algorithm decides to stop. The stopping criterion will be explained in Section 5.3. Figure 7 shows the final discretization tree. The resulting histogram is the one shown at the beginning of this section in Figure 2. In the final tree each leaf node represents a bin of the histogram. Each internal node represents a cut.

5.3 The Stopping Criterion

The third and last part of our algorithm is the stopping criterion. Based on the likelihood on the training data, the algorithm would not stop cutting until all subranges contain a single value (i.e. it would overfit). The stopping criterion sets the maximal number of cut points and prevents overfitting.

We use the 10-fold cross-validated log-likelihood to find an appropriate number. We start with zero and increase the maximal number of cut points in increments of one. This can be implemented efficiently: to find k cut points, one can use the division into $k - 1$ cut points and add one more. By default the algorithm iterates up to $N - 1$ as the maximal number of cut points (i.e. the cross-validated log-likelihood is computed for all trees with 1 up to $N - 1$ cut points). For each of the $N - 1$ iterations the algorithm computes the average log-likelihood over the test folds and from this the number of splits that exhibits the maximum value is chosen.

In our above example the cross-validated log-likelihood curve has its maximum at four cut points and therefore four cuts have been performed. Note that this method involves growing a density estimation tree eleven times: first once for each of the ten training folds, and finally for the full dataset based on the chosen number of cut points. The time complexity of the discretization algorithm is $O(N log N)$.

5.4 A Problem: Small Cuts

The log-likelihood criterion used to find the cut point in a range can be unstable. Sometimes a cut is found at the border of a subrange that contains very few instances and is very small. This can lead to a very high density value and the criterion decides to cut.

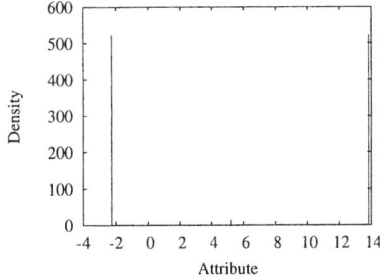

Fig. 8. Distorted histogram due to small cuts

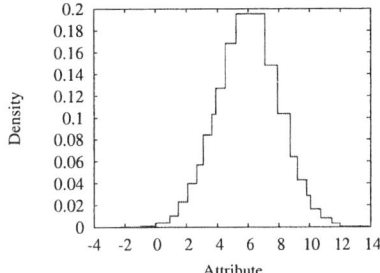

Fig. 9. Small cuts eliminated with heuristic

Hence we implemented a heuristic that avoids these small cuts in most cases. More specifically, we disallow cuts that are smaller than 0.1 percent of the whole range of the data and set the minimum number of instances to $\sqrt{0.1 * N}$. Figure 8 shows a strongly distorted histogram of a normal density that is due to two small cuts that have very high density. This was created by our method without using the heuristic. The same dataset is used in Figure 9, where the small cuts have been avoided using the heuristic.

6 Evaluation

We evaluated the TUBE discretization method using numeric attributes from 21 UCI datasets [7]. The algorithm works on univariate numeric data, and thus the numeric attributes of the UCI datasets have been extracted and converted into 464 one-attribute datasets.

A surprising finding was that many of these numeric attributes have a low uniqueness in their values. Low uniqueness means that they have many instances with the same value. Table 1 lists the number of attributes sorted into columns according to their level of uniqueness (e.g. $[0-20)$ means that the percentage of unique values is between 0 and 20). The table also shows the UCI datasets the attributes have been extracted from and the number of instances.

Table 1. 464 numeric attributes from UCI datasets and their levels of uniqueness

Dataset	[0-20)	[20-40)	[40-60)	[60-80)	[80-100]	num inst
anneal	6	-	-	-	-	898
arrythmia	182	7	14	3	-	452
autos	13	-	-	1	1	205
balance-scale	4	-	-	-	-	625
winsconsin-breast-cancer	9	-	-	-	-	699
horse-colic	7	-	-	-	-	368
german-credit	6	-	-	-	1	1000
ecoli	7	-	-	-	-	336
glass	3	3	2	1	-	214
heart-statlog	12	1	-	-	-	270
hepatitis	4	1	1	-	-	155
hypothyroid	7	-	-	-	-	3772
ionosphere	2	-	2	31	-	351
iris	4	-	-	-	-	150
labor	8	-	-	-	-	57
lymphography	3	-	-	-	-	148
segment	14	3	-	2	-	2310
sick	7	-	-	-	-	3772
sonar	-	7	4	-	46	208
vehicle	17	1	-	5	-	846
vowel	-	-	4	8	-	990
Sum	315	23	27	51	48	
In percent	68	5	6	11	10	

These datasets are used to test how well TUBE discretization estimates the true density. The density estimates that are generated are evaluated using 10x10-fold cross-validation, measuring the log-likelihood on the test data. Note that this "outer" cross-validation was performed in addition to the "inner" cross-validation used to select the number of cut points.

Our new discretization method (TUBE) is compared against equal-width discretization with 10 bins (EW-10), equal-width with cross-validation for the number of bins (EWcvB), equal-width with cross-validation for the origin of the bins and the number of bins (EWcvBO), and equal-frequency discretization with 10 bins (EF-10). The equal-frequency method could not produce useful models for the attributes with uniqueness lower than 20 and has therefore been left out in that category. TUBE, EWcvB and EWcvBO were all run with the maximum bin number set to 100.

6.1 Evaluating the Fit to the True Distribution

Table 2 lists the summary of the comparison. Each value in the table is the percentage of all attributes in that uniqueness category for which TUBE was

Table 2. Comparison of the density estimation results. Result of paired t-test based on cross-validated log-likelihood.

	EW-10	EWcvB	EWcvBO	EF-10
(0-20)				
TUBE significantly better	99	100	100	-
TUBE equal	1	0	0	-
TUBE significantly worse	0	0	0	-
[20-40)				
TUBE significantly better	48	43	43	48
TUBE equal	52	57	57	52
TUBE significantly worse	0	0	0	0
[40-60)				
TUBE significantly better	8	8	8	37
TUBE equal	92	92	92	63
TUBE significantly worse	0	0	0	0
[60-80)				
TUBE significantly better	53	56	56	67
TUBE equal	44	40	42	30
TUBE significantly worse	3	3	2	3
[80-100]				
TUBE significantly better	13	17	15	13
TUBE equal	85	81	81	85
TUBE significantly worse	2	2	4	2
Total				
TUBE significantly better	76	77	77	43
TUBE equal	23	22	22	55
TUBE significantly worse	1	1	1	2

significantly better, equal or worse respectively based on the corrected resampled t-test [8]. In almost all cases our method is at least as good as the other methods and shows especially good results in cases with low uniqueness and some cases of high uniqueness. We have analyzed the corresponding attributes and they show that TUBE is generally better when attributes exhibit discontinuities in their distributions.

It is difficult to split the datasets precisely into attributes with continuous distributions and attributes with discontinuous distributions. Datasets below 20 percent uniqueness can be considered discontinuous but there are some datasets in the higher uniqueness category that showed discontinuities.

Attributes with low uniqueness exhibit discontinuous distributions of different kinds. Some of the attributes are very discrete and have only integer values (e.g. vehicle-9) or a low precision (e.g.iris-4), some have irregularly distributed data spikes (e.g. segment-7) and some have data spikes in regular intervals (e.g. balance-scale-1). In the category of (0-20) uniqueness TUBE outperforms all other methods on almost all of the datasets.

In the category [60-80) half of the attributes have a distribution that is a mixture between continuous data and discrete data (most of the ionosphere at-

Table 3. Comparison of the number of bins

	EW-10	EWcvB	EWcvBO	EF-10
(0-20)				
TUBE significantly fewer	14	62	62	-
TUBE equal	2	8	7	-
TUBE significantly more	84	30	31	-
[20-40)				
TUBE significantly fewer	31	13	26	31
TUBE equal	4	30	17	4
TUBE significantly more	65	57	57	65
[40-60)				
TUBE significantly fewer	29	46	54	29
TUBE equal	38	42	38	38
TUBE significantly more	33	12	8	33
[60-80)				
TUBE significantly fewer	44	94	97	44
TUBE equal	14	6	3	14
TUBE significantly more	42	0	0	42
[80-100]				
TUBE significantly fewer	96	85	92	96
TUBE equal	2	15	8	2
TUBE significantly more	2	0	0	2
Total				
TUBE significantly fewer	29	65	68	56
TUBE equal	5	12	9	12
TUBE significantly more	66	23	23	32

tributes in this category have a mixed distribution). TUBE's density estimation was better for all these attributes.

6.2 Comparing the Number of Bins

Table 3 shows a comparison of the number of bins generated by the different methods. A smaller number of bins gives histograms that are easier to understand and analyze. In the category 80 percent and higher the TUBE discretization generates a significantly smaller number of bins than the other methods. Whenever it produces more bins this appears to result in a better fit to the data.

7 Conclusion

TUBE discretization provides a good algorithm for density estimation with histograms. The density estimation of very discontinuous data is often difficult. Our results show that TUBE outperforms equal-width and equal-frequency discretization on discontinuous attributes. It finds and can represent "spikes" in the density function that are caused by discrete data (many instances with the same value) and can reliably detect empty areas present in the value range.

On truly continuous data the method provides a discretization that represents the data as well as the other methods but with fewer bins and hence gives a clearer picture of areas of different density. A possible application of our method would be density estimation for classification in Naive Bayes [9]. We plan to investigate this application in future work.

References

1. Eibe Frank and Ian H. Witten. Making better use of global discretization. *Proc. of the Sixteenth International Conference on Machine Learning*, pages 115–123, 1999.
2. B.W.Silverman. *Density Estimation for Statistics and Data Analysis*. Chapman and Hall, 1986.
3. D.W.Scott. *Multivariate Density Estimation*. John Wiley & Sons, 1992.
4. Ian H. Witten and Eibe Frank. *Data Mining: Practical Machine Learning Tools and Techniques with Java Implementations*. Morgan Kaufmann, 2000.
5. P. Smyth. Model selection for probabilistic clustering using cross-validated likelihood. *Staistics and Computing*, pages 63–72, 2000.
6. J.R.Quinlan. *C4.5 Programs for Machine Learning*. Morgan Kaufmann, 1993.
7. C.L. Blake S. Hettich and C.J. Merz. UCI repository of machine learning databases, 1998.
8. C. Nadeau and Y. Bengio. Inference for the generalization error. *Machine Learning*, 52:239–281, 2003.
9. Y. Yang and G. I. Webb. Proportional k-interval discretization for naive-bayes classifiers. In *Proceedings of the 12th European Conference on Machine Learning (ECML)*, pages 159–173, Tokyo, 2001.

Weighted Average Pointwise Mutual Information for Feature Selection in Text Categorization

Karl-Michael Schneider

Department of General Linguistics, University of Passau,
94030 Passau, Germany
schneide@phil.uni-passau.de

Abstract. Mutual information is a common feature score in feature selection for text categorization. Mutual information suffers from two theoretical problems: It assumes independent word variables, and longer documents are given higher weights in the estimation of the feature scores, which is in contrast to common evaluation measures that do not distinguish between long and short documents. We propose a variant of mutual information, called *Weighted Average Pointwise Mutual Information* (WAPMI) that avoids both problems. We provide theoretical as well as extensive empirical evidence in favor of WAPMI. Furthermore, we show that WAPMI has a nice property that other feature metrics lack, namely it allows to select the best feature set size automatically by maximizing an objective function, which can be done using a simple heuristic without resorting to costly methods like EM and model selection.

1 Introduction

Automatic text categorization, i.e. the assignment of text documents to predefined categories, is an important task in many NLP applications. The common *bag of words* approach results in a document space with very high dimensionality. In order to speed up parameter estimation and classification and to improve the classifier performance, it is common to use feature selection to reduce the dimensionality of the document space. This is typically done using a filtering approach [1] in which each feature is assigned a score based on an independent evaluation, and the features are then ranked according to their scores, and the N highest ranked features are selected, where N is the desired vocabulary size. Wrapper methods, which use the classifier directly to evaluate different feature subsets [1], are not commonly used for text classification because of the high dimensionality of the feature space that makes searching for the best feature subset intractable.

Mutual Information (MI) is an information-theoretic measure that is often used to evaluate features. It measures the amount of information that the value of a feature in a document (e.g. the presence or absence of a word) gives about the class of the document. Feature selection studies have obtained good results with MI [2]. However, there are two problems associated with the use of MI for feature ranking: First, MI treats each feature as an independent random variable. This is a problem because words in a text are not independent. Second, classifiers based on generative models, such as Naive

Bayes [3], estimate class-conditional probability distributions over words from training data. In the multinomial Naive Bayes model [3,4] this is done by concatenating the training documents in each class to one long document and estimating the distribution of words in this long document. This gives larger weights to longer documents. However, in classifier evaluation, all (test) documents have equal weight irrespective of their length—that is, there is a mismatch between classifier training and evaluation.

This paper proposes a variant of MI, *Weighted Average Pointwise Mutual Information* (WAPMI) that avoids both aforementioned problems. We present theoretical (using an information-theoretic argument that links WAPMI to multinomial Naive Bayes) and empirical evidence (through extensive experimentation) in favor of WAPMI. WAPMI improves the performance of multinomial Naive Bayes over MI on a variety of standard benchmark corpora. It also outperforms several other standard metrics for feature ranking.

In addition, WAPMI has a very nice property compared to other metrics, including MI: It allows to determine the (theoretically) best feature set size by maximizing an objective function. This can be done using a simple heuristic by applying a general, data-independent threshold to the feature scores, without the need to resort to computationally intensive methods like EM and model selection. Other feature metrics only evaluate the relative usefulness, and it is not entirely clear how they could be used to define an objective function for feature selection.

We demonstrate the effectiveness of this general thresholding method in our experiments. On some datasets (notably those that are commonly regarded "easy" classification tasks) we obtain smaller feature sets and better performance, while on "difficult" datasets (i.e. large datasets with great variability in the vocabulary) WAPMI selects larger feature sets than other metrics while outperforming them.

The paper is structured as follows. In Sect. 2 we review the probabilistic framework of multinomial Naive Bayes. In Sect. 3 we define weighted average pointwise mutual information and motivate its use for feature ranking. We also discuss its relation to distributional clustering. The experimental setup is described in Sect. 4, and Sect. 4 presents our experiments and the results. Section 5 finishes with some conclusions.

2 Naive Bayes

Naive Bayes is a simple probabilistic classifier that is widely used for text classification [3,4]. Despite this independence assumption, Naive Bayes performs surprisingly well on text classification problems [5].

Let $C = \{c_1, \ldots, c_{|C|}\}$ denote the set of possible classes of documents, and let $V = \{w_1, \ldots, w_{|V|}\}$ be a vocabulary. The multinomial Naive Bayes classifier assumes that a document d is drawn from a multinomial distribution by $|d|$ independent trials on a random variable $W \in V$ with class-conditional distribution $p(w_t|c_j)$ (where $|d|$ denotes document length):

$$p(d|c_j) = p(|d|)|d|! \prod_{t=1}^{|V|} \frac{p(w_t|c_j)^{x_t}}{x_t!}$$

x_t is the number of times W yields w_t, i.e. the number of times the word w_t occurs in d. The parameters $p(w_t|c_j)$ are usually estimated from training documents using maximum likelihood with Laplace smoothing to avoid zero probabilities:

$$\hat{p}(w_t|c_j) = \frac{1 + n(c_j, w_t)}{|V| + n(c_j)}$$

where $n(c_j, w_t)$ is the number of occurrences of w_t in the training documents in c_j and $n(c_j)$ is the total number of word occurrences in c_j.

The posterior probability of the class given the document is given by Bayes' rule:

$$p(c_j|d) = \frac{p(c_j)p(d|c_j)}{p(d)}$$

where $p(d)$ is the total probability of d:

$$p(d) = \sum_{j=1}^{|C|} p(c_j) p(d|c_j)$$

The class priors $p(c_j)$ are estimated from training documents as the fraction of documents in class c_j. Given a document, the Naive Bayes classifier selects the class with the highest posterior probability (we can omit those parts that do not depend on the class in the maximization):

$$c^*(d) = \arg\max_{c_j} p(c_j) p(d|c_j) \tag{1}$$

3 Weighted Average Pointwise Mutual Information

3.1 Defining Weighted Average Pointwise Mutual Information

Mutual Information is a measure of the information that one random variable gives about the value of another random variable [6]. Let W be a random variable that ranges over the vocabulary V, and let C be random variable that ranges over classes. The mutual information between W and C is defined as:

$$I(W; C) = \sum_{t=1}^{|V|} \sum_{j=1}^{|C|} p(w_t, c_j) \log \frac{p(w_t|c_j)}{p(w_t)} \tag{2}$$

The term $\log \frac{p(w_t|c_j)}{p(w_t)}$ is called *pointwise mutual information* [7].[1] Note that mutual information can be written as a weighted sum of Kullback-Leibler (KL) divergences. The KL-divergence between two probability distributions p and q is defined as $D(p\|q) = \sum_x p(x) \log \frac{p(x)}{q(x)}$ [6]. Thus (2) can be written as the weighted average KL-divergence

---

[1] In [2] this is called *information gain*, and the term *mutual information* is used as a synonym for pointwise mutual information.

between the class-conditional distribution of words and the global (unconditioned) distribution in the entire corpus:

$$I(W;C) = \sum_{j=1}^{|C|} p(c_j) D(p(W|c_j) \| p(W))$$

To rank features we would like a measure for each feature. A common method is to define new binary random variables, W_t, for each word that indicate whether the next word in a document is w_t (or some other word) [3,8]: $p(W_t = 1) = p(W = w_t)$. Then the MI-score for w_t is given by:

$$MI(w_t) := I(W_t;C) = \sum_{j=1}^{|C|} \sum_{x=0,1} p(W_t = x, c_j) \log \frac{p(W_t = x | c_j)}{p(W_t = x)} \quad (3)$$

The problem with (3) is that it treats W_t as an independent random variable, but in fact $\sum_{t=1}^{|V|} p(W_t = 1) = 1$! To avoid this independence assumption, we consider (2) as a sum over word scores, where the score for w_t is the pointwise mutual information with the class, averaged over all classes:

$$PMI(w_t) := \sum_{j=1}^{|C|} p(w_t, c_j) \log \frac{p(w_t | c_j)}{p(w_t)} \quad (4)$$

The problem with (4) is that it treats all training documents in one class as one big document (because of the way the class-conditional probabilities are estimated). Thus, if there is variation in the document lengths, (4) is dominated by the longer documents. To avoid this problem, we replace the weight $p(w_t, c_j)$ with a term that is a weighted average of the document-conditional probabilities $p(w_t|d_i) = n(w_t, d_i)/|d_i|$ where $n(w_t, d_i)$ is the number of times w_t occurs in d_i and $|d_i|$ is the length of d_i.[2] Thus weighted average pointwise mutual information is defined as:

$$WAPMI(w_t) := \sum_{j=1}^{|C|} \sum_{d_i \in c_j} \alpha_i p(w_t | d_i) \log \frac{p(w_t | c_j)}{p(w_t)} \quad (5)$$

We consider several alternatives for the weights α_i, which can be associated with different measures for classifier evaluation:

- $\alpha_i = p(c_j) \cdot |d_i| / \sum_{d_i \in c_j} |d_i|$. This gives each document a weight proportional to its lengths and yields (4).
- $\alpha_i = 1 / \sum_{j=1}^{|C|} |c_j|$. This gives equal weight to all documents. This corresponds to an evaluation measure that counts each misclassified document as the same error, i.e. classification accuracy.
- $\alpha_i = 1/(|c_j| \cdot |C|)$ where $d_i \in c_j$. This gives equal weight to the classes by normalizing for class size, i.e. documents from smaller categories receive higher weights. This compensates for the dominance of larger categories in classifier evaluation.

---
[2] Note that any word that does not occur in d_i has zero probability.

By summing (5) over all words we obtain the total weighted average pointwise mutual information between the word variable W and the class variable C:

$$WAPMI(W;C) := \sum_{t=1}^{|V|} \sum_{j=1}^{|C|} \sum_{d_i \in c_j} \alpha_i p(w_t|d_i) \log \frac{p(w_t|c_j)}{p(w_t)} \quad (6)$$

In the following subsections we provide theoretical evidence that total WAPMI could be used as an objective function, and the goal of feature selection is to maximize that objective function.

3.2 Relation to Distributional Clustering

Note that (6) can be written as a weighted sum of the difference between (i) the KL-divergence of the document-conditional distribution from the corpus distribution and (ii) the KL-divergence of the document-conditional distribution from the class-conditional distribution:

$$\sum_{j=1}^{|C|} \sum_{d_i \in c_j} \alpha_i \left[D(p(W|d_i) \| p(W)) - D(p(W|d_i) \| p(W|c_j)) \right] \quad (7)$$

This can be interpreted as an estimate of how similar the documents in one class are and how dissimilar documents of different classes are. From a clustering perspective we can say that (7) is large if the documents that belong to the same class form tight clusters, with wide separation between the clusters. Interpreting text categorization as an information retrieval task (i.e. regarding classes as queries) this is a desirable property that has been argued to improve document retrieval performance in the vector space model [9].

In distributional clustering the goal is to cluster similar objects (e.g. documents) together so as to maximize the value of an objective function that measures the quality of the clustering [10]. Below we argue that maximizing (7) is expected to improve the accuracy of the multinomial Naive Bayes classifier. Thus we can regard total weighted average pointwise mutual information as an objective function (since it is a function of the entire training corpus). However, in contrast to clustering, we do not change the clusters (which correspond to the classes in the training corpus and which we consider to be fixed). Instead our goal is to improve the clustering by changing the document representation (i.e. by using a subset of the features).

3.3 Relation to Multinomial Naive Bayes

We can use (7) to get an estimate of the expected performance of Naive Bayes on the training set (and by generalization also on a test set, if the test documents are draw from the same distribution). We manipulate the Naive Bayes classifier (1) in an information theoretic framework using the fact that a document defines a probability distribution over words. We define the distance of a document, d_i, from a class, c_j, as the KL-divergence between the document-conditional word distribution and the class-conditional distribution. Naive Bayes can then be written in the following form by taking logarithms, dividing by the length of d_i and adding the entropy of d_i, $H(p(W|d_i)) = -\sum_t p(w_t|d_i) \log p(w_t|d_i)$ [10]:

$$c^*(d_i) = \arg\min_{c_j} \left[D(p(W|d_i) \| p(W|c_j)) - \frac{1}{|d_i|} \log p(c_j) \right] \qquad (8)$$

Note that the modifications in (8) do not change the classification of documents. Assuming equal class priors, Naive Bayes can thus be interpreted as selecting the class which has the least distance from the document. Taking into account the arguments from the previous subsection, maximizing the total weighted average pointwise mutual information (6) would thus increase the probability that each document is nearer to its true class than to any other class, and would therefore be classified correctly by multinomial Naive Bayes.

3.4 Using WAPMI as an Objective Function for Feature Selection

Taking into account the arguments in the previous subsections, the best feature set would be one that maximizes the total WAPMI (6). Note that the WAPMI score (5) can be negative, which suggests the following simple heuristic for maximizing total WAPMI: Simply select all words with a positive WAPMI score and removing all other words. This is equivalent to applying a threshold of θ = 0 to the WAPMI score. We examine this empirically in Sect. 4. In contrast, mutual information is always non-negative (and almost always positive), and it is not entirely clear how mutual information could be used as an objective function in feature selection.

Note that the above heuristic is only an approximation. In fact, feature selection isn't entirely well-defined in multinomial Naive Bayes, since we are not only pruning the model but the data too! Pruning the vocabulary changes the distribution of the remaining words. An alternative would be to not greedily discard words but perform several iterations and recompute the objective function after each iteration until convergence. We tried this, but there was almost no difference. In most cases, convergence occurred after only two or three iterations, with only a few additional words removed after the first round.

4 Experiments

4.1 Datasets and Procedures

We perform experiments on five text categorization datasets, described in Table 1. The 20 Newsgroups dataset[3] consists of Usenet articles distributed evenly in 20 different newsgroups that make up the classes [11]. We remove newsgroup headers and binary attachments and use only words consisting of alphabetic characters as tokens, after converting to lower case and mapping numbers, URLs and email addresses to special tokens.

The WebKB dataset and the 7 Sectors dataset are both available from the WebKB project [12].[4] WebKB contains web pages gathered from computer science departments and categorized in six classes plus one *other* class. We use only the four most populous classes *course, faculty, project* and *student*. The 7 Sectors data consists of web pages

---
[3] http://people.csail.mit.edu/people/jrennie/20Newsgroups/
[4] http://www.cs.cmu.edu/afs/cs.cmu.edu/project/theo-11/www/wwkb/

Table 1. Corpus statistics. The last two columns show the number of documents in the smallest and biggest categories, respectively.

Dataset	Classes	Vocabulary	Documents	Smallest	Largest
20 Newsgroups	20	94,897	19,997	997	1,000
WebKB	4	41,015	4,199	504	1,641
7 Sectors	48	42,110	4,582	39	105
Reuters-10 (train)	10	22,430	6,490	181	2,877
Reuters-10 (test)	10	13,849	2,545	56	1,087
Reuters-90 (train)	90	24,719	7,770	1	2,877
Reuters-90 (test)	90	15,660	3,019	1	1,087

from different companies divided into a hierarchy of classes. We use the flattened version of the data. We strip all HTML tags and use only words and numbers as tokens, after converting to lower case and mapping numbers and other expressions to special tokens.

The Reuters-21578 dataset[5] consists of Reuters news articles belonging to zero or more topic classes. We use the ModApte split [13] and produce two versions of the corpus. Reuters-10 uses only the 10 largest topics. On average, each document belongs to 1.105 topic classes. Reuters-90 uses all 90 topics that have at least one document in the training and test set, with an average of 1.235 topics per document.

Except on Reuters, all experiments are performed using cross-validation. We follow the methodology in [3]. For 20 Newsgroups and 7 Sectors, we split the data into five parts of equal size and with equal class distribution. For WebKB we produce ten train/test splits using stratified random sampling with 70% training and 30% test data. We report average classification accuracy across trials.

For the Reuters experiments we build a binary classifier for each topic, using the documents belonging to each topic as positive examples and all other documents as negative examples. Following the standard methodology with multi-label datasets, we ignore the classification decision of the classifier and use the classification scores to rank the documents. We then report precision/recall breakeven points averaged over all topics (called "macroaverage"). Instead of the Naive Bayes posterior probabilities, which tend to produce extreme values with growing document length due to the Naive Bayes independence assumption and are not comparable across documents, we use the normalized KL-divergence based classification scores described in [12].

4.2 Quality of Selected Features

We compare our WAPMI scoring function against three other scoring functions: Mutual Information [3], Chi-squared [2] and Bi-normal separation [14]. We evaluate the quality of the selected features by varying the number of selected features. We use WAPMI with equal weighting for all documents (we also experimented with equal class weights but found no statistically significant difference). Table 2 shows the top 20 words in the entire 20 Newsgroups corpus according to Mutual Information and WAPMI.

Figure 1 shows classification accuracy on the three datasets. As can be seen, the WAPMI scoring function yields higher classification accuracy, although on WebKB

---
[5] http://www.daviddlewis.com/resources/testcollections/reuters21578/

Table 2. 20 words with highest MI (left) and WAPMI score (right) in the 20 Newsgroups corpus

MI	Word	MI	Word	WAPMI	Word	WAPMI	Word
0.02833	ax	0.00174	g	0.00221	rainbowthreedigit	0.00073	rainbowdigits
0.01555	rainbowonedigit	0.00168	w	0.00179	sale	0.00070	mac
0.00387	rainbowdigits	0.00161	m	0.00150	rainbowtwodigit	0.00068	clipper
0.00374	rainbowtwodigit	0.00155	u	0.00140	windows	0.00067	taggedemail
0.00336	x	0.00144	v	0.00129	x	0.00067	card
0.00222	q	0.00143	of	0.00091	car	0.00066	thanks
0.00188	rainbowthreedigit	0.00124	god	0.00089	god	0.00065	team
0.00182	f	0.00119	r	0.00087	game	0.00064	he
0.00181	max	0.00109	p	0.00083	drive	0.00064	i
0.00175	the	0.00104	that	0.00074	bike	0.00064	space

the difference is statistically significant only for up to 2,000 words. In general, the improvement seems to be higher on smaller vocabulary sizes.

The class distribution is highly skewed in the Reuters datasets. The largest category (earn) has 2,877 documents in the training set, while the smallest category in Reuters-10 (corn) has 181 documents in the training set. In Reuters-90 there are 29 categories with less than 10 documents in the training set.

For the Reuters experiments we use two versions of WAPMI: with equal weights for all documents (WAPMI1), and with equal class weights (WAPMI2) (cf. Sect. 3.1), which deemphasizes the impact of the larger classes. Figure 2 shows the results on the Reuters datasets with 10 and 90 categories. We report macroaveraged precision/recall breakeven, which gives equal weight to the performance on each category. WAPMI with equal weights on documents does not perform better than the other metrics, except for very small vocabularies on Reuters-90. However, when the weights are set such that documents from smaller categories receive higher weights (WAPMI2), WAPMI clearly outperforms the other feature scoring methods.

4.3 Global Thresholding

In addition to the experiments with varying numbers of features we also examined the possibility of using a global thresholding strategy, with a fixed threshold that is applied to all datasets. We are interested in the sensitivity of the various feature scoring functions to the difficulty of the classification task. In general, the Naive Bayes classifier performs better with large vocabularies, but the optimal vocabulary size depends on the dataset. For instance, the 20 Newsgroups dataset requires a larger vocabulary for optimal classification accuracy than the other datasets [3].

For Mutual Information, Chi-squared and Bi-normal separation we select a threshold that yields relatively good performance on all datasets. For WAPMI we use the theoretically best threshold 0. For all datasets except 20 Newsgroups we use both variants with equal weights on documents (WAPMI1) and on classes (WAPMI2). For 20 Newsgroups WAPMI1 and WAPMI2 are the same because all classes have the same number of documents.

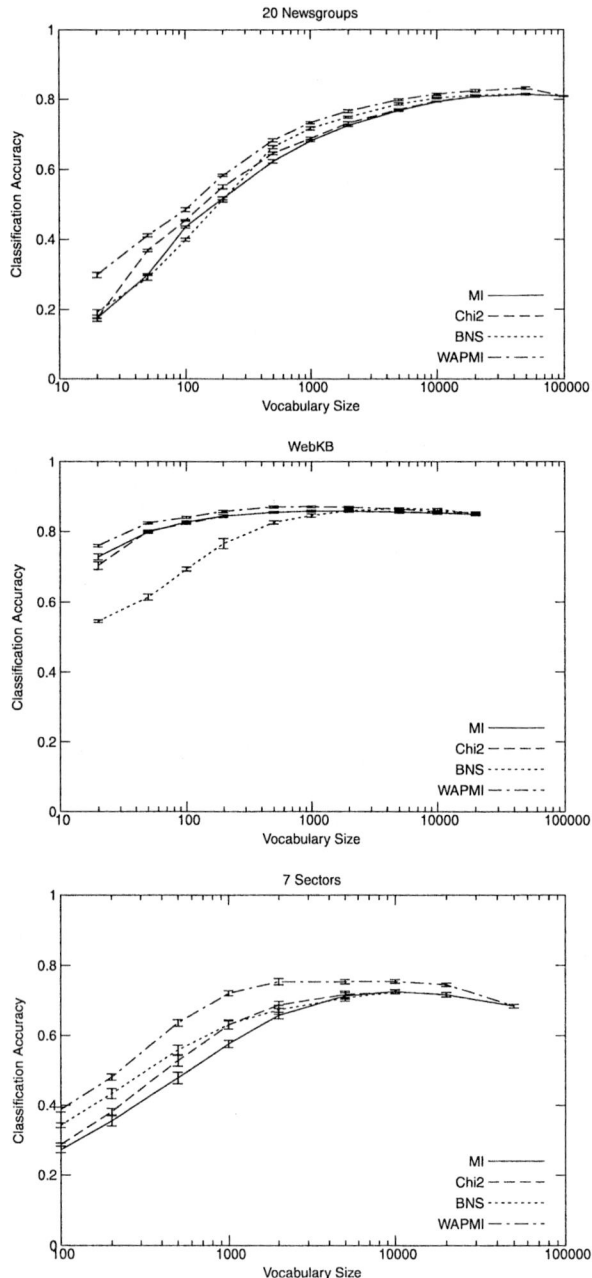

Fig. 1. Classification accuracy on 20 Newsgroups (top), WebKB (middle) and 7 Sectors (bottom). Curves show small error bars twice the width of the standard error of the mean. Differences between WAPMI and the other metrics are statistically significant (at the 95% confidence level using a two-tailed paired t-test) at the following vocabulary sizes: on 20 Newsgroups from 20 to 50,000 words; on WebKB from 20 to 2,000 words; on 7 Sectors from 100 to 20,000 words.

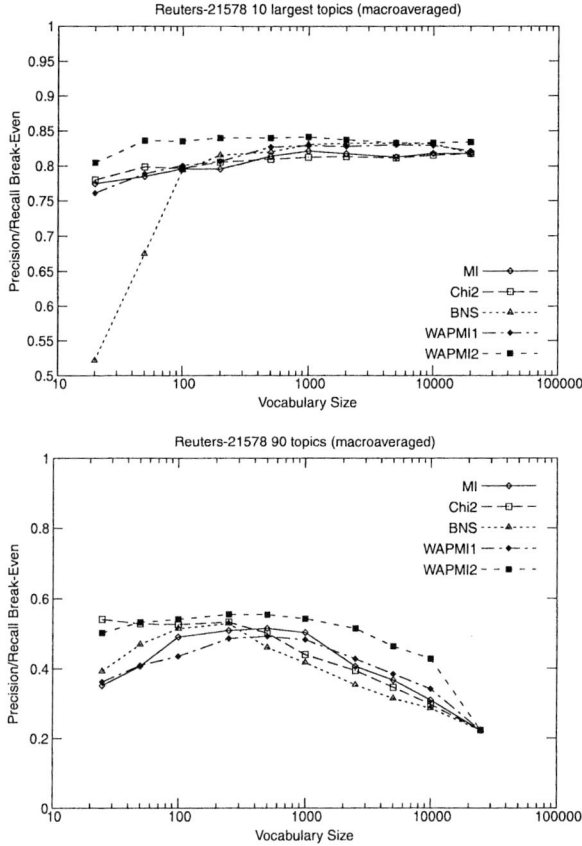

Fig. 2. Macroaveraged precision/recall breakeven on the Reuters datasets with 10 (top) and 90 (bottom) topic classes. WAPMI1 gives equal weight to documents, while WAPMI2 gives equal weight to classes.

Table 3 shows the results. For each dataset and each scoring function we report the number of features and the classification performance at the selected threshold. In addition we show the classification performance at the full vocabulary (i.e. with no feature selection).

We make two observations in Table 3. First, WAPMI is always among the top performers, although its performance is significantly better only on 20 Newsgroups and Reuters. Mutual Information performs significantly worse than the other metrics on 7 Sectors. Secondly and more importantly, the number of features selected by WAPMI seems to reflect the difficulty of the datasets better than for the other scoring methods. For 20 Newsgroups, which requires many features, WAPMI1 selects more features than any other method, while it still omits some features which results in an improvement of 2 percentage points compared to the full vocabulary. In contrast, the WAPMI scores select considerably less features on the Reuters datasets than the other methods, with better results.

Table 3. Global thresholding results. Shown are the number of selected words at the predefined threshold, classification performance, and standard deviation where applicable. Statistically significant differences (at $p = 0.95$ using a two-tailed paired t-test) are printed in boldface. For Reuters, macroaveraged precision/recall breakeven points are shown.

	20 Newsgroups			WebKB			7 Sectors		
	Words	Acc	SDev	Words	Acc	SDev	Words	Acc	SDev
Chi^2=0.1	65,194	81.35%	0.36%	32,712	84.79%	0.99%	15,147	72.29%	1.19%
MI=10^{-7}	77,694	81.13%	0.37%	32,776	84.79%	1.01%	37,474	**68.32%**	1.17%
BNS=0.05	62,777	81.42%	0.26%	32,550	84.78%	0.99%	8,545	72.27%	1.78%
WAPMI1=0	85,870	**82.92%**	0.72%	32,091	85.00%	0.96%	37,422	73.12%	0.57%
WAPMI2=0				32,278	85.06%	1.03%	37,428	73.14%	1.01%
Full	86,019	80.97%	0.29%	32,873	84.80%	0.99%	37,474	68.32%	1.17%

	Reuters-10		Reuters-90	
	Words	P/R	Words	P/R
Chi^2=0.1	18,861	81.72%	23,395	22.30%
MI=10^{-7}	18,014	81.72%	22,571	22.57%
BNS=0.05	20,086	81.76%	23,778	22.26%
WAPMI1=0	7,617	82.47%	3,066	**44.58%**
WAPMI2=0	10,610	**83.17%**	20,762	38.97%
Full	22,430	81.61%	24,719	22.28%

5 Conclusions

This paper proposes weighted average pointwise mutual information (WAPMI) as a replacement for mutual information to rank features for feature selection in text categorization. Experiments on a number of standard benchmark datasets show that WAPMI outperforms several other feature scoring metrics, including mutual information, Chi-squared and Bi-normal separation. An important property of WAPMI is that the feature set size (i.e. the number of selected features) can be set automatically, depending on the complexity and difficulty of the dataset, by using a simple constant-threshold heuristics that maximizes an objective function and does not require EM or model selection.

WAPMI contains weights that can be set to account for skewed class distributions, which we used in our experiments with the Reuters dataset and obtained improved classification performance. It is not entirely clear how this could be done with other metrics.

We have used WAPMI with the multinomial Naive Bayes classifier, but future work should deal with other classification models, e.g. support vector machines. A general open problem is that feature selection for multinomial Naive Bayes is not entirely well-defined, thus we are actually approximating feature selection. More work is required to better understand how feature selection affects the class-conditional distributions.

Acknowledgments

The author would like to thank the anonymous reviewers for their detailed comments and suggestions that helped to improve the paper.

References

1. John, G.H., Kohavi, R., Pfleger, K.: Irrelevant features and the subset selection problem. In Cohen, W.W., Hirsh, H., eds.: Machine Learning: Proceedings of the Eleventh International Conference, San Francisco, CA, Morgan Kaufmann Publishers (1994) 121–129
2. Yang, Y., Pedersen, J.O.: A comparative study on feature selection in text categorization. In: Proc. 14th International Conference on Machine Learning (ICML-97). (1997) 412–420
3. McCallum, A., Nigam, K.: A comparison of event models for Naive Bayes text classification. In: Learning for Text Categorization: Papers from the AAAI Workshop, AAAI Press (1998) 41–48 Technical Report WS-98-05.
4. Eyheramendy, S., Lewis, D.D., Madigan, D.: On the Naive Bayes model for text categorization. In Bishop, C.M., Frey, B.J., eds.: AI & Statistics 2003: Proceedings of the Ninth International Workshop on Artificial Intelligence and Statistics. (2003) 332–339
5. Friedman, J.H.: On bias, variance, 0/1-loss, and the curse-of-dimensionality. Data Mining and Knowledge Discovery **1** (1997) 55–77
6. Cover, T.M., Thomas, J.A.: Elements of Information Theory. John Wiley, New York (1991)
7. Church, K.W., Hanks, P.: Word association norms, mutual information, and lexicography. Computational Linguistics **16** (1990) 22–29
8. Rennie, J.D.M.: Improving multi-class text classification with Naive Bayes. Master's thesis, Massachusetts Institute of Technology (2001)
9. Salton, G., Wong, A., Yang, C.S.: A vector space model for automatic indexing. Communications of the ACM **18** (1975) 613–620
10. Dhillon, I.S., Mallela, S., Kumar, R.: A divisive information-theoretic feature clustering algorithm for text classification. Journal of Machine Learning Research **3** (2003) 1265–1287
11. Lang, K.: NewsWeeder: Learning to filter netnews. In: Proc. 12th International Conference on Machine Learning (ICML-95), Morgan Kaufmann (1995) 331–339
12. Craven, M., DiPasquo, D., Freitag, D., McCallum, A., Mitchell, T., Nigam, K., Slattery, S.: Learning to construct knowledge bases from the World Wide Web. Artificial Intelligence **118** (2000) 69–113
13. Apté, C., Damerau, F., Weiss, S.M.: Towards language independent automated learning of text categorization models. In: Proc. 17th ACM SIGIR Conference on Research and Development in Information Retrieval (SIGIR '94). (1994) 23–30
14. Forman, G.: An extensive empirical study of feature selection metrics for text classification. Journal of Machine Learning Research **3** (2003) 1289–1305

Non-stationary Environment Compensation Using Sequential EM Algorithm for Robust Speech Recognition[*]

Haifeng Shen, Jun Guo[1], Gang Liu[1], and Qunxia Li[2]

[1] Beijing University of Posts and Telecommunications, 100876, Beijing, China
shen_hai_feng@126.com, guojun@bupt.edu.cn, lg@pris.edu.cn
[2] University of Science and Technology Beijing, 100083, Beijing, China
kellylqx@163.com

Abstract. The paper presents a non-stationary environment compensation using sequential EM estimation for tracking the complicated environment. All of the noisy features used in the recognition system are effectively compensated. The speech corruption in the log domain such as the 24 log-filterbank coefficients and the log-energy feature can be modeled as a nonlinear model. For efficient estimating noise parameter using the subsequent sequential Expectation-Maximization (EM) algorithm, the nonlinear environment model is linearized by the truncated first-order vector Taylor series (VTS) approximation. Due to the cepstral features are nearly independence, we train the clean speech using cepstral features and the log-energy feature, and then obtain a diagonal Gaussian mixture model in the log domain by taking inverse discrete cosine transform (IDCT). The experiments are conducted on the large vocabulary continuous speech recognition (LVCSR) system. Results demonstrate that it achieves attractive improvements when compared with CMN (cepstral mean normalization) and the batch-EM based compensation approach.

1 Introduction

The recognition performance will be severely degraded in the acoustic-distorted environments due to mismatches between the training and the test environments. The test utterances represent specific conditions such as specific speakers, specific speaking styles, specific noisy conditions, which generally are not included in the training data set and usually differ from the training conditions. There are many compensation approaches for reducing the influences of these mismatches on the speech. CMN (cepstral mean normalization), with the merits of inexpensive computation load and good recognition performance, can remove the cepstral mean from all vectors with the cepstral mean calculated separately from each sentence assuming that the average cepstral mean in the training and testing environments are equal to each other. The data-driven approach such that SNR-Dependent Cepstral Normalization (SDCN), Fixed Codeword-Dependent Cepstral Normalization (FCDCN) [1], needs a "stereo"

---

[*] This research was sponsored by NSFC (National Natural Science Foundation of China) under Grant No.60475007, the Foundation of China Education Ministry for Century Spanning Talent and BUPT Education Foundation.

database that contains time-aligned samples of speech which had been simultaneously recorded in both the training and the reprehensive test environments. The cepstral features of the incoming speech are compensated by direct comparison. The problem of the data-driven approach is that the stereo data recorded in a specific test environment is not suitable for another real environment. Moreover, this kind of the approaches is really complicated in recording the "stereo" databases and not effective when dealing with the non-stationary environment. Recently, the model-based approach becomes the most attractive technique [2]-[10]. The acoustic-distorted environment is modeled as an explicit model. For effectively modeling the statistical distribution of the noisy observation and estimating the environment parameters, the environment model is postprocessed to achieve the compact model. For instance, by employing the truncated first order vector Taylor series (VTS) approximation [2] [3] [5] and statistical linear approximation (SLA) [6]-[8], the nonlinear model is linearized. It is proven that such environment approximation approaches achieve the considerable performance on speech recognition. Furthermore, based on maximum likelihood estimation (ML) [7] or maximum a posteriori estimation (MAP) criterion [3], the noise parameter is iteratively updated to the real value using EM algorithm, generally, using the batch-EM algorithm. It is clear that the batch-EM algorithm can be carried out assuming that the environment is stationary, that is, the noise statistics is iteratively updated by computing the posteriori probabilities of all of the incoming speech frames. Although the batch-EM algorithm also improves the recognition performance in the non-stationary environment, this improvement is rather limited, especially in the high time-varying environment. The sequential EM algorithm [8]-[10], can deal with this problem and can improve recognition performance considerably compared with the batch EM environment compensation, especially in the time-varying environment.

In this paper, we present a non-stationary environment compensation based on sequential EM algorithm. Generally speaking, most of state-of-the-art speech recognition systems use the Mel frequency cepstral coefficients (MFCCs) and the log-energy feature as the acoustic vector. It is well known that the log-energy feature also makes a significant contribution for improving the recognition performance. Because the cepstral coefficients can be obtained from the log-filterbank coefficients by taking DCT transform, a number of the papers in literature [2]-[6], [8] [10] deal with the cepstral coefficients or the log-filterbank coefficients for making the feature robust against the noise environments. But it is clear that if the log-energy feature isn't well compensated, the system also can deteriorate the system performance, especially in the condition with a large mount of noise. Therefore, in this paper, we compensate all of the log-filterbank coefficients and the log-energy feature. Then taking DCT transform and corresponding dynamic features computation, the compensated cepstal coefficients and the log-energy feature plus the first and second differentials are obtained. For effectively estimating the environment parameter, the environment in the log domain can be modeled as a nonlinear model and linearized using the truncated first-order VTS approximation. It is noticeable that the clean speech model has a severe influence on the recognition performance. Generally, the clean speech model is modeled as the diagonal Gaussian mixture distribution for the effectiveness of the subsequent environment parameter estimation, also for decreasing the huge computation load. Due to the aforementioned nearly independence in the cepstral domain, our approach to this is based on combination of all cepstral coefficients and the log-energy feature. Then the diagonal clean model in the log domain is obtained by taking inverse DCT transform on the cepstral statistics of the trained model. Based on initializing the truncated first-order VTS coefficients by employing the current estimated noise parameter and the next noisy speech frame, we update the next frame noise parameter by using sequential EM algorithm until the last noisy frame. The experiments are conducted on the large vocabulary continuous speech recognition (LVCSR) system. Results demonstrate that the environment compensation using the sequential EM algorithm improves recognition performance considerably compared with the batch EM environment compensation, especially in the time-varying environment. After introducing the forgetting factor for tracking the non-stationary time-varying environment, the performance of the speech recognition can further be improved in the non-stationary environment. The rest of the paper is organized as follows. The next section briefly describes the environment model approximation and accordingly investigates the statistical characteristics of the noisy speech. In section 3, we present sequential EM algorithm for noise parameter estimation. The experimental results are given in section 4 and some conclusions are drawn in section 5.

2 Environment Model Approximation

As seen in the appendix, due to the noise is additive in the linear spectral domain, the speech corruption will be nonlinear in the log spectral domain. In addition, the log-energy feature has the same corruption form as those of the log filterbank coefficients. So we can describe the corruption of these features in the noisy environment jointly. Denote the noisy feature, the clean feature and the noise in the log domain by y, x and n. The corruption is well represented as

$$y = x + \log(1 + \exp(n - x)) = x + f(x,n). \qquad (1)$$

We assume the clean speech is modeled as a Gaussian mixture model:

$$p(x) = \sum_{j=1}^{M} p_j N(x; \mu_{xj}, \Sigma_{xj}), \qquad (2)$$

in which M denotes the number of mixture components, p_j, μ_{xj} and Σ_{xj} denote the mixture coefficient, the mean vector and the diagonal covariance matrix for the j th mixture component, respectively. In our system, we first train the clean cepstral coefficients and the log-energy feature to obtain Gaussian mixture model. Then taking inverse DCT transform on these cepstral probability statistics, Gaussian mixture model in the log domain can be derived. We assume the noise is a Gaussian and statistically independent from the clean speech. The probability distribution of the noisy speech, unfortunately, is not the Gaussian mixture model due to the nonlinear relationship between the noisy speech and the clean speech described in Eq.(1). To simplify the distribution of the noisy speech and efficient noise estimation using sequential EM algorithm in the following step, we employ the truncated first-order VTS expansion to linearize the nonlinearity $f(x,n)$ in Eq.(1) around the vector points (μ_{xj}, n_0). This gives the linearized model in the j th mixture component:

$$y = A_j x + B_j n + C_j, \tag{3}$$

where

$$\begin{cases} A_j = 1 + \nabla_x f(\mu_{xj}, n_0) \\ B_j = \nabla_n f(\mu_{xj}, n_0) \\ C_j = f(\mu_{xj}, n_0) - \nabla_x f(\mu_{xj}, n_0)\mu_{xj} - \nabla_n f(\mu_{xj}, n_0)n_0 \end{cases}, \tag{4}$$

and the gradients $\nabla_x f(\mu_{xj}, n_0)$ and $\nabla_n f(\mu_{xj}, n_0)$ have the following close form:

$$\begin{cases} \nabla_x f(\mu_{xj}, n_0) = diag\left(\frac{1}{1+\exp\{n_0 - \mu_{xj}\}}\right). \\ \nabla_n f(\mu_{xj}, n_0) = 1 - \nabla_x f(\mu_{xj}, n_0) \end{cases} \tag{5}$$

3 Noise Estimation Using Sequential EM Algorithm

Assuming that the noise is a single Gaussian distribution with mean vector n_t and covariance matrix Σ_n in each instant time t, we can see that the distribution of the noisy speech is a Gaussian mixture model by applying the first-order VTS approximation. In this paper, for simplicity, we are only interested in the noise mean estimation in each frame. The covariance matrix of each frame is set with equal value and can be estimated from silence frames. Given the acoustic-distorted feature sequence $Y_{t+1} = \{y_1, y_2, \cdots, y_{t+1}\}$ and the previous noise estimate sequence $\Lambda_{nt} = \{\hat{n}_0, \hat{n}_1, \cdots, \hat{n}_t\}$ in which $\hat{n}_0$ is the initial parameter estimate and $\hat{n}_t$ is the noise estimate at time t, the noise $\hat{n}_{t+1}$ at time $t+1$ can be obtained under ML criterion:

$$\hat{n}_{t+1} = \arg\max_{n_{t+1}} \{\log P(Y_{t+1}, J_{t+1} \mid n_{t+1}, \Lambda_{nt})\}, \tag{6}$$

where $J_{t+1} = \{j_1, j_2, \cdots, j_{t+1}\}$ is the a set of the mixture components up to time $t+1$.

In general, it is not easy to estimate instant noise parameter. In this section, we use the sequential EM algorithm to iteratively estimate the different instant noise. At each iteration, the likelihood in Eq.(6) are increase until convergence. The auxiliary function is given below

$$Q(\hat{n}_{t+1} \mid n_{t+1}, \Lambda_{nt}) = E\{\log P(Y_{t+1}, J_{t+1} \mid \hat{n}_{t+1}, \Lambda_{nt}) \mid Y_{t+1}, n_{t+1}, \Lambda_{nt}\}, \tag{7}$$

where n_{t+1} is the initial value needed to know beforehand. In the slow time-varying acoustic-distorted environment, the value n_{t+1} can be approximated using the previous estimate $\hat{n}_t$, then the above equation can be compactly written as

$$Q(\hat{n}_{t+1} \mid \Lambda_{nt}) \approx E\{\log P(Y_{t+1}, J_{t+1} \mid \hat{n}_{t+1}, \Lambda_{nt}) \mid Y_{t+1}, \Lambda_{nt}\} \tag{8}$$

$$\propto -\sum_{\tau=1}^{t+1}\sum_{j=1}^{M} p(j_\tau = j \mid y_\tau, \hat{n}_{\tau-1})\{y_\tau - \hat{\mu}_{y_\tau, j}(\hat{n}_\tau)\}' \Sigma_{y_\tau, j}^{-1}\{y_\tau - \hat{\mu}_{y_\tau, j}(\hat{n}_\tau)\},$$

where

$$\begin{cases} \hat{\mu}_{y_\tau, j}(\hat{n}_\tau) = A_j(\hat{n}_{\tau-1})\mu_{xj} + B_j(\hat{n}_{\tau-1})\hat{n}_\tau + C_j(\hat{n}_{\tau-1}) \\ \Sigma_{y_\tau, j} = A_j(\hat{n}_{\tau-1})\Sigma_{xj}A_j'(\hat{n}_{\tau-1}) + B_j(\hat{n}_{\tau-1})\Sigma_n B_j'(\hat{n}_{\tau-1}) \end{cases} \tag{9}$$

where the coefficients $A_j(\cdot)$, $B_j(\cdot)$ and $C_j(\cdot)$ are the functions of the noise paramter $\hat{n}_{\tau-1}$. That is, the nonlinear function $f(x, \hat{n}_\tau)$ in Eq.(1) is approximated around the vector point $(\mu_{xj}, \hat{n}_{\tau-1})$ by using vector Taylor expansion.

The posteriori probability $p(j_\tau = j \mid y_\tau, \hat{n}_{\tau-1})$ in Eq.(8) can be computed as

$$p(j_\tau = j \mid y_\tau, \hat{n}_{\tau-1}) = \frac{p_j N(y_\tau; \hat{\mu}_{y_\tau, j}, \Sigma_{y_\tau, j})}{\sum_{j=1}^{M} p_j N(y_\tau; \hat{\mu}_{y_\tau, j}, \Sigma_{y_\tau, j})}, \tag{10}$$

where $\hat{\mu}_{y_\tau, j} = \mu_{xj} + f(\mu_{xj}, \hat{n}_{\tau-1})$.

In the non-stationary environment, the history observation data is not useful or not really important to current noise estimation. We can add the different weights according to their contributions on current noise estimation. The different weights can be added by introducing the forgetting factor ρ where ρ is a non-negative constant with value less than 1, thus, Eq.(8) can be rewritten as

$$Q(\hat{n}_{t+1} \mid \Lambda_{nt}) = -\sum_{\tau=1}^{t+1}\rho^{t+1-\tau}\cdot\left\{\sum_{j=1}^{M} p(j_\tau = j \mid y_\tau, \hat{n}_{\tau-1})\{y_\tau - \hat{\mu}_{y_\tau, j}(\hat{n}_\tau)\}' \Sigma_{y_\tau, j}^{-1}\{y_\tau - \hat{\mu}_{y_\tau, j}(\hat{n}_\tau)\}\right\}. \tag{11}$$

By Taylor series expansion to the above auxiliary function, choosing the truncated second order items, and maximizing the approximated items with respect to the noise parameter, the noise at time $t+1$ can be estimated [8]-[11]

$$\hat{n}_{t+1} = \hat{n}_t + \gamma\cdot\{K_{t+1}(\hat{n}_t)\}^{-1} S_{t+1}(\hat{n}_t), \tag{12}$$

where the disturbing factor γ is a non-negative constant with value greater than 0, the Fisher information matrix $K_{t+1}(\hat{n}_t)$ and the score vector $S_{t+1}(\hat{n}_t)$ are defined as following

$$K_{t+1}(\hat{n}_t) = -\frac{\partial^2 Q(n \mid \Lambda_{nt})}{\partial^2 n}\bigg|_{n=\hat{n}_t} = \sum_{\tau=1}^{t+1}\rho^{t+1-\tau}\sum_{j=1}^{M} p(j_\tau = j \mid y_\tau, \hat{n}_{\tau-1})\cdot B_j'(\hat{n}_{\tau-1})\Sigma_{y_\tau, j}^{-1} B_j(\hat{n}_{\tau-1})$$
$$= \rho\cdot K_t + \sum_{j=1}^{M} p(j_{t+1} = j \mid y_{t+1}, \hat{n}_t) B_j'(\hat{n}_t)\Sigma_{y_{t+1}, j}^{-1} B_j(\hat{n}_t), \tag{13}$$

$$S_{t+1}(\hat{n}_t) = \frac{\partial Q(n \mid \Lambda_{nt})}{\partial n}\bigg|_{n=\hat{n}_t} = \sum_{j=1}^{M} p(j_{t+1} = j \mid y_{t+1}, \hat{n}_t) B_j(\hat{n}_t)' \Sigma_{y_{t+1}, j}^{-1}\{y_t - \hat{\mu}_{y_t}(\hat{n}_t)\}. \tag{14}$$

4 Experimental Results

A continuous hidden Markov model (HMM)-based speech recognition system is used in the recognition experiments for examining the presented approach. The utterances of 82 speakers (41 males and 41 females) from the mandarin Chinese corpus provided by the 863 plan (China High-Tech Development Plan[12]) are trained for triphone-based HMM acoustic models, where each triphone unit was modeled as a three-emitting-state left-right topology with a mixture of 16 Gaussian per state and diagonal covariance matrices. The utterances of 9 speakers from the clean corpus are used for subsequent artificial contamination with different noise class.

In order to extract Mel frequency cepstral coefficients (MFCCs) from the 16Hz noisy speech data, we use a power spectrum which is calculated every 10ms on a 25ms Hmming window with pre-emphasis coefficient 0.97, then take a mel-scaled triangular filterbank and logarithmic computation and accordingly obtain the Mel-scaled 24 log-fiterbank coefficients. After transforming them into the cepstral domain with DCT transform, we obtain the first 12 cepstral coefficients (excluding the zero coefficients). The log-energy feature in each frame is computed after taking Hmming windowing. Accordingly, 39 dimensional features consisting of the 12 cepstral coefficients, the log-energy feature coefficient and their time derivatives are computed.

In our feature compensation paradigm, for modeling the clean speech, we extract a set of 24 MFCCs and one log-energy feature from the clean speech data for training and obtain a mixture of 128 Gaussian distributions. Then the mean vector of each mixture component in the Mel-scaled log spectral domain is obtained using inverse cosine transformation matrix. The covariance matrix is computed also from the cepstral domain using the inverse cosine transformation matrix and its transpose. By ignoring the off-diagonal elements in the covariance matrices assuming that the different coefficients are statistically independent, we obtain the diagonal covariance matrices in the log domain. With the developed sequential EM algorithm, the 24 dimensional log-filterbank features and a log-energy feature are compensated. With DCT transform and delta and delta-delta regression equations, the static coefficients (12 MFCCs plus the log-energy feature) and the corresponding dynamic coefficients (13 delta coefficients and 13 delta-delta coefficients) are computed.

In order to test the validity of the feature compensation algorithm, a number of experiments have been performed. They include the baseline without compensation, compensation with CMN (cepstral mean normalization), batch-EM estimation and sequential EM estimation. The forenamed three approaches are titled as "baseline", "CMN" and "batch-EM", respectively in Table 1 and Table 2. In the sequential EM estimation, to investigate the behavior in the non-stationary environment, we get three forms: "Seq-0.90", "Seq-0.95" and "Seq-1.00" according to the different forgetting value ρ with 0.90, 0.95 and 1.00. And we add the stationary white noise and the non-stationary babble noise from NoiseX92 [13] to the test set according to different SNR varying from 0dB to 20dB. It is observed from Table 1 that, the sequential estimation gives considerable performances, compared with "baseline", "CMN" and "Batch-EM". For example, in the 5dB white noisy condition, "baseline" only achieves 2.54% recognition rate, "CMN" achieves 10.98% recognition rate, and "Batch-EM" achieves 17.00% recognition rate. The sequential estimation with the forgetting factor ρ set to 0.90, 0.95 and 1.00 gives 18.93%, 18.97% and 18.91% recognition rates and achieves

1.93%, 1.97%, and 1.91% improvements over that by "Batch-EM", respectively. As a whole, the sequential estimation with different forgetting factor value achieves 0.77%, 0.75%, and 0.82% improvements over that by "Batch-EM", respectively. It is clear that the presented approach is very effective in the stationary noisy condition.

Table 1. Recognition rates in the white noisy environment (%)

SNR	0dB	5dB	10dB	15dB	20dB	Avg.
baseline	0.32	2.54	11.14	30.00	56.04	20.01
CMN	3.31	10.98	29.99	36.94	61.65	28.57
Batch-EM	5.51	17.00	39.37	62.36	77.12	40.27
Seq-0.90	4.99	18.93	39.99	63.19	78.10	41.04
Seq-0.95	4.99	18.97	40.19	62.85	78.12	41.02
Seq-1.00	5.02	18.91	40.38	63.02	78.14	41.09

Table 2. Recognition rates in the babble noisy environment (%)

SNR	0dB	5dB	10dB	15dB	20dB	Avg.
baseline	3.87	24.61	54.84	62.98	80.23	45.31
CMN	11.16	32.38	55.95	71.04	80.31	50.17
Batch-EM	15.86	39.25	61.78	75.23	81.09	54.64
Seq-0.90	17.72	40.71	62.56	75.53	81.07	55.52
Seq-0.95	17.50	40.70	62.49	75.50	81.21	55.48
Seq-1.00	17.44	40.31	62.60	75.51	81.46	55.46

To test the validity of the sequential estimation in non-stationary conditions, we further test the babble noise in different SNR levels. It is observed in Table 2 that, using "baseline", performance degradation is not obvious in the high SNR condition, such as in the 20dB condition. But when the noise amount increases, recognition performance quickly deteriorates with only 3.87% recognition rate in the 0dB condition. With "CMN" and "Batch-EM", the phenomena can be relatively restrained. However, they still have the main limitations to cope with the non-stationary environments. Although compensation is applied to reduce the mismatch among the clean acoustic model and the test set, they remain a minor mismatch which they don't obtain the best performance at all of non-stationary noisy conditions. With the sequential estimation algorithm, it can further reduce the mismatch and can improve the system performance in most of noisy conditions, especially in low SNR conditions. For example, in 5dB condition, the sequential EM algorithm with different forgetting factor achieves 1.46%, 1.45% and 1.06% improvements in comparison to "Batch-EM", respectively.

From Table 1 and Table 2, we also observe that the sequential estimation averagely provides slight improvement when the forgetting factor ρ is 1.00 over that of ρ is 0.9 or 0.95 for the white noise. But we notice that it averagely provides slight improvement when ρ is 0.90 over that of ρ is 0.95 or 1.00 for the babble noise. The cause of this behavior is that the white noise is the stationary noise and the babble noise is the non-stationary noise. For the white noisy condition, it is clear that the history data is very useful to noise estimation. With the reasonable forgetting factor, the presented approach can ignore the history data which is effective for computing the current noise parameter in the non-stationary condition. Due to the babble noise is a slow time-varying noise, the forgetting factor can be set with a high value relatively. For the highly time-varying conditions, ρ can be a low value to reasonably track the non-stationary characteristics.

5 Conclusions

We have presented an approach to environment compensation for robust speech recognition based on a sequential EM algorithm. The algorithm compensates entirely all of the features to deal with the environment corruption. The corruption causing distortion in the speech signal in the log domain can be modeled a nonlinear function and linearized by the truncated first-order VTS approximation. Furthermore, all of the clean cepstral coefficients and the log-energy feature are trained and postprocessed by taking corresponding inverse DCT transform to obtain a reasonable Gaussian mixture model in the log domain. They give a reasonable basis for the subsequent speech recognition. Experiment results show that the algorithm presented provides improvements of about 20% in the white noise and about 10% in the babble noise when compared with the performances under distortion environments. Moreover, the performance of speech recognition system by using sequential EM algorithm achieves considerable improvement compared with the traditional batch-EM algorithm. In the future work, we will investigate the relationship of the forgetting factor with the degree of the non-stationary characteristics and the noise class.

References

1. Stern, R.M., Raj, B., Moreno, P.J.: Compensation for Environmental Degradation in Automatic Speech Recognition. In: Proc. ESCA-NATO Tutorial Research Workshop Robust Speech Recognition for Unknown Communication Channels(1997)33–42
2. Moreno, P.J., Raj, B., Stern, R.M.: A Vector Taylor Series Approach for Environment-Independent Speech Recognition. In: Proceedings of IEEE(1995)733-736
3. Haifeng, S., Jun, G., Gang, L., and Qunxia, L.: Environment Compensation Based on Maximum a Posteriori Estimation for Improved Speech Recognition. Accepted in The Mexican International Conference on Artificial Intelligence (MICAI)(2005)
4. Raj, B., Gouvea, E.B., Moreno, P.J., Stern, R.M.: Cepstral Compensation by Polynomial Approximation for Environment-Independent Speech Recognition. In: Proceedings of Int. Conf. Spoken Language Processing, Philadelphia(1996)2340-2343
5. Kim, N.S., Kim, D.Y., Byung, K.G., Kim S.R.: Application of VTS to Environment Compensation with Noise Statistics. In: ESCA Workshop on Robust Speech Recognition. Pont-a-Mousson, France(1997)99-102
6. Kim, N.S.: Statistical Linear Approximation for Environment Compensation. IEEE Signal Processing Letters, 1(1998)8-10

7. Haifeng, S., Gang, L., Jun, G., and Qunxia, L.: Two-Domain Feature Compensation for Robust Speech Recognition. In: Wang, J., Liao, X., and Yi, Z. (eds.): Advance in Neural Network- ISNN 2005. Lecture Notes in Computer Science 3497, Springer-Verlag, Berlin Heidelberg New York(2005)351–356
8. Kim, N.S.: Nonstationary Environment Compensation Based on Sequential Estimation. IEEE Signal Processing Letters, 3(1998)8-10
9. Zhao, Y., Wang, S., Yen, K.C.: Recursive Estimation of Time-Varying Environments for Robust Speech Recognition. In: Proceedings of IEEE(2001)225-228
10. Deng, L., Droppo, J., Acero, A.: Recursive Noise Estimation Using Iterative Stochastic Approximation for Stereo-Based Robust Speech Recognition. In: Proceedings of IEEE (2002)81-84
11. Krishnamurthy, V., Moore, J.B.: "Online Estimation of Hidden Markov Model Parameters Based on the Kullback-Leibler Information Measure. IEEE Trans. Sig. Proc, 8 (1993)2557-2573
12. Zu, Y. Q.: Issues in the Scientific Design of the Continuous Speech Database. Available: http://www.cass.net.cn/chinese/s18_yys/yuyin/report/report_1998.htm.
13. Varga, A., Steenneken, H. J. M., Tomilson, M., Jones, D.: The NOISEX–92 Study on the Effect of Additive Noise on Automatic Speech Recognition. Tech. Rep. DRA Speech Research Unit(1992)

Appendix

If we only consider the additive noise, the corruption in the signal domain is shown as following

$$y_t = x_t + n_t, \qquad (15)$$

where y_t denotes the noisy sample, x_t for the clean sample, n_t for the additive noise.

Generally we assume that x_t and n_t are statistically independent. If we transform the above relation into the power spectral domain, the corruption can be expressed as:

$$Y(\omega) = X(\omega) + N(\omega), \qquad (16)$$

where $Y(\omega)$, $X(\omega)$ and $N(\omega)$ represent the power spectrum of the noisy speech, clean speech and additive noise, respectively. If we take a logarithmic computation on both sides of Eq.(16),

$$\begin{aligned}\log\{Y(\omega)\} &= \log\{X(\omega) + N(\omega)\} \\ &= \log\{X(\omega)\} + \log\left\{1 + \frac{N(\omega)}{X(\omega)}\right\} \\ &= \log\{X(\omega)\} + \log\{1 + \exp\{\log\{N(\omega)\} - \log\{X(\omega)\}\}\}.\end{aligned} \qquad (17)$$

Let $y = \log\{Y(\omega)\}$, $x = \log\{X(\omega)\}$ and $n = \log\{N(\omega)\}$, we have [2]

$$y = x + \log(1 + \exp(n - x)) = x + f(x,n), \qquad (18)$$

where y, x and n are respectively the noisy speech, the clean speech and the noise in the log spectral domain. From Eq.(18), For each log-filterbank bin, it is noticeable that the corruption becomes a complex nonlinear contamination procedure.

Now we describe the log-energy feature contamination procedure. In order to attenuate the discontinuities at the window, we generally use the Hmming window before extracting the feature coefficients. The energy of one frame after taking Hmming windowing on speech can be written as

$$E_y = \sum_{l=1}^{L} [h(y_l)]^2 , \qquad (19)$$

where L denotes the number of samples in each frame, E_y is the noisy energy, $h(\cdot)$ represents operation with Hmming windowing. Due to the clean speech and the noise are statistical independent and $h(\cdot)$ is a linear computation, the above equation can be rewritten as

$$\begin{aligned} E_y &= \sum_{l=1}^{L} \{h(x_l + n_l)\}^2 = \sum_{l=1}^{L} \{h(x_l) + h(n_l)\}^2 \\ &= \sum_{l=1}^{L} \{h(x_l)\}^2 + \sum_{l=1}^{L} \{h(n_l)\}^2 = E_x + E_n, \end{aligned} \qquad (20)$$

where E_x and E_n are respectively the clean energy and the noise energy in one frame. If we take a logarithmic transformation on Eq.(20), the corruption of the log-energy feature is

$$y_e = x_e + \log(1 + \exp(n_e - x_e)) , \qquad (21)$$

in which y_e, x_e and n_e are respectively the noisy log-energy feature, the clean log-energy feature and the noise, $y_e = \log(E_y)$, $x_e = \log(E_x)$, $n_e = \log(E_n)$.

As seen in Eq.(18) and Eq.(21), the corruptions of the log-filtebank coefficients and the log-energy feature have the same functional form.

Hybrid Cost-Sensitive Decision Tree

Shengli Sheng and Charles X. Ling

Department of Computer Science, The University of Western Ontario,
London, Ontario N6A 5B7, Canada
{cling, ssheng}@ csd.uwo.ca

Abstract. Cost-sensitive decision tree and cost-sensitive naïve Bayes are both new cost-sensitive learning models proposed recently to minimize the total cost of test and misclassifications. Each of them has its advantages and disadvantages. In this paper, we propose a novel cost-sensitive learning model, a hybrid cost-sensitive decision tree, called DTNB, to reduce the minimum total cost, which integrates the advantages of cost-sensitive decision tree and of the cost-sensitive naïve Bayes together. We empirically evaluate it over various test strategies, and our experiments show that our DTNB outperforms cost-sensitive decision and the cost-sensitive naïve Bayes significantly in minimizing the total cost of tests and misclassification based on the same sequential test strategies, and single batch strategies.

1 Introduction

Inductive learning techniques have had great success in building classifiers and classifying test examples into classes with a high accuracy or low error rate. However, in many real-world applications, lowering misclassification error is not the goal as "errors" can cost very differently. This type of learning is called cost-sensitive learning. Turney [14] surveys a whole range of costs in cost-sensitive learning, among which two types of costs are most important: misclassification costs and test costs. For example, in a binary classification task, the cost of false positive (FP) and the cost of false negative (FN) are often very different. In addition, attributes (tests) may have different costs, and acquiring values of attributes also incurs costs. The goal of learning is to minimize the sum of the misclassification costs and the test costs.

Tasks involving both misclassification and test costs are abundant in real-world applications. For example, when building a model for medical diagnosis from the training data, we must consider the cost of tests (such as blood tests, X-ray, etc.) and the cost of misclassifications (errors in the diagnosis). Further, when a doctor sees a new patient (a test example), tests are normally ordered, at a cost to the patient or his/her insurance company. To better diagnose or predict the disease of the patient (i.e., reducing the misclassification cost). Doctors must balance the trade-off between potential misclassification costs and test costs to determinate which tests should be ordered, and at what order, to reduce the expected total cost. A case study on heart disease is given in the paper.

In this paper, we propose a new cost-sensitive learning model, DTNB, which integrates the advantages of the cost-sensitive decision tree and the cost-sensitive naïve Bayes, both of which minimize the total cost of misclassifications and tests.

DTNB uses the cost-sensitive decision tree to collect the required tests for test examples, and uses the cost-sensitive naïve Bayes to classify. For a test example, after the required tests are collected according to the cost-sensitive decision tree, the tests are performed with a cost and their results are available. Then the cost-sensitive naïve Bayes built on all the training data is applied to classify the test example. The naïve Bayes model can make use of the known values which do not appear in the path which the test example follows to go down to a leaf in the cost-sensitive decision tree. Thus, we can expect that the cost-sensitive DTNB can achieve lower total cost than the cost-sensitive decision tree and the cost-sensitive naïve Bayes do alone.

The rest of paper is organized as follows. We first review the related work in Section 2. Then we describe our new cost-sensitive learning model, DTNB, to reduce the minimum total cost of tests and misclassifications in Section 3. In Section 4, we present empirical experiments. The paper concludes with discussion and some directions for the future work.

2 Review of Previous Work

Cost-sensitive learning has received extensive attentions in recent years. Turney [14] analyzes a variety of costs in machine learning, such as misclassification costs, test costs, active learning costs, computation cost, human-computer interaction cost, etc. Two types of costs are singled out as the most important in machine learning: misclassification costs and test costs, and test costs are normally considered in conjunction with misclassification costs. Much work has been done in considering non-uniform misclassification costs (alone), such as [4, 5, 7]. Those works can often used to solve problem of learning with very imbalanced datasets [3]. Some previous work, such as [10, 12], consider the test cost alone without incorporating misclassification cost. As pointed out by [14] it is obviously an oversight. As far as we know, the only work considering both misclassification and test costs includes [13, 15, 9, 2]. We discuss these works in detail below.

In [15], the cost-sensitive learning problem is cast as a Markov Decision Process (MDP), and an optimal solution is given as a search in a state space for optimal policies. While related to our work, their research adopts an optimal search strategy, which may incur very high computational cost to conduct the search. In contrast, we adopt the local search similar to [11] using a polynomial time algorithm to build a new decision trees, and our test strategies are also polynomial to the tree size. (Greiner et al. 2002) studied the theoretical aspects of active learning with test costs using a PAC learning framework, which models how to use a budget to collect the relevant information for the real-world applications with no actual data at beginning. Our algorithm builds a model from history data to minimize the total cost of misclassification and tests for a new case with missing values. Turney [13] presented a system called ICET, which uses a genetic algorithm to build a decision tree to minimize the cost of tests and misclassification. Our algorithm essentially adopts the same decision-tree building framework as in [11], and it is expected to be more efficient than Turney's genetic algorithm based approach.

Ling et al. [9] propose a cost-sensitive decision tree learning program that minimizes the total cost of tests and misclassifications. They also propose several test strategies, and compare their results to C4.5. However, for a test example, the cost-sensitive decision tree ignores the information supplied by the known attributes which do not appear in the path which the test example follows to go down to a leaf in the cost-sensitive decision tree. Chai et al. [2] propose a cost-sensitive naïve Bayes based algorithm, called CSNB, which searches for minimal total cost of tests and misclassifications. They also propose a sequential test strategy and a single batch test strategy. However, the cost-sensitive naïve Bayes does not learn the general attribute structure (such as the tree structure) but only probability tables from training data. The test sequence for each test example is less comprehensible.

Our model, DTNB, combines the advantages of cost-sensitive decision tree and naïve Bayes. It utilizes the structure of the cost-sensitive decision tree to collect the beneficiary tests for a test example and makes use of the information in the known attributes which are ignored by the cost-sensitive decision tree to reduce the misclassification cost. We expect that our DTNB outperform cost-sensitive decision tree and cost-sensitive naïve Bayes alone in terms of the total cost of tests and misclassification.

The new cost-sensitive model, DTNB, is composed of decision tree and naïve Bayes, but it is much different from NBTree [8] proposed by Kohavi. First of all, NBTree is not a cost-sensitive learning model. The learning algorithm of NBTree is similar to C4.5 [Qui93]. DTNB is a cost-sensitive learning to minimize the total cost of tests and misclassification. Secondly, in NBTree, a naïve Bayes is constructed for each leaf using the data associated with the leaf. However, DTNB only constructs one naïve Bayes using all the training data. This naïve Bayes acts as a hidden node at each node (including the leaves) of the cost-sensitive decision tree. The details of difference between NBTree and DTNB are explained in Section 3.

3 The New Cost-Sensitive Learning - DTNB

We assume that we are given a set of training data (with possible missing attribute values), the misclassification costs, and test costs for each attribute. We propose a novel cost-sensitive learning model, DTNB, which combines the advantages of cost-sensitive decision tree and naïve Bayes. The rationale of DTNB is based on our observations. We note that cost-sensitive decision tree has the ability of learning a general structure, and the structure of the tree plays an important role for collecting the most beneficiary unknown values. However, the decision tree ignores the original known values which do not appear in the tree for classify a test example. In non-cost-sensitive learning, this is one reasonable feature of decision tree. But in cost-sensitive learning, any value is available with a certain cost. We do not want waste any available information. Naturally, making use of all known values can reduce the total cost. The information of the known attributes which do not appear in the path through which the test example goes down to a leaf of the tree is useful for cost-sensitive classification to reduce the misclassification cost. Fortunately, cost-sensitive naïve Bayes indeed utilizes all known attributes for misclassification, but it does not have a structure learning ability to help determine which tests and in what order should be done for unknown attributes.

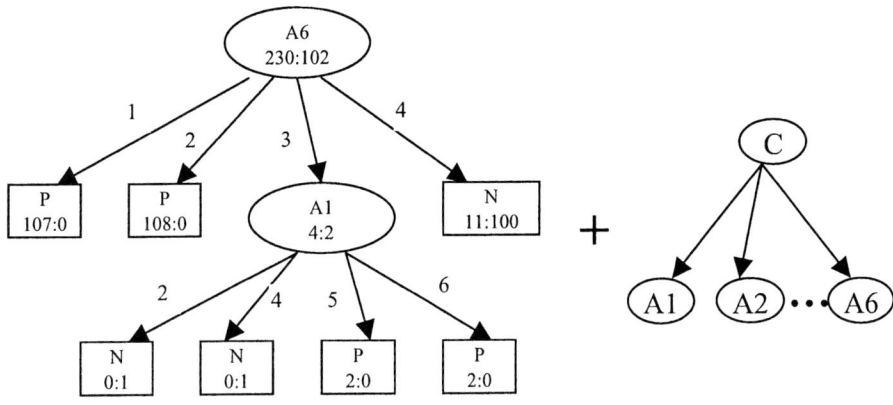

Fig. 1. An example of cost-sensitive DTNB

In order to overcome these drawbacks and combine those advantages in the two cost-sensitive models, we propose a novel cost-sensitive learning model, which integrates cost-sensitive decision tree with cost-sensitive naïve Bayes, called DTNB. Figure 1 shows the structure of an example of the novel cost-sensitive learning model DTNB. We can see DTNB is an integration model with two parts. The left part is a cost-sensitive decision tree which is used for finding the required tests for each testing example. Besides the cost-sensitive tree, DTNB also contains a naïve Bayes (right part), which is for classification.

First of all, DTNB builds a cost-sensitive decision tree, given a set of training data, the misclassification costs, and test costs for each attribute. The building procedure is similar to C4.5. Instead of using entropy based splitting criteria, we use the *expected total misclassification cost* to select an attribute for splitting. This gives a more accurate choice for attribute selection. That is, an attribute may be selected as a root node of a decision tree if the sum of the test cost and the expected misclassification costs of all branches is the minimum among other attributes, and is less than that of the root. For a subset of examples with tp positive examples and tn negative examples, if $C_P = tp \times TP + tn \times FP$ is the total misclassification cost of being a positive leaf, and $C_N = tn \times TN + tp \times FN$ is the total misclassification cost of being a negative leaf, then the probability of being positive is estimated by the relative cost of C_P and C_N; the smaller the cost, the larger the probability (as minimum cost is sought). Thus, the probability of being positive is: $1 - \dfrac{C_P}{C_P + C_N} = \dfrac{C_N}{C_P + C_N}$. The expected misclassification cost of being positive is: $E_P = \dfrac{C_N}{C_P + C_N} \times C_P$. Similarly, the probability of being a negative leaf is $\dfrac{C_P}{C_P + C_N}$; and the expected misclassification cost of being negative is: $E_N = \dfrac{C_P}{C_P + C_N} \times C_N$. Therefore, without splitting, the expected total misclassification cost of a given set of examples is: $E = E_P + E_N = \dfrac{2 \times C_P \times C_N}{C_P + C_N}$. If an attribute A has l branches, then the expected total misclassification cost after splitting on A is: $E_A = 2 \times \sum_{i=1}^{l} \dfrac{C_{P_i} \times C_{N_i}}{C_{P_i} + C_{N_i}}$. Thus, ($E - E_A - T_C$) is the expected cost reduction splitting on A, where T_C is the total test cost for all examples on A. It is easy to find out which attribute has the smallest expected total cost (the sum of the test cost and the expected misclassification cost), and if it is smaller than the one without split (if so, it is worth to split). With the expected total misclassification cost described above as the splitting criterion, the lazy-tree learning algorithm is shown in Figure 2.

Simultaneously, we build a cost sensitive naïve Bayes. Note that this model is built on all the training data, and for all nodes in the tree. However, NBTree [Koh96] treats the segmentation of decision tree as an advantage. It builds a naïve Bayes at each leaf of the decision tree. And the naïve Bayes constructed for a leaf uses only the data associated with the leaf. However, as the tree grows, the training data are split into the lower level nodes. Finally, there are very little data in the leaves. The classification based on these leaves is far less accurate, so that the misclassification cost goes higher. This is reason that NBTree is proposed for larger dataset. However, without larger dataset assumption DTNB overcomes the shortcoming of segmentation of decision tree by constructing only one naïve Bayes using all the training data. This naïve Bayes acts as a hidden model at each node (including the leaves) of the cost-sensitive decision tree. The hidden model is only for classification. Thus, DTNB does not utilize the data which go down into a leaf of the tree to classify a testing example which drops into this leaf. It classifies the test example by the only hidden cost-sensitive naïve Bayes.

DTNB only builds one general naïve Bayes from all the training data. Whereas, the posterior probabilities of a test example e are computed from the known attributes and the tested unknown attributes. The unknown attributes which are not selected to perform testing are not concerned. With the posterior probabilities, if $FN \times P(+|e) > FP \times P(-|e)$, this test example is classified as negative, otherwise, as positive. A misclassification cost may be incurred if the prediction of the test example is wrong. Thus, for each test example, not only the attributes appearing on the tree, but also the known attributes can be fully used to make correct classification, so that the total misclassification cost can be reduced, as any known value is worthy of a certain cost. But for the cost-sensitive decision tree, it is possible some known attributes are not used to split the training data, so that they become useless for the classification. DTNB makes use of all known attributes, as well as the available values of the collected unknown attributes at certain test costs.

CSDT(*Examples, Attributes, TestCosts*)
1. Create a *root* node for the tree
2. If all examples are positive, return the single-node tree, with *label* = +
3. If all examples are negative, return the single-node tree, with *label* = -
4. If attributes is empty, return the single-node tree, with label assigned according to *min (E_P, E_N)*
5. Otherwise Begin
   a. If *maximum cost reduction* < *0* return the single-node tree, with label assigned according to *min (E_P, E_N)*
   b. *A* is an attribute which produces maximum cost reduction among all the remaining attributes
   c. Assign the attribute *A* as the tree *root*
   d. For each possible value vi of the attribute *A*
      i. Add a new branch below root, corresponding to the test $A=v_i$
      ii. Segment the training examples into each branch *Example_v_i*
      iii. If no examples in a branch, add a leaf node in this branch, with label assigned according to *min (E_P, E_N)*
      iv. Else add a subtree below this branch, CSDT(*examples_v_i, Attributes-A, TestCosts*)
6. End
7. Return *root*

Fig. 2. Algorithm of cost-sensitive decision tree

In the naïve Bayes model of DTNB, the Laplace Correction is applied. That is, $p(a \mid +) = \dfrac{N_a + 1}{N + m}$, where *Na* is the number of instances whose attribute $A_1=a$, *N* is the number of instances whose class is +, and m is the number of classes.

After DTNB is built, for each testing example, there are two steps to find the minimum total cost of tests and misclassifications. The first step is to utilize the tree structure of the cost-sensitive decision tree to collect a set of tests which need be performed according to a certain strategy (there are several strategies explained in Section 4). The total test cost is accumulated in the step. After the set of tests are done, the values of the unknown attributes in the test example are available. It automatically goes to the second step, where the cost-sensitive naïve Bayes model is used to classify the test example into a certain class. The naïve Bayes uses not only the unknown attributes tested but also all known attributes. If it is classified incorrectly, there is misclassification cost. We empirically evaluate it over various test strategies in next section.

4 Experiments

We evaluate the performance of DTNB on two categories of test strategies: Sequential Test, and Single Batch Test. For a given test example with unknown attributes, the

Sequential Test can request only one test at a time, and wait for the test result to decide which attribute to be tested next, or if a final prediction is made. The Single Batch Test, on the other hand, can request one set (batch) of one or many tests to be done simultaneously before a final prediction is made.

4.1 DTNB's Optimal Sequential Test

Recall that Sequential Test allows one test to be performed (at a cost) each time before the next test is determined, until a final prediction is made. Ling, et al. [9] described a simple strategy called *Optimal Sequential Test* (or OST in short) that directly utilizes the decision tree built to guide the sequence of tests to be performed in the following way: when the test example is classified by the tree, and is stopped by an attribute whose value is unknown, a test of that attribute is made at a cost. This process continues until the test case reaches a leaf of the tree. According to the leaf reached, a prediction is made, which may incur a misclassification cost if the prediction is wrong. Clearly the time complexity of OST is only linear to the depth of the tree.

One weakness with this approach is that it ignores some known attributes which do not appear in the path through which a test example goes down to a leaf. However, these attributes can be useful for reducing the misclassification cost. Like the OST, We also propose an Optimal Sequential Test strategy for DTNB (section 3), called DNOST in short. It has the similar process as OST. The only difference is that the class prediction which is not made by the leaf it reached, but the naïve Bayesian classification model in DTNB. This strategy utilizes the tree structure to collect the most useful tests for a test example. And it also utilizes the entire original known attributes in the test example with the unknown attributes tested to predict the class of the test example. We can expect DNOST outperforms OST.

Table 1. Datasets used in the experiments

	No. of Attributes	No. of Examples	Class dist. (N/P)
Ecoli	6	332	230/102
Breast	9	683	444/239
Heart	8	161	98/163
Thyroid	24	2000	1762/238
Australia	15	653	296/357
Tic-tac-toe	9	958	332/626
Mushroom	21	8124	4208/3916
Kr-vs-kp	36	3196	1527/1669
Voting	16	232	108/124
Cars	6	446	328/118

Comparing Sequential Test Strategies. To compare various sequential test strategies, we choose 10 real-world datasets which are listed in Table 1, from the UCI Machine Learning Repository [1]. The datasets are first discretized using the minimal entropy method [6]. These datasets are chosen because they are binary class, have at least some discrete attributes, and have a good number of examples. Each dataset is split into two parts: the training set (60%) and the test set (40%). Unlike the case study of heart disease, the detailed test costs and group information [13] of these datasets are unknown. To make the comparison possible, we simply choose randomly the test costs of all attributes to be some values between 0 and 100. This is reasonable because we compare the relative performance of all test strategies under the same chosen costs. To make the comparisons straightforward, we set up the same misclassification costs 200/600 (200 for false positive and 600 for false negative). For test examples, a certain ratio of attributes (0.2, 0.4, 0.6, 0.8, and 1) are randomly selected and marked as unknown to simulate test cases with various degrees of missing values.

In this section, we compare our DNOST with the other two sequential test strategies available, OST, and CSNB [2] on 10 real-world datasets to see which one is better (having a smaller total cost). Note that DNOST and OST use the same decision tree to collect beneficiary tests. However, DNOST uses DTNB's naïve Bayes for classification, while OST uses the leaves of tree to classify test examples. CSNB follows the same test strategy: determine next test based on the previous test result. However, it is based on the naïve Bayes only. In all, all of them are based on the same test strategy, but they are applied different cost-sensitive learning models. That is, their performances directly stand for the performances of different learning models. We repeat this process 25 times, and the average total costs for the 10 datasets are plotted in Figure 3.

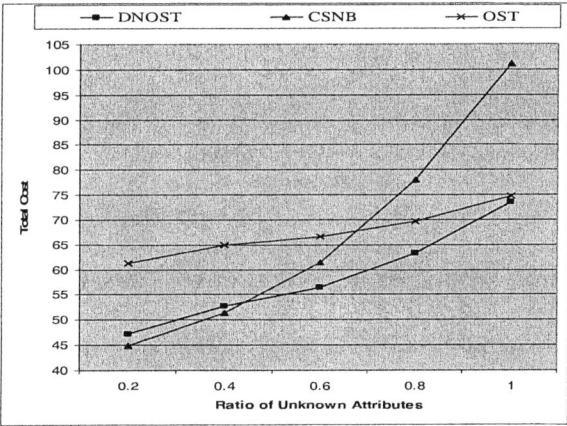

Fig. 3. The total cost of our new Sequential Test Strategy DNOST compared to previous strategies (OST and CSNB)

We can make several interesting conclusions. First, DNOST performs the best among the three sequential test strategies. When the unknown attribute ratio is higher, the difference between DNOST and CSNB becomes bigger. However, DNOST is gradually close to OST when the unknown ratio is increased. When the unknown ratio is lower, the difference between DNOST and OST is bigger, as more known attributes are utilized in DTNB, but they are ignored in cost-sensitive decision tree. Second, the results proof our expectation which DTNB integrates the advantage of the decision tree and the naïve Bayes and overcomes their defects. When the unknown ratio is lower, there are more known attributes ignored by OST, so that OST performs worse, whereas DNOST and CSNB perform better and are closer, as they make use of the known values. When the unknown ratio is higher, there are less known attributes ignored by OST and both DNOST and OST utilize the tree structure to collect the most beneficial tests, so that they perform better and are close to each other.

4.2 Single Batch Test Strategies

The Sequential Test Strategies have to wait for the result of each test to determine which test will be the next one. Waiting not only costs much time, but also increases the pressure and affects the life quality of patients in medical diagnosis. In manufacturing diagnoses, it delays the progress of engineering. Even in some particular situations, for example, emergence, we have to make decisions as soon as possible. In medical emergence, doctors normally order one set of tests (at a cost) to be done at once. This is the case of the Single Batch Test.

In [9] a very simple heuristic is described. The basic idea is that when a test example is classified by a minimum-cost tree and is stopped by the first attribute whose value is unknown in the test case, all unknown attributes under and including

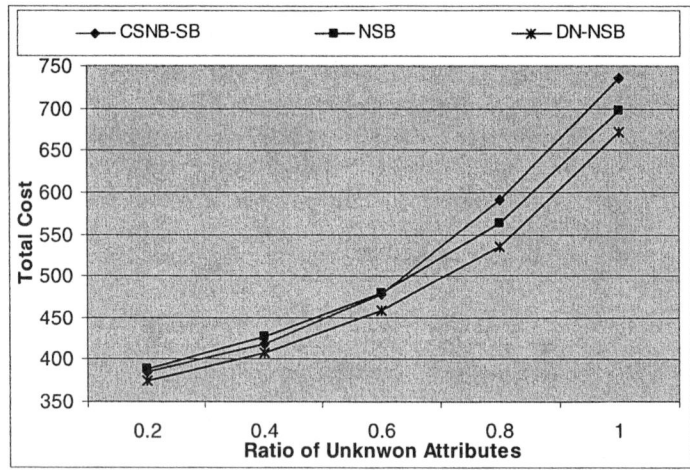

Fig. 4. The total cost of our new Single Batch Test Strategies DN-NSB compared to their previous strategies (NSB and CSNB-SB)

this first attribute would be tested, as a single batch. Clearly, this strategy would have exactly the same misclassification cost as the Optimal Sequential Test, but the total test cost is higher as extra tests are performed. This strategy is called Naïve Single Batch (NSB).

The weakness of NSB is that it ignores some known attributes which do not appear in the path through which a test example goes down to a leaf after the tests are performed. However, these attributes can be useful for reducing the misclassification cost. Like the NSB, we apply the similar process on DTNB. The only difference is the class prediction which is not made by the leaf a test example reached after the tests are performed, but by the naïve Bayes classification model. We call this process DTNB's Naïve Single Batch Test (or DN-NSB in short).

Comparing Single Batch Test Strategies. We use the same experiment procedure on the same 10 datasets used in Section 4.1 (see Table 1) to compare various Single Batch Test strategies including CSNB-SB [2]. The only change is the misclassification costs, which are set to 2000/6000 (2000 for false positive and 6000 for false negative). The misclassification costs are set to be larger so the trees will be larger and the batch effect is more evident. Note that DN-NSB and NSB use the same decision tree to collect beneficiary tests. However, DN-NSB uses DTNB's naïve Bayes for classification, while NSB uses the leaves of tree to classify test examples. CSNB follows the same test strategy: request one set (batch) of one or many tests to be done simultaneously before a final prediction is made. However, it is based on the naïve Bayes only. In all, all of them are based on the same test strategy, but they are applied to different cost-sensitive learning models. That is, their performances directly stand for the performances of different learning models. The total costs for the 10 datasets are compared and plotted in Figure 4.

We can make several interesting conclusions. First, the single batch test strategy (DN-NSB) based on DTNB outperforms others on any unknown ratio. CSNB-SB outperforms NSB when the unknown ratio is higher, but it is worse than NSB when the unknown ratio goes down. Second, the results again proof our expectation which DTNB integrates the advantage of the decision tree and the naïve Bayes and overcomes their defects. When the unknown ratio is lower, there are more known attributes ignored by NSB, so that NSB performs worse. DN-NSB and CSNB-SB perform better, as they make use of the known values. When the unknown ratio is higher, there are less known attributes ignored by NSB and both DN-NSB and NSB utilize the tree structure to collect the most beneficiary tests, so that they perform better.

5 Conclusion and Future Work

In this paper, we present a hybrid decision tree learning algorithm, which integrate with naïve Bayes, to minimize the total cost of misclassifications and tests. We evaluate the performance (in terms of the total cost) empirically, compared to previous methods using decision tree and naïve Bayes alone. The results show that our novel learning algorithm, DTNB, performs significantly better than the decision tree learning and the naïve Bayes learning alone.

In our future work we plan to design smart single batch test strategies. We also plan to incorporate other types of costs in our hybrid decision tree learning DTNB and test strategies.

References

1. Blake, C.L., and Merz, C.J., *UCI Repository of machine learning databases (website)*. Irvine, CA: University of California, Department of Information and Computer Science (1998).
2. Chai, X., Deng, L., Yang, Q., and Ling,C.X., Test-Cost Sensitive Naïve Bayesian Classification. *In Proceedings of the Fourth IEEE International Conference on Data Mining*. Brighton, UK : IEEE Computer Society Press (2004).
3. Chawla,N.V., Japkowicz, N., and Kolcz, A. eds., *Special Issue on Learning from Imbalanced Datasets*. SIGKDD, 6(1): ACM Press (2004).
4. Domingos, P., MetaCost: A General Method for Making Classifiers Cost-Sensitive. *In Proceedings of the Fifth International Conference on Knowledge Discovery and Data Mining*, 155-164. San Diego, CA: ACM Press (1999).
5. Elkan, C., The Foundations of Cost-Sensitive Learning. *In Proceedings of the Seventeenth International Joint Conference of Artificial Intelligence*, 973-978. Seattle, Washington: Morgan Kaufmann (2001).
6. Fayyad, U.M., and Irani, K.B., Multi-interval discretization of continuous-valued attributes for classification learning. *In Proceedings of the 13th International Joint Conference on Artificial Intelligence*, 1022-1027. France: Morgan Kaufmann (1993).
7. Ting, K.M., Inducing Cost-Sensitive Trees via Instance Weighting. *In Proceedings of the Second European Symposium on Principles of Data Mining and Knowledge Discovery*, 23-26. Springer-Verlag (1998).
8. Kohavi, R., Scaling up the accuracy of Naïve-Bayes Classifier: a Decision-Tree Hybrid. *In Proceeding of the Second International Conference on Knowledge Discovery and Data Mining* (KDD96). AAAI Press (1996) 202-207.
9. Ling, C.X., Yang, Q., Wang, J., and Zhang, S., Decision Trees with Minimal Costs. *In Proceedings of the Twenty-First International Conference on Machine Learning*, Banff, Alberta: Morgan Kaufmann (2004).
10. Nunez, M., The use of background knowledge in decision tree induction. *Machine learning*, 6:231-250 (1991).
11. Quinlan, J.R. eds., *C4.5: Programs for Machine Learning*. Morgan Kaufmann (1993).
12. Tan, M., Cost-sensitive learning of classification knowledge and its applications in robotics. *Machine Learning Journal*, 13:7-33 (1993).
13. Turney, P.D., Cost-Sensitive Classification: Empirical Evaluation of a Hybrid Genetic Decision Tree Induction Algorithm. *Journal of Artificial Intelligence Research* 2:369-409 (1995).
14. Turney, P.D., Types of cost in inductive concept learning. *In Proceedings of the Workshop on Cost-Sensitive Learning at the Seventeenth International Conference on Machine Learning*, Stanford University, California (2000).
15. Zubek, V.B., and Dietterich, T., Pruning improves heuristic search for cost-sensitive learning. *In Proceedings of the Nineteenth International Conference of Machine Learning*, 27-35, Sydney, Australia: Morgan Kaufmann (2002).

Characterization of Novel HIV Drug Resistance Mutations Using Clustering, Multidimensional Scaling and SVM-Based Feature Ranking

Tobias Sing[1], Valentina Svicher[2], Niko Beerenwinkel[3],
Francesca Ceccherini-Silberstein[2], Martin Däumer[4], Rolf Kaiser[4],
Hauke Walter[5], Klaus Korn[5], Daniel Hoffmann[6], Mark Oette[7],
Jürgen K. Rockstroh[8], Gert Fätkenheuer[4],
Carlo-Federico Perno[2], and Thomas Lengauer[1]

[1] Max Planck Institute for Informatics, Saarbrücken, Germany*
[2] University of Rome "Tor Vergata", Italy
[3] University of California, Berkeley, CA, USA
[4] University of Cologne, Germany
[5] University of Erlangen-Nürnberg, Germany
[6] Center for Advanced European Studies and Research, Bonn, Germany
[7] University of Düsseldorf, Germany
[8] University of Bonn, Germany

Abstract. We present a case study on the discovery of clinically relevant domain knowledge in the field of HIV drug resistance. Novel mutations in the HIV genome associated with treatment failure were identified by mining a relational clinical database. Hierarchical cluster analysis suggests that two of these mutations form a novel mutational complex, while all others are involved in known resistance-conferring evolutionary pathways. The clustering is shown to be highly stable in a bootstrap procedure. Multidimensional scaling in mutation space indicates that certain mutations can occur within multiple pathways. Feature ranking based on support vector machines and matched genotype-phenotype pairs comprehensively reproduces current domain knowledge. Moreover, it indicates a prominent role of novel mutations in determining phenotypic resistance and in resensitization effects. These effects may be exploited deliberately to reopen lost treatment options. Together, these findings provide valuable insight into the interpretation of genotypic resistance tests.

Keywords: HIV, clustering, multidimensional scaling, support vector machines, feature ranking.

1 Introduction

1.1 Background: HIV Combination Therapy and Drug Resistance

Human immunodeficiency virus HIV-1 is the causative agent of the acquired immunodeficiciency syndrome AIDS, a disease in which persistent virus-induced

---

* This work was conducted in the context of the European Union Network of Excellence BioSapiens (grant no. LHSG-CT-2003-503265). T.S. would like to thank Oliver Sander for the lively and stimulating discussions on the topic of this paper.

depletion of helper T cells leads to immune failure and death due to opportunistic infections. While to date there is no cure for HIV infection, the introduction of highly active antiretroviral therapy (HAART), in which three to six antiretroviral drugs are administered in combination, has significantly improved life quality and survival time of patients. However, incomplete suppression of HIV replication by current drugs, combined with high mutation and replication rates of HIV ultimately results in the selection of viral populations carrying resistance-conferring mutations in their genomes. The fixation of these strains in the population eventually leads to therapy failure, upon which a new combination of drugs has to be chosen as next-line regimen.

1.2 Motivation: Evidence for Additional Resistance-Associated Mutations and Mutational Clusters

To date, the decision for follow-up drug combinations in patients failing therapy is routinely based on sequencing the relevant genomic region of the viral population harbored by the individual. The sequence is then analyzed to identify the presence of resistance-associated mutations for each of the 19 drugs currently available for anti-HIV therapy, by using mutation lists annually updated by the International AIDS Society (IAS) [1] or other panels of human experts.

The situation is complicated by the fact that resistance mutations do not accumulate independently from each other. Rather, they are loosely time-ordered along mutational pathways, leading to distinct mutational complexes or clusters.[1] Rational therapy planning is severely compromised by our limited understanding of these effects. Increasing evidence on additional mutations involved in the development of drug resistance [2,3], besides those listed by the IAS, provides the incentive for our present study.

1.3 Outline

We describe an approach towards the discovery and characterization of novel mutations associated with therapy failure from a large relational database, and their evolutionary and phenotypic characterization using supervised and unsupervised statistical learning methods. We focus on resistance against seven drugs from the class of nucleoside reverse transcriptase inhibitors (NRTIs), which target an HIV protein called reverse transcriptase (RT). This enzyme is responsible for translating the RNA genome of HIV back to DNA prior to its integration into the human genome. NRTIs are analogues of the natural building blocks of DNA, but lack a group essential for chain elongation. Thus, incorporation of a nucleoside analogue during DNA polymerization terminates the chain elongation process.

The knowledge discovery process described in this paper combines heterogeneous data from three different virological centers. To allow for integrated

---

[1] Throughout this paper, the words *complex*, *cluster*, and *pathway* are used interchangeably.

analysis, these data are stored in a relational database, whose structure is outlined in section 2. Systematic mining for mutations with differing propensities in NRTI-treated and untreated patients, respectively, as detailed in section 3, leads to the identification of 14 novel mutations associated with therapy failure. In section 4, we propose an approach towards characterizing the covariation structure of novel mutations and their association into complexes using hierarchical clustering and multidimensional scaling. Stability results are provided using a bootstrap method. Feature ranking based on support vector machines, described in section 5, allows for assessing the actual phenotypic impact of novel mutations. In section 6, we conclude by summarizing our approach, related work, and open problems.

2 The *Arevir* Database for Managing Multi-center HIV/AIDS Data

This study is based on multi-center virological data, including HIV genomic sequences from over 2500 patients, *in vitro* measurements of drug resistance [4], and clinical data such as viral load and helper T cell counts. Our relational HIV database *Arevir*, implemented in MySQL and Perl, and in use and ongoing development since 2002, provides an appropriate platform to address the challenges of data management and integration. The Arevir database schema is grouped into different modules, each consisting of a few tables, corresponding to information on patients, therapies, sequences, isolates, and specific (predicted or measured) isolate properties. Registered users can perform queries or enter data directly through a web interface. Upload of new data triggers the execution of several scripts, including programs for sequence alignment. To ensure privacy, connection between client and server is established via an SSH-tunneled Virtual Network Computing (VNC) client.[2]

3 Mining for Novel Mutations

Our approach towards identifying mutations associated with NRTI therapy is based on the assumption that these should occur with differential frequencies in treatment-naive subjects and in patients failing therapy, respectively.

Thus, mining for novel mutations was based on contrasting the frequency of the wild-type residue with that of a specific mutation in 551 isolates from drug-naive patients and 1355 isolates from patients under therapy failure, at RT positions 1–320 [5]. Chi-square tests were performed for all pairs of wild-type and mutant residues to determine mutations for which the null hypothesis that amino acid choice is independent from the patient population can be rejected. Correction for multiple testing was performed using the Benjamini-Hochberg method [6] at a false discovery rate of 0.05.

---

[2] Computational analyses are performed on completely anonymized data, retaining only patient identifiers instead of full names.

This procedure revealed 14 novel mutations significantly associated with NRTI treatment, in addition to those previously described in [1]: K43E/Q/N, E203D/K, H208Y, D218E[3] were virtually absent in therapy-naives (< 0.5%), while K20R, V35M, T39A, K122E, and G196E were already present in the naive population with a frequency of > 2.5% but showed significant increase in treated patients. Surprisingly, mutations I50V and R83K showed significant decrease in the treated population as compared to therapy-naives.

4 Identifying Mutational Clusters

In this section we describe an unsupervised learning approach towards characterizing the covariation structure of a set of mutations and its application to the newly discovered mutations. Mutational complexes can give rise to distinct physical resistance mechanisms, but can also reflect different ways to achieve the same resistance mechanism. Indeed, the two most prominent complexes associated with NRTI resistance, the nucleoside analogue mutations (NAMs), groups 1 and 2, consisting of mutations M41L/L210W/T215Y and K70R/K219Q/D67N, respectively, both confer resistance via an identical mechanism, called primer unblocking. On the other hand, the multi-NRTI resistance complex with Q151M as the main mutation mediates a different physical mechanism in which recognition of chemically modified versions of the DNA building blocks is improved to avoid unintended integration. In essence, to appreciate the evolutionary role of novel mutations it is important to identify whether they aggregate with one of these complexes or whether they form novel clusters, possibly reflecting additional resistance mechanisms. This analysis was performed focusing on 1355 isolates from patients failing therapy.

4.1 Pairwise Covariation Patterns

Patterns of pairwise interactions among mutations associated with NRTI treatment were identified from the database using Fisher's exact test. Specifically, for each pair of mutations co-occurrence frequencies for mutated and corresponding wild-type residues were contrasted in a 2-way contingency table, from which the test statistic was computed.

A visual summary of these pairwise comparisons, part of which is shown in Fig. 1, immediately reveals the classical mutational clusters described above. It is also apparent that no significant interactions are formed between the Q151M complex and mutations from the NAM clusters, suggesting that resistance evolution along the former pathway is largely independent from the other complexes and that different pathways may act simultaneously on a sequence, at least if they mediate different physical resistance mechanisms.

In contrast, significant interactions take place across the two NAM complexes. Antagonistic interactions between the core NAM 1 mutations L210W / M41L /

---

[3] We use the syntax axb to denote amino acid substitutions in RT, where a is the most frequent amino acid in virus from untreated patients and b the mutated residue.

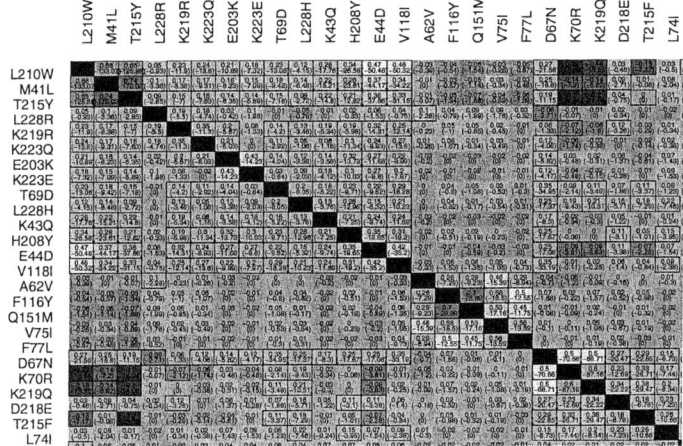

Fig. 1. Pairwise ϕ correlation coefficients between mutations (part view), with red indicating maximal observed positive covariation and blue maximal observed negative covariation. Boxes indicate pairs whose covariation behavior deviates significantly from the independence assumption, according to Fisher's exact test and correction for multiple testing using the Benjamini-Hochberg method at a false discovery rate of 0.01. The classical mutational complexes introduced in section 4 form distinct clusters, from left to right: NAM 1, Q151M multi-NRTI, NAM 2.

T215Y and NAM 2 mutations K70R and K219Q might indicate negative effects of simultaneous evolution along these two pathways, which both contribute to the primer unblocking mechanism.

4.2 Clustering Mutations

Dendrograms obtained from hierarchical clustering allow for a more detailed analysis of mutation covariation structure. The similarity between pairs of mutations was assessed using the ϕ (Matthews) correlation coefficient, as a measure of association between two binary random variables, with 1 and -1 representing maximal positive and negative association, respectively. This similarity measure was transformed into a dissimilarity δ by mapping $\phi = 1$ to $\delta = 0$ and $\phi = -1$ to $\delta = 1$, with linear interpolation in between. Since it is impossible to obtain adequate dissimilarity estimates for pairs of mutations at a single position from cross-sectional data, [4] these were treated as missing values in our approach. The resulting partial dissimilarity matrix was taken as the basis for average linkage hierarchical agglomerative clustering.[5]

The dendrogram in Fig. 2 reveals that most novel mutations group within the NAM 1 cluster (T215Y/M41L/L210W), except for D218E and F214L, which

---

[4] Such mutation pairs never co-occur in a sequence.
[5] In average linkage with missing values, the distance between clusters is simply the average of the *defined* distances.

aggregrate to NAM 2. Interestingly, mutations R83K and I50V, which occur more frequently in naive than in treated patients appear to form a novel outgroup.

To assess the stability of the dendrogram, 100 bootstrapped samples of RT sequences were drawn from the original 1355 sequences. Distance calculation and hierarchical clustering were performed for each of these samples as described above. Then, for each subtree of the dendrogram in Fig. 2, the fraction of bootstrap runs was counted in which the *set* of mutations defined by the subtree occurred as a subtree, without additional mutations. [6]

The four edge weights next to the root of the dendrogram show that the reported association of mutations D218E and F214L with NAM 2 is indeed highly stable across resampled data subsets, as is the grouping of other novel mutations with NAM 1, and the outgroup status of R83K and I50V. Bootstrap values for the lower dendrogram levels have been omitted for the sake of clarity; they range from 0.35 to 0.99, reflecting considerable variability of intra-cluster accumulation order. Finally, the core NAM 1 and NAM 2 mutations, respectively, are again grouped together with maximal confidence.

4.3 Multidimensional Scaling in Mutation Space

As can be seen in Fig. 1, certain mutations interact positively with mutations from both NAM pathways – an effect which might be missed in a dendrogram representation, and which can be visualized, at least to some extent, using multidimensional scaling (MDS).

The goal in MDS is, given a distance matrix D between entities, to find an embedding of these entities in $\mathbb{R}^n$ (here $n = 2$), such that the distances D' induced by the embedding match those provided in the matrix optimally, defined via minimizing a particular "stress" function. Our embedding is based on the Sammon stress function [7],

$$E(D, D') = \frac{1}{\sum_{i \neq j} D_{ij}} \sum_{i \neq j} \frac{(D_{ij} - D'_{ij})^2}{D_{ij}}, \qquad (1)$$

which puts emphasis on reproducing small distances accurately. As in clustering, mutation pairs at a single position are excluded from the computation of the stress function, to avoid undue distortions.

The optimal Sammon embedding for the mutation distance matrix derived from pairwise ϕ values is shown in Fig. 3. Note that due to the non-metricity of this matrix, which violates the triangle inequality, such an embedding cannot be expected to preserve all original distances accurately, Still, the MDS plot supports the main conclusions from section 4.2, such as to the structure of the classical NAM complexes, the outgroup status of R83K and I50V, and the exclusive propensity of certain mutations, such as K43E/Q or F214L, to a unique

---

[6] Thus, in computing confidence values increasingly closer to the root, topology of included subtrees is deliberately ignored (otherwise, values would be monotonically decreasing from leaves to the root).

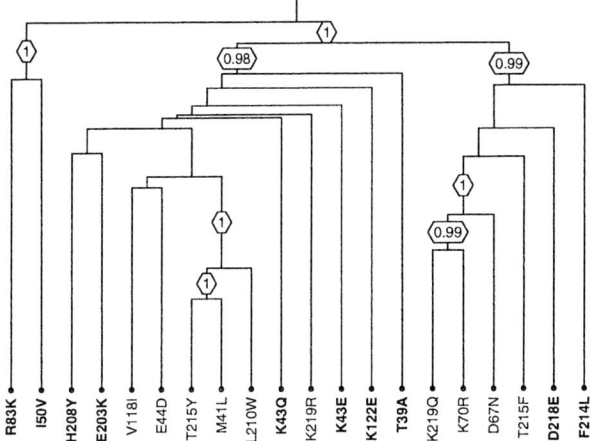

Fig. 2. Dendrogram, as obtained from average linkage hierarchical clustering, showing the clear propensity of novel mutations to cluster within one of the classical NAM complexes T215Y/M41L/L210W and K219Q/K70R/D67N, or in the case of R83K and I50V, to a distinct outgroup. Bootstrap values which are not relevant for our discussion have been removed for the sake of clarity. Distances between mutations at a single position are treated as missing values in the clustering procedure. Remarkably, such pairs of mutations can show differential clustering behavior, as is apparent in the case of K219Q/R and T215F/Y.

pathway. In addition, the plot also suggests a role in both NAM pathways for several mutations, such as H208Y, D67N, or K20R.

5 Phenotypic Characterization of Novel Mutations Using SVM-Based Feature Ranking

The analyses described above allowed us to associate novel mutations with treatment failure and to group them into distinct mutational complexes. In this section we address the question whether novel mutations contribute directly to increased resistance or merely exert compensatory functions in removing catalytic deficiencies induced by the main resistance-conferring mutations. We do so by analyzing their role in classification models for predicting phenotypic drug resistance.

Resistance of a given HIV strain against a certain drug can be measured *in vitro* by comparing the replicative capacity of the mutant strain with that of a non-resistant reference strain, at increasing drug concentrations [4]. The result of this comparison is summarized in a scalar *resistance factor*. On the basis of 650 matched genotype-phenotype pairs for each drug, we have built predictive models, using decision trees [8], and support vector machine classification and regression. These models are implemented in a publically available web server

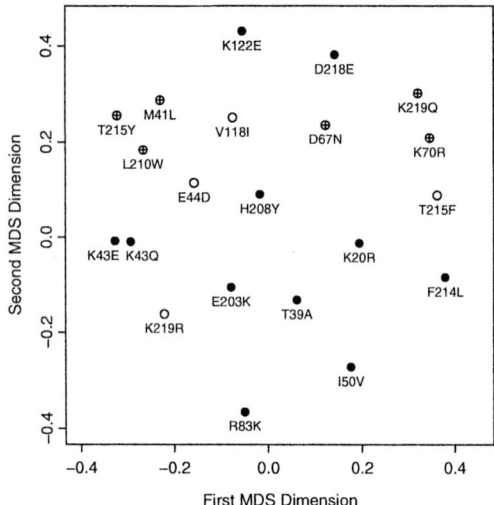

Fig. 3. Multidimensional scaling plot of novel (shown in black) and classical mutations (in white; main NAMs indicated by a cross), showing a two-dimensional embedding which optimally (according to Sammon's stress function) preserves the distances among the mutations, as derived from the ϕ correlation coefficient. Distances between mutations at a single position were treated as missing values.

called *geno2pheno* [9] (http://www.geno2pheno.org), which has been used over 36000 times since December 2000.

While support vector machines are widely considered as the state-of-the-art in prediction performance, there is a common attitude that these models are difficult to interpret and suffer from "the same disadvantage as neural networks, viz. that they yield black-box models" [10]. In fact, a substantial set of techniques is available for feature ranking with SVMs (e.g. [11]), by removing features or destroying their information through permutation, and even for extracting rule sets from SVMs.

In our case, using the linear kernel $k(x,y) = \langle x, y \rangle$ (standard nonlinear kernels did not significantly improve accuracy), feature ranking is particularly straightforward. Due to the bilinearity of the scalar product, the SVM decision function can be written as a linear model,

$$f(x) = \sum_i y_i \alpha_i k(x_i, x) + b = \langle \sum_i y_i \alpha_i x_i, x \rangle + b, \qquad (2)$$

allowing for direct assessment of the model weights.

Figure 4 shows the result of this SVM-based feature ranking for zidovudine (ZDV), one of the seven NRTIs. All mutations associated with resistance to ZDV in the current resistance update provided by the International AIDS Society [1]

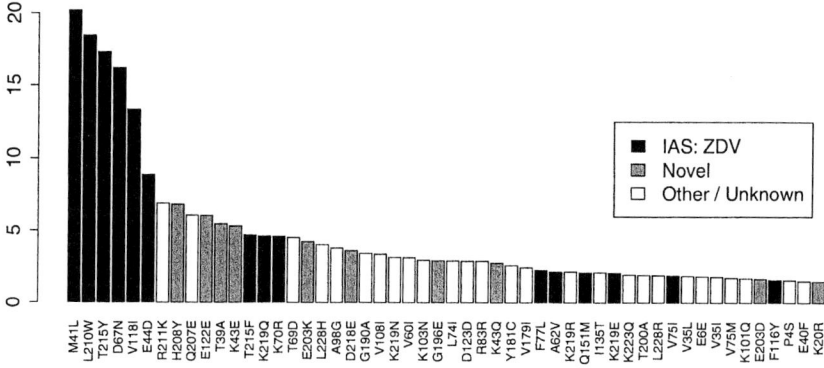

Fig. 4. Major mutations conferring resistance to zidovudine (ZDV), as obtained from SVM-based ranking of 5001 mutations. Bar heights indicate z-score-normalized feature weights (for example, mutation M41L is more than 20 standard deviations above the mean feature weight). Mutations associated with ZDV resistance by the International AIDS Society are shown in black; novel mutations identified from frequency comparisons in treated and untreated patients are shown in grey.

appear in the top 50 of 5001 features (250 positions, 20 amino acids each, plus 1 indicator for an insertion), with the first six positions exclusively occupied by classical NAM mutations (shown in black). This observation provides evidence that our models have adequately captured established domain knowledge as contributed by human experts. Remarkably, when investigating the role of novel mutations (shown in grey) in the model, we find that many of them are prominently involved in determining ZDV resistance, ranking even before several of the classical ZDV mutations.

These findings generalize to the whole NRTI drug class, as is obvious from table 1, which shows the ranks of novel mutations in the individual drug models. Table 1 also reveals some striking and unexpected differences among mutations. For example, various results suggest a close relationship of mutations H208Y and E203K, which form a tight cluster in the dendrogram, show up as neighbors in the multidimensional scaling plot, and exhibit similar rank profiles – with the notable exception of their differential impact on ddC resistance.

This surprising difference and other effects are more readily appreciated in Fig. 5, which shows the weights associated with novel mutations in the individual SVM drug models (after drug-wise z-score weight normalization for improved comparability). Indeed, increased resistance against ZDV, 3TC, and ABC upon appearance of E203K seems to coincide with *resensitization* (i.e. increased susceptibility) towards ddC. A similar, even more extreme effect can be observed in the case of T39A, for which increased resistance against ZDV and TDF again contrasts with increased ddC susceptibility. R83K shows dual behavior: increased d4T resistance and increased ZDV susceptibility. The presence of I50V is associated with increased susceptibility against all NRTIs, explaining its decreased frequency in treated patients.

Table 1. Ranks of novel mutations in SVM models for seven NRTIs, with rank 1 indicating maximal contribution to resistance, and rank 5001 maximal contribution to susceptibility. The classical mutation M184V is shown here for comparison, due to its particularly strong resensitization effect. The clinical (but not virological) relevance of results concerning ddC is limited by the limited popularity of this drug.

	ZDV	ddI	ddC	d4T	3TC	ABC	TDF
R83K	4972	3722	718	79	4973	539	154
I50V	4910	803	4702	4855	4736	4818	4899
H208Y	8	16	170	9	114	20	65
E203K	17	271	4963	103	8	19	103
K43Q	30	121	72	684	19	32	18
K43E	12	19	641	10	107	49	10
K122E	10	21	37	45	72	72	774
T39A	11	3814	4882	528	169	4017	50
D218E	20	22	103	50	25	13	659
F214L	119	898	4019	735	128	303	4844
M184V	67	2	1	4971	1	1	4994

Related effects have attracted considerable recent interest due to their possible benefits in reopening lost treatment options [12]. Arguably the most pronounced behavior can be seen in the classical mutation M184V (table 1), known to confer high-level resistance to 3TC but inducing d4T and TDF resensitization. SVM-based feature ranking reproduces this effect in a most striking manner: For ddI, ddC, 3TC, and ABC, M184V turns out to be the top resistance mutation, with contributions of 11.2,15.4,42.0, and 20.8 standard deviations above the mean. In contrast, the same mutation appears to be one of the major contributors of increased susceptibility towards d4T and TDF, 3.5 and 8.2 standard deviations *below* the mean, respectively.

6 Discussion

We have presented a case study on mining a multi-center HIV database using supervised and unsupervised methods. Previously undescribed mutations could be associated with resistance towards the drug class of nucleoside reverse transcriptase inhibitors and grouped into mutational clusters. SVM-based feature ranking on an independent data set suggests a direct contribution of novel mutations to phenotypic resistance and an involvement in resensitization effects which might be exploited in the design of antiretroviral combination therapies.

Mutation Screening. Novel mutations were found by position-wise comparisons, leaving inter-residue effects aside. It is conceivable that additional sets of mutations related to therapy failure, whose effect is too weak to discern in isolation, could be identified using other methods, such as discriminating item set miners. In fact, we have recently proposed an approach towards mining discriminating item sets, in which an overall rule weight in a mixture model of rules

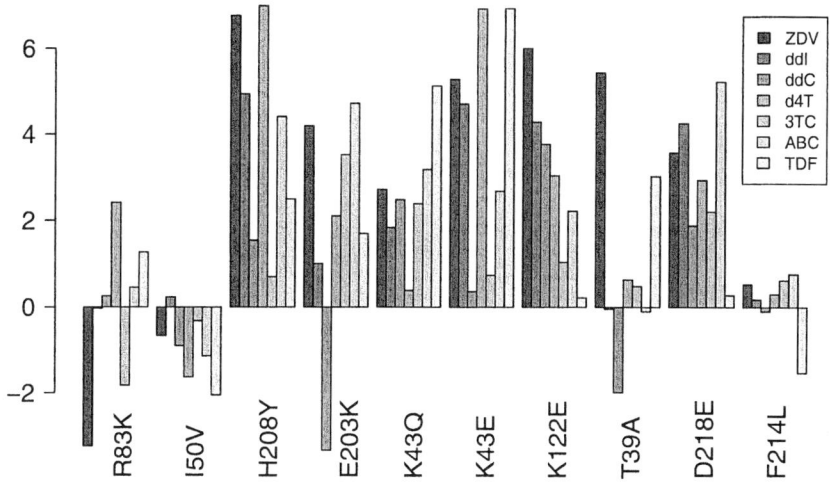

Fig. 5. Weights of novel mutations (after z-score normalization) in SVM models for seven NRTIs. For example, mutation E203K contributes significantly to ZDV resistance, while increasing susceptibility towards ddC.

is modulated by the genomic background in which a rule matches [13]. Further work will have to explore the possible benefits of using such strategies in the present context.

Covariation Versus Evolution. Dendrograms and MDS analyses describe the association of mutations into mutational complexes, but refrain from explicit statements on the accumulation order of mutations. Other approaches, most notably mutagenetic tree models [14], are explicitly tailored towards elucidating HIV evolutionary pathways from cross-sectional data as those used in our study. However, while novel mutations exhibit distinct clustering behavior, the actual order of their accumulation seems to be relatively flexible, challenging the applicability of such evolutionary models in this setting.

SVM-based Versus Correlation-Based Feature Ranking. To date, feature ranking is performed mostly using simple correlation methods, in which features are assessed in their performance to discriminate between classes *individually*, e.g. by using *mutual information*. However, as detailed in [11], feature ranking with correlation methods suffers from the implicit orthogonality assumptions that are made, in that feature weights are computed from information on a single feature in isolation, without taking into account mutual information between features. In contrast, statistical learning models such as support vector machines are inherently multivariate. Thus, their feature ranking is much less prone to be misguided by inter-feature dependencies than simple correlation methods. Further analysis of the feature rankings induced by different methods can provide valuable insights into their particular strenghts and weaknesses and suggest novel strategies for combining models from different model classes.

References

1. Johnson, V.A., Brun-Vezinet, F., Clotet, B., Conway, B., Kuritzkes, D.R., Pillay, D., Schapiro, J., Telenti, A., Richman, D.: Update of the Drug Resistance Mutations in HIV-1: 2005. Top HIV Med **13** (2005) 51–7
2. Gonzales, M.J., Wu, T.D., Taylor, J., Belitskaya, I., Kantor, R., Israelski, D., Chou, S., Zolopa, A.R., Fessel, W.J., Shafer, R.W.: Extended spectrum of HIV-1 reverse transcriptase mutations in patients receiving multiple nucleoside analog inhibitors. AIDS **17** (2003) 791–9
3. Svicher, V., Ceccherini-Silberstein, F., Erba, F., Santoro, M., Gori, C., Bellocchi, M., Giannella, S., Trotta, M., d'Arminio Monforte, A., Antinori, A., Perno, C.: Novel human immunodeficiency virus type 1 protease mutations potentially involved in resistance to protease inhibitors. Antimicrob. Agents Chemother. **49** (2005) 2015–25
4. Walter, H., Schmidt, B., Korn, K., Vandamme, A.M., Harrer, T., Uberla, K.: Rapid, phenotypic HIV-1 drug sensitivity assay for protease and reverse transcriptase inhibitors. J Clin Virol **13** (1999) 71–80
5. Svicher, V., Ceccherini-Silberstein, F., Sing, T., Santoro, M., Beerenwinkel, N., Rodriguez, F., Forbici, F., d'Arminio Monforte, A., Antinori, A., Perno, C.: Additional mutations in HIV-1 reverse transcriptase are involved in the highly ordered regulation of NRTI resistance. In: Proc. 3rd Europ. HIV Drug Resistance Workshop. (2005) Abstract 63
6. Benjamini, Y., Hochberg, Y.: Controlling the false discovery rate: A practical and powerful approach to multiple testing. Journal of the Royal Statistical Society (Series B) **57** (1995) 289–300
7. Sammon, J.: A non-linear mapping for data structure analysis. IEEE Trans. Comput. **C-18** (1969) 401–409
8. Beerenwinkel, N., Schmidt, B., Walter, H., Kaiser, R., Lengauer, T., Hoffmann, D., Korn, K., Selbig, J.: Diversity and complexity of HIV-1 drug resistance: a bioinformatics approach to predicting phenotype from genotype. Proc Natl Acad Sci U S A **99** (2002) 8271–6
9. Beerenwinkel, N., Däumer, M., Oette, M., Korn, K., Hoffmann, D., Kaiser, R., Lengauer, T., Selbig, J., Walter, H.: Geno2pheno: Estimating phenotypic drug resistance from HIV-1 genotypes. Nucleic Acids Res **31** (2003) 3850–5
10. Lucas, P.: Bayesian analysis, pattern analysis, and data mining in health care. Curr Opin Crit Care **10** (2004) 399–403
11. Guyon, I., Weston, J., Barnhill, S., Vapnik, V.: Gene selection for cancer classification using support vector machines. Machine Learning **46** (2002) 389–422
12. Wang, K., Samudrala, R., Mittler, J.E.: HIV-1 genotypic drug-resistance interpretation algorithms need to include hypersusceptibility-associated mutations. J Infect Dis **190** (2004) 2055–6
13. Sing, T., Beerenwinkel, N., Lengauer, T.: Learning mixtures of localized rules by maximizing the area under the ROC curve. In José Hernández-Orallo, et al., ed.: 1st International Workshop on ROC Analysis in Artificial Intelligence, Valencia, Spain (2004) 89–96
14. Beerenwinkel, N., Däumer, M., Sing, T., Rahnenführer, J., Lengauer, T., Selbig, J., Hoffmann, D., Kaiser, R.: Estimating HIV Evolutionary Pathways and the Genetic Barrier to Drug Resistance. J Infect Dis **191** (2005) 1953–60

Object Identification with Attribute-Mediated Dependences

Parag Singla and Pedro Domingos

Department of Computer Science and Engineering,
University of Washington,
Seattle, WA 98195-2350, USA
{parag, pedrod}@cs.washington.edu

Abstract. Object identification is the problem of determining whether different observations correspond to the same object. It occurs in a wide variety of fields, including vision, natural language, citation matching, and information integration. Traditionally, the problem is solved separately for each pair of observations, followed by transitive closure. We propose solving it collectively, performing simultaneous inference for all candidate match pairs, and allowing information to propagate from one candidate match to another via the attributes they have in common. Our formulation is based on conditional random fields, and allows an optimal solution to be found in polynomial time using a graph cut algorithm. Parameters are learned using a voted perceptron algorithm. Experiments on real and synthetic datasets show that this approach outperforms the standard one.

1 Introduction

In many domains, the objects of interest are not uniquely identified, and the problem arises of determining which observations correspond to the same object. For example, in vision we may need to determine whether two similar shapes appearing at different times in a video stream are in fact the same object. In natural language processing and information extraction, a key task is determining which noun phrases are co-referent (i.e., refer to the same entity). When creating a bibliographic database from reference lists in papers, we need to determine which citations refer to the same papers in order to avoid duplication. When merging multiple databases, a problem of keen interest to many large scientific projects, businesses, and government agencies, we need to determine which records represent the same entity and should therefore be merged. This problem, originally defined by Newcombe et al. [14] and placed on a firm statistical footing by Fellegi and Sunter [7], is known by the name of object identification, record linkage, de-duplication, merge/purge, identity uncertainty, hardening soft information sources, co-reference resolution, and others. There is a large literature on it, including Winkler [21], Hernandez and Stolfo [9], Cohen et al. [4], Monge and Elkan [13], Cohen and Richman [5], Sarawagi and Bhamidipaty [17], Tejada et al. [20], Bilenko and Mooney [3], etc. Most approaches are variants of the original Fellegi-Sunter model, in which object identification is viewed as a classification problem: given a vector of similarity scores between the attributes of two observations, classify it as "Match" or "Non-match." A separate match decision is made for each candidate pair, followed by transitive closure to eliminate inconsistencies. Typically, a logistic regression model is used [1].

Making match decisions separately ignores that information gleaned from one match decision may be useful in others. For example, if we find that a paper appearing in *Proc. PKDD-04* is the same as a paper appearing in *Proc. 8th PKDD*, this implies that these two strings refer to the same venue, which in turn can help match other pairs of PKDD papers. In this paper, we propose an approach that accomplishes this propagation of information. It is based on conditional random fields, which are discriminatively trained, undirected graphical models [10]. Our formulation allows us to find the globally optimal match in polynomial time using a graph cut algorithm. The parameters of the model are learned using a voted perceptron [6].

Recently, Pasula et al. [15] proposed an approach to the citation matching problem that has collective inference features. This approach is based on directed graphical models, uses a different representation of the matching problem, also includes parsing of the references into fields, and is quite complex. It is a generative rather than discriminative approach, requiring modeling of all dependences among all variables, and the learning and inference tasks are correspondingly more difficult. A collective discriminative approach has been proposed by McCallum and Wellner [12], but the only inference it performs across candidate pairs is the transitive closure that is traditionally done as a post-processing step. Bhattacharya and Getoor [2] proposed an *ad hoc* approach to matching authors taking into account the citations they appear in. Our model can be viewed as a form of relational Markov network [18], except that it involves the creation of new nodes for match pairs, and consequently cannot be directly created by queries to the databases of interest. Max-margin Markov networks [19] can also be viewed as collective discriminative models, and applying their type of margin-maximizing training to our model is an interesting direction for future research.

We first describe in detail our approach, which we call the collective model. We then report experimental results on real and semi-artificial datasets, which illustrate the advantages of our model relative to the standard Fellegi-Sunter one.

2 Collective Model

Using the original database-oriented nomenclature, the input to the problem is a database of records (set of observations), with each record being a tuple of fields (attributes). We now describe the graphical structure of our model, its parameterization, and inference and learning algorithms for it.

2.1 Model Structure

Consider a database relation $R = \{r_1, r_2, \ldots, r_n\}$, where r_i is the ith record in the relation. Let $F = \{F^1, F^2, \ldots, F^m\}$ denote the set of fields in the relation. For each field F^k, we have a set FV^k of corresponding field values appearing in the relation, $FV^k = \{f_1^k, f_2^k, \ldots, f_{l_k}^k\}$. We will use the notation $r_i.F^k$ to refer to the value of kth field of record r_i. The goal is to determine, for each pair of records (r_i, r_j), whether they refer to the same underlying entity. Our graphical model contains three types of nodes:

Record-match nodes. The model contains a Boolean node R_{ij} for each pairwise question of the form: "Is record r_i the same as record r_j?"

Field-match nodes. The model contains a Boolean node F_{xy}^k for each pairwise question of the form: "Do field values f_x^k and f_y^k represent the same underlying property?" For example, for the venue field in a bibliography database, the model might contain a node for the question: "Do the strings 'Proc. PKDD-04' and 'Proc. 8th PKDD' represent the same venue?"

Field-similarity nodes. For pair of field values $f_x^k, f_y^k \in FV^k$, the model contains a node S_{xy}^k whose domain is the [0, 1] interval. This node encodes how similar the two field values are, according to a pre-defined similarity measure. For example, for textual fields this could be the TF/IDF score [16]. Since their values are computed directly from the data, we will also call these nodes *evidence nodes*.

Because of the symmetric nature of their semantics, R_{ij}, F_{xy}^k and S_{xy}^k represent the same nodes as R_{ji}, F_{yx}^k and S_{yx}^k, respectively.

The structure of the model is as follows. Each record-match node R_{ij} is connected by an edge to each corresponding field-match node $F_{xy}^k, 1 \leq k \leq m$. Formally, R_{ij} is connected to F_{xy}^k iff $r_i.F^k = f_x^k$ and $r_j.F^k = f_y^k$. Each field-match node F_{xy}^k is in turn connected to the corresponding field-similarity node S_{xy}^k. Each record-match node R_{ij} is also directly connected to the corresponding field-similarity node S_{xy}^k. In general, a field-match node will be linked to many record-match nodes, as the same pair of field values can be shared by many record pairs. This sharing lies at the heart of our model. The field-match nodes allow information to propagate from one candidate record pair to another. Notice that merging the evidence nodes corresponding to the same field value pairs, without introducing field-match nodes, would not work. This is because evidence nodes have known values at inference time, rendering the record-match nodes independent and reducing our approach to the standard one. Figure 1(a) shows a four-record bibliography database, and 1(b) shows the corresponding graphical representation for the candidate pairs (b_1, b_2) and (b_3, b_4). Note how dependences flow through the shared field-match node corresponding to the venue field. Inferring that b_1 and b_2 refer to the same underlying paper will lead to the inference that the corresponding venue strings "Proc. PKDD-04" and "Proc. 8th PKDD" refer to the same underlying venue, which in turn might provide sufficient evidence to merge b_3 and b_4. In general, our model can capture complex interactions between candidate pair decisions, potentially leading to better object identification.

One limitation of the model is that it makes a global decision on whether two fields are the same, which may not always be appropriate. For example, "J. Doe" may sometimes be the same as "Jane Doe," and sometimes the same as "Julia Doe." In this case the model will tend to choose whichever match is most prevalent. This simplifies inference and learning, and in many domains will not sigificantly affect overall performance. Nevertheless, relaxing it is an item for future work.

2.2 Conditional Random Fields

Conditional random fields, introduced by Lafferty et al. [10], define the conditional probability of a set of output variables $\mathbf{Y}$ given a set of input or evidence variables $\mathbf{X}$. Formally,

$$P(\mathbf{y}|\mathbf{x}) = \frac{1}{Z_\mathbf{x}} \sum_{c \in C} \exp \sum_l \lambda_{lc} f_{lc}(y_c, x_c) \qquad (1)$$

where C is the set of cliques in the graph, x_c and y_c denote the subset of variables participating in clique c, and $Z_\mathbf{x}$ is a normalization factor. f_{lc}, known as a feature function, is a function of variables involved in clique c, and λ_{lc} is the corresponding weight. In many domains, rather than having different parameters (feature weights) for each clique in the graph, the parameters of a conditional random field are tied across repeating clique patterns in the graph, called clique templates [18]. The probability distribution can then be specified as

$$P(\mathbf{y}|\mathbf{x}) = \frac{1}{Z_\mathbf{x}} \sum_{t \in T} \sum_{c \in C_t} \exp \sum_l \lambda_{lt} f_{lt}(y_c, x_c) \qquad (2)$$

where T is the set of all the templates, C_t is the set of cliques which satisfy template t, and f_{lt} and λ_{lt} are respectively a feature function and a feature weight, pertaining to template t.

2.3 Model Parameters

Our model has a singleton clique for each record-match node and one for each field-match node, a two-way clique for each edge linking a record-match node to a field-match node, a two-way clique for each edge linking a record-match node to a field-similarity node, and a two-way clique between each field-match node and the corresponding field-similarity node. The parameters for all cliques of the same type are tied; there is a template for the singleton record-match cliques, one for each type of singleton field-match clique (e.g., in a bibliography

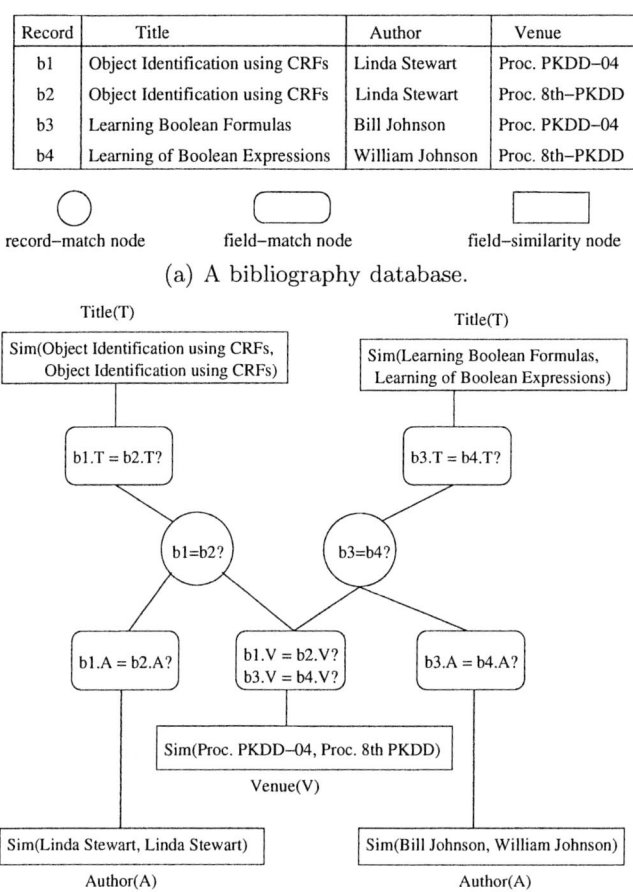

Fig. 1. Example of collective object identification. For clarity, we have omitted the edges linking the record-match nodes to the corresponding field-similarity nodes.

database, one for author fields, one for title fields, one for venue fields, etc.), and so on. The probability of a particular assignment **r** to the record-match and field-match nodes, given that the field-similarity (evidence) node values are **s**, is

$$P(\mathbf{r}|\mathbf{s}) = \frac{1}{Z_\mathbf{s}} \exp \sum_{i,j} \left[\sum_l \lambda_l f_l(r_{ij}) + \sum_k \left(\sum_l \phi_{kl} f_l(r_{ij}.F^k) + \sum_l \gamma_{kl} g_l(r_{ij}, r_{ij}.F^k) \right. \right.$$
$$\left. \left. + \sum_l \eta_{kl} h_l(r_{ij}, r_{ij}.S^k) + \sum_l \delta_{kl} h_l(r_{ij}.F^k, r_{ij}.S^k) \right) \right] \qquad (3)$$

where (i, j) ranges over all candidate pairs and k ranges over all fields. $r_{ij}.F^k$ and $r_{ij}.S^k$ refer to the kth field-match node and field-similarity node, respectively, for the record pair (r_i, r_j). λ_l and ϕ_{kl} denote the feature weights for singleton cliques. γ_{kl} denotes the feature weights for a two-way clique between a record-match node and a field-match node. η_{kl} and δ_{kl} denote the feature weights for a two-way clique between a Boolean node (record-match node or field-match node, respectively) and a field-similarity node. Cliques have one feature per possible state. Singleton cliques thus have two (redundant) features: $f_0(x) = 1$ if $x = 0$, and $f_0(x) = 0$ otherwise; $f_1(x) = 1$ if $x = 1$, and $f_1(x) = 0$ otherwise. Two-way cliques involving Boolean variables have four features: $g_0(x,y) = 1$ if $(x,y) = (0,0)$; $g_1(x,y) = 1$ if $(x,y) = (0,1)$; $g_2(x,y) = 1$ if $(x,y) = (1,0)$; $g_3(x,y) = 1$ if $(x,y) = (1,1)$; each of these features is zero in all other states. Two-way cliques between a Boolean node (record-match node or field-match node) q and a field-similarity node s have two features, defined as follows: $h_0(q,s) = 1 - s$ if $q = 0$, and $h_0(q,s) = 0$ otherwise; $h_1(q,s) = s$ if $q = 1$, and $h_1(q,s) = 0$ otherwise. This captures the fact that, the more similar two field values are, the more likely they are to match.

Notice that a particular field-match node appears in Equation 3 once for each pair of records containing the corresponding field values. This reflects the fact that that node is effectively the result of merging the field-match nodes from each of the individual record-match decisions.

2.4 Inference and Learning

Inference in our model corresponds to finding the configuration $\mathbf{r}^*$ of non-evidence nodes that maximizes $P(\mathbf{r}^*|\mathbf{s})$. For random fields where maximum clique size is two and all non-evidence nodes are Boolean, this problem can be reduced to a graph min-cut problem, provided certain constraints on the parameters are satisfied [8]. Our model is of this form, and it can be shown that satisfying the following constraints suffices for the min-cut reduction to hold: $\gamma_{k0} + \gamma_{k3} - \gamma_{k1} - \gamma_{k2} \geq 0$, $\forall k, 1 \leq k \leq m$, where the $\gamma_{kl}, 0 \leq l \leq 3$, are the parameters of the clique template for edges linking record-match nodes to field-match nodes of type F^k (see Equation 3).[1] This essentially corresponds to requiring that nodes be positively correlated, which should be true in this application. Our learning algorithm ensures that the learned parameters satisfy these constraints. Since min-cut can be solved exactly in polynomial time, we have a polynomial-time exact inference algorithm for our model.

Learning involves finding maximum-likelihood parameters from data. The partial derivative of the log-likelihood L (see Equation 3) with respect to the parameter γ_{kl} is

$$\frac{\partial L}{\partial \gamma_{kl}} = \sum_{i,j} g_l(r_{ij}, r_{ij}.F^k) - \sum_{\mathbf{r}'} P_\Lambda(\mathbf{r}'|\mathbf{s}) \sum_{i,j} g_l(r'_{ij}, r'_{ij}.F^k) \quad (4)$$

where $\mathbf{r}'$ varies over all possible configurations of the non-evidence nodes in the graph, and $P_\Lambda(\mathbf{r}'|\mathbf{s})$ denotes the probability distribution according to the current

---

[1] The constraint mentioned in Greig et al. [8] translates to $\gamma_{k0}, \gamma_{k3} \geq 0$, $\gamma_{k1}, \gamma_{k2} \leq 0$, which is a more restrictive version of the constraint above.

set of parameters. In words, the derivative of the log-likelihood with respect to a parameter is the difference between the empirical and expected counts of the corresponding feature, with the expectation taken according to the current model. The other components of the gradient are found analogously. To satisfy the constraint $\gamma_{k0} + \gamma_{k3} - \gamma_{k1} - \gamma_{k2} \geq 0$, we perform the following re-parameterization: $\gamma_{k0} = f(\beta_1) + \beta_2$, $\gamma_{k1} = f(\beta_1) - \beta_2$, $\gamma_{k2} = -f(\beta_3) + \beta_4$, $\gamma_{k3} = -f(\beta_3) - \beta_4$, where $f(x) = log(1+e^x)$. We then learn the β parameters using the appropriate transformation of Equation 4. The second term in this equation involves the expectation over an exponential number of configurations, and its computation is intractable. We use a voted perceptron algorithm [6], which approximates this expectation by the feature counts of the most likely configuration, which we find using our polynomial-time inference algorithm with the current parameters. The final parameters are the average of the ones learned during each iteration of the algorithm. Notice that, because parameters are learned at the template level, we are able to propagate information through field values that did not appear in the training data.

2.5 Combined Model

Combining models is often a simple way to improve accuracy. We combine the standard and collective models using logistic regression. For each record-match node in the training set, we form a data point with the outputs of the two models as predictors, and the true value of the node as the response variable. We then apply logistic regression to this dataset. Notice that this still yields a conditional random field.

3 Experiments

We performed experiments on real and semi-artificial datasets, comparing the performance of (a) the standard Fellegi-Sunter model using logistic regression, (b) the collective model, and (c) the combined model. If we consider every possible pair of records for a match, the potential number of matches is $O(n^2)$, which is a very large number even for datasets of moderate size. Therefore, we used the technique of first clustering the dataset into possibly-overlapping *canopies* using an inexpensive distance metric, as described by McCallum et al. [11], and then applying our inference and learning algorithms only to record pairs which fall in the same canopy. This reduced the number of potential matches to at most the order 1% of all possible matches. In our experiments we used this technique with all the three models being compared. The field-similarity nodes were computed using cosine similarity with TF/IDF [16].

3.1 Real-World Data

Cora. The hand-labeled Cora dataset is provided by McCallum[2] and has previously been used by Bilenko and Mooney [3] and others. This dataset is a collec-

---
[2] www.cs.umass.edu/~mccallum/data/cora-refs.tar.gz

Table 1. Experimental results on the Cora dataset (performance measured in %)

	Citation Matching					
Model	Before transitive closure			After transitive closure		
	F-measure	Recall	Precision	F-measure	Recall	Precision
Standard	86.9	89.7	85.3	84.7	98.3	75.5
Collective	87.4	91.2	85.1	88.9	96.3	83.3
Combined	85.8	86.1	87.1	89.0	94.9	84.5
	Author Matching					
Model	Before transitive closure			After transitive closure		
	F-measure	Recall	Precision	F-measure	Recall	Precision
Standard	79.2	65.8	100	89.5	81.1	100
Collective	90.4	99.8	83.1	90.1	100	82.6
Combined	88.7	99.7	80.1	88.6	99.7	80.2
	Venue Matching					
Model	Before transitive closure			After transitive closure		
	F-measure	Recall	Precision	F-measure	Recall	Precision
Standard	48.6	36.0	75.4	59.0	70.3	51.6
Collective	67.0	62.2	77.4	74.8	90.0	66.7
Combined	86.5	85.7	88.7	82.0	96.5	72.0

tion of 1295 different citations to computer science of research papers from the Cora Computer Science Research Paper Engine. The original dataset contains only unsegmented citation strings. Bilenko and Mooney [3] segmented each citation into fields (author, venue, title, publisher, year, etc.) using an information extraction system. We used this processed version of Cora. We further cleaned it up by correcting some labels. This cleaned version contains references to 132 different research papers. We used only the three most informative fields: author, title and venue (with venue including conferences, journals, workshops, etc.). We compared the performance of the algorithms for the task of de-duplicating citations, authors and venues.[3] For training and testing purposes, we hand-labeled the field pairs. The labeled data contains references to 50 authors and 103 venues. We carried out five runs of two-fold cross-validation, and report the average F-measure, recall and precision on post-canopy record match decisions. (To avoid contamination of test data by training data, we ensured that no true set of matching records was split between folds.) Next, we took the transitive closure over the matches produced by each model as a post-processing step to remove any inconsistent decisions. Table 1 shows the results obtained before and after this step. The combined model is the best-performing one for de-duplicating citations and venues. The collective model is the best one for de-duplicating authors. Transitive closure has a variable effect on the performance, depending upon the algorithm and the de-duplication task (i.e. citations, authors, venues).

---

[3] For the standard model, TFIDF similarity scores were used as the match probabilities for de-duplicating the fields (i.e. authors and venues).

Table 2. Experimental results on the BibServ dataset (performance measured in %)

Model	Citation Matching					
	Before transitive closure			After transitive closure		
	F-measure	Recall	Precision	F-measure	Recall	Precision
Standard	82.7	99.8	70.7	68.5	100.0	52.1
Collective	82.8	100.0	70.7	73.6	99.5	58.4
Combined	85.6	99.8	75.0	76.0	99.5	61.5

We also generated precision/recall curves on Cora for de-duplicating citations, and the collective model dominated throughout. [4]

BibServ. BibServ.org is a publicly available repository of about half a million pre-segmented citations. It is the result of merging citation databases donated by its users, CiteSeer, and DBLP. We experimented on the user-donated subset of BibServ, which contains 21,805 citations. As before, we used the author, title and venue fields. After forming canopies, we obtained about 58,000 match pair decisions. We applied the three models to these pairs, using the parameters learned on Cora (Training on BibServ was not possible because of the unavailability of labeled data.). We then hand-labeled 100 random pairs on which at least one model disagreed with the others, and 100 random pairs on which they all agreed. From these, we extrapolated the (approximate) results that would be obtained by hand-labeling the entire dataset.[5] Table 2 shows the results obtained for de-duplicating citations before and after transitive closure. All the models have close to 100% recall on the BibServ data. The combined model yields the best precision, resulting in the overall best F-measure. Transitive closure hurts all models, with the standard model being the worst hit. This is attributable to the fact that BibServ is much noisier and broader than Cora; the parameters learned on Cora produce an excess of matches on BibServ, and transitive closer compounds this. Collective inference, however, makes the model more resistant to this effect.

Summary. These experiments show that the collective and the combined models are able to exploit the flow of information across candidate pairs to make better predictions. The best combined model outperforms the best standard model in F-measure by 2% on de-duplicating citations in Cora, 27.5% on de-duplicating venues in Cora and 3% on de-duplicating citations in BibServ. On de-duplicating authors in Cora, the best collective model outperforms the best standard model by 0.9%.

3.2 Semi-artificial Data

To further observe the behavior of the algorithms, we generated variants of the Cora dataset by taking distinct field values from the original dataset and

---

[4] For the collective model, the match probabilities needed to generate precision/recall curves were computed using Gibbs sampling starting from the graph cut solution.

[5] Notice that the quality of this approximation does not depend on the size of the database.

randomly combining them to generate distinct papers. This allowed us to control various factors like the number of clusters, level of distortion, etc., and observe how these factors affect the performance of our algorithms. To generate the semi-artificial dataset, we created eight distorted duplicates of each field value taken from the Cora dataset. The number of distortions within each duplicate was chosen according to a binomial distribution whose "probability of success" parameter we varied in our experiments; a single Bernoulli trial corresponds to the distortion of a single word in the original string. The total number of records was kept constant at 1000 in all the experiments with semi-artificial data. To generate the records in the dataset, we first decided the number of clusters, and then created duplicate records for each cluster by randomly choosing the duplicates for each field value in the cluster. The results reported are over the task of de-duplicating citations, were obtained by performing five runs of two-fold cross-validation on this data, and are before transitive closure.[6]

The first set of experiments compared the relative performance of the models as we varied the number of clusters from 50 to 400, with the first two cluster sizes being 50 and 100 and then varying the size at an interval of 100. The binomial distortion parameter was kept at 0.4. Figures 2(a), 2(c) and 2(e) show the results. The F-measure (Figure 2(a)) drops as the number of clusters is increased, but the collective model always outperforms the standard model. The recall curve (Figure 2(c)) shows similar behavior. Precision (Figure 2(e)) appears to drop with increasing number of clusters, with collective model outperforming the standard model throughout.

The second set of experiments compared the relative performance of the models as we varied the level of distortion from 0 to 1, at intervals of 0.2. (0 means no distortion, and 1 means that every word in the string is distorted.) The number of clusters in the dataset was kept constant at 100. Figures 2(b), 2(d) and 2(f) show the results. As expected, the F-measure (Figure 2(b)) drops as the level of distortion in the data increases, with the collective model dominating between the distortion levels of 0.2 to 0.6. The two models seem to perform equally well at other distortion levels. The recall curve (Figure 2(c)) shows similar behavior. Precision (Figure 2(e)) seems to fluctuate with increasing distortion, with the collective model dominating throughout.

Overall, the collective model clearly dominates the standard model over a broad range of the number of clusters and level of distortion in the data.

4 Conclusion and Future Work

Determining which observations correspond to the same object is a key problem in information integration, citation matching, natural language, vision, and other areas. It is traditionally solved by making a separate decision for each pair of observations. In this paper, we proposed a collective approach, where information is propagated among related decisions via the attribute values they have

---

[6] For clarity, we have not shown the curves for the combined model, which are similar to the collective model's.

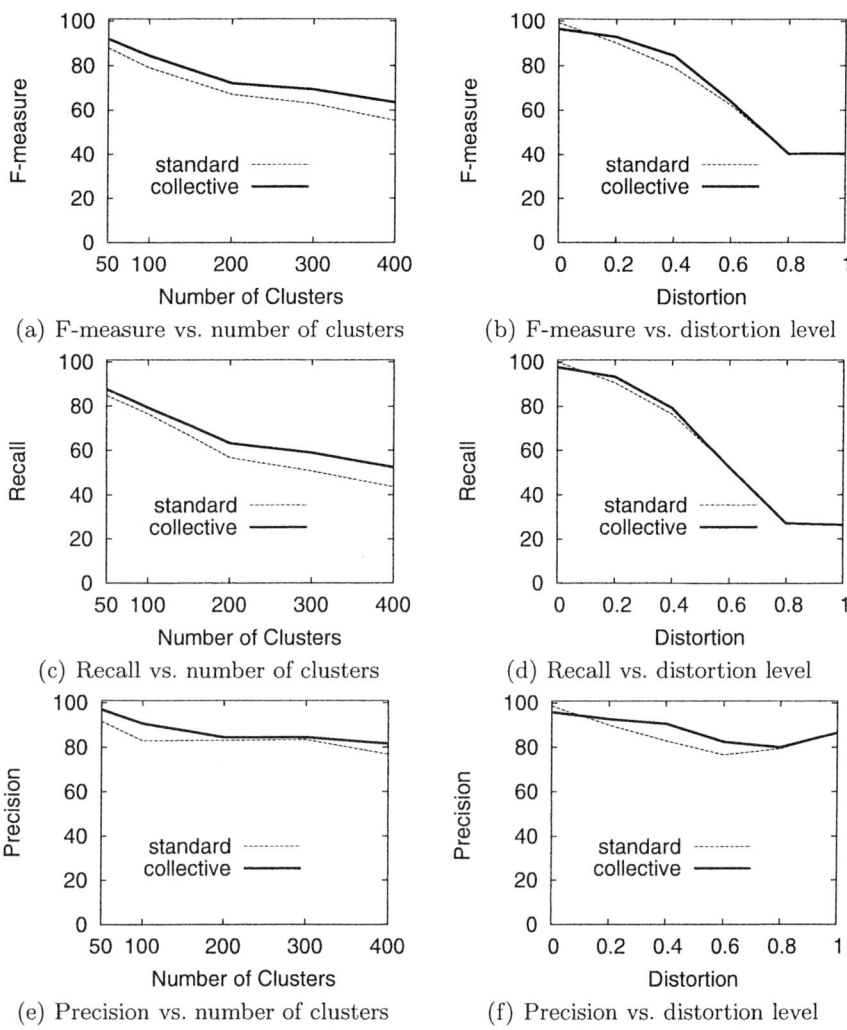

Fig. 2. Experimental results on semi-artificial data

in common. In our experiments, this produced better results than the standard method. Directions for future work include enriching the model with more complex dependences (which will entail moving to approximate inference), using it to deduplicate multiple types of objects at once, etc.

Acknowledgments

This research was partly supported by ONR grant N00014-02-1-0408 and by a Sloan Fellowship awarded to the second author.

References

1. A. Agresti. *Categorical Data Analysis*. Wiley, 1990.
2. I. Bhattacharya and L. Getoor. Iterative record linkage for cleaning and integration. In *Proc. SIGMOD-04 DMKD Wkshp.*, 2004.
3. M. Bilenko and R. Mooney. Adaptive duplicate detection using learnable string similarity measures. In *Proc. KDD-03*, pages 7–12, 2003.
4. W. Cohen, H. Kautz, and D. McAllester. Hardening soft information sources. In *Proc. KDD-00*, pages 255–259, 2000.
5. W. Cohen and J. Richman. Learning to match and cluster large high-dimensional data sets for data integration. In *Proc. KDD-02*, pages 475–480, 2002.
6. M. Collins. Discriminative training methods for hidden Markov models: Theory and experiments with perceptron algorithms. In *EMNLP-02*, pages 1–8, 2002.
7. I. Fellegi and A. Sunter. A theory for record linkage. *J. American Statistical Association*, 64:1183–1210, 1969.
8. D. Greig, B. Porteous, and A. Seheult. Exact maximum a posteriori estimation for binary images. *J. Royal Statistical Society B*, 51:271–279, 1989.
9. M. Hernandez and S. Stolfo. The merge/purge problem for large databases. In *Proc. SIGMOD-95*, pages 127–138, 1995.
10. J. Lafferty, A. McCallum, and F. Pereira. Conditional random fields: Probabilistic models for segmenting and labeling sequence data. In *Proc. ICML-01*, pages 282–289, 2001.
11. A. McCallum, K. Nigam, and L. Ungar. Efficient clustering of high-dimensional data sets with application to reference matching. In *Proc. KDD-00*, pages 169–178, 2000.
12. A. McCallum and B. Wellner. Conditional models of identity uncertainty with application to noun coreference. In *Adv. NIPS 17*, pages 905–912, 2005.
13. A. Monge and C. Elkan. An efficient domain-independent algorithm for detecting approximately duplicate database records. In *Proc. SIGMOD-97 DMKD Wkshp.*, 1997.
14. H. Newcombe, J. Kennedy, S. Axford, and A. James. Automatic linkage of vital records. *Science*, 130:954–959, 1959.
15. H. Pasula, B. Marthi, B. Milch, S. Russell, and I. Shpitser. Identity uncertainty and citation matching. In *Adv. NIPS 15*, pages 1401–1408, 2003.
16. G. Salton and M. McGill. *Introduction to Modern Information Retrieval*. McGraw-Hill, 1983.
17. S. Sarawagi and A. Bhamidipaty. Interactive deduplication using active learning. In *Proc. KDD-02*, pages 269–278, 2002.
18. B. Taskar, P. Abbeel, and D. Koller. Discriminative probabilistic models for relational data. In *Proc. UAI-02*, pages 485–492, 2002.
19. B. Taskar, C. Guestrin, B. Milch, and D. Koller. Max-margin Markov networks. In *Adv. NIPS 16*, 2004.
20. S. Tejada, C. Knoblock, and S. Minton. Learning domain-independent string transformation weights for high accuracy object identification. In *Proc. KDD-02*, pages 350–359, 2002.
21. W. Winkler. The state of record linkage and current research problems. Technical report, Statistical Research Division, U.S. Census Bureau, 1999.

Weka4WS: A WSRF-Enabled Weka Toolkit for Distributed Data Mining on Grids

Domenico Talia, Paolo Trunfio, and Oreste Verta

DEIS, University of Calabria,
Via P. Bucci 41c, 87036 Rende, Italy
{talia, trunfio}@deis.unical.it

Abstract. This paper presents Weka4WS, a framework that extends the Weka toolkit for supporting distributed data mining on Grid environments. Weka4WS adopts the emerging Web Services Resource Framework (WSRF) for accessing remote data mining algorithms and managing distributed computations. The Weka4WS user interface is a modified Weka Explorer environment that supports the execution of both local and remote data mining tasks. On every computing node, a WSRF-compliant Web Service is used to expose all the data mining algorithms provided by the Weka library. The paper describes the design and the implementation of Weka4WS using a first release of the WSRF library. To evaluate the efficiency of the proposed system, a performance analysis of Weka4WS for executing distributed data mining tasks in different network scenarios is presented.

1 Introduction

Complex business and scientific applications require access to distributed resources (e.g., computers, databases, networks, etc.). Grids have been designed to support applications that can benefit from high performance, distribution, collaboration, data sharing and complex interaction of autonomous and geographically dispersed resources. Since computational Grids emerged as effective infrastructures for distributed high-performance computing and data processing, a few Grid-based KDD systems has been proposed [1,2,3,4]. By exploiting a service-oriented approach, data-intensive and knowledge discovery applications can be developed by exploiting the Grid technology to deliver high performance and manage data and knowledge distribution. As critical for scalable knowledge discovery, our focus here is on distributed data mining services beginning to allow distributed teams or virtual organizations accessing and mining data in a high-level, standard and reliable way.

This paper presents *Weka4WS*, a framework that extends the widely used Weka toolkit [5] for supporting distributed data mining on Grid environments. Weka provides a large collection of machine learning algorithms written in Java for data pre-processing, classification, clustering, association rules, and visualization, which can be invoked through a common *Graphical User Interface (GUI)*. In Weka, the overall data mining process takes place on a single machine, since the algorithms can be executed only locally. The goal of Weka4WS is to extend Weka to support remote execution of the data mining algorithms. In such a way, distributed data mining tasks can be executed on decentralized Grid nodes by exploiting data distribution and improving application performance.

In Weka4WS, the data-preprocessing and visualization phases are still executed locally, whereas data mining algorithms for classification, clustering and association rules can be also executed on remote Grid resources. To enable remote invocation, each data mining algorithm provided by the Weka library is exposed as a Web Service, which can be easily deployed on the available Grid nodes. Thus, Weka4WS also extends the Weka GUI to enable the invocation of the data mining algorithms that are exposed as Web Services on remote machines. To achieve integration and interoperability with standard Grid environments, Weka4WS has been designed and developed by using the emerging *Web Services Resource Framework (WSRF)* [6] as enabling technology.

WSRF is a family of technical specification concerned with the creation, addressing, inspection, and lifetime management of *stateful resources*. The framework codifies the relationship between Web Services and stateful resources in terms of the *implied resource pattern*. A stateful resource that participates in the implied resource pattern is termed *WS-Resource*. WSRF describes the WS-Resource definition and association with the description of a Web Service interface, and describes how to make the properties of a WS-Resource accessible through a Web Service interface.

Initial work on WSRF has been performed by the Globus Alliance and IBM, with the goal of integrating previous work on the so-called *Open Grid Services Architecture (OGSA)* [7] with new Web Services mechanisms and standards. The Globus Alliance recently released the Globus Toolkit 4 (GT4) [8], which provides an open source implementation of the WSRF library and incorporates services implemented according to the WSRF specifications. The Weka4WS prototype described in this paper has been developed by using the Java WSRF library provided by a development release of Globus Toolkit 4 (Globus Toolkit 3.9.2 Core version).

The paper describes the design, implementation and performance evalution of Weka4WS. To evaluate the efficiency of the proposed system, a performance analysis of Weka4WS executing distributed data mining tasks in different network scenarios is presented. The remainder of the paper is organized as follows. Section 2 describes the architecture and the implementation of the Weka4WS framework. Section 3 presents a performance analysis of the Weka4WS prototype. Section 4 discusses related work. Finally, Section 5 concludes the paper.

2 The Weka4WS Framework

Figure 1 shows the general architecture of the Weka4WS framework that includes three kinds of nodes: *storage nodes*, which store the datasets to be mined; *computing nodes*, on which the remote data mining tasks are executed; *user nodes*, which are the local machines of the users.

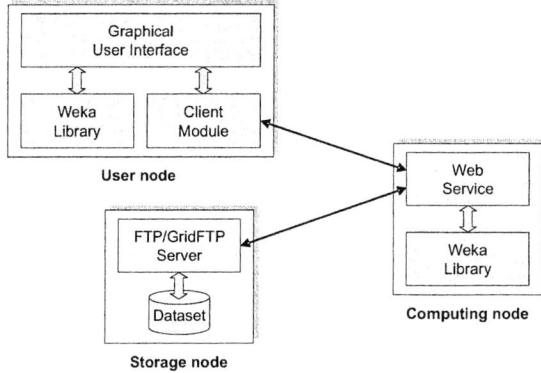

Fig. 1. The general architecture of the Weka4WS framework

User nodes include three components: *Graphical User Interface* (*GUI*), *Client Module* (*CM*), and *Weka Library* (*WL*). The GUI is an extended Weka Explorer environment that supports the execution of both local and remote data mining tasks. Local tasks are executed by directly invoking the local WL, whereas remote tasks are executed through the CM, which operates as an intermediary between the GUI and Web Services on remote computing nodes.

Figure 2 shows a snapshot of the current GUI implementation. As highlighted in the figure, a "Remote" pane has been added to the original Weka Explorer environment. This pane provides a list of the remote Web Services that can

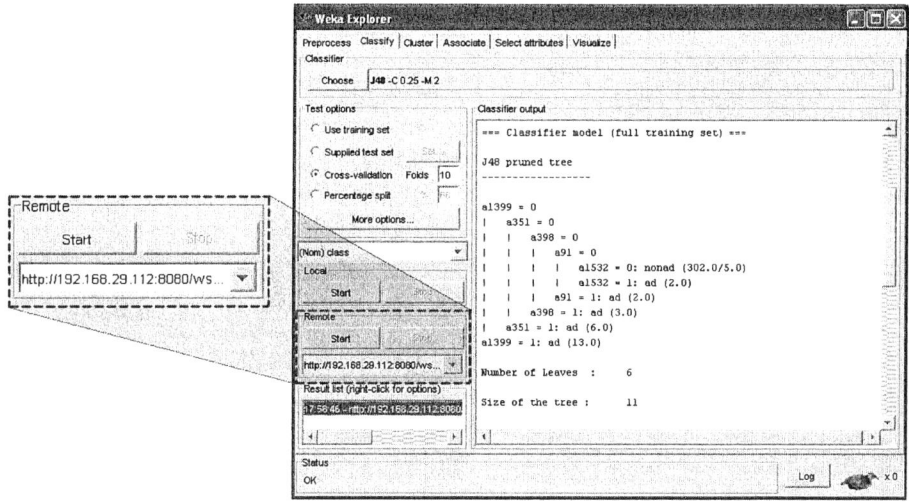

Fig. 2. The Graphical User Interface: a "Remote" pane has been added to the original Weka Explorer to start remote data mining tasks

be invoked, and two buttons to start and stop the data mining task on the selected Web Service. Through the GUI a user can both: *i*) start the execution locally by using the "Local" pane; *ii*) start the execution remotely by using the "Remote" pane. Each task in the GUI is managed by an independent thread. Therefore, a user can start multiple data mining tasks in parallel on different Web Services, this way taking full advantage of the distributed Grid environment for implementing parallel and distributed data mining tasks. Whenever the output of a data mining task has been received from a remote computing node, it is visualized in the standard "output" pane (on the right of Figure 2).

Computing nodes include two components: a *Web Service (WS)* and the *Weka Library (WL)*. The WS is a WSRF-compliant Web Service that exposes the data mining algorithms provided by the underlying WL. Therefore, requests to the WS are executed by invoking the corresponding WL algorithms.

Finally, storage nodes provide access to data to be mined. To this end, an *FTP server* or a *GridFTP server* [9] is executed on each storage node. The dataset to be mined can be locally available on a computing node, or downloaded to a computing node in response to an explicit request of the corresponding WS.

2.1 Web Service Operations

Table 1 shows the operations provided by each Web Service in the Weka4WS framework.

Table 1. Operations provided by each Web Service in the Weka4WS framework

Operation	Description
createResource	Creates a new WS-Resource.
subscribe	Subscribes to notifications about resource properties changes.
destroy	Explicitly requests destruction of a WS-Resource.
classification	Submits the execution of a classification task.
clustering	Submits the execution of a clustering task.
associationRules	Submits the execution of an association rules task.

The first three operations are related to WSRF-specific invocation mechanisms (described below), whereas the last three operations - `classification`, `clustering` and `associationRules` - are used to require the execution of a specific data mining task. In particular, the `classification` operation provides access to the complete set of classifiers in the Weka Library (currently, 71 algorithms). The `clustering` and `association rules` operations expose all the clustering and association rules algorithms provided by the Weka Library (5 and 2 algorithms, respectively).

To improve concurrency the data mining operations are invoked in an asynchronous way, i.e., the client submits the execution in a non-blocking mode, and results will be notified to the client whenever they have been computed.

Table 2. Input parameters of the Web Service data mining operations

Operation	Parameter	Description
classification	algorithm	Name of the classification algorithm to be used.
	arguments	Arguments to be passed to the algorithm.
	testOptions	Options to be used during the testing phase.
	classIndex	Index of the attribute to use as the class.
	dataSet	URL of the dataset to be mined.
clustering	algorithm	Name of the clustering algorithm.
	arguments	Algorithm arguments.
	testOptions	Testing phase options.
	selectedAttrs	Indexes of the selected attributes.
	classIndex	Index of the class w.r.t. evaluate clusters.
	dataSet	URL of the dataset to be mined.
associationRules	algorithm	Name of the association rules algorithm.
	arguments	Algorithm arguments.
	dataSet	URL of the dataset to be mined.

Table 2 lists the input parameters of the Web Service data mining operations. Three parameters, in particular, are required in the invocation of all the data mining operations: `algorithm`, `arguments`, and `dataSet`. The `algorithm` argument specifies the name of the Java class in the Weka Library to be invoked (e.g., *"weka.classifiers.trees.J48"*). The `arguments` parameter specifies a sequence of arguments to be passed to the algorithm (e.g., *"-C 0.25 -M 2"*). Finally, the `dataSet` parameter specifies the URL of the dataset to be mined (e.g., *"gsiftp://hostname/path/ad.arff"*).

2.2 Task Execution

This section describes the steps that are performed to execute a data mining task on a remote Web Service in the Weka4WS framework.

Figure 3 shows a *Client Module (CM)* that interacts with a remote *Web Service (WS)* to execute a data mining task. In particular, this example assumes that the CM is requesting the execution of a clustering analysis on a dataset local to the user node, which then acts also as a storage node. Notice that this is a worst case since in several scenarios the dataset is available on the remote computing node. In order to perform this task, the following steps are executed (see Figure 3):

1. **Resource creation.** The CM invokes the `createResource` operation, which creates a new WS-Resource used to maintain the state of the subsequent clustering computation. The state is stored as *properties* of the resource. In particular, a "clustering model" property is used to store the result of the clustering computation. The WS returns the *endpoint reference (EPR)* of the created resource. The EPR is unique within the WS, and distinguishes

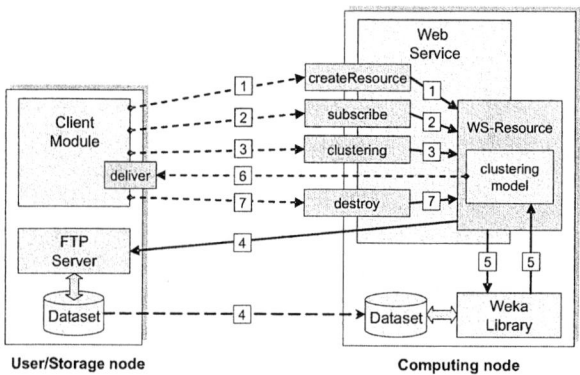

Fig. 3. Execution of a data mining task on a remote Web Service

this resource from all other resources in that service. Subsequent requests from the CM will be directed to the resource identified by that EPR.

2. **Notification subscription.** The CM invokes the `subscribe` operation, which subscribes for notifications about changes that will occur to the "clustering model" resource property. Whenever this property will change its value (i.e., whenever the model has been computed), the CM will receive a notification containing that value, which represents the result of the computation.

3. **Task submission.** The CM invokes the `clustering` operation to require the execution of the clustering analysis. This operation receives six parameters as shown in Table 2, among which the name of the clustering algorithm and the URL of the dataset to be mined. The operation is invoked in an asynchronous way, i.e., the client may proceed its execution without waiting for the completion of the operation.

4. **Dataset download.** Since in this example the dataset is assumed not available on the computing node, the WS downloads the dataset to be mined from the URL specified in the `clustering` invocation. The download request is directed to an FTP server running on the user node. Note that different protocols could be used, such as HTTP or GridFTP, as mentioned before.

5. **Data mining.** After the dataset has been downloaded to the computing node, the clustering analysis is started by invoking the appropriate Java class in the Weka Library. The execution is handled within the WS-Resource created on Step 1, and the result of the computation (i.e., the inferred model) is stored in the "clustering model" property.

6. **Results notification.** Whenever the "clustering model" property has been changed, its new value is notified to the CM, by invoking its implicit `deliver` operation. This mechanism allows for the asynchronous delivery of the execution results whenever they are generated.

7. **Resource destruction.** The CM invokes the `destroy` operation, which destroys the WS-Resource created on Step 1.

The next section presents a performance analysis of the execution mechanisms described above.

3 Performance Analysis

To evaluate the efficiency of the proposed system, we carried out a performance analysis of Weka4WS for executing a typical data mining task in different network scenarios. In particular, we evaluated the execution times of the different steps needed to perform the overall data mining task, as described at the end of the previous section. The main goal of our analysis is to evaluate the overhead introduced by the WSRF mechanisms with respect to the overall execution time.

For our analysis we used the *census* dataset from the UCI repository [10]. Through random sampling we extracted from it ten datasets, containing a number of instances ranging from 1700 to 17000, with a size ranging from 0.5 to 5 MB. We used Weka4WS to perform a clustering analysis on each of these datasets. In particular, we used the *Expectation Maximization (EM)* clustering algorithm, using 10 as the number of clusters to be identified on each dataset.

The clustering analysis on each dataset was executed in two network scenarios:

- **LAN**: the computing node N_c and the user/storage node N_u are connected by a LAN network, with an average bandwidth of 94.4 Mbps and an average round-trip time (RTT) of 1.4 ms. Both N_c and N_u machines are Pentium4 2.4 GHz with 1 GB RAM.
- **WAN**: the computing node N_c and the user/storage node N_u are connected by a WAN network, with an average bandwidth of 213 kbps and an average RTT of 19 ms. N_c is an Pentium4 2.4 GHz with 1 GB RAM, whereas N_u is an Athlon 2.14 GHz with 512 MB RAM.

For each dataset size and network scanario we run 20 independent executions. The values reported in the following graphs are computed as an average of the values measured in the 20 executions.

Figure 4 represents the execution times of the different steps of the clustering task in the LAN scenario for a dataset size ranging from 0.5 to 5 MB. As shown in the figure, the execution times of the WSRF-specific steps are independent from the dataset size, namely: *resource creation* (1698 ms, on the average), *notification subscription* (275 ms), *task submission* (342 ms), *results notification* (1354 ms), and *resource destruction* (214 ms).

On the contrary, the execution times of the *dataset download* and *data mining* steps are proportional to the dataset size. In particular, the execution time of the *dataset download* ranges from 218 ms for 0.5 MB to 665 ms for 5 MB, while the *data mining* execution time ranges from 107474 ms for the dataset of 0.5 MB, to 1026584 ms for the dataset of 5 MB. The total execution time ranges from 111798 ms for the dataset of 0.5 MB, to 1031209 ms for the dataset of 5 MB. Note that in Figure 4 the lines representing the total execution time and

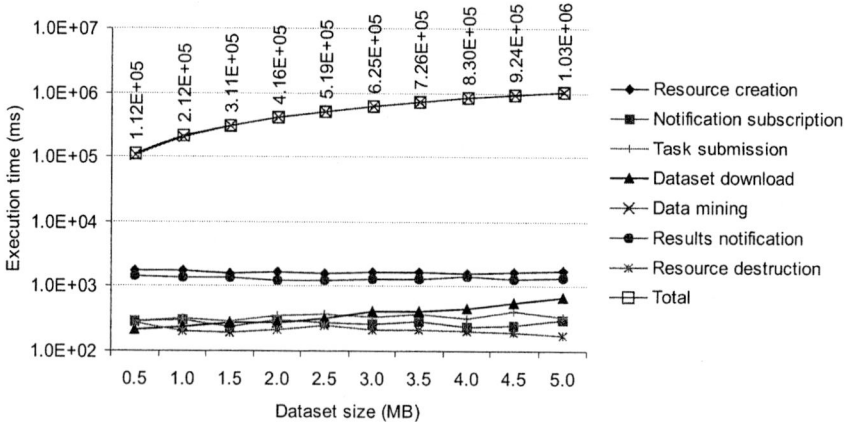

Fig. 4. Execution times of the different steps of the clustering task in the LAN scenario

the *data mining* execution time appear coincident, because the *data mining* step takes from 96% to 99% of the total execution time, as discussed below.

Figure 5 represents the execution times of the different steps in the WAN scenario. The execution times of the WSRF-specific steps are similar to those measured in the LAN scenario. The only significant difference is the execution time of the *results notification* step, which passes from an average of 1354 ms in the LAN scenario to an average of 2790 ms in the WAN scenario, due to additional time needed to transfer the clustering model through a low-speed network. For the same reason, the transfer of the dataset to be mined requires an execution time significantly greater than the one measured in the LAN scenario. In particular, the execution time of the *dataset download* step in the WAN scenario ranges from 14638 ms for 0.5 MB to 132463 ms for 5 MB.

The *data mining* execution time is similar to that measured in the LAN scenario, since the clustering analysis is executed on an identical computing node, as mentioned before. Mainly due to the additional time required by the *dataset download* step, the total execution time is greater than the one measured in the LAN scenario, ranging from 130488 ms for the dataset of 0.5 MB to 1182723 ms for the dataset of 5 MB. Like Figure 4, the line of the total execution time is very close to the line of the *data mining* execution time.

To better highlight the overhead introduced by the WSRF mechanisms and the distributed scenario, Figure 6 and Figure 7 show the percentage of the execution times of the *data mining*, *dataset download*, and the other steps (i.e., *resource creation*, *notification subscription*, *task submission*, *results notification*, *resource destruction*), with respect to the total execution time in the LAN and WAN scenarios.

In the LAN scenario (see Figure 6) the *data mining* step represents from 96.13% to 99.55% of the total execution time, the *dataset download* ranges from 0.19% to 0.06%, and the other steps range from 3.67% to 0.38%. Notice that the *data mining* time corresponds to total Weka execution time on a single node.

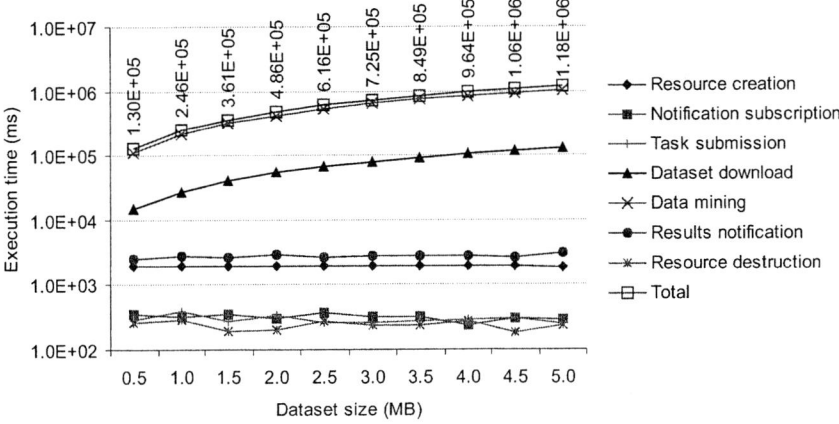

Fig. 5. Execution times of the different steps of the clustering task in the WAN scenario

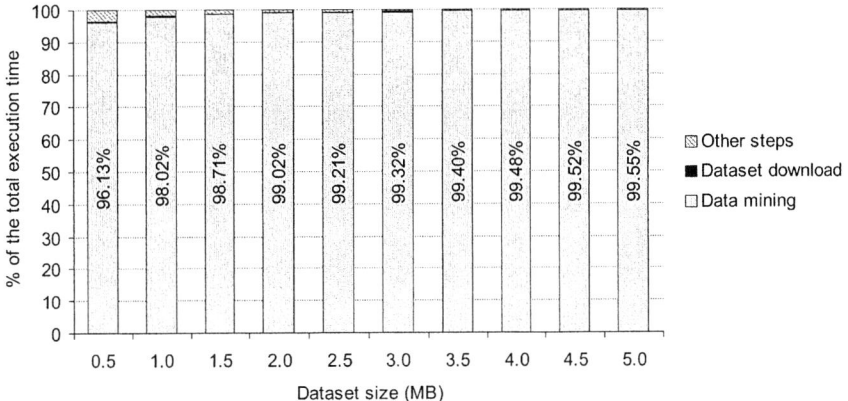

Fig. 6. Percentage of the execution times of the different steps in the LAN scenario

In the WAN scenario (see Figure 7) the *data mining* step represents from 84.62% to 88.32% of the total execution time, the *dataset download* ranges from 11.22% to 11.20%, while the other steps range from 4.16% to 0.48%.

We can observe that in the LAN scenario neither the *dataset download* nor the other steps represent a significant overhead with respect to the total execution time. In the WAN scenario, on the contrary, the *dataset download* is a critical step that can significantly affect the overall execution time. For this reason, the use of high-performance file transfer protocols such as GridFTP can be of great importance.

The performance analysis discussed above demonstrates the efficiency of the WSRF mechanisms as a means to execute data mining tasks on remote machines. By exploiting such mechanisms, Weka4WS can provide an effective way to perform compute-intensive distributed data analysis on a large-scale Grid environment.

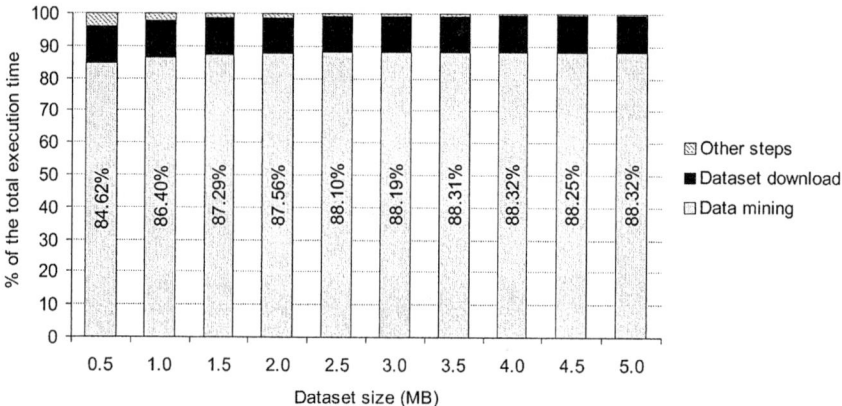

Fig. 7. Percentage of the execution times of the different steps in the WAN scenario

4 Related Work

The idea of adapting the Weka toolkit to a Grid environment has been recently explored, although none of the proposed systems makes use of WSRF as enabling technology.

Grid Weka [11] modifies the Weka toolkit to enable the use of multiple computational resources when performing data analysis. In this system, a set of data mining tasks can be distributed across several machines in an ad-hoc environment. Tasks that can be executed using Grid Weka include: building a classifier on a remote machine, labelling a dataset using a previously built classifier, testing a classifier on a dataset, and cross-validation. Even if Grid Weka provides a way to use multiple resources to execute distributed data mining tasks, it has been designed to work within an ad-hoc environment, which does not constitute a Grid per se. In particular, the invocation of remote resources in the Weka Grid framework is not service-oriented, and makes use of ad-hoc solutions that do not take into considerations fundamental Grid aspects (e.g., interoperability, security, etc.). On the contrary, Weka4WS exposes all its functionalities as WSRF-compliant Web Services, which enable important benefits, such as dynamic service discovery and composition, standard support for authorization and cryptography, and so on.

FAEHIM (Federated Analysis Environment for Heterogeneous Intelligent Mining) [12] is a Web Services-based toolkit for supporting distributed data mining. This toolkit consists of a set of data mining services, a set of tools to interact with these services, and a workflow system used to assemble these services and tools. The Triana problem solving environment [13] is used as the workflow system. Data mining services are exposed as Web Services to enable an easy integration with other third party services, allowing data mining algorithms to be embedded within existing applications. Most of the Web Services in FAEHIM are derived from the Weka library. All the data mining algorithms available in Weka were converted into a set of Web Services. In particular, a general "Classifier Web Service" has been implemented to act as a wrapper for a complete set of classifiers in Weka, a "Clustering Web Service" has been used to wrap a variety of clustering algorithms, and so on. This service-oriented approach is similar to that adopted in Weka4WS. However, in Weka4WS standard WSRF mechanisms are used for managing remote tasks execution and asynchronous results notification (which is missing in that system), and all the algorithms are exposed on every node as a single WSRF-compliant Web Service to facilitate the deployment in a large Grid environment.

WekaG [14] is another adaptation of the Weka toolkit to a Grid environment. WekaG is based on a client/server architecture. The server side defines a set of *Grid Services* that implement the functionalities of the different algorithms and phases of the data mining process. A WekaG client is responsible for communicating with Grid Services and offering the interface to users. A prototype that implements the capabilities of the *Apriori* algorithm has been developed using Globus Toolkit 3. In this prototype an *Apriori Grid Service* has been developed to produce association rules from a dataset, while GridFTP is used for the deployment of the files to the Grid Service node. WekaG shares this service-orientation with Weka4WS. However, the WekaG prototype provides access to only one data mining algorithm (Apriori), whereas Weka4WS currently provides access to 78 different Weka algorithms through a single Web Service interface. Moreover, WekaG uses the old Grid Service technology [15], which - differently from WSRF - is largely incompatible with current Web Service and Grid computing standards.

5 Conclusions

Weka4WS adopts the emerging Web Services Resource Framework (WSRF) for accessing remote data mining algorithms and composing distributed KDD applications.

The paper described the design and the implementation of Weka4WS using a first release of the WSRF library. To evaluate the efficiency of the proposed system, a performance analysis of Weka4WS for executing a distributed data mining task in different network scenarios has been also discussed.

The experimental results demonstrate the efficiency of the WSRF mechanisms as a means to execute data mining tasks on remote resources. By exploiting such mechanisms, Weka4WS can provide an effective way to perform compute-intensive distributed data analysis on large-scale Grids. The Weka4WS software prototype will be made available to the research community.

Acknowledgements

This research work is carried out under the FP6 Network of Excellence CoreGRID funded by the European Commission (Contract IST-2002-004265). This work has been also supported by the Italian MIUR FIRB Grid.it project RBNE01KNFP on High Performance Grid Platforms and Tools.

References

1. Curcin, V., Ghanem, M., Guo, Y., Kohler, M., Rowe, A., Syed, J., Wendel, P.: Discovery Net: Towards a Grid of Knowledge Discovery. 8th Int. Conf. on Knowledge Discovery and Data Mining (2002).
2. Brezany, P., Hofer, J., Tjoa, A. M., Woehrer, A.: Towards an open service architecture for data mining on the grid. Conf. on Database and Expert Systems Applications (2003).
3. Skillicorn, D., Talia, D.: Mining Large Data Sets on Grids: Issues and Prospects. Computing and Informatics, vol. 21 n. 4 (2002) 347-362.
4. Cannataro, M., Talia, D.: The Knowledge Grid. Communications of the ACM, vol. 46 n. 1 (2003) 89-93.
5. Witten, H., Frank, E.: Data Mining: Practical machine learning tools with Java implementations. Morgan Kaufmann (2000).
6. Czajkowski, K. et al: The WS-Resource Framework Version 1.0 (2004). http://www-106.ibm.com/developerworks/library/ws-resource/ws-wsrf.pdf.
7. Foster, I., Kesselman, C., Nick, J., Tuecke, S.: The Physiology of the Grid. In: Berman, F., Fox, G., A. Hey, A. (Eds.), Grid Computing: Making the Global Infrastructure a Reality, Wiley (2003) 217-249.
8. Foster, I.: A Globus Primer (2005). http://www.globus.org/primer.
9. Allcock, B., Bresnahan, J., Kettimuthu, R., Link, M., Dumitrescu, C., Raicu, I., Foster, I.: The Globus Striped GridFTP Framework and Server. Conf. on Supercomputing (SC'05) (2005).
10. The UCI Machine Learning Repository. http://www.ics.uci.edu/~mlearn/MLRepository.html.
11. Khoussainov, R., Zuo, X., Kushmerick, N.: Grid-enabled Weka: A Toolkit for Machine Learning on the Grid. ERCIM News, n. 59 (2004).
12. Shaikh Ali, A., Rana, O. F., Taylor, I. J.: Web Services Composition for Distributed Data Mining. Workshop on Web and Grid Services for Scientific Data Analysis (2005).
13. The Triana Problem Solving Environment. http://www.trianacode.org.
14. Prez, M. S., Sanchez, A, Herrero, P, Robles, V., Pea. J. M.: Adapting the Weka Data Mining Toolkit to a Grid based environment. 3rd Atlantic Web Intelligence Conf. (2005).
15. Tuecke, S. et al.: Open Grid Services Infrastructure (OGSI) Version 1.0 (2003). http://www-unix.globus.org/toolkit/draft-ggf-ogsi-gridservice-33_2003-06-27.pdf.

Using Inductive Logic Programming for Predicting Protein-Protein Interactions from Multiple Genomic Data

Tuan Nam Tran, Kenji Satou, and Tu Bao Ho

School of Knowledge Science,
Japan Advanced Institute of Science and Technology,
1-1 Asahidai Nomi Ishikawa 923-1292, Japan
{tt-nam, ken, bao}@jaist.ac.jp

Abstract. Protein-protein interactions play an important role in many fundamental biological processes. Computational approaches for predicting protein-protein interactions are essential to infer the functions of unknown proteins, and to validate the results obtained of experimental methods on protein-protein interactions. We have developed an approach using Inductive Logic Programming (ILP) for protein-protein interaction prediction by exploiting multiple genomic data including protein-protein interaction data, SWISS-PROT database, cell cycle expression data, Gene Ontology, and InterPro database. The proposed approach demonstrates a promising result in terms of obtaining high sensitivity/specificity and comprehensible rules that are useful for predicting novel protein-protein interactions. We have also applied our method to a number of protein-protein interaction data, demonstrating an improvement on the expression profile reliability (EPR) index.

1 Introduction

The interaction between proteins is fundamental to a broad spectrum of biological functions, including regulation of metabolic pathways, immunologic recognition, DNA replication, progression through the cell cycle, and protein synthesis. Therefore, mapping the organism-wide protein-protein interaction network plays an important role in functional inference of the unknown proteins. With the development of genomic technology, new experimental methods have vastly increased the number of protein-protein interactions for various organisms. An enormous amount of protein-protein interaction data have been obtained recently for yeast and other organisms using high-throughput experimental approaches such as yeast two-hybrid [12], affinity purification and mass spectrometry [2], phage display [22]. However, a potential difficulty with these kinds of data is a prevalence of false positive (interactions that are seen in an experiment but never occur in the cell or are not physiologically relevant) and false negatives (interactions that are not detected but do occur in the cell). As such, the prediction of protein-protein interactions using computational approaches can be used to validate the results of high-throughput interaction screens and used to complement the experimental approaches.

There have been a number of studies using computational approaches applied to predicting interactions. Bock and Gough [3] applied a Support Vector Machine learning system to predict directly protein-protein interactions from primary structure and associated data. Jansen *et al.* [13] used a Bayesian networks approach for integrating weakly predictive genomic features into reliable predictions of protein-protein interactions. A different approach is based on interacting domain pairs, attempting to understand protein-protein interactions at the domain level. Sprinzak and Margalit [23] proposed the AM (Association Method) for computing the score for each domain pair. Deng *et al.* [9] estimated the probabilities of interactions between every pair of domains using an EM algorithm, using the inferred domain-domain interactions to predict interactions between proteins. The major drawback of this approach is that there are currently no efficient experimental methods for detecting domain-domain interactions. Also, in [11], Grigoriev demonstrated that there is a significant relationship between gene expression and protein interactions on the proteome scale, finding that the mean correlation coefficients of gene expression profiles between interacting proteins are higher than those between random protein pairs.

In this paper, we present an approach for predicting genome-wide protein-protein interactions in yeast using the ILP system Aleph [1], a successor to Progol [16]. Unlike the other work, our approach is able to exploit the relationships among features of multiple genomic data, and to induce rules that give possible insight into the binding mechanism of the protein-protein interactions. Concerning rule-based methods using protein-protein interaction data, Oyama *et al.* [21] applied Association Rule Mining to extracting rules from protein-protein interaction data, however, the goal of this work is descriptive while our aim is to generate rules for predictive purposes.

2 ILP and Bioinformatics

Inductive Logic Programming (ILP) is the area of AI which is built on a foundation laid by research in machine learning and computational logic. ILP deals with the induction of hypothesized predicate definitions from examples and background knowledge. Logic programs are used as a single representation for examples, background knowledge and hypotheses. ILP is differentiated from most other forms of Machine Learning (ML) both by its use of an expressive representation language and its ability to make use of logically encoded background knowledge. This has allowed successful applications of ILP in areas such as molecular biology and natural language which both have rich sources of background knowledge and both benefit from the use of an expressive concept representation languages [17].

It is considered that one of the most important application domains for machine learning in general is bioinformatics. There have been many ILP systems that are successfully applied to various problems in bioinformatics. ILP is particular suitable for bioinformatics tasks because of its ability to take into account background knowledge and work directly with structured data. The ILP system GOLEM [18] was used to model the structure activity relationships of trimethoprim analogues binding to dihydrofolate reductase [14]. A study of discriminating molecules with positive mutagenicity from those with negative mutagenicity [15] has been conducted using Progol [16], another ILP system. ILP has also been applied to many other tasks in bioinformatics, such as protein secondary structure prediction [19] and protein fold recognition [26].

3 Using ILP for Predicting Protein-Protein Interactions

In this section, we present an algorithm for discovering rules using ILP. We use a multi-relational data mining approach to discover rules from multiple genomic data concerning protein-protein interactions. At present, we are using five kinds of genomic data:

1. **SWISS-PROT** [5], containing description of the function of a protein, its domains structure, post-translational modifications, variants, and so on.
2. **MIPS** [4], containing highly accurate protein interaction data for yeast.

Algorithm 1 Discovering rules for protein-protein interactions

Require:
  Set of protein interacting pairs $I = \{(p_i, p_j)\}$, $p_i \in P$, $p_j \in P$, where P is the set of proteins occurred
  Number of negative examples N
  Multiple genomic data used for extracting background knowledge ($S^{SWISS-PROT}, S^{MIPS}, S^{expression}, S^{GO}, S^{InterPro}$)

Ensure: Set of rules R for protein-protein interaction prediction

1: $R := \emptyset$, $S_{pos} := I$
2: Extract protein annotation information concerning each p of P from $S^{SWISS-PROT}$
3: Extract protein information concerning each p of P from S^{MIPS}
4: Call GENERATE-NEGATIVES for artificially generating N negative examples
5: Extract the expression correlation coefficients from $S^{expression}$ for every protein pairs (p_k, p_l), where $p_k \in P, p_l \in P$.
6: Extract all is_a and part_of relations (g_1, g_2), $g_1 \in G_P, g_2 \in G_P$, where G_P is the set of GO terms associated with P
7: Extract all relations between InterPro domains and GO terms $(d_{InterPro}, g)$ from $S^{InterPro}$, $d_{InterPro} \in D_P^{InterPro}, g \in G_P$, where $D_P^{InterPro}$ is the set of InterPro domains associated with P
8: Select a positive example at random
9: Saturate it to find the most specific clause that entails this example
10: Do top-down search for selecting the best clause c and add c to R
11: Remove covered positive examples
12: If there remain positive examples, go to step 8
13: **return** R

3. **Gene expression data** [24], containing the correlation of mRNA amounts with temporal profiles during the cell cycle.
4. **Gene Ontology (GO)** [20], containing the relations between GO terms.
5. **InterPro** [7], containing the relations between InterPro domains and their corresponding GO terms.

Our algorithm 1 consists of two main parts. The first part (step 1 to 7) is concerned with generating negative examples and extracting background knowledge from multiple genomic data. The second part (step 8 to 12) deals with inducing rules given the lists of positive, negative examples and background knowledge using Aleph [1]. Aleph is an ILP system that uses a top-down ILP covering algorithm, taking as input background information in the form of predicates, a list of modes declaring how these predicates can be chained together, and a designation of one predicate as the head predicate to be learned. Aleph is able to use a variety of search methods to find good clauses, such as the standard methods of breadth-first search, depth-first search, iterative beam search, as well as heuristic methods requiring an evaluation function. We use the default evaluation function *coverage* (the number of positive and negative examples covered by the clause) in our work.

Algorithm 2 GENERATE-NEGATIVES

Require:
  Number of negative examples N and S^{MIPS}
Ensure: Set of negative examples S_{neg} consisting of N protein pairs

1: $n := 0$, $S_{neg} := \emptyset$
2: **repeat**
3:   Select an arbitrary pair (p_k, p_l), where $p_k \in P, p_l \in P$
4:   Find the sets of subcellular location L_k and L_l of p_k and p_l from S^{MIPS}
5:   **if** $L_k \cap L_l = \emptyset$ **then**
6:     Add (p_k, p_l) to S_{neg}
7:     $n := n + 1$
8:   **endif**
9: **until** $n = N$
10: **return** S_{neg}

In this paper, we want to learn the following target predicate

`interact(Protein, Protein)`: the instances of this relation represent the interaction between two proteins.

For background knowledge, we shortly denote all predicates used by each genomic data. Note that Aleph uses *mode declarations* to build the bottom clause, and there are three types of variables: (1) an input variable (+), (2) an output variable (−), and (3) a constant term (#). Table 1 shows the list of predicates used as background knowledge for each genomic data.

4 Experiments

4.1 Data Preparation

We used the core data of the Yeast Interacting Proteins Database provided by Ito [6] as positive examples. Ito et al. [12] conducted comprehensive analysis using their system to examine two-hybrid interactions in all possible combinations between the 6000 proteins of the budding yeast *Saccharomyces cerevisiae*. Among 4,549 interactions detected using yeast-hybrid analysis, the "core" data consist of 841 interactions with more than two IST hits[1], accounting for 18.6% of the whole data. Note that the core data used in this paper is a subset of protein-protein interactions of MIPS [4] database, which is considered as the gold-standard for positive examples in [13]. A negatives gold-standard is defined similar to [13] in which negative examples are synthesized from lists of proteins in separate subcellular compartments.

We employ our approach to predict protein-protein interactions. We used the core data of Ito data set [6] mentioned above as positive examples, selecting at random 1000 protein pairs whose elements are in separate subcellular compartments as negative examples. Each interaction in the interaction data originally shows a pair of bait and prey ORF (Open Reading Frame)[2] some of which are not found in SWISS-PROT database. After removing all interactions in which either bait ORF or prey ORF is not found in SWISS-PROT, we obtained 602 interacting pairs from the original 841 pairs.

4.2 Analysis of Sensitivity/Specificity

To validate our proposed method, we conducted a 10-fold cross-validation test, comparing cross-validated sensitivity and specificity with those obtained by using AM [23] and SVM method. The AM method calculates a score d_{kl} to each domain pair (D_k, D_l) as the number of interacting protein pairs containing (D_k, D_l) divided by the number of protein pairs containing (D_k, D_l).

In the approach of predicting protein-protein interactions based on domain-domain interactions, it can be assumed that domain-domain interactions are independent and two proteins interact if at least one domain pairs of these two proteins interact. Therefore, the probability p_{ij} that two proteins P_i and P_j interact can be calculated as

$$p_{ij} = 1 - \prod_{D_k \in P_i, D_l \in P_j} (1 - d_{kl})$$

We implemented the AM and SVM methods in order to compare with our proposed method. We used the PFAM domains extracted from SWISS-PROT and superdomains, i.e. proteins without any domain information. The probability threshold is set to 0.05 for the simplicity of comparison. For SVM method, we

---

[1] IST hit means how many times the corresponding interaction was observed. The higher IST number, the much more reliable the corresponding interaction is.
[2] ORF is a series of codons which can be translated into a protein.

Table 1. Predicates used as background knowledge in various genomic data

Genomic data	Background Knowledge
SWISS-PROT	haskw(+Protein,#Keyword): A protein contains a keyword
	hasft(+Protein,#Feature): A protein contains a feature
	ec(+Protein,#EC): An enzyme code for a protein
	pfam(+Protein,-PFAM_Domain) A protein contains a Pfam domain
	interpro(+Protein,-InterPro_Domain) A protein contains a InterPro domain
	pir(+Protein,-PIR_Domain) A protein contains a Pir domain
	prosite(+Protein,-PROSITE_Domain) A protein contains a Prosite domain
	go(+Protein,-GO_Term) A protein contains a GO term
MIPS	subcellular_location(+Protein,#Subcellular_Structure) Relation between proteins and the subcellular structures in which they are found.
	function_category(+Protein,#Function_Category) A protein which is categorized to a certain function category
	protein_category(+Protein,#Protein_Category) A protein which is categorized to a certain protein category
	phenotype_category(+Protein,#Phenotype_Category) A protein which is categorized to a certain phenotype category
	complex_category(+Protein,#Complex_Category) A protein which is categorized to a certain complex category
Gene expression	correlation(+Protein,+Protein,-Expression) Expression correlation coefficient between two proteins
GO	is_a(+GO_Term,-GO_Term) is_a relation between two GO terms
	part_of(+GO_Term,-GO_Term) part_of relation between two GO terms
InterPro	interpro2go(+InterPro_Domain,-GO_Term) Mapping of InterPro entries to GO

used SVM^{light} [25] for learning, and used the same set of PFAM domains and superdomains as used in AM method. The linear kernel with default value of the parameters was used. For Aleph, we selected $minpos = 2$ and $noise = 0$, i.e. the lower bound on the number of positive examples to be covered by an acceptable clause is 2, and there are no negative examples allowed to be covered by an acceptable clause. We also used the default evaluation function *coverage* which is defined as $P - N$, where P, N are the number of positive and negative examples covered by the clause.

Table 2 shows the performance of Aleph compared with AM and SVM methods. The sensitivity of a test is described as the proportion of true positives it detects of all the positives, measuring how accurately it identifies positives. On the other hand, the specificity of a test is the proportion of true negatives it detects of all the negatives, thus is a measure of how accurately it identifies negatives. It can be seen from this Table that the proposed method showed a considerably high sensitivity and specificity given a certain number of negative examples. The number of negative examples should be chosen neither too large nor too small to avoid the imbalanced learning problem. At present, we did not compare our approach with EM method [9] in which they obtained 42.5% specificity and 77.6% sensitivity using the combined Uetz and Ito protein-protein interaction data.

Table 2. Performance of Aleph compared with AM and SVM methods. The sensitivity and specificity are obtained for each randomly chosen set of negative examples. The last column demonstrates the number of rules obtained using our proposed method with the minimum positive cover is set to 2.

# Neg	Sensitivity			Specificity			# Rules
	AM	SVM	Aleph	AM	SVM	Aleph	
100	0.70	**0.99**	0.90	**0.46**	0.01	0.44	27
500	0.68	0.54	**0.79**	0.42	0.61	**0.84**	63
1000	0.71	0.32	**0.73**	0.39	0.88	**0.93**	62
2000	**0.69**	0.26	**0.69**	0.38	0.95	**0.96**	58
4000	**0.69**	0.15	0.68	0.39	0.98	**0.99**	68

4.3 Rule Analysis

Figure 1 demonstrates a number of selective rules obtained when providing 602 positive examples and 1000 randomly chosen negative examples. Those rules are manually ranked using the difference between positive and negative coverages. It can be seen that although some of rules can be obtained using other propositional learning methods, some other rules can only be obtained using ILP. Rule 1 supports the approach using domain-domain interactions, demonstrating that two proteins interact if they share a common PFAM domain (81 cases covered among a total of 602 positive examples). Some rules obtained also match the result reported in [11] that the mean correlation coefficients of gene expression profiles between interacting proteins are higher than those between random protein pairs.

Using the Gene Ontology Term Finder tool [10], we also searched for significant GO terms, or parents of the GO terms used to describe the pair of protein interaction of each positive example covered by those rules in Figure 1. As a result, it can be found that rule 5, 6, 10, 12, 13, 14 are relevant with very high confidence, rule 7, 8, 9, 11 are relevant with lower confidence, and rule 15 is irrelevant.

4.4 Assessment of the Reliability Using EPR Index

Since high-throughput experimental methods may produce false positives, it is essential to assess the reliability of protein-protein interaction data obtained. Deane et al. [8] proposed the expression profile reliability (EPR) index to assess the reliability of measurement of protein interaction. The EPR index estimates the biologically relevant fraction of protein interactions detected in a high throughput screen. For each given data, we retrieved all protein pairs that classified as positive. Table 3 shows the EPR index calculated using the original and our proposed method for a number of well-known protein-protein interaction data. It can be seen that the EPR index of our method is higher than the original one, demonstrating the validity of the proposed method.

Rule 1 [Pos cover = 81 Neg cover = 0]
　　$interact(A, B) :- pfam(B, C), pfam(A, C).$
Rule 2 [Pos cover = 61 Neg cover = 0]
　　$interact(A, B) :- go(B, C), go(A, C), is\_a(C, D).$
Rule 3 [Pos cover = 51 Neg cover = 0]
　　$interact(A, B) :- interpro(B, C), interpro(A, C), interpro2go(C, D).$
Rule 4 [Pos cover = 15 Neg cover = 0]
　　$interact(A, B) :- go(B, C), go(A, C),$
　　$hasft(A, domain\_coiled\_coil\_potential).$
Rule 5 [Pos cover = 8 Neg cover = 0]
　　$interact(A, B) :- go(B, C), go(A, C),$
　　$complex\_category(A, intracellular\_transport\_complexes).$
Rule 6 [Pos cover = 6 Neg cover = 0]
　　$interact(A, B) :- subcellular\_location(B, nucleus),$
　　$function\_category(A, cell\_cycle\_and\_dna\_processing),$
　　$phenotype\_category(B, cell\_morphology\_and\_organelle\_mutants).$
Rule 7 [Pos cover = 6 Neg cover = 0]
　　$interact(A, B) :- pfam(A, C), subcellular\_location(B, er),$
　　$haskw(B, autophagy).$
Rule 8 [Pos cover = 5 Neg cover = 0]
　　$interact(A, B) :- phenotype\_category(B, conditional\_phenotypes),$
　　$hasft(A, domain\_rna\_binding\_rrm).$
Rule 9 [Pos cover = 5 Neg cover = 0]
　　$interact(A, B) :- correlation(B, A, C), gteq(C, 0.241974),$
　　$hasft(A, domain\_rna\_binding\_rrm).$
Rule 10 [Pos cover = 4 Neg cover = 0]
　　$interact(A, B) :- pfam(A, C), haskw(B, direct\_protein\_sequencing),$
　　$hasft(B, domain\_histone\_fold).$
Rule 11 [Pos cover = 4 Neg cover = 0]
　　$interact(A, B) :- correlation(A, B, C), gteq(C, 0.236007),$
　　$hasft(A, domain\_poly\_gln).$
Rule 12 [Pos cover = 4 Neg cover = 0]
　　$interact(A, B) :- protein\_category(A, gtp-binding\_proteins),$
　　$correlation(A, B, C), gteq(C, 0.144137).$
Rule 13 [Pos cover = 4 Neg cover = 0]
　　$interact(A, B) :- function\_category(B, cell\_fate),$
　　$hasft(B, transmem\_potential), hasft(A, transmem\_potential).$
Rule 14 [Pos cover = 3 Neg cover = 0]
　　$interact(A, B) :- subcellular\_location(B, integral\_membrane),$
　　$correlation(A, B, C), gteq(C, 0.46332).$
Rule 15 [Pos cover = 2 Neg cover = 0]
　　$interact(A, B) :- correlation(B, A, C), gteq(C, 0.599716),$
　　$haskw(A, cell\_division).$

Fig. 1. Some rules obtained with $minpos = 2$. For example, rule 14 means that protein A will interact with protein B if protein B is located in the integral membrane of the cell, and the expression correlation coefficient between protein A and protein B is greater than 0.46332.

Table 3. Evaluated the proposed method using EPR index. The number of interactions after preprocessing means the number of interactions obtained after removing all interactions in which either bait ORF or prey ORF it not found in SWISS-PROT.

Data	Number of interactions			EPR index	
	Original	After preprocessing	Proposed	Original	Proposed
Ito	4549	3174	1925	0.1910 ± 0.0306	**0.2900** ± 0.0481
Uetz	1474	1109	738	0.4450 ± 0.0588	**0.5290** ± 0.0860
Ito+Uetz	5827	4126	2567	0.2380 ± 0.0287	**0.3170** ± 0.0431
MIPS	14146	10894	7080	0.5950 ± 0.0337	**0.6870** ± 0.0420
DIP	15409	12152	8674	0.4180 ± 0.0260	**0.5830** ± 0.0374

5 Conclusions and Future Work

We have presented an approach using ILP to predict protein-protein interactions. The experimental results demonstrate that our proposed method can produce comprehensible rules, and at the same time, showing a considerably high performance compared with other work on protein-protein interaction prediction. In future work, we would like to investigate further about the biological significance of novel protein-protein interactions obtained by our method, and apply the ILP approach to other important tasks, such as predicting protein functions and subcellular locations using protein-protein interaction data. We are also investigating to exploit the GO structures as background knowledge, rather than using the occurence of a single GO term as in the current work.

Acknowledgements

This research is supported by the grant-in-aid for scientific research on priority areas (C) "Genome Information Science" from the Japanese Ministry of Education, Culture, Sports, Science and Technology. The authors would like to thank JST BIRD (Institute for Bioinformatics Research and Development) for all the support during the period of this work.

References

1. Aleph A. Srinivasan. http://web.comlab.ox.ac.uk/oucl/research/areas/machlearn/Aleph/aleph_toc.html.
2. A. Bauer and B. Kuster. Affinity purification-mass spectrometry: Powerful tools for the characterization of protein complexes. *Eur. J. Biochem.*, 270(4):570–578, 2003.
3. J. R. Bock and D. A. Gough. Predicting protein-protein interactions from primary structure. *Bioinformatics*, 17(5):455–460, 2001.
4. Comprehensive Yeast Genome Database. http://mips.gsf.de/genre/proj/yeast/index.jsp.

5. SWISS-PROT database. http://www.expasy.ch/sprot.
6. Yeast Interacting Proteins Database. http://genome.c.kanazawa-u.ac.jp/Y2H/.
7. InterPro database concerning protein families and domains. http://www.ebi.ac.uk/interpro/.
8. C. M. Deane, L. Salwinski, I. Xenarios, and D. Eisenberg. Protein interactions: Two methods for assessment of the reliability of high-throughput observations. *Mol. Cell. Prot.*, 1:349–356, 2002.
9. M. Deng, S. Mehta, F. Sun, and T. Chen. Inferring domain-domain interactions from protein-protein interactions. *Genome Res.*, 12(10):1540–1548, 2002.
10. SGD Gene Ontology Term Finder. http://db.yeastgenome.org/cgi-bin/GO/goTermFinder.
11. A. Grigoriev. A relationship between gene expression and protein interactions on the proteome scale: analysis of the bacteriophage t7 and the yeast *saccharomyces cerevisiae*. *Nucleic Acids Res.*, 29(17):3513–3519, 2001.
12. T. Ito, T. Chiba, R. Ozawa, M. Yoshida, M. Hattori, and Y. Sakaki. A comprehensive two-hybrid analysis to explore the yeast protein interactome. In *Proc. Natl. Acad. Sci. USA 98*, pages 4569–4574, 2001.
13. R. Jansen, H. Yu, D. Greenbaum, Y. Kluger, N. J. Krogan, S. Chung, A. Emili, M. Snyder, J. F. Greenblatt, and M. Gerstein. A bayesian networks approach for predicting protein-protein interactions from genomic data. *Science*, 302(5644):449–453, 2003.
14. R. King, S. Muggleton, R. A. Lewis, and M. J. Sternberg. Drug design by machine learning: the use of inductive logic programming to model the structure-activity relationships of trimethoprim analogues binding to dihydrofolate reductase. In *Proc. Natl. Acad. Sci.*, pages 11322–11326, 1992.
15. R. King, S. Muggleton, A. Srinivasan, and M. J. Sternberg. Structure-activity relationships derived by machine learning: the use of atoms and their bond connectives to predict mutagenicity by inductive logic programming. In *Proc. Natl. Acad. Sci.*, pages 438–442, 1996.
16. S. Muggleton. Inverse entailment and progol. *New Generation Computing*, 13:245–286, 1995.
17. S. Muggleton. Inductive logic programming: Issues, results and the challenge of learning language in logic. *Artificial Intelligence*, 114:283–296, 1999.
18. S. Muggleton and C. Feng. Efficient induction of logic programs. In *Proceedings of the First Conference on Algorithmic Learning Theory*, 1990.
19. S. Muggleton, R. King, and M. Sternberg. Protein secondary structure prediction using logic-based machine learning. *Protein Engineering*, 5(7):647–657, 1992.
20. Gene Ontology. http://www.geneontology.org/.
21. T. Oyama, K. Kitano, K. Satou, and T. Ito. Extracting of knowledge on protein-protein interaction by association rule discovery. *Bioinformatics*, 18(5):705–714, 2002.
22. G. P. Smith. Filamentous fusion phage: Novel expression vectors that display cloned antigens on the vision surface. *Science*, 228(4705):1315–1317, 1985.
23. E. Sprinzak and H. Margalit. Correlated sequence-signatures as markets of protein-protein interaction. *J. Mol. Biol.*, 311:681–692, 2001.
24. Yale Gerstein Lab Supplementary data. http://networks.gersteinlab.org/genome/intint/supplementary.htm.
25. SVM^{light} T. Joachim. http://svmlight.joachims.org.
26. M. Turcotte, S. Muggleton, and M. J. Sternberg. Protein fold recognition. In *International Workshop on Inductive Logic Programming (ILP-98)*, C. D. Page (Ed.), 1998.

ISOLLE: Locally Linear Embedding with Geodesic Distance

Claudio Varini[1,2], Andreas Degenhard[2], and Tim Nattkemper[1]

[1] Applied Neuroinformatics Group, Faculty of Technology,
University of Bielefeld, Bielefeld, Germany
{cvarini, tnattkem}@techfak.uni-bielefeld.de
[2] Condensed Matter Theory Group, Faculty of Physics,
University of Bielefeld, Bielefeld, Germany
adegenha@physik.uni-bielefeld.de

Abstract. Locally Linear Embedding (LLE) has recently been proposed as a method for dimensional reduction of high-dimensional nonlinear data sets. In LLE each data point is reconstructed from a linear combination of its n nearest neighbors, which are typically found using the Euclidean Distance. We propose an extension of LLE which consists in performing the search for the neighbors with respect to the geodesic distance (ISOLLE). In this study we show that the usage of this metric can lead to a more accurate preservation of the data structure. The proposed approach is validated on both real-world and synthetic data.

1 Introduction

The analysis of complex high-dimensional data structures is essential in many real-world applications, including medical high resolution time-series data. Algorithms for dimensional data reduction are particularly useful for discerning the information contained in a high-dimensional data structure.

In recent years several methods for the analysis of nonlinear data sets have been proposed, including Locally Linear Embedding (LLE) [1]. LLE has already been successfully applied to many problems, including face recognition [2], prediction of membrane protein types [3] and the analysis of micro array data [4]. The algorithm assumes linearity in the local area centered on each data point. Each area is mathematically characterized by a set of coefficients (weights) which correlate the particular data point with its n nearest neighbors. The aggregation of all areas can be intuitively thought as an assemblage of linear patches which approximates the nonlinear data structure. The high-dimensional data is then projected into a lower-dimensional space while preserving the coefficients between neighboring data points.

The number of neighbors n strongly influences the accuracy of the linear approximation of nonlinear data. Specifically, the smaller n, the smaller the area, the more faithful is the linear approximation. However, if these areas are disjoint, LLE can fail to detect the global data structure [5]. Disjoint areas can be obtained especially when the data is sparse or spread among multiple clusters.

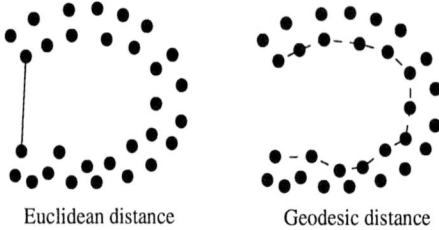

Euclidean distance   Geodesic distance

Fig. 1. The short circuit induced by Euclidean distance is shown on the left. In case the number of neighbors n is set to a relative high value, the two points in figure can be treated as neighbors, although they are on the opposite parts of the horseshoe. This may cause LLE to fail to detect the real global structure of the data. On the right are shown the benefits of the geodesic distance. In this case the two points are not neighbors, as they are faraway according to the geodesic distance.

To address this problem, in [6] it is proposed to search for the $n/2$ nearest and $n/2$ furthest neighbors of each data point. Another approach is given in [7], where the authors suggest to connect the disjoint manifold or interpolating the embeddings of some samples.

In general, for larger values of n the linear areas are more likely to overlap. The number of neighbors n therefore needs to be sufficiently high to satisfy this condition. On the other hand, as the neighbors search is typically conducted using the Euclidean distance, this may lead a data point to have neighbors which are instead very distant as one considers the intrinsic geometry of the data. More intuitively, one can imagine this fact as a short circuit (see Fig. 1). The presence of short circuits is undesirable, as they can cause LLE to misinterpret the actual data structure.

To address the above outlined problems occurring to LLE when employed with a high number of neighbors, we propose the usage of LLE with geodesic distance (ISOLLE). More specifically, the n nearest neighbors are searched with respect to the geodesic distance. This metric has already been employed in other methods for nonlinear dimensional data reduction such as Isomap [8], Curvilinear Distance Analysis [9] and Self-organizing Maps [10]. The geodesic distance between two data points can be intuitively thought as their distance along the contour of an object (see Fig. 1 right). For example let us consider the distance between Paris and New York. Their geodesic distance is the distance along the curvature of the Earth. Their Euclidean distance instead is the length of the straight line connecting the two cities which is below the level of the ground. Points faraway from each other, as measured by the geodesic distance, may appear deceptively close in the high-dimensional input space as measured by the Euclidean distance.

In this work we demonstrate that the employment of the geodesic distance can lower the probability to create short circuits during the neighbors search, thereby allowing for a more accurate dimensional reduction. Our approach to investigate the performances of ISOLLE as compared to conventional LLE is basically twofold. Firstly, we perform the analysis on synthetic data, namely a three-dimensional swissroll which was also used in [1] and [8]. By this phantom data set we illustrate the difference between both techniques. Secondly, we analyze a complex, medical real-world data set acquired using dynamic contrast-enhanced magnetic resonance imaging (DCE-MRI). DCE-MRI involves the repeated imaging of a region of interest, in our case the female breast with tumor lesions, after the administration of a contrast agent, yielding a high-dimensional spatio-temporal data structure.

Both data sets are reduced to two dimensions using different values of the number of neighbors n. The dimensional reduction of the swissroll is evaluated qualitatively, while the analysis of the tumor data set requires a statistical approach because of the complexity of the data. Specifically for this purpose we consider the percentage of nearest points in the original space that are preserved as nearest neighbors in the dimensional reduced space, and the stress induced by the dimensional reduction. In addition, in the final part the running times of LLE and ISOLLE are compared.

2 Locally Linear Embedding (LLE)

The LLE algorithm is based on three steps involving standard methods of linear algebra. Its input comprises N D-dimensional vectors $\{\mathbf{X}_i\}$. The first step consists in searching for the n nearest neighbors of each data point.

Once the neighbors are determined, by minimizing the following error function (step 2)

$$\Psi(W) = \sum_{i=1}^{N} |\mathbf{X}_i - \sum_{j=1}^{n} W_{ij}\mathbf{X}_j|^2 \qquad (1)$$

subject to the constraint $\sum_{j=1}^{n} W_{ij} = 1$, one obtains the weights $\{W_{ij}\}$ that best allow to reconstruct each data point from its neighbors. With the above constraints, Eq. (1) can be simplified to a linear system and the weights can be computed in closed form as follows: given a particular data point $\mathbf{X}_i$ with n-nearest neighbors $\mathbf{X}_j$ and reconstruction weights W_j that sum to one, we can write the reconstruction error as

$$\Psi(W) = \sum_{i=1}^{N} |\mathbf{X}_i - \sum_{j=1}^{n} W_j\mathbf{X}_j|^2 = \sum_{jk} W_j W_k C_{jk}. \qquad (2)$$

In the second identity, the term

$$C_{jk} = (\mathbf{X}_i - \mathbf{X}_j) \cdot (\mathbf{X}_i - \mathbf{X}_k) \qquad (3)$$

is the local covariance matrix. The weights which minimize the error function of Eq. (1) are given by:

$$W_j = \frac{\sum_k C_{jk}^{-1}}{\sum_{lm} C_{lm}^{-1}}, \; l, m \in \{1, .., n\}. \qquad (4)$$

In some cases, for example if the number of neighbors is greater than the input dimension ($n > D$), it arises that the matrix C is singular or nearly singular and the solution of Eq. (2) is not unique. In this case the matrix C must be conditioned by adding a small multiple of the identity matrix [11]:

$$C_{ij} \leftarrow C_{ij} + \delta_{ij}\Gamma \tag{5}$$

where Γ is defined as

$$\Gamma = \frac{\text{Tr}(C)}{n}\Delta^2. \tag{6}$$

The term Δ is a correction parameter set by the user and its value must be much smaller than 1.

The third and last step of the LLE algorithm consists in mapping each data point $\mathbf{X}_i$ to a low dimensional vector $\mathbf{Y}_i$, such that the following embedding error function is minimized:

$$\Phi(Y) = \sum_{i=1}^{N} |\mathbf{Y}_i - \sum_{j=1}^{n} W_{ij}\mathbf{Y}_j|^2 \tag{7}$$

under the conditions $\frac{1}{N}\sum_{i=1}^{N} \mathbf{Y}_i \mathbf{Y}_i^T = I$ and $\sum_{i=1}^{N} \mathbf{Y}_i = 0$, which provide a unique solution. Note that the weights are kept constant in order to preserve the local neighborhood of each data point. The most straightforward method for computing the M-dimensional coordinates is to find the bottom $M+1$ eigenvectors of the sparse matrix

$$S = (I - W)^T (I - W). \tag{8}$$

These eigenvectors are associated with the $M+1$ smallest eigenvalues of S. The bottom eigenvector is related to the smallest eigenvalue whose value is closest to zero. This eigenvector is the unit vector with all equal components and is discarded.

3 The ISOLLE Algorithm

The ISOLLE algorithm differs from LLE only in the first step, i.e. the neighbors search. More specifically, ISOLLE computes the n nearest neighbors of each data point according to the geodesic distance. For this purpose we employ a small variation of Dijkstra's algorithm [12]. Given a graph, this algorithm computes the shortest paths from a particular node to all remaining nodes. In our case we restrict the computation to the n shortest paths.

In practice, the process of finding the geodesic neighbors is composed of two phases. The first phase consists in constructing a weighted graph G over the data set where neighboring data points are connected. In principle, any similarity measure d_E can be adopted to determine neighboring relations, and probably the Euclidean distance is the most common choice. Two points are neighbors if are closer than a fixed distance ϵ (ϵ-graph), or one is the K nearest point of the other (K-graph). These relations between neighbors are represented by edges of weights $d_E(\mathbf{X}_i, \mathbf{X}_j)$ [8].

In the second phase the n nearest neighbors of each data point are found according to the geodesic distance computed by Dijkstra's algorithm. This algorithm begins at a specific node (source vertex) and extends outward within the graph until all the vertices have been reached (in our case only the n nearest nodes). Dijkstra's algorithm creates labels associated with vertices. These labels represent the distance (cost) from the source vertex to that particular vertex. Within the graph, there exists two kinds of labels: temporary and permanent. The temporary labels are given to vertices that have not been reached. The value given to these temporary labels can vary. Permanent labels are given to vertices that have been reached and their distance (cost) to the source vertex is known. The value given to these labels is the distance (cost) of that vertex to the source vertex. For any given vertex, there must be a permanent label or a temporary label, but not both. An animated example of Dijkstra's algorithm can be seen at [13]. Both steps of the neighboring search are detailed in the following:

Construct the neighborhood graph: define the graph G over all data points by connecting points $\mathbf{X}_i$ and $\mathbf{X}_j$ if (as measured by $d_E(\mathbf{X}_i, \mathbf{X}_j)$) they are closer than ϵ, or if $\mathbf{X}_i$ is one of the K nearest neighbors of $\mathbf{X}_j$. Set edge lengths equal to $d_E(\mathbf{X}_i, \mathbf{X}_j)$.

Compute n nearest points with Dijkstra's algorithm: given a graph G=(V,E) where V is a set of vertices and E a set of edges, Dijkstra algorithm keeps two sets of vertices:

S —the set of vertices whose shortest paths from the source vertex have already been determined. These vertices have a permanent label

V-S —the remaining vertices. These have a temporary label

The other data structures needed are:

$\mathbf{X}_0$ —initial beginning vertex (source vertex)

N —number of vertices in G

$\mathbf{D}$ —array of estimates of shortest path to $\mathbf{X}_0$.

The basic mode of operation of Dijkstra's algorithm is:

1 S={$\mathbf{X}_0$}
2 For i=1 to N
   D[i]=E[$\mathbf{X}_0$,i]
3 For i=1 to N-1
   Choose a vertex w in V-S such that D[w] is minimum and add it to S
   For each vertex v in V-S
   D[v]=min(D[v],D[w]+E[w,v])

The construction of graph G requires a further parameter (ϵ or K) to be set by the user. In [8] it is pointed out the scale-invariant parameter K is typically easier to set than ϵ, but may yield misleading results when the local dimensionality varies across the data set. A sensible way to set this parameter can be to choose the minimal value such that all the pairwise geodesic distances are finite.

4 Data Sets

The performances of ISOLLE and LLE are tested on two data sets whose numbers of points are displayed in table 1. The first data set is a three-dimensional synthetic distributions, namely a swissroll (Fig. 2(a)). The second is a real-world data set comprising the signal intensity values obtained by dynamic contrast-enhanced magnetic resonance imaging (DCE-MRI) on female breast with tumor.

The DCE-MRI technique consists in acquiring a sequence of images (six in our case) from a region of interest (the female breast in our case), whose movement is carefully restricted, over time in order to monitor the dynamic of a previous injected contrast agent within the tissue. As a result, tissue types with higher level of vascularity have enhanced values of signal intensity, proportionally to the amount of absorbed contrast agent. After the acquisition of the images, a time-series of intensity values is correlated with each voxel (see Fig. 2(b)). As benign and malignant tumor tissues are expected to differ in the level of vascularization, the respective contrast characteristics are expected to exhibit different behaviors.

The tumor data set in this work comprises the time-series relative to six benign and six malignant cancerous lesions which were labeled and pathologically analyzed by an expert physician. The discrimination between benign and malignant lesions in DCE-MRI is a particularly challenging and delicate problem in light of the relatively high rate of false positive cases characterizing this imaging technique, as published in the literature [14]. In this study the time-series associated with the voxel of each tumor is treated as a data point in a six-dimensional signal space. It is therefore interesting to project this six-dimensional space in two dimensions in order to visualize how benign and malignant data differs from each other.

5 Method for Comparing ISOLLE and LLE

The difference between LLE and ISOLLE are illustrated by considering the three-dimensional swissroll. At first we visualize the neighbors graphs obtained by LLE

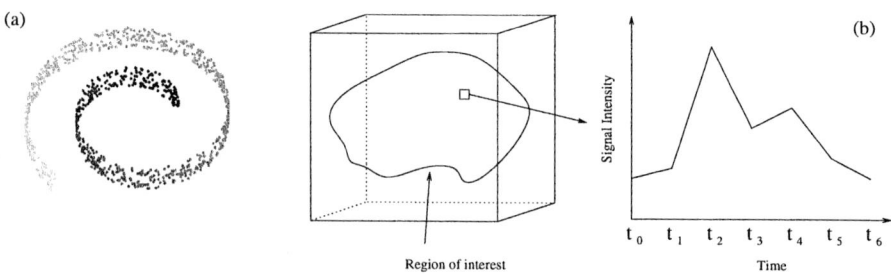

Fig. 2. (a) Three-dimensional swissroll data set. (b) In DCE-MRI, a time-series of MR signal intensity values is associated with each voxel.

Table 1. Data sets investigated in this study

Data set	Number of points	Dimension
Swissroll	1000	3
DCE-MRI breast tumor data	2449	6

and ISOLLE with different values of n in order to highlight the advantages of the geodesic distance. Each graph is obtained by connecting each data point with its n neighbors by an edge (note that this is not the graph G used to compute the geodesic distances and described in section 3). This allows us to check for the possible presence of short circuits induced by the neighbors search. The effects of these short circuits are then qualitatively evaluated by visualizing the respective two-dimensional projections.

The evaluation of the dimensional reduction of the tumor data set requires a statistical analysis, as the output can not be predicted a priori because of the complexity and multi-dimensional nature of the data, and consequently it is not possible to visually evaluate the accuracy of the low-dimensional projection.

The quality of the tumor data embeddings is estimated by means of two numerical quantities, namely neighborhood preservation (NP) and stress (ST). The first quantity is given by the average percentage of neighbors which are preserved after the dimensional reduction. It is defined as

$$\text{NP} = \frac{1}{V} \sum_{i=1}^{V} p_t(\mathbf{X}_i) \qquad (9)$$

where $p_t(\mathbf{X}_i)$ is the percentage of the t-nearest neighbors of point $\mathbf{X}_i$ in the original space which are preserved in the low-dimensional space. For example, if only 25% of its t-nearest neighbors are preserved in the embedding, then $\text{pt}(\mathbf{X}_i)$ will equal 0.25. In this work we use $t = 5$. A high value of NP (close to 1) denotes a good preservation of the local relations between data points in the low-dimensional space.

Stress reflects the preservation of the global structure of the original data set in the embedding. More specifically, it quantifies the overall deviation (i. e. the extent to which they differ) between the distances in the original and embedded space [15]. Let $\mathbf{X}_i$ and $\mathbf{X}_j$ be two data points; their distance in the original and in the embedding space are indicated by $d(\mathbf{X}_i, \mathbf{X}_j)$ and $\delta(\mathbf{X}_i, \mathbf{X}_j)$, respectively. Stress is typically defined in terms of variance as

$$\text{ST} = \frac{\sum_{\mathbf{X}_i, \mathbf{X}_j} (\delta(\mathbf{X}_i, \mathbf{X}_j) - d(\mathbf{X}_i, \mathbf{X}_j))^2}{\sum_{\mathbf{X}_i, \mathbf{X}_j} d(\mathbf{X}_i, \mathbf{X}_j)^2}. \qquad (10)$$

Prior to the computation of the value of stress, both the original and embedded coordinates are scaled to [0,1] in order to allow for a correct comparison between different embeddings. Low values of stress (close to 0) reflect a good preservation of the original pairwise distances in the low dimensional space.

6 Experiments

Graph G is computed by setting ϵ to the minimal possible value such that all the pairwise distances are finite. These values empirically found for each data set are: ϵ(swissroll)=5; ϵ(tumor data)=90.

The two data sets are reduced to two dimensions by LLE and ISOLLE with the number of neighbors n varying between 5 and 40.

7 Results and Discussion

In Fig. 3 one can see the neighbors graphs of the swissroll. It is obvious that already with $n = 15$ LLE with Euclidean distance meets some short circuit effects in the neighbors search. With $n = 40$ the number of short circuits increases noticeably. By contrast, the graphs relative to ISOLLE do not present short circuit effects, even when the number of neighbors n equals 40. This shows that the usage of the geodesic distance can drastically reduce the number of short circuits.

Possible effects of these short circuits on the two-dimensional projection of the swissroll data set can be seen in Fig. 4. Here it is clear that LLE fails to preserve the global structure of the data with $n = 15$ and in particular $n = 40$, as in both cases the darkest points are mapped close to brighter points. On the contrary, ISOLLE can correctly unfold the swissroll in all three cases, and the structure of the data is clearly preserved. In particular, the ISOLLE projection is also accurate with $n = 40$, while the respective LLE projection results completely incorrect.

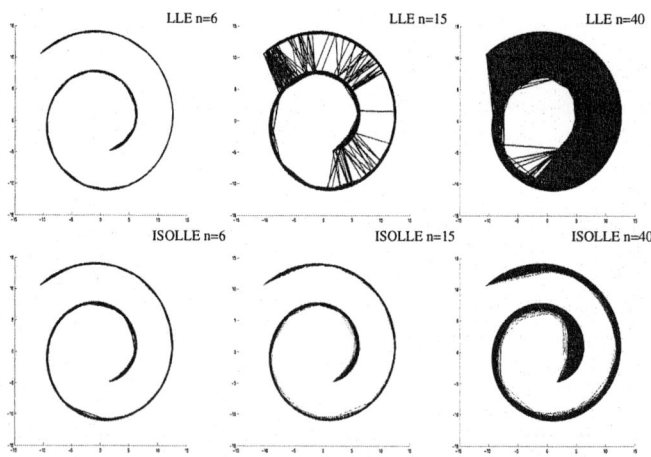

Fig. 3. Neighbors graphs of the swissroll data set. In the LLE graph with $n = 15$ there are already short circuits. Their number considerably increases with $n = 40$. Conversely, in all the ISOLLE graphs there are no short circuits.

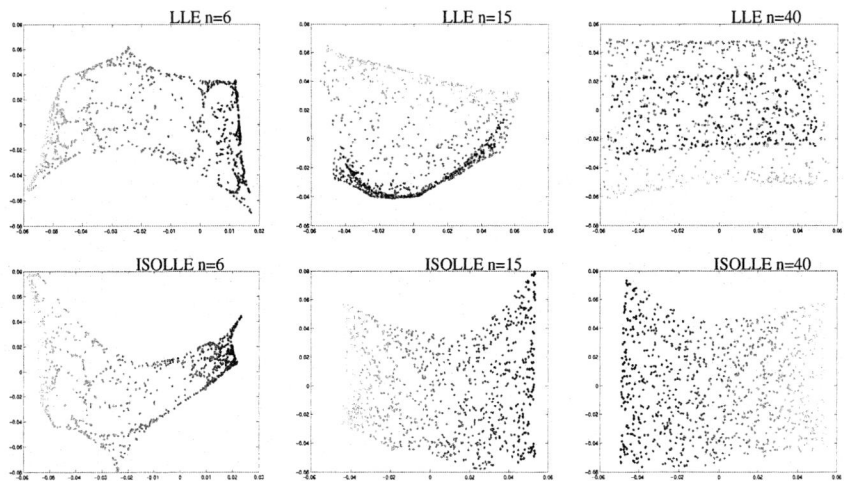

Fig. 4. Two-dimensional reductions of the swissroll data set. While LLE fails to preserve the structure of the swissroll with $n \geq 15$, ISOLLE yields a good projection of the data in all cases.

The evaluation of dimensional reduction of the tumor data set is conducted by taking into account the neighborhood preservation and stress measures. Their average values with the respective variances computed with respect to n comprised between 5 and 40 are displayed in table 2. The projections by ISOLLE result better with respect to both quantities. Indeed, the average ST value is lower than the one by LLE, suggesting that ISOLLE better preserves the metric of the tumor data. The higher value of the average NP by ISOLLE gives evidence that this algorithm also leads to a better preservation of the topology of the data. Two scatter plots of the DCE-MRI breast data embeddings obtained by LLE and ISOLLE with $n = 20$ are shown in Fig. 5. Interestingly, the benign cluster in the projection by ISOLLE appears more localized and compact than in the projection by LLE. Moreover, benign and malignant data overlap more in the projection by LLE. This indicates that ISOLLE can better separate benign from malignant data and this is of considerable value from the medical point of view. In addition, the compactness of the benign cluster in the ISOLLE projection shows that benign tumors are rather homogeneous, while the malignant ones are more heterogeneous, in agreement with the clinical experience of physicians [16].

Table 2. Average and variance values of stress (ST) and neighborhood preservation(NP) computed for the tumor data set

ST(LLE)	ST(ISOLLE)	NP(LLE)	NP(ISOLLE)
0.454±0.034	**0.337 ± 0.025**	0.081±0.001	**0.115 ± 0.002**

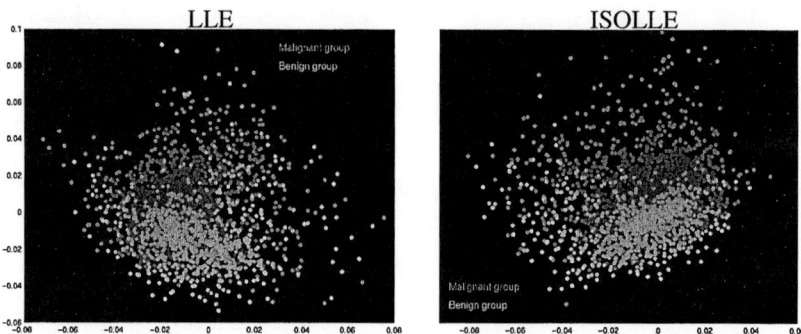

Fig. 5. Two scatter plots of the two-dimensional embeddings of the DCE-MRI breast data set obtained by LLE and ISOLLE. In both cases n equals 20. Note that the benign and malignant clusters overlap much less in the ISOLLE embedding. In particular, here the benign cluster is more compact and localized.

Table 3. Table of the running times in seconds

n	Swissroll		DCE-MRI	
	LLE	ISOLLE	LLE	ISOLLE
10	0.20	2.66	1.26	16.55
20	0.23	6.32	1.39	39.24
30	0.27	11.25	1.62	69.31
40	0.30	17.29	1.75	106.37

Finally, we compare the performances of LLE and ISOLLE is terms of running time. Both algorithms were run with different n on a Pentium IV 2.8 GHz. The respective values of running times are shown in table 3. The ISOLLE algorithm involves a larger computation time and the divergence of speed becomes more marked as n increases. The higher computational time of ISOLLE is somewhat expected as the algorithm requires a further step as compared to LLE, i. e. the construction of the neighborhood graph over all data points.

In general, the usage of ISOLLE should be preferred to LLE in particular when n needs to be relatively high (for example in case of sparse or clustered data) and in turn short circuits are more likely to occur. One way to determine if a certain data set requires a relatively high value of n is to perform an analysis of the smallest eigenvalues of matrix S from eq. (8). Specifically, in standard conditions matrix S has only one eigenvalue close to 0. However, if n is so small that the linear areas are disjoint, then matrix S will have more than one close-to-zero eigenvalue [5]. Therefore, the minimum n for which S has only one eigenvalue close to 0 can be taken into account in order to evaluate which algorithm is more suited for the data analysis.

8 Conclusions

In this study we propose a new approach to the neighbor search in LLE based on the geodesic distance. Its usage can reduce the number of short circuits considerably, thereby improving the preservation of the data structure. We show this by investigating the neighbors graphs obtained by LLE and ISOLLE on a synthetic three-dimensional swissroll. The ISOLLE graphs do not exhibit short circuits, even when the number of neighbors is high. By contrast, the standard neighbors search with Euclidean distance in LLE causes many short circuits. As a consequence, ISOLLE can detect the intrinsic two-dimensional structure of the swissroll with both small and large values of the number of neighbors n. Conversely, LLE fails to unfold the swissroll with $n \geq 15$.

Regarding the dimensional reduction of the tumor data set, our results clearly show that ISOLLE significantly outperforms LLE in terms of both stress and neighborhood preservation. In addition, ISOLLE appears to better distinguish between benign and malignant lesions.

Experiments concerning the running times revealed that ISOLLE is slower than LLE and this becomes more noticeable as n increases.

In conclusion, ISOLLE exhibits a superior ability to project the investigated data sets into a two-dimensional space while preserving the original data structure but at the cost of a larger running time.

Future work will include the comparison of the performances of LLE and ISOLLE with respect to data sets having different density distributions, with particular regard to sparse data.

References

1. Roweis, S. T. and Saul, L. K. Nonlinear Dimensionality Reduction by Locally Linear Embedding. *Science*, 290:2323–2326, 2000.
2. Zhang, J., Shen, H. and Zhou, Z. H. Feature Dimension Reduction for Microarray Data Analysis using Locally Linear Embedding. In *SINOBIOMETRICS*, pages 296–304, 2004.
3. Wang, M., Yang, J., Xu, Z. and Chou, K. SLLE for predicting Membrane Proteins Types. *Journal of Theoretical Biology*, 232:7–15, 2005.
4. Chao, S. and Lihui, C. Feature Dimension Reduction for Microarray Data Analysis using Locally Linear Embedding. In *3rd Asia-Pacific Bioinformatic Conference*, 2004.
5. Polito, M. and Perona, P. Grouping and Dimensionality Reduction by Locally Linear Embedding. In *Neural Information Processing Systems NIPS*, 2001.
6. Vlachos, M. et al. Non-Linear Dimensionality Reduction Techniques for Classification and Visualization. In *8th ACM SIGKDD International Conference on Knowledge Discovery and Data Mining*, 2002.
7. Hadid, A. and Pietikäinen, M. Efficient Locally Linear Embeddings of Imperfect Manifolds. In *MLDM*, 2003.
8. Tenenbaum, J. B., de Silva, V. and Langford, J. C. A Global Geometric Framework for Nonlinear Dimensionality Reduction . *Science*, 290:2319–2322, 2000.

9. Lee, J.A., Lendasse, A., Donckers, N. and Verleysen, M. A robust nonlinear Projection Method. In *Proceedings of ESANN 2000*, pages 13–20, Bruges, Belgium, 2000.
10. Wu, Y.X. and Takatsuka, M. The geodesic self-organizing Map and its Error Analysis. In *Australian Computer Science Conference*, volume 38, pages 343–352, Newcastle, Australia, 2005.
11. Lawrence, J. B., and Roweis, S. T. Think Globally, Fit Locally: Unsupervised Learning of Low Dimensional Manifolds. *Journal of Machine Learning Research*, 4:119–155, 2003.
12. Dijkstra, E. W. A Note on two Problems in Connection with Graphs. *Numer. Math*, 1:269–271, 1959.
13. http://www.cs.sunysb.edu/ skiena/combinatorica/animations/dijkstra.html.
14. Kelcz, F., Furman-Haran, E., Grobgeld, D. et al. Clinical Testing of High-Spatial-Resolution Parametric Contrast Enhanced MR Imaging of the Breast. *AJR*, 179:1485–1492, 2002.
15. Hjaltason, G. R., and Samet, H. Properties of Embeddings Methods for Similarity Searching in Metric Spaces. *IEEE Transactions on Pattern Analysis and Machine Intelligence*, 25(5):530–549, 2003.
16. Furman-Haran, E., Grobgeld, D., Kelcz, F. et al. Critical Role of Spatial Resolution in Dynamic Contrast-Enhanced Breast MRI. *Journal of Magnetic Resonance Imaging*, 13:862–867, 2001.

Active Sampling for Knowledge Discovery from Biomedical Data

Sriharsha Veeramachaneni[1], Francesca Demichelis[1,2],
Emanuele Olivetti[1], and Paolo Avesani[1]

[1] SRA Division, ITC-IRST, Trento, Italy 38050
[2] Department of Pathology, Brigham and Women's Hospital,
Harvard Medical School, Boston, USA
{sriharsha, michelis, olivetti, avesani}@itc.it

Abstract. We describe work aimed at cost-constrained knowledge discovery in the biomedical domain. To improve the diagnostic/prognostic models of cancer, new biomarkers are studied by researchers that might provide predictive information. Biological samples from monitored patients are selected and analyzed for determining the predictive power of the biomarker. During the process of biomarker evaluation, portions of the samples are consumed, limiting the number of measurements that can be performed. The biological samples obtained from carefully monitored patients, that are well annotated with pathological information, are a valuable resource that must be conserved. We present an active sampling algorithm derived from statistical first principles to incrementally choose the samples that are most informative in estimating the efficacy of the candidate biomarker. We provide empirical evidence on real biomedical data that our active sampling algorithm requires significantly fewer samples than random sampling to ascertain the efficacy of the new biomarker.

1 Introduction

In the biomedical domain, the acquisition of data is often expensive. The cost constraints generally limit the amount of data that is available for analysis and knowledge discovery. We present a methodology for intelligent incremental data acquisition for performing knowledge discovery under cost constraints.

In biological domains molecular tests, called biomarkers, conducted on biological samples (e.g. tumor tissue samples) that provides predictive information to pre-existing clinical data are studied. The development of molecular biomarkers for clinical application is a long process that must go through many phases starting from early discovery phases to more formalized clinical trials. This process involves the analysis of biological samples obtained from a large population of patients in a retrospective manner. The biological samples that need to be properly preserved are collected together with corresponding clinical data over time and are therefore very valuable. There is therefore a need to carefully optimize the use of the samples while studying new biomarkers. We address the issue of cost-constrained biomarker evaluation for developing diagnostic/prognostic models for cancer.

In general the acquisition of new data can be performed automatically by querying the environment by choosing the queries that are most likely to provide 'useful' information. This learning paradigm called *active learning*, where the learner is endowed with the ability to choose the data to be acquired, has been shown to yield comparable accuracy with significantly fewer data [1,6,9,11]. Traditionally active learning methods have been applied for training classifiers in the presence of unlabeled data, where the class labels of carefully chosen samples are queried. These techniques are suitable for situations where the class labels are considerably more expensive to obtain than the feature representation of the examples. Moreover the queries are chosen in order to learn the classifier accurately (with low cost).

In contrast, for our problem new biomarkers are tested on biological samples from patients who are labeled according to their disease and survival status. Moreover for each patient we have additional information such as grade of the disease, tumor dimensions and lymphonode status. That is, the samples are class labeled as well as described by some previous features. The goal is to choose the new feature (the biomarker) among many that is most correlated with the class label given the previous features. Since the cost involved in the evaluation of all the biomarkers on all the available data is prohibitive, we need to *actively* choose the the samples on which the new features (biomarkers) are tested. Therefore our objective at every sampling step is to choose the query (the sample on which the new feature is measured) so as to learn the efficacy of the biomarker most accurately.

Although the general theory of active learning has been studied in statistics in the area of optimal experimentation [7,10,13], it has seldom been applied to problems in knowledge discovery. The reason being the difficulty in resolving various practical issues such as finding good approximations to the theory and learning with sampling bias (which is a side-effect of active sampling).

In Section 2 we provide an overview of the process of identifying and evaluating biomarkers with a description of the resources required. In Section 3 we present an abstract formulation of the problem and outline a solution. We then describe the dataset obtained from Tissue Microarray analysis and will provide experimental evidence for the efficacy of our solution. We conclude with a discussion of the insights gained and directions for future work.

2 Cancer Biomarker Evaluation

Current models for cancer characterization, that lead to diagnostic/prognostic models, mainly involve the histological parameters (such as grade of the disease, tumor dimensions, lymphonode status) and biochemical parameters (such as the estrogen receptor). The diagnostic models used for clinical cancer care are not yet definitive and the prognostic and therapy response models do not accurately predict patient outcome and follow up. For example, for lung cancer, individuals affected by the same disease and equally treated demonstrate different treatment responses, evidencing that still unknown tumor subclasses (different istotypes) exist. This incomplete view results, at times, in the unnecessary over-treatment of patients, that is some patients do not benefit from the treatment they undertake. Diagnostic and prognostic models used in clinical cancer care can be improved by embedding new biomedical knowledge. The ultimate goal is to improve the diagnostic and prognostic ability of the pathologists and clinicians leading to better decisions about treatment and care.

Ongoing research in the study and characterization of cancer is aimed at the refinement of the current diagnostic and prognostic models. As disease development and progression are governed by gene and protein behaviour, new biomarkers found to be associated with patient diagnosis or prognosis are investigated. The identification of new potential biomarkers is recently driven by high throughput technologies, called Microarrays. They enable the identification of genes that provide information with a potential impact on understanding disease development and progression [5].

Although the initial high throughput discovery techniques are rapid, they often only provide qualitative data. Promising genes are further analyzed by using other experimental approaches (focusing on DNA, RNA or proteins), to test specific hypotheses. Usually a well characterized dataset of tumor samples from a retrospective population of patients is identified and the experimental process of analyzing one biomarker (feature) on one sample at a time is conducted. These analyses are usually based on the comparison between 1)non-diseased (e.g. normal) and diseased (e.g. tumors) biological samples, 2)between diseased samples pharmacologically treated and untreated at variable time points or 3)between samples of different diseases. The efficacy of specific biomakers can for example be determined based on their discriminative power in distinguishing between patients with poor or good prognosis, meaning patients with short or long overall survival respectively or cancer recurring or not recurring. This process can be time consuming, depending on the type of experimental technique which is adopted.

More importantly, well annotated tissue samples are very precious. Monitoring the status of patients over years, even decades, and store them so that to be useful for studies is not trivial and requires organizational efforts. It is not uncommon, for example, that patients who undertook the treatment at a hospital, will be monitored during the follow up period in another hospital and even in another country. Therefore keeping track of their status may become quite difficult. When the biomarker is tested on a biological sample, a portion of the sample is consumed, implying that each sample can be used for only a finite number of experiments. This motivates the need to develop an active sampling approach to conserve the valuable biological sample resource.

3 Active Measurement of Feature Values

We first present an abstract statement of the problem and show how it fits the real problem in the biomedical domain. We will then derive an active sampling algorithm as a solution. Let us consider a finite set of monitored pattern instances (or subjects) $T = \{t_i\}_{i=1,...,N}$. Let the random variable corresponding to the class label be denoted by **c** taking values in $\mathcal{C}$. The random variables[1] **x** and **y** correspond to features (or attributes) that can be measured on any instance, taking on values in $\mathcal{X}$ and $\mathcal{Y}$ respectively. The class label and the feature value **x** are known for every instance in T. In other words the random vector $\mathbf{s} = (\mathbf{c}, \mathbf{x})$ is instantiated on all the instances. However, initially, feature value **y** has not been acquired on any instance. Let the probability distribution over $\mathcal{C} \times \mathcal{X} \times \mathcal{Y}$ be parameterized by $\theta \in \Theta$. It is required to learn a concept g (which is a function of θ) on $\mathcal{C} \times \mathcal{X} \times \mathcal{Y}$ accurately by minimizing the number of instances on which **y** is measured (or probed).

In this formulation the class label **c** represents the diagnostic class, overall survival or relapse free survival status or the response to the treatment of the particular patient, the previous feature vector **x** represents the histological parameters of the tumor sample or biomarkers previously determined to be effective and the feature **y** represents the expression level of the candidate biomarker that is being evaluated. The concept **g** to be learned represents the efficacy of the new feature, or equivalently the error rate (denoted by **e**) of the classifier $\phi : \mathcal{X} \times \mathcal{Y} \to \mathcal{C}$ operating on the full feature space. Since the goal is to save the valuable resource we would like to measure as few samples as possible with the new biomarker before deciding whether it provides any useful information.

3.1 Active Sampling to Minimize Predicted Mean Squared Error

The active sampling approach is to iteratively choose the *best* $s = (c, x)$ where to probe the feature value **y**. Then an instance in the dataset with the particular s is chosen on which **y** is probed. Let us assume that after k iterations of this sampling process we obtained the dataset of measurements $T_k = \{s_1 \to y_1, s_2 \to y_2, \ldots, s_k \to y_k\}$. We will now describe how the choice of *best* s can be performed from a Bayesian statistical viewpoint.

Let us assume that the estimate of the concept is the Bayes minimum mean square error (MMSE) estimate. That is, given data T, the estimate of the concept is given by $\hat{g}(T) = E[\mathbf{g}|T]$. The predicted mean squared error (MSE) of the estimate of g at step $k+1$, if s were to be probed for the value of **y**, is given by

$$\text{MSE}(s)_{k+1} = \int \int (E[\mathbf{g}|T_k, s \to y] - g)^2 p(g|T_k, s \to y) p(s \to y|T_k) dg\, dy$$

$$= \int \int (E[\mathbf{g}|T_k, s \to y] - g)^2 p(g, s \to y|T_k) dg\, dy \quad (1)$$

---

[1] In general **x** and **y** can be random feature vectors of different lengths.

Note that the MSE is averaged over all the possible values of **y**, with the probabilities given by conditioning over all the data we have seen thus far. Now the best s to probe is the one that yields the lowest predicted mean squared error[2].

Lemma 1. *To select the sample s to minimize the predicted mean squared error in Equation 1, we can equivalently maximize the squared difference between the Bayes estimates of the concept before and and after s is probed, averaged over the possible outcomes. That is*

$$B(s) = \operatorname*{argmin}_{s} MSE(s)_{k+1}$$
$$= \operatorname*{argmax}_{s} \int_{\mathcal{Y}} (E[\mathbf{g}|T_k, s \to y] - E[\mathbf{g}|T_k])^2 p(s \to y|T_k) dy$$

The proof is outlined in [12]. The result implies that the most informative s to sample at is the one where the sampling would lead to the most change from the current estimate of the concept in the expected sense. In most problems, however, it is difficult to estimate the concept using a Bayes MMSE approach. Therefore we relax this constraint to approximate the objective function as

$$B(s) = \int_{\mathcal{Y}} (\hat{g}(T_k, s \to y) - \hat{g}(T_k))^2 p(s \to y|T_k) dy \qquad (2)$$

where $\hat{g}(T)$ is any estimate of the concept **g** from data T, that is appropriate for the problem. Our active sampling method based on this benefit criterion is called the *Maximum Average Change* (MAC) sampling algorithm[3]. An approach such as ours where the best instance to sample is decided based upon the benefit at the next step is called *myopic* active learning. An alternative would be reason about the best sample based upon the benefit it will have after many sampling steps. This alternative, however, can be computationally very expensive.

For feature selection we need to evaluate the benefit of adding each candidate feature **y** to the current set of features **x**. Therefore the concept **g** to be learned is the error rate (denoted by **e**) of the classifier $\phi : \mathcal{X} \times \mathcal{Y} \to \mathcal{C}$ operating on the full feature space. Although the above derivation of the active learning strategy is problem and classifier independent, we implemented the active feature measurement algorithm for the estimation of the error rate of a *Bayesian maximum a posteriori* classifier with discrete valued features and class labels. As opposed to the *Naive Bayes* classifier, we do not assume that the features **x** and **y** are class-conditionally independent. We will now briefly describe how the active sampling algorithm was implemented and provide the equations for the estimation of the probability distribution and for the error rate of the classifier.

---

[2] In case of non-uniform costs a different objective function that uses both the sampling cost and the MSE should be optimized.

[3] In [12] we have shown that MAC heuristic incurs significantly lower sampling cost than the heuristics proposed by Lizotte et al. [4] and Zheng and Padmanabhan [14] for similar problems.

3.2 Implementation

All probability distributions are multinomial whose parameters are estimated from data using Bayes MMSE estimators under uniform Dirichlet priors. Due to the difficulty in obtaining the exact Bayes MMSE estimate of the error rate, we approximate it by the error rate computed from the Bayes estimate of the distribution $p(c, x, y)$ over $\mathcal{C} \times \mathcal{X} \times \mathcal{Y}$.

We will now describe how the estimation of the joint probability is performed and present the formulae for the computation of the classifier and its error rate. At a given iteration of the active sampling process some of the instances have feature value **y** missing. Moreover because of the active sampling the missing values are not uniformly distributed. In [3] MacKay asserts that the biases introduced in the induced concept because of non-random sampling can be avoided by taking into account how we gathered the data. Therefore to construct the estimator $\hat{p}(c, x, y)$ over $\mathcal{C} \times \mathcal{X} \times \mathcal{Y}$ it is necessary to consider the sampling process. Since all the examples in the database are completely described with respect to **c** and **x** we already have the density $p(c, x)$. In addition, at any iteration of the active sampling algorithm there is an incomplete database with **y** values missing non-uniformly across various configurations of $(\mathbf{c}, \mathbf{x})$. However for each (c, x) the samples for y are independent and identically distributed. We incorporate this information in the estimator of the probability density from incomplete data T as follows. We first calculate

$$\hat{p}_T(y|c, x) = \frac{n_{c,x,y} + 1}{\sum_y n_{c,x,y} + |\mathcal{Y}|} \quad (3)$$

where $n_{c,x,y}$ is the number of instances of the particular combination of (c, x, y) among all the completely described instances in T. Note that $\hat{p}_T(y|c, x)$ is the same as $p(s \rightarrow y|T)$ used in the equations above. Now the probability density over $\mathcal{C} \times \mathcal{X} \times \mathcal{Y}$ is calculated as

$$\hat{p}_T(c, x, y) = \hat{p}_T(y|c, x) \times p(c, x) \quad (4)$$

Once we have the estimate $\hat{p}_T(c, x, y)$ all other quantities can be computed easily and in particular the estimate of the error rate $\hat{e}(T)$ is computed as follows.

$$\hat{e}(T) = 1 - \sum_{\mathcal{X} \times \mathcal{Y}} \hat{p}_T(\hat{\phi}(x, y), x, y) \quad (5)$$

where $\hat{\phi}_T$ is the *Bayes maximum a posteriori* classifier learned from data T given by

$$\hat{\phi}_T(x, y) = \underset{c \in \mathcal{C}}{\operatorname{argmax}} \frac{\hat{p}_T(c, x, y)}{\sum_c \hat{p}_T(c, x, y)} \quad (6)$$

For a given amount of budget and candidate feature **y**, the MAC active sampling algorithm to learn the utility of (i.e., the error rate given) the feature is given in pseudocode below.

Algorithm : ACTIVESAMPLINGFORERRORRATE(*DataSet*,**y**, *Budget*)

$cost \leftarrow 0$;
$ErrorRate \leftarrow EstimateErrorRate(DataSet)$ **comment:** cf. Equation 5
while $(cost < Budget)$
   for each $s \in \mathcal{C} \times \mathcal{X}$
     $B[s] \leftarrow 0$;
     for each $y \in \mathcal{Y}$
       $p(y|s) \leftarrow CalcConditionalProb(DataSet)$
           comment: cf. Equation 3
       $AugmentedDataSet \leftarrow AddSample(DataSet, (s \rightarrow y))$
       $PredErrorRate \leftarrow EstimateErrorRate(AugmentedDataSet)$
       $B[s] \leftarrow B[s] + (PredErrorRate - ErrorRate)^2 \times p(y|s)$
     end
   end
   $BestSample \leftarrow RandomChooseSample(\underset{s}{\operatorname{argmax}} B[s])$
   comment: Randomly select an incomplete sample among samples with max Benefit
   $DataSet \leftarrow AddSample(DataSet, (BestSample \rightarrow ExtractY(BestSample)))$
         comment: Measure **y** on *BestSample* and update *DataSet*
   $ErrorRate \leftarrow EstimateErrorRate(DataSet)$
   $cost \leftarrow cost + SamplingCost$
end
return $(ErrorRate)$

4 Dataset for Experimentation

To provide evidence that our method is effective in reducing costs in the biomedical domain we experimentally evaluated our method on a breast cancer Tissue Microarray dataset[4]. Although the biomarker evaluation problem is not relevant for this particular dataset we use it to demonstrate the utility of our approach.

The dataset was acquired using the recently developed technique of Tissue Microarray [8] that improves the in-situ experimentation process by enabling the placement of hundreds of samples on the same glass slide. Core tissue biopsies are carefully selected in morphologically representative areas of original samples and then arrayed into a new "recipient" paraffin block, in an ordered array allowing for high-throughput in situ experiments. A TMA block contains up to 600 hundred tissue biopsies. TMA approach has a dramatic advantage over the conventional approach in performing in situ experiments on standard glass slides by allowing the simultaneous staining of hundreds of tissue samples from as many patients, ensuring experimental standardization and conserving the

---

[4] The data used for experimentation was collected by the Department of Histopathology and the Division of Medical Oncology, St. Chiara Hospital, Trento, Italy. Tissue Microarray experiments were conducted at the Department of Histopathology [2].

limited tissue resource, which is vital given the increasing number of candidate genes that need to be explored.

For each patient there is a record that described by clinical, histological and biomarkers information. The entire dataset consisted of 400 records defined by 11 features. Each of the clinical features is described by a binary status value and a time value. Some of the records have missing values. The data are described by the following features:

Clinical Features

1. the status of the patient (binary, dead/alive) after a certain amount of time (in months, integer from 1 to 160)
2. the presence/absence of tumor relapse (binary value) after a certain amount of time (in months, integer from 1 to 160 months)

Histological Features

3. diagnosis of tumor type made by pathologists (nominal, 14 values)
4. pathologist's evaluation of metastatic lymphonodes (integer valued)
5. pathologist's evaluation of morphology (called grading, ordinal, 4 values)

Biomarkers Features (manually measured by experts in TMA)

6. Percentage of nuclei expressing ER (estrogen receptor) marker.
7. Percentage of nuclei expressing PGR (progesterone receptor) marker.
8. Score value (combination of colour intensity and percentage of stained area measurements) of P53 (tumor suppressor protein) maker in cells nuclei.
9. Score value (combination of colour intensity and percentage of stained area measurements) of cerbB marker in cells membrane.

The learning task defined on this dataset is the prediction of the status of the patient (dead/alive or relapse) given some previous knowledge (histological information or known biomarkers). The goal is to choose the new biomarker which can be used along with the histological features that provides accurate prediction. The experiments address the issue of learning which additional feature has to be sampled.

The dataset was preprocessed as follows. Continuous features have been discretized to reduce the level of detail and to narrow the configuration space for the sampling problem. Feature values have been discretized encoded into binary variables according to the convention suggested by experts in the domain.

We designed 10 experiments corresponding to different learning situations. The experiments differ in the choice of attribute for the class label (**c**), the attributes used as the previous features (**x**) and the feature used as the new candidate feature (**y**). The various configurations are shown below.

	Class Label (C)	Known Features (X)	New Feature (Y)	Size (#)
I	dead/alive	all histological information	PGR	160
II	dead/alive	all histological information	P53	164
III	dead/alive	all histological information	ER	152
IV	dead/alive	all histological information	cerbB	170
V	relapse	all histological information	PGR	157
VI	relapse	all histological information	P53	161
VII	relapse	all histological information	ER	149
VIII	relapse	all histological information	cerbB	167
IX	dead/alive	PGR, P53, ER	cerbB	196
X	relapse	PGR, P53, ER	cerbB	198

For the empirical evaluation we performed an additional preprocessing step of removing all the records with missing values for each experiment separately. For this reason the sizes of datasets used for different experiments are different.

5 Experiments

For each of the 10 experimental configurations described above, the random and MAC sampling schemes are compared for different number of acquired samples. The evaluation metric is computed as follows. For each choice of **x** and **y** we calculated the expected error rate e_F of a maximum a posteriori classifier trained on the entire database (i.e., with all the values of **c**, **x** and **y** known). Then for a given sample size L we sampled **y** values on L samples from the database (either by MAC or random sampling) and calculate the predicted error rate e_L for each method. We then computed the root mean square difference between e_L and e_F over several runs of the sampling scheme. Under the assumption of unit cost for feature value acquisition the rms difference will measure the efficacy of a sampling scheme in estimating the error rate of a classifier trained on both **x** and **y** as a function of the number of feature values acquired.

To relate the evaluation measure used to the biological problem, we note that e_F can be viewed as the true error rate of the classifier that uses the new biomarker **y** and e_L as the estimate of the error rate after sampling **y** on L samples. Since our goal is to predict the error rate accurately minimizing L, we can measure the effectiveness of our sampling algorithm by the rms difference between e_F and e_L.

For each experiment we plotted the rms value against the number of samples probed which are shown in Figure 1. In each plot, to compare the MAC sampling scheme to the random method for cost effectiveness, we must compare the number of feature values sampled for a required rms error.

In some plots the rms value starts at zero before increasing to a maximum value. This happens in situations where the new feature **y** adds no new information for predicting the class **c** given the previous features **x**. Therefore in the beginning of the sampling process, the estimated error rate is just the error rate obtained by using **x** which is the actual value. As more samples are added the

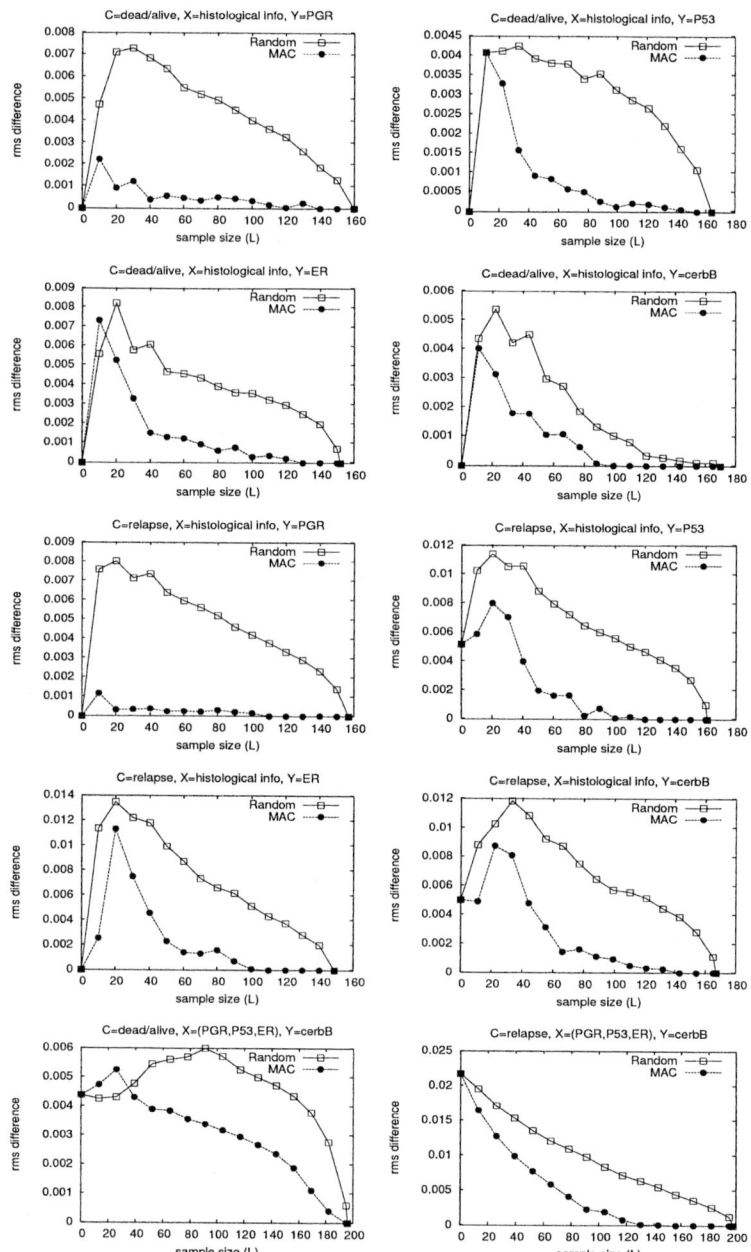

Fig. 1. Plots of the rms difference between the 'true' error rate of the classifier operating on the full feature space and that estimated from the acquired samples for all three sampling schemes as a function of the number of samples acquired. The features chosen for **c**, **x** and **y** are also indicated. The rms value is computed from 500 runs of the sampling experiment for each configuration. The error bars are smaller than the markers used in the plots.

estimate becomes optimistic predicting an error rate lower than the actual value until sufficiently large number of samples are added to learn the new feature is useless.

We observe from the plots that our MAC active sampling algorithm is significantly better in reducing the number of samples needed for error rate estimation than random sampling. For example in the top-left subplot, we observe that to obtain an error rate estimate with an rms error of 0.001, MAC algorithm needs less than 40 samples as opposed to about 150 for random. This behaviour is observed in most of the experimental configurations.

In terms of the biomedical problem, this implies that using the MAC active sampling, we can evaluate a higher number of biomarkers on the same amount of bio-sample resource than using the standard random sampling method.

6 Conclusions and Future Work

We presented a preliminary solution to the problem of evaluating new biomarkers that enable better characterization of cancer while conserving the limited amount of well annotated biological samples. We showed experimentally that our active sampling algorithm holds promise in accurately evaluating the efficacy of the new biomarker with significantly fewer samples tested. This allows for the evaluation of more biomarkers using the biological samples at hand. One way to exploit the efficacy of the active sampling algorithm is to initially test the biomarkers on only a limited number of samples to discard uninformative biomarkers and proceed to test the remaining ones exhaustively.

There are still several open problems that need to be addressed. With the Tissue Microarray method several tissue samples are stained with the biomarker simultaneously. Therefore we need to extend our sampling algorithm by reasoning about incrementally sampling not one but a batch of samples. Although we can follow the same analysis presented above to derive an algorithm for the batch case, the algorithm would be computationally expensive. Therefore we need to make further approximations to derive a feasible algorithm. Since the Tissue Microarray process is complicated in the preparation of the slides, the costs involved are not uniform. We intend to develop active sampling algorithms for more general cost models. We also intend to investigate active sampling with other classification schemes. Our goal is to develop a fully functioning active sampling decision support system that can be employed for biomarker evaluation in biomolecular laboratories.

References

1. D.A.Cohn, Z.Ghahramani, and M.I.Jordan. Active learning with statistical models. In G. Tesauro, D. Touretzky, and T. Leen, editors, *Advances in Neural Information Processing Systems*, volume 7, pages 705–712. MIT Press, 1995.
2. F. Demichelis, A. Sboner, M. Barbareschi, and R. Dell'Anna. Tmaboost: an integrated system for comprehensive management of tissue microarray data. *IEEE Trans. Inf. Technol. Biomed.* In Press.

3. D.J.C.MacKay. Information-based objective functions for active data selection. *Neural Computation*, 4(4):590–604, 1992.
4. D.Lizotte, O.Madani, and R.Greiner. Budgeted learning of naive-bayes classifiers. In *Proceedings of the 19th Annual Conference on Uncertainty in Artificial Intelligence (UAI-03)*, pages 378–385, San Francisco, CA, 2003. Morgan Kaufmann.
5. T.R. Golub, D.K. Slonim, P. Tamayo, C. Huard, M. Gaasenbeek, J.P. Mesirov, H. Coller, M.L. Loh, J.R. Downing, M.A. Caligiuri, C.D. Bloomfield, and E.S. Lander. Molecular classification of cancer: class discovery and class prediction by gene expression monitoring. *Science*, 286(5439):531–537, 1999.
6. H.S.Seung, M.Opper, and H.Sompolinsky. Query by committee. In *Proceedings of the fifth annual workshop on Computational learning theory*, pages 287–294. ACM Press, 1992.
7. K.Chaloner and I.Verdinelli. Bayesian experimental design: A review. *Statistical Science*, 10:273–304, 1995.
8. J. Kononen, L.Bubendorf, A.Kallioniemi, M.Barlund, P.Schraml, S.Leighton, J.Torhorst, M.Mihatsch, G.Seuter, and O.P.Kallioniemi. Tissue microarrays for high-throughput molecular profiling of tumor specimens. *Nature Medicine*, 4(7):844–847, 1998.
9. M.Saar-Tsechansky and F.Provost. Active Sampling for Class Probability Estimation and Ranking. In *Proc. 7th International Joint Conference on Artificial Intelligence*, pages 911–920, 2001.
10. P.Sebastiani and H.P.Wynn. Maximum entropy sampling and optimal Bayesian experimental design. *Journal of Royal Statistical Society*, pages 145–157, 2000.
11. S.Tong and D.Koller. Support vector machine active learning with applications to text classification. In *Proceedings of the International Conference on Machine Learning*, pages 999–1006, 2000.
12. S.Veeramachaneni, E.Olivetti, and P.Avesani. Active feature sampling for low cost feature evaluation. Technical report, ITC-irst, 2005.
13. V.V.Fedorov. *Theory of optimal experiments*. Academic Press, New York, 1972.
14. Z.Zheng and B.Padmanabhan. On active learning for data acquisition. In *Proceedings of the International Conference on Datamining*, pages 562–570, 2002.

A Multi-metric Index for Euclidean and Periodic Matching

Michail Vlachos[1], Zografoula Vagena[2],
Vittorio Castelli[1], and Philip S. Yu[1]

[1] IBM. T.J. Watson Research Center
[2] University of California, Riverside

Abstract. In many classification and data-mining applications the user does not know a priori which distance measure is the most appropriate for the task at hand without examining the produced results. Also, in several cases, different distance functions can provide diverse but equally intuitive results (according to the specific focus of each measure). In order to address the above issues, we elaborate on the construction of a hybrid index structure that supports query-by-example on shape and structural distance measures, therefore lending enhanced exploratory power to the system user. The shape distance measure that the index supports is the ubiquitous Euclidean distance, while the structural distance measure that we utilize is based on important periodic features extracted from a sequence. This new measure is phase-invariant and can provide flexible sequence characterizations, loosely resembling the Dynamic Time Warping, requiring only a fraction of the computational cost of the latter. Exploiting the relationship between the Euclidean and periodic measure, the new hybrid index allows for powerful query processing, enabling the efficient answering of kNN queries on both measures in a single index scan. We envision that our system can provide a basis for fast tracking of correlated time-delayed events, with applications in data visualization, financial market analysis, machine monitoring/diagnostics and gene expression data analysis.

1 Introduction

Even though many time-series distance functions have been proposed in the data-mining community, none of them has received the almost catholic acceptance that the Euclidean distance enjoys. The Euclidean norm can be considered as the most rudimentary *shape-matching* distance measure, but it has been shown to outperform many complex measures in a variery of clustering/classification tasks [3], while having only a fraction of the computational and logical complexity of the competing measures.

Lately however, time-series researchers are also starting to acknowledge certain limitations of shape matching distance measures, and therefore we are gradually experiencing a shift to more *structural* measures of similarity. These new structural measures can greatly enhance our ability to assess the inherent similarity between time sequences and tend to be more coherent with theories governing the human perception and cognition. Recent work quantifying structurally the similarity between sequences, may take into consideration a variety of features, such as change-point-detection [2], sequence burstiness [7], ARIMA or ARMA generative models [9], and sequence compressibility [4].

In many cases though, there is no clear indication whether a shape or a structural measure is best suited for a particular application. In the presence of a heterogeneous dataset, specific queries might be tackled better using different measures. The distance selection task becomes even more challenging, if we consider that different distance measures can sometimes also provide diverse but equally intuitive search results.

In an effort to mitigate the distance selection dilemma, we present an index structure that can answer multi-metric queries based on both shape and structure, allowing the end user to contrast answer sets, explore and organize more effectively the resulting query matches. The proposed indexing scheme seamlessly blends the Euclidean norm with a structural *periodic* measure. Periodic distance functions were recently presented in [8] and have been shown to perform very effectively for many classes of datasets (i.e., ECG data, machine diagnostics, etc). However, in the original paper no indexing scheme had been proposed. Recognizing that the periodic measure can easily (and cost-effectively) identify time-shifted versions of the query sequence (therefore loosely resembling Time-Warping), we exploit the relationship between the euclidean and the periodic measure in the frequency domain, in order to design an index that supports query-by-example on both metrics. By intelligently organizing the extracted sequence features and multiplexing the euclidean and periodic search we can return the k-NN matches of both measures in a *single* index scan. Both result sets are presented to the user, expanding the possibilities of interactive data exploration, providing more clues as to the appropriate distance function.

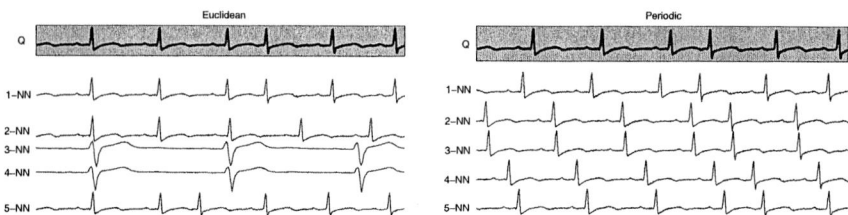

Fig. 1. 5-NN euclidean and periodic matches on an ECG dataset

A sample output of the proposed index for a database of ECG data is shown in Fig. 1. For the specific query, all instances returned by the periodic measure belong to the same class of sequences and correspond to time-shifted variations of the query sequence. The 1,2,5-Nearest-Neighbor (NN) matches of the Euclidean metric can also be considered similar to the query, however the 3-NN and 4-NN would be characterized as spurious matches by a human. The purpose of this (rather simplistic) example, is to emphasize that in many cases multiple measures are necessary, since each metric can harvest a different subset of answers.

Even though other multi-metric distances have been presented in [10,6] (but for different sets of distance functions), queries needed to be issued multiple times for retrieving the results of the different measures. Therefore, the presented index has two distinct advantages:

- It supports concurrent euclidean and periodic matching, returning both sets of Nearest-Neighbor matches in a *single* index scan. So it allows for both rigid matching (euclidean distance), or more flexible periodic matching, by identifying arbitrary time shifts of a query (periodic measure).
- Performance is not compromised, but is in fact improved (compared to the dual index approach) due to the reduced index size and the intelligent tree traversal.

Given the above characteristics, we expect that the new index structure can provide necessary building blocks for constructing powerful 'all-in-one' tools, within the scope of applications such as decision support, analysis of causal data relationships and data visualization.

2 Background

The periodic measure and the hybrid index that we will describe later operate in the frequency domain, therefore we will succinctly describe important concepts from harmonic analysis.

2.1 Frequency Analysis

A discrete-time signal $\mathbf{x} = [x_0, \ldots, x_{N-1}]$ of length N can be thought of as a period of a periodic signal and represented in terms of its Fourier-series coefficients $\{X_k\}_{k=0}^{N-1}$ by

$$x_n = \frac{1}{\sqrt{N}} \sum_{k=0}^{N-1} X_k e^{2\pi j(k/N)n}, \qquad n = 0, \ldots, N-1,$$

where $j = \sqrt{-1}$ is the imaginary unit. The coefficient X_k is defined by

$$X_k = \rho_k e^{j\theta_k} = \frac{1}{\sqrt{N}} \sum_{n=0}^{N-1} x_n e^{-2\pi j(k/N)n}, \qquad k = 0, \ldots, N-1,$$

and corresponds to the frequency $f_k = k/N$. Here ρ_k and θ_k are respectively the magnitude and the phase of X_k. Parseval's theorem states that the energy $\mathcal{P}$ of the signal computed in the frequency domain is equal to the energy computed in the Fourier domain:

$$\mathcal{P}(\mathbf{x}) = \|\mathbf{x}\|^2 = \sum_{k=0}^{N-1} x_k^2 = \mathcal{P}(\mathbf{X}) = \|\mathbf{X}\|^2 = \sum_{k=0}^{N-1} \|X_k\|^2.$$

Many operations are substantially more efficient in the frequency domain than in the time domain. The use of frequency-domain operations is often appealing thanks to the existence of the efficient Fast Fourier Transform, which has computational complexity of $O(N \log N)$.

3 Distance Functions

3.1 Euclidean Distance

Let $\mathbf{x}$ and $\mathbf{y}$ be two time sequences of length N having Discrete Fourier Transform $\mathbf{X}$ and $\mathbf{Y}$, respectively. The Euclidean distance $d(\mathbf{x}, \mathbf{y})$ between $\mathbf{x}$ and $\mathbf{y}$ (i.e., the ℓ_2 norm of $\mathbf{x} - \mathbf{y}$) is defined by $d(\mathbf{x}, \mathbf{y}) = \sqrt{(\mathbf{x} - \mathbf{y}) \cdot (\mathbf{x} - \mathbf{y})} = \sqrt{\sum_{k=1}^{N} |x_k - y_k|^2}$, where $\cdot$ denotes the inner product. Parseval's Theorem ensures that $d(\mathbf{x}, \mathbf{y}) = d(\mathbf{X}, \mathbf{Y})$. We can decompose the Euclidean distance into the sum of the magnitude distance and a non-negative term involving both magnitudes and phases:

$$[d(\mathbf{x}, \mathbf{y})]^2 = \sum_{k=0}^{N-1} \|x_k - y_k\|^2 = \sum_{k=0}^{N-1} \|\rho_k e^{j\theta_k} - \tau_k e^{j\phi_k}\|^2$$

$$\stackrel{(a)}{=} \sum_{k=0}^{N-1} (\rho_k \cos(\theta_k) - \tau_k \cos(\phi_k))^2 + (\rho_k \sin(\theta_k) - \tau_k \sin(\phi_k))^2$$

$$\stackrel{(b)}{=} \sum_{k=0}^{N-1} \rho_k^2 + \tau_k^2 - 2\rho_k \tau_k \left(\sin\theta_k \sin\phi_k + \cos\theta_k \cos\phi_k \right)$$

$$\stackrel{(c)}{=} \sum_{k=0}^{N-1} (\rho_k - \tau_k)^2 + 2 \sum_{k=1}^{N} \rho_k \tau_k \left[1 - \cos(\theta_k - \phi_k) \right], \tag{1}$$

where (a) is the Pythagorean theorem, (b) follows from algebraic manipulations and elementary trigonometric identities, and (c) follows by adding and subtracting $2\rho_k \tau_k$ to (b), collecting terms, and using an elementary trigonometric identity. Having expressed the Euclidean distance using magnitude and phase terms, we explore its connection with a periodic measure in the following section.

3.2 Periodic Measure

We present a distance measure that can quantify the structural similarity of sequences based on *periodic* features extracted from them. The periodic measure was discussed, together with applications, in [8] and is explicated here for completeness of presentation. In this work we make the connection with euclidean distance in the frequency domain and show how to combine both in an efficient index.

The introduction of periodic measures is motivated by the inability of shape-based measures such as the Euclidean to capture accurately the rudimentary human notion of similarity between two signals. For example two sequences that are identical except for a small time shift should be considered similar in a variety of applications (Fig. 2), in spite of the potentially large euclidean distance between them. Therefore, the periodic measure loosely resembles time-warping measures, requiring only linear computational complexity, thus rendering it very suitable for large data-mining tasks.

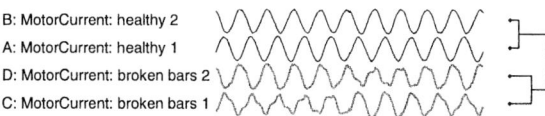

Fig. 2. Dendrogram on 4 sequences using a periodic measure

3.3 Periodic Distance (pDist)

To assess the periodic similarity of two sequences we examine the difference of their harmonic content. We define the periodic distance between two sequences **x** and **y**, with Fourier transforms **X** and **Y**, respectively as the euclidean distance between their magnitude vectors:

$$[pDist(\mathbf{X}, \mathbf{Y})]^2 = \sum_{k=1}^{N} (\rho_k - \tau_k)^2.$$

Notice that the omission of the phase information renders the new similarity measure shift-invariant in the time domain, allowing for global time-shifting in $O(n)$ time (another alternative would be the use of using Time-Warping with $O(n^2)$ complexity).

In order to meaningfully compare the spectral power distribution of two sequences in the database, we normalize them to contain the same amount of energy by studentizing them (thus producing zero-mean, unit-energy sequences):

$$\hat{x}(n) = \frac{x(n) - \frac{1}{N}\sum_{i=1}^{N} x(i)}{\sqrt{\sum_{i=1}^{N}(x(n) - \frac{1}{N}\sum_{i=1}^{N} x(i))^2}}, \quad n = 1, \ldots, N$$

For the rest of the paper we will assume that all database sequences are studentized (whether we are considering euclidean or periodic distance). We also remark a property of the periodic distance that will be useful to provide more effective traversal of our indexing structure.

Lemma 1. *The periodic distance is a lower bound to the Euclidean distance:* $pDist(\mathbf{X}, \mathbf{Y}) \leq d(\mathbf{X}, \mathbf{Y})$.

This lemma is proved by noting that the first sum on the RHS Equation 1 is the periodic distance, and that the second sum is non-negative.

4 Lower Bounding and Coefficient Selection

In order to efficiently incorporate a distance measure with an indexing structure, one needs to: (i) compress a sequence (dimensionality reduction) (ii) provide a lower bounding function of the original distance using the compressed object. We will show how both of these can be achieved in an effective way.

After a sequence is transformed in the frequency domain, it can be compressed by recording only a small subset of its coefficients. Therefore, $\{X_k\}_{k=0}^{k=N-1}$

$\leadsto \{X_k\}_{k \in S}$, $S \subset \{0, \ldots, N-1\}$, $|S| \ll N$. It is straightforward to show that the euclidean or periodic distance on the compressed vectors will lower bound the original distances, because they are a sum of positive numbers:

$$d(\mathbf{X}_k, \mathbf{Y}_k)_{k \in S} \leq d(\mathbf{X}, \mathbf{Y})$$

$$pDist(\mathbf{X}_k, \mathbf{Y}_k)_{k \in S} = \sum_{k \in S} (\rho_k - \tau_k)^2 \leq pDist(\mathbf{X}, \mathbf{Y})$$

The majority of data-mining work has adapted the selection of the first k coefficients for sequence compression [5], which can provide effective approximation of signals with low frequency content (e.g., stock price movement). Recent work also suggested picking the k coefficients for each sequence that preserve most of its energy [7]. High energy coefficients can provide effective sequence reconstruction, but are not necessarily suitable for data retrieval purposes (i.e. incur the least number of accesses to the original data). Consider a dataset where the k coefficients with the highest energy are the same for all sequences, and the sequences only differ in the low energy coefficients (i.e. there are some small but distinct nuances between each sequence). In this case, one needs to select the coefficients that will *discriminate better* the sequences. Generally speaking, we can capture more effectively the data differences by recording those coefficients that account for most of the data variation. With this observation in mind, we will record the k coefficients that depict that largest variance:

$$arg \max_k var(X_k^{(j)})_{j=1 \ldots m}$$

where X_k^j denotes the kth coefficient of sequence j. We compare the performance of various coefficient selection schemes with a comprehensive experiment on 40 datasets (each containing 1000 sequences of length 1024), obtained from the UCR time-series archive [1]. We perform a 1-NN leave-one-out search and estimate the pruning power of each method as given by the ratio: *(examined objects)/(total objects)*. The methods for coefficient selection that we consider are: (i) first k coefficients, (ii) coefficients with maximum energy (iii) coefficients with maximum variance.

The results attest to the superiority of the 'max-variance' method for Nearest-Neighbor retrieval, depicting an improvement in 21 out of the 40 datasets. The average improvement over the next best method is 17.17% (with a maximum of > 80% in the *darwin* dataset). Eight datasets do not exhibit any change at all, which is observed because the set of coefficients selected by the 3 methods were the same. Finally, 11 datasets depict a deterioration in the k-NN performance which however is almost negligible (never exceeding -0.5%), with the average negative improvement being -0.13%. For the remainder of the paper, we will assume that the coefficients selected from each time-series are the ones exhibiting the largest variance.

---

[1] http://www.cs.ucr.edu/~eamonn/TSDMA/

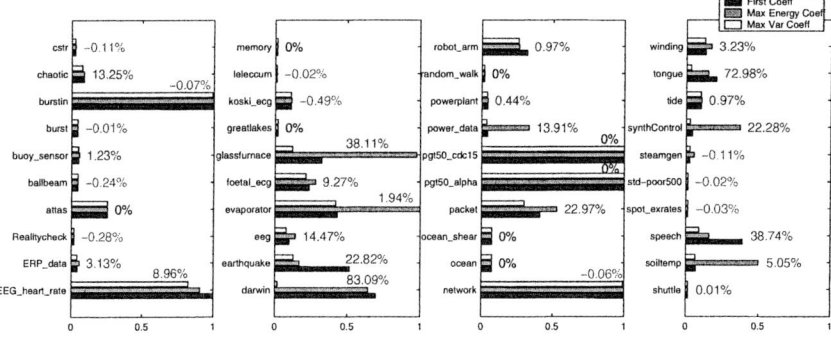

Fig. 3. Comparison of coefficient selection (smaller numbers are better). Improvement of *max-variance* method vs second best is reported next to each performance bar.

5 Index for Euclidean and Periodic Distance

Instead of constructing a different index structure for each measure, we can exploit the common representation of the euclidean and periodic measure in the magnitude/phase space, as well as their lower bounding relationship, for designing a metric index structure that can simultaneously answer queries on both measures. Our index structure borrows ideas from the family of metric index structures [1], recursively partitioning the search space into disjoint regions, based on the relative distance between objects. Our indexing approach has three important differences from generic metric trees: (i) only compressed sequences are stored within the index nodes, reducing the index space requirements, (ii) the index uses a different distance measure for data partitioning on each alternating level of the tree, (iii) a novel tree traversal is presented, that can answer euclidean and periodic queries in a single index scan.

5.1 MM-Tree Structure

We introduce a hybrid metric structure, the MM-Tree (Multi-Metric Tree). Similar to VP-trees, each node of the index contains a reference point (or vantage point), which is used to partition the points associated with this node into two distinct and equal sized clusters. Vantage points are chosen to be the sequences with the highest variance of distances to the remaining objects. The distances of the node objects to the reference point (sequence) are calculated, distances are sorted and the median distance μ is identified. Subsequently, any sequence associated with the examined node, is assigned to a *left* or a *right subtree*, depending on whether its distance from the vantage point is smaller or larger than the median distance. The index tree is constructed by recursively performing this operation for all subtrees.

The unique structure of the MM-Tree derives from the fact that it uses a *different* distance function to partition objects at each alternating tree level.

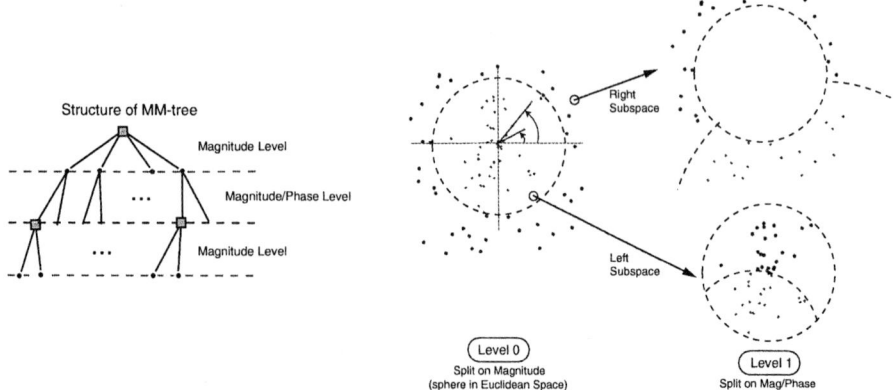

Fig. 4. MM-tree structure and space partitioning. Dotted circles/arcs indicate median distance μ.

Even depth levels (root is zero) are partitioned using the periodic distance, while odd levels utilize the euclidean (Fig. 4). We follow this construction for providing a good partitioning on both distance measures, since usage of a single distance function during the tree construction would have impacted the search on the other domain (potentially leading to search on both left and right subtrees).

Unlike other metric indexing structures, intermediate tree nodes contain the *compressed* representation of a vantage point (in addition to the median distance μ). For example, in the tree of Figure 4 the vantage points of nodes at even depth are represented by the magnitudes of the preserved coefficients (and are called "P-nodes"), while those of nodes at odd depth are represented by both magnitude and phase (and are called "E-nodes"). Finally, leaf nodes contain both magnitude and phase information of the compressed data sequences.

5.2 Multiplexing Search for Periodic and Euclidean Distances

We now describe how the MM-Tree can efficiently multiplex searches in the euclidean and periodic spaces and simultaneously return the nearest-neighbors in both domains. The key idea of the search algorithm is to identify, in a single index traversal, the union of the necessary index nodes for both queries. In figure 5 we provide a pseudocode of the multiplexed search.

The combined search employs two sorted lists, $BEST_p$ and $BEST_e$, that maintain the current k closest points using periodic and euclidean distances, respectively. The algorithm also records a state, depicting whether a visited node is marked for search in the euclidean, or in the periodic domain, or both. The root node is marked for search in both domains.

Searching a P-node node. If the node is marked for search only in the euclidean domain, both subtrees are searched in the euclidean domain only. Otherwise, the algorithm computes the lower bound $LB_p(q, v)$ of the periodic distance

```
/* perform 1-NN search for query sequence Q */
1NNSearch(Q) {
   // farthest results are in Best_P[0] and Best_E[0]
   Best_P = new Sorted_List(); // Modified by searchLeaf_Periodic
   Best_E = new Sorted_List(); // Modified by searchLeaf_Euclidean
   search_Node(Q, ROOT, TRUE);
}

search_Node(Q, NODE, searchPeriodic) {
   if (NODE.isLeaf) {
      search_Leaf(Q, NODE, searchPeriodic);
   } else {
      search_Inner_Node(Q, NODE, searchPeriodic);
   }
}

search_Inner_Node(Q, NODE, searchPeriodic) {
   add_Point_To_Queue(PQ, vantagePoint, searchPeriodic);
   if (NODE.E_NODE) { /* E-Node */
      if (searchPeriodic) {
         search_Inner_Node(Q, NODE.LEFT, searchPeriodic);
      } else { /* only search in euclidean space */
         if (LowerBoundEuclidean(Q, vantagePoint) - Best_E[0] < median)
            search_Inner_Node(Q, NODE.LEFT, searchPeriodic);
      }
   } else { /* P-Node */
      if (searchPeriodic) {
         if (LowerBoundPeriodic(Q, vantagePoint) - Best_P[0] < median)
            search_Inner_Node(Q, NODE.LEFT, searchPeriodic);
         else
            search_Inner_Node(Q, NODE.LEFT, FALSE);
      } else { /* only search in euclidean space */
         search_Inner_Node(Q, NODE.LEFT, searchPeriodic);
      }
   }
   search_Inner_Node(Q, NODE.RIGHT, searchPeriodic);
}

search_Leaf(Q, NODE, searchPeriodic) {
   if (searchPeriodic) search_Leaf_Periodic(Q, NODE);   // update Best_P
   search_Leaf_Euclidean(Q, NODE); // update Best_E
}
```

Fig. 5. Multiplexing euclidean and periodic search on the MM-Tree

between the vantage point of the node and the query sequence. Let r_p be the periodic distance to the farthest entry in $BEST_p$ to the query. Noting that:

$$median < LB_p(q,v) - r_p \Rightarrow median < pDist(q,v) - r_p,$$

where $pDist(q,v)$ is the periodic distance of the corresponding uncompressed sequences, we conclude that the algorithm should search the left subtree only in the euclidean domain (but not in the periodic) if $median < LB_p(q,v) - r_p$.

Searching an E-node node. If the node is marked for search in both domains, all subtrees are searched using both measures. Otherwise, the algorithm computes the lower bound $LB_e(q,v)$ of the euclidean distance between the vantage point of the node and the query sequence. Let r_e be the euclidean distance of the farthest entry in $BEST_e$ to the query. Then, if $LB_e(q,v) - r_e > median$ the left subspace is discarded. It is also important to note that, for both types of nodes, since the vantage point is in compressed form and we use lower bounds to distances, we do not have sufficient information to discard the right subtree, unless we load the uncompressed representation of the vantage point.

A global priority queue PQ, whose priority is defined by the lower bounds of the periodic distances, is employed, in order to efficiently identify the query results. In particular, whenever the compressed representation of a data sequence v is accessed, the lower bound of the periodic distance $LB_p(q,v)$ between v and the query sequence is computed and the pair $(LB_p(q,s),s)$ is pushed into PQ. When a sequence s is popped from the PQ, the associated lower bound of the periodic distance $LB_p(q,v)$ is compared against the current $BEST_p[0]$ and $BEST_e[0]$ values. If $LB_p(q,v)$ is larger than both of those values, the sequence is discarded. However, if it is smaller than $BEST_e[0]$, the lower bound of the euclidean distance $LB_e(q,v)$ is computed and if it is larger than the $BEST_e[0]$ value, the sequence can still be safely discarded. In all other cases, the uncompressed sequence is loaded from disk and the actual periodic and euclidean distances are computed to determine whether it belongs to any of the $BEST_p$ or $BEST_e$ lists.

6 Experiments

We demonstrate the performance and meaningfulness of results of the MM-tree for answering simultaneously queries on both euclidean and periodic distance measures. The experiments reveal that the new index offers better response time and reduced storage requirements compared to the alternative approach of using two dedicated indices, one for each distance measure.

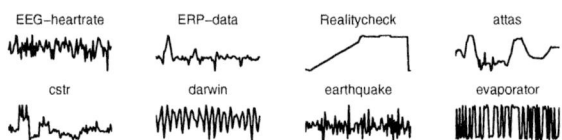

Fig. 6. Sample from the Mixed-Bag dataset

For our experiments we used the same mixture of 40 datasets that was utilized in the coefficient selection section, a sample of which is depicted in figure 6 (*MIXEDBAG* dataset). We used this dataset to create larger datasets with increasing data cardinalities of 4000, 8000, 16000 and 32000 sequences, in order to quantify the index storage requirements and its scalability. All used datasets can be obtained by emailing the first author.

6.1 Matching Results

We depict the meaningfulness of results returned by the MM-tree index, when searching in both euclidean and periodic spaces. Using the *MIXEDBAG* dataset we retrieve the 5-NN of various queries and the results are plotted in Figure 7. It is immediately apparent that the periodic measure always returns sequences with great structural affinity (i.e., belong to the same dataset). The euclidean measure

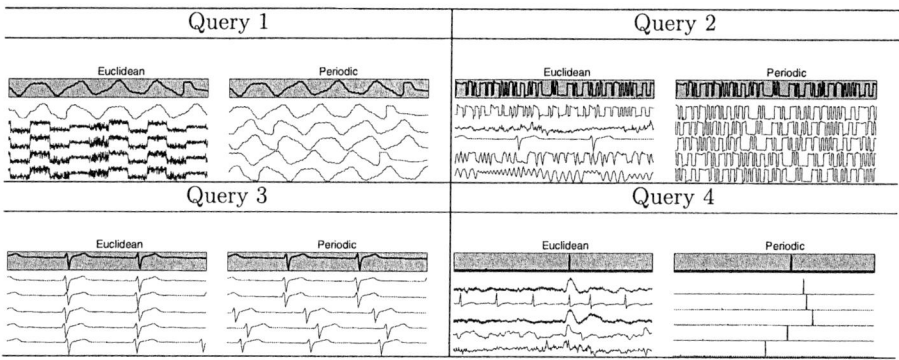

Fig. 7. 5-NN euclidean & periodic matches using the MM-tree (*MIXEDBAG* dataset)

returns meaningful results only when the database contains sequences that are very similar to the query (queries 1 & 3). In such cases, the periodic measure can meaningfully augment the result set of the purely euclidean matches, by retrieving time-shifted variations of the query. In the cases where there are no direct matches to the query (queries 2 & 4), the euclidean measure simply returns spurious matches, while the periodic measure can easily discover instances of the query that belong in the same class of sequence shapes.

6.2 Index Size

The MM-tree presents also the additional advantage of having reduced space requirements, compared to the alternative of maintaining 2 separate indices. Construction of two index structures (one on magnitude and the other on magnitude and phase) results in higher space occupancy, because the magnitude component of each preserved coefficient is stored twice. This is better illustrated in Figure 8, where we plot the total size occupied by the proposed MM-tree, as well as the total disk size occupied by two dedicated metric trees. As expected MM-tree only requires 2/3 of the space of the dual index approach. Moreover, as shown in the next section, the information compaction that takes place during the MM-tree construction, can lead to a significant performance boost of this new hybrid index structure.

6.3 Index Performance

Finally, we evaluate the performance of the multi-query index traversal on the MM-tree, which returns euclidean and periodic matches in a single scan. In Figure 9 we illustrate the performance gain that is realized by this novel tree traversal, for various coefficient cardinalities and kNN index searches, as captured by metrics such as the pruning power (*examined sequences/total sequences*) and the running time. The results compare the MM-tree with the dual index approach (i.e. total cost of executing one euclidean and one periodic query, each

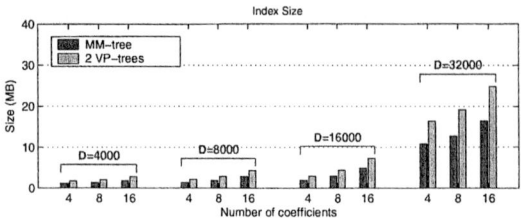

Fig. 8. Index size of MM-tree vs two index structures (euclidean & periodic)

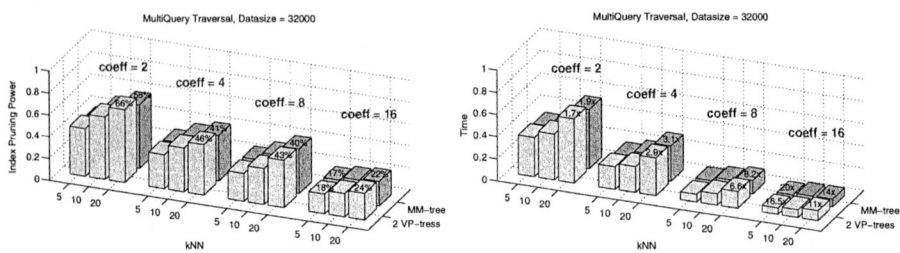

Fig. 9. Performance charts: MM-tree vs 2 VP-trees. Improvement over sequential scan is reported on top of each bar. *(Left)* Pruning power, *(Right)* Running time.

on a dedicated metric VP-tree index). Performance comparisons are conducted with VP-trees, since they have been shown to have superior performance than other metric structures, as well as R-trees [1]. The index performance charts are reported as a fraction of the cost incurred by the sequential scan of the data for the same operation. For sequential scan, the data are traversed just once while maintaining 2 priority queues, each one holding the kNN neighbors of the specific distance function. From the graph it is apparent that the performance of the MM-tree supersedes the dual index execution. The greatest performance margin is observed when retaining the largest number of coefficients per sequence, where the speedup of the MM-tree can be 20 faster than the sequential scan, while the individual metric trees are 16.5 times faster.

This final experiment displays the full potential of the proposed hybrid structure. The MM-tree due to its unique structure outperforms the dedicated metric tree structures when answering both distance queries at the same time, because it can collect the results of both distance measures in a single tree traversal.

7 Conclusion

We have presented a hybrid index structure that can efficiently multiplex queries on euclidean and periodic spaces. The new index allocates disk space more judiciously compared to two dedicated index structures, and its unique structure allows for more effective tree traversal, returning k-NN matches on two distance measures in the single index scan. We hope that our system can provide the necessary building blocks for constructing powerful 'all-in-one' tools, within the scope of applications such as decision support, analysis of causal data relationships and data visualization.

References

1. A. Fu, P. Chan, Y.-L. Cheung, and Y. S. Moon. Dynamic VP-Tree Indexing for N-Nearest Neighbor Search Given Pair-Wise Distances. *The VLDB Journal*, 2000.
2. T. Ide and K. Inoue. Knowledge discovery from heterogeneous dynamic systems using change-point correlations. In *Proc. of SDM*, 2005.
3. E. Keogh and S. Kasetty. On the need for time series data mining benchmarks: A survey and empirical demonstration. In *Proc. of SIGKDD*, 2002.
4. E. Keogh, S. Lonardi, and A. Ratanamahatana. Towards parameter-free data mining. In *Proc. of SIGKDD*, 2004.
5. D. Rafiei and A. Mendelzon. On Similarity-Based Queries for Time Series Data. *In Proc. of FODO*, 1998.
6. M. Vlachos, M. Hadjieleftheriou, D. Gunopulos, and E. Keogh. Indexing Multi-Dimensional Time-Series with Support for Multiple Distance Measures. In *Proc. of SIGKDD*, 2003.
7. M. Vlachos, C. Meek, Z. Vagena, and D. Gunopulos. Identification of Similarities, Periodicities & Bursts for Online Search Queries. In *Proc. of SIGMOD*, 2004.
8. M. Vlachos, P. Yu, and V. Castelli. On periodicity detection and structural periodic similarity. In *SIAM Datamining*, 2005.
9. Y. Xiong and D.-Y. Yeung. Time series clustering with arma mixtures. In *Pattern Recognition 37(8)*, pages 1675–1689, 2004.
10. B.-K. Yi and C. Faloutsos. Fast Time Sequence Indexing for Arbitrary Lp Norms. In *Proceedings of VLDB, Cairo Egypt*, Sept. 2000.

Fast Burst Correlation of Financial Data

Michail Vlachos, Kun-Lung Wu, Shyh-Kwei Chen, and Philip S. Yu

IBM. T.J. Watson Research Center,
19 Skyline Dr, Hawthorne, NY

Abstract. We examine the problem of monitoring and identification of correlated burst patterns in multi-stream time series databases. Our methodology is comprised of two steps: a burst detection part, followed by a burst indexing step. The burst detection scheme imposes a variable threshold on the examined data and takes advantage of the skewed distribution that is typically encountered in many applications. The indexing step utilizes a memory-based interval index for effectively identifying the overlapping burst regions. While the focus of this work is on financial data, the proposed methods and data-structures can find applications for anomaly or novelty detection in telecommunications and network traffic, as well as in medical data. Finally, we manifest the real-time response of our burst indexing technique, and demonstrate the usefulness of the approach for correlating surprising volume trading events at the NY stock exchange.

1 Introduction

"Panta rhei", said Heraklitos; everything is 'in flux'. The truth of this famous aphorism by the ancient Greek philosopher is so much more valid today. People need to make decisions about financial, personal or inter-personal matters based on the observations of various factoring parameters. Therefore, since everything is in constant flow, monitoring the volatility/variability of important measurements over time, becomes a critical determinant in any decision making process.

When dealing with time sequences, or time-series data, one important indicator of change is the presence of 'burstiness', which suggests that more events of importance are happening within the same time frame. Therefore, the identification of bursts can provide useful insights about an imminent change in the monitoring quantity, allowing the system analyst or individual to act upon a timely and informed decision.

Monitoring and modeling of burst behavior is significant in many areas; first and foremost, in *computer networks* it is generally recognized that network traffic can be bursty in various time-scales [9,6]. Detection of bursts is therefore inherently important for identification of network bottlenecks or for *intrusion detection*, since an excessive amount of incoming packets may be a valid indication that a network system is under attack [13]. Additionally, for applications such as *fraud detection* it is very critical to efficiently recognize any anomalous activity (typically in the form of over-utilization of resources). For example, burst

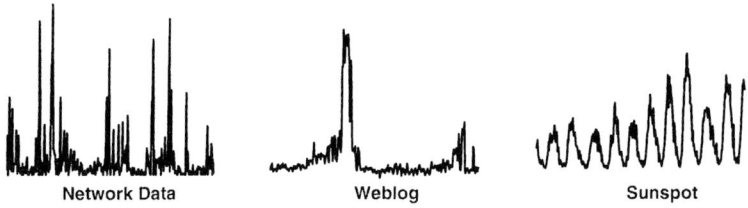

Fig. 1. Burst examples in time-series data

detection techniques can be fruitfully utilized for spotting suspicious activities in large stock trading volumes [10] or for identification of fraudulent phone activity [12]. Finally, in *epidemiology and bio-terrorism*, scientists are interested in the early detection of a disease outbreak. This may be indicated by the discovery of a sudden increase in the number of illnesses or visits to the doctor within a certain geographic area [16,17].

Many recent works address the problem of burst detection [19,7]. However, in many disciplines, more effective knowledge discovery can be achieved by identifying *correlated* bursts when monitoring multiple data sources. From a data-mining perspective, this task is more exciting and challenging, since it involves the identification of burst 'clusters' and it can also aid the discovery of causal chains of burst events, which possibly occur across multiple data streams. Instances of the above problems can be encountered in many financial and stock market applications, e.g., for triggering fraud alarms. Finally, burst correlation can be applicable for the discovery and measurement of gene coexpression (in this field, burst appears under the term 'up-regulation'), which holds substantial biological significance, since it can provide insight into functionally related groups of genes and proteins [5].

Addressing the above issues, this paper presents a complete framework for effective multi-stream burst correlation. Similar to [15], we represent detected bursts as a time interval of their occurrence. We provide a new burst detection scheme, which is tailored for skewed distributions, such as the financial data that we examine here. Additionally, we introduce a memory-based index structure for identification of overlapping bursts. The new index structure is based on the idea of *containment-encoded intervals* (CEI's), which were originally used for performing stabbing queries [18]. Building on the idea of encoded time intervals, we develop a new search algorithm that can efficiently answer overlapping range queries. Moreover, we develop an approach to incrementally maintain the index as more recent data values are added. Using this new index structure we can achieve more than 3 orders of magnitude better search performance for solving the problem of burst overlap computation, compared to the B+tree solution proposed in [15]. Below we summarize the main contributions of this paper:

1. We elaborate on a flexible and robust method of burst extraction on skewed distributions.

2. We present a memory-based index structure that can store the identified burst regions of a sequence and perform very effective overlap estimation of burst regions.
3. Finally, we depict the real-time response of the proposed index and we demonstrate the intuitiveness of the matching results on financial stock data at the NYSE.

2 Problem Formulation

Let us consider a database $\mathcal{D}$, containing m time-series of the form $S = s_1 \ldots s_n$, $s_i \in \mathbb{R}$. Fundamental is also the notion of a *burst interval* $b = [t^{start}, t^{end})$, representing a time-span of a detected burst, with an inclusive left endpoint and an exclusive right endpoint, where t^{start}, t^{end} are integers and $t^{start} < t^{end}$.

Between two burst intervals q, b one can define a time overlap operator $\cap$, such that:

$$q \cap b = \begin{cases} 0 & \text{if } t_q^{end} \leq t_b^{start} \\ 0 & \text{if } t_q^{start} \geq t_b^{e} \\ min(t_q^{end}, t_b^{end}) - max(t_q^{start}, t_b^{start}) & \text{otherwise} \end{cases}$$

We dissect the burst correlation problem into the following steps:

(i) Burst identification on sequences residing in a database $\mathcal{D}$. The burst detection process will return for each sequence S a set of burst intervals $B^S = \{b_1, \ldots, b_k\}$, with a different value of k for every sequence. The set containing all burst intervals of database $\mathcal{D}$, is denoted as $B^{\mathcal{D}}$.

(ii) Organization of $B^{\mathcal{D}}$ in a *CEI-Overlap* index $\mathcal{I}$.

(iii) Discovery of overlapping bursts with a query Q given index $\mathcal{I}$, where Q is also a set of burst intervals: $Q = \{q_1, \ldots q_l\}$. The output of the index will be a set of intervals $V = \{v_1, \ldots, v_r\}, v_j \in B^{\mathcal{D}}$ such that:

$$\sum_i \sum_j q_i \cap v_j \neq 0$$

(iv) Return of top-k matches [*optional*]. This step involves the ranking of the returned sequences based on the degree of overlap, between their respective burst intervals and the query intervals. Since this step is merely a sorting of the result set, we do not elaborate any further on this for the remaining of the paper.

3 Burst Detection

The burst detection process involves the identification of time regions in a sequence, which exhibit over-expression of a certain feature. In our setting, we consider the actual value of a sequence S as an indication of a burst. That is, if $s_i > \tau$, then time i is marked as a burst. The determination of the threshold τ depends on the distributional characteristics of the dataset. Assuming a gaussian data distribution τ could be set as the mean value μ plus 3 times the standard deviation.

In this work we focus on financial data, therefore we first examine the distribution of their values. In Figure 2 we depict the volume distribution of traded shares for two stocks (period 2001-2004). Similar shapes were observed for the majority of stocks. We notice a highly skewed distribution that is also typically encountered in many streaming applications [1]. We capture the shape of this distribution using an exponential model, because of its simplicity and intuitiveness of the produced results. The CDF of the exponential distribution of a random variable $\mathbf{X}$ is given by:

$$P(\mathbf{X} > x) = e^{-\lambda x}$$

where the mean value μ of $\mathbf{X}$ is $\frac{1}{\lambda}$. Solving for x, after elementary calculations we derive at the following:

$$x = -\mu \cdot ln(P) = -\frac{\sum_{i=1}^{n} s_i \cdot \ln(P)}{n}$$

In order to calculate the critical threshold above which all values are considered as bursts, we estimate the value of x by looking at the tail of the distribution, hence setting P to a very small probability, i.e. 10^{-4}. Figure 2 depicts the threshold value and the discovered bursts on two stock volume sequences.

Notice that the computed threshold is amenable to incremental computation in the case of streaming time-series (either for a sliding or aggregate window), because it only involves the maintenance of the running sum of the sequence values.

However, setting a global threshold might introduce a bias when the range of values changes drastically within the examined window, i.e. when there is a 'concept drift' [8,4]. Therefore, one can compute a variable threshold, dividing the examined data into overlapping partitions. The distribution in each partition still remains highly skewed and can be estimated by the exponential distribution, due to the self similar nature of financial data [11,14]. An example of the modified threshold (for the second stock of Fig. 2) is shown in Figure 3, where the length of the partition is 200 and the overlap is 100. At the overlapping part, the threshold is set as the average threshold calculated by the 2 consecutive windows. We observe that in this case we can also detect the smaller burst patterns that were overshadowed by the high threshold value of the whole window (notice that a similar algorithm can be utilized for streaming sequences).

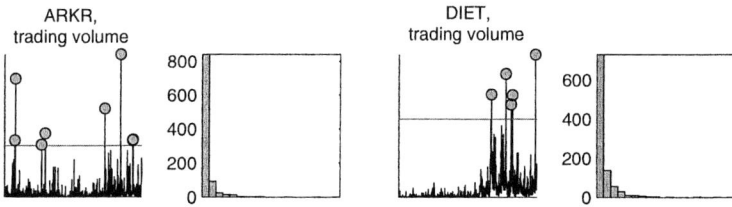

Fig. 2. Examples of the value distributions of stock trading volumes

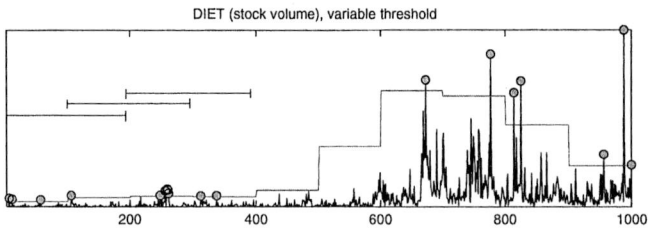

Fig. 3. Variable threshold using overlapping subwindows

After the bursts of a sequence are marked, each identified burst is transcribed into a burst record. Consecutive burst points are compacted into a *burst interval*, represented by its start and end position in time, such as [m,n), $m < n$. Burst points at time m are therefore represented by an interval [m,m+1). In what follows, we will explicate how these burst regions can be organized into an efficient index structure.

4 Index Structure

For the fast identification of overlapping burst intervals[1], we adapt the notion of containment-encoded-intervals (CEI's), which were originally utilized for answering stabbing queries [18] (CEI-Stab). In this work we present the CEI-Overlap index, which shares a similar structure with CEI-Stab. We introduce a new efficient search technique for identifying overlapping bursts regions. Moreover, we present an effective approach for handling the nonstop progress of time.

4.1 Building a CEI-Overlap index

There are two kinds of intervals in CEI-Overlap indexing: (a) burst intervals and (b) virtual construct intervals. Burst intervals are identified as described in section 3. The notion of virtual construct intervals is also introduced for facilitating the decomposition and numbering of burst intervals, in addition to enabling an effective search operation. As noted before, burst intervals are represented by their start and end position in time and the *query search regions* are also expressed similarly.

Fig. 4 shows an example of containment-encoded intervals and their local ID labeling. Assume that the burst intervals to be indexed cover a time-span between [0, r)[2]. First, this range is partitioned into r/L segments of length L, denoted as S_i, where $i = 0, 1, \cdots, (r/L - 1)$, $L = 2^k$, and k is an integer. Note that r is assumed to be a multiple of L. In general, the longer the average length of burst regions is, the larger L should be [18]. Segment S_i contains time interval

---

[1] For the remainder of the paper, "burst regions" and "burst intervals" will be used interchangeably.

[2] Section 4.3 will describe how to handle the issue of choosing an appropriate r as time continues to advance nonstop.

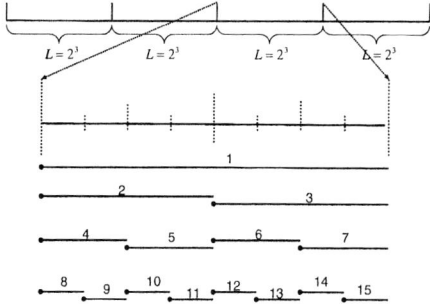

Fig. 4. Example of containment-encoded intervals and their ID labeling

$[iL, (i+1)L)$. Segment boundaries can be treated as *guiding posts*. Then, $2L - 1$ CEI's are defined for each segment as follows: (a) Define one CEI of length L, containing the entire segment; (b) Recursively define 2 CEI's by dividing a CEI into 2 halves until the length is one. For example, there are one CEI of length 8, 2 CEI's of length 4, 4 CEI's of length 2 and 8 CEI's of length one in Fig. 4.

These $2L - 1$ CEI's are defined to have containment relationships among them. Every unit-length CEI is contained by a CEI of size 2, which is in turn contained by a CEI of size 4, ... and so on. The labeling of CEI's is encoded with containment relationships. The ID of a CEI has two parts: the segment ID and the local ID. The local ID assignment follows the labeling of a perfect binary tree. The global unique ID for a CEI in segment S_i, where $i = 0, 1, \cdots, (r/L) - 1$, is simply computed as $l + 2iL$, where l is the local ID. The local ID of the parent of a CEI with local ID l is $\lfloor l/2 \rfloor$, and it can be efficiently computed by a logical right shift by 1 bit.

To insert a burst interval, it is first decomposed into one or more CEI's, then its ID is inserted into the ID lists associated with the decomposed CEI's. The CEI index maintains a set of burst ID lists, one for each CEI. Fig. 5 shows an example of a CEI-Overlap index. It shows the decomposition of four burst intervals: b_1, b_2, b_3 and b_4 within a specific segment containing CEI's of $c_1, \cdots, c_7$. b_1 completely covers the segment, and its ID is inserted into c_1. b_2 lies within the segment and is decomposed into c_5 and c_6, the largest CEI's that can be used for decomposition. b_3 also resides within the segment, but its right endpoint coincides with a guiding post. As a result, we can use c_3, instead of c_6 and c_7 for decomposition. Similarly, c_2 is used to decompose b_4. Burst IDs are inserted into the ID lists associated with the decomposed CEI's.

4.2 Identification of Overlapping Burst Regions

To identify overlapping burst regions, we must first find the overlapping CEI's. One simple approach is to divide the input interval into multiple unit-sized CEI's and perform a point search for each of the unit-sized CEI's using the CEI-Stab search algorithm. However, replicate elimination is required to remove redundant

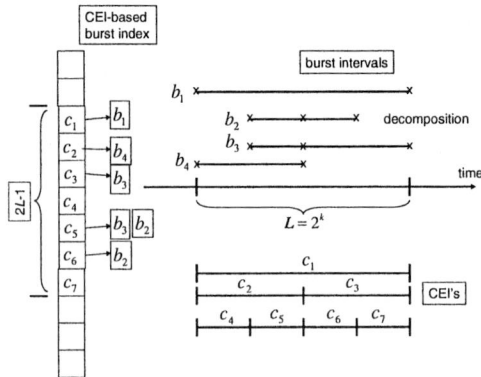

Fig. 5. Example of CEI-Overlap indexing

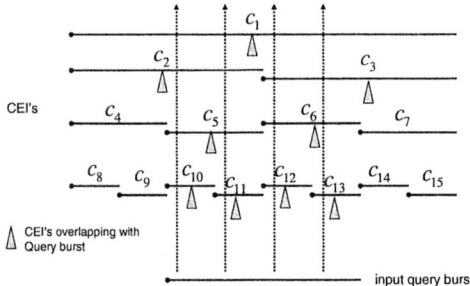

Fig. 6. Example of finding CEI's overlapping with an input interval

overlapping CEI's. Fig. 6 shows an example of identifying CEI's overlapping with an input interval. There are 9 unique overlapping CEI's. Using the point search algorithm of the CEI-Stab index [18], there will be 16 overlapping CEI's, 4 from each upward-pointing dotted arrow. Seven of them are replicates. There are 4 replicates of c_1, and two duplicates each of c_2, c_3, c_5 and c_6, respectively, if we use the point search algorithm of CEI-Stab for searching overlap CEI's.

Eliminating redundant CEI's slows down search time. In this paper, we develop a new search algorithm for CEI-Overlap that does not involve replicate elimination. Fig. 7 shows the pseudo code for systematically identifying all the overlapping bursts for an input region $[x, y)$, where x and y are integers, $x < y$ and $[x, y)$ resides within two consecutive guiding posts (other cases will be discussed later).

First, we compute the segment ID $i = \lfloor x/L \rfloor$. Then, the local IDs of the leftmost unit-sized CEI, $l_1 = x - iL + L$, and the rightmost unit-sized CEI, $l_2 = (y - 1) - iL + L$, that overlap with $[x, y)$ are computed. From l_1 and l_2, we can systematically locate all the CEI's overlapping with the input interval. Any CEI's whose local ID is between l_1 and l_2 also overlaps with the input. We then move up one level to the parents of l_1 and l_2. This process repeats until

```
Search ([x,y)) { // [x,y) resides between two consecutive guiding posts
  i = ⌊x/L⌋; // segment ID
  l₁ = x - iL + L; // leftmost unit-sized CEI overlapping with [x,y)
  l₂ = (y - 1) - iL + L; // rightmost unit-sized CEI overlapping with [x,y)
  for (j = 0; j ≤ k; j = j + 1) {
    for (l = l₁; l ≤ l₂; l = l + 1) {
      c = 2iL + l; // global ID of an overlap CEI
      if (IDList[c] ≠ NULL) { output(IDList[c]); }
      l₁ = l₁/2; // local ID of parent of l₁
      l₂ = l₂/2; // local ID of parent of l₂
    }
  }
}
```

Fig. 7. Pseudo code for searching overlap bursts

$l_1 = l_2 = c_1$. Each overlapping CEI is examined only once. Hence, no duplicate elimination is needed. Fig. 6 shows the identification of overlapping CEI's, from which the overlapping bursts can easily be found via the CEI index.

Now we discuss the cases where the input interval does not reside within two consecutive segment boundaries. Similar to the decomposition process, the input interval can be divided along the segment boundaries. Any remnant can use the search algorithm described in Fig. 7. The full segment, if any, has all the $2L - 1$ CEI's within that segment as the overlapping CEI's.

In contrast to CEI-Stab [18], there might be duplicate burst IDs in the search results of CEI-Overlap. Note that, even though the search algorithm of CEI-Overlap has no duplicate in overlapping CEI's, it might return duplicates in overlapping burst ID's. This is because a burst can be decomposed into one or more CEI's and more than one of them can overlap with an input interval. To efficiently eliminate these duplicates, the burst ID lists are maintained so that the IDs are sorted within individual ID lists. During search, instead of reporting all the burst IDs within each overlapping CEI one CEI at a time, we first locate all the overlapping CEI's. Then, the multiple ID lists associated with these CEI's are merged to report the search result. During the merge process, duplicates can be efficiently eliminated.

4.3 Incrementally Maintaining the Index

Since time continues to advance nonstop, no matter what initial $[0, r)$ is chosen, current time will exceed at some point the maximal range r. Selecting a large r to cover a time-span deep in the future is not a good approach because the index storage cost will increase [18]. A better approach is to choose an r larger than the maximum window of burst regions at the moment, and to keep two indexes in memory, similar to the double-buffering concept. More specifically, we start with $[0, r)$. When time passes r, we create another index for $[r, 2r)$. When time passes $2r$, we create an index for $[2r, 3r)$, but the index for $[0, r)$ will be likely not needed any more and can be discarded or flushed into disk. Using this approach no false dismissals are introduced, since any burst interval covering two regions can be divided along the region boundary and indexed or searched accordingly.

5 Experiments

We evaluate 3 parameters of the burst correlation scheme: (i) the quality of results (is the burst correlation useful?), (ii) the index response time (how fast can we obtain the results?), (iii) indexing scheme comparison (how much better is it than other approaches?).

5.1 Meaningfulness of Results

Our first task is to assess the quality of results obtained through the burst correlation technique. To this end, we search for burst patterns in stock trading volumes during the days before and after the 9/11 attack, with the intention of examining our hypothesis that financial and/or travel related companies might have been affected by the events. We utilize historical stock data obtained from finance.yahoo.com totaling 4793 stocks of length 1000, that cover the period between 2001-2004 (STOCK dataset). We use the trading *volume* of each stock as the input for the burst detection algorithm. Our burst query range is set for the dates 9/7/2001 - 9/20/2001, while we should note that the stock market did not operate for the dates between 9/11 and 9/16. Figures 8, 9, 10 illustrate examples of several affected stocks. The graphs display the volume demand of the respective stocks, while on the top right we also enclose the stock price movement for the whole month of September (the price during the search range is depicted in thicker line style). Stocks like *'Priceline'* or *'Skywest'* which are related to traveling, experience a significant increase in selling demand, which leads to share depreciation when the stock market re-opens on Sep. 17. At the same time, the stock price of *'NICE Systems'* (a provider of air traffic control equipment) depicts a 25% increase in value. More examples of stocks with bursty trends in the stock demand within the requested time frame are presented in Table 1.

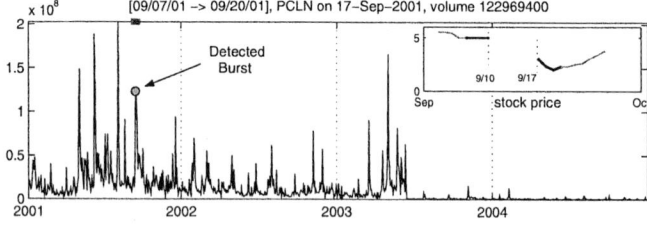

Fig. 8. Volume trading for the `Priceline` stock. We notice a large selling tendency, which results in a drop in the share price.

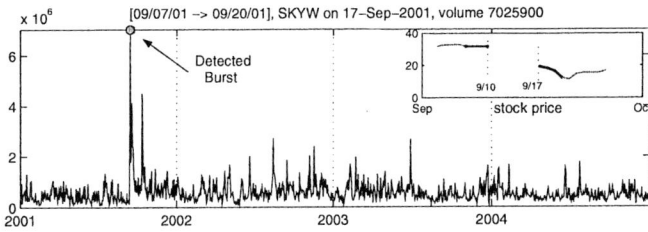

Fig. 9. Volume trading for the Skywest stock

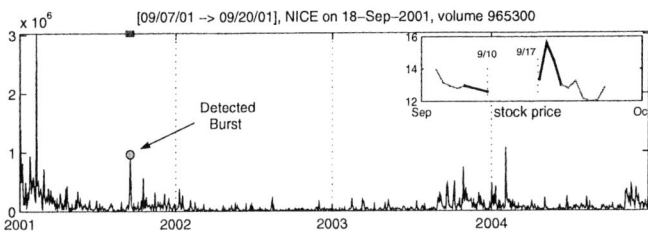

Fig. 10. Volume trading for the stock of Nice Systems (provider of air traffic control systems). In this case, the high stock demand results in an increase of the share price.

Table 1. Some of the stocks that exhibited high trading volume after the events of 9/11/2001

Symbol	Name (Description)	Price
LIFE	Lifeline Systems (Medical Emergency Response)	1.5% ↓
MRCY	Mercury Computer Systems	48% ↑
MAIR	Mair Holdings (Airline Subsidiary)	36% ↓
NICE	NICE Systems (Air traffic Control Systems)	25% ↑
PCLN	Priceline	60% ↓
PRCS	Praecis Pharmaceuticals	60% ↓
SKYW	Skywest Inc	61 % ↓
STNR	Steiner Leisure (Spa & Fitness Services)	51 % ↓

5.2 Index Response Time

We compare the performance of the new CEI-Overlap indexing scheme with the B+tree approach proposed in [15]. Both approaches rely on memory based indexes, so here we report the time required to identify overlapping burst regions for a number of burst query ranges. CEI-based indexing has been shown to outperform other interval indexing schemes for the stabbing case [18], such as the 'Interval Skip Lists' [3] and R-trees [2], therefore due to space limitations we refrain from reporting such comparisons in this version of the paper.

Because for this experiment the STOCK dataset is quite small, we generate a larger artificial dataset that simulates the burst ranges returned by a typical burst detection algorithm. The dataset contains 250,000 burst ranges, at various

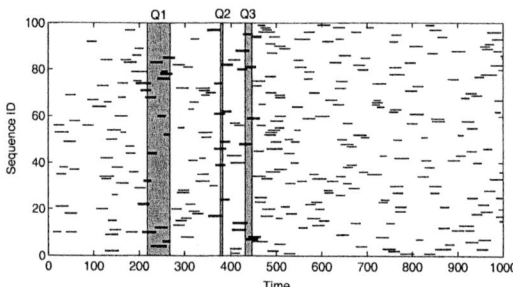

Fig. 11. Artificial dataset and example of 3 burst range queries

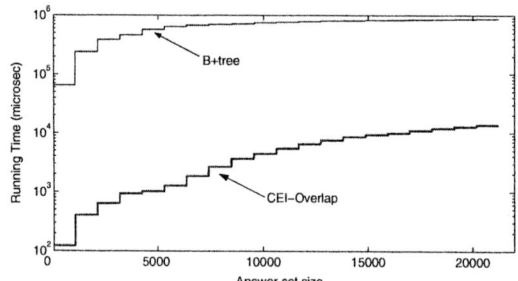

Fig. 12. B+Tree vs CEI-Overlap runtime (log plot)

positions and covering different time-spans. A small sample of this dataset (corresponding to the bursts of 100 'virtual' sequences), along with 3 query ranges, is depicted in Fig. 11. On both the CEI-Overlap and the B+tree we probed 5000 query ranges that cover different positions and ranges.

Intuitively, the cost of the search operation is proportional to the number of burst intervals that overlap with a given query. Therefore, we order the running time of each query based on the size of the answer set (more overlaps suggest longer running time). We create a histogram of the running time by dividing the range of the answer set into 20 bins and in Fig. 12 we plot the average running time of all the results that ended in the same histogram bin. The results indicate the superior performance of the CEI-based index, which is approximately 3 orders of magnitude faster than the competing B+tree approach. We should also notice that the running time is reported in $\mu secs$, which demonstrates the real-time search performance of the proposed indexing scheme.

6 Conclusion

We have presented a complete framework for efficient correlation of bursts. The effectiveness of our scheme is attributed not only to the effective burst detection but also to the efficient memory-based index. The index hierarchically organizes important burst features of time sequences (in the form of 'burst segments') and can subsequently perform very efficient overlap computation of the discovered burst regions. We have demonstrated the enhanced response time of the proposed indexing scheme, and presented interesting burst correlations that we mined from financial data. Encouraged by the excellent responsiveness and scalability of the index, in the immediate future we plan to investigate the applicability of the indexing structure under high data rates and particularly for the online burst detection and correlation of data-streams.

References

1. G. Cormode and S. Muthukrishnan. Summarizing and Mining Skewed Data Streams. In *Proc. of SDM*, pages 44–55, 2005.
2. A. Guttman. R-trees: A dynamic index structure for spatial searching. In *Proc. of ACM SIGMOD*, pages 47–57, 1984.
3. E. Hanson and T. Johnson. Selection predicate indexing for active databases using interval skip lists. *Information Systems*, 21(3):269–298, 1996.
4. M. Harries and K. Horn. Detecting Concept Drift in Financial Time Series Prediction. In *8th Australian Joint Conf. on Artif. Intelligence*, pages 91–98, 1995.
5. L. J. Heyer, S. Kruglyak, and S. Yooseph. Exploring expression data: identification and analysis of coexpressed genes. In *Genome Research, 9:11*, 1999.
6. H.Jiang and C. Dovrolis. Why is the Internet traffic bursty in short (sub-RTT) time scales? In *Proc. of ACM SIGMETRICS*, pages 241–252, 2005.
7. J. Kleinberg. Bursty and Hierarchical Structure in Streams. In *Proc. 8th ACM SIGKDD*, pages 91–101, 2002.
8. M. Lazarescu, S. Venkatesh, and H. H. Bui. Using Multiple Windows to Track Concept Drift. In *Intelligent Data Analysis Journal, Vol 8(1)*, 2004.
9. W. E. Leland, M. S. Taqqu, W. Willinger, and D. V. Wilson. On the Self-Similar Nature of Ethernet Traffic. In *Proc. of ACM SIGCOMM*, pages 183–193, 1993.
10. A. Lerner and D. Shasha. The Virtues and Challenges of Ad Hoc + Streams Querying in Finance. In *IEEE Data Engineering Bulletin*, pages 49–56, 2003.
11. T. Lux. Long-term Stochastic Dependence in Financial Prices: Evidence from the German Stock Market. In *Applied Economics Letters Vol. 3*, pages 701–706, 1996.
12. T. M. Nguyen and A. M. Tjoa. Grid-based Mobile Phone Fraud Detection System. In *Proc. of PAKM*, 2004.
13. Steven L. Scott. A Bayesian Paradigm for Designing Intrusion Detection Systems. In *Computational Statistics and Data Analysis. (special issue on Computer Security) 45*, pages 69–83, 2004.
14. A. Turiel and C. Perez-Vicente. Multifractal geometry in stock market time series. In *Physica A, vol.322*, pages 629–649, 2003.
15. M. Vlachos, C. Meek, Z. Vagena, and D. Gunopulos. Identification of Similarities, Periodicities & Bursts for Online Search Queries. In *Proc. of SIGMOD*, 2004.
16. M.-A. Widdowson, A. Bosman, E. van Straten, M. Tinga, S. Chaves, L. van Eerden, and W. van Pelt. Automated, laboratory-based system using the Internet for disease outbreak detection, the Netherlands. In *Emerg Infect Dis 9*, 2003.
17. W.-K. Wong, A. Moore, G. Cooper, and M. Wagner. WSARE: What's Strange About Recent Events? In *Journal of Urban Health 80*, pages 66–75, 2003.
18. K.-L. Wu, S.-K. Chen, and P. S. Yu. Interval query indexing for efficient stream processing. In *Proc. of ACM CIKM*, pages 88–97, 2004.
19. Y. Zhu and D. Shasha. Efficient elastic burst detection in data streams. In *Proc. of SIGKDD*, pages 336–345, 2003.

A Propositional Approach to Textual Case Indexing

Nirmalie Wiratunga[1], Rob Lothian[1], Sutanu Chakraborti[1], and Ivan Koychev[2]

[1] School of Computing,
The Robert Gordon University,
Aberdeen AB25 1HG, Scotland, UK
{nw, rml, sc}@comp.rgu.ac.uk
[2] Institute of Mathematics and Informatics,
Bulgarian Academy of Science,
Sofia - 1113, Bulgaria
ikoychev@math.bas.bg

Abstract. Problem solving with experiences that are recorded in text form requires a mapping from text to structured cases, so that case comparison can provide informed feedback for reasoning. One of the challenges is to acquire an indexing vocabulary to describe cases. We explore the use of machine learning and statistical techniques to automate aspects of this acquisition task. A propositional semantic indexing tool, PSI, which forms its indexing vocabulary from new features extracted as logical combinations of existing keywords, is presented. We propose that such logical combinations correspond more closely to natural concepts and are more transparent than linear combinations. Experiments show PSI-derived case representations to have superior retrieval performance to the original keyword-based representations. PSI also has comparable performance to Latent Semantic Indexing, a popular dimensionality reduction technique for text, which unlike PSI generates linear combinations of the original features.

1 Introduction

Discovery of new features is an important pre-processing step for textual data. This process is commonly referred to as feature extraction, to distinguish it from feature selection, where no new features are created [18]. Feature selection and feature extraction share the aim of forming better dimensions to represent the data. Historically, there has been more research work carried out in feature selection [9,20,16] than in extraction for text pre-processing applied to text retrieval and text classification tasks. However, combinations of features are better able to tackle the ambiguities in text (e.g. synonyms and polysemys) that often plague feature selection approaches. Typically, feature extraction approaches generate linear combinations of the original features. The strong focus on classification effectiveness alone has increasingly justified these approaches, even though their black-box nature is not ideal for user interaction. This argument applies even more strongly to combinations of features using algebraic or higher mathematical functions. When feature extraction is applied to tasks such as help desk systems, medical or law document management, email management or even Spam filtering, there is often a need for user interaction to guide retrieval or to support incremental query elaboration. The primary communication mode between system and user has the extracted features as vocabulary. Hence, these features should be transparent as well as providing good dimensions for classification.

The need for features that aid user interaction is particularly strong in the field of Case-Based Reasoning (CBR), where transparency is an important element during retrieval and reuse of solutions to similar, previously solved problems. This view is enforced by research presented at a mixed initiative CBR workshop [2]. The indexing vocabulary of a CBR system refers to the set of features that are used to describe past experiences to be represented as cases in the case base. Vocabulary acquisition is generally a demanding knowledge engineering task, even more so when experiences are captured in text form. Analysis of text typically begins by identifying keywords with which an indexing vocabulary is formed at the keyword level [14]. It is here that there is an obvious opportunity to apply feature extraction for index vocabulary acquisition with a view to learning transparent and effective textual case representations.

The focus of this paper is extraction of features to automate acquisition of index vocabulary for knowledge reuse. Techniques presented in this paper are suited for applications where past experiences are captured in free text form and are pre-classified according to the types of problems they solve. We present a Propositional Semantic Indexing (PSI) tool, which extracts interpretable features that are logical combinations of keywords. We propose that such logical combinations correspond more closely to natural concepts and are more transparent than linear combinations. PSI employs boosting combined with rule mining to encourage learning of non-overlapping (or orthogonal) sets of propositional clauses. A similarity metric is introduced so that textual cases can be compared based on similarity between extracted logical clauses. Interpretability of these logical constructs creates new avenues for user interaction and naturally leads to the discovery of knowledge. PSI's feature extraction approach is compared with the popular dimensionality reduction technique Latent Semantic Indexing (LSI), which uses singular value decomposition to extract orthogonal features that are linear combinations of keywords [7]. Case representations that employ PSI's logical expressions are more comprehensible to domain experts and end-users compared to LSI's linear keyword combinations. Ideally we wish to achieve this expressiveness without significant loss in retrieval effectiveness.

We first establish our terminology for feature selection and extraction, before describing how PSI extracts features as logical combinations. We then describe LSI, highlighting the problem of interpretability with linear combinations. Finally we show that PSI's approach achieves comparable retrieval performance yet remains expressive.

2 Feature Selection and Extraction

Consider the hypothetical example in Figure 1 where the task is to weed out Spam from legitimate email related to AI. To assist with future message filtering these messages must be mapped onto a set of cases before they can be reused. We will refer to the set of all labelled documents (cases) as $\mathcal{D}$. The keyword-vector representation is commonly used to represent a document d by considering the presence or absence of words [17]. Essentially the set of features are the set of words $\mathcal{W}$ (e.g. "conference", "intelligent"). Accordingly a document d is represented as a pair $(\mathbf{x}, y)$, where $\mathbf{x} = (x_1, \ldots, x_{|\mathcal{W}|})$ is

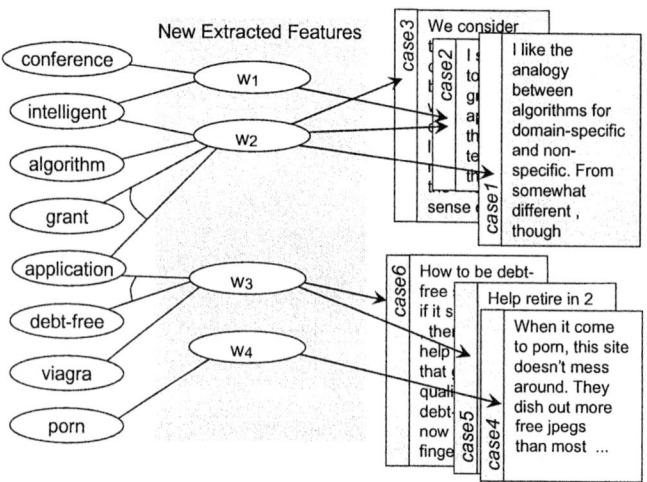

Fig. 1. Features as logical keyword combinations

a binary valued feature vector corresponding to the presence or absence of words in $\mathcal{W}$; and y is d's class label.

Feature selection reduces $|\mathcal{W}|$ to a smaller feature subset size m [20]. Information Gain (IG) is often used for this purpose, where m features with highest IG are retained and the new binary-valued feature vector $\mathbf{x}'$ is formed with the reduced word vocabulary set $\mathcal{W}'$, where $\mathcal{W}' \subset \mathcal{W}$ and $|\mathcal{W}'| \ll |\mathcal{W}|$. The new representation of document d with $\mathcal{W}'$ is a pair $(\mathbf{x}', y)$. Selection using IG is the base-line algorithm in this paper and is referred to as BASE. An obvious shortcoming of BASE is that it fails to ensure selection of non-redundant keywords. Ideally we want $\mathbf{x}'$ to contain features that are representative but also orthogonal. A more serious weakness is that BASE's one-to-one feature-word correspondence operates at a lexical level, ignoring underlying semantics.

Figure 1 illustrates a proof tree showing how new features can be extracted to capture keyword relationships using propositional disjunctive normal form clauses (DNF clauses). When keyword relationships are modelled, ambiguities in text can be resolved to some extent. For instance "grant" and "application" capture semantics akin to legitimate messages, while the same keyword "application" in conjunction with "debt-free" suggests Spam messages.

Feature extraction, like selection, also reduces $|\mathcal{W}|$ to a smaller feature subset size m. However unlike selected features, extracted features no longer correspond to presence or absence of single words. Therefore, with extracted features the new representation of document d is $(\mathbf{x}'', y)$, but $\mathcal{W}'' \not\subset \mathcal{W}$. When extracted features are logical combinations of keywords as in Figure 1, then a new feature $w'' \in \mathcal{W}''$, represents a propositional clause. For example the new feature w_2'' represents the clause: "intelligent" $\vee$ "algorithm" $\vee$ ("grant" $\wedge$ "application").

3 Propositional Semantic Indexing (PSI)

PSI discovers and captures underlying semantics in the form of propositional clauses. PSI's approach is two-fold. Firstly, decision stumps are selected by IG and refined by association rule mining, which discovers sets of Horn clause rules. Secondly, a boosting process encourages selection of non-redundant stumps. The PSI feature extraction algorithm and the instantiation of extracted features appear at the end of this section after a description of the main steps.

3.1 Decision Stump Guided Extraction

A decision stump is a one-level decision tree [12]. In PSI, a stump is initially formed using a single keyword, which is selected to maximise IG. An example decision stump formed with "conference" in its decision node appears in Figure 2. It partitions documents into leaves, based on whether or not "conference" appears in them. For instance 70 documents contain the word "conference" and just 5 of these are Spam (i.e. +5). It is not uncommon for documents containing "conference" to still be semantically similar to those not containing it. So documents containing "workshop" without "conference" in the right leaf are still contextually similar to those containing "conference" in the left leaf. A generalised decision node has the desired effect of bringing such semantically related documents closer [19]. Generalisation refines the decision node formed with a single feature w', to an extracted feature w'', containing a propositional clause. Typically a refined node results in an improved split (see right stump in Figure 2).

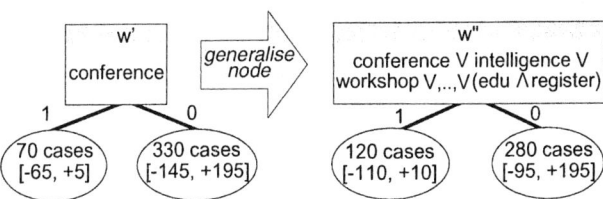

Fig. 2. Node Generalisation

A propositional clause is formed by adding disjuncts to an initial clause containing just the selected feature w'. Each disjunct is a conjunction of one or more keyword co-occurrences with similar contextual meaning to that of w'. An exhaustive search for disjuncts will invariably be impractical. Fortunately the search space can be pruned by using w' as a handle over this space. Instead of generating and evaluating all disjuncts, we generate propositional Horn clause rules that conclude w' given other logical keyword combinations.

3.2 Growing Clauses from Rules

Examples of five association rules concluding in "conference" (i.e. w') appear in Figure 3. These rules are of the form H ← B, where the rule body B is a conjunction of

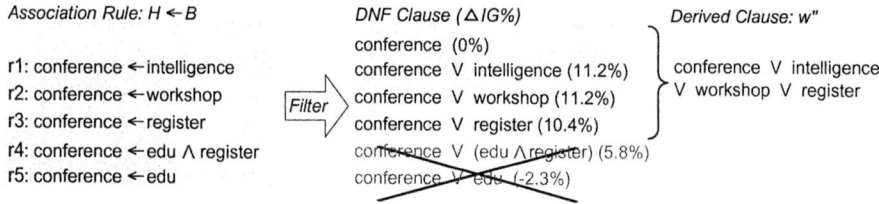

Fig. 3. Growing clauses from selected rules

keywords, and the rule head H is a single keyword. These conjunctions are keyword combinations that have been found to co-occur with the head keyword. Rule bodies are a good source of disjuncts with which to grow our DNF clause w'', which initially contains only the selected keyword "conference". However, an informed selection strategy is necessary to identify those disjuncts that are good descriptors of underlying semantics.

The contribution of each disjunct to clause growth is measured by comparing IG of w'' with and without the disjunct (rule body) included in the DNF clause. Disjuncts that fail to improve IG are filtered out by the *gain-filter*. Those remaining are passed onto *gen-filter* where any specialised forms of a disjunct with a lower IG compared to any one of its generalised forms are filtered out. The DNF clauses in Figure 3 show how each rule is converted into a potential DNF clause (difference in IG, used for filtering appear in brackets). The final DNF clause derived once the filtering step is completed is: "conference" ∨ "intelligence" ∨ "workshop" ∨ "register". We use the Apriori [1] association rule learner to generate feature extraction rules that conclude a selected w'. Apriori typically generates many rules, but the filters are able to identify useful rules.

3.3 Feature Extraction with Boosting

PSI's iterative approach to feature extraction employs boosted decision stumps (see Figure 4). The number of features to be extracted is determined by *vocabulary_size*. The general idea of boosting is to iteratively generate several (weak) learners, with each learner biased by the training error in the previous iteration [10]. This bias is expressed by modifying weights associated with documents. When boosted stumps are used for feature selection the new document distribution discourages selection of a redundant feature given the previously selected feature [6]. Here, with extracted features, unlike with single keyword-based features, we need to discourage discovery of an overlapping clause given the previously discovered clause. We achieve this by updating document weights in PSI according to the error of the decision stump created with the new extracted feature, w'', instead of w'.

3.4 Feature Instantiation

Once PSI has extracted new features, textual cases are mapped to a new representation. For a new feature w_i'', let $S_i = \bigvee_j s_{ij}$, be its propositional clause, where $s_{ij} = \bigwedge_k x_{ijk}$

```
W" = ∅; n = |D|; vocabulary_size = m
Algorithm: PSI
    Repeat
        initialise document weights to 1 / n
        w_j = feature with highest IG
        W = W \ w_j
        w''_j = GROWCLAUSE(w_j, W)
        W" = W" ∪ w''_j
        stump = CREATESTUMP(w''_j)
        err = error(stump)
        update document weights using err
    Until ( |W"| = vocabulary_size)
    Return W"
```

Fig. 4. Feature Extraction with PSI

is the jth conjunction in this clause. The new representation of document $d = (\mathbf{x''}, y)$ is obtained by:

$$x''_i = \sum_j \text{gain\_inc}(s_{ij}) * \text{infer}(s_{ij})$$

here gain_inc returns the increase in gain achieved by s_{ij} when growing $\mathcal{S}_i$. Whether or not s_{ij} can be inferred (satisfied) from a document's initial representation $d = (\mathbf{x}, y)$ (i.e. using all features in $\mathcal{W}$) is determined by:

$$\text{infer}(s_{ij}) = \begin{cases} 1 & \text{if } (\bigwedge_k x_{ijk}) = \text{True} \\ 0 & \text{otherwise} \end{cases}$$

The PSI-derived representation enables case comparison at a semantic (or conceptual) level, because instantiated features now capture the degree to which each clause is satisfied by documents. In other words, satisfaction of the same disjunct will contribute more towards similarity than satisfaction of different disjuncts in the same clause.

4 Latent Semantic Indexing (LSI)

LSI is an established method of feature extraction and dimension reduction. The matrix whose columns are the document vectors $\mathbf{x}_1, \ldots, \mathbf{x}_{|D|}$, known as the term-document matrix, constitutes a vector space representation of the document collection. In LSI, the term-document matrix is subjected to singular value decomposition (SVD). The SVD extracts an orthogonal basis for this space, consisting of new features that are linear combinations of the original features (keywords). Crucially, these new features are ranked according to their importance. It is assumed that the m highest-ranked features contain the true semantic structure of the document collection and the remaining features, which are considered to be noise, are discarded. Any value of m less than the

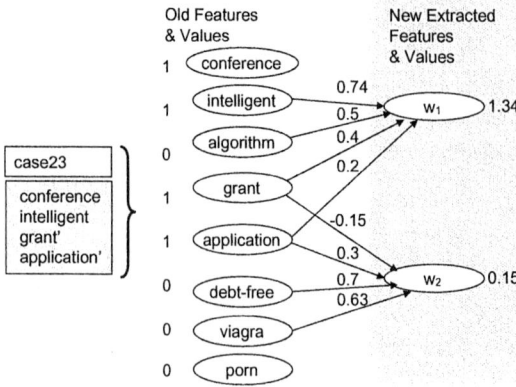

Fig. 5. Feature extraction using LSI

rank of the term-document matrix can be used, but good values will be much smaller than this rank and, hence, much smaller than either $|W|$ or $|D|$. These features form a new lower-dimensional representation which has frequently been found to improve performance in information retrieval and classification tasks [18,22]. Full technical details can be found in the original paper [7].

Figure 5 shows a hypothetical example from the AI-Spam domain (cf. Figure 1). The first extracted feature is a combination of "intelligent", "algorithm", "grant" and "application". Any document containing most of these is likely to be legitimate, so high values of this feature indicate non-Spam. The second feature has positive weights for "application", "debt-free" and "viagra" and a negative weight for "grant". A high value for this feature is likely to indicate Spam. The new features are orthogonal, despite having two keywords in common. The first feature has positive weights for both "grant" and "application", whereas the second has a negative weight for "grant". This shows how the modifying effect of "grant" on "application" might manifest itself in a LSI-derived representation. With a high score for the first extracted feature and a low score for the second, the incoming test case is likely to be classified as legitimate email.

LSI extracted features are linear combinations of typically very large numbers of keywords. In practice this can be in the order of hundreds/thousands of keywords, unlike in our illustrative example involving just 8 keywords. Consequently, it is difficult to interpret these features in a meaningful way. In contrast, a feature extracted by PSI combines far fewer keywords and its logical description of underlying semantics is easier to interpret. A further difference is that, although both PSI and LSI exploit word-word co-occurrences to discover and preserve underlying semantics, PSI also draws on word-class co-occurrences while LSI does not naturally exploit this information.

5 Evaluation

The goodness of case representations derived by BASE, LSI and PSI in terms of retrieval performance is compared on a retrieve-only CBR system, where the weighted majority vote from the 3 best matching cases are re-used to classify the test case. A modified case similarity metric is used so that similarity due to absence of words (or words in linear combinations or in clauses) is treated as less important compared to their presence [19].

Experiments were conducted on 6 datasets; 4 involving email routing tasks and 2 involving Spam filtering. Various groups from the 20Newsgroups corpus of 20 Usenet groups [13], with 1000 postings (of discussions, queries, comments etc.) per group, form the routing datasets: SCIENCE (4 science related groups); REC (4 recreation related groups); HW (2 hardware problem discussion groups, one on Mac, the other on PC); and RELPOL (2 groups, one concerning religion, the other politics in the middle-east). Of the 2 Spam filtering datasets: USREMAIL [8] contains 1000 personal emails of which 50% are Spam; and LINGSPAM [16] contains 2893 messages from a linguistics mailing list of which 27% are Spam.

Equal-sized disjoint train-test splits were formed. Each split contains 20% of the dataset and also preserves the class distribution of the original corpus. All text was pre-processed by removing stop words (common words) and punctuation. Remaining words were stemmed to form W, where $|W|$ varies from approximately 1,000 in USREMAIL to 20,000 in LINGSPAM. Generally, with both routing and filtering tasks, the overall aim is to assign incoming messages into appropriate groups. Hence, test set accuracy was chosen as the primary measure of the effectiveness of the case representation as a facilitator of case comparison. For each test corpus and each method, the accuracy (averaged over 15 trials) was computed for representations with 20, 40, 60, 80, 100 and 120 features.

Paired t-tests were used to find improvements by LSI and PSI compared to BASE (one-tailed test) and differences between LSI and PSI (two-tailed test), both at the 95% significance level. Precision[1] is an important measure when comparing Spam filters, because it penalises error due to false positives (Legitimate → Spam). Hence, for the Spam filtering datasets, precision was tested as well as accuracy.

5.1 Results

Accuracy results in Figure 6 shows that BASE performs poorly with only 20 features, but gets closer to the superior PSI when more features are added. PSI's performance is normally good with 20 features and is robust to the feature subset size compared to both BASE and LSI. LSI clearly performs better for smaller sizes. This motivated an investigation of LSI with fewer than 20 features. We found that 10-feature LSI consistently outperforms 20-feature LSI and is close to optimal. Consequently, 10-feature LSI was used for the significance testing, in order to give a more realistic comparison with the other methods.

Table 5.1 compares performance of BASE and PSI (both 20 features) and LSI (10 features). Where LSI or PSI are significantly better than BASE, the results are in bold. Where LSI and PSI are significantly different, the better result is starred. It can be seen that LSI is significantly better than BASE on 6 of 8 measures and to PSI on 3. PSI is better than BASE on 7 measures and better than LSI on 2. We conclude that the 20-dimensional representations extracted by PSI have comparable effectiveness to the 10-dimensional representations extracted by LSI. Generally, BASE needs a much larger

---

[1] Precision = TP/(TP+FP) where TP is no. of true positives and FP is no. of false positives.

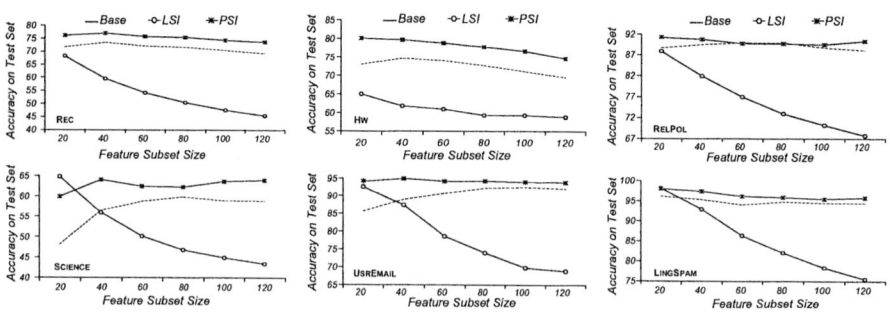

Fig. 6. Accuracy results for datasets

Table 1. Summary of significance testing for feature subset size 20 (10 for LSI)

Algo.	Routing: Accuracy				Filtering: Accuracy (Precision)	
	REC	HW	RELPOL	SCIENCE	USREMAIL	LINGSPAM
BASE	71.7	73.0	88.7	48.1	85.7 (89.5)	94.2 (92.0)
LSI	*78.7	65.5	90.4	*71.8	93.9 (*96.8)	96.8 (89.0)
PSI	76.2	*80.1	91.2	59.9	94.1 (95.2)	95.8 (*92.1)

indexing vocabulary to achieve comparable performance. PSI works well with a small vocabulary of features, which are more expressive than LSI's linear combinations.

5.2 Interpretability

Figure 7 provides a high-level view of sample features extracted by PSI in the form of logical combinations (for 3 of the datasets). It is interesting to compare the differences in extracted combinations (edges), the contribution of keywords (ovals) to different extracted features (boxes) and the number of keywords used to form conjunctions (usually not more than 3). We see a mass of interconnected nodes with the HW dataset on which PSI's performance was far superior to that of LSI. Closer examination of this data set shows that there are many keywords that are polysemous given the two classes. For instance "drive" is applicable both to Macs and PCs but combined with "vlb" indicates PC while with "syquest" indicates Mac. Unlike HW, the multi-class SCIENCE dataset contains several disjoint graphs each relating to a class, suggesting that these concepts are easily separable. Accuracy results show LSI to be a clear winner on SCIENCE. This further supports our observation that LSI operates best only in the absence of class-specific polysemous relationships. Finally, features extracted from LINGSPAM in figure 7 show that the majority of new features are single keywords rather than logical combinations. This explains BASE's good performance on LINGSPAM.

An obvious advantage of interpretability is knowledge discovery. Consider the SCIENCE tree, here, without the extracted clauses indicating that "msg", "food" and "chinese" are linked through "diet", one would not understand the meaning in context of a

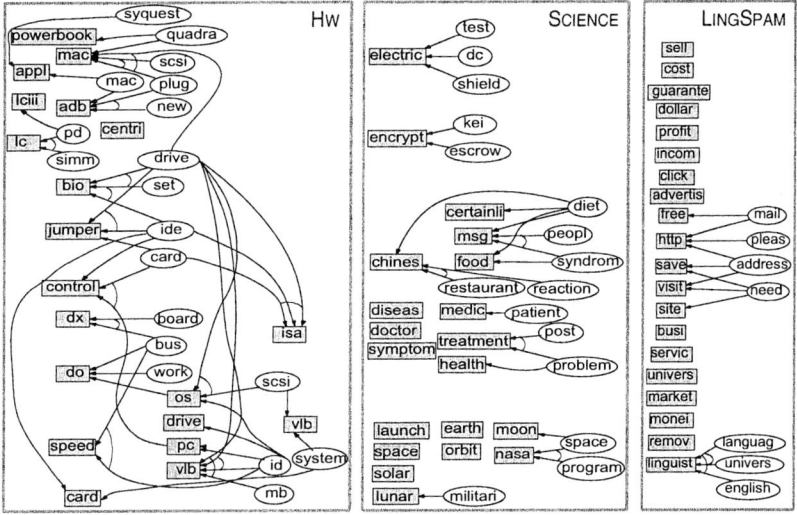

Fig. 7. Logical combinations extracted from datasets

term such as "msg". Such proof trees, which are automatically generated by PSI, highlight relations between keywords; this knowledge can further aid with glossary generation, often a demanding manual task (e.g. FALLQ [14]). Case authoring is a further task that can benefit from PSI generated trees. For example disjoint graphs involving "electric" and "encrypt" with many fewer keyword associations may suggest the need for case creation or discovery in that area. From a retrieval standpoint, PSI generated features can be exploited to facilitate query elaboration, within incremental retrieval systems. The main benefit to the user would be the ability to tailor the expanded query by deactivating disjuncts to suit retrieval needs. Since clauses are granular and can be broken down into semantically rich constituents, retrieval systems can gather statistics of which clauses worked well in the past, based on user interaction; this is difficult over linear combinations of features mined by LSI.

6 Related Work

Feature extraction is an important area of research particularly when dealing with textual data. In Textual-CBR (TCBR) research the SMILE system provides useful insight into how machine learning and statistical techniques can be employed to reason with legal documents [3]. As in PSI, single keywords in decision nodes are augmented with other keywords of similar context. Unlike PSI, these keywords are obtained by looking up a manually created domain-specific thesaurus. Although PSI grows clauses by analysing keyword co-occurrence patterns, they can just as easily be grown using existing domain-specific knowledge. A more recent interest in TCBR involves extraction of features in the form of predicates. The FACIT framework involving semi-automated index vocabulary acquisition addresses this challenge but also highlights the need for reliance on deep syntactic parsing and the acquisition of a generative lexicon which warrants significant manual intervention [11].

In text classification and text mining research, there is much evidence to show that analysis of keyword relationships and modelling them as rules is a successful strategy for text retrieval. A good example is RIPPER [5], which adopts complex optimisation heuristics to learn propositional clauses for classification. A RIPPER rule is a Horn clause rule that concludes a class. In contrast, PSI's propositional clauses form features that can easily be exploited by CBR systems to enable case comparison at a semantic level. Such comparisons can also be facilitated with the FEATUREMINE [21] algorithm, which also employs association rule mining to create new features based on keyword co-occurrences. FEATUREMINE generates all possible pair-wise keyword co-occurrences converting only those that pass a significance test into new features. What is unique about PSI's approach is that firstly, search for associations is guided by an initial feature selection step, secondly, associations remaining after an informed filtering step are used to grow clauses, and, crucially, boosting is employed to encourage growing of non-overlapping clauses. The main advantage of PSI's approach is that instead of textual case similarity based solely on instantiated feature value comparisons (as in FEATUREMINE), PSI's clauses enable more fine-grained similarity comparisons. Like PSI, WHIRL [4] also integrates rules resulting in a more fine-grained similarity computation over text. However these rules are manually acquired.

The use of automated rule learning in an Information Extraction (IE) setting is demonstrated by TEXTRISE, where mined rules predict text for slots based on information extracted over other slots [15]. The vocabulary is thus limited to template slot fillers. In contrast PSI does not assume knowledge of case structures and is potentially more useful in unconstrained domains.

7 Conclusion

A novel contribution of this paper is the acquisition of an indexing vocabulary in the form of expressive clauses, and a case representation that captures the degree to which each clause is satisfied by documents. The propositional semantic indexing tool, PSI, introduced in the paper, enables text comparison at a semantic, instead of a lexical, level. Experiments show that PSI's retrieval performance is significantly better than that of retrieval over keyword-based representations. Comparison of PSI-derived representations with the popular LSI-derived representations generally shows comparable retrieval performance. However in the presence of class-specific polysemous relationships PSI is the clear winner. These results are very encouraging, because, although features extracted by LSI are rich mathematical descriptors of the underlying semantics in the domain, unlike PSI, they lack interpretability. We note that PSI's reliance on class knowledge inevitably restricts its range of applicability. Accordingly, future research will seek to develop an unsupervised version of PSI.

Acknowledgements

We thank Susan Craw for helpful discussions on this work.

References

1. R. Agrawal, H. Mannila, R. Srikant, H. Toivonen, and A. I. Verkamo. Fast discovery of association rules. In *Advances in KD and DM*, pages 307–327, 1995. AAAI/MIT.
2. D. Aha, editor. *Mixed-Initiatives Workshop at 6th ECCBR*, 2002. Springer.
3. S. Bruninghaus and K. D. Ashley. Bootstrapping case base development with annotated case summaries. In *Proc of the 2nd ICCBR*, pages 59–73, 1999. Springer.
4. W. W. Cohen. Providing database-like access to the web using queries based on textual similarity. In *Proc of the Int Conf on Management of Data*, pages 558–560, 1998.
5. W. W. Cohen and Y. Singer. Context-sensitive learning methods for text categorisation. *ACM Transactions in Information Systems*, 17(2):141–173, 1999.
6. S. Das. Filters, wrappers and a boosting based hybrid for feature selection. In *Proc of the 18th ICML*, pages 74–81, 2001. Morgan Kaufmann.
7. S. C. Deerwester, S. T. Dumais, T. K. Landauer, G. W. Furnas, and R. A. Harshman. Indexing by latent semantic analysis. *Journal of the American Society of Information Science*, 41(6):391–407, 1990.
8. S. J. Delany and P. Cunningham. An analysis of case-base editing in a spam filtering system. In *Proc of the 7th ECCBR*, pages 128–141, 2004. Springer.
9. G. Forman and I. Cohen. Learning with Little: Comparison of Classifiers Given Little Training. In *Proc of the 8th European Conf on PKDD*. pages 161–172, 2004.
10. Y. Freund and R. Schapire. Experiments with a new boosting algorithm. In *Proc of the 13th ICML*, pages 148–156, 1996.
11. K. M. Gupta and D. W. Aha. Towards acquiring case indexing taxonomies from text. In *Proc of the 17th Int FLAIRS Conference*, pages 307–315, 2004. AAAI press.
12. W. Iba and P. Langley. Induction of one-level decision trees. In *Proc of the 9th Int Workshop on Machine Learning*, pages 233–240, 1992.
13. T. Joachims. A probabilistic analysis of the Rocchio algorithm with TFIDF. Technical report, Carnegie Mellon University CMU-CS-96-118, 1996.
14. M. Lenz. Defining knowledge layers for textual CBR. In *Proc of the 4th European Workshop on CBR*, pages 298–309, 1998. Springer.
15. U. Y. Nahm and R. J. Mooney. Mining soft-matching rules from textual data. In *Proc of the 17th IJCAI*, pages 979–984, 2001.
16. G. Sakkis, I. Androutsopoulos, G. Paliouras, V. Karkaletsis, C. Spyropoulos, and P. Stamatopoulos. A memory-based approach to anti-spam filtering for mailing lists. *Information Retrieval*, 6:49–73, 2003.
17. G. Salton and M. J. McGill. *An introduction to modern IR*. 1983, McGraw-Hill.
18. F. Sebastiani. ML in automated text categorisation. *ACM Computing surveys*, 34:1–47, 2002.
19. N. Wiratunga, I. Koychev, and S. Massie. Feature selection and generalisation for textual retrieval. In *Proc of the 7th ECCBR*, pages 806–820, 2004. Springer.
20. Y. Yang and J. O. Pedersen. A comparative study on feature selection in text categorisation. In *Proc of the 14th ICML*, pages 412–420, 1997. Springer.
21. S. Zelikovitz. Mining for features to improve classification. In *Proc of Machine Learning, Models, Technologies and Applications*, 2003.
22. S. Zelikovitz and H. Hirsh. Using LSI for text classification in the presence of background text. In *Proc of the 10th Int Conf on Information and KM*, 2001.

A Quantitative Comparison of the Subgraph Miners MoFa, gSpan, FFSM, and Gaston

Marc Wörlein, Thorsten Meinl, Ingrid Fischer, and Michael Philippsen

University of Erlangen-Nuremberg, Computer Science Department 2,
Martensstr. 3, 91058 Erlangen, Germany
simawoer@stud.informatik.uni-erlangen.de
{meinl, idfische, philippsen}@cs.fau.de

Abstract. Several new miners for frequent subgraphs have been published recently. Whereas new approaches are presented in detail, the quantitative evaluations are often of limited value: only the performance on a small set of graph databases is discussed and the new algorithm is often only compared to a single competitor based on an executable. It remains unclear, how the algorithms work on bigger/other graph databases and which of their distinctive features is best suited for which database. We have re-implemented the subgraph miners MoFa, gSpan, FFSM, and Gaston within a common code base and with the same level of programming expertise and optimization effort. This paper presents the results of a comparative benchmarking that ran the algorithms on a comprehensive set of graph databases.

1 Introduction

Mining of frequent subgraphs in graph databases is an important challenge, especially in its most important application area "cheminformatics" where frequent molecular fragments help finding new drugs. Subgraph mining is more challenging than frequent itemset mining, since instead of bit vectors (i.e., frequent itemsets) arbitrary graph structures must be generated and matched. Since graph isomorphism testing is a hard problem [3], fragment miners are exponential in runtime and/or memory consumption. For a general overview see [1].

The naive fragment miner starts from the empty graph and recursively generates all possible refinements/fragment extensions by adding edges and nodes to already generated fragments. For each new possible fragment, it then performs a subgraph isomorphism test conceptually on each of the graphs in the graph database to determine if that fragment appears frequently (i.e., if it has enough *support*). Since a new refinement can only appear in those graphs that already hold the original fragment, the miner keeps appearance lists to restrict isomorphism testing to the graphs in these lists.

All possible graph fragments of a graph database form a lattice, see Fig. 1 for an example with just one graph. The empty graph ∗ is given at the top, the final graph at the bottom of the picture. During the search this lattice will be pruned at infrequent fragments since their refinements will appear even more rarely.[1] Efficient fragment miners have to solve three main subproblems.

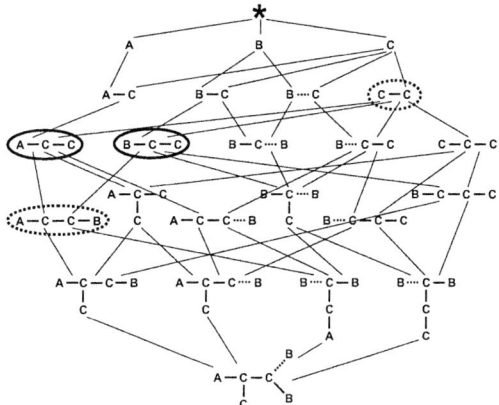

* is the empty fragment. Each graph is subgraph of all its descendants in the lattice. Subgraphs on one level have the same number of edges.

The dashed C-C-fragment is the common core of the two circled fragments. The new subgraph A-C-C-B can be generated by taking this core and adding the two edges A- and B- that only appear in one of the subgraphs.

Fig. 1. The complete subgraph lattice of the graph shown at the bottom

(A) **Purposive refinement.** Mining gets faster if instead of *all potential* refinements only those are created that might *appear* in the database. Two basic approaches exist: on the one hand two graphs can be joined from the previous level of the lattice that share a common core (Fig. 1). Although this may create some subgraphs that do not appear in the database, the appearance list of the refinement (i.e., the intersection of both preceeding appearance lists) is quickly checked. On the other hand, an existing subgraph can be extended by an edge and a node (or only an edge if cycles are closed). The node to be extended and the extension must be chosen carefully based on the appearance list.

(B) **Efficient enumeration.** Generated duplicates of the fragments have to be filtered out. One possibility are isomorphism tests on the database, which are costly. Hence, miners that generate less isomorphic refinements are faster. Using a canonical graph representation, some time is saved by detecting these duplicates before isomorphism testing on the database.

(C) **Focused isomorphism testing.** Known approaches either use efficient subgraph isomorphism tests, e.g. Nauty [3], or they trade time versus storage and keep embeddings. An embedding is a mapping of the nodes and edges of a fragment to the corresponding nodes and edges in the graph it occurs in. When counting the support of fragments, excessive isomorphism tests are necessary. It has to be clarified whether embedding lists lead to better results compared to isomorphism tests.

Early fragment miners generated refinements in a breadth first way, e.g., SUB-DUE [6] (incomplete beam-search), AGM [7], and FSG [8]. Depth first search

---
[1] Similar to the *frequency antimonotone principle* in frequent itemset mining [4,5].

(dfs) approaches need less memory to store appearance lists because the number of lists that have to be stored in memory is proportional to the *depth* of the lattice (i.e. the size of the biggest graph) whereas it is proportional to its *width* (i.e. the maximal number of subgraphs in one level) in breadth first searches. The dfs-algorithms MoFa [9], gSpan [10], FFSM [11], and Gaston [12] attack the subproblems (A–C) quite differently. But since it is difficult to prove that a solution is better than another, the authors usually select a few databases and present benchmarks that demonstrate that their proposed solution works better than a competitor based on executables. Since different authors use different databases there is no general picture. It is unknown which of the solutions to (A–C) perform best under which conditions. To make things worse, sometimes only executables of the algorithms are available. Hence, measurements are skewed by use of different programming language and by exploitation of varying compiler optimization technology, etc.

In this paper we present an unbiased and detailed comparison of the four fragment miners MoFa, gSpan, FFSM and Gaston. We implemented them all from scratch using a common graph framework, i.e. all use the same graph data structures. In section 2 we briefly characterize how these algorithms solve subproblems (A–C). Section 3 contains the main body of this paper: the detailed experimental evaluation of the four contestants.

2 Distinctive Ideas of MoFa, gSpan, FFSM, and Gaston

All four fragment miners work on general, undirected graphs with labeled nodes and edges. They all are restricted to finding connected subgraphs and traverse the lattice as mentioned before in depth-first order.

MoFa (Mo̲lecule F̲ragment Miner, by Borgelt and Berthold in 2002 [9]) has been targeted towards molecular databases, but it can also be used for arbitrary graphs. MoFa stores all embeddings (both nodes and edges). Extension is restricted to those fragments, that actually appear in the database. Isomorphism tests in the database can cheaply be done by testing whether an embedding can be refined in the same way. MoFa uses a fragment-local numbering scheme to reduce the number of refinements generated from a fragment: MoFa counts the nodes of a fragment according to the sequence in which they have been added. When a fragment is extended at node n, later refinements may only occur at n or at nodes bigger than n. Moreover, all extensions that grow from the same node n are ordered according to increasing node and edge labels. Although this local ordering helps, MoFa still generates many isomorphic fragments and then uses standard isomorphism testing to prune duplicates.

gSpan (graph-based S̲ubstructure patter̲n, by Yan and Han in 2002 [10]) uses a canonical representation for graphs, called dfs-code. A dfs-traversal of a graph defines an order in which the edges are visited. The concatenation of edge representations in that order is the graph's dfs-code. Refinement generation is restricted by gSpan in two ways: First, fragments can only be extended at nodes that lie on the *rightmost path* of the dfs-tree. Secondly, fragment generation is guided by occurrence in the appearance lists. Since these two pruning rules cannot fully prevent isomorphic fragment generation, gSpan computes the canonical (lexicographically smallest) dfs-code for each refinement by means of a series of permutations. Refinements with non-minimal dfs-code can be pruned. Since instead of embeddings, gSpan only stores appearance lists for each fragment, explicit subgraph isomorphism testing must be done on all graphs in these appearance lists.

FFSM (Fast Frequent Subgraph Mining, by Huan, Wang, and Prins in 2003 [11]) represents graphs as triangle matrices (node labels on the diagonal, edge labels elsewhere). The matrix-code is the concatenation of all its entries, left to right and line by line. Based on lexicographic ordering, isomorphic graphs have the same canonical code (CAM – Canonical Adjacency Matrix). When FFSM joins two matrices of fragments to generate refinements, only at most two new structures result. FFSM also needs a restricted extension operation: a new edge-node pair may only be added to *the last node* of a CAM. After refinement generation, FFSM permutes matrix lines to check whether a generated matrix is in canonical form. If not, it can be pruned. FFSM stores embeddings to avoid explicit subgraph isomorphism testing. However, FFSM only stores the matching nodes, edges are ignored. This helps speeding up the join and extension operations since the embedding lists of new fragments can be calculated by set operations on the nodes.

Gaston (GrAph/Sequence/Tree extractiON, by Nijssen and Kok 2004 [12]) stores all embeddings, to generate only refinements that actually appear and to achieve fast isomorphism testing. The main insight is that there are efficient ways to enumerate paths and (non-cyclic) trees. By considering fragments that are paths or trees first, and by only proceeding to general graphs with cycles at the end, a large fraction of the work can be done efficiently. Only in that last phase, Gaston faces the NP-completeness of the subgraph isomorphism problem. Gaston defines a global order on cycle-closing edges and only generates those cycles that are "larger" than the last one. Duplicate detection is done in two phases: hashing to pre-sort and a graph isomorphism test for final duplicate detection.

For gSPan and MoFa several extensions exist that are described in section 3.5.

3 The Comparison

In the following sections we compare the four algorithms based on an analysis of the main computational parts on detailed experiments and on some special features of the algorithms.

3.1 Setup of Experiments

The tests were all done on 64bit Linux systems because of the huge memory requirements of some algorithms on the bigger datasets. Because of the lengthy tests we used several machines: Most experiments were run on a Dual-Itanium 2 PC running at 1.3 GHz with 10GB of RAM. Here we used IBM's Java Virtual

Machine (JVM) 1.4.2 because it produced the best runtime results for all algorithms.[2] The maximal heap space available to the JVM was set to 8GB to avoid swapping influences. For the memory tests we used the SUN JVM[3] as the IBM JVM showed garbage collector artifacts. The test on varying database sizes was carried out on an SGI Altix 3700 system[4] with Itanium 2 processors at 1.3 GHz. There only BEA Weblogic's JVM 1.4.2 was available.[5] The maximum heap was set to 14GB. Except for the database size experiments we aborted tests that ran longer than four hours.

We chose Java as programming language because this kept the implementation work at a bearable level. This may of course not lead to astonishingly fast execution times but the relative performance of the algorithms should not be affected significantly. Also the algorithms could be run on 64bit systems with no changes at all, which was important for some experiments.

Because the main application area of frequent subgraphs miners are molecular datasets, experiments were done on the databases described in Fig. 2. The IC93 dataset [13] is used to find out how the algorithms behave if the number of found fragments and the fragments itself get large. At a minimum support value of 4% the largest frequent fragment has 22 bonds, the number of fragments is 37,727. Typically all molecules of the HIV assay from 1999[6] are used for performance evaluations. The complete NCI database[7] is used to determine how the algorithms scale with increasing database size. The found fragments will very likely have no chemical meaning because the molecules in the dataset are very diverse.

Dataset	# molecules	average size # edges	largest molecule # edges	# node labels
IC93	1,283	28	81	10
HIV	42,689	27	234	58
NCI	237,771	22	276	78
PTE	337	26	213	66
CAN2DA99	32,557	28	236	69
HIV CA	423	42	196	21
HIV CM	1,083	34	234	27

Fig. 2. The molecular datasets used for testing and their sizes. There are always four edge labels in molecules.

Only to retest performance comparisons from [14,15] the PTE database[8], the DTP Human Tumor Cell Line Screen (dataset CAN2DA99)[9] and parts of the HIV dataset containing only the confirmend moderately active molecules (HIV CM) and the confirmed active molecules (HIV CA) were used. Except for the

---

[2] http://www-128.ibm.com/developerworks/java/jdk/index.html
[3] http://java.sun.com/
[4] http://www.sgi.com/products/servers/altix/index.html
[5] http://www.bea.com/framework.jsp/content/products/jrockit/
[6] http://dtp.nci.nih.gov/docs/aids/aids_data.html
[7] http://cactus.nci.nih.gov/ncidb2/download.html
[8] See [16] and http://web.comlab.ox.uk/oucl/research/areas/machlearn/PTE/. The dataset we used was provided by Siegfried Nijssen.
[9] http://dtp.nci.nih.gov/docs/cancer/cancer_data.html

CAN2DA99 these datasets are rather small compared to the complete HIV or the NCI dataset.

3.2 Hotspots

Section 2 has summarized how the four algorithms solve subproblems (A-C) and which tasks need to be done. Hence, we first show the runtime distribution by percentage for each task and each algorithm, a measurement, that was not done before in the literature. We used Quest's JProbe on a Profiler[10] on the PC with the SUN JVM for monitoring a run on the IC93 dataset with a minimum support of 5%, see Fig. 3. Using a profiler slows down the runtime a lot, so we took the biggest databases, that are manageable for this experiment: IC93 and HIV CA + HIV CM.

	IC93				HIV CA+CM			
	MoFa	gSpan	FFSM	Gaston	MoFa	gSpan	FFSM	Gaston
Duplicate filtering/pruning	11.3%	3.1%	0.1%	1.8%	12.3%	1.4%	0.2%	1.0%
Support computation	9.3%	62.9%	3.7%		9.6%	70.7%	3.3%	
Embedding list calculations	19.1%	-	60.4%	87.8%	18.1%	-	62.7%	95.9%
Extending of subgraphs	29.9%	17.3%	10.2%		31.1%	14.9%	8.1%	
Joining of subgraphs	-	-	0.1%	-	-	-	0.1%	-

Fig. 3. The table shows the main parts of the subgraph mining process and how much time (relative to the total runtime) each of the four algorithms spends for them

Filtering/pruning duplicates plays only a minor role in the whole subgraph mining process (0.1% - 12.3% of the total runtime). For MoFa, the time contains both the graph isomorphism tests for already generated graphs, and the deletion of extensions that do not comply with the structural pruning rules.

Support Computation or Embedding list calculation is where the algorithms spend most of their time. Using embedding lists (MoFa and FFSM) leads to low numbers in support computation, but calculating them is expensive. Although MoFa's 19.1% for IC93 seem faster than FFSM's 60.4% for IC93, both algorithms have spent about the same number of seconds in this task. If no embedding lists are used (gSpan), expensive subgraph isomorphism tests are necessary. For Gaston, it is impossible to separate runtimes for support computation, embedding list calculation and the extension of fragments. The 87.8% for IC93 includes Gaston's ability to uniquely generate paths and trees.

Extending or joining subgraphs takes about the same time in MoFa, gSpan and FFSM. Joining is only done by FFSM and is very cheap compared to the extension process.

The number for HIV CA + CM do not differ much from the numbers measured for IC93.

---
[10] http://www.quest.com/jprobe/

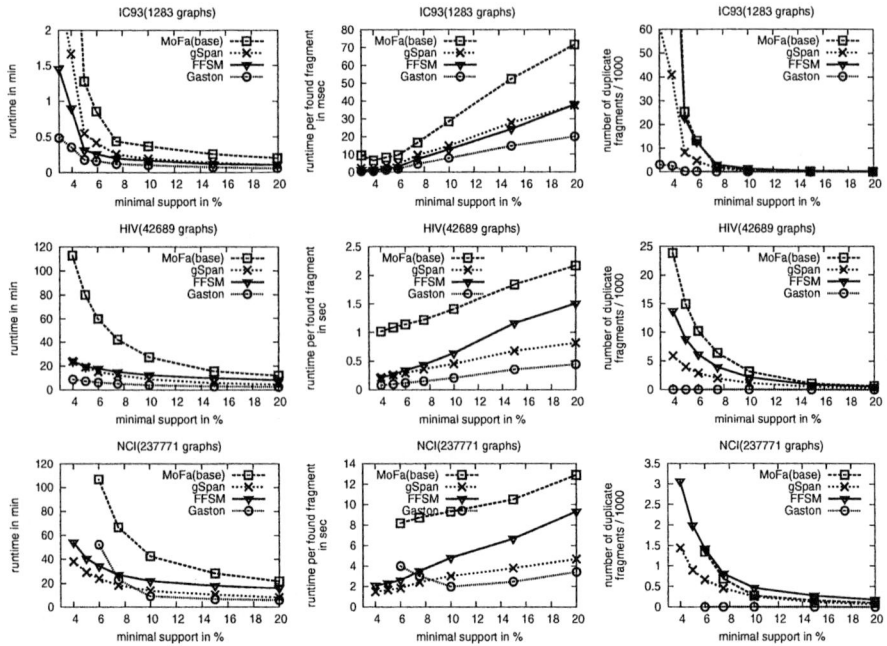

Fig. 4. Total runtime, runtime per found fragment, and the number of found duplicates for the three datasets IC93, HIV, and NCI measured for varying minimum support

3.3 Tests on Molecular Databases

First we retested successfully the results published in [14,15] (PC, IBM JVM) to prove that our implementations can compete with the original implementations and provide qualitatively the same results as given in the literature.

Second we recorded for each algorithm the total time needed at varying support values, the time needed per found frequent fragment and the number of found duplicates (which have to be filtered out by the algorithm in some way) for the IC93, HIV and NCI dataset. A comparison of MoFa[11], gSpan, FFSM and Gaston based on these databases was never published before. Figure 4 shows the results. The first obvious conclusion is the exponential rise in runtime with lower support values (left column). This is not very surprising as the number of fragments found also increases exponentially. Therefore, the runtime per found fragment (second column) is more interesting. For all datasets it shrinks with lower support values which can be explained by the cheaper frequency determination and calculation of embedding lists. The runtime per graph rises for Gaston on the NCI dataset for low support values. This is a memory problem as NCI is

---

[11] As for MoFa several extensions (closed fragments, ring mining, fuzzy chains) exist we did not use in our experiments, this algorithm is marked as *MoFa base* in the pictures, see section 3.5.

the largest database and Gaston needs the most memory of all algorithms, see Fig. 5 on the left.

There is a more or less clear runtime ranking among the four algorithms: MoFa is always the slowest. On the big datasets, FFSM is the second slowest algorithm, only on IC93 it is faster than gSpan. The result of this IC93 test equals the test result in [15]. The likely reason why gSpan is so slow on the IC93 dataset is the growing number of subgraph isomorphism tests gSpan has to do at these low support value (more than 37,000). All other algorithms use embedding list which speed up these tests especially for large fragments. On the large datasets, however, gSpan is faster than FFSM. Gaston is the fastest of all algorithms except at lower support values on the complete NCI dataset. A reason for this may be the amount of bookkeeping because of the large number of embeddings. Also a slowdown because of more frequent garbage collections may be the cause. The fragments found are however rather small so that gSpan gets by with cheap tests.

The number of found duplicates (right column) gives an insight into the power of the fragment refinement mechanisms. Also different pruning techniques minimize the number of duplicates, but as shown in section 3.2 they are not as relevant. Gaston wins as it does not produce duplicates for non-cyclic graphs. On the other hand FFSM's and MoFa's extension methods and pruning rules seem to be the weakest. For FFSM a look at relative time spend in filtering out these duplicates (see table 3), which is only 0.1% like Gaston, indicates that the canonical representation is very efficient.

Next the memory consumption at varying support values was recorded based on the SUN JVM. We frequently called the garbage collector and recorded the maximum heap size. This does not necessarily give the exact value of the memory consumption, but is a very good approximation. Because it slows down the runtime dramatically only the values for the HIV dataset were recorded. As can be seen in Fig. 5, gSpan needs the least memory as it does not use embedding lists. Although MoFa stores both edges and nodes in the embedding lists whereas FFSM only stores the nodes, MoFa still needs less memory. This is because MoFa only needs to store in each node of the search tree the embeddings of one subgraph, while in FFSM a search tree node consists of many subgraphs together with their embeddings. Gaston needs the most memory because embedding lists for a new fragment are built based on the embedding lists of the parent. Extensions to the parent's embedding list are stored with the children. Therefore, the size of the embedding lists does also depend on the number of children a fragment has. This results in the rise of the curve for low support values.

Finally the scalability of the algorithms for increasing database size was tested (Bea JVM, Altix), see Fig. 5, right. The complete NCI database was split into 119 pieces of 2,000 randomly selected molecules. For 5% support we have tested the performance for various subsets of the NCI database, each subset consisting of a growing number of these pieces. An obvious conclusion is, that all algorithms scale linearly with the database size, but with different factors. The surprising result is, that in this test Gaston is always slower than gSpan

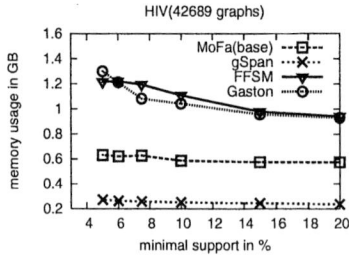 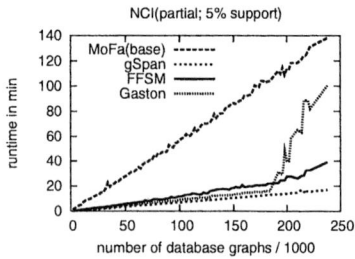

Fig. 5. Memory usage on the HIV database and the runtime in dependence of the database size on the complete NCI database

which was not the case in all other tests. We have performed some tests to be convinced that this is not an artefact of the different JVMs. Instead it seems that the uncommon memory architecture of the SGI Altix system penalizes memory intensive algorithms like Gaston. This also explains the raise in runtime for Gaston for larger databases. Testing Gaston with the IBM JVM on the Itanium on the same subsets of the NCI database did not result in this steep rise of the runtime curve.

3.4 Tests on Artificial Graph Databases

Real-world datasets are never "random". For example typical characteristics of molecular databases are certain distribution of labels, distinct cycles, and low node degrees. Although artificial generated graph databases seem to be a way to do general graph comparisons, there are several obstacles. The main problem is, that even with some fixed parameters randomly generated graph databases can be very different from each other and cause a wide spectrum of runtimes. By considering only the average or median of these results, no valid conclusion can be drawn.

Nevertheless we did some experiments with synthetical databases by our own graph generator. The most interesting test was done with graphs of varying edge densities as molecules mostly have a low edge density. We took a fixed number of 2000 graphs with an average of 50 nodes (ranging from 1 to 100) and 10 uniformly distributed different nodes and edge labels. Then we increased the number of edges in the graphs, starting at an edge density of 10% up to 40% (which means that the graphs contain e.g. $\frac{0.1 \cdot (\#nodes)^2}{2}$ edges). The minimum support was set to 10%. Figure 6 shows the runtime per graph in the left diagram and the total number of discovered fragments in the right one. Except MoFa, all other algorithms show a slight increase in the runtime which seems not to be strongly correlated to the number of found graphs. MoFa however shows a steep increase in the runtime. One reason is the number of discovered fragments: each new fragment has to be checked against all others to find out if it has already been found. The other algorithms rely on canonical representations and the test for duplicates is independent of the number of already discovered structures.

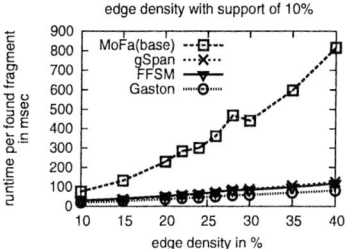 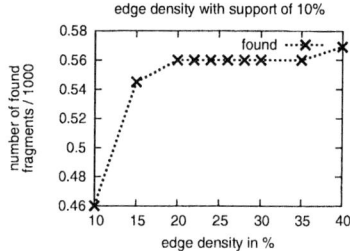

Fig. 6. Runtime and number of frequent subgraphs on synthetic datasets with varying edge density

3.5 Special Features and Possible Extensions

Some of the presented algorithms have special extensions not taken into account for this comparison, but which might improve the performance of the algorithms. One example are *closed subgraphs*. A subgraph is said to be closed if there is no bigger supergraph containing it that occurs in the same transactions of the database. Unclosed subgraphs can easily be filtered out after the search (and partly during the search), but for gSpan and MoFa there exist special extensions that prune branches of the search tree if only closed subgraphs are to be found [17,18]. This speeds up the search considerably (on some datasets for gSpan a speedup of a factor of 10 is reported, for MoFa the runtime is almost halved).

Another issue is the search in directed graphs. FFSM strongly relies on the triangle matrices, that cannot be used for directed graphs. Gaston's rules for uniquely constructing all paths and trees cannot be used for directed graphs without major changes. It is e.g. unclear how a spanning tree can be constructed in a directed graph. MoFa is capable of finding directed frequent subgraphs and also for gSpan only minor changes should be necessary.

Another topic of interest is the search for unconnected subgraphs. An example are molecules in which a certain part of the fragment must be present but the rest of the fragment is not known yet. MoFa can start the search with an unconnected *seed* instead of the empty graph. It is unclear how seeds can be combined with any of the other three algorithms.

For MoFa there also exists an extension for molecular databases that treats rings as single entities [19]. This not only dramatically reduces the number of search tree nodes but also avoids the reporting of fragments with open ring systems that normally make no sense for the biochemists. Another addition enables MoFa to find fragments with carbon chains of varying lengths [20], because this length is not important for biochemical reactions.

4 Conclusions

After re-implementing and testing four famous subgraph mining algorithms, the following conclusions can be drawn:

- Contrary to common belief embedding lists do not considerably speed up the search for frequent fragments. Even though gSpan does not use them, it is competitive to Gaston and FFSM. Only if the fragments become large (like in the IC93 dataset), gSpan falls off. On the other hand, embedding lists can cause problems if not enough memory is available or if the memory throughput is not high enough.
- The power of the pruning strategies to avoid duplicates is not the most important factor. The generation of candidates and support/embedding lists computations are much more critical.
- Using canonical representations for detecting duplicates is more efficient than doing explicit graph isomorphism test. Even better is the complete avoidance of duplicate fragment generation like Gaston does (at least for non-cyclic fragments).
- All algorithms scale linearly with the database size though with different factors.
- Depending on the used Java Virtual Machine results can sometimes differ. This problem can not be solved by the algorithms themself.
- Pure performance is not everything. Although MoFa is the slowest algorithm in all tests it offers much more functionality than the other miners for molecular databases and biochemical questions.

It is not yet clear, where the development of frequent subgraph mining will lead in the future. Possible directions are distributed or parallel search to overcome memory and performance limits. Exploring new application areas is expected to lead to new insights.

References

1. Fischer, I., Meinl, T.: Subgraph Mining. In Wang, J., ed.: Encyclopedia of Data Warehousing and Mining. Idea Group Reference, Hershey, PA, USA (2005)
2. Washio, T., Motoda, H.: State of the Art of Graph-based Data Mining. SIGKDD Explorations Newsletter **5** (2003) 59–68
3. McKay, B.: Practical graph isomorphism. Congressus Numerantium **30** (1981)
4. Agrawal, R., Imielinski, T., Swami, A.N.: Mining Association Rules between Sets of Items in Large Databases. In Buneman, P., Jajodia, S., eds.: Proc. 1993 ACM SIGMOD Int'l Conf. on Management of Data, Washington, D.C., USA, ACM Press (1993) 207–216
5. Zaki, M.J., Parthasarathy, S., Ogihara, M., Li, W.: New Algorithms for Fast Discovery of Association Rules. In Heckerman, D., Mannila, H., Pregibon, D., Uthurusamy, R., Park, M., eds.: In 3rd Int'l Conf. on Knowledge Discovery and Data Mining, AAAI Press (1997) 283–296
6. Cook, D.J., Holder, L.B.: Substructure Discovery Using Minimum Description Length and Background Knowledge. J. of Artificial Intelligence Research **1** (1994) 231–255
7. Inokuchi, A., Washio, T., Motoda, H.: An apriori-based algorithm for mining frequent substructures from graph data. In: PKDD '00: Proceedings of the 4th European Conference on Principles of Data Mining and Knowledge Discovery, London, UK, Springer (2000) 13–23

8. Kuramochi, M., Karypis, G.: Frequent subgraph discovery. In: Proceedings of the IEEE Intl. Conf. on Data Mining ICDM, Piscataway, NJ, USA, IEEE Press (2001) 313–320
9. Borgelt, C., Berthold, M.R.: Mining Molecular Fragments: Finding Relevant Substructures of Molecules. In: Proc. IEEE Int'l Conf. on Data Mining ICDM, Maebashi City, Japan (2002) 51–58
10. Yan, X., Han, J.: gSpan: Graph–Based Substructure Pattern Mining. In: Proc. IEEE Int'l Conf. on Data Mining ICDM, Maebashi City, Japan (2002) 721–723
11. Huan, J., Wang, W., Prins, J.: Efficient mining of frequent subgraphs in the presence of isomorphism. In: Proceedings of the 3rd IEEE Intl. Conf. on Data Mining ICDM, Piscataway, NJ, USA, IEEE Press (2003) 549–552
12. Nijssen, S., Kok, J.N.: Frequent Graph Mining and its Application to Molecular Databases. In Thissen, W., Wieringa, P., Pantic, M., Ludema, M., eds.: Proc. of the 2004 IEEE Conf. on Systems, Man and Cybernetics, SMC 2004, Den Haag, The Netherlands (2004) 4571 – 4577
13. Institute of Scientific Information, Inc. (ISI): Index chemicus - subset from 1993 (1993)
14. Nijssen, S., Kok, J.N.: A quickstart in frequent structure mining can make a difference. Technical report, Leiden Institute of Advanced Computer Science, Leiden University (2004)
15. Huan, J., Wang, W., Prins, J.: Efficient mining of frequent subgraphs in the presence of isomorphism. Technical report, Department of Computer Science at the University of North Carolina, Chapel Hill (2003)
16. Srinivasan, A., King, R.D., Muggleton, S.H., Sternberg, M.: The predictive toxicology evaluation challenge. In: Proceedings of the Fifteenth International Joint Conference on Artificial Intelligence (IJCAI-97). Morgan-Kaufmann (1997) 1–6
17. Yan, X., Han, J.: Closegraph: Mining Closed Frequent Graph Patterns. In: Proc. of the 9th ACM SIGKDD Int'l Conf. on Knowledge Discovery and Data Mining, Washington, DC, USA, ACM Press (2003) 286–295
18. Meinl, T., Borgelt, C., Berthold, M.R.: Discriminative Closed Fragment Mining and Pefect Extensions in MoFa. In Onaindia, E., Staab, S., eds.: STAIRS 2004 - Proc. of the Second Starting AI Researchers' Symp. Volume 109 of Frontiers in Artificial Intelligence and Applications., Valencia, Spain, IOS Press (2004) 3–14
19. Hofer, H., Borgelt, C., Berthold, M.R.: Large Scale Mining of Molecular Fragments with Wildcards. In: Advances in Intelligent Data Analysis. Number 2810 in Lecture Notes in Computer Science, Springer (2003) 380–389
20. Meinl, T., Borgelt, C., Berthold, M.R.: Mining Fragments with Fuzzy Chains in Molecular Databases. In Kok, J.N., Washio, T., eds.: Proc. of the Workshop W7 on Mining Graphs, Trees and Sequences (MGTS '04), Pisa, Italy (2004) 49–60

Efficient Classification from Multiple Heterogeneous Databases*

Xiaoxin Yin and Jiawei Han

University of Illinois at Urbana-Champaign, Urbana, IL 61801, USA
{xyin1, hanj}@uiuc.edu

Abstract. With the fast expansion of computer networks, it is inevitable to study data mining on heterogeneous databases. In this paper we propose *MDBM*, an accurate and efficient approach for classification on multiple heterogeneous databases. We propose a regression-based method for predicting the usefulness of inter-database links that serve as bridges for information transfer, because such links are automatically detected and may or may not be useful or even valid. Because of the high cost of inter-database communication, MDBM employs a new strategy for cross-database classification, which finds and performs actions with high benefit-to-cost ratios. The experiments show that MDBM achieves high accuracy in cross-database classification, with much higher efficiency than previous approaches.

1 Introduction

The rapid growth of the number of data sources on the internet has brought great need for computation over multiple data sources, especially knowledge discovery from multiple data sources. For example, biologists need databases of genes, proteins, and microarrays in their research; a credit card company needs data from a credit bureau for building models for handling applications. Data integration approaches [5,11] may be used to overcome the heterogeneity problem. However, perfect integration of heterogeneous data sources is a very challenging problem, and it is often impossible to migrate one whole database to another site. In contrast, distributed data mining [3,7,8,12,13] aims at discovering knowledge from a dataset that is stored at different sites. But they focus on a homogeneous dataset (a single table or a set of transactions) that is distributed to multiple sites, thus are unable to handle heterogeneous relational databases.

In this paper we study the problem of *cross-database classification*, which aims at building accurate classifiers based on multiple heterogeneous databases, because a single database often contains insufficient information for a classification task. For example, Yahoo shopping may want to build a model for predicting customers' behaviors (as in Figure 1), and thus needs important information

---

* The work was supported in part by the U.S. National Science Foundation NSF IIS-02-09199/IIS-03-08215. Any opinions, findings, and conclusions or recommendations expressed in this paper are those of the authors and do not necessarily reflect the views of the funding agencies.

Efficient Classification from Multiple Heterogeneous Databases 405

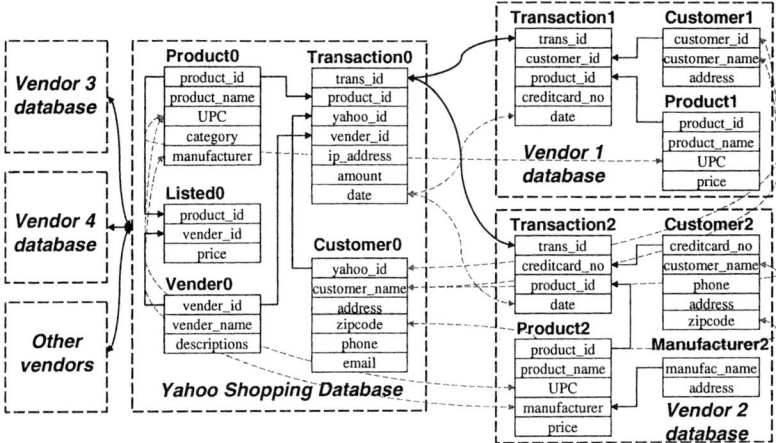

Fig. 1. Databases of Yahoo shopping and venders

from databases of different vendors. In this example the *Customer0* relation in the Yahoo shopping database is called *target relation*, whose tuples are *target tuples*. The goal of cross-database classification is to build an accurate classifier for predicting the class labels of target tuples.

There are two major challenges in cross-database classification. The first is the **data heterogeneity problem**. To transfer information across heterogeneous databases, one must detect *inter-database links*, which are links between matched attributes (as in Figure 1) and can serve as bridges for information transfer. There are many studies on this issue, such as schema mapping [5] and mining database structures [4]. However, some links detected may be vague and sometimes connect unrelated objects. For example, *Customer0.name* → *Customers1.name* may connect different persons with same name, and *Customer0.zipcode* → *Customer1.zipcode* may lead to an explosive number of joined tuples. The second challenge is the **efficiency problem**. It is often expensive to transfer information between two databases, which may be far from each other physically. Thus we must be able to build accurate cross-database classifiers with as low inter-database communication cost as possible. In this paper we propose *MDBM* (Multi-Database Miner), an efficient and accurate approach for classification across multiple heterogeneous databases.

The first contribution of this paper is to propose an approach for predicting the usefulness of links. As mentioned above, some links can lead to useful features, while some others may be useless and only add burdens to the classification procedure. We define the usefulness of a link as the maximum information gain of any feature generated by propagating information through this link. We propose a regression-based approach for building a model to predict usefulness of links based on properties of links. Our experiments show that this approach achieves reasonably high prediction accuracy.

Our second contribution is *economical classification*. As many approaches on relational (or first-order) classification [1,9,10,14], MDBM also uses rule-based classification. All previous approaches build rules by searching for predicates (or literals) with highest information gain (or Foil gain), in order to build accurate rules. Although this strategy is effective in single databases, it may lead to high inter-DB communication cost in multi-database classification. With the prediction model for gainfulness of links, MDBM can predict the gain and cost of each action of searching for predicates. The strategy of *economical classification* always selects the action with highest gain-to-cost ratio, i.e., the action of lowest price per unit of gain. It can achieve same total gain with much lower cost. Our experiments show that MDBM achieves as high accuracy as previous approaches, but is much more efficient in both running time and inter-DB communication.

The rest of the paper is organized as follows. We discuss related work in Section 2. Section 3 describes the approach for building prediction models for usefulness of links. We describe the strategy of economical cross-database classification in Section 4. We present empirical evaluation in Section 5, and conclude this study in Section 6.

2 Related Work

The traditional way of mining multiple databases is to first integrate the databases [5,11], then apply data mining algorithms. However, it is often hard to integrate heterogeneous databases or to migrate one whole database to another site, because of both efficiency and privacy concerns. Thus in multi-database mining we need efficient approaches that can produce good mining results with low inter-database communication cost.

Distributed data mining received much attention in the last several years, which aims at discovering knowledge from a dataset that is distributed at different sites. There are two types of distributed data: (1) horizontally partitioned data, in which data about different objects with same attributes are owned by different sites; (2) vertically partitioned data, in which different attributes of the same set of objects are stored at different sites. Either way of distribution divides the rows or columns of a table into different parts. Distributed data mining approaches for horizontally partitioned data include meta-learning [3] that merges models built from different sites, and privacy preserving techniques including decision tree [8] and association rule mining [7]. Those for vertically partitioned data include association rule mining [12] and k-means clustering [13]. Distributed data mining works on a well-formatted data table stored at different sites. It is fundamentally different from cross-database data mining, which works on multiple heterogeneous databases, each containing a set of interconnected relations.

There are many studies on relational (or first-order) classification [1,9,10,14], which aims at building accurate classifiers in relational databases. Such algorithms search among different relations for useful predicates, by transferring information across relations. They either build rules by adding good literals (or predicates), or build decision trees recursively. Such approaches have proven to be efficient and accurate in single-database scenarios. However, in multi-database classification, they may have high inter-database communication cost, because they only focus on finding gainful literals but not on how much data needs to be transferred. MDBM follows their main philosophy of classification (rule-based, greedy classification), but adopts a new strategy called *economical classification* which can achieve as high accuracy with much lower cost.

3 Predicting Usefulness of Links

3.1 Propagating Information Across Databases

In [4] an efficient approach is proposed to identify joinable attributes in a relational database. Its main idea is to compute the set resemblance of sets of values of different attributes, and it achieves good scalability with a sampling technique. MDBM uses this approach to find all joinable attributes across databases. For two attributes A_1 and A_2 in different databases, if a significant portion (at least 25%) of values of A_1 are joinable to A_2, or those of A_2 are joinable to A_1, then MDBM assumes there is a link between A_1 and A_2. This approach has the limitation that it can only detect simple links. A recent schema matching approach [5] that can detect complex links (e.g., "$firstname + lastname \rightarrow name$") can also be easily integrated into MDBM.

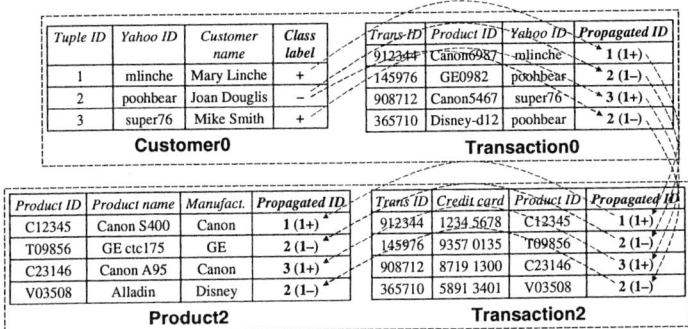

Fig. 2. Example of Tuple ID Propagation

During cross-database mining, large amounts of data needs to be exchanged across databases frequently, and we need an approach that transfers minimum required information to enable effective data mining. In [14] an approach called *Tuple ID Propagation* is proposed, which propagates the unique IDs of target tuples and their class labels across different relations. Tuple ID Propagation is a method for virtually joining relations, and the propagated IDs can be used to identify useful features in different relations. As shown in Figure 2, the IDs can be propagated freely across different relations and databases. As shown in Figure 1, there are usually a large number of inter-database links. Some links serve as good bridges for cross-database mining, such as links of trans_id. While some other links are weak or even incorrect, such as links of zipcode and date.

3.2 Gainfulness of Links

As previous approaches of relational classification [1,9,10,14], MDBM uses rule-based classification. A rule consists of a list of predicates and a class label. Suppose the class label is whether a customer will buy photo printers. An example rule is "[$Customer0 \rightarrow Transaction0$, $Transaction0.amount \geq 500$], [$Transaction0 \rightarrow Product0$, $Product0.category=$'digital camera'] $\Rightarrow$ +". It contains two predicates: the first one is that the customer buys some product of at least \$500, and the second one is that this product is a digital camera. As in [10,14], we use *Foil gain*, a variant of information gain, to measure the usefulness of a predicate. Foil gain measures how many bits can be saved in representing class labels of positive tuples, by adding a predicate to the current rule.

Definition 1 (Foil gain). *For a rule r, we use $P(r)$ and $N(r)$ to denote the numbers of positive and negative target tuples satisfying r. We use $r+p$ to denote the rule constructed by appending predicate p to r. Suppose the current rule is $\hat{r}$. The Foil gain of predicate p is defined as*

$$Foil\_gain(p) = P(\hat{r}+p) \cdot \left[\log \frac{P(\hat{r}+p)}{P(\hat{r}+p) + N(\hat{r}+p)} - \log \frac{P(\hat{r})}{P(\hat{r}) + N(\hat{r})} \right] \quad (1)$$

A link is considered to be a useful one if it brings significant Foil gain, and vice versa. To build a model for predicting the gainfulness of links, we need to first define the gainfulness of links in a predictable way. This definition must indicate the potential gain we can get from a link, but should not be significantly affected by the problem settings (e.g. usage of different classification goals) other than the properties of the link itself.

The definition of Foil gain mainly depends on two factors that vary greatly for different classification goals, even on same dataset. If there are a large number of positive target tuples, the Foil gain of each link is likely to be large. If the number of positive tuples is very small compared to that of negative tuples, then the entropy difference for each positive tuple is large, and Foil gain is likely to be large. Although these factors are not related to the links, they may affect their Foil gain greatly. Therefore, we eliminate the influences of these factors in the definition of gainfulness of links. We define the *gainfulness of a link* as the maximum Foil gain we get from it, divided by the number of positive target tuples, and the maximum possible entropy gain for each positive tuple, as follows.

Definition 2 (gainfulness of link). *Suppose there are P positive target tuples and N negative ones. Suppose p_l is the predicate with highest Foil gain that is found by propagating through link l. The gainfulness of l is defined as*

$$gainfulness(l) = \frac{Foilgain(p_l)}{P \cdot (-\log \frac{P}{P+N})} \quad (2)$$

3.3 Building Prediction Model

In order to build a model for predicting gainfulness of links, we need to select a good set of properties of links that are related to their gainfulness. The first property of a link is the type of its source and destination attributes. Each attribute can be a *key*, a *foreign-key*, or a *semi-key* (an attribute that can almost distinguish every tuple in a relation). Links between other attributes are not considered because they seldom convey strong semantic relationships.

Besides the types of links, the following three properties are selected: *coverage*, *fan-out*, and *correlation*. For a link $l = R_1.A \to R_2.B$, they are defined as follows. The *coverage* of link l is the proportion of tuples in R_1 that are joinable with R_2 via l. Propagating information through a link with high coverage is likely to generate predicates covering many positive tuples. The *fan-out* of link l is the average number of tuples in R_2 joinable with each tuple in R_1 via l. Low fan-out usually indicates stronger relationships between linked objects. The *correlation* of link l is the maximum information gain of using any attribute of R_2 to predict the value of any attribute of R_1[1]. It indicates whether link l brings correlation between some attributes of R_1 and R_2. For example, the link *Product0.UPC* $\to$ *Product2.UPC* has high correlation because *category* of *Product0* can be predicted by *manufacturer* and some specifications of *Product2*.

The coverage, fan-out, and correlation of each link can be computed when searching for matching attributes between different databases. These properties can be roughly computed by sampling techniques in an efficient way.

Based on the properties of links, we use regression techniques to predict their gainfulness. Regression is a well studied field, with many mature approaches such as linear or non-linear regression, support vector machines, and neural networks. We finally choose neural networks [6], because it has high scalability and accuracy, and can model arbitrary functions. A neural network learns to predict values by keeping adapting itself when training examples are fed into it.

We perform multi-relational classification on some datasets to get properties and gainfulness of links, in order to get training data and build models. Our experiments show that these models achieve reasonably high accuracy when predicting for gainfulness of links on other datasets.

4 Economical Cross-Database Classification

4.1 Classification Algorithm

The procedure of rule-based classification consists of a series of actions of searching for gainful predicates. It keeps performing the following action: *propagating information across a link between two relations, and searching for good predicates based on propagated information*. For each action, there is a certain cost (of inter-database communication, computation, etc.), and a certain benefit (in predicting class labels of target tuples). The goal of economical cross-database classification is to achieve high accuracy, with as low cost as possible.

For example, the estimated costs and benefits of four actions are shown in Figure 3. The main philosophy of economical classification is to always select the "cheapest" action, i.e., the action with highest benefit-to-cost ratio (the second

---
[1] Numerical attributes are discretized when computing correlation.

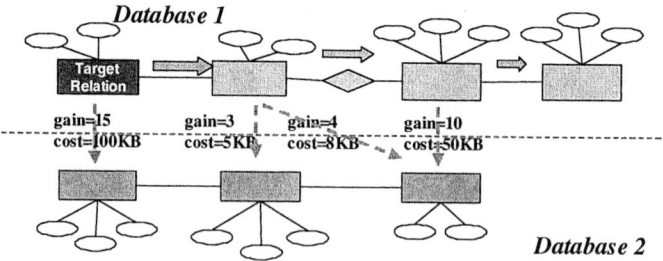

Fig. 3. Economical cross-database classification

one in Figure 3). In a cross-database classification process, the most gainful action is usually not the cheapest one, and vice versa. By selecting the cheapest actions instead of the most gainful ones, MDBM can achieve similar total gain with a much lower "average price", thus achieves high accuracy with low cost.

In general MDBM uses the sequential covering algorithm as in [14] to build rules. At each step of searching for gainful predicates, instead of evaluating all possible predicates as in [14], it uses the strategy of economical cross-database classification to select a gainful predicate. At each step, there is a set of candidate links for propagation, each having an estimated benefit-to-cost ratio. MDBM conducts the action with highest benefit-to-cost ratio. If the real benefit of this action mostly meets our expectation, MDBM stops and moves to the next step. If the real benefit is much lower than estimation, and there is another action with higher estimated benefit-to-cost ratio, then MDBM will conduct that action.

The benefit of a propagation is defined as the maximum Foil gain of any feature found by this propagation, which can be estimated by the prediction model. Suppose there are P positive and N negative tuples satisfying the current rule. The estimated maximum Foil gain of propagation through a link l is

$$est\_Foilgain(l) = gainfulness(l) \cdot P \cdot \left(-\log \frac{P}{P+N}\right) \qquad (3)$$

We use the communication overhead of a propagation as its cost, which can be estimated by the properties of link l and statistics of the source relation R_s of propagation. Suppose there are $|R_s|$ tuples in R_s, and each tuple is associated with I tuple IDs on average. [2]

$$est\_cost(l) = l.coverage \cdot |R_s| \cdot I \qquad (4)$$

Now we describe the MDBM classification algorithm, which follows the main principles of previous relational classification approaches [10,14]. MDBM builds a set of rules for each class. For a certain class, it builds rules one by one, and

---

[2] Because each propagation leads to some computational cost, the estimated cost of a propagation is set to MIN_COST if it is less than this. This threshold prevents MDBM from selecting many extremely cheap actions with very low gain.

removes all positive tuples that are correctly classified by each rule, until more than a proportion of $(1-\epsilon)$ of positive tuples are covered by any rule. To build a rule, it keeps searching for gainful predicates and adding them to the current rule. At each step, MDBM considers all links from the target relation or any relation used in the current rule. MDBM also utilizes some idea of beam search [15]. Suppose it builds a rule "$p_1, p_2 \Rightarrow +$", and this rule only covers a small portion of the positive tuples covered by p_1, then MDBM will try to build another rule based on those uncovered tuples satisfying p_1. By using the idea of beam search, MDBM tries to utilize all Foil gain of p_1, which saves some inter-database communication cost compared with starting from another empty rule.

4.2 Analysis of the Search Strategy

The strategy of previous rule-based classification algorithms [10,14] is to try every possible action at each step and select the most gainful one. While our strategy is to select the cheapest action at each step. Using cheap actions will lead to the generation of predicates and rules with less Foil gain. For our strategy to be effective, we need to prove that *many cheap actions can achieve similar classification accuracy as a smaller number of "expensive" actions, if their total gain are similar.*

Theorem 1. *Suppose a rule set S contains L rules $r_1, r_2, \ldots, r_L$. Each rule r_i covers p_i positive and n_i negative tuples, which are not covered by previous rules $(r_1, \ldots, r_{i-1})$. For another rule r that covers $(\sum_{i=1}^{L} p_i)$ positive and $(\sum_{i=1}^{L} n_i)$ negative tuples,*

$$\sum_{i=1}^{L} Foil\_gain(r_i) \leq Foil\_gain(r).$$

Corollary 1. *Suppose a rule set S contains L rules $r_1, r_2, \ldots, r_L$. Each rule r_i covers p_i positive and n_i negative tuples, which are not covered by previous rules $(r_1, \ldots, r_{i-1})$. If S has higher total Foil gain than a single rule r, then r either covers less positive tuples or more negative tuples than S.*

Theorem 1 and Corollary 1 show that, (1) if a rule set S covers identical numbers of positive and negative tuples as any single rule r, S will have less total gain, and (2) if S has total Foil gain of g, then for any single rule r with Foil gain less than g, r must cover either less positive or more negative examples than S. Although it cannot be strictly proven, we believe that in most cases if a rule set S has higher total gain than a rule r, S will have higher classification accuracy or at least cover more tuples with similar accuracy.

As mentioned before, in cross-database classification we want to achieve high classification accuracy with as low inter-database communication cost as possible. Let us compare MDBM with an existing rule-based multi-relational classification approach (e.g., [10] and [14]). MDBM always selects actions with high gain-to-cost ratios. Thus if both approaches build rule sets with similar total gains, MDBM will usually be much more efficient. On the other hand, our experiments show that MDBM achieves similar accuracies as the approach in [14], which means that MDBM probably builds a rule set with less total gain (according to Corollary 1), and is thus more efficient. The efficiency and accuracy of MDBM is also verified in our experiments.

Although MDBM usually builds more rules, it uses the same thresholds to control the complexity of each rule (by limiting the length of rule and minimum Foil gain of each predicate). Therefore, MDBM will not build overly complex rules, and overfitting is not a big concern.

5 Empirical Evaluation

We perform comprehensive experiments on both synthetic and real databases. The experiments are run on a 2.4GHz Pentium 4 PC with 1GB memory, running Windows XP Pro. The algorithms are implemented with Visual Studio.Net. The following parameters are used in MDBM: MIN_COST=0.5KB, MIN_GAIN= 6.0, and $\epsilon = 0.1$. MDBM is compared with CrossMine [14], a recent approach for relational classification that is order of magnitude more efficient than previous approaches. We keep the implementation details and parameters of CrossMine, and reimplement it to make it capable of performing cross-database classification. We use the code of neural networks at
http://www-2.cs.cmu.edu/afs/cs.cmu.edu/user/mitchell/ftp/faces.html.

5.1 Experiments on Predicting Gainfulness of Links

We perform experiments on three real datasets to test the accuracy and efficiency of MDBM. The first one is CS Dept + DBLP dataset. CS Dept dataset[3] was collected from the web sources of Dept. of CS, UIUC. It contains eight relations: *Student, Advise, Professor, Registration, OpenCourse, Course, WorkIn*, and *ResearchGroup*. DBLP dataset is retrieved from DBLP web site and contains three relations: *Author, Publication*, and *Publish*. The target relation is *Student*, and the class labels are their research areas, which are inferred from their research groups, advisors, and recent publications.

The second dataset is Loan application + Bank dataset. This is from the financial dataset used in PKDD CUP 99, and is split into two datasets. One of them contains information about loan applications and has three relations: *Loan, Account*, and *District*. The other is about bank transactions and records and has five relations: *Client, Disposition, Card, Order*, and *Transaction*. The target relation is *Loan*. It stores the loan applications and their results (approved or not), which are used as class labels.

The third dataset is Movie + People dataset. It is from the Movies dataset in UCI KDD archive, and is split into two databases. One of them contains information of people and has three relations: *Actor, Director*, and *Studio*. The other contains information about movies and has four relations: *Movie, MovieCategory* (a movie may belong to multiple categories), *Cast*, and *Award*.

---

[3] http://dm1.cs.uiuc.edu/csuiuc_dataset/

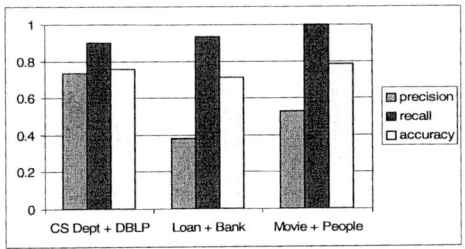

Fig. 4. Accuracy of predicting gainfulness of links

The target relation is *Director* and the class label is whether a director is old or new (whether she started her career before or after 1970). All temporal information is removed from *Director* relation before training.

We first test the accuracy of predicting gainfulness of links. Cross-validation is used in this experiment as well as others, which means that a model is built based on the links from two datasets, and is used to predict the gainfulness of links in the third dataset. In this experiment a link is considered as gainful if its gainfulness is greater than 0.25. The precision, recall, and accuracy of prediction on each dataset is shown in Figure 4. Recall is more important than precision because important features may be missed if a gainful link is predicted as gainless, but it does not hurt much to predict a gainless link as gainful. It can be seen that we achieve high recall and overall accuracy for predicting gainfulness of links. This training process only takes about one second.

5.2 Experiments on Classification Accuracy

For each of the three datasets, we compare the accuracy, running time, and inter-database communication of three approaches: (1) *Single-DB CrossMine*—the CrossMine algorithm for single database; (2) *Multi-DB CrossMine*—the Cross-Mine algorithm that is able to propagate information and search for features across databases; (3) *MDBM*: our cross-database classification algorithm.

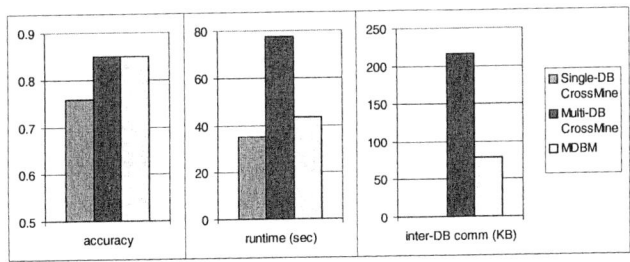

Fig. 5. Accuracy, runtime, and inter-DB communication on CS Dept + DBLP dataset

The results on CS Dept + DBLP dataset are shown in Figure 5. It can be seen that using multi-database information can significantly increase classification accuracy. MDBM achieves much higher efficiency in both running time and inter-database communication, which shows the effectiveness of our approach.

The results on Loan application + Bank dataset are shown in Figure 6. One can see that both Multi-DB CrossMine and MDBM achieve high accuracy, and MDBM is much more efficient in inter-DB communication and running time.

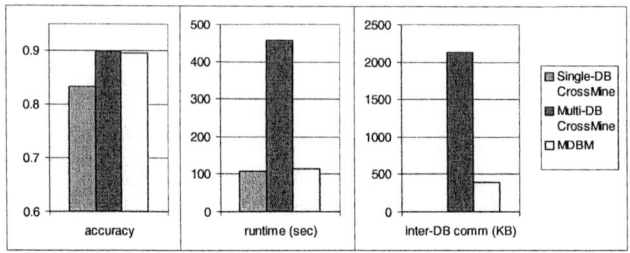

Fig. 6. Accuracy, runtime, and inter-DB communication on Loan + Bank dataset

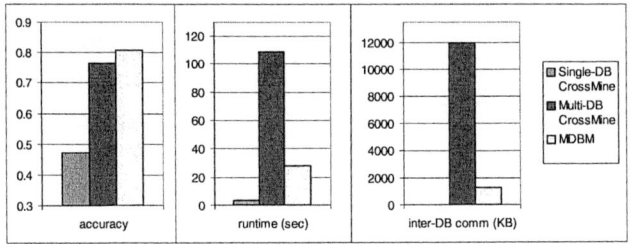

Fig. 7. Accuracy, runtime, and inter-DB communication on Movie + People dataset

The results on Movie + People dataset are shown in Figure 7. It can be seen that MDBM achieves higher accuracy than Multi-DB CrossMine. Again MDBM is much more efficient in inter-database communication (about 10% of that of Multi-DB CrossMine) and running time. Single-DB CrossMine runs fast because it cannot generate any meaningful rules.

5.3 Experiments on Scalability

We test the scalability of MDBM w.r.t. number of databases and number of tuples on synthetic datasets. We use the data generator for CrossMine [14], which can randomly generate a relational database with $|R|$ relations, each having N tuples on average. The target tuples are generated according to a set of randomly generated rules that involve different relations. After a dataset is generated, we randomly partition it into several databases, and use database structuring tool to identify inter-database links.

We first test the scalability of MDBM and Multi-DB CrossMine w.r.t. the number of databases. Five datasets are generated, with number of databases being one to five. Each database has five relations, and the expected number of tuples in each relation is 1000. The accuracy, runtime and inter-database communication of two algorithms are shown in Figure 8. It can be seen that their accuracies are close, but MDBM achieves much higher efficiency and scalability than CrossMine, especially in inter-database communication.

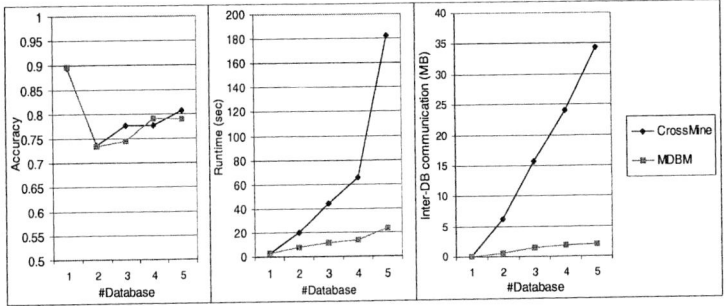

Fig. 8. Scalability w.r.t. number of databases

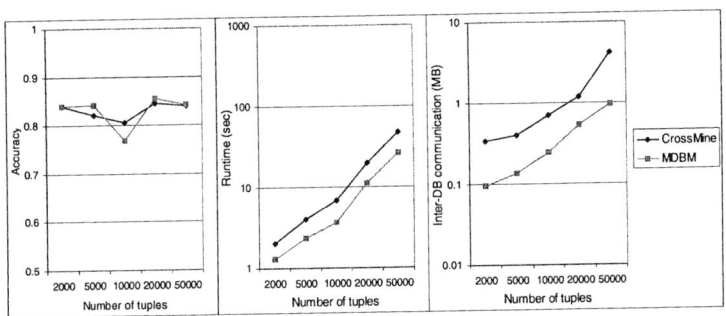

Fig. 9. Scalability w.r.t. number of tuples

We also test the scalability of MDBM and Multi-DB CrossMine w.r.t. the number of tuples. Five datasets are generated with identical schemas, each having two databases and five relations in each database. The expected number of tuples in each relation grow from 200 to 5,000. The accuracy, runtime and inter-database communication of two algorithms are shown in Figure 9. It can be seen that both algorithms are linear scalable in runtime and inter-database communication, and MDBM is much more efficient than CrossMine.

6 Conclusions

In this paper we present MDBM, a new approach for cross-database classification. MDBM can perform accurate classification with data stored in multiple heterogeneous databases, with low inter-database communication. It builds a prediction model for usefulness of links from cross-database mining processes on available datasets that can guide the mining tasks. To achieve high classification accuracy with as low cost as possible, MDBM adopts an economical strategy for cross-database mining, which selects actions with high gain-to-cost ratio. It is shown by experiments that MDBM achieves both high accuracy and high efficiency (especially in inter-database communication) on classification tasks on both real and synthetic datasets.

References

1. H. Blockeel, L.D. Raedt. Top-down induction of logical decision trees. *Artificial Intelligence*, 1998.
2. P. Clark and R. Boswell. Rule induction with CN2: Some recent improvements. In *European Working Session on Learning*, 1991.
3. D. W. Cheung, V. T. Ng, A. W. Fu, Y. Fu. Efficient Mining of Association Rules in Distributed Databases. *TKDE*, 1996.
4. T. Dasu, T. Johnson, S. Muthukrishnan, V. Shkapenyuk. Mining Database Structure; Or, How to Build a Data Quality Browser. *SIGMOD*, 2002.
5. R. Dhamankar, Y. Lee, A. Doan, A. Halevy, P. Domingos. iMAP: Discovering Complex Semantic Matches between Database Schemas. *SIGMOD*, 2004.
6. J. Hertz, R. Palmer, A. Krogh. Introduction to the Theory of Neural Computation. Addison-Wesley, 1991.
7. M. Kantarcioglu, C. Clifton. Privacy-preserving Distributed Mining of Association Rules on Horizontally Partitioned Data. *TKDE*, 2004.
8. Y. Lindell, B. Pinkas. Privacy Preserving Data Mining. *CRYPTO*, 2000.
9. S. Muggleton. Inverse entailment and progol. In *New Generation Computing, Special issue on Inductive Logic Programming*, 1995.
10. J. R. Quinlan and R. M. Cameron-Jones. FOIL: A midterm report. In *European Conf. Machine Learning*, 1993.
11. E. Rahm, P.A. Bernstein. A Survey of Approaches to Automatic Schema Matching. *VLDB Journal*, 2001.
12. J. Vaidya, C. Clifton. Privacy Preserving Association Rule Mining in Vertically Partitioned Data. *KDD*, 2002.
13. J. Vaidya, C. Clifton. Privacy-Preserving K-Means Clustering over Vertically Partitioned Data *KDD*, 2003.
14. X. Yin, J. Han, J. Yang, P.S. Yu. CrossMine: Efficient Classification Across Multiple Database Relations. *ICDE*, 2004.
15. W. Zhang. Search techniques. *Handbook of data mining and knowledge discovery*, Oxford University Press, 2002.

A Probabilistic Clustering-Projection Model for Discrete Data

Shipeng Yu[1,2], Kai Yu[2], Volker Tresp[2], and Hans-Peter Kriegel[1]

[1] Institute for Computer Science, University of Munich, Germany
[2] Siemens Corporate Technology, Munich, Germany

Abstract. For discrete co-occurrence data like documents and words, calculating optimal projections and clustering are two different but related tasks. The goal of projection is to find a low-dimensional latent space for words, and clustering aims at grouping documents based on their feature representations. In general projection and clustering are studied independently, but they both represent the intrinsic structure of data and should reinforce each other. In this paper we introduce a probabilistic clustering-projection (PCP) model for discrete data, where they are both represented in a unified framework. Clustering is seen to be performed in the projected space, and projection explicitly considers clustering structure. Iterating the two operations turns out to be exactly the variational EM algorithm under Bayesian model inference, and thus is guaranteed to improve the data likelihood. The model is evaluated on two text data sets, both showing very encouraging results.

1 Introduction

Modelling discrete data is a fundamental problem in machine learning, pattern recognition and statistics. The data is usually represented as a large (and normally sparse) matrix, where each entry is an integer and characterizes the relationship between corresponding row and column. For example in document modelling, the "bag-of-words" methods represent each document as a row vector of occurrences of each word, ignoring any internal structure and word order. This is taken as the working example in this paper, but the proposed model is generally applicable to other discrete data.

Data projection and clustering are two important tasks and have been widely applied in data mining and machine learning (e.g., principal component analysis (PCA) and k-means [1]). Projection is also referred as feature mapping that aims to find a new representation of data, which is low-dimensional and physically meaningful. On the other hand, clustering tries to group similar data patterns together, and thus uncovers the structure of data. Traditionally these two methods are studied separately and mainly on continuous data. However in this paper we investigate them on *discrete* data and treat them *jointly*.

Projection on discrete data differs from the case on continuous space, where, for example, the most popular technology PCA tries to find the orthogonal dimensions (or factors) that explains the *covariance* of data dimensions. However, one cannot make the same orthogonal assumption on the low-dimensional factors of discrete data and put the interests on the covariance anymore. Instead, it is desired to find the *independent* latent factors that explain the *co-occurrence* of dimensions (e.g., words). In text modelling, if we refer the factors as topics, the projection actually represent each document as a data point in a low-dimensional topic space, where a co-occurrence factor actually suggests more or less a cluster of words (i.e., a group of words often occurring together). Intuitively, if the projected topic space is informative enough, it should also be highly indicative to reveal the clustering structure of documents. On the other hand, a truly discovered clustering structure reflects the shared topics within document clusters and the distinguished topics across document clusters, and thus can offer evidence for the projection side. Therefore, it is highly desired to consider the two problems in a unified model.

In this paper a novel probabilistic clustering-projection (PCP) model is proposed, to jointly handle the projection and clustering for discrete data. The projection of words is explicitly formulated with a matrix of model parameters. Document clustering is then incorporated using a mixture model on the projected space, and we model each mixture component as a multinomial over the latent topics. In this sense this is a *clustering model using projected features* for documents if the projection matrix is given, and a *projection model with structured data* for words if the clustering structure is known. A nice property of the model is that we can perform clustering and projection *iteratively*, incorporating new information on one side to the updating of the other. We will show that they are corresponding to a Bayesian variational EM algorithm that improves the data likelihood iteratively.

This paper is organized as follows. The next section reviews related work. Section 3 introduces the PCP model and explicitly points out the clustering and projection effects. In Section 4 we present inference and learning algorithm. Then Section 5 presents experimental results and Section 6 concludes the paper.

2 Related Work

PCA is perhaps the most well-known projection technique, and has its counterpart in information retrieval called latent semantic indexing [4]. For discrete data, an important related work is probabilistic latent semantic indexing (pLSI) [7] which directly models latent topics. PLSI can be treated as a projection model, since each latent topic assigns probabilities to a set of words and thus a document, represented as a bag of words, can be treated as generated from a mixture of multiple topics. However, the model is not built for clustering and, as pointed by Blei et al. [2], it is not a proper generative model, since it treats document IDs as random variables and thus cannot generalize to new documents. Latent Dirichlet allocation (LDA) [2] generalizes pLSI by treating the topic mixture parameters (i.e., a multinomial over topics) as variables drawn from a Dirichlet distribution. This model is a well-defined generative model and performs much better than pLSI, but the clustering effect is still missing. On the other side, document clustering has been intensively investigated and the most popular method is probably partition-based algorithms like k-means (see, e.g., [1]). Non-negative matrix factorization (NMF) [11] is another candidate and is shown to obtain good results in [13].

Despite that plenty of work has been done in either clustering or projection, the importance of considering both in a single framework has been noticed only recently, e.g., [6] and [12]. Both works are concerned about document clustering and projection on continuous data, while lacking the probabilistic interpretations to the connections among documents, clusters and factors. Buntine et al. [3] noticed this problem for discrete data and pointed out that the multinomial PCA model (or discrete PCA) takes clustering and projection as two extreme cases. Another closely related work is the so-called two-sided clustering, like [8] and [5], which aims to clustering words and documents simultaneously. In [5] it is implicitly assumed a one-to-one correspondence between the two sides of clusters. [8] is a probabilistic model for discrete data, but it has similar problems as in pLSI and not generalizable to new documents.

3 The PCP Model

We consider a corpus $\mathcal{D}$ containing D documents, with vocabulary $\mathcal{V}$ having V words. Following the notation in [2], each document d is a sequence of N_d words that is denoted by $\mathbf{w}_d = \{w_{d,1}, \ldots, w_{d,N_d}\}$, where $w_{d,n}$ is a variable for the nth word in $\mathbf{w}_d$ and denotes the index of the corresponding word in $\mathcal{V}$.

To simplify explanations, we use "clusters" for components in document clustering structure and "topics" for projected space for words. Let M denote the number of clusters and K the dimensionality of topics. Roman letters d, m, k, n, j are indices for documents, clusters, topics, words in $\mathbf{w}_d$, and words in $\mathcal{V}$. They are up to D, M, K, N_d, V, respectively. Letter i is reserved for temporary index.

3.1 The Probabilistic Model

The PCP model is a generative model for a document corpus. Figure 1 (left) illustrates the sampling process in an informal way. To generate one document d, we first choose a cluster from the M clusters. For the mth cluster, the cluster center is denoted as θ_m and defines a topic mixture over the topic space. Therefore θ_m is a K-dimensional vector and satisfies $\theta_{m,k} \geq 0$, $\sum_{k=1}^{K} \theta_{m,k} = 1$ for all $m = 1, \ldots, M$. The probability of choosing a specific cluster m for document d is denoted as π_m, and $\boldsymbol{\pi} := \{\pi_1, \ldots, \pi_M\}$ satisfies $\pi_m \geq 0$, $\sum_{m=1}^{M} \pi_m = 1$.

When document d chooses cluster m, it defines a document-specific topic mixture θ_d, which is obtained exactly from the cluster center θ_m. Note that everything is discrete and two documents belonging to the same cluster will have the same topic mixtures. Words are then sampled independently given topic mixture θ_d, in the same way as in LDA. Each word $w_{d,n}$ is generated by first choosing a topic $z_{d,n}$ given the topic mixture, and then sampling the word given the projection β. β is the $K \times V$ matrix where $\beta_{k,j}$ specifies the probability

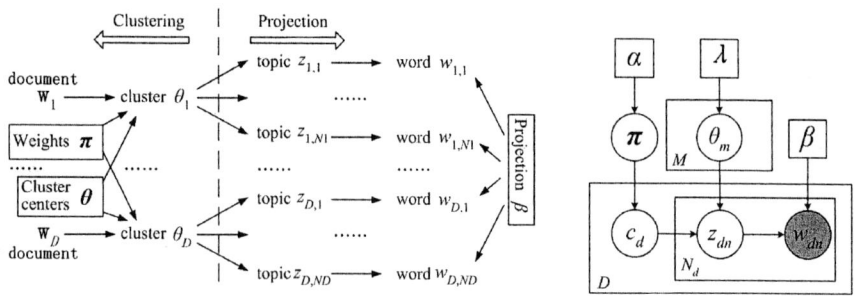

Fig. 1. Informal sampling process (left) and plate model (right) for the PCP model. In the left figure, dark arrows show dependencies between entities and the dashed line separates the clustering and projection effects. In the plate model, rectangle means independent sampling, and hidden variables and model parameters are denoted as circles and squares, respectively. Observed quantities are marked in black.

of generating word j given topic k, $\beta_{k,j} = p(w^j = 1|z^k = 1)$. Therefore each row $\beta_{k,:}$ defines a multinomial distribution for all words over topic k and satisfies $\beta_{k,j} \geq 0$, $\sum_{j=1}^{V} \beta_{k,j} = 1$.

To complete the model, we put a Dirichlet prior $\text{Dir}(\boldsymbol{\lambda})$ for all the cluster centers $\theta_1, \ldots, \theta_M$, and a symmetric Dirichlet prior $\text{Dir}(\alpha/M, \ldots, \alpha/M)$ for the mixing weights $\boldsymbol{\pi}$. Note that they are sampled only once for the whole corpus.

Finally we obtain the probabilistic model formally illustrated in Figure 1 (right), using standard plate model. c_d takes value $\{1, \ldots, M\}$ and acts as the indicator variable saying which cluster document d takes on out of the M clusters. All the model parameters are $\alpha, \boldsymbol{\lambda}, \beta$ and amount to $1 + M + K \times (V - 1)$. The following procedure describes the sampling process for the whole corpus:

1. Choose model parameter $\alpha, \boldsymbol{\lambda}, \beta$;
2. For the mth cluster, choose $\theta_m \sim \text{Dir}(\boldsymbol{\lambda}), m = 1, \ldots, M$;
3. Choose the mixing weight $\boldsymbol{\pi} \sim \text{Dir}(\alpha/M, \ldots, \alpha/M)$;
4. For each document $\mathbf{w}_d$:
   (a) Choose a cluster m with mixing weights $\boldsymbol{\pi}$, and obtain $\theta_d = \theta_m$;
   (b) For each of the N_d words $w_{d,n}$:
      i. Choose a topic $z_{d,n} \sim \text{Mult}(\theta_d)$;
      ii. Choose a word $w_{d,n} \sim \text{Mult}(\beta_{z_{d,n},:})$.

Denote $\boldsymbol{\theta}$ as the set of M cluster centers $\{\theta_1, \ldots, \theta_M\}$, the likelihood of the corpus $\mathcal{D}$ can be written as

$$\mathcal{L}(\mathcal{D}; \alpha, \boldsymbol{\lambda}, \beta) = \int_{\boldsymbol{\pi}} \int_{\boldsymbol{\theta}} \prod_{d=1}^{D} p(\mathbf{w}_d | \boldsymbol{\theta}, \boldsymbol{\pi}; \beta) dP(\boldsymbol{\theta}; \boldsymbol{\lambda}) \, dP(\boldsymbol{\pi}; \alpha), \quad (1)$$

where $p(\boldsymbol{\theta}; \boldsymbol{\lambda}) = \prod_{m=1}^{M} p(\theta_m; \boldsymbol{\lambda})$, and the likelihood of document d is a mixture:

$$p(\mathbf{w}_d | \boldsymbol{\theta}, \boldsymbol{\pi}; \beta) = \sum_{c_d=1}^{M} p(\mathbf{w}_d | \boldsymbol{\theta}, c_d; \beta) p(c_d | \boldsymbol{\pi}). \quad (2)$$

Given mixture component c_d, likelihood term $p(\mathbf{w}_d|\boldsymbol{\theta}, c_d; \beta)$ is then given by

$$p(\mathbf{w}_d|\theta_{c_d}; \beta) = \prod_{n=1}^{N_d} \sum_{z_{d,n}=1}^{K} p(w_{d,n}|z_{d,n}; \beta) p(z_{d,n}|\theta_{c_d}). \quad (3)$$

3.2 PCP as a Clustering Model

As can be seen from (2) and (3), PCP is a clustering model when the projection β is assumed known. The essential terms now are the probabilities of clusters $p(m|\boldsymbol{\pi}) = \pi_m$, probabilistic clustering assignment for documents $p(\mathbf{w}_d|\theta_m; \beta)$, and cluster centers θ_m, for $m = 1, \ldots, M$. Note from (3) that cluster centers θ_m are not modelled directly with words like $p(w|\theta_m)$, but with topics, $p(z|\theta_m)$. This means we are not clustering documents in word space, but in *topic space*. This is analogous to clustering continuous data on the latent space found by PCA [6], and K is exactly the dimensionality of this space. To obtain the probability that document d belongs to cluster m, we project each word into topic space, and then calculate the distance to cluster center θ_m by considering all the words in $\mathbf{w}_d$. This explains (3) from perspective of clustering.

To improve generalization and avoid overfitting, we put priors to θ_m and $\boldsymbol{\pi}$ and treat them as hidden variables, as usually done in mixture modelling. The prior distributions are chosen to be Dirichlet that is *conjugate* to multinomial. This will make model inference and learning much easier (see Section 4).

3.3 PCP as a Projection Model

A projection model aims to learn projection β, mapping words to topics. As can be seen from (3), the topics are not modelled directly with documents $\mathbf{w}_d$, but with cluster centers θ_m. Therefore if clustering structure is already known, PCP will learn β by using the richer information contained in cluster centers, not just individual documents. In this sense, PCP can be explained as a *projection model with structured data* and is very attractive because clustered documents are supposed to contain less noise and coarser granularity. This will make the projection more accurate and faster.

As a projection model, PCP is more general than pLSI because document likelihood (3) is well defined and generalizable to new documents. Although LDA uses similar equation as (3), the topic mixture θ_d is only sampled for current document and no inter-similarity of documents is directly modelled. Documents can only exchange information via the hyperparameter for θ_d's, and thus its effect to β is only implicit. On the contrary, PCP directly models similarity of documents and incorporate all information to learn β.

As discussed in [2], projection β can be smoothed by putting a common prior to all the rows. If only the *maximum a posteriori* (MAP) estimate of β is considered, the effect of smoothing turns out to add a common factor to each entry of β before normalization each row. This is also straightforward in PCP model and we will not discuss it in detail for simplicity. In the experiments we will use this smoothing technique.

4 Inference and Learning

In this section we consider model inference and learning. As seen from Figure 1, for inference we need to calculate the *a posteriori* distribution of latent variables

$$\hat{p}(\boldsymbol{\pi}, \boldsymbol{\theta}, \mathbf{c}, \mathbf{z}) := p(\boldsymbol{\pi}, \boldsymbol{\theta}, \mathbf{c}, \mathbf{z} | \mathcal{D}, \alpha, \boldsymbol{\lambda}, \beta),$$

including both effects of clustering and projection. Here for simplicity we denote $\boldsymbol{\pi}, \boldsymbol{\theta}, \mathbf{c}, \mathbf{z}$ as groups of $\pi_m, \theta_m, c_d, z_{d,n}$, respectively. This requires to compute (1), where the integral is however analytically infeasible. A straightforward Gibbs sampling method can be derived, but it turns out to be very slow and inapplicable to high dimensional discrete data like text, since for each word we have to sample a latent variable z. Therefore in this section we suggest an efficient *variational method* by introducing variational parameters for latent variables [9]. Then we can maximize the data likelihood by iteratively updating these parameters and obtain a *variational EM* algorithm until convergence. The interesting thing is that this algorithm is equivalent to performing clustering and projection iteratively, which we will discuss in detail.

4.1 Variational EM Algorithm

The idea of variational EM algorithm is to propose a joint distribution $q(\boldsymbol{\pi}, \boldsymbol{\theta}, \mathbf{c}, \mathbf{z})$ for latent variables conditioned on some free parameters, and then enforce q to approximate the *a posteriori* distributions of interests by minimizing the KL-divergence $D_{\text{KL}}(q\|\hat{p})$ with respect to those free parameters. We propose a variational distribution q over latent variables as the following

$$q(\boldsymbol{\pi}, \boldsymbol{\theta}, \mathbf{c}, \mathbf{z} | \boldsymbol{\eta}, \boldsymbol{\gamma}, \boldsymbol{\psi}, \boldsymbol{\phi}) = q(\boldsymbol{\pi}|\boldsymbol{\eta}) \prod_{m=1}^{M} q(\theta_m | \gamma_m) \prod_{d=1}^{D} q(c_d | \psi_d) \prod_{n=1}^{N_d} q(z_{d,n} | \phi_{d,n}), \quad (4)$$

where $\boldsymbol{\eta}, \boldsymbol{\gamma}, \boldsymbol{\psi}, \boldsymbol{\phi}$ are groups of *variational parameters*, each tailoring the variational *a posteriori* distribution to each latent variable. In particular, $\boldsymbol{\eta}$ specifies an M-dim. Dirichlet for $\boldsymbol{\pi}$, γ_m specifies a K-dim. Dirichlet for distinct θ_m, ψ_d specifies an M-dim. multinomial for indicator c_d of document d, and $\phi_{d,n}$ specifies a K-dim. multinomial over latent topics for word $w_{d,n}$. It turns out that minimization of the KL-divergence is equivalent to maximization of a lower bound of the log likelihood $\ln p(\mathcal{D}|\alpha, \boldsymbol{\lambda}, \beta)$, derived by applying Jensen's inequality [9]:

$$\mathcal{L}_q(\mathcal{D}) = \mathbb{E}_q[\ln p(\boldsymbol{\pi}|\alpha)] + \sum_{m=1}^{M} \mathbb{E}_q[\ln p(\theta_m|\boldsymbol{\lambda})] + \sum_{d=1}^{D} \mathbb{E}_q[\ln p(c_d|\boldsymbol{\pi})]$$

$$+ \sum_{d=1}^{D} \sum_{n=1}^{N_d} \mathbb{E}_q[\ln p(w_{d,n}|z_{d,n}, \beta) p(z_{d,n}|\boldsymbol{\theta}, c_d)] - \mathbb{E}_q[\ln q(\boldsymbol{\pi}, \boldsymbol{\theta}, \mathbf{c}, \mathbf{z})]. \quad (5)$$

The optimum is found by setting the partial derivatives with respect to each variational and model parameter to be zero, which corresponds to the variational E-step and M-step, respectively. In the following we separate these equations into two parts and interpret them from the perspective of clustering and projection, respectively.

4.2 Updates for Clustering

As we mentioned in Section 3.2, the specific variables for clustering are document-cluster assignments c_d, cluster centers θ_m, and cluster probabilities $\boldsymbol{\pi}$. It turns out that their corresponding variational parameters are updated as follows:

$$\psi_{d,m} \propto \exp\left\{\sum_{k=1}^{K}\left[\left(\Psi(\gamma_{m,k}) - \Psi(\sum_{i=1}^{K}\gamma_{m,i})\right)\sum_{n=1}^{N_d}\phi_{d,n,k}\right] + \Psi(\eta_m) - \Psi(\sum_{i=1}^{M}\eta_i)\right\}, \quad (6)$$

$$\gamma_{m,k} = \sum_{d=1}^{D}\psi_{d,m}\sum_{n=1}^{N_d}\phi_{d,n,k} + \lambda_k, \qquad \eta_m = \sum_{d=1}^{D}\psi_{d,m} + \frac{\alpha}{M}, \quad (7)$$

where $\Psi(\cdot)$ is the digamma function, the first derivative of the log Gamma function. $\psi_{d,m}$ are the *a posteriori* probabilities $p(c_d = m)$ that document d belongs to cluster m, and define a *soft cluster assignment* for each document. $\gamma_{m,k}$ characterize the cluster centers θ_m and are basically the kth coordinate of θ_m on the topic space. Finally η_m control the mixing weights for clusters and define the probability of cluster m. $\phi_{d,n,k}$ are the variational parameters that measure the *a posteriori* probability that word $w_{d,n}$ in document d is sampled from topic k. They are related to projection of words and assumed fixed at the moment.

These equations seem to be complicated and awful, but they turn out to be quite intuitive and just follow the standard clustering procedure. In particular,

- $\psi_{d,m}$ is seen from (6) to be a multiplication of two factors p_1 and p_2, where p_1 includes the γ terms in the exponential and p_2 the η terms. Since η_m controls the probability of cluster m, p_2 acts as a *prior term* for $\psi_{d,m}$; p_1 can be seen as the *likelihood term*, because it explicitly measures the probability of generating $\mathbf{w}_d$ from cluster m by calculating the inner product of projected features and cluster centers. Therefore, (6) directly follows from Bayes' rule, and a normalization term is needed to ensure $\sum_{m=1}^{M}\psi_{d,m} = 1$.
- $\gamma_{m,k}$ is updated by summing over the *prior position* λ_k and the *empirical location*, the weighted sum of projected documents that belong to cluster k.
- Similar to $\gamma_{m,k}$, η_k is empirically updated by summing over the *belongingnesses* of all documents to cluster k. α/M acts as a prior or a *smoothing term*, shared by all the clusters.

Since these parameters are coupled, clustering is done by iteratively updating (6) and (7). Note that the words are incorporated into the clustering process *only* via the projected features $\sum_{n=1}^{N_d}\phi_{d,n,k}$. This means that the clustering is performed not in word space, but in the more informative topic space.

4.3 Updates for Projection

If ψ, γ, η are fixed, projection parameters ϕ and β are updated as:

$$\phi_{d,n,k} \propto \beta_{k,w_{d,n}}\exp\left\{\sum_{m=1}^{M}\psi_{d,m}\left[\Psi(\gamma_{m,k}) - \Psi(\sum_{i=1}^{K}\gamma_{m,i})\right]\right\}, \quad (8)$$

$$\beta_{k,j} \propto \sum_{d=1}^{D}\sum_{n=1}^{N_d}\phi_{d,n,k}\delta_j(w_{d,n}), \quad (9)$$

where $\delta_j(w_{d,n}) = 1$ if $w_{d,n}$ takes word index j, and 0 otherwise. Please recall that $\phi_{d,n,k}$ is the *a posteriori* probability that word $w_{d,n}$ is sampled from topic k, and $\beta_{k,j}$ measures the probability of generating word j from topic k. Normalization terms are needed to ensure $\sum_{k=1}^{K} \phi_{d,n,k} = 1$ and $\sum_{j=1}^{V} \beta_{k,j} = 1$, respectively. Update (9) for $\beta_{k,j}$ is quite intuitive, since we just sum up all the documents that word j occurs, weighted by their generating probabilities from topic k. For update of $\phi_{d,n,k}$ in (8), $\beta_{k,w_{d,n}}$ is the probability that topic k generates word $w_{d,n}$ and is thus the *likelihood* term; the rest exponential term defines the *prior*, i.e., the probability that document d selects topic k. This is calculated by taking into account the clustering structure and summing over all cluster centers with corresponding soft weights. Therefore, the projection model is learned via clusters of documents, not simply individual ones. Finally we iterate (8) and (9) until convergence to obtain the optimal projection.

4.4 Discussion

As guaranteed by variational EM algorithm, iteratively performing the given clustering and projection operations will improve the data likelihood monotonically until convergence, where a local maxima is obtained. The convergence is usually very fast, and it would be beneficial to initialize the algorithm using some simple projection models like pLSI.

The remaining parameters α and λ control the mixing weights π and cluster centers θ_m *a priori*, and they can also be learned by setting their partial derivatives to zero. However, there are no analytical updates for them and we have to use computational methods like Newton-Raphson method as in [2].

The PCP model can also be seen as a Bayesian generalization of the TTMM model [10], where π and θ_m are directly optimized using EM. Treating them as variables instead of parameters would bring more flexibility and reduce the impact of overfitting. We summarize the PCP algorithm in the following table:

Table 1. The PCP Algorithm

1. Initialize model parameters α, λ and β. Choose $M > 0$ and $K > 0$. Choose initial values for $\phi_{d,n,k}, \gamma_{m,k}$ and η_k.
2. **Clustering**: Calculate the projection term $\sum_{n=1}^{N_d} \phi_{d,n,k}$ for each document d and iterate the following steps until convergence:
   (a) Update cluster assignments $\psi_{d,m}$ by (6);
   (b) Update cluster centers $\gamma_{m,k}$ and mixing weights η_k by (7).
3. **Projection**: Calculate the clustering term $\sum_{m=1}^{M} \psi_{d,m} \left[\Psi(\gamma_{m,k}) - \Psi(\sum_{i=1}^{K} \gamma_{m,i}) \right]$ for each document d and iterate the following steps until convergence:
   (a) Update word projections $\phi_{d,n,k}$ by (8);
   (b) Update projection matrix β by (9).
4. Update α and λ if necessary.
5. Calculate the lower bound (5) and go to Step 2 if not converged.

5 Empirical Study

In this section we illustrate experimental results for the PCP model. In particular we compare it with other models in the following three perspectives:

- **Document Modelling**: How good is the generalization in PCP model?
- **Word Projection**: Is the projection really improved in PCP model?
- **Document Clustering**: Will the clustering be better in PCP model?

We will make comparisons based on two text data sets. The first one is Reuters-21578, and we select all the documents that belong to the five categories *money-fx, interest, ship, acq* and *grain*. After removing stop words, stemming and picking up all the words that occur at least in 5 documents, we finally obtain 3948 documents with 7665 words. The second data set consists of four groups taken from 20Newsgroup, i.e., *autos, motorcycles, baseball* and *hockey*. Each group has 1000 documents, and after the same preprocessing we get 3888 documents with 8396 words. In the following we use "Reuters" and "Newsgroup" to denote these two data sets, respectively. Before giving the main results, we illustrate one case study for better understanding of the algorithm.

5.1 Case Study

We run the PCP model on the Newsgroup data set, and set topic number $K = 50$ and cluster number $M = 20$. α is set to 1 and λ is set with each entry being $1/K$. Other initializations are chosen randomly. The algorithm runs until the improvement on $\mathcal{L}_q(\mathcal{D})$ is less than 0.01% and converges after 10 steps.

Figure 2 illustrates part of the results. In (a) 10 topics are shown with 10 words that have highest assigned probabilities in β. The topics are seen to be very meaningful and each defines one projection for all the words. For instance, topic 5 is about "bike", and 1, 7, 9 are all talking about "car" but with different subtopics: 1 is about general stuffs of car; 7 and 9 are specifying car systems and purchases, respectively. Besides finding topic 6 that covers general terms for "hockey", we even find two topics that specify the hockey teams in US (4) and Canada (8). These topics provide the building blocks for document clustering.

Figure 2(b) gives the 4 cluster centers that have highest probabilities after learning. They define topic mixtures over the whole 50 topics, and for illustration we only show the given 10 topics as in (a). Darker color means higher weight. It is easily seen that they are corresponding to the 4 categories *autos, motorcycles, baseball* and *hockey*, respectively. If we sort all the documents with their true labels, we obtain the document-cluster assignment matrix as shown in Figure 2(c). Documents that belong to different categories are clearly separated.

5.2 Document Modelling

In this subsection we investigate the generalization of PCP model. We compare PCP with pLSI and LDA on the two data sets, where 90% of the data are used

1	2	3	4	5	6	7	8	9	10
car	ball	game	gm	bike	team	car	pit	car	team
engin	runner	basebal	rochest	clutch	hockei	tire	det	price	year
ford	hit	gant	ahl	back	nhl	brake	bo	dealer	win
problem	base	pitch	st	gear	leagu	drive	tor	year	morri
mustang	write	umpir	john	front	game	radar	chi	model	cub
good	fly	time	adirondack	shift	season	oil	nyi	insur	game
probe	rule	call	baltimor	car	citi	detector	van	articl	write
write	articl	strike	moncton	time	year	system	la	write	jai
ve	left	write	hockei	work	star	engin	stl	cost	won
sound	time	hirschbeck	utica	problem	minnesota	spe	buf	sell	clemen

(a)

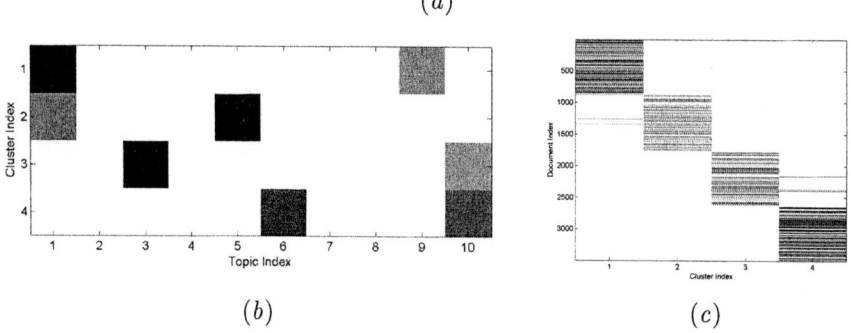

(b) (c)

Fig. 2. A case study of PCP model on Newsgroup data. (a) shows 10 topics and 10 associated words for each topic with highest generating probabilities. (b) shows 4 clusters and the topic mixture on the 10 topics. Darker color means higher value. (c) gives the assignments to the 4 clusters for all the documents.

for training and the rest 10% are held out for testing. The comparison metric is *perplexity*, which is conventionally used in language modelling and defined as $\text{Perp}(\mathcal{D}_{\text{test}}) = \exp(-\ln p(\mathcal{D}_{\text{test}})/\sum_d |\mathbf{w}_d|)$, where $|\mathbf{w}_d|$ is the length of document d. A lower perplexity score indicates better generalization performance.

We follow the formula in [2] to calculate perplexity for pLSI. For PCP model, we take the similar approach as in LDA, i.e., we run the variational inference and calculate the lower bound (5) as the likelihood term. M is set to be the number of training documents for initialization. As suggested in [2], a smoothing term for β is used and optimized for LDA and PCP. All the three models are trained until the improvement is less than 0.01%. We compare all three algorithms using different K's, and the results are shown in Table 2. PCP outperforms both pLSI and LDA in all the runs, which indicates that the model fits the data better.

5.3 Word Projection

All the three models pLSI, LDA and PCP can be seen as projection models and learn the mapping β. To compare the quality, we train a support vector machine (SVM) on the low-dimensional representations of these models and measure the

Table 2. Perplexity comparison for pLSI, LDA and PCP on Reuters and Newsgroup

K	Reuters						Newsgroup					
	5	10	20	30	40	50	5	10	20	30	40	50
pLSI	1995	1422	1226	1131	1128	1103	2171	2018	1943	1868	1867	1924
LDA	1143	892	678	599	562	533	2083	1933	1782	1674	1550	1513
PCP	1076	882	670	592	555	527	2039	1871	1752	1643	1524	1493

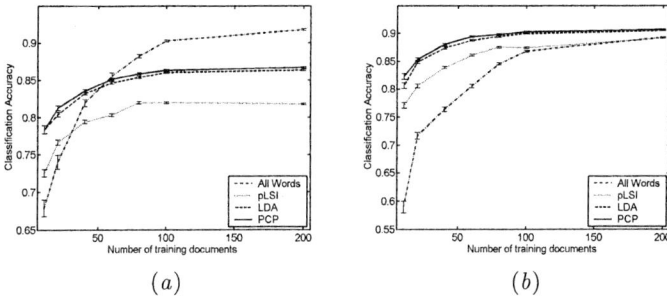

Fig. 3. Classification results on Reuters (a) and Newsgroup (b)

classification rate. For pLSI, the projection for document d is calculated as the *a posteriori* probability of latent topics conditioned on d, $p(z|d)$. This can be computed using Bayes' rule as $p(z|d) \propto p(d|z)p(z)$. In LDA it is calculated as the *a posteriori* Dirichlet parameters for d in the variational E-step [2]. In PCP model this is simply the projection term $\sum_{n=1}^{N_d} \phi_{d,n,k}$ which is used in clustering.

We train a 10-topic model on the two data sets and then train a SVM for each category. Note that we are reducing the feature space by 99.8%. In the experiments we gradually improve the number of training data from 10 to 200 (half positive and half negative) and randomize 50 times. The performance averaged over all categories is shown in Figure 3 with mean and standard deviation. It is seen that PCP obtains better results and learns a better word projection.

5.4 Document Clustering

In our last experiment we demonstrate the performance of PCP model on document clustering. For comparison we implement the original version of NMF algorithm [11] which can be shown as a variant of pLSI, and a k-means algorithm that uses the learned features by LDA. For NMF we tune its parameter to get best performance. The k-means and PCP algorithms are run with the true cluster number, and we tune the dimensionality K to get best performance.

The experiments are run on both two data sets. The true cluster number is 5 for Reuters and 4 for Newsgroup. For comparison we use the normalized mutual information [13], which is just the mutual information divided by the

Table 3. Comparison of clustering using different methods

	NMF	LDA+k-means	PCP
Reuters	0.246	0.331	**0.418**
Newsgroup	0.522	0.504	**0.622**

maximal entropy of the two cluster sets. The results are given in Table 3, and it can be seen that PCP performs the best on both data sets. This means iterating clustering and projection can obtain better clustering structure for documents.

6 Conclusions

This paper proposes a probabilistic clustering-projection model for discrete co-occurrence data, which unifies clustering and projection in one probabilistic model. Iteratively updating the two operations turns out to be the variational inference and learning under Bayesian treatments. Experiments on two text data sets show promising performance for the proposed model.

References

1. C. M. Bishop. *Neural Networks for Pattern Recognition*. Oxford University Press, 1995.
2. D. M. Blei, A. Y. Ng, and M. I. Jordan. Latent Dirichlet Allocation. *Journal of Machine Learning Research*, 3:993–1022, 2003.
3. W. Buntine and S. Perttu. Is multinomial PCA multi-faceted clustering or dimensionality reduction? In *Proceedings of the 9th International Workshop on Artificial Intelligence and Statistics*, pages 300–307, 2003.
4. S. C. Deerwester, S. T. Dumais, T. K. Landauer, G. W. Furnas, and R. A. Harshman. Indexing by latent semantic analysis. *Journal of the American Society of Information Science*, 41(6):391–407, 1990.
5. I. S. Dhillon. Co-clustering documents and words using bipartite spectral graph partitioning. In *SIGKDD*, pages 269–274, 2001.
6. C. Ding, X. He, H. Zha, and H. D. Simon. Adaptive dimension reduction for clustering high dimensional data. In *ICDM*, pages 147–154, 2002.
7. T. Hofmann. Probabilistic Latent Semantic Indexing. In *Proceedings of the 22nd Annual ACM SIGIR Conference*, pages 50–57, Berkeley, California, August 1999.
8. T. Hofmann and J. Puzicha. Statistical models for co-occurrence data. Technical Report AIM-1625, 1998.
9. M. I. Jordan, Z. Ghahramani, T. Jaakkola, and L. K. Saul. An introduction to variational methods for graphical models. *Machine Learning*, 37(2):183–233, 1999.
10. M. Keller and S. Bengio. Theme Topic Mixture Model: A Graphical Model for Document Representation. January 2004.
11. D. D. Lee and H. S. Seung. Learning the parts of objects with nonnegative matrix factorization. *Nature*, 401:788–791, Oct. 1999.
12. T. Li, S. Ma, and M. Ogihara. Document clustering via adaptive subspace iteration. In *Proceedings of SIGIR*, 2004.
13. W. Xu, X. Liu, and Y. Gong. Document clustering based on non-negative matrix factorization. In *Proceedings of SIGIR*, pages 267–273, 2003.

Collaborative Filtering on Data Streams

Jorge Mario Barajas and Xue Li

School of Information Technology and Electrical Engineering,
The University of Queensland, Brisbane, Australia
s4071254@student.uq.edu.au, xueli@itee.uq.edu.au

Abstract. Collaborate Filtering is one of the most popular recommendation algorithms. Most Collaborative Filtering algorithms work with a static set of data. This paper introduces a novel approach to providing recommendations using Collaborative Filtering when user rating is received over an incoming data stream. In an incoming stream there are massive amounts of data arriving rapidly making it impossible to save all the records for later analysis. By dynamically building a decision tree for every item as data arrive, the incoming data stream is used effectively although an inevitable trade off between accuracy and amount of memory used is introduced. By adding a simple personalization step using a hierarchy of the items, it is possible to improve the predicted ratings made by each decision tree and generate recommendations in real-time. Empirical studies with the dynamically built decision trees show that the personalization step improves the overall predicted accuracy.

1 Introduction

Nowadays an individual may have access to so many sources of information that is difficult to find the interesting information. The task of a Recommender System [1] is to find interesting items among the vast sea of information. Collaborative Filtering [2] is a recommender system technology that works by building a database of Users-Items with the ratings and then using this database to make recommendations to a user, based on the rating on an item referred to as the Target Item. The whole idea of using users' rating database is to automate the natural social process where people rely on other people's recommendations, this phenomenon is also know as "word of mouth". Collaborative Filtering algorithms can be grouped into two categories: memory-based and model-based [3]. In memory-based algorithms [4], the entire Users-Items database is used to generate a prediction while in model-based approaches [5] a model of user ratings is first developed to make predictions.

One fundamental problem of Collaborative Filtering is managing a big Users-Items database and using it effectively to make predictions about ratings on target items. Any popular online store receives thousands of orders and visits per hour, millions each month. Using all this information to provide quick and quality recommendations using the immense data collected about past users can be challenging. With an incoming data stream it is not possible to build a static

Users-Items database because there are massive volumes of records arriving continuously at a rapid rate. Generating recommendations over a data stream has the added constraints that the algorithm gets only one look of the data, there is a limit on the number of records it can store, and the recommendations must be made in real-time.

The rest of the document is organized as follows: the next section briefly describes the literature related to collaborative filtering algorithms that aim at giving recommendations in real-time. In section 3 the proposed approach for handling an incoming stream of ratings is described. Section 4 describes the experiments conducted and section 5 finishes with conclusions.

2 Related Work

Linden et al [6] proposed the item-to-item collaborative filtering algorithm that scales to massive datasets and provides recommendations in real-time by recording items occurring together. However, the similarities among items are calculated off-line. The off-line batch process is followed by most collaborative filtering algorithms to provide recommendations in real-time, but this produces an outdated model where the quality of the recommendations is low. Online algorithms in Collaborative Filtering [7] are more suitable for handling an incoming data stream since these are fast, incremental and there is no need to store all the previously seen examples. The first online algorithm applied to collaborative filtering was the Weighted Majority Prediction (WMP) [8], Delgado et al [9] extended this approach for multi-valued ratings. Papagelis et al [10] developed a method to incrementally update similarities among users, and Domingos et al [11] proposed VFDT, a system that allows the building of decision trees dynamically to mine data streams.

3 Proposed Approach

The goal of data stream processing is to mine patterns, process queries and compute different statistics on data streams in real-time [12]. The proposed approach attempts to deal with the prediction problem of Collaborative Filtering when the ratings are received over a continuous data stream. The prediction problem in Collaborative Filtering over a data stream is defined as follows. Having a list of items I and a set of online users U, an incoming stream S of opinions $\langle U_i, \{I_j, ..., I_k\}, \{O_j, ..., O_k\}\rangle$ is received, where U_i identifies the i-th online user, $\{I_j, ..., I_k\} \subset I$ is the set of items rated by U_i, and $\{O_j, ..., O_k\}$ are the opinions of user U_i on items $\{I_j, ..., I_k\}$. The task is to predict in real-time, the opinion O_b of the online user U_i on a target item I_b where $I_b \notin \{I_j, ..., I_k\}$.

The main idea of the proposed approach is to build a decision tree for every item by assuming they are all related to each other, using a very fast algorithm to handle the rapid incoming stream of ratings effectively and then personalizing the predicted rating made by the decision tree by scaling it up or down depending on the hierarchy of items liked by the user. The scaling of the decision trees predicted value with the items' hierarchy deals with the inevitable hit in accuracy that is introduced when building the decision trees over the incoming stream. The next sections describe the proposed approach in detail.

3.1 Building Decision Trees Dynamically

Observing the layout of the Users-Ratings database, the prediction of the rating for a target item by the active user can be seen as a classification problem [13] where an attribute value is found instead of trying to come up with a number between the ratings's range. The classification problem view can be applied easily to both scaled or binary ratings. With 0-5 rating values, each item has 6 classes and with binary ratings only 2 ('Disliked' and 'Liked'). Having N items, the classification problem involves $N - 1$ attributes (the rest of the items in the database) and the class is the target's item rating.

For every item it is necessary to find other items that have stronger predictive capacity so they can be used in the prediction of the item's rating. If someone purchases a cereal he would eventually need milk, or if someone purchases a lamp he would need a bulb. The idea is to find these items that are strongly associated with the target item and use their ratings to predict the target's rating. Decision trees are useful for this task, but with an incoming data stream they need to be built dynamically since it is not possible to load all the examples into memory. The VFDT learning system [11] allows the building of a decision tree on a data stream by considering a small subset of cases to find the best attribute to make a split decision using information gain. By using VFDT, it is possible to build a decision tree for every item in the database when the ratings are received over a data stream as shown in figure 1. Nevertheless, the dynamic building of decision trees for each item brings a trade-off between the accuracy and the amount of memory used to store the records statistics.

Time	Coke	Pepsi	Beer	Wine	Lettuce	Tomato	Onion	Broccoli
10:51:11	'1'	'4'	'4'	'2'	'2'	'2'	'2'	'5'
10:51:12	'3'	'2'	'2'	'2'	?	'1'	'5'	'2'
10:51:13	'1'	?	'3'	'1'	'4'	'4'	?	'2'
10:51:14	'1'	?	'2'	'3'	'2'	'3'	?	'2'
10:51:15	'3'	'1'	'2'	?	'2'	?	'3'	?
10:51:16	'1'	?	'3'	'1'	'3'	'2'	?	'4'
10:51:17	'4'	'4'	?	?	'3'	'3'	'3'	?

Fig. 1. Building decision trees for every item in the database dynamically

3.2 The Hierarchy of Items

Once the prediction of a rating for an active user has been made, using the dynamically built decision trees, the predicted rating can be improved. Each decision tree provides a predicted rating that has been proven to be true with the general statistical properties of the past examples, no true personalization of the results has been done. An accuracy hit has also been added to the decision trees by building them dynamically. There is still the issue of having to operate on a data stream where there is little or no information at all about past users to use in this personalization step. Here is where some information known or derived about the items can be used online to improve the prediction made by the decision tree.

The hierarchy of items [14] can be used to make certain assumptions about the users' interests. One possible way to start understanding a user's interests is when he starts pointing out some of the items that will be interesting to him. If a hierarchy of the items liked by the user is built, this hierarchy can be used to look for items that belong to categories contained in this hierarchy and provide recommendations for that user. If an item was predicted as disliked by the active user by the item's decision tree and this item belongs to one of the categories contained in the active user's hierarchy of liked items, it is possible to argue that the decision tree is not completely applicable to this user and the rating should be scaled up.

There are some important aspects to take into account when using the hierarchy of items to scale up or down the predicted rating given by the dynamically built decision trees. Scaling up all the ratings of items belonging to the built hierarchy that were marked as disliked by the decision tree can be too aggressive. A user might go to purchase groceries at a supermarket where the taxonomy of items is varied and complex, his list of groceries will include many items of many taxonomic categories but he will probably only be interested in a few products that belong to his favorite category (e.g. Cereals) and he usually purchases a lot of items belonging to this single category. Is in this case where it makes sense to scale up all the ratings of the items predicted as disliked belonging to the user's favorite category or set of favorite categories so he can receive recommendations about items he hasn't seen but belong to the favorite category of the active user.

4 Experiments and Results

A series of experiments were conducted to examine the prediction ability of the decision trees built dynamically for each item and the improving capability of using a hierarchy of items. In particular, for the decision trees it was important to first examine which attribute values definition had a better performance for building dynamically the decision trees and how these performed when exposed to different number of attributes. VFDT [11] was the algorithm used to build the decision trees for each item. Finally, the improving ability of using a hierarchical taxonomy of items to change the predictions given by each decision tree was evaluated.

4.1 Experiments Setup

The proposed approach was evaluated with EachMovie [1]. EachMovie is a dataset of movie ratings made publicly available by Digital Equipment Corporation (DEC)

---

[1] http://research.compaq.com/SRC/eachmovie/

for Collaborative Filtering research. The dataset contains the numeric ratings for 1628 movies given by 72916 users. Each numeric rating is a score from 0.0 to 1.0 with 0.2 spaces that maps linearly to the zero-to-five star rating were a five star rating stands for a 'liked it very much' and a 0 star rating for a 'Didn't or won't like it'. Each vote is accompanied by a time stamp. For the experiments a subset of 500 movies with the greatest number of votes was first extracted. The entries were ordered by the time stamp to simulate an incoming data stream. The evaluation metrics used were the Mean Absolute Error (MAE) and the Mean Squared Error (MSE) which are shown in equations 1 and 2 and where p_i and q_i is the predicted-rating pair for item i, and N is the total number of items. These two metrics were measured for the whole database and for each movie.

$$MAE = \frac{\sum_{i=0}^{N} |p_i - q_i|}{N} \quad (1)$$

$$MSE = \frac{\sum_{i=0}^{N} |p_i - r_i|^2}{N} \quad (2)$$

4.2 Experiment 1: Decision Trees' Attributes Evaluation

In our first experiment, the goal was to find out which was the best way to build the decision trees for every item in terms of the definition of the attributes' values. The evaluation metrics were calculated with the distance between the predicted rating class and the real rating given by the user. Only 50 Movies with the greatest number of votes were used. VFDT's parameters were initialized to nominal values: $\tau = 0.055$, $\delta = 0.0000001$ and $n_{min} = 250$. The results are shown in figure 2.

The decision trees built with 7 cardinal valued attributes outperformed the others built with different attribute values. In this setting, every unknown rating

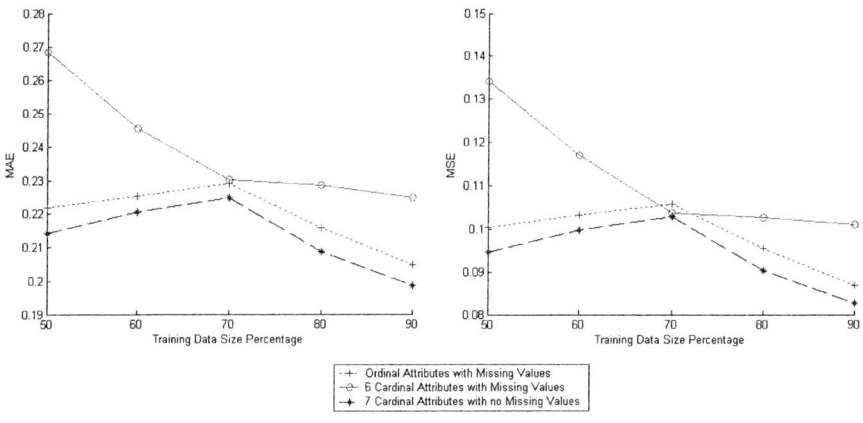

Fig. 2. Dynamically built decision trees performance with different types of attributes

was assigned to a new 'Undefined' class so there weren't any missing values. The decision trees built with continuous attributes also performed well compared with the decision trees built with 6 valued attributes and missing values.

4.3 Experiment 2: Number of Attributes for Decision Trees

To see how the decision trees behaved when exposed to more attributes and more data sparsity, movies with the greatest number of votes were added from the unused subset. These experiments used 80 percent of the dataset to build the decision trees, and the other 20 percent was used to test them measuring the MAE and MSE as in the first experiment. VFDT's parameters were set to $\tau = 0.1$, $\delta = 0.0000001$ and $n_{min} = 250$. Results are shown in table 1. The decision trees performed relatively stable when exposed to more data. The more attributes are available, the lower the MAE for the best performing decision tree. This can be explained by the fact that for building each decision tree there are more items so there is a bigger chance of finding items with stronger predictive capacity and the target item's predicted rating is more accurate. On the other hand, the worst performing decision trees decreased their accuracy with more attributes because some items didn't have enough examples to build an accurate decision tree.

4.4 Experiment 3: Using the Items' Hierarchy

In order to improve the decision's trees results using the item's hierarchy and face the accuracy trade off by dynamically building the decision trees, the genre classification about each movie was used. In the EachMovie dataset, each movie belongs to one or more genres, to deal with this, the set of genres of each movie were treated as one category; that is, if a movie belonged to the genres action and drama, a category called 'Action and Drama' was created and all the movies belonging to only these two genres were added to the category. The settings of experiment 4.3 were used and a hierarchy of items was built with the items liked by the user: those that had a rating greater or equal to 4 stars. If the target item didn't belong to any of the genres contained in the hierarchy of items, the decision trees' predicted rating were decreased by one if it was liked, if the movie

Table 1. Errors of the dynamically built decision trees

Movies	MAE	MSE	MAE Best DT	MAE Worst DT	MSE Best DT	MSE Worst DT
100	0.2062	0.0896	0.1089	0.5401	0.0319	0.4130
200	0.2166	0.0984	0.0676	0.5401	0.0370	0.4130
300	0.2224	0.1043	0.0676	0.5581	0.0291	0.4327
400	0.2238	0.1062	0.0676	0.5687	0.0336	0.4327
500	0.2287	0.1111	0.0676	0.7526	0.0336	0.6071

belonged to the most popular set of genres in the hierarchy of liked items by the user, the decision trees' predicted rating was increased by one if it was disliked (less than or equal to '3'), otherwise the decision trees' predicted rating was left unchanged. Results are shown in table 2.

Table 2. Hierarchy Personalization Errors

Movies	MAE	MSE	MAE Best DT	MAE Worst DT	MSE Best DT	MSE Worst DT
100	0.2039	0.0818	0.1074	0.5051	0.0296	0.3608
200	0.2127	0.0891	0.0692	0.4801	0.0316	0.3224
300	0.2185	0.0946	0.0668	0.5437	0.0302	0.3866
400	0.2203	0.0965	0.0721	0.5424	0.0331	0.3853
500	0.2242	0.1002	0.0709	0.7522	0.0331	0.6065

The use of the liked items hierarchy per user effectively improved the decision trees' results. Each decision tree was built dynamically so there was an inevitable tradeoff in the accuracy of the predicted rating and the amount of memory used. As we can see from table 4 that the MSE was lowered, meaning that the big errors produced by the decision trees were reduced. In all cases the performance of prediction, with the worst-case errors, is improved.

5 Conclusions

This paper introduced a novel approach to providing recommendations with Collaborative Filtering over an incoming data stream containing user rating over items. By dynamically building decision trees for every item, it is possible to deal with an incoming data stream and use all received data to generate recommendations. A hierarchy of items was used to improve the personalization of each generated decision tree. Moreover, by using different types of hierarchies of items, the same set of decision trees may behave differently for the different mappings between the items and their categories. So, different item hierarchies can lead to different predictions. For example a hierarchy of items might be a taxonomy that focuses on understanding of user profile. Then the recommendation will be based on user tastes. While another hierarchy could group popular and unpopular items that will lead to some novel recommendations.

The results of our experiments have shown that our proposed approach of building the multiple decision trees on the fly for the real-time recommendation is effective and efficient.

Acknowledgements

This work is partially founded by the Australian ARC Large Grant DP0558879.

References

1. Resnick, P., Varian, H.R.: Recommender systems. Commun. ACM **40** (1997) 56–58
2. Goldberg, D., Nichols, D., Oki, B.M., Terry, D.: Using collaborative filtering to weave an information tapestry. Commun. ACM **35** (1992) 61–70
3. Breese, J.S., Heckerman, D., Kadie, C.: Empirical analysis of predictive algorithms for collaborative filtering. In: Fourteenth Annual Conference on Uncertainty in Artificial Intelligence. (1998) 43–52
4. Sarwar, B.M., Karypis, G., Konstan, J.A., Reidl, J.: Item-based collaborative filtering recommendation algorithms. In: World Wide Web. (2001) 285–295
5. Hofmann, T.: Collaborative filtering via gaussian probabilistic latent semantic analysis. In: SIGIR '03: Proceedings of the 26th annual international ACM SIGIR conference on Research and development in informaion retrieval, New York, NY, USA, ACM Press (2003) 259–266
6. Linden, G., Smith, B., York, J.: Industry report: Amazon.com recommendations: Item-to-item collaborative filtering. IEEE Distributed Systems Online **4** (2003)
7. Calderón-Benavides, M.L., González-Caro, C.N., de J. Pérez-Alcázar, J., García-Díaz, J.C., Delgado, J.: A comparison of several predictive algorithms for collaborative filtering on multi-valued ratings. In: SAC '04: Proceedings of the 2004 ACM symposium on Applied computing, New York, NY, USA, ACM Press (2004) 1033–1039
8. Nakamura, A., Abe, N.: Collaborative filtering using weighted majority prediction algorithms. In: ICML '98: Proceedings of the Fifteenth International Conference on Machine Learning, San Francisco, CA, USA, Morgan Kaufmann Publishers Inc. (1998) 395–403
9. Delgado, J., Ishii, N.: Memory-based weighted-majority prediction for recommender systems. In: Proceedings of the ACM SIGIR-99. (1999)
10. Papagelis, M., Rousidis, I., Plexousakis, D., Theoharopoulos, E.: Incremental collaborative filtering for highly-scalable recommendation algorithms. In: International Symposium on Methodologies of Intelligent Systems (ISMIS'05). (2005)
11. Domingos, P., Hulten, G.: Mining high-speed data streams. In: Knowledge Discovery and Data Mining. (2000) 71–80
12. Garofalakis, M.N., Gehrke, J.: Querying and mining data streams: You only get one look. In: VLDB. (2002)
13. Basu, C., Hirsh, H., Cohen, W.W.: Recommendation as classification: Using social and content-based information in recommendation. In: AAAI/IAAI. (1998) 714–720
14. Ganesan, P., Garcia-Molina, H., Widom, J.: Exploiting hierarchical domain structure to compute similarity. ACM Trans. Inf. Syst. **21** (2003) 64–93

The Relation of Closed Itemset Mining, Complete Pruning Strategies and Item Ordering in Apriori-Based FIM Algorithms

Ferenc Bodon[1],* and Lars Schmidt-Thieme[2]

[1] Department of Computer Science and Information Theory,
Budapest University of Technology and Economics
bodon@cs.bme.hu
[2] Computer Based New Media Group (CGNM),
Albert-Ludwigs-Universität Freiburg
lst@informatik.uni-freiburg.de

Abstract. In this paper we investigate the relationship between closed itemset mining, the complete pruning technique and item ordering in the Apriori algorithm. We claim, that when proper item order is used, complete pruning does not necessarily speed up Apriori, and in databases with certain characteristics, pruning increases run time significantly. We also show that if complete pruning is applied, then an intersection-based technique not only results in a faster algorithm, but we get free closed-itemset selection concerning both memory consumption and run-time.

1 Introduction

Frequent itemset mining (FIM) is a popular and practical research field of data mining. Techniques and algorithms developed here are used in the discovery of association rules, sequential patterns, episode rules, frequent trees and subgraphs, and classification rules. The set of *frequent closed itemsets* (FC) is an important subset of the frequent itemsets (F) because it offers a compact representation of F. This means that FC contains fewer elements, and from FC we can completely determine the frequent itemsets [1].

Over 170 FIM and FCIM algorithms have been proposed in the last decade, each claiming to outperform its existing rivals [2]. Thanks to some comparisons from independent authors (the FIMI competitions [2] are regarded to be the most important), the chaos seems to be settling. The most successful algorithms are Apriori [3], ECLAT [4][5], FP-growth [6] and variants of these. Adaptations of these algorithms used to extract closed itemsets are also the most popular and most efficient FCIM algorithms.

Apriori is regarded to be the first FIM algorithm that can cope with large datasets. One of the most important surprises of the FIM competition was that

---

* This work was supported in part by OTKA Grants T42481, T42706, TS-044733 of the Hungarian National Science Fund, NKFP-2/0017/2002 project Data Riddle and by a Madame Curie Fellowship (IHP Contract nr. HPMT-CT-2001-00251).

this algorithm is competitive regarding run time (particularly at high support thresholds), and its memory need was outstandingly low in many cases. Moreover, the resulting closed extension, Apriori-Close [1], is the best algorithm for certain sets of test. An inherent feature of Apriori is *complete pruning*, which only allows the generation of candidates that possess only frequent subsets. Due to complete pruning Apriori never generated more candidates than those algorithms which traverse the itemset space in a depth-first manner (DFS algorithms), as do Eclat and FP-growth. Complete pruning in Apriori is considered to be so essential, that the frequent pattern mining community has accepted it as a rule of thumb.

In this paper, we investigate the efficiency of complete pruning, and draw the surprising conclusion, that this technique is not as necessary as once believed. If the database has a certain characteristic, then pruning may even slow down Apriori. We also show, that the efficiency of pruning depends on the item ordering used during the algorithm.

We also investigate the connection between pruning and closed-itemset selection. By presenting a novel pruning strategy, we will show that closed-itemset mining comes for free. In Apriori-Close, this does not hold because closed itemset selection is merged into the phase where infrequent candidates are removed, and requires many scans of the data structure which stores the frequent itemsets. This can be saved by applying our new pruning strategy.

2 Problem Statement

Frequent itemset mining came from efforts to discover useful patterns in customers' transaction databases. A customers' transaction database is a sequence of transactions ($\mathcal{T} = \langle t_1, \ldots, t_n \rangle$), where each transaction is an itemset ($t_i \subseteq \mathcal{I}$). An itemset with k elements is called a k-itemset. The *support* of an itemset X in $\mathcal{T}$, denoted as $supp_\mathcal{T}(X)$, is the number of transactions containing X, i.e. $supp_\mathcal{T}(X) = |\{t_j : X \subseteq t_j\}|$. An itemset is *frequent* if its support is greater than a *support threshold*, originally denoted by min_supp. The frequent itemset mining problem is to discover all frequent itemsets in a given transaction database.

Itemset I is *closed* if no proper superset of I exists that has the same support as I. The set of closed itemsets is a compact representation of the frequent itemsets. All frequent itemsets together with their supports can be generated if only the closed itemsets and their supports are known. In some databases the number of closed itemsets is much smaller than the number of frequent sets, thus it is an important data mining task to determine FCI.

The concepts of *negative border* and *order-based negative border* play an important role in our contributions. Let F be the set of frequent itemsets, and $\prec$ a total order on the elements of $2^\mathcal{I}$. The negative border of F is the set of itemsets, whose elements are infrequent, but all their proper subsets are frequent (formally: $NB(F) = \{I | I \notin F, \forall I' \subset I, I' \in F\}$). The order-based negative border (denoted by $NB^\prec(F)$) is a superset of $NB(F)$. An itemset I is element of $NB^\prec(F)$, if I is not frequent, but the two smallest ($|I| - 1$)-subsets of I

are frequent. Here "smallest" is understood with respect to a fixed ordering of items. For example, if $\mathcal{I} = \{A, B, C\}$ and $F = \{\emptyset, A, B, C, AB, AC\}$ then $NB(F) = \{BC\}$ and $NB^{\prec}(F) = \{BC, ABC\}$ if $\prec$ is the alphabetic order.

In the rest of the paper the ascending and descending order according to supports of the items are denoted by $\prec_D$ and $\prec_A$ respectively.

The Apriori algorithm plays a central role in frequent itemset mining. Although it is one of the oldest algorithms, the intensive researches that polished its data structure and implementation specific issues [7] [8] have raised it to a competitive algorithm which outperforms the newest DFS algorithms in some cases [2]. We assume that the reader is familiar with the Apriori algorithm. In this paper we concentrate on its candidate generation method.

3 Candidate Generation of Apriori

To understand our claims we also need to understand the main data structure of Apriori, i.e. the *trie* (also called *prefix-tree*). The data structure trie was originally introduced by de la Briandais [9] and Fredkin to store and efficiently retrieve words (i.e. sequence of letters) of a dictionary. In the FIM setting the alphabet is the set of items, and the itemsets are converted to sequences by a predefined order. A trie is a rooted, (downward) directed tree. The root is defined to be at depth 0, and a node at depth d can point to nodes at depth $d+1$. A pointer is also called *edge* or *link*, which is labeled by an item. If node u points to node v, then we call u the *parent* of v, and v is a *child node* of u. Nodes with no child are called *leaves*.

Every leaf ℓ represents an itemset which is the union of the letters in the path from the root to ℓ. Note that if the first k letters are the same in two words, then the first k steps on their paths are the same as well. In the rest of the paper the node that represents itemset I is referred to node I. For more details about the usage of the trie data structure in Apriori the reader is referred to [7][8].

In Apriori's candidate generation phase we generate $(\ell + 1)$-itemset candidates. Itemset I becomes a candidate if all proper subsets of I are frequent. The trie that stores the frequent items, supports this method. Each itemset that fulfills the complete pruning requirement can be obtained by taking the union of the representations of two sibling nodes. In the so called *simple pruning* we go through all nodes at depth $\ell-1$, take the pairwise union of the children and then check all subsets of the union if they are frequent. Two straightforward modifications can be applied to reduce unnecessary work. On one hand, we do not check those subsets that are obtained by removing the last and the one before the last elements of the union. On the other hand, the prune check is terminated as soon as a subset is infrequent, i.e. not contained in the trie.

3.1 Pruning by Intersection

A problem with the simple pruning method is that it unnecessarily travels some part of the trie many times. We illustrate this by an example. Let $ABCD$,

$ABCE$, $ABCF$, $ABCG$ be the four frequent 4-itemsets. When we check the subsets of potential candidates $ABCDE$, $ABCDF$, $ABCDG$ then we travel through nodes ABD, ACD and BCD three times. This gets even worse if we take into consideration all potential candidates that stem from node ABC. We travel to each subset of ABC 6 times.

To save these superfluous traversals we propose an *intersection-based pruning* method. We denote by u the current leaf that has to be extended, the depth of u by ℓ, the parent of u by P and the label that is on the edge from P to u by i. To generate new children of u, we do the following. First determine the nodes that represent all the $(\ell - 2)$-element subsets of the $(\ell - 1)$-prefix. Let us denote these nodes by $v_1, v_2, \ldots, v_{\ell-1}$. Then find the child v'_j of each v_j that is pointed by an edge with label i. If there exists a v_j that has no edge with label i (due to the dead-end branch removal), then the extension of u is terminated and the candidate generation continues with the extension of u's sibling (or with the next leaf, if u does not have any siblings). The complete pruning requirement is equivalent to the condition that only those labels can be on an edge that starts from u, which are labels of an edge starting from v'_j and labels of one starting from P. This has to be fulfilled for each v'_j, consequently, the labels of the new edges are exactly the intersection of labels starting from v'_j and P nodes.

The siblings of u have the same prefix as u, hence, in generating of the children of siblings, we can use the same $v_1, v_2, \ldots, v_{\ell-1}$ nodes. It is enough to find their children with the proper label (the new v'_j nodes) and to make the intersection of the labels of edges that starts from the prefix and the new $v'_1, v'_2, \ldots, v'_{\ell-1}$. This is the real advantage of this method. The $(\ell - 2)$-subset nodes of the prefix are reused, hence the paths representing the subsets are traversed only once, instead of $\binom{n}{2}$, where n is the number of the children of the prefix.

As an illustrative example let us assume that the trie that is obtained after removing infrequent itemsets of size 4 is depicted in Fig. 1.

To get the children of node $ABCD$ that fulfill complete pruning requirement (all subsets are frequent), we find the nodes that represent the 2-subsets of the prefix (ABC). These nodes are denoted by v_1, v_2, v_3. Next we find their children that are reached by edges with label D. These children are denoted by v'_1, v'_2

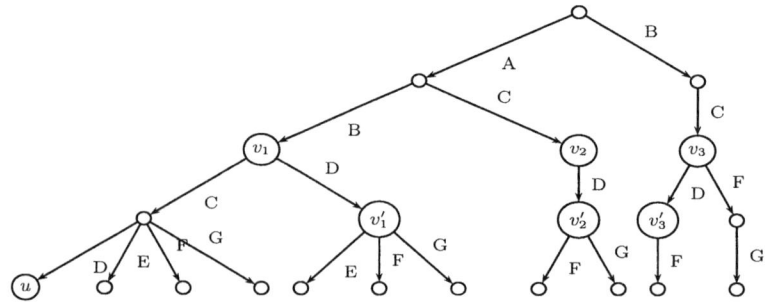

Fig. 1. Example: intersection-based pruning

and v'_3 in the trie. The intersection of the label sets associated to the children of the prefix, v'_1, v'_2 and v'_3 is: $\{D,E,F,G\} \cap \{E,F,G\} \cap \{F,G\} \cap \{F\} = \{F\}$, hence only one child will be added to node $ABCD$, and F will be the label of this new edge.

3.2 Closed Itemset Selection

Closed itemsets can be retrieved from frequent itemsets by post-processing, but it results in a faster solution, if the selection is pushed into the FIM algorithm. In Apriori-Close the infrequent candidate deletion is extended by a step, where the subsets of the frequent candidate are checked. By default all subsets are marked as closed, which is changed if the subsets' support equals to the candidate's actually examined. Consequently, in Apriori-Close all subsets of the frequent candidates are generated, which mean many travels in the trie.

These superfluous travels are avoided if the closed itemset filtering is done in the candidate generation phase and intersection-based pruning is applied. In this method the subsets are already determined, hence checking support equivalence does not require any extra travels.

4 Item Ordering and the Pruning Efficiency

Our previous work has [8] shown that the order of items used in the trie to convert itemsets to sequences greatly affects both run-time and memory need of the Apriori. Next we show that the efficacy of complete pruning is also influenced by this factor.

The advantage of the pruning is to reduce the number of candidates. The number of candidates in Apriori equals to the number of frequent itemsets plus the number of infrequent candidates, i.e. the negative border of the frequent itemsets. If pruning is not used then the number of infrequent candidates becomes the size of the order-based negative border, where the order corresponds to the order used in the trie. It follows, that if we want to decrease the redundant work (i.e determining a support of the infrequent candidates) then we have to use the order that results in the smallest order-based negative border. This comes into play in all DFS algorithms, so we already know the answer: the ascending order according to supports achieves in most cases the best result. This is again a rule of thumb, that works well on real and synthetic datasets. The statement cannot be proven unless the distribution of the items is known and the independence of the items is assumed.

The disadvantage of the pruning strategy is simple: we have to traverse some part of the trie to decide if all subsets are frequent or not. Obviously this needs some time.

Here we state that pruning is not necessarily an important part of Apriori. This statement is supported by the following observation, that applies in most cases:
$$|NB^{\prec_A}(F) \setminus NB(F)| \ll |F|.$$

The left-hand side of the inequality gives the number of infrequent itemsets that are not candidates in the original Apriori, but are candidates in Apriori-NOPRUNE. So the left-hand side is proportional to the extra work to be done by omitting pruning. On the other hand, $|F|$ is proportional to the extra work done with pruning. Candidate generation with pruning checks all the subsets of each element of F, while Apriori-NOPRUNE does not. The outcomes of the two approaches are the same for frequent itemsets, but the pruning-based solution determines the outcome with much more work (i.e traverses the trie many times).

Although the above inequality holds for most cases, this does not imply that pruning is unnecessary, and slows down Apriori. The extra work is just proportional to the formulas above. Extra work caused by omitting pruning means determining the support of some candidates, which is affected by many factors, such as the size of these candidates, the number of transactions, the number of elements in the transactions, and the length of matching prefixes in the transaction. The extra work caused by pruning comes in a form of redundant traversals of the tree during checking the subsets.

As soon as pruning strategy is omitted, Apriori can be further tuned by merging the candidate generation and the infrequent node deletion phases. After removing the infrequent children of a node, we extend each child the same way as we would do in candidate generation. This way we spare an entire traversal of the trie.

5 Experiments

All tests were carried out on 16 public "benchmark" databases, which can be downloaded from the FIMI repository[1]. Results would require too much space, hence only the most typical ones are shown below. All results, all programs as well as the test scripts can be downloaded from http://www.cs.bme.hu/~bodon/en/fim/test.html. For Apriori implementation we have used an improved version of our code, that took part in the FIMI'04 competition, and reached many times outstanding results concerning memory requirement. Due to the improvements it is now a true rival of the best Apriori implementation [7] and outperforms it in many cases. The code can be downloaded from http://fim.informatik.uni-freiburg.de.

Comparing just pruning techniques, Apriori-IBP (Apriori that uses intersection-based pruning) was always faster than Apriori-SP (simple pruning), however, the differences were insignificant in many cases. The intersection-based pruning was 25% - 100% faster than the original solution at databases BMS-WebView-1, BMS-WebView-2, T10I5N1KP5KC0.25D200K.

It is not so easy to declare a winner in the competition of Apriori-IBP and Apriori-NOPRUNE. Apriori-NOPRUNE was faster in 85% of the tests, however in most cases the difference was under 10%. Using low support threshold six measurements showed significant differences. In the case of BMS-WebView-1 and

---

[1] http://fimi.cs.helsinki.fi/data/

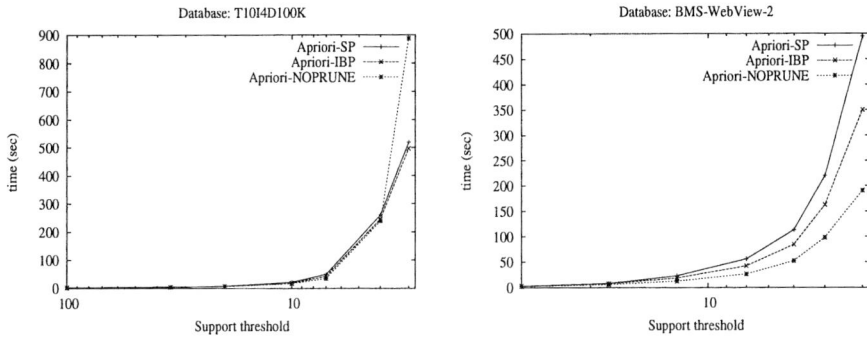

Fig. 2. Candidate generation with different pruning strategies

BMS-WebView-2 Apriori-NOPRUNE was fast twofold faster than Apriori-IBP, but in T10I4D100K the contrary was true. Figure 2 shows the run-times of these cases, and the run-time of a typical result (Apriori-IBP slightly faster than Apriori-SP; Apriori-NOPRUNE is 10%-20% faster than Apriori-IBP).

To understand why Apriori-IBP was the faster in the first case and why Apriori-NOPRUNE in the second, we have to examine the number of candidates generated by the two algorithms and the number of frequent itemsets. These data are summarized in the next table (for the sake of better readability the numbers of itemsets are divided by 1000).

Table 1. Number of frequent itemsets and number of candidates

database	min_{supp}	$|F|$	$|NB(F)|$	$|NB^{\prec_A}|$	$|NB^{\prec_D}|$	$\dfrac{|NB^{\prec_A}| - |NB|}{|F|}$
T10I4D100K	3	5 947	39 404	92 636	166 461	8.95
BMS-WebView-2	4	60 083	3 341	9 789	197 576	0.11

The data support our hypothesis. The ratio of number of frequent itemset to the difference of the two negative borders greatly determines pruning efficiency. Obviously, if the number of extra candidates is insignificant compared to the number of frequent itemsets, then pruning slows down Apriori. The table also shows the importance of the proper order. In the case of T10I4D100K the $|NB^{\prec_D}|$ is so large that the algorithm did not fit in the 2GB of main memory. This does not occur at BMS-WebView-2, but the number of infrequent candidates was 20-fold more compared to the ascending order based solution.

6 Conclusions

In this paper, we have proposed an intersection-based pruning strategy that outperforms the classic candidate-generation method. The other advantage of the method is that closed-itemset selection comes for free. Since the new candidate-generation method does not affect any other part of the algorithm, it can also be applied in Apriori-Close to obtain an improved version.

The major contribution of the paper is the investigation of the pruning efficiency in Apriori. We claim that, if ascending order is used, then pruning does not necessarily speed-up the algorithm, and if $(|NB^{\prec_A}(F)| - |NB(F)|)/|F|$ is small, then the run-time increases in most cases. Note that this conclusion does not only affect Apriori and its variants, but also all those Apriori modifications that discover other type of frequent patterns, like sequences, episodes, boolean formulas, trees or graphs. Since in such cases subpattern inclusion check is more complicated (for example in the case of labeled graphs this requires a graph isomorphism test) the difference can be more significant, and thus needs to be investigated.

References

1. Pasquier, N., Bastide, Y., Taouil, R., Lakhal, L.: Pruning closed itemset lattices for association rules. In: Proceedings of the 14th BDA French Conference on Advanced Databases, Hammamet, Tunisie (1998) 177–196
2. Goethals, B., Zaki, M.J.: Advances in frequent itemset mining implementations: Introduction to fimi03. In: Proceedings of the IEEE ICDM Workshop on Frequent Itemset Mining Implementations (FIMI'03). Volume 90 of CEUR Workshop Proceedings., Melbourne, Florida, USA (2003)
3. Agrawal, R., Srikant, R.: Fast algorithms for mining association rules. In Bocca, J.B., Jarke, M., Zaniolo, C., eds.: Proceedings of the 20th International Conference on Very Large Data Bases (VLDB), Chile, Morgan Kaufmann (1994) 487–499
4. Zaki, M.J., Parthasarathy, S., Ogihara, M., Li, W.: New algorithms for fast discovery of association rules. In Heckerman, D., Mannila, H., Pregibon, D., Uthurusamy, R., Park, M., eds.: Proceedings of the 3rd International Conference on Knowledge Discovery and Data Mining, California, USA, AAAI Press (1997) 283–296
5. Schmidt-Thieme, L.: Algorithmic features of eclat. In: Proceedings of the IEEE ICDM Workshop on Frequent Itemset Mining Implementations (FIMI'04). Volume 126 of CEUR Workshop Proceedings., Brighton, UK (2004)
6. Han, J., Pei, J., Yin, Y.: Mining frequent patterns without candidate generation. In: Proceedings of the 2000 ACM SIGMOD international conference on Management of data, Dallas, Texas, United States, ACM Press (2000) 1–12
7. Borgelt, C.: Efficient implementations of apriori and eclat. In: Proceedings of the IEEE ICDM Workshop on Frequent Itemset Mining Implementations (FIMI'03). Volume 90 of CEUR Workshop Proceedings., Melbourne, Florida, USA (2003)
8. Bodon, F.: Surprising results of trie-based fim algorithms. In: Proceedings of the IEEE ICDM Workshop on Frequent Itemset Mining Implementations (FIMI'04). Volume 126 of CEUR Workshop Proceedings., Brighton, UK (2004)
9. de la Briandais, R.: File searching using variable-length keys. In: Proceedings of the Western Joint Computer Conference. (1959) 295–298

Community Mining from Multi-relational Networks*

Deng Cai[1], Zheng Shao[1], Xiaofei He[2], Xifeng Yan[1], and Jiawei Han[1]

[1] Computer Science Department, University of Illinois at Urbana Champaign
{dengcai2, zshao1, xyan, hanj}@cs.uiuc.edu
[2] Computer Science Department, University of Chicago
xiaofei@cs.uchicago.edu

Abstract. Social network analysis has attracted much attention in recent years. Community mining is one of the major directions in social network analysis. Most of the existing methods on community mining assume that there is only one kind of relation in the network, and moreover, the mining results are independent of the users' needs or preferences. However, in reality, there exist multiple, heterogeneous social networks, each representing a particular kind of relationship, and each kind of relationship may play a distinct role in a particular task. In this paper, we systematically analyze the problem of mining hidden communities on heterogeneous social networks. Based on the observation that different relations have different importance with respect to a certain query, we propose a new method for learning an optimal linear combination of these relations which can best meet the user's expectation. With the obtained relation, better performance can be achieved for community mining.

1 Introduction

With the fast growing Internet and the World Wide Web, Web communities and Web-based social networks are flourishing, and more and more research efforts have been put on Social Network Analysis (SNA) [1][2]. A social network is modeled by a graph, where the nodes represent individuals, and an edge between nodes indicates that a direct relationship between the individuals. Some typical problems in SNA include discovering groups of individuals sharing the same properties [3] and evaluating the importance of individuals [4][5]. In a typical social network, there always exist various relationships between individuals, such as friendships, business relationships, and common interest relationships.

Most of the existing algorithms on social network analysis assume that there is only one single social network, representing a relatively homogenous relationship (such as Web page linkage). In real social networks, there always exist

---

* The work was supported in part by the U.S. National Science Foundation NSF IIS-02-09199/IIS-03-08215. Any opinions, findings, and conclusions or recommendations expressed in this paper are those of the authors and do not necessarily reflect the views of the funding agencies.

various kinds of relations. Each relation can be treated as a **relation network**. Such kind of social network can be called *multi-relational social network* or *heterogeneous social network*, and in this paper the two terms will be used interchangeably depending on the context. These relations play different roles in different tasks. To find a community with certain properties, we first need to identify which relation plays an important role in such a community. Moreover, such relation might not exist explicitly, we might need to first discover such a hidden relation before finding the community on such a relation network.

Such a problems can be modeled mathematically as relation selection and extraction in *multi-relational* social network analysis. The problem of relation extraction can be simply stated as follows: *In a heterogeneous social network, based on some labeled examples (e.g., provided by a user as queries), how to evaluate the importance of different relations? Also, how to get a combination of the existing relations which can best match the relation of labeled examples?* In this paper, we propose an algorithm for relation extraction and selection. The basic idea of our algorithm is to *model this problem as an optimization problem*. Specifically, we characterize each relation by a graph with a *weight matrix*. Each element in the matrix reflects the relation strength between the two corresponding objects. Our algorithm aims at finding a linear combination of these weight matrices that can best approximate the weight matrix associated with the labeled examples. The obtained combination can better meet user's desire. Consequently, it leads to better performance on community mining.

The rest of this paper is organized as follows. Section 2 presents our algorithm for relation extraction. The experimental results on the DBLP data set are presented in Section 3. Finally, we provide some concluding remarks and suggestions for future work in Section 4.

2   Relation Extraction

In this section, we begin with a detailed analysis of the relation extraction problem followed by the algorithm.

2.1   The Problem

A typical social network likely contains multiple relations. Different relations can be modeled by different graphs. These different graphs reflect the relationship of the objects from different views. For the problems of community mining, these different relation graphs can provide us with different communities.

As an example, the network in Figure 1 may form three different relations. Suppose a user requires the four colored objects belong to the same community. Then we have:

1. Clearly, these three relations have different importance in reflecting the user's information need. As can be seen, the relation (a) is the most important one, and the relation (b) the second. The relation (c) can be seen as noise in reflecting the user's information need.

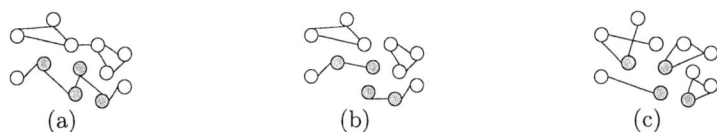

Fig. 1. There are three relations in the network. The four colored objects are required to belong to the same community, according to a user query.

2. In the traditional social network analysis, people do not distinguish these relations. The different relations are equally treated. So, they are simply combined together for describing the structure between objects. Unfortunately, in this example, the relation (c) has a negative effect for this purpose. However, if we combine these relations according to their importance, the relation (c) can be easily excluded, and the relation (a) and (b) will be used to discover the community structure, which is consistent with the user's requirement.
3. In the above analysis, the relationship between two objects is considered as a boolean one. The problem becomes much harder if each edge is assigned with a real value weight which indicates to what degree the two objects are related to each other. In such situation, an optimal combination of these relations according to the user's information need cannot be easily obtained.

Different from Figure 1, a user might submit a more complex query in some situations. Take Figure 2 as another example. The relations in the network are the same as those in Figure 1. However, the user example (prior knowledge) changes. The two objects with lighter color and the two with darker color should belong to different communities. In this situation, the importance of these three relations changes. The relation (b) becomes the most important, and the relation (a) becomes the useless (and even negative) one.

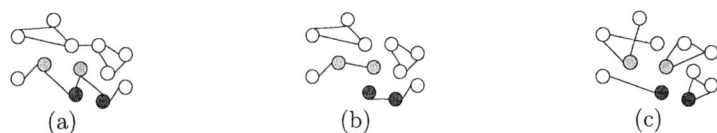

Fig. 2. Among the three relations in the network, the two objects with lighter color and the two with darker color should belong to different communities, as user required

As we can see, in multi-relational social network, community mining should be dependent on the user's example (or information need). A user's query can be very flexible. Since previous community mining techniques only focus on *single* relational network and are independent of the user's query, they cannot cope with such a complex situation.

In this paper, we focus on the relation extraction problem in multi-relational social network. The community mining based on the extracted relation graph is more likely to meet the user's information need. For relation extraction, it can be either linear or nonlinear. Due to the consideration that in real world applications it is almost impossible for a user to provide sufficient information, nonlinear techniques tend to be unstable and may cause over-fitting problems. Therefore, here we only focus on linear techniques.

This problem of relation extraction can be mathematically defined as follows. Given a set of objects and a set of relations which can be represented by a set of graphs $G_i(V, E_i)$, $i = 1, \ldots, n$, where n is the number of relations, V is the set of nodes (objects), and E_i is the set of edges with respect to the i-th relation. The weights on the edges can be naturally defined according to the relation strength of two objects. We use M_i to denote the weight matrix associated with G_i, $i = 1, \ldots, n$. Suppose there exists a hidden relation represented by a graph $\widehat{G}(V, \widehat{E})$, and $\widehat{M}$ denotes the weight matrix associated with $\widehat{G}$. Given a set of labeled objects $X = [\mathbf{x}_1, \cdots, \mathbf{x}_m]$ and $\mathbf{y} = [y_1, \cdots, y_m]$ where y_j is the label of $\mathbf{x}_j$ (Such labeled objects indicate partial information of the hidden relation $\widehat{G}$), find a linear combination of the weight matrices which can give the best estimation of the hidden matrix $\widehat{M}$.

2.2 A Regression-Based Algorithm

The basic idea of our algorithm is trying to find an combined relation which makes the relationship between the intra-community examples as tight as possible and at the same time the relationship between the inter-community examples as loose as possible.

For each relation, we can normalize it to make the biggest strength (weight on the edge) be 1. Thus we construct the target relation between the labeled objects as follows:

$$\widetilde{M}_{ij} = \begin{cases} 1, & \text{example } i \text{ and example } j \text{ have the same label}; \\ 0, & \text{otherwise.} \end{cases}$$

where $\widetilde{M}$ is a $m \times m$ matrix and $\widetilde{M}_{ij}$ indicates the relationship between examples i and j. Once the target relation matrix is built, we aim at finding a linear combination of the existing relations to optimally approximate the target relation in the sense of L_2 norm. Sometimes, a user is uncertain if two objects belong to the same community and can only provide the possibility that two objects belong to the same community. In such case, we can define $\widetilde{M}$ as follows.

$$\widetilde{M}_{ij} = \text{Prob}(\mathbf{x}_i \text{ and } \mathbf{x}_j \text{ belong to the same community})$$

Let $\mathbf{a} = [a_1, a_2, \cdots, a_n]^T \in R^n$ denote the combination coefficients for different relations. The approximation problem can be characterized by solving the following optimization problem:

$$\mathbf{a}^{opt} = \arg\min_{\mathbf{a}} \|\widetilde{M} - \sum_{i=1}^{n} a_i M_i\|^2 \tag{1}$$

This can be written as a vector form. Since the matrix $M_{m\times m}$ is symmetric, we can use a $m(m-1)/2$ dimensional vector **v** to represent it. The problem (1) is equivalent to:

$$\mathbf{a}^{opt} = \arg\min_{\mathbf{a}} \|\tilde{\mathbf{v}} - \sum_{i=1}^{n} a_i \mathbf{v}_i\|^2 \qquad (2)$$

Equation (2) is actually a linear regression problem [6]. From this point of view, the relation extraction problem is interpreted as a prediction problem. Once the combination coefficients are computed, the hidden relation strength between any object pair can be predicted. There are many efficient algorithms in the literature to solve such a regression problem [7].

The objective function (2) models the relation extraction problem as an unconstrained linear regression problem. One of the advantages of the unconstrained linear regression is that, it has a close form solution and is easy to compute. However, researches on linear regression problem show that in many cases, such unconstrained least squares solution might not be a satisfactory solution and the coefficient shrinkage technique should be applied based on the following two reasons [6].

1. *Prediction accuracy*: The least-squares estimates often have low bias but large variance [6]. The overall relationship prediction accuracy can sometimes be improved by shrinking or setting some coefficients to zero. By doing so we sacrifice a little bit of bias to reduce the variance of the predicted relation strength, and hence may improve the overall relationship prediction accuracy.
2. *Interpretation*: With a large number of explicit (base) relation matrices and corresponding coefficients, we often would like to determine a smaller subset that exhibit the strongest effects. In order to get the "big picture", we are willing to sacrifice some of the small details.

Our technical report [8] provides an example to explain such consideration. This problem can be solved by using some coefficient shrinkage techniques [6].

Thus, for each relation network, we normalize all the weights on the edges in the range [0, 1]. And, we put a constraint $\sum_{i=1}^{n} a_i^2 \leq 1$ on the objective function (2). Finally, our algorithm tries to solve the following minimization problem,

$$\mathbf{a}^{opt} = \arg\min_{\mathbf{a}} \|\tilde{\mathbf{v}} - \sum_{i=1}^{n} a_i \mathbf{e}_i\|^2$$
$$\text{subject to } \sum_{i=1}^{n} a_i^2 \leq 1 \qquad (3)$$

Such a constrained regression is called *Ridge Regression* [6] and can be solved by some numerical methods [7]. When we use such constrained relation extraction, the coefficients of the extracted relation for the above example are 1, 0, 0, 0. This shows that our constrained relation extraction can really solve the problem. For more details on our relation extraction algorithm, please refer to [8].

3 Mining Hidden Networks on the DBLP Data

In this part, we present our experimental results based on DBLP (Digital Bibliography & Library Project) data. The DBLP server (http://dblp.uni-trier.de/) provides bibliographic information on major computer science journals and proceedings. It indexes more than 500000 articles and more than 1000 different conferences (by May 2004).

Taking the authors in DBLP as objects, there naturally exist multiple relations between them. Authors publish paper in difference conferences. If we treat that authors publish paper(s) in the same conference as one kind of relation, these 1000 conferences provide us 1000 different relations. Given some examples (e.g., a group of authors), our experiment is to study how to extract a new relation using such examples and find all the other groups in the relation. The extracted relation can be interpreted as the groups of authors that share a certain kind of similar interests.

3.1 Data Preparation and Graph Generation

The DBLP server provides all the data in the XML format as well a simple DTD. We extracted the information of author, paper and conference.

We generate different kinds of graphs (social networks) based on the extracted information. For each proceeding, we construct a graph with researchers as the nodes, which is called *proceeding graph* thereafter. If two researchers have paper(s) in this proceeding, the edge between the two corresponding nodes is set to 1. Otherwise, it is set to 0. For each conference, we add up the proceeding graphs of the same conference over years, which is called *conference graph* thereafter. Finally, we choose the top 70 conference graphs based on the number of distinct authors in that conference.

Every conference graph reflects the relationship between the researchers pertaining to a certain research area. Generally, if two researchers are connected by an edge in the conference graph, they may share the same research interests.

For each graph, we normalize the edge weight by dividing the maximum weight in the whole graph. The resulting weight has a range [0, 1]. The greater the weight is, the stronger the relation is.

3.2 Experiment Results

In this experiment, we provide the system with some queries (some groups of researchers) to examine if our algorithm can capture the hidden relation between the researchers. We only show one example in this paper, please refer to [8] for more example queries.

Experiment 1. In the first case, there are two queries provided by the user.

1. Philip S. Yu, Rakesh Agrawal, Hans-Peter Kriegel, Padhraic Smyth, Bing Liu, Pedro Domingos.

2. Philip S. Yu, Rakesh Agrawal, Hans-Peter Kriegel, Hector Garcia-Molina, David J. DeWitt, Michael Stonebraker.

Both of the two queries contain 6 researchers. The first three researchers are the same in the two queries.

Table 1. Coefficients of different conference graphs for two queries (sorted on the coefficients)

Query 1		Query 2	
Conference	Coefficient	Conference	Coefficient
KDD	0.949	SIGMOD	0.690
SIGMOD	0.192	ICDE	0.515
ICDE	0.189	VLDB	0.460
VLDB	0.148	KDD	0.215

Table 1 shows the coefficients of the extracted relation for the two queries. KDD is a data mining conference, and high weight on the KDD graph indicates the common interest on data mining. On the other hand, SIGMOD, VLDB and ICDE are three database conferences. High weights on these conference graphs indicate the common interest on database area. The extracted relation for query 1 has KDD graph with weighting 1, which tells us that the researchers in query 1 share common interest on data mining. For query 2, the extracted relation tells us those researchers share common interest on database.

Table 2. Researchers' activities in conferences

Researcher	KDD	ICDE	SIGMOD	VLDB
Philip S. Yu	7	15	10	11
Rakesh Agrawal	6	10	13	15
Hans-Peter Kriegel	7	9	11	8
Padhraic Smyth	10	1	0	0
Bing Liu	8	1	0	0
Pedro Domingos	8	0	2	0
Hector Garcia-Molina	0	15	12	12
David J. DeWitt	1	4	20	16
Michael Stonebraker	0	12	19	15

Table 3. Combined Coefficients

Conference Name	Coefficient
SIGMOD	0.586
KDD	0.497
ICDE	0.488
VLDB	0.414

While we examine the publication of these researchers on these four conferences as listed in Table 2, we clearly see the extracted relation really captures the semantic relation between the researchers in the queries.

Furthermore, with the extracted relation graph, we applied the community mining algorithm *threshold cut* [8] and obtained the corresponding communities. For each query, we list one example community below:

1. Community for query 1: Alexander Tuzhilin, Bing Liu, Charu C. Aggarwal, Dennis Shasha, Eamonn J. Keogh, ...
2. Community for query 2: Alfons Kemper, Amr El Abbadi, Beng Chin Ooi, Bernhard Seeger, Christos Faloutsos, ...

Let us see what will happen if we only submit the first three names in one query. The extracted relation is shown in Table 3. The extracted relation really captures the two areas (data mining and dababase) in which these researchers are interested.

4 Conclusions

Different from most social network analysis studies, we assume that there exist multiple, heterogeneous social networks, and the sophisticated combinations of such heterogeneous social networks may generate important new relationships that may better fit user's information need. Therefore, our approach to social network analysis and community mining represents a major shift in methodology from the traditional one, a shift from single-network, user-independent analysis to multi-network, user-dependant, and query-based analysis. Our argument for such a shift is clear: multiple, heterogeneous social networks are ubiquitous in the real world and they usually *jointly* affect people's social activities.

Based on such a philosophy, we worked out a new methodology and a new algorithm for relation extraction. With such query-dependent relation extraction and community mining, fine and subtle semantics are captured effectively. It is expected that the query-based relation extraction and community mining would give rise to a lot of potential new applications in social network analysis.

References

1. Milgram, S.: The small world problem. Psychology Today **2** (1967) 60–67
2. Wasserman, S., Faust, K.: Social Network Analysis: Methods and Applications. Cambridge University Press, Cambridge, UK (1994)
3. Schwartz, M.F., Wood, D.C.M.: Discovering shared interests using graph analysis. Communications of the ACM **36** (1993) 78–89
4. Kautz, H., Selman, B., Milewski, A.: Agent amplified communication. In: Proceedings of AAAI-96. (1996) 3–9
5. Domingos, P., Richardson, M.: Mining the network value of customers. In: Proceedings of the seventh ACM SIGKDD international conference on Knowledge discovery and data mining, ACM Press (2001) 57–66
6. Hastie, T., Tibshirani, R., Friedman, J.H.: The Elements of Statistical Learning. Springer-Verlag (2001)
7. Bjorck, A.: Numerical Methods for Least Squares Problems. SIAM (1996)
8. Cai, D., Shao, Z., He, X., Yan, X., Han, J.: Mining hidden community in heterogeneous social networks. Technical report, Computer Science Department, UIUC (UIUCDCS-R-2005-2538, May, 2005)

Evaluating the Correlation Between Objective Rule Interestingness Measures and Real Human Interest

Deborah R Carvalho[1,3], Alex A. Freitas[2], and Nelson Ebecken[3]

[1] Universidade Tuiuti do Paraná (UTP), Brazil
deborah@utp.br
[2] Computing Laboratory University of Kent, CT2 7NF, UK
A.A.Freitas@kent.ac.uk
[3] COPPE/ Universidade Federal do Rio de Janeiro, Brazil
nelson@ntt.ufrj.br

Abstract. In the last few years, the data mining community has proposed a number of objective rule interestingness measures to select the most interesting rules, out of a large set of discovered rules. However, it should be recalled that objective measures are just an *estimate* of the true degree of interestingness of a rule to the user, the so-called real human interest. The latter is inherently subjective. Hence, it is not clear how effective, in practice, objective measures are. More precisely, the central question investigated in this paper is: "how effective objective rule interestingness measures are, in the sense of being a good estimate of the true, subjective degree of interestingness of a rule to the user?" This question is investigated by extensive experiments with 11 objective rule interestingness measures across eight real-world data sets.

1 Introduction

Data mining essentially consists of extracting *interesting* knowledge from real-world data sets. However, there is no consensus on how the interestingness of discovered knowledge should be measured. Indeed, most of the data mining literature still avoids this thorny problem and implicitly interprets "interesting" as meaning just "accurate" and sometimes also "comprehensible". Although accuracy and comprehensibility are certainly important, they are not enough to measure the real, *subjective* interestingness of discovered knowledge *to the user*. Consider, e.g., the classic example of the following rule: IF (patient is pregnant) THEN (patient is female). This rule is very accurate and comprehensible, but it is *not* interesting, since it represents an obvious pattern. As a real-world example, [8] reports that less than 1% of the discovered rules were found to be interesting to medical experts. It is also possible that a rule be interesting to the user even though it is not very accurate. For instance, in [9] rules with an accuracy around 40%-60% represented novel knowledge that gave new insights to medical doctors. Hence, there is a clear motivation to investigate the relationship between rule interestingness measures and the subjective interestingness of rules to the user – *an under-explored topic in the literature*.

Rule interestingness measures can be classified into two broad groups: user-driven (subjective) and data-driven (objective) measures. User-driven measures are based on comparing discovered rules with the previous knowledge or believes of the user. A rule is considered interesting, or novel, to the extent that it is different from the user's previous knowledge or believes. User-driven measures have the advantage of being a direct measure of the user's interest in a rule, but they have a twofold disadvantage. First, they require, as input, a specification of the user's believes or previous knowledge – a very time-consuming task to the user. Second, they are strongly domain-dependent and user-dependent. To avoid these drawbacks, the literature has proposed more than 40 data-driven rule interestingness measures [5], [7], [3]. These measures estimate the degree of interestingness of a rule to the user in a user-independent, domain-independent fashion, and so are much more generic. Data-driven measures have, however, the disadvantage of being an indirect *estimate* of the true degree of interestingness of a rule to the user, which is an inherently *subjective* interestingness.

This begs a question rarely addressed in the literature: *how effective data-driven rule interestingness measures are, in the sense of being a good estimate of the true, subjective degree of interestingness of a rule to the user?* The vast majority of works on data-driven rule interestingness measures ignore this question because they do not even show the rules to the user. A notable exception is the interesting work of [5], which investigates the effectiveness of approximately 40 data-driven rule interestingness measures, by comparing their values with the subjective values of the user's interest – what they called *real human interest*. Measuring real human interest involves showing the rules to the user and ask her/him to assign a subjective interestingness score to each rule. Therefore, real human interest should not be confused with the above-mentioned user-driven rule interestingness measures.

This paper follows the same general line of research. We investigate the effectiveness of 11 data-driven rule interestingness measures, by comparing them with the user's subjective real human interest. Although we investigate a smaller number of rule interestingness measures, this paper extends the work of [5] by presenting results for eight data sets, whereas [5] did experiments with just one medical data set, a limitation from the point of view of generality of the results.

2 Objective (Data-Driven) Rule Interestingness Measures

This work involves 11 objective rule interestingness measures – all of them used to evaluate classification rules. Due to space limitations we mention here a brief definition of each of those measures – which are discussed in more detail in the literature. The measures defined by formulas (1)–(8) [5], [7] are based on the coverage and accuracy of a rule. Their formulas are expressed using a notation where A denotes the rule antecedent; C denotes the rule consequent (class); $P(A)$ denotes the probability of A – i.e., the number of examples satisfying A divided by the total number of examples; $P(C)$ denotes the probability of C; "$\neg A$" and "$\neg C$" denote the logical negation of A and C. The measures defined by formulas (9)-(11) [2] use the same notation of A

and C to denote a rule's antecedent and consequent, but they also involve heuristic principles based on variables other than a rule's coverage and accuracy.

The Attribute Surprisingness measure – formula (9) – is based on the idea that the degree of surprisingness of an attribute is estimated as the inverse of its information gain. The rationale for this measure is that the occurrence of an attribute with a high information gain in a rule will not tend to be surprising to the user, since users often know the most relevant attributes for classification. However, the occurrence of an attribute with a low information gain in a rule tends to be more surprising, because this kind of attribute is usually considered little relevant for classification. In formula (9), A_i denotes the attribute in the i-th condition of the rule antecedent A, m is the number of conditions in A, and #classes is the number of classes.

$$\Phi\text{-Coefficiente} = (P(A,C)-P(A)P(C))/\sqrt{P(A)P(C)(1-P(A))(1-P(C))} \tag{1}$$

$$\text{Odds Ratio} = P(A,C)P(\neg A, \neg C)/P(A, \neg C)P(\neg A, C) \tag{2}$$

$$\text{Kappa} = (P(A,C)+P(\neg A, \neg C)-P(A)P(C)-P(\neg A)P(\neg C)) \tag{3}$$
$$/ (1-P(A)P(C)-P(\neg A)P(\neg C))$$

$$\text{Interest} = P(A,C)/(P(A)*P(C)) \tag{4}$$

$$\text{Cosine} = P(A,C) / \sqrt{(P(A)*P(C))} \tag{5}$$

$$\text{Piatetsky-Shapiro's} = P(A,C)-P(A)P(C) \tag{6}$$

$$\text{Collective Strength} = ((P(A,C)+P(\neg A, \neg C))/(P(A)P(C)+P(\neg A)P(\neg C))) * \tag{7}$$
$$((1-P(A)P(C) - P(\neg A)P(\neg C))/(1-P(A,C)-P(\neg A, \neg C))$$

$$\text{Jaccard} = P(A,C) / (P(A)+ P(C) - P(A,C)) \tag{8}$$

$$\text{Attribute Surprisingness} = 1 - ((\sum_{i=1}^{m} \text{InfoGain}(A_i) / m) / \log_2(\text{\#classes})) \tag{9}$$

$$\text{MinGen} = N / m \tag{10}$$

$$\text{InfoChange-ADT} = I^{AB1} - I^{ABo} \tag{11.1}$$

$$I^{ABo} = (- \Pr(X|AB) \log_2 \Pr(X|AB) + (- \Pr(\neg X |AB) \log_2 \Pr(\neg X |AB))) \tag{11.2}$$

$$I^{AB1} = - \Pr(X|AB) [\log_2 \Pr(X|A) + \log_2 \Pr(X|B)] \tag{11.3}$$
$$- \Pr(\neg X |AB) [\log_2 \Pr(\neg X|A)+ \log_2 \Pr(\neg X|B)]$$

The MinGen measure – formula 10 – considers the minimum generalizations of the current rule r and counts how many of those generalized rules predict a class different from the original rule r. Let m be the number of conditions (attribute-value pairs) in the antecedent of rule r. Then rule r has m minimum generalizations. The k-th minimum generalization of r, $k=1,...m$, is obtained by removing the k-th condition from r. Let C be the class predicted by the original rule r (i.e., the majority class among the examples covered by the antecedent of r) and C_k be the class predict by the k-th minimum generalization of r (i.e., the majority class of the examples covered by the antecedent of the k-th minimum generalization of r). The system compares C with each C_k, $k=1,..., m$, and N is defined as the number of times where C is different from C_k.

InfoChange-ADT (Adapted for Decision Trees) is a variation of the InfoChange measure proposed by [4]. Let $A \rightarrow C$ be a common sense rule and $A, B \rightarrow \neg C$ be an exception rule. The original InfoChange measure computes the interestingness of an exception rule based on the amount of change in information relative to common sense rules. In formulas (11.1), (11.2) and (11.3), I^{ABo} denotes the number of bits required to describe the specific rule $AB \rightarrow C$ in the absence of knowledge represented by the generalized rules $A \rightarrow C$ and $B \rightarrow C$, whereas I^{AB1} is the corresponding number of bits when the relationship between C and AB is rather described by the two rules $A \rightarrow C$ and $B \rightarrow C$. One limitation of the original InfoChange measure is that it requires the existence of a pair of exception and common sense rules, which is never the case when converting a decision tree into a set of rules – since the derived rules have mutually exclusive coverage. In order to avoid this limitation and make InfoChange useful in our experiments, the new version InfoChange-ADT is introduced in this paper, as follows. A path from the root to a leaf node corresponds to an exception rule. The common sense rule for that exception rule is produced by removing the condition associated with the parent node of the leaf node. This produces a common sense rule which is "the minimum generalization" of the exception rule. Even with this modification, InfoChange-ADT still has the limitation that its value cannot always be computed, because sometimes the minimum generalization of an exception rule predicts the same class as the exception rule, violating the conditions for using this measure.

For all the 11 rule interestingness measures previously discussed, the higher the value of the measure, the more interesting the rule is estimated to be.

3 Data Sets and Experimental Methodology

In order to evaluate the correlation between objective rule interestingness measures and real, subjective human interest, we performed experiments with 8 data sets. Public domain data sets from the UCI data repository are not appropriate for our experiments, simply because we do not have access to any user who is an expert in those data sets. Hence, we had to obtain real-world data sets where an expert was available to subjectively evaluate the interestingness of the discovered rules. Due to the difficult of finding available real-world data and expert users, our current experiments involved only one user for each data set. This reduces the generality of the results in each data set, but note that the overall evaluation of each rule interestingness measure is (as discussed later) averaged over 8 data sets and over 9 rules for each data set, i.e. each of the 11 measures is evaluated over 72 rule-user pairs. The 8 data sets are summarized in Table 1. Next, we describe the five steps of our experimental methodology.

Table 1. Characteristics of data sets used in the experiments

Data Set	Nature of Data	# Examp.	# Attrib.
CNPq1	Researchers' productivity (# publications), data from the Brazilian Research Council (CNPq)	5690	23
ITU	Patients in Intensive Care Unit	7451	41
UFPR-CS	Students' performance in comp. sci. admiss. exam	1181	48
UFPR-IM	Students' performance in info. manag. admis. exam	235	48
UTP-CS	Comp. Sci. students' end of registration	693	11
Curitiba	Census data for the city of Curitiba, Brazil	843	43
Londrina	Census data for the city of Londrina, Brazil	4115	42
Rio Branco	Census data for city of Rio Branco do Ivai, Brazil	223	43

Step 1 – Discovery of classification rules using several algorithms
We applied, to each data set, 5 different classification algorithms. Three of them are decision-tree induction algorithms (variants of C4.5 [6]), and two are genetic algorithms (GA) that discover classification rules. In the case of the decision tree algorithms, each path from the root to a leaf node was converted into an IF-THEN classification rule as usual [6]. A more detailed description of the 5 algorithms can be found in [1], where they are referred to as default C4.5, C4.5 without pruning, "double C4.5", "Small-GA", "Large-GA". The Rule Interestingness (RI) measures were applied to each of the discovered rules (after all the classification algorithms were run), regardless of which classification algorithm generated that rule.

Step 2 – Ranking all rules based on objective rule interestingness measures
For each data set, all classification rules discovered by the 5 algorithms are ranked based on the values of the 11 objective RI measures, as follows. First, for each rule, the value of each of the 11 RI measures is computed. Second, for each RI measure, all discovered rules are ranked according to the value of that measure. I.e., the rule with the best value of that RI measure is assigned the rank number 1, the second best rule assigned the rank number 2, and so. This produces 11 different rankings for the discovered rules, i.e., one ranking for each RI measure. Third, we compute an *average* ranking over the 11 rankings, by assigning to each rule a rank number which is the *average* of the 11 rank numbers originally associated with that rule. This average rank number is then used for the selection of rules in the next step.

Step 3 – Selection of the rules to be shown to the user
Table 2 shows, for each data set, the total number of rules discovered by all the 5 algorithms applied to that data set. It is infeasible to show a large number of discovered rules to the user. Hence, we asked each user to evaluate the subjective degree of interestingness of just 9 rules out of all rules discovered by all algorithms. The set of 9 rules showed to the user consisted of: (a) the three rules with the lowest rank number (i.e., rules with rank 1, 2, 3, which were the three most interesting rules according to the objective RI measures); (b) the three rules with the rank number closest to the median rank (e.g., if there are 15 rules, the three median ranks would be 7, 8, 9); and (c) the three rules with the highest rank number (least interesting rules). The selection of rules with the lowest, median and highest rank numbers creates three distinct groups of rules which ideally should have very different user-specified interestingness scores. The correlation measure calculated over such a broad range of different

objective ranks is more reliable than the correlation measure that would be obtained if we selected instead 9 rules with very similar objective ranks.

Step 4 – Subjective evaluation of rule interestingness by the user

For each data set, the 9 rules selected in step 3 were shown to the user, who assigned a subjective degree of interestingness to each rule. The user-specified score can take on three values, viz.: <1> – the rule is not interesting, because it represents a relationship known by the user; <2> – the rule is somewhat interesting, i.e., it contributes a little to increase the knowledge of the user; <3> – the rule is truly interesting, i.e., it represents novel knowledge, previously unknown by the user.

Step 5 – Correlation between objective and subjective rule interestingness

We measured the correlation between the rank number of the selected rules – based on the *objective* RI measures – and the *subjective* RI scores – <1>, <2>, <3> – assigned by the user to those rules. As a measure of correlation we use the Pearson coefficient of linear correlation, with a value in [–1...+1], computed using SPSS.

Table 2. Total number of discovered rules for each data set

Data Set:	CNPq1	ITU	UFPR-CS	UFPR-IM	UTP-CS	Curitiba	Londrina	Rio Branco do Ivai
# Rules:	20,253	6,190	1,345	232	2,370	1,792	1,261	486

4 Results

Table 3 shows, for each data set, the correlation between each objective RI measure and the corresponding subjective RI score assigned by the user. These correlations are shown in columns 2 through 9 in Table 3, where each column corresponds to a data set. To interpret these correlations, recall that the lower the objective rank number the more interesting the rule is *estimated to be*, according to the objective RI measure; and the higher the user's subjective score the more interesting the rule *is to the user*. Hence, an ideal objective RI measure should behave as follows. When a rule is assigned the best possible subjective score (<3>) by the user, the RI measure should assign a low rank number to the rule. Conversely, when a rule is assigned the worst possible subjective score (<1>) by the user, the RI measure should assign a high rank number to the rule. Therefore, the closer the correlation value is to –1 the more effective the corresponding objective RI measure is in *estimating the true degree of interestingness of a rule to the user*. In general a correlation value ≤ –0.6 can be considered a strong negative correlation, which means the objective RI measure is quite effective in estimating the real human interest in a rule. Hence, in Table 3 all correlation values ≤ –0.6 are shown in bold.

In columns 2 through 9 of Table 3, the values between brackets denote the ranking of the RI measures for each data set (column). That is, for each data set, the first rank (1) is assigned to the smallest (closest to –1) value of correlation in that column, the second rank (2) is assigned to the second smallest value of correlation, etc. Finally, the last column of Table 3 contains the average rank number for each RI measure –

i.e., the arithmetic average of all the rank numbers for the RI measure across all the data sets. The numbers after the symbol "±" are standard deviations.

Two cells in Table 3 contain the symbol "N/A" (not applicable), rather than a correlation value. This means that SPSS was not able to compute the correlation in question because the user's subjective RI scores were constant for the rules evaluated by the user. This occurred when only a few rules were shown to the user. In general each correlation was computed considering 9 rules selected shown to the user, as explained earlier. However, in a few cases the value of a given objective RI measure could not be computed for most selected rules, and in this case the rules without a value for an objective RI measure were not considered in the calculation of the correlation for that measure. For instance, the N/A symbol in the cell for InfoChange-ADT and data set UFPR-CS is explained by the fact that only 2 out of the 9 selected rules were assigned a value of that objective RI measure, and those two rules had the same subjective RI score assigned by the user.

Table 3. Correlations between objective rule interestingness measures and real human interest; and ranking of objective rule interestingness measures based on these correlations

Rule interestingness measure	Data Set							Avg. Rank
	ITU	UFPR-CS	UTP-CS	Curitiba	UFPR-IM	Londrina	CNPq1	Rio Bran
Φ-Coefficient	**-0.63** (1)	**-0.91** (4)	**-0.69** (7)	-0.17 (5)	**-0.97** (2)	0.01 (4)	**-0.48** (4)	0.45 (10)	4.63 ±2.8
Infochange-ADT (*)	-0.18 (10)	N/A	-0.17 (11)	**-0.70** (1)	**-1.00** (1)	-0.54 (2)	0.15 (8)	**-1.00** (1)	4.86 ±4.6
Kappa	-0.44 (6)	**-0.94** (3)	**-0.74** (5)	-0.12 (6)	**-0.87** (4)	0.12 (5)	-0.18 (7)	**-0.56** (3)	4.88 ±1.5
Cosine	**-0.55** (3)	**-0.79** (6)	**-0.93** (2)	-0.49 (2)	**-0.81** (7)	0.37 (8)	**-0.64** (1)	0.79 (11)	5.00 ±3.6
Piatesky Shapiro	-0.45 (5)	**-0.95** (1)	**-0.68** (8)	-0.09 (9)	**-0.87** (5)	0.19 (7)	**-0.49** (3)	-0.55 (4)	5.25 ±2.7
Interest	-0.40 (8)	**-0.77** (7)	**-0.85** (3)	-0.44 (3)	**-0.87** (6)	**-0.61** (1)	0.28 (9)	-0.22 (7)	5.50 ±2.8
Collective Strength	-0.44 (7)	**-0.94** (2)	**-0.66** (9)	-0.10 (7)	**-0.88** (3)	0.19 (6)	0.35 (10)	-0.56 (2)	5.75 ±3.1
Jaccard	-0.49 (4)	**-0.69** (8)	**-0.93** (1)	-0.10 (8)	-0.30 (9)	0.41 (9)	-0.45 (5)	-0.52 (5)	6.13 ±2.9
Odds Ratio	-0.59 (2)	**-0.91** (5)	**-0.85** (4)	-0.28 (4)	N/A	0.48 (10)	0.43 (11)	0.19 (9)	6.43 ±3.5
MinGen	-0.36 (9)	**-0.60** (9)	**-0.71** (6)	0.00 (10)	0.36 (10)	-0.22 (3)	-0.53 (2)	-0.23 (6)	6.88 ±3.1
Attsurp	0.42 (11)	-0.46 (10)	-0.54 (10)	0.63 (11)	**-0.62** (8)	0.59 (11)	-0.37 (6)	-0.10 (8)	9.38 ±1.9

(*) Although InfoChange-ADT obtained the second best rank overall, it was not possible to compute the value of this measure for many discovered rules (see text).

As shown in Table 3, the strength of the correlation between an objective RI measure and the user's subjective RI score is quite dependent on the data set. In three data sets – namely UFPR-CS, UTP-CS and UFPR-IM – the vast majority of the objective RI measures were quite effective, having a strong correlation ($\leq$ –0.6, shown in bold) with the user's true degree of interestingness in the rules. On the other hand, in each of the other five data sets there was just one objective RI measure that was effective, and in most cases the effective measure (with correlation value shown in bold) was different for different data sets. Correlation values that are very strong ($\leq$ –0.9) are rarer in Table 3, but they are found for five RI measures in the UFPR-CS data set, and for one or two RI measures in three other data sets.

Consider now the average rank number of each measure shown in the last column of Table 3. The RI measures are actually in increasing order of rank number, so that, overall, across the eight data sets, the most effective RI measure was the Φ-Coefficient, with an average rank of 4.63. However, taking into account the standard deviations, there is no significant difference between the average rank of Φ-Coefficient and the average rank of the majority of the measures. The only measure which performed significantly worse than Φ-Coefficient was Attribute Surprisingness, the last in the average ranking.

There is, however, an important caveat in the interpretation of the average ranking of InfoChange-ADT. As explained earlier, there are several rules where the value of this RI measure cannot be computed. More precisely, out of the 9 rules selected to be shown to the user for each data set, the number of rules with a value for InfoChange-ADT varied from 2 to 5 across different data sets. This means that the average rank assigned to InfoChange-ADT is less reliable than the average rank assigned to other measures, because the former was calculated from a considerably smaller number of samples (rules). In particular, the correlation value of InfoChange-ADT was –1 (the best possible value) in two data sets, viz. UFPR-IM and Rio Branco, and in both data sets only 2 out of the 9 selected rules had a value for InfoChange-ADT.

5 Conclusions and Future Research

The central question investigated in this paper was: "how effective objective rule interestingness measures are, in the sense of being a good estimate of the true, subjective degree of interestingness of a rule to the user?" This question was investigated by measuring the correlation between each of 11 objective rule interestingness measures and real human interest in rules discovered from 8 different data sets. Overall, 31 out of the 88 (11 $\times$ 8) correlation values can be considered strong (correlation $\geq$ 60%). This indicates that objective rule interestingness measures were effective (in the sense of being good estimators of real human interest) in just 35.2% (31 / 88) of the cases. There was no clear "winner" among the objective measures – the correlation values associated with each measure varied considerably across the 8 data sets.

A research direction would be to try to predict which objective rule interestingness measure would be most correlated with real human interest for a given target data set, or to predict the real human interest in a rule using a combination of results from different objective measures. This could be done, in principle, using a meta-learning framework, mining data from previously-computed values of the correlation between objective interestingness measures and subjective human interest for a number of rules that have been previously evaluated by a given user.

References

[1] Carvalho, D.R.; Freitas, A.A. Evaluating Six Candidate Solutions for the Small-Disjunct Problem and Choosing the Best Solution via Meta Learning. *AI Review* 24(1), 61-98, 2005

[2] Carvalho, D.R.; Freitas, A.A.; Ebecken, N.F. (2003) A Critical Review of Rule Surprisingness Measures. Proc. 2003 Int. Conf. on Data Mining, 545-556. WIT Press.

[3] Hilderman, R.J.; Hamilton H.J. *Knowledge Discovery Measures of Interest.* Kluwer, 2001.

[4] Hussain, F.; Liu, H.; Lu, H. Exception Rule Mining with a Relative Interestingness Measure. PAKDD-2000, LNAI 1805, 86-96. Springer-Verlag.

[5] Ohsaki, M., Kitaguchi, S., Okamoto, K., Yokoi, H. Yamaguchi, T. Evaluation of rule interestingness measures with a clinical dataset on hepatitis. *Knowledge Discovery in Databases: PKDD 2004, LNAI 3202*, 362-373. Springer-Verlag, 2004

[6] Quinlan, J.R. *C4.5: programs for machine learning.* Morgan Kaufmann. 1993.

[7] Tan, P.N.; Kumar, V. and Srivastava, J. Selecting the right interestingness measure for association patterns. *Proc. ACM SIGKDD KDD-2002.* ACM Press, 2002

[8] Tsumoto, S. Clinical knowledge discovery in hospital information systems. *Principles of Data Mining and Knowledge Discover, PKDD-2000*, 652-656. Springer-Verlag, 2000.

[9] Wong, M.L. and Leung, K.S. *Data mining using grammar-based genetic programming and applications.* Kluwer, 2000.

A Kernel Based Method for Discovering Market Segments in Beef Meat

Jorge Díez[1], Juan José del Coz[1], Carlos Sañudo[2],
Pere Albertí[3], and Antonio Bahamonde[1]

[1] Artificial Intelligence Center, University of Oviedo at Gijón (Asturias), Spain
{jdiez, juanjo, antonio}@aic.uniovi.es
www.aic.uniovi.es
[2] Facultad de Veterinaria, University of Zaragoza, Zaragoza (Aragón), Spain
csanudo@unizar.es
[3] Service of Agriculture and Food Science Research, Zaragoza (Aragón), Spain
palberti@aragon.es

Abstract. In this paper we propose a method for learning the reasons why groups of consumers prefer some food products instead of others. We emphasize the role of groups given that, from a practical point of view, they may represent market segments that demand different products. Our method starts representing people's preferences in a metric space; there we are able to define a kernel based similarity function that allows a clustering algorithm to discover significant groups of consumers with homogeneous tastes. Finally in each cluster, we learn, with a SVM, a function that explains the tastes of the consumers grouped in the cluster. To illustrate our method, a real case of consumers of beef meat was studied. The panel was formed by 171 people who rated 303 samples of meat from 101 animals with 3 different aging periods.

1 Introduction

Consumer preferences for food products address the strategies of industries and breeders, and should be carefully considered when export and commercial policies are designed. In this paper we present a method to deal with data collected from panels of consumers in order to discover groups with differentiated tastes; these groups may constitute significant market segments that demand different kinds of food products. Additionally, our approach studies the factors that could contribute to the success or failure of food products in each segment.

From a conceptual point of view, consumer panels are made up of untrained consumers; these are asked to rate their degree of acceptance or satisfaction about the tested products on a scale. The aim is to be able to relate product descriptions (human and mechanical) with consumer preferences. Nevertheless, the Market is not interested in tastes of individual consumers, the purpose of marketing studies of sensorial data is to discover, if there exist widespread ways to appreciate food products that can be considered as market segments. These segments can be seen as *clusters* of consumers with similar tastes. In this paper, we will show that the similarity of preference criteria of consumers can be computed in a high dimension space; for this purpose, we present here a kernel-based method. To illustrate our method, we used a data set that collects the ratings of a panel of beef meat consumers. The panel studied was formed by 171 people rating samples of 303 different kinds of beef meat [1] from different breeds, live weights, and aging periods.

2 Description of the General Approach

The main assumption behind the approach presented in this paper is that we are able to map people's preferences into a metric space in such a way that we can assume some kind of continuity. A first attempt to provide such a mapping would consist in associating, to each consumer, the vector of his or her ratings, taking the set of samples as indexes. However, this is not a wise option since ratings have only a relative meaning, and therefore they cannot assume an absolute role. There is a kind of *batch effect*: a product will obtain a higher/lower rating when it is assessed together with other products that are clearly worse/better. In fact, if we try to deal with sensory data as a regression problem, we will fail [2]; due to this batch effect, the ratings have no numerical meaning: they are only a relative way to express preferences between products of the same session.

To overcome this, instead of ratings, we can assign to each product its ordinal position in the ranking of preferences. Unfortunately, this is not always possible given that, in general, the size of the sample of food prevents panelists from testing all products. Hence, we cannot ask our panelists to spend long periods rating the whole set of food samples. Typically, each consumer only participates in one or a small number of testing sessions, usually in the same day. Notice that tasting a large sample of food may be physically impossible, or the number of tests performed would damage the sensory capacity of consumers. The consequence is that consumers' rankings are not comparable because they deal with different sets of products. Thus, in this case we will codify people preferences by the weighting vector of a linear function (called *preference* or *ranking function*) in a high dimensional space: the space of features where we represent the descriptions of food products. Then, the similarity is defined by means of the kernel attached to the representation map.

Once we have people preferences represented in a metric space, and we have defined a similarity function, then we use a clustering algorithm. Finally, we only need to explain the meaning and implications of each cluster in the context of the food products. For this purpose, we will learn a preference or ranking function from the union of preference judgments expressed by the member of the cluster; this will provide the consensus assessment function of the cluster.

3 Description of the Beef Meat Experiment

To illustrate our method we used a database described in [1]. The data collects the sensory ratings of a panel of beef meat consumers about three aspects: flavour, tenderness, and acceptability.

For this experience, more than 100 animals of 7 Spanish breeds were slaughtered to obtain two kinds of carcasses: lights, from animals with a live weight around 300–350 kg (light); and heavies, from animals at 530–560 kg. The set of animals was uniformly distributed by breeds and weights. Additionally, to test the influence of aging in consumers' appreciation, each piece of meat was prepared with 3 aging periods, 1, 7, and 21 days. On the other hand, the 7 breeds used constitute a wide representation of beef cattle. These breeds can be divided into four types: double muscled (DM, one breed), fast growth (FG, two breeds), dual purpose (DP, one breed), and unimproved rustic type (UR, three breeds). In Table 1 for each breed, we show the average percentages of fats, muscle and bone.

Table 1. Carcass compositions of 7 Spanish beef breeds used in the experiment

Breed		Fat %		Bone	Muscle	Intramuscular
Name	Type	inter-muscular	subcutaneous	%	%	fat %
Asturiana Valles	DM	4.77	0.89	16.00	78.34	0.90
Avileña	UR	13.17	3.53	19.25	64.05	2.28
Morucha	UR	12.46	3.46	19.28	64.80	2.10
Parda Alpina	DP	9.65	2.32	20.86	67.17	1.82
Pirenaica	FG	9.02	3.01	17.33	70.63	1.48
Retinta	UR	14.16	4.75	20.89	60.20	2.13
Rubia Gallega	FG	5.73	1.20	16.56	76.52	1.12

Each kind of meat was also described by a panel of 11 trained experts who rate 12 traits of products such as fibrosis, flavor, odor, etc.. In this paper, we considered the average rate of each trait. The characterization of meat samples was completed with 6 physical features describing its texture.

4 Vectorial Representation of Preference Criteria

As was explained above, in order to compare the preference criteria of consumers we need to state a common language. We cannot use for this purpose the ratings assigned by consumers to food products, since they have rated, in general, different sets of samples. Then we are going to induce a reasonable extension of the preferences expressed by each consumer to obtain a function able to capture the pairwise orderings, not the rates. Then we will manage to define similarities in the space of those functions.

Although there are other approaches to learn preferences, we will follow [3, 4, 5]. Then we will try to induce a real *preference*, *ranking*, or *utility function* f from the input space of object descriptions, say $\mathbf{R}^d$, in such a way that it maximizes the probability of having f(**v**) > f(**u**) whenever **v** is preferable to **u**; we call such pairs, *preference judgments*. This functional approach can start from a set of objects endowed with a (usually ordinal) rating, as in regression; but essentially, we only need a collection of preference judgments.

When we have a set of ratings given by a consumer c, we most take into account the session where the ratings have been assessed [6, 7], as was explained in section 2. Thus, for each session we include in the set of preference judgments, PJ_c, the pairs (**v**, **u**) whenever consumer c assessed to sample represented by **v** a higher rating than to the sample represented by **u**. In order to induce the ranking function, as in [3], we look for a function $F_c: \mathbf{R}^d \times \mathbf{R}^d \to \mathbf{R}$ such that

$$\forall \mathbf{x}, \mathbf{y} \in \mathbf{R}^d, \; F_c(\mathbf{x}, \mathbf{y}) > 0 \Leftrightarrow F_c(\mathbf{x}, \mathbf{0}) > F_c(\mathbf{y}, \mathbf{0}) \quad (1)$$

Notice that the right hand side of (1) establishes an ordering of functional expressions of a generic couple ($\mathbf{x}$, $\mathbf{y}$) of objects representations. This suggests the definition

$$f_c: \mathbf{R}^d \to \mathbf{R}, \; f_c(\mathbf{x}) = F_c(\mathbf{x}, \mathbf{0}) \quad (2)$$

The idea is then to obtain ranking functions f_c from functions like F_c, as in (2), when F_c fulfils (1). Thus, given the set of preference judgments PJ_c, we can specify F_c by means of the constraints

$$\forall \; (\mathbf{v}, \mathbf{u}) \in PJ_c, \; F_c(\mathbf{v}, \mathbf{u}) > 0 \text{ and } F_c(\mathbf{u}, \mathbf{v}) < 0 \quad (3)$$

Therefore, PJ_c gives rise to a set of binary classification training set to induce F_c

$$E_c = \{(\mathbf{v}, \mathbf{u}, +1), (\mathbf{u}, \mathbf{v}, -1): (\mathbf{v}, \mathbf{u}) \in PJ_c\} \quad (4)$$

Nevertheless, a separating function for E_c does not necessarily fulfill (1). Thus, we need an additional constraint. So, if we represent each object description $\mathbf{x}$ in a higher dimensional feature space by means of $\phi(\mathbf{x})$, then we can represent pairs ($\mathbf{x}$, $\mathbf{y}$) by $\phi(\mathbf{x}) - \phi(\mathbf{y})$. Hence, a classification SVM can induce from E_c a function of the form:

$$F_c(\mathbf{x}, \mathbf{y}) = \sum_{s \in S(c)} \alpha_s z_s \langle \phi(\mathbf{x}_s^{(1)}) - \phi(\mathbf{x}_s^{(2)}), \phi(\mathbf{x}) - \phi(\mathbf{y}) \rangle \quad (5)$$

where $\langle \mathbf{x}, \mathbf{y} \rangle$ stands for the inner product of vectors $\mathbf{x}$ and $\mathbf{y}$; $S(c)$ is the set of support vectors, notice that they are formed by two d-dimensional vectors $(\mathbf{x}_s^{(1)}, \mathbf{x}_s^{(2)})$, while the scalars z_s represent the class $+1$ or -1. Trivially, F_c fulfils the condition (1). Let us remark that if k is a kernel function, defined as the inner product of two objects represented in the feature space, that is, $k(\mathbf{x}, \mathbf{y}) = \langle \phi(\mathbf{x}), \phi(\mathbf{y}) \rangle$, then the kernel function used to induce F_c is

$$K(\mathbf{x}_1, \mathbf{x}_2, \mathbf{x}_3, \mathbf{x}_4) = k(\mathbf{x}_1, \mathbf{x}_3) - k(\mathbf{x}_1, \mathbf{x}_4) - k(\mathbf{x}_2, \mathbf{x}_3) + k(\mathbf{x}_2, \mathbf{x}_4) \quad (6)$$

Usually it is employed a linear or a simple polynomial kernel; that is, $k(\mathbf{x}, \mathbf{y}) = \langle \mathbf{x}, \mathbf{y} \rangle$, or $k(\mathbf{x}, \mathbf{y}) = (\langle \mathbf{x}, \mathbf{y} \rangle + 1)^g$, with $g = 2$.

Once we have a function F_c for a consumer c fulfilling (1), then, using (2), a ranking or preference or utility function f_c is given (but for an irrelevant constant) by

$$f_c(\mathbf{x}) = \sum_{s \in S(c)} \alpha_s z_s \langle \phi(\mathbf{x}_s^{(1)}) - \phi(\mathbf{x}_s^{(2)}), \phi(\mathbf{x}) \rangle = \sum_{s \in S(c)} \alpha_s z_s \left(k(\mathbf{x}_s^{(1)}, \mathbf{x}) - k(\mathbf{x}_s^{(2)}, \mathbf{x}) \right) \quad (7)$$

Therefore, f_c can be represented by the weight vector $\mathbf{w}^c$ in the higher dimensional space of features such that

$$f_c(\mathbf{x}) = \langle \mathbf{w}^c, \phi(\mathbf{x}) \rangle, \quad \mathbf{w}^c = \sum_{s \in S(c)} \alpha_s z_s \left(\phi(\mathbf{x}_s^{(1)}) - \phi(\mathbf{x}_s^{(2)}) \right) \quad (8)$$

Notice that (8) defines the ranking of an object represented by a vector $\mathbf{x}$. This is not an absolute value; its importance is the relative position that gives to $\mathbf{x}$ against to other objects $\mathbf{y}$ in the *competition* for gaining the appreciation of consumer c. Now we only need to define the distance of consumers' preferences. Given that preferences are codified by those weighting vectors, we define the similarity of the preferences of consumer c and c' by the cosine of their weighting vectors. In symbols,

$$\text{similarity}(\mathbf{w}^c, \mathbf{w}^{c'}) = \cos(\mathbf{w}^c, \mathbf{w}^{c'}) = \frac{\langle \mathbf{w}^c, \mathbf{w}^{c'} \rangle}{\|\mathbf{w}^c\| * \|\mathbf{w}^{c'}\|} \quad (9)$$

Given that this definition uses scalar products instead of coordinates of weighting vectors, we can easily rewrite (10) in terms of the kernels used in the previous derivations. The essential equality is:

$$\langle \mathbf{w}^c, \mathbf{w}^{c'} \rangle = \sum_{s \in S(c)} \sum_{l \in S(c')} \alpha_s \alpha_l z_s z_l \langle \phi(\mathbf{x}_s^{(1)}) - \phi(\mathbf{x}_s^{(2)}), \phi(\mathbf{x}_l^{(1)}) - \phi(\mathbf{x}_l^{(2)}) \rangle \quad (10)$$

$$= \sum_{s \in S(c)} \sum_{l \in S(c')} \alpha_s \alpha_l z_s z_l K(\mathbf{x}_s^{(1)}, \mathbf{x}_s^{(2)}, \mathbf{x}_l^{(1)}, \mathbf{x}_l^{(2)})$$

5 Clustering Consumers with Homogeneous Tastes

In the previous section we have associated one data point for each consumer in the space of preference criteria represented by ranking or preference functions. Moreover, we have defined a reasonable similarity measure for preference criteria; now we proceed to look for clusters of consumers with homogeneous tastes. For this purpose, we applied a nonparametric pairwise algorithm [8].

Let $S = (s_{ij})$ be a square matrix where s_{ij} stands for the similarity between data points i and j; in our case, data points are the vectorial representation of the preference criteria of consumers, and similarities are given by equation (9). Then, matrix S is transformed iteratively, following a two step procedure that converges to a two values matrix (1 and 0), yielding a bipartition of the data set into two clusters. Then, recursively, the partition mechanism is applied to each of the resulting clusters represented by their corresponding submatrices. To guarantee that only meaningful splits take places, in [8] the authors provide a cross validation method that measures an index that can be read as a significance level; we will only accept splits which level is above 0.90.

The first step normalizes the columns of S using the L_∞ norm; then the proximities are re-estimated using the Jensen-Shannon divergence. The idea is to formalize that two preference criteria are close (after these two steps) if they were both similar and dissimilar to analogous sets of criteria before the transformation.

6 Experimental Results

In this section, we report the outputs obtained with the database of beef meat consumers. In order to consider significant opinions, we first selected those people involved in our consumers' panel whose ratings gave rise to at least 30 preference judgments; these yielded us to consider a set of 171 panelists that tested from 9 to 14 samples of meat of 101 different animals. The total amount of different samples was 303, since the meat from each animal was prepared with 3 different aging periods: 1, 7, and 21 days. Then the opinions of our panelists can be estimated inducing a preference or ranking function as was explained in section 4. Notice that only such functions can be used in order to compare the preferences of different consumers; in general, two arbi-

Table 2. For clusters of acceptance and tenderness datasets, this table reports the number of preference judgments (PJ), percentage of disagreements, and classification errors achieved into clusters with their own ranking or preference function, and using the function of the other cluster

Dataset	cluster	PJ	disagreements %	classification errors using function own %	other %
acceptance	left	1927	16.19	19.20	50.96
	right	2150	17.07	21.12	54.95
tenderness	left	2487	15.96	19.38	61.98
	right	2432	15.21	19.59	61.06

trary consumers have not tested samples of the same animal prepared with the same aging. However, it is possible to compare the preference functions of any couple of consumers as vectors in a high dimension space following the kernel based method of section 4.

The clustering algorithm [8] returns the trees depicted in Figure 1. Split nodes achieved a confidence level of 91% for tenderness dataset, and 97% for acceptance. The leaves of these trees and the dataset of flavor reached lower confidence levels, and therefore they were rejected.

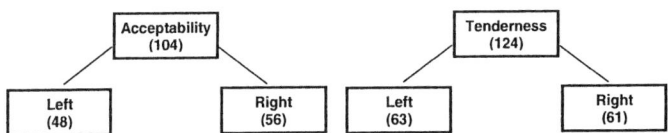

Fig. 1. Trace of the clustering algorithm. In each node we report the number of consumers

The job of clustering is to compute groups with minimal intra-group and maximal inter-group distances or differences. In our case, the relevance of clusters can be estimated, in part, by the coherence of consumers included into the same cluster, which can be measured by the classification error of the SVM used to compute the ranking or preference function of each cluster. Let us notice that the union of preference judgments of the members of the same cluster has some disagreements; if for each pair of samples we choose the most frequent relative ordering, then about 16% of preference pairs of each cluster express a particular disagreement with the majority opinion of the cluster, see Table 2. However, every preference judgment is included in the training set of each cluster; this sums more than 2000 preference judgments, what means (see equation 4 in section 4) more than 4000 training instances for the corresponding classification sets. When we use a polynomial kernel of degree 2, the errors range from 19.20% to 21.12%; we used this kernel following [2, 6, 7]. Nevertheless, if we apply the induced classification function of each cluster to the other one, then the errors rise to more than 50% in the case of acceptance, and more than 60% in the case of tenderness. Notice that in both cases we are ranking the same samples and these errors can be understood as the probability of reversing the order given by one of such clusters when we use the criteria of the other one. Therefore, 50% of error means a random classification, and over that threshold means that ranking criteria is approaching the exactly opposite, see Table 2.

In general, it is well known that meat qualities are mainly the result of a set of complex factors. In this study, we are interested in knowing if there are different groups of people who prefer some breeds to others. To gain insight into the meaning of the preference criteria of each cluster, we used the ranking or preference functions to order the samples of meat; then we assessed 10 points to those samples included in the first decile, 9 to the second decile, and so on. Graphical representations of the average points obtained by each breed are shown in Figure 2; notice that the average score of all samples is 5.5. The results are quite the same if we use quartiles instead of deciles or any other division of the relative rankings of each cluster.

In the acceptance dataset (Fig. 2 left), let us emphasize the opposite role played by Retinta and Asturiana breeds: they were first and last (or almost last) in each cluster alternatively. In [6, 7] we used Boolean attributes to include the breed in the description of each sample, and then Retinta and Asturiana were found to be the most relevant Boolean features in order to explain consumer's acceptance of meat. Additionally, these two breeds have significant differences in carcass composition (see Table 1). Notice that Asturiana breed is the only double muscled breed of the sample, and then it has the lowest values in percentages of subcutaneous and inter-muscular fat, and bone; while Retinta is the unimproved rustic breed with the highest percentages of fat and bone. Therefore, there are some reasons so as to assign opposite ratings to samples of these two breeds, although, in general, the final acceptance scorings rely on a complex set of features.

In tenderness dataset (Fig. 2 right), meat from Pirenaica and Retinta breeds are the tenderest for people in left cluster, however they are ranked in low positions in right cluster. We can say exactly the opposite of meat from Asturiana and Parda breeds. Again, Asturiana and Retinta breeds play opposites roles in each cluster.

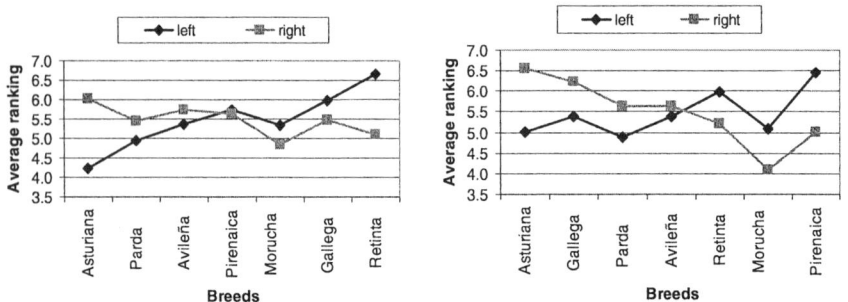

Fig. 2. Average ranking scores for each breed. Acceptance (left). Tenderness (right)

Acknowledgments

We would like to thank: the authors of Spider [9], a MatLab toolbox that includes kernel based algorithms; Thorsten Joachims for his SVM^{light} [10]. Those systems were used in the experiments reported in this paper; and INIA (Instituto Nacional de Investigación y Tecnología Agraria y Alimentaria of Spain) and Breeders Associations for the financial (Grant SC-97019) and technical support.

References

1. Sañudo, C.; Macie, E.S.; Olleta, J.L.; Villarroel, M.; Panea, B.; Albertí, P.. The effects of slaughter weight, breed type and ageing time on beef meat quality using two different texture devices. Meat Science, 66 (2004), 925–932
2. Díez, J.; Bayón, G. F.; Quevedo, J. R.; del Coz, J. J.; Luaces, O.; Alonso, J.; Bahamonde, A.. Discovering relevancies in very difficult regression problems: applications to sensory data analysis. Proceedings of the European Conference on Artificial Intelligence (ECAI 2004), 993-994
3. Herbrich, R.; Graepel, T.; and Obermayer, K.: Large margin rank boundaries for ordinal regression. In A. Smola, P. Bartlett, B. Scholkopf, and D. Schuurmans, editors, Advances in Large Margin Classifiers, 115–132. MIT Press, Cambridge, MA, (2000)
4. Joachims, T.: Optimizing search engines using clickthrough data. In: Proceedings of the ACM Conference on Knowledge Discovery and Data Mining (KDD) (2002)
5. Bahamonde, A.; Bayón, G. F.; Díez, J.; Quevedo, J. R.; Luaces, O.; del Coz, J. J.; Alonso, J.; Goyache, F.. Feature subset selection for learning preferences: a case study. Proceedings of the 21st International Conference on Machine Learning, (ICML 2004), 49-56
6. Del Coz, J. J.; Bayón, G. F.; Díez, J.; Luaces, O.; Bahamonde, A.; Sañudo, C.. Trait selection for assessing beef meat quality using non-linear SVM. Proceedings of the Eighteenth Annual Conference on Neural Information Processing Systems (NIPS 2004), **321-328**
7. Luaces, O.; Bayón, G.F.; Quevedo, J.R.; Díez, J.; del Coz, J.J.; Bahamonde, A.. Analyzing sensory data using non-linear preference learning with feature subset selection. Proceedings of the 15th European Conference of Machine Learning, (ECML 2004), 286-297
8. Dubnov, S.; El-Yaniv, R.; Gdalyahu, Y.; Schneidman, E.; Tishby, N.; Yona, G.. A New Nonparametric Pairwise Clustering Algorithm Based on Iterative Estimation of Distance Profiles. Machine Learning, 47 (2002), 35–61
9. Weston, J.; Elisseeff, A.; BakIr, G.; Sinz, F.: SPIDER: object-orientated machine learning library. http://www.kyb.tuebingen.mpg.de/bs/people/spider/
10. Joachims, T.. Making large-Scale SVM Learning Practical. Advances in Kernel Methods - Support Vector Learning, B. Schölkopf and C. Burges and A. Smola (ed.), MIT-Press, (1999)

Corpus-Based Neural Network Method for Explaining Unknown Words by WordNet Senses

Bálint Gábor, Viktor Gyenes, and András Lőrincz*

Eötvös Loránd University, Pázmány P. sétány 1/C, Budapest 1117
{gbalint, gyenesvi}@inf.elte.hu, andras.lorincz@elte.hu
http://nipg.inf.elte.hu/

Abstract. This paper introduces an unsupervised algorithm that collects senses contained in WordNet to explain words, whose meaning is unknown, but plenty of documents are available that contain the word in that unknown sense. Based on the widely accepted idea that the meaning of a word is characterized by its context, a neural network architecture was designed to reconstruct the meaning of the unknown word. The connections of the network were derived from word co-occurrences and word-sense statistics. The method was tested on 80 TOEFL synonym questions, from which 63 questions were answered correctly. This is comparable to other methods tested on the same questions, but using a larger corpus or richer lexical database. The approach was found robust against details of the architecture.

1 Introduction

The Internet is an immensely large database; large amount of domain specific text can be found. Intelligent tools are being developed to determine the meaning of documents, and manually created lexical databases are intended to provide help for such tools. However, manually assembled lexical databases are unable to cover specific, emerging subjects, thus documents may contain words of unknown meanings; words that are not contained in the lexical databases, or the contained meanings do not fit into the context found in the documents. However, the meaning of these words can often be inferred from their contexts of usage. The aim of our method is to explain words that are unknown to a human or machine reader, but are contained in many documents in the same sense.

To achieve this goal, our method looks for WordNet senses that are semantically close to the unknown meaning of the word. We rely on the common practice of measuring the similarity of words based on their contextual features, and designed a neural network architecture by means of three databases as sources of information. The first is WordNet[1], where the words are grouped into synonym sets, called *synsets*. The second source of information that our method exploits is SemCor[2], which is a corpus tagged with WordNet senses. SemCor was used

---

* Corresponding author
[1] http://www.cogsci.princeton.edu/~wn/
[2] http://www.cs.unt.edu/~rada/downloads/semcor/semcor2.0.tar.gz

to obtain information on the statistical distributions of the senses of the words. The third database used is the British National Corpus (BNC[3]), a collection of English texts of 100 million words. Our assumption is that the meaning of a word is similar to the meaning of a *sense*, if they appear in the same context, where we define context by the *senses* that they often co-occur with[4].

Testing of any new method requires a controllable benchmark problem. Our method was evaluated on 80 synonym questions from the Test of English as a Foreign Language (TOEFL[5]). The system scored 78.75% (63 correct answers). Many other studies had also chosen this TOEFL benchmark problem: Landauer and Dumais's Latent Semantic Analysis (LSA) [1] is based on co-occurrences in a corpus, and it provides generalization capabilities. It was able to answer 64.4% of the questions correctly. Turney's Pointwise Mutual Information Information Retrieval (PMI-IR) algorithm [2] performed 73.25% on the same set of questions. This is also a co-occurrence based corpus method, which examines noun enumerations. It uses the whole web as a corpus and exploits AltaVista's special query operator, the *NEAR* operator. Terra and Clarke [3] compared several statistical co-occurrence based similarity measures on a one terabyte web corpus, and scored 81.25%. Jarmasz and Szapakovicz constructed a thesaurus-based method [4], which performed 78.75% on these questions. They utilized Roget's Thesaurus to calculate path lengths in the semantic relations graph between two words, from which a semantic similarity measure could be derived.

This paper is organized as follows: Section 2 details the neural network method. Section 3 describes the various test cases and presents the results. Discussion is provided in Section 4, conclusions are drawn in Section 5.

2 Methods

As it was already mentioned, our method exploits three databases (Fig. 1(A)). BNC is used to obtain word co-occurrence statistics. The following estimations are also required: given a synset, how frequently is one of its words used to express that sense, and, on the other hand, if a word is used, how frequently is it used in one of its senses. Database SemCor was used to obtain these statistics. Since all these pieces of information are needed, we only used words and synsets which occur in SemCor at least once. This means 23141 words and 22012 synsets. Later, we extended these sets by adding the trivial synsets, which contained single words. Then we could experiment with 54572 words and 53443 synsets.

2.1 Co-occurrence Measures

The aim of our method is to find semantically close synsets to an unknown word. Two words that occur in similar contexts can be considered as similar in

---

[3] http://www.natcorp.ox.ac.uk/
[4] The words *sense* and *synset* is used interchangeably in this paper.
[5] http://www.ets.org/

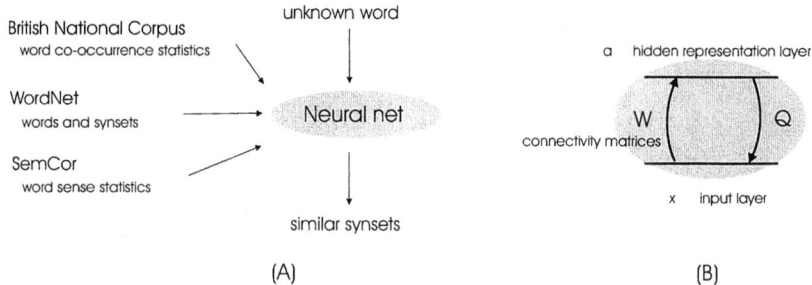

Fig. 1. Databases and reconstruction network architecture (A): Scheme of our method. (B): Basic computational architecture. Input layer (x) and hidden representation (a) are connected by bottom-up (W) and top-down (Q) matrices.

meaning. Therefore, we need to express the measures of co-occurrence between words and synsets; i.e. values indicating how often words and synsets co-occur.

The probability that word w_1 occurs near word w_2 can be estimated as follows: $P(w_1|w_2) = \frac{f(w_1,w_2)}{f(w_2)}$, where $f(w_1,w_2)$ is the number of times w_1 and w_2 co-occur in a 5 wide context window[6] and $f(w)$ is the frequency of word w. We say that w_1 and w_2 are *near words* of each other, if both $P(w_1|w_2)$ and $P(w_2|w_1)$ are high, meaning that w_1 and w_2 are likely to co-occur. The following measure derived from mutual co-occurrences expresses this idea: $N(w_1, w_2) = min(P(w_1|w_2), P(w_2|w_1))$. It is expected that this co-occurrence measure describes the contexts of the words. Given a word w, we call the *near word list* of w is the 100 words w_i for which the $N(w, w_i)$ values are the highest. This near word list can be represented as a *feature vector* of the word, the entries of the vector are the $N(w, w_i)$ values. Then the co-occurrence information about the words can be summarized in a quadratic and symmetric matrix N_W, where the ith row of the matrix is the feature vector of the ith word: $N_W(i,j) = N(w_i, w_j)$, where w_j is the jth word in our vocabulary.

In SemCor, every occurrence of a word is tagged with a WordNet synset that expresses the meaning of the actual occurrence of the word. By counting these tags we can compute the desired probabilities. The probability that for a given word w, the expressed sense is s, can be estimated as $P(s|w) = \frac{f(w,s)}{f(w)}$, where $f(w, s)$ is the frequency of word w in sense s and $f(w)$ is the frequency of word w in any of its senses. We also need the probability that a given sense s is expressed by word w, which can be estimated as: $P(w|s) = \frac{f(w,s)}{f(s)}$, where $f(s)$ is the frequency of sense s, whichever word it is expressed by. These probabilities can also be summarized in matrix forms, denoted by S_W and W_S: $S_W(i,j) = P(s_i|w_j)$ and $W_S(i,j) = P(w_i|s_j)$.

Using the measures introduced, a co-occurrence measure between synsets can be derived. The idea is the following: given a synset s, the *near synsets* of s are the synsets of the near words of the words expressing s. This idea is

---

[6] Increasing the context window by a factor of 2 had no significant effects.

expressed by the appropriate concatenation of the three matrices introduced above: $N_S = S_W N_W W_S$.

2.2 Reconstruction Networks

The basic reconstruction network model has two neuron layers. Connections bridge these layers. The lower layer is the input layer of the network, and the upper layer is called the hidden or internal representation layer (Fig. 1(B)). The network reconstructs its input by optimizing the hidden representation. For this reason, we call it reconstruction network. Formally, the following quadratic cost function is involved [5]:

$$J(a) = ||x - Qa||_2^2 ,\qquad(1)$$

where Q is the connectivity matrix, x is the input vector, a is the hidden representation vector. The columns of the connectivity matrix can be thought of as basis vectors, which must be linearly combined with the appropriate coefficients so that the combination falls close to the input. The optimization can either be solved directly

$$a = (Q^T Q)^{-1} Q^T x ,\qquad(2)$$

or iteratively

$$\Delta a \propto W(x - Qa) ,\qquad(3)$$

where $W = Q^T$, which can be derived from the negative gradient of cost function (1). The form of (3) is more general than required by (1) but it still suitable as long as WQ is positive definite. Both methods have advantages and disadvantages. Directly solving the optimization returns the exact solution, but might require a considerable amount of memory, while the iterative solution requires less resources, but is computationally intensive.

The reconstruction network described above shall be called 'one-tier' network. We designed both 'one-tier' and 'two-tier' networks for the word-sense reconstruction. The two-tier network has two one-tiers on the top of each other. The internal representation layer of the first tier serves as input for the second tier. There are differences between the two architectures in computation speed and in numerical precision.

The feature vector representation of the context of the unknown word serves as input for the network. In the hidden layer, the neurons represent the candidate synsets. In the one-tier network (Fig. 2(A)), the top-down and bottom-up matrices are defined as follows: $Q = W_S N_S$ and $W = N_S^T S_W$[7]. Thus, in the one-tier method, hidden synset activities are (a) transformed to near synset activities and then (b) the activities of the near words are generated. An illustrative iteration is depicted in Fig. 2(B): The activities in the topmost layer change during the reconstruction of the input word *frog*. It can be seen that only a few activities become high, others remain small. Note the horizontal scale: there were about 23,000 neurons in the topmost layer in this iteration.

---

[7] However, we found that $\tilde{Q} = N_W W_S$ and $\tilde{W} = \tilde{Q}^T = W_S^T N_W^T$ are simpler, express the same relations and converge faster, thus these were used in the computations.

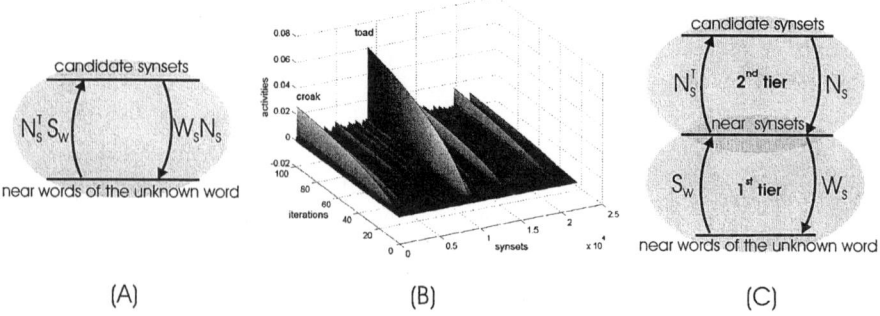

Fig. 2. One-tier and two-tier networks (A) and (C): one and two tier networks. The input of the ith node of the lowest layer is $N(w, w_i)$, where w is the unknown word. $S_W(i,j) = P(s_i|w_j)$, $W_S(i,j) = P(w_i|s_j)$, $N_S = S_W N_W W_S$, where $N_W(i,j) = N(w_i, w_j)$ for all i, j. The *result* of the computation is the synset represented by the highest activity unit of the top layer after running the network. (B): Convergence of the iterative approximation. Input word is *frog*. High activity nodes correspond to sense *toad* and *croak*.

In the 'two-tier' network (Fig. 2(C)) the two steps of the transformation are separated. The nodes in the intermediate layer correspond to the near synsets of the unknown word. The bottom-up and top-down matrices are the S_W and the W_S matrices, respectively. In the second tier the connectivity matrix is N_S in both directions, which contains the synset near values. This captures the idea mentioned in the introduction; the meaning of the unknown word is similar to the meaning of a synset if they have the same near *synsets*.

3 Tests and Results

We tested our method on 80 TOEFL synonym questions, each question consisted of a *question word* (for example *grin*) and four *candidate answer words*, (for example: *exercise, rest, joke, smile*). The task was to find the candidate word that was the most similar in meaning to the question word (*smile*).

To meet our goals, we considered the question word the unknown word. We simulated the situation of the question word being *unknown* by erasing all information about the question word and its meaning from the SemCor statistics and WordNet synsets. After running the network for the context of the question word as input, we examined the activities corresponding to the synsets of the candidate words, and assigned a value to each candidate word equal to the highest activity of the synsets of the candidate word in the upper layer of the network. The candidate word with the highest value was the chosen answer.

In the one-tier network and in the second tier of the two-tier network, the huge number of connections between the nodes required the application of the less precise iterative method. However, the first tier of the two-tier network could be optimized directly, because the connectivity matrices were very sparse.

In some cases we have additional – top-down (TD) – information about the unknown word, for example its part-of-speech or the candidate answer words, this reduces the set of candidate synsets. The implementation of this filtering is simple in our system, since we can simply leave out the unnecessary synsets.

In order to test whether or not using synsets increases the efficiency of the system, we constructed a one tier *control* network which used only words. The input of the network was the same, however, the nodes in the upper layer corresponded to words. The connectivity matrix between the two layers is N_W, as defined in 2.1. The activities of the upper layer were examined; words having similar context as the input word were returned.

The number of correct answers in the various cases can be seen in Table 1. The best result, 63 correct answers (78.75%), was produced with part-of-speech constraint with the 23141 word data set and also without TD constraint with the 54572 word data set, utilizing the one-tier network. However, the best first iterations were achieved by the two-tier network. It can be seen, that iteration has improved the precision in almost all the cases. The control network started from 53 and 54 correct answers for the two word sets and reached 60 and 59 correct answers, respectively, by iterations. These results are considerably smaller than those of without the synsets, which supports our starting assumptions. We should note that if information related to the question words are not deleted from the database, then the number of correct answers is 68, which *amounts to 85%*.

Table 1. Results *No constr*: No constraint is applied. *PoS*: Synsets in the upper layer correspond to the part-of-speech of the word. *Candidate*: Only the synsets of the candidate words are used. *Control*: Single tier control network. *Direct*: non-iterative solution.

23141 words	No constr		PoS		Candidate		Control
	1 tier	2 tiers	1 tier	2 tiers	1 tier	2 tiers	1 tier
1 iter	55	58	55	58	55	58	53
10 iter	60	57	61	59	55	58	56
100 iter	61	60	63	58	58	56	60
direct	-	-	-	-	57	55	-
54572 words	No constr		PoS		Candidate		Control
	1 tier	2 tiers	1 tier	2 tiers	1 tier	2 tiers	1 tier
1 iter	56	59	56	59	56	59	54
10 iter	62	60	59	58	56	59	56
100 iter	63	61	62	58	57	56	59
direct	-	-	-	-	57	57	-

By examining the activities corresponding to the candidate answer words, decision points can be incorporated. Then the system may deny to answer a question, if the answer is uncertain. We could improve precision but the number of answers decreased considerably. Still, this property should be useful in multiple expert schemes, where experts may be responsible for different domains.

4 Discussion

Compared to the other methods, LSA performs relatively poorly (64.5%). However, the original intention of LSA was not to serve as an efficient TOEFL solver, but to model human memory. LSA reads the dictionary (the text database of the experiment) and runs the singular value decomposition only once and without knowing anything about the questions beforehand. After this procedure LSA can immediately answer the questions. While the first phase in LSA models a person's general learning process, this second phase imitates how someone solves questions without relying on any external aid [6]. By contrast, many other methods are allowed to use their databases after they have observed the questions.

Our method resembles LSA. Alike LSA, it works by the optimization of reconstruction using hidden variables over Euclidean norm. We also build a kind of memory model (the connectivity matrices of the neural network) before the questions are observed. When the questions are observed, the answer can be produced by running the network. Alike to LSA, our method was not developed for solving TOEFL questions, but for explaining unknown words. Considering this, our comparably high score (78.75%) is promising. True though, the original network incorporates information contained in WordNet and SemCor, however, the strength of the approach is shown by the *control network*, which did not use any lexical information, and gave 60 correct answers (i.e., 75%).

The Hyperspace Analogue to Language (HAL) model [7] works in high dimensions alike to our method. According to the HAL model, the strength of a term–term association is inversely proportional to the Euclidean distance between the context and the target words. Alike to HAL, our method makes use of the whole table of co-occurrences. This seems important; the larger table gave better result for us. Our method combines the advantages of LSA and HAL: it makes use of all information like HAL and adopts hidden variables like LSA.

We also included other information, the uncertainty of the answer, that goes beyond the statistics of co-occurrences. It may be worth noting here that our approach can be generalized to hidden, overcomplete, and sparse representations [8]. Such non-linear generalizations can go beyond simple computational advantages when additional *example based information* [9] or *supervisory training* are to be included.

A recent paper on meaning discovery using Google queries [10] thoroughly details the development of semantic distances between words. The method uses first order co-occurrence counts (Google page counts) to determine the semantic distance of two words. The article describes a semantic distance called Normalized Google Distance (NGD), derived from the same formula that we use to calculate the co-occurrence measure of two words. However, in our case, the formula was used to examine second order co-occurrences instead of first order co-occurrences. We conducted two studies with NGD. First, we solved the 80 TOEFL synonym questions using NGD as described in the paper; we measured the distance of the question word and each candidate word, and chosen the one with the smallest distance. Depending on the database we used to collect word frequencies, the results were different, however, surprisingly low: 30 correct answers (37.5%) when BNC was used, and 40 correct answers (50.0%) when Google was used to return the page counts needed for NGD. In the other study we used our neural network method based on NGD instead of our co-occurrence measure. Results in this case were almost identical to the original setting, when we used our own co-occurrence measure, indicating the robustness of our solution against these details.

5 Conclusions

We have studied neural network architectures for explaining unknown words by known senses, senses that are contained in our lexical databases. We tested the method on TOEFL synonym questions. It was found that the networked solution provided good results, and was found robust against the details. At the cost of decreasing recall, the precision of the system can be improved. These features make our method attractive for various circumstances.

Acknowledgments

We are grateful to Prof. Landauer for providing us the TOEFL examples. We thank one of the referees for calling our attention to Burgess' HAL model.

References

[1] Landauer, T.K., Dumais, S.T.: A solution to Plato's problem: The latent semantic analysis theory of acquisition, induction and representation of knowledge. Psychol. Rev. **104** (1997) 211–240
[2] Turney, P.: Mining the Web for synonyms: PMI-IR versus LSA on TOEFL. In: ECML Proceedings, Freiburg, Germany (2001) 491–502
[3] Terra, E., Clarke, C.L.A.: Choosing the word most typical in context using a lexical co-occurrence network. In: Proc. of Conf. on Human Language Technol. and North American Chapter of Assoc. of Comput. Linguistics. (2003) 244–251
[4] Jarmasz, M., Szpakowicz, S.: Roget's Thesaurus and semantic similarity. In: Proc. of the Int. Conf. on Recent Advances in Natural Language Proc. (RANLP-03). (2003)
[5] Haykin, S.: Neural Networks: A comprehensive foundation. Prentice Hall, New Jersey, USA (1999)
[6] Landauer, T.K. (personal communication)
[7] Burgess, C.: From simple associations to the building blocks of language: Modeling meaning in memory with the HAL model. Behav. Res. Methods, Instr. and Comps. **30** (1998) 188–198
[8] Olshausen, B.A.: Learning linear, sparse factorial codes. A.I. Memo 1580, MIT AI Lab (1996) C.B.C.L. Paper No. 138.
[9] Szatmáry, B., Szirtes, G., Lőrincz, A., Eggert, J., Körner, E.: Robust hierarchical image representation using non-negative matrix factorization with sparse code shrinkage preprocessing. Pattern Anal .and Appl. **6** (2003) 194–200
[10] Cilibrasi, R., Vitanyi, P.: Automatic meaning discovery using google. arXiv:cs.CL/0412098 v2 (2005)

Segment and Combine Approach for Non-parametric Time-Series Classification

Pierre Geurts† and Louis Wehenkel

University of Liège, Department of Electrical Engineering and Computer Science,
Sart-Tilman B28, B4000 Liège, Belgium,
† Postdoctoral researcher, FNRS, Belgium
{p.geurts,l.wehenkel}@ulg.ac.be

Abstract. This paper presents a novel, generic, scalable, autonomous, and flexible supervised learning algorithm for the classification of multivariate and variable length time series. The essential ingredients of the algorithm are randomization, segmentation of time-series, decision tree ensemble based learning of subseries classifiers, combination of subseries classification by voting, and cross-validation based temporal resolution adaptation. Experiments are carried out with this method on 10 synthetic and real-world datasets. They highlight the good behavior of the algorithm on a large diversity of problems. Our results are also highly competitive with existing approaches from the literature.

1 Learning to Classify Time-Series

Time-series classification is an important problem from the viewpoint of its multitudinous applications. Specific applications concern the non intrusive monitoring and diagnosis of processes and biological systems, for example to decide whether the system is in a healthy operating condition on the basis of measurements of various signals. Other relevant applications concern speech recognition and behavior analysis, in particular biometrics and fraud detection.

From the viewpoint of machine learning, a time-series classification problem is basically a supervised learning problem, with temporally structured input variables. Among the practical problems faced while trying to apply classical (propositional) learning algorithms to this class of problems, the main one is to transform the non-standard input representation into a fixed number of scalar attributes which can be managed by a propositional base learner and at the same time retain information about the temporal properties of the original data.

One approach to solve this problem is to define a (possibly very large) collection of temporal predicates which can be applied to each time-series in order to compute (logical or numerical) features which can then be used as input representation for any base learner (e.g. [7,8,10,11]). This feature extraction step can also be incorporated directly into the learning algorithm [1,2,14]. Another approach is to define a distance or similarity measure between time-series that takes into account temporal specific peculiarities (e.g. invariance with respect to time or amplitude rescaling) and then to use this distance measure in combination with nearest neighbors or other kernel-based methods [12,13]. A potential advantage of these approaches is the possibility to bias the representation by exploiting prior problem specific knowledge. At the same time, this problem specific modeling step makes the application of machine learning non autonomous.

The approach investigated in this paper aims at developing a fully generic and off-the-shelf time-series classification method. More precisely, the proposed algorithm relies on a generic pre-processing stage which extracts from the time-series a number of randomly selected subseries, all of the same length, which are labeled with the class of the time-series from which they were taken. Then a generic supervised learning method is applied to the sample of subseries, so as to derive a subseries classifier. Finally, a new time-series is classified by aggregating the predictions of all its subseries of the said size. The method is combined with a ten-fold cross-validation wrapper in order to adjust automatically the size of the subseries to a given dataset. As base learners, we use tree-based methods because of their scalability and autonomy.

Section 2 presents and motivates the proposed algorithmic framework of segmentation and combination of time-series data and Section 3 presents an empirical evaluation of the algorithm on a diverse set of time-series classification tasks. Further details about this study may be found in [4].

2 Segment and Combine

Notations. A time-series is originally represented as a discrete time finite duration real-valued vector signal. The different components of the vector signal are called temporal attributes in what follows. The number of time-steps for a given temporal attribute is called its *duration*. We suppose that all temporal attributes of a given time-series have the same duration. On the other hand, the durations of different time-series of a given problem (or dataset) are not assumed to be identical. A given time series is related to a particular observation (or object). A learning sample (or dataset) is a set (ordering is considered irrelevant at this level) of N preclassified time-series denoted by $LS_N = \left\{ \left(\boldsymbol{a}(t^{d(o)}, o), c(o)\right) \big| o = 1, \ldots, N \right\}$, where o denotes an observation, $d(o) \in \mathbb{N}$ stands for the duration of the time-series, $c(o)$ refers to the class associated to the time-series, and

$$\boldsymbol{a}(t^{d(o)}, o) = (a_1(t^{d(o)}, o), \ldots, a_n(t^{d(o)}, o))', a_i(t^{d(o)}, o) = (a_i(1, o), \ldots, a_i(d(o), o))',$$

represents the vector of n real-valued temporal attributes of duration $d(o)$.

The objective of the time-series classification problem is to derive from LS_N a classification rule $\hat{c}(\boldsymbol{a}(t^{d(o)}, o))$ which predicts output classes of an unseen time-series $\boldsymbol{a}(t^{d(o)}, o)$ as accurately as possible.

Training a subseries classifier. In its training stage, the segment and combine algorithm uses a propositional base learner to yield a subseries classifier from LS_N in the following way:

Subseries sampling. For $i = 1, \ldots, N_s$ choose $o_i \in \{1, \ldots, N\}$ randomly, then choose a subseries offset $t_i \in \{0, \ldots, d(o_i) - \ell\}$ randomly, and create a scalar attribute vector

$$a_{t_i}^\ell(o_i) = (a_1(t_i + 1, o_i), \ldots, a_1(t_i + \ell, o_i), \ldots, a_n(t_i + 1, o_i), \ldots, a_n(t_i + \ell, o_i))$$

concatenating the values of all n temporal attributes over the time interval $t_i + 1, \ldots, t_i + \ell$. Collect the samples in a training set of subseries

$$LS_{N_s}^\ell = \{(a_{t_i}^\ell(o_i), c(o_i)) \mid i = 1, \ldots, N_s\}.$$

Classifier training. Use the base learner on $LS_{N_s}^\ell$ to build a subseries classifier. This "classifier" is supposed to return a class-probability vector $P_c^\ell(a^\ell)$.

Notice that when N_s is greater than the total number of subseries of length ℓ, no sampling is done and $LS_{N_s}^\ell$ is taken as the set of all subseries.

Classifying a time-series by votes on its subseries. For a new time-series $a(t^{d(o)}, o)$, extract systematically all its subseries of length ℓ, $a_i^\ell(o), \forall i \in \{0, \ldots, d(o) - \ell\}$, and classify it according to

$$\hat{c}(a(t^{d(o)}, o)) \triangleq \arg\max_c \left\{ \sum_{i=0}^{d(o)-\ell} P_c^\ell(a_i^\ell(o)) \right\}.$$

Note that if the base learner returns 0/1 class indicators, the aggregation step merely selects the class receiving the largest number of votes.

Tuning the subseries length ℓ. In addition to the choice of base learner discussed below, the sole parameters of the above method are the number of subseries N_s and the subseries length ℓ. In practice, the larger N_s, the higher the accuracy. Hence, the choice of the value of N_s is only dictated by computational constraints. On the other hand, the subseries length ℓ should be adapted to the temporal resolution of the problem. Small values of ℓ force the algorithm to focus on local (shift-invariant) patterns in the original time-series while larger values of ℓ amount to considering the time-series more globally. In our method, we determine this length automatically by trying out a set of candidate values $\ell_i \leq \min_{o \in LS_N} d(o)$, estimating for each ℓ_i the error-rate by ten-fold cross-validation over LS_N, and selecting the value ℓ_* yielding the lowest error rate estimate.

Base learners. In principle, any propositional base learner (SVM, kNN, MLP etc.) could be used in the above approach. However, for scalability reasons, we recommend to use decision trees or ensembles of decision trees. In the trials in the next section we will compare the results obtained with two different tree-based methods, namely single unpruned CART trees and ensembles of extremely randomized trees. The extremely randomized trees algorithm (Extra-Trees) is

Table 1. Summary of datasets

Dataset	Src.	N_d	n	c	$\underline{d}-\overline{d}$	Protocol	Best	Ref	$\{\ell_i\},\{s_i\}$
CBF	1	798	1	3	128	10-fold cv	0.00	[7]	1,2,4,8,16,32,64,96,128
CC	2	600	1	6	60	10-fold cv	0.83	[1]	1,2,5,10,20,30
CBF-tr	1	5000	1	3	128	10-fold cv	–		1,2,4,8,16,32,64,96,128
Two-pat	1	5000	1	3	128	10-fold cv	–		1,2,4,8,16,32,64,96,128
TTest	1	999	3	3	81-121	10-fold cv	0.50	[7]	3,5,10,20,40,60
Trace	3	1600	4	16	268-394	holdout 800	0.83	[1]	10,25,50,100,150,200,250
Auslan-s	2	200	8	10	32-101	10-fold cv	1.50	[1]	1,2,5,10,20,30
Auslan-b	5	2566	22	95	45-136	holdout 1000	2.10	[7]	1,2,5,10,20,30,40
JV	2	640	8	10	7-29	holdout 270	3.80	[8]	2,3,5,7
ECG	4	200	2	2	39-152	10-fold cv	–		1,2,5,10,20,30,39

[1] http://www.montefiore.ulg.ac.be/~geurts/thesis.html [2] [5] [3] http://www2.ife.no
[4] http://www-2.cs.cmu.edu/~bobski/pubs/tr01108.html
[5] http://waleed.web.cse.unsw.edu.au/new/phd.html

described in details in [4]. It grows a tree by selecting the best split from a small set of candidate random splits (both attribute and cut-point are randomized). This method allows to reduce strongly variance without increasing bias too much. It is also significantly faster in the training stage than bagging or boosting which search for optimal attribute and cut-points at each node.

Notice that because the segment and combine approach has some intrinsic variance reduction capability, it is generally counterproductive to prune single trees in this context. For the same reason, the number of trees in the tree ensemble methods can be chosen reasonably small (25 in our experiments).

3 Empirical Analysis

3.1 Benchmark Problems

Experiments are carried out on 10 problems. For the sake of brevity, we only report in Table 1 the main properties of the 10 datasets. We refer the interested reader to [4] and the references therein for more details. The second column gives the (web) source of the dataset. The next four columns give the number N_d of time-series in the dataset, the number of temporal attributes n of each time-series, the number of classes c, and the range of values of the duration $d(o)$; the seventh column specifies our protocol to derive error rates; the eighth and ninth columns give respectively the best published error rate (with identical or comparable protocol to ours) and the corresponding reference; the last column gives the trial values used for the parameters ℓ and s. The first six problems are artificial problems specifically designed for the validation of time-series classification methods, while the last four problems correspond to real world problems.

3.2 Accuracy Results

Accuracy results on each problem are gathered in Table 2. In order to assess the interest of the segment and combine approach, we compare it with a simple

Table 2. Error rates (in %) and optimal values of s and ℓ

Dataset	Temporal normalization				Segment&Combine ($N_s = 10000$)			
	ST		ET		ST		ET	
	Err%	s^*	Err%	s^*	Err%	ℓ^*	Err%	ℓ^*
CBF	4.26	24.0 ± 8.0	**0.38**	27.2 ± 7.3	1.25	92.8 ± 9.6	0.75	96.0 ± 0.0
CC	3.33	21.0 ± 17.0	0.67	41.0 ± 14.5	0.50	35.0 ± 6.7	**0.33**	37.0 ± 4.6
CBF-tr	13.28	30.4 ± 13.3	2.51	30.4 ± 4.8	**1.63**	41.6 ± 14.7	1.88	57.6 ± 31.4
Two-pat	25.12	8.0 ± 0.0	14.37	36.8 ± 46.1	2.00	96.0 ± 0.0	**0.37**	96.0 ± 0.0
TTest	18.42	40.0 ± 0.0	13.61	40.0 ± 0.0	3.00	80.0 ± 0.0	**0.80**	80.0 ± 0.0
Trace	50.13	50	40.62	50	8.25	250	**5.00**	250
Auslan-s	19.00	5.5 ± 1.5	4.50	10.2 ± 4.0	5.00	17.0 ± 7.8	**1.00**	13.0 ± 4.6
Auslan-b	22.82	10	**4.51**	10	18.40	40.0 ± 0.0	5.16	40.0 ± 0.0
JV	16.49	2	4.59	2	8.11	3	**4.05**	3
ECG	25.00	18.5 ± 10.0	**15.50**	19.0 ± 9.4	25.50	29.8 ± 6.0	24.00	32.4 ± 8.5

normalization technique [2,6], which aims at transforming a time-series into a vector of fixed dimensionality of scalar numerical attributes: the time interval of each object is divided into s equal-length segments and the average values of all temporal attributes along these segments are computed, yielding a new vector of $n \cdot s$ attributes which are used as inputs to the base learner. The two approaches are combined with single decision trees (ST) and ensembles of 25 Extra-Trees (ET) as base learners. The best result in each row is highlighted.

For the segment and combine method, we randomly extracted 10,000 subseries. The optimal values of the parameters ℓ and s are searched among the candidate values reported in the last column of Table 1. When the testing protocol is holdout, the parameters are adjusted by 10-fold cross-validation on the learning sample only; when the testing protocol is 10-fold cross-validation, the adjustment of these parameters is made for each of the ten folds by an internal 10-fold cross-validation. In this latter case average values and standard deviations of the parameters s^* and ℓ^* over the (external) testing folds are provided.

From these results we first observe that "Segment and Combine" with Extra-Trees (ET) yields the best results on six out of ten problems. On three other problems (CBF, CBR-tr, Auslan-b) its accuracy is close to the best one. Only on the ECG problem, the results obtained are somewhat disappointing with respect to the normalization approach. On the other hand, it is clear that the combination of the normalization technique with single trees (ST) is systematically (much) less accurate than the other variants.

We also observe that, both for "normalization" and "segment and combine", the Extra-Trees always give significantly better results than single trees.[1] On the other hand, the improvement resulting from the segment and combine method is stronger for single decision trees than for Extra-Trees. Indeed, error rates of the former are reduced in average by 65% while error rates of the latter are

---

[1] There is only one exception, namely CBF-tr where the ST method is slightly better than ET in the case of "segment and combine".

only reduced by 30%. Actually, with "segment and combine", single trees and Extra-Trees are close to each other in terms of accuracy on several problems, while they are not with "normalization". This can be explained by the intrinsic variance reduction effect of the segment and combine method, which is due to the virtual increase of the learning sample size and the averaging step and somewhat mitigates the effect variance reduction techniques like ensemble methods (see [3] for a discussion of bias and variance of the segment and combine method).

From the values of ℓ^* in the last column of Table 2, it is clear that the optimal ℓ^* is a problem dependent parameter. Indeed, with respect to the average duration of the time-series this optimal values ranges from 17% (on JV) to 80% (on TTest). This highlights the usefulness of the automatic tuning by cross-validation of ℓ^* as well as the capacity of the segment and combine approach to adapt itself to variable temporal resolutions.

A comparison of the results of the last two columns of Table 2 with the eighth column of Table 1, shows that the segment and combine method with Extra-Trees is actually quite competitive with the best published results. Indeed, on CBF, CC, TTest, Auslan-s, and JV, its results are very close to the best published ones.[2] Since on Trace, and to a lesser extent on Auslan-b, the results were less good, we ran a side-experiment to see if there is room for improvement. On Trace we were able (with Extra-Trees and $N_s = 15000$) to reduce the error rate from 5.00% to 0.875% by first resampling the time series into a fixed number of 268 time points. The same approach with 40 time points decreased also the error rate on Auslan-b from 5.16% to 3.94%.

3.3 Interpretability

Let us illustrate the possibility to extract interpretable information from the subseries classifiers. Actually, these classifiers provide for each time point a vector estimating the class-probabilities of subseries centered at this point. Hence, subseries that correspond to a high probability of a certain subset of classes can be considered as typical patterns of this subset of classes.

Figure 1 shows for example, in the top part, two temporal attributes for three instances of the Trace problem respectively of classes 1, 3, and 5, and in the bottom part the evolution of the probabilities of these three classes as predicted for subseries (of length $\ell = 50$) as they move progressively from left to right on the time axis. The Class 3 signal (top middle) differs from the Class 1 signal (top left) only in the occurrence of a small sinusoidal pattern in one of the attribute (around $t = 200$); on the other hand, Class 1 and 3 differ from Class 5 (top right) in the occurrence of a sharp peak in the other attribute (around $t = 75$ and $t = 100$ respectively). From the probability plots we see that, for $t \leq 50$ the three classes are equally likely, but at the time where the peak appears ($t \in [60 - 70]$) the probability of Class 5 decreases for the two

---

[2] Note that on CBF, CC, TTest, and Auslan-s, our test protocols are not strictly identical to those published since we could not use the same ten folds. This may be sufficient to explain small differences with respect to results from the literature.

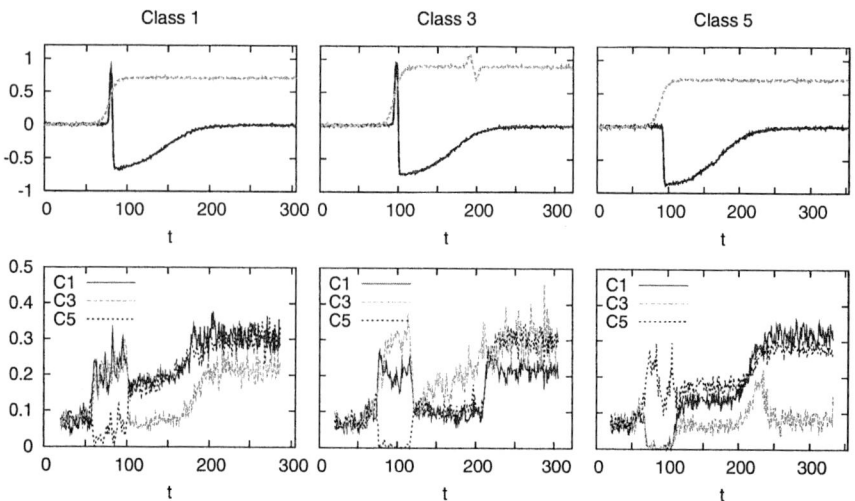

Fig. 1. Interpretability of "Segment and Combine" (Trace dataset, $N_s = 10000$, ET)

first series (where a peak appears) and increases for the right-most series (where no peak appears). Subsequently, around $t = 170$, the subseries in the middle instance start to detect the sinusoidal pattern, which translates into an increase of the probability of Class 3, while for the two other time-series Classes 1 and 5 become equally likely and Class 3 relatively less. Notice that the voting scheme used to classify the whole time-series from its subseries amounts to integrating these curves along the time axis and deciding on the most likely class once all subseries have been incorporated. This suggests that, once a subseries classifier has been trained, the segment combine approach can be used in real-time in order to classify signals through time.

4 Conclusion

In this paper, we have proposed a new generic and non-parametric method for time-series classification which randomly extracts subseries of a given length from time-series, induces a subseries classifier from this sample, and classifies time-series by averaging the prediction over its subseries. The subseries length is automatically adapted by the algorithm to the temporal resolution of the problem. This algorithm has been validated on 10 benchmark problems, where it yielded results competitive with state-of-the-art algorithms from the literature. Given the diversity of benchmark problems and conceptual simplicity of our algorithm, this is a very promising result. Furthermore, the possibility to extract interpretable information from time-series has been highlighted.

There are several possible extensions of our work, such as more sophisticated aggregation schemes and multi-scale subseries extraction. These would allow to handle problems with more complex temporally related characteristic patterns of variable time-scale. We have also suggested that the method could be used for real-time time-series classification, by adjusting the voting scheme.

The approach presented here for time-series is essentially identical to the work reported in [9] for image classification. Similar ideas could also be exploited to yield generic approaches for the classification of texts or biological sequences. Although these latter problems have different structural properties, we believe that the flexibility of the approach makes it possible to adjust it to these contexts in a straightforward manner.

References

1. J. Alonso González and J. J. Rodríguez Diez. Boosting interval-based literals: Variable length and early classification. In M. Last, A. Kandel, and H. Bunke, editors, *Data mining in time series databases*. World Scientific, June 2004.
2. P. Geurts. Pattern extraction for time-series classification. In L. de Raedt and A. Siebes, editors, *Proceedings of PKDD 2001, 5th European Conference on Principles of Data Mining and Knowledge Discovery*, LNAI 2168, pages 115–127, Freiburg, September 2001. Springer-Verlag.
3. P. Geurts. *Contributions to decision tree induction: bias/variance tradeoff and time series classification*. PhD thesis, University of Liège, Belgium, May 2002.
4. P. Geurts and L. Wehenkel. Segment and combine approach for non-parametric time-series classification. Technical report, University of Liège, 2005.
5. S. Hettich and S. D. Bay. The UCI KDD archive, 1999. Irvine, CA: University of California, Department of Information and Computer Science. http://kdd.ics.uci.edu.
6. M. W. Kadous. Learning comprehensible descriptions of multivariate time series. In *Proceedings of the Sixteenth International Conference on Machine Learning, ICML'99*, pages 454–463, Bled, Slovenia, 1999.
7. M. W. Kadous and C. Sammut. Classification of multivariate time series and structured data using contructive induction. *Machine learning*, 58(1-2):179–216, February/March 2005.
8. M. Kudo, J. Toyama, and M. Shimbo. Multidimensional curve classification using passing-through regions. *Pattern Recognition Letters*, 20(11-13):1103–1111, 1999.
9. R. Marée, P. Geurts, J. Piater, and L. Wehenkel. Random subwindows for robust image classification. In *Proceedings of the IEEE International Conference on Computer Vision and Pattern Recognition (CVPR 2005)*, 2005.
10. I. Mierswa and K. Morik. Automatic feature extraction for classifying audio data. *Machine Learning*, 58(1-2):127–149, February/March 2005.
11. R. T. Olszewski. *Generalized feature extraction for structural pattern recognition in time-series data*. PhD thesis, Carnegie Mellon University, Pittsburgh, PA, 2001.
12. C.A. Ratanamahatana and E. Keogh. Making time-series classification more accurate using learned constraints. In *Proceedings of SIAM*, 2004.
13. H. Shimodaira, K.I. Noma, M. Nakai, and S. Sagayama. Dynamic time-alignment kernel in support vector machine. In *Advances in Neural Information Processing Systems 14, NIPS2001*, volume 2, pages 921–928, December 2001.
14. Y. Yamada, E. Suzuki, H. Yokoi, and K. Takabayashi. Decision-tree induction from time-series data based on standard-example split test. In *Proceedings of the 20th International Conference on Machine Learning (ICML-2003)*, 2003.

Producing Accurate Interpretable Clusters from High-Dimensional Data

Derek Greene and Pádraig Cunningham

University of Dublin, Trinity College,
Dublin 2, Ireland
{derek.greene, padraig.cunningham}@cs.tcd.ie

Abstract. The primary goal of cluster analysis is to produce clusters that accurately reflect the natural groupings in the data. A second objective is to identify features that are descriptive of the clusters. In addition to these requirements, we often wish to allow objects to be associated with more than one cluster. In this paper we present a technique, based on the spectral co-clustering model, that is effective in meeting these objectives. Our evaluation on a range of text clustering problems shows that the proposed method yields accuracy superior to that afforded by existing techniques, while producing cluster descriptions that are amenable to human interpretation.

1 Introduction

The unsupervised grouping of documents, a frequently performed task in information retrieval systems, can be viewed as having two fundamental goals. Firstly, we seek to identify a set of clusters that accurately reflects the topics present in a corpus. A second objective that is often overlooked is the provision of information to facilitate human interpretation of the clustering solution.

The primary choice of representation for text mining procedures has been the *vector space model*. However, corpora modelled in this way are generally characterised by their sparse high-dimensional nature. Traditional clustering algorithms are susceptible to the well-known problem of the *curse of dimensionality*, which refers to the degradation in algorithm performance as the number of features increases. Consequently, these methods will often fail to identify coherent clusters when applied to text data due to the presence of many irrelevant or redundant terms. In addition, the inherent sparseness of the data can further impair an algorithm's ability to correctly uncover the data's underlying structure.

To overcome these issues, a variety of techniques have been proposed to project high-dimensional data to a lower-dimensional representation in order to minimise the effects of sparseness and irrelevant features. In particular, dimension reduction methods based on spectral analysis have been frequently applied to improve the accuracy of document clustering algorithms, due to their ability to uncover the latent relationships in a corpus. However, from the perspective of domain users, the production of clear, unambiguous descriptions of cluster content is also highly important. A simple but effective means of achieving this goal is to generate weights signifying the relevance of the terms in the corpus vocabulary to each cluster, from which a set of cluster labels can be derived. The provision of document weights can also help the end-user to gain an insight into a clustering solution. In particular, when a document is assigned to a cluster, one may wish to quantify the confidence of the assignment. Additionally, the use of soft clusters allows us to represent cases where a given document relates to more than one topic.

In this paper, we introduce a co-clustering technique, based on spectral analysis, that provides interpretable membership weights for both terms and documents. Furthermore, we show that by applying an iterative matrix factorisation scheme, we can produce a refined clustering that affords improved accuracy and interpretability. We compare our algorithms with existing methods on a range of datasets, and briefly discuss the generation of useful cluster descriptions. Note that an extended version of this paper is available as a technical report with the same title [1].

2 Matrix Decomposition Methods

In this section, we present a brief summary of two existing dimension reduction methods that have been previously applied to document clustering. To describe the algorithms discussed in the remainder of the paper, we let $\mathbf{A}$ denote the $m \times n$ term-document matrix of a corpus of n documents, each of which is represented by an m-dimensional feature vector. We assume that k is an input parameter indicating the desired number of clusters.

2.1 Spectral Co-clustering

Spectral clustering methods have been widely shown to provide an effective means of producing disjoint partitions across a range of domains [2,3]. In simple terms, these algorithms analyse the spectral decomposition of a matrix representing a dataset in order to uncover its underlying structure. The reduced space, constructed from the leading eigenvectors or singular vectors of the matrix, can be viewed as a set of semantic variables taking positive or negative values.

A novel approach for simultaneously clustering documents and terms was suggested by Dhillon [4], where the co-clustering problem was formulated as the approximation of the optimal normalised cut of a weighted bipartite graph. It was shown that a relaxed solution may be obtained by computing the *singular value decomposition* (SVD) of the normalised matrix $\mathbf{A_n} = \mathbf{D_1}^{-1/2}\mathbf{A}\mathbf{D_2}^{-1/2}$, where $[D_1]_{ii} = \sum_{j=1}^{n} A_{ij}$ and $[D_2]_{jj} = \sum_{i=1}^{m} A_{ij}$ are diagonal matrices. A reduced representation $\mathbf{Z}$ is then constructed from the the left and right singular vectors of $\mathbf{A_n}$, corresponding to the $\log_2 k$ largest non-trivial singular values. Viewing the matrix $\mathbf{Z}$ as a l-dimensional geometric embedding of the original data, the k-means algorithm is applied in this space to produce a disjoint co-clustering.

2.2 Non-negative Matrix Factorisation (NMF)

Non-Negative Matrix Factorisation (NMF) [5] has recently been identified as a practical approach for reducing the dimensionality of non-negative matrices such as the term-document representation of a text corpus [6]. Unlike spectral decomposition, NMF is constrained to produce non-negative factors, providing an interpretable clustering without the requirement for further processing.

Given the term-document matrix $\mathbf{A}$, NMF generates a rank-k approximation of the corpus in the form of the product of two non-negative matrices $\mathbf{A} \approx \mathbf{UV}^T$. The factor $\mathbf{U}$ is a $m \times k$ matrix consisting of k basis vectors, which can be viewed as a set of semantic variables corresponding to the topics in the data, while $\mathbf{V}$ is a $n \times k$ matrix of coefficients describing the contribution of the documents to each topic. The factors are determined by a given objective function that seeks to minimise the error of the reconstruction of $\mathbf{A}$ by the approximation $\mathbf{UV}^T$.

3 Soft Spectral Clustering

In this section, we discuss the problem of inducing membership weights from a disjoint partition, and we propose an intuitive method to produce soft clusters based on the spectral co-clustering model.

For the task of generating feature weights from a hard clustering, a common approach is to derive values from each cluster's centroid vector [7]. However, an analogous technique for spectral clustering is not useful due to the presence of negative values in centroid vectors formed in the reduced space. Another possibility is to formulate a document's membership weights as a function of the document's similarity to each cluster centroid [8].

The success of spectral clustering methods has been attributed to the truncation of the eigenbasis, which has the effect of amplifying the association between points that are highly similar, while simultaneously attenuating the association of points that are dissimilar [3]. However, while this process has been shown to improve the ability of a post-processing algorithm to identify cohesive clusters, the truncation of the decomposition of $\mathbf{A}$ to $k \ll m$ singular vectors introduces a distortion that makes the extraction of natural membership weights problematic. As a consequence, we observe that directly employing embedded term-centroid similarity values as membership weights will not provide intuitive cluster labels.

3.1 Inducing Soft Clusters

As a starting point, we construct a reduced space based on the spectral co-clustering model described in Section 2.1. However, we choose to form the embedding $\mathbf{Z}$ from the leading k singular vectors, as truncating the eigenbasis to a smaller number of dimensions may lead to an inaccurate clustering [2]. By applying the classical k-means algorithm using the cosine similarity measure, we generate k disjoint subsets of the points in the embedded geometric space. We represent this clustering as the $(m+n) \times k$ partition matrix $\mathbf{P} = [P_1, \ldots, P_k]$, where P_i is a binary membership indicator for the i-th cluster. We denote the k centroids of the clustering by $\{\mu_1, \ldots, \mu_k\}$.

As the spectral co-clustering strategy is based on the principle of the duality of clustering documents and terms [4], we argue that we can induce a soft clustering of terms from the partition of documents in $\mathbf{Z}$ and a soft clustering of documents from the partition of terms. Note that the matrix $\mathbf{P}$ has the structure:

$$\mathbf{P} = \begin{bmatrix} \mathbf{P_1} \\ \mathbf{P_2} \end{bmatrix}$$

where $\mathbf{P_1}$ and $\mathbf{P_2}$ indicate the assignment of terms and documents respectively. An intuitive approach to producing term weights is to apply the transformation $\mathbf{A}^T\hat{\mathbf{P}}_1$, where $\hat{\mathbf{P}}_1$ denotes the matrix $\mathbf{P}_1$ with columns normalised to unit length. This effectively projects the centroids of the partition of documents in $\mathbf{Z}$ to the original feature space. Similarly, to derive document-cluster association weights $\mathbf{V}$, we can apply the transformation $\mathbf{A}\hat{\mathbf{P}}_2$, thereby projecting the embedded term cluster centroids to the original data.

However, the above approach will not reflect the existence of boundary points lying between multiple clusters or outlying points that may be equally distant from all centroids. To overcome this problem, we propose the projection of the centroid-similarity values from the embedded clustering to the original data. Due to the presence of negative values in $\mathbf{Z}$, these similarities will lie in the range $[-1, 1]$. We rescale the values to the interval $[0, 1]$ and normalise the k columns to unit length, representing them by the matrix $\mathbf{S}$ as defined by:

$$S_{ij} = \frac{1 + cos(z_i, \mu_j)}{2}, \quad S_{ij} \leftarrow \frac{S_{ij}}{\sum_l S_{lj}} \quad (1)$$

As with the partition matrix of the embedded clustering, one may divide $\mathbf{S}$ into two submatrices, where $\mathbf{S_1}$ corresponds to the $m \times k$ term-centroid similarity matrix and $\mathbf{S_2}$ corresponds to the $n \times k$ document-centroid similarity matrix:

$$\mathbf{S} = \begin{bmatrix} \mathbf{S_1} \\ \mathbf{S_2} \end{bmatrix}$$

By applying the projections $\mathbf{A}^T\mathbf{S_1}$ and $\mathbf{A}\mathbf{S_2}$, we generate membership weights that capture both the affinity between points in the embedded space and the raw term-frequency values of the original dataset.

3.2 Soft Spectral Co-clustering (SSC) Algorithm

Motivated by the duality of the co-clustering model, we now present a spectral clustering algorithm with soft assignment of terms and documents that employs a combination of the transformation methods described in the previous section. We formulate the output of the algorithm as a pair of matrices $(\mathbf{U}, \mathbf{V})$, representing the term-cluster and document-cluster membership functions respectively. As a document membership function, we select the projection $\mathbf{A}^T\mathbf{S_1}$, on the basis that the use of similarity values extracts more information from the embedded clustering than purely considering the binary values in **P**. We observe that this generally leads to a more accurate clustering, particularly on datasets where the natural groups overlap.

The requirements for a term membership function differ considerably from those of a document membership function, where accuracy is the primary consideration. As the production of useful cluster descriptions is a central objective of our work, we seek to generate a set of weights that results in the assignment of high values to relevant features and low values to irrelevant features. Consequently, we select the projection $\mathbf{A}\hat{\mathbf{P}}_2$ as previous work has shown that centroid vectors can provide a summarisation of the important concepts present in a cluster [7]. Our choice is also motivated by the observation that the binary indicators in $\hat{\mathbf{P}}_1$ result in sparse discriminative weight vectors, whereas the projection based on $\mathbf{S}_2$ leads to term weights such that the highest ranking words tend to be highly similar across all clusters. We now summarise the complete procedure, which we refer to as the *Soft Spectral Co-clustering* (SSC) algorithm:

1. Compute the k largest singular vectors of $\mathbf{A}_n$ to produce the truncated factors $\mathbf{U_k} = (u_1, \ldots, u_k)$ and $\mathbf{V_k} = (v_1, \ldots, v_k)$.
2. Construct the embedded space $\mathbf{Z}$ by scaling and stacking $\mathbf{U_k}$ and $\mathbf{V_k}$:

$$\mathbf{Z} = \begin{bmatrix} \mathbf{D_1}^{-1/2}\mathbf{U_k} \\ \mathbf{D_2}^{-1/2}\mathbf{V_k} \end{bmatrix}$$

3. Apply the k-means algorithm with cosine similarity to $\mathbf{Z}$ to produce a disjoint co-clustering, from which the matrices $\mathbf{S_1}$ and $\hat{\mathbf{P}}_1$ are computed.
4. Form soft clusters by applying the projections $\mathbf{U} = \mathbf{A}\hat{\mathbf{P}}_2$ and $\mathbf{V} = \mathbf{A}^T\mathbf{S_1}$.

We employ a cluster initialisation strategy similar to that proposed in [2], where each centroid is chosen to be as close as possible to $90°$ from the previously selected centroids. However, rather than nominating the first centroid at random, we suggest that accurate deterministic results may be produced by selecting the most centrally located data point in the embedded space.

4 Refined Soft Spectral Clustering

We now present a novel technique for document clustering by dimension reduction that builds upon the co-clustering techniques described in Section 3.

The dimensions of the space produced by spectral decomposition are constrained to be orthogonal. However, as text corpora will typically contain documents relating to multiple topics, the underlying semantic variables in the data will rarely be orthogonal. The limitations of spectral techniques to effectively identify overlapping clusters has motivated other techniques such as NMF, where each document may be represented as an additive combination of topics [6]. However, the standard approach of initialising the factors $(\mathbf{U}, \mathbf{V})$ with random values can lead to convergence to a range of solutions of varying quality. We argue that initial factors, produced using the soft cluster induction techniques discussed previously, can provide a set of well-separated "core" clusters. By subsequently applying matrix factorisation with non-negativity constraints to the membership matrices, we can effectively uncover overlaps between clusters.

4.1 Refined Soft Spectral Co-clustering (RSSC) Algorithm

In the SSC algorithm described in Section 3.1, our choice of projection for the construction of the term membership matrix was motivated by the goal of producing human-interpretable weights. However, the projection $\mathbf{AS_2}$ retains additional information from the embedded clustering, while simultaneously considering the original term frequencies in $\mathbf{A}$. Consequently, we apply soft spectral co-clustering as described previously, but we select $\mathbf{U} = \mathbf{A}^T\mathbf{S_1}$ and $\mathbf{V} = \mathbf{AS_2}$ as our initial pair of factors.

We refine the weights in $\mathbf{U}$ and $\mathbf{V}$ by iteratively updating these factors in order to minimise the divergence or entropy between the original term-document matrix $\mathbf{A}$ and the approximation $\mathbf{UV}^T$ as expressed by

$$D(\mathbf{A}\|\mathbf{UV}^T) = \sum_{i=1}^{m}\sum_{j=1}^{n}\left(A_{ij}\log\frac{A_{ij}}{[\mathbf{UV}^T]_{ij}} - A_{ij} + [\mathbf{UV}^T]_{ij}\right) \quad (2)$$

This function can be shown to reduce to the Kullback-Leibler divergence measure when both $\mathbf{A}$ and $\mathbf{UV}^T$ sum to 1. To compute the factors a diagonally scaled gradient descent optimisation scheme is applied in the form of a pair of multiplicative update rules that converge to a local minimum. We summarise the Refined Soft Spectral Co-clustering (RSSC) algorithm as follows:

1. Decompose $\mathbf{A_n}$ and construct the embedded space $\mathbf{Z}$ as described previously.
2. Apply k-means to the rows of $\mathbf{Z}$ to produce a disjoint clustering, from which $\mathbf{S_1}$ and $\mathbf{S_2}$ are constructed.
3. Generate the initial factors $\mathbf{U} = \mathbf{A}^T\mathbf{S_1}$ and $\mathbf{V} = \mathbf{AS_2}$.
4. Update $\mathbf{V}$ using the rule

$$v_{ij} \leftarrow v_{ij}\left[\left(\frac{A_{ij}}{[\mathbf{UV}^T]_{ij}}\right)^T \mathbf{U}\right]_{ij} \quad (3)$$

5. Update $\mathbf{U}$ using the rule

$$u_{ij} \leftarrow u_{ij}\left[\frac{A_{ij}}{[\mathbf{UV}^T]_{ij}}\mathbf{V}\right]_{ij}, \quad u_{ij} \leftarrow u_{ij}\frac{U_{ij}}{\sum_{l=1}^{m}U_{lj}} \quad (4)$$

6. Repeat from step 4 until convergence.

To provide a clearer insight into the basis vectors, we subsequently apply a normalisation so that the Euclidean length of each column of $\mathbf{U}$ is of unit length and we scale the factor $\mathbf{V}$ accordingly as suggested in [6].

5 Experimental Evaluation

In our experiments we compared the accuracy of the SSC and RSSC algorithms to that of spectral co-clustering (CC) based on k singular vectors and NMF using the divergence objective function given in (2) and random initialisation. Choosing the number of clusters k is a difficult model-selection problem which lies beyond the scope of this paper. For the purpose of our experiments we set k to correspond to the number of annotated classes in the data.

The experimental evaluation was conducted on a diverse selection of datasets, which differ in their dimensions, complexity and degree of cluster overlap. For a full discussion of the datasets used in our experiments, consult [1]. To pre-process these datasets, we applied standard stop-word removal and stemming techniques. We subsequently excluded terms occurring in less than three documents. No further feature selection or term normalisation was performed.

5.1 Results

To compare algorithm accuracy, we apply the *normalised mutual information* (NMI) external validation measure proposed in [9]. We elected to evaluate hard clusterings due to the disjoint nature of the annotated classes for the datasets under consideration, and to provide a means of comparing the non-probabilistic document weights generated by our techniques with the output of the spectral co-clustering algorithm. Therefore, we induce a hard clustering from $\mathbf{V}$ by assigning the i-th document to the j-th cluster if $j = \arg\max_j (v_{ij})$.

Table 1 summarises the experimental results for all datasets as averaged across 20 trials. In general, the quality of the clusters produced by the SSC algorithm was at least comparable to that afforded by the spectral co-clustering method described in [4]. By virtue of their ability to perform well in the presence of overlapping clusters, both the NMF and RSSC methods generally produced clusterings that were superior to those generated using only spectral analysis. However, the RSSC algorithm's use of spectral information to seed well-separated "core clusters" for subsequent refinement leads to a higher level of accuracy on most datasets. When applied to larger datasets, we observe that the NMF and CC methods exhibit considerable variance in the quality of the clusters that they

Table 1. Performance comparison based on NMI

Dataset	CC	NMF	SSC	RSSC
bbc	0.78	0.80	0.82	**0.86**
bbcsport	0.64	0.69	0.65	**0.70**
classic2	0.29	0.34	0.46	**0.79**
classic3	0.92	**0.93**	0.92	**0.93**
classic	0.63	0.70	0.62	**0.87**
ng17-19	0.39	0.36	0.45	**0.50**

Dataset	CC	NMF	SSC	RSSC
ng3	0.68	0.78	0.70	**0.84**
re0	0.33	0.39	0.35	**0.40**
re1	0.39	0.42	0.41	**0.43**
reviews	0.34	0.53	0.40	**0.57**
tr31	0.38	0.54	0.51	**0.65**
tr41	0.58	0.60	**0.67**	**0.67**

produce, whereas the deterministic nature of the initialisation strategy employed by the newly proposed algorithms leads to stable solutions.

5.2 Cluster Labels

Given the term membership weights produced by the SSC and RSSC algorithms, a natural approach to generating a set of labels for each cluster is to select the terms with the highest values from each column of the matrix **U**. Due to space restrictions, we only provide a sample of the labels selected for clusters produced by the RSSC algorithm on the *bbc* dataset in Table 2, where the natural categories are: business, politics, sport, entertainment and technology.

Table 2. Labels produced by RSSC algorithm for *bbc* dataset

Cluster	Top 7 Terms
C1	company, market, firm, bank, sales, prices, economy
C2	government, labour, party, election, election, people, minister
C3	game, play, win, players, england, club, match
C4	film, best, awards, music, star, show, actor
C5	people, technology, mobile, phone, game, service, users

6 Concluding Remarks

In this paper, we described a method based on spectral analysis that can yield stable interpretable clusters in sparse high-dimensional spaces. Subsequently, we introduced a novel approach to achieve a more accurate clustering by applying a constrained matrix factorisation scheme to refine an initial solution produced using spectral techniques. Evaluations conducted on a variety of text corpora demonstrate that this method can lead to the improved identification of overlapping clusters, while simultaneously producing document and term weights that are amenable to human interpretation.

References

1. Greene, D., Cunningham, P.: Producing accurate interpretable clusters from high-dimensional data. Technical Report CS-2005-42, Trinity College Dublin (2005)
2. Ng, A., Jordan, M., Weiss, Y.: On spectral clustering: Analysis and an algorithm. In: Proc. Advances in Neural Information Processing. (2001)
3. Brand, M., Huang, K.: A unifying theorem for spectral embedding and clustering. In: Proc. 9th Int. Workshop on AI and Statistics. (2003)
4. Dhillon, I.S.: Co-clustering documents and words using bipartite spectral graph partitioning. In: Knowledge Discovery and Data Mining. (2001) 269–274
5. Lee, D.D., Seung, H.S.: Learning the parts of objects by non-negative matrix factorization. Nature **401** (1999) 788–91
6. Xu, W., Liu, X., Gong, Y.: Document clustering based on non-negative matrix factorization. In: Proc. 26th Int. ACM SIGIR. (2003) 267–273

7. Dhillon, I.S., Modha, D.S.: Concept decompositions for large sparse text data using clustering. Machine Learning **42** (2001) 143–175
8. Zhao, Y., Karypis, G.: Soft clustering criterion functions for partitional document clustering: a summary of results. In: Proc. 13th ACM Conf. on Information and Knowledge Management. (2004) 246–247
9. Strehl, A., Ghosh, J.: Cluster ensembles - a knowledge reuse framework for combining multiple partitions. JMLR **3** (2002) 583–617

Stress-Testing Hoeffding Trees

Geoffrey Holmes, Richard Kirkby, and Bernhard Pfahringer

Department of Computer Science,
University of Waikato,
Hamilton, New Zealand
{geoff, rkirkby, bernhard}@cs.waikato.ac.nz

Abstract. Hoeffding trees are state-of-the-art in classification for data streams. They perform prediction by choosing the majority class at each leaf. Their predictive accuracy can be increased by adding Naive Bayes models at the leaves of the trees. By stress-testing these two prediction methods using noise and more complex concepts and an order of magnitude more instances than in previous studies, we discover situations where the Naive Bayes method outperforms the standard Hoeffding tree initially but is eventually overtaken. The reason for this crossover is determined and a hybrid adaptive method is proposed that generally outperforms the two original prediction methods for both simple and complex concepts as well as under noise.

1 Introduction

The Hoeffding tree induction algorithm [2] has proven to be one of the best methods for data stream classification. Standard Hoeffding trees use the majority class at each leaf for prediction. Previous work [3] has shown that adding so-called functional (or Naive Bayes) leaves to Hoeffding trees for both synthetically generated streams, and real datasets of sufficient size, as well as in the presence of noise outperforms standard Hoeffding trees.

In this paper we use an experimental evaluation methodology for data streams, where every single instance in the stream is used for both learning and testing. Using this methodology and an order of magnitude more data we discover situations were the standard Hoeffding tree unexpectedly outperforms its Naive Bayes counterpart. We investigate the cause and propose modifications to the original algorithm. An empirical investigation compares these modifications to both the standard Hoeffding tree and its Naive Bayes variant, and shows that one of the possible modifications is very robust across all combinations of concept complexity and noise. The modifications *only* concern the method of prediction. The standard Hoeffding tree learning algorithm is used in all cases.

The paper is arranged as follows. Section 2 contains an evaluation of Hoeffding trees with an unexpected result. Section 3 proposes several solutions to address the problem, and Section 4 evaluates and discusses them. Finally Section 5 concludes the paper.

2 Examining Hoeffding Trees

Data streams present unique opportunities for evaluation, due to the volume of data available and the any-time property of the algorithms under examination. We consider a method of evaluation that exploits this property whilst maximizing use of the data. This is achieved by using every instance as a testing example on the current model before using it to train the model, incrementally updating statistics at each point.

The particular implementation of Hoeffding Tree induction discussed in this paper uses information gain as the split criterion, the original VFDT Hoeffding bound formulation [2] to determine when to split (using parameters $\delta = 10^{-6}$, $\tau = 5\%$, and $n_{min} = 300$), and handles numeric attributes by Gaussian approximation (throughout **ht** refers to this algorithm and **htnb** the same algorithm with Naive Bayes prediction at the leaves).

Our first analysis looks at the difference in accuracy between **ht** and **htnb**. We start with data generated by a randomly constructed decision tree consisting of 10 nominal attributes with 5 values each, 10 numeric attributes, 2 classes, a tree depth of 5, with leaves starting at level 3 and a 0.15 chance of leaves thereafter (the final tree had 741 nodes, 509 of which were leaves)—which we shall refer to as the simple random tree. Note that for all graphs in this paper we have averaged over 10 runs to eliminate order effects.

Figure 1 shows the result of evaluating over 10 million instances with no noise present. As in previous studies, it is clear that **htnb** gives an improvement in classification accuracy, with both variants performing well, reaching around 99% accuracy in the long run.

Figure 2 shows the impact that noise has. 10% noise was introduced to the data with uniform randomness. A different picture emerges—**htnb** looks better initially but somewhere before 2 million instances the graphs cross over and in the long run **htnb** is worse.

Next, the tree generator is adjusted to produce a complex random tree—50 nominal attributes with 5 values each, 50 numeric attributes, 2 classes, a tree

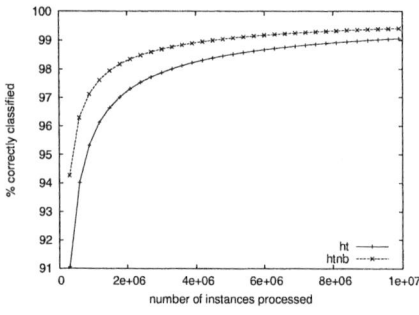
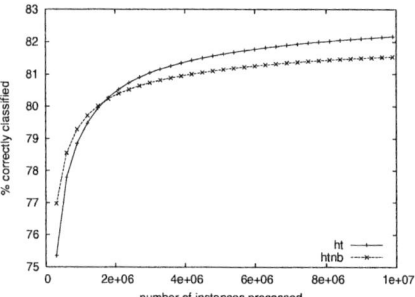

Fig. 1. Simple random tree generator with no noise

Fig. 2. Simple random tree generator with 10% noise

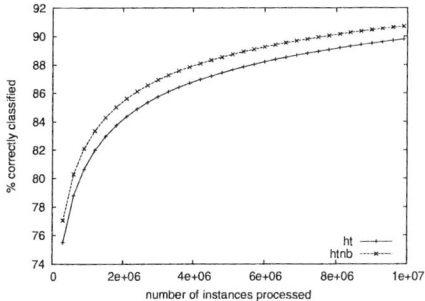 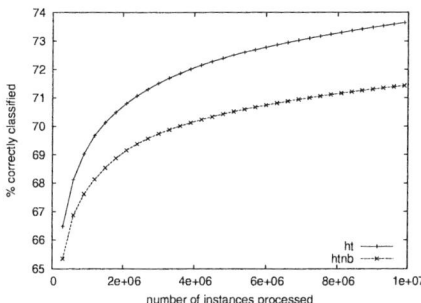

Fig. 3. Complex random tree generator with no noise

Fig. 4. Complex random tree generator with 10% noise

depth of 10, with leaves starting at level 5 and a 0.15 chance of leaves thereafter (the final tree had 127,837 nodes, 90,259 of which were leaves).

Figures 3 and 4 show the learning curves resulting from the complex random tree data, both clean and noisy. The same pattern is observed, only this time **htnb** is worse from the outset.

Testing the simple tree with other noise levels gives result similar to Figure 2, at 5% noise the crossover occurs later, and at 20% it occurs earlier than the 10% case. On the complex tree data the gap in Figure 4 widens as the noise increases.

3 Possible Solutions to the Problem

The evaluation uncovered cases where **htnb** is less accurate than **ht**. This behaviour appears when noise is introduced, and becomes more pronounced when the concept is more difficult to learn. The problem shows up on other datasets also, though sometimes it is less apparent due to the millions of instances that are needed before crossover occurs.

We speculate that the problem is due to small disjuncts in the tree in combination with noisy data. The leaves see only an insufficient number of examples, and the ones they do see are noisy, causing **htnb** to be less accurate than **ht**.

Experiments with severely limited maximum tree sizes supported this hypothesis: **htnb** eventually outperformed **ht** once tree growth was artificially stopped.

If the problem with **htnb** under noise is due to the models being unreliable in their early stages, then there are several ways the problem could be solved.

One solution has been proposed by Gama who in [3] suggests the use of a short term memory cache housing some of the recently seen instances. A problem with this solution is determining a sufficient size for the cache. As the tree grows in complexity, fewer of the instances in the cache will be applicable to the new leaves deep in the tree (we refer to this as **htnb-stm**x where x is the cache size).

Another idea is to inherit information from the parent once a split has been chosen. For an attribute split, it is possible to approximate the distribution of values resulting from the split. For the other attributes, less information is known about the result of the new split, but we can assume that the distribution is the same as in the parent. This approximation may be grossly incorrect, but at least it gives the model a starting point rather than starting with no information (**htnbp**).

A potentially serious problem with this approach is that if the statistics used to make split choices are primed then the split decisions will be altered, having an impact on the tree structure. As will be shown in Section 4, this generally has a detrimental effect on accuracy.

A solution to this is to maintain a separate model per leaf that is used for prediction purposes only, and leave the split decision statistics untouched. This effectively doubles the storage requirement per leaf (**htnbps**).

An adaptive solution is to see how often the Naive Bayes leaves make classification errors compared with choosing the majority class, using Naive Bayes leaves only when their measured accuracy is higher. The data stream setting affords us the ability to do this as we can monitor performance on unseen instances in the same way that the overall evaluation is performed (**htnba**).

The method works by performing a Naive Bayes prediction per training instance, comparing its prediction with the majority class. Counts are stored to measure how many times the Naive Bayes prediction gets the true class correct as compared to the majority class. When performing a prediction on a test instance, the leaf will only return a Naive Bayes prediction if it has been more accurate overall than the majority class, otherwise it resorts to a majority class prediction.

To complete the experimentation, we added priming and model separation to **htnba**, these are referred to as **htnbap** and **htnbaps** respectively.

Table 1 summarizes the costs associated with the candidates, beyond that needed for **htnb**. The costs associated with the adaptive choice are minor—a few extra counts and a single comparison per prediction. The Naive Bayes prediction per training instance is a cost that can be shared with the evaluation mechanism. The costs associated with maintaining a separate prediction model are the greatest—effectively doubling the storage and update time per leaf. As splitting is a much less frequent operation than anything else, higher splitting costs usually do not have much impact on the overall total cost.

4 Results and Discussion

Figures 5 to 8 show the result of the various prediction strategies on the simple and complex tree data, with and without noise.

Figure 5 shows that the two methods of priming without separate models (**htnbp** and **htnbap**) do worse than **ht**. All other Naive Bayes methods outperform **ht** by roughly equal amounts. Introducing noise in Figure 6 sees **htnb** doing worst overall, even worse than **htnbp** and **htnbap**. The other methods (besides those using short term memory) successfully overcome the problem with little between them.

Table 1. Additional space/time costs beyond **htnb** requirements

	space per tree	space per leaf	time per training instance	time per test instance	time per split
htnb-stm$x$	cache of x instances		cache update		pass instances to leaves + NB updates
htnbp					distribution estimation
htnbps		NB model	NB update		distribution estimation
htnba		error count	NB prediction count update	decide MC or NB	
htnbap		error count	NB prediction count update	decide MC or NB	distribution estimation
htnbaps		error count NB model	NB prediction count update NB update	decide MC or NB	distribution estimation

Figure 7 explores the case of a more complex but still noise-free concept. Results are similar to the simple tree case (Figure 5). There is not much separation within the group of methods that outperform **ht**. Once again both **htnbp** and **htnbap** perform worse than **ht**.

Adding noise to more complex concepts results in the greatest separation between the techniques. Figure 8 shows the short term memory solution to be unsatisfactory. A short term memory of 1000 instances hardly does better than **htnb**, which is the worst performer. Increasing the cache size to 10000 instances does little to improve the situation.

Figure 8 demonstrates the superiority of the adaptive method. **htnbp** and **htnbps** fall short of **ht**, while all of the adaptive methods do better. The best performing method is also the most costly one (**htnbaps**), with the less expensive **htnba** not far behind.

To investigate whether these findings hold more generally, experiments on two additional UCI datasets [1] were conducted. The LED dataset is a synthetic generator allowing us to generate the desired 10 million instances. The particular configuration used of the LED generator produced 24 binary attributes, 10 classes, and 10% noise.

The results in Figure 9 exhibit slightly different looking curves. The majority of the methods hover around 26% error which is known to be the optimal Bayes error for this problem. The exceptions are **ht** (which has a much slower learning curve without the aid of Naive Bayes leaves), and **htnb**. The failure of **htnb** shows that the problem extends beyond tree generated data. The success of the others show that the problem can be alleviated.

These results contradict those reported by Gama et al. [3,4], whose conclusion was that Naive Bayes leaves are always better on the LED data. They used a

Table 2. Final accuracies achieved on tree generators

	simple tree no noise	simple tree 10% noise	complex tree no noise	complex tree 10% noise
ht	99.056 ± 0.033	82.167 ± 0.031	89.793 ± 0.168	73.644 ± 0.151
htnb	99.411 ± 0.026	81.533 ± 0.021	90.723 ± 0.153	71.425 ± 0.118
htnb-stm1k	99.407 ± 0.027	81.544 ± 0.019	90.768 ± 0.150	71.527 ± 0.108
htnb-stm10k	99.409 ± 0.025	81.593 ± 0.018	91.008 ± 0.153	71.658 ± 0.085
htnbp	97.989 ± 0.058	81.853 ± 0.042	88.326 ± 0.209	73.029 ± 0.121
htnbps	99.376 ± 0.028	82.456 ± 0.023	90.598 ± 0.153	73.063 ± 0.124
htnba	99.408 ± 0.027	82.510 ± 0.024	90.874 ± 0.153	74.089 ± 0.141
htnbap	98.033 ± 0.057	81.938 ± 0.040	88.609 ± 0.211	73.675 ± 0.127
htnbaps	99.375 ± 0.028	82.545 ± 0.024	90.935 ± 0.148	74.249 ± 0.134

Table 3. Final accuracies achieved on other datasets

	LED	Covertype
ht	72.851 ± 0.031	66.832 ± 0.163
htnb	71.645 ± 0.013	69.064 ± 0.135
htnbp	73.928 ± 0.005	68.476 ± 0.040
htnbps	73.799 ± 0.041	69.049 ± 0.145
htnba	73.935 ± 0.005	70.998 ± 0.087
htnbap	73.961 ± 0.004	71.388 ± 0.037
htnbaps	73.996 ± 0.005	71.054 ± 0.095

hold out test set and quoted the final accuracy attained. The largest training set used in their work was 1.5 million instances. In our experiment the problem does not occur until about 4 million instances.

Finally, the algorithms were tested on real data using the Forest Covertype dataset. This consists of 581,012 instances, 10 numeric attributes, 44 binary attributes and 7 classes. To do 10 runs over this data the instances were randomly permuted 10 different ways. In Figure 10 we see three distinct groups. The worst performer is **ht**. The next group consists of the non-adaptive methods **htnb**, **htnbp** and **htnbps**. The group of best performers are the adaptive ones. This result demonstrates that even in cases where **htnb** is not obviously underperforming, adding the adaptive modification can enhance performance.

Tables 2 and 3 show the final accuracies achieved along with the standard error for all of the graphs displayed in Figures 5 through 10.

Overall these results support the conclusion that priming the leaf models without using a separate model per leaf results in poor performance. Without the separate model, the split decisions are altered in such a way that the tree is less accurate. Inclusion of the separate model improves the situation (at a cost), but it appears not as helpful as using the adaptive method.

Our experiments demonstrate that **htnba** provides a good compromise between accuracy and cost. In some cases it did slightly worse than **htnbaps**, but the difference does not justify the extra cost. The adaptive approach of **htnba** has a relatively low overhead, meaning it can be justified over **ht**, and especially over **htnb**, in all but the most extreme resource-bounded situations.

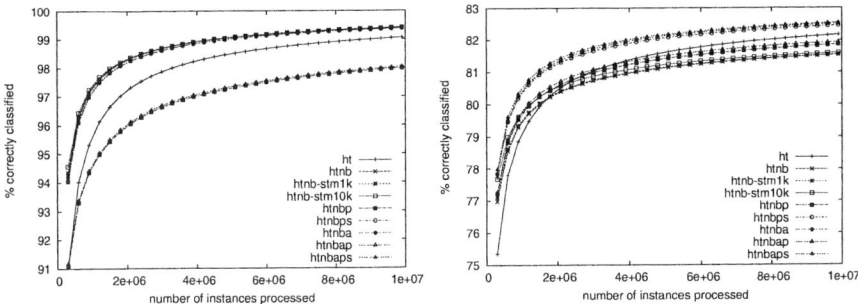

Fig. 5. Simple random tree generator with no noise

Fig. 6. Simple random tree generator with 10% noise

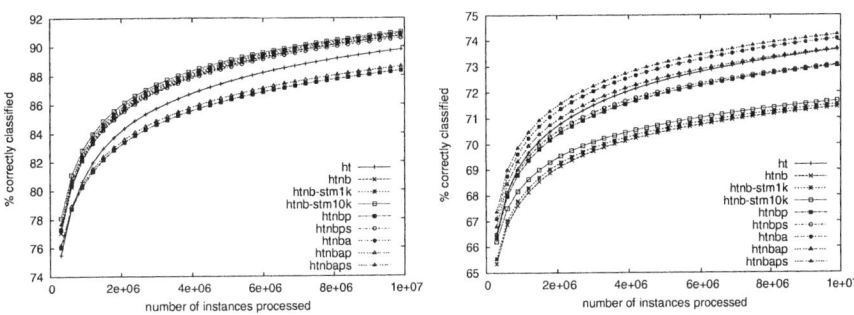

Fig. 7. Complex random tree generator with no noise

Fig. 8. Complex random tree generator with 10% noise

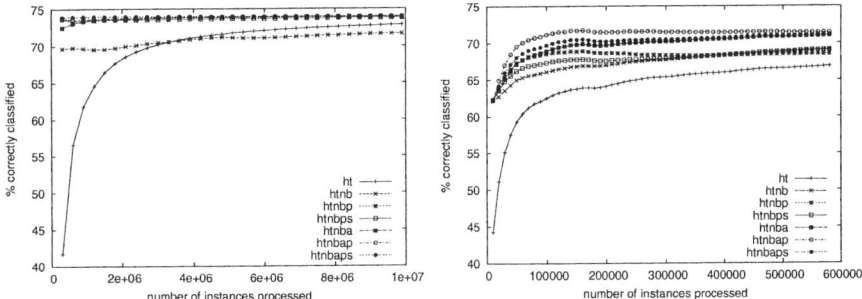

Fig. 9. LED generator

Fig. 10. Covertype

5 Conclusions

By using noise and synthetically generating large and complex concepts we have performed experiments that stress-test Hoeffding trees. Our focus has been on the two common prediction methods used in association with these trees: majority class prediction and Naive Bayes leaf prediction. In this experimental environment we uncovered an unexpected problem with Naive Bayes leaf prediction. Multiple improvements to the shortcomings of this method were invented and empirically evaluated. The best solution adaptively decides when the Naive Bayes leaves are accurate enough to be trusted. This adaptive method only imposes a minor additional cost on the algorithm, yet seems to almost guarantee equal or better accuracy than a simple majority class prediction.

References

1. C.L. Blake and C.J. Merz. UCI repository of machine learning databases, 1998.
2. Pedro Domingos and Geoff Hulten. Mining high-speed data streams. In *Knowledge Discovery and Data Mining*, pages 71–80, 2000.
3. Joao Gama, Pedro Medas, and Ricardo Rocha. Forest trees for on-line data. In *SAC '04: Proceedings of the 2004 ACM symposium on Applied computing*, pages 632–636, New York, NY, USA, 2004. ACM Press.
4. Joao Gama, Ricardo Rocha, and Pedro Medas. Accurate decision trees for mining high-speed data streams. In *KDD '03: Proceedings of the ninth ACM SIGKDD international conference on Knowledge discovery and data mining*, pages 523–528, New York, NY, USA, 2003. ACM Press.

Rank Measures for Ordering

Jin Huang and Charles X. Ling

Department of Computer Science,
The University of Western Ontario,
London, Ontario, Canada N6A 5B7
{jhuang33, cling}@csd.uwo.ca

Abstract. Many data mining applications require a ranking, rather than a mere classification, of cases. Examples of these applications are widespread, including Internet search engines (ranking of pages returned) and customer relationship management (ranking of profitable customers). However, little theoretical foundation and practical guideline have been established to assess the merits of different rank measures for ordering. In this paper, we first review several general criteria to judge the merits of different single-number measures. Then we propose a novel rank measure, and compare the commonly used rank measures and our new one according to the criteria. This leads to a preference order for these rank measures. We conduct experiments on real-world datasets to confirm the preference order. The results of the paper will be very useful in evaluating and comparing rank algorithms.

1 Introduction

Ranking of cases is an increasingly important way to describe the result of many data mining and other science and engineering applications. For example, the result of document search in information retrieval and Internet search is typically a ranking of the results in the order of match. This leaves two issues to be addressed. First, given two orders of cases, how do we design or choose a measure to determine which order is better? Second, given two different rank measures of ordering, how do we tell which rank measure is more desirable?

In previous research, the issue of determining which order is better is usually addressed using accuracy and its variants, such as recall and F-measures, which are typically used in information retrieval. More recently, AUC (Area Under Curve) of the ROC (Receiver Operating Characteristics) has gained an increasing acceptance in comparing learning algorithms [1] and constructing learning models [2,3]. Bradley [4] experimentally compared popular machine learning algorithms using both accuracy and AUC, and found that AUC exhibits several desirable properties when compared to the accuracy.

However, accuracy is traditionally designed to judge the merits of classification results, and AUC is simply used as a replacement of accuracy without much reasoning for why it is a better measure, especially for the case of ordering. The main reason for this lack of understanding is that up to now, there has been no theoretical study on whether any of these measures work better than others, or whether there are even better measures in existence.

In this paper, we first review our previous work [5] on general criteria to compare two arbitrary single-number measures (see Section 2.1). Then we compare six rank measures for ordering using our general criteria. Our contributions in this part consist of a novel measure for the performance of ordering (Section 2.4), and a preference order discovered for these measures (Section 3.1). The experiments on real-world datasets confirm our analysis, which show that better rank measures are more sensitive in comparing rank algorithms (see Section 3.2).

2 Rank Measures for Ordering

In this section, we first review the criteria proposed in our previous work to compare two arbitrary measures. We then review five commonly used rank measures, and propose one new rank measure, OAUC. Then based on the comparison criteria, we will make a detailed comparison among these measures, which leads to a preference order of the six rank measures. Finally, we perform experiments with real-world data to confirm our conclusions on the preference order. The conclusions of the paper are significant for future machine learning and data mining applications involving ranking and ordering.

2.1 Review of Formal Criteria for Comparing Measures

In [5] the *degree of consistency* and *degree of discriminancy* of two measures are proposed and defined. The degree of consistency between two measures f and g, denoted as $\mathbf{C_{f,g}}$, is simply the fraction (probability) that two measures are consistent over some distribution of the instance space. Two measures are consistent when comparing two objects a and b, if f stipulates that a is better than b, g also stipulates that a is better than b. [5] define that two measures f and g are *consistent* iff the degree of consistency $\mathbf{C_{f,g}} > 0.5$. That is, f and g are consistent if they agree with each other on over half of the cases.

The *degree of discriminancy* of f over g, denoted as $\mathbf{D_{f/g}}$, is defined as the ratio of cases where f can tell the difference but g cannot, over the cases where g can tell the difference but f cannot. [5] define that a measure f is *more discriminant* (or *finer*) than g iff $D_{f/g} > 1$. That is, f is finer than g if there are more cases where f can tell the difference but g cannot, than g can tell the difference but f cannot.

2.2 Notation of Ordering

We will use some simple notations to represent ordering throughout this paper. Without loss of generality, for n examples to be ordered, we use the actual ordering position of each example as the label to represent this example in the ordered list. For example, suppose that the label of the actual highest ranked example is n, the label of the actual second highest ranked example is $n-1$, etc. We assume the examples are ordered incrementally from left to right. Then the *true-order list* is $l = 1, 2, \ldots, n$. For any ordered list generated by an ordering algorithm, it is a permutation of l. We use $\pi(l)$ to denote the ordered list generated by ordering algorithm π. $\pi(l)$ can be written as $a_1, a_2, \ldots, a_n$, where a_i is the actual ordering position of the example that is ranked ith in $\pi(l)$.

Table 1. An example of ordered lists

l	1	2	3	4	5	6	7	8
	a_1	a_2	a_3	a_4	a_5	a_6	a_7	a_8
$\pi(l)$	3	6	8	1	4	2	5	7

Table 1 gives an instance of ordered lists with eight examples. In this table, l is the true-order list and $\pi(l)$ is the ordered list generated by an ordering algorithm π. In $\pi(l)$ from left to right are the values of a_i. We can find that $a_1 = 3$, $a_2 = 6$, ..., $a_8 = 7$.

2.3 Previous Rank Measures for Ordering

We first review five most commonly used rank measures. Later we will invent a new rank measure which we will evaluate among the rest.

We call some of the rank measures "true-order" rank measures, because to obtain the evaluation values, we must know the true order of the original lists. Some other rank measures, however, are not true-order rank measures. They do not need the true order to obtain evaluation values; instead, only a "rough" ordering is sufficient. For example, accuracy and AUC are not true-order rank measures. As long as we know the true classification, we can calculate their values. In a sense, positive examples can be regarded as "the upper half", and negative examples are the "lower half" in an ordering, and such a rough ordering is sufficient to obtain AUC and accuracy.

1. **Euclidean Distance (ED)**
   If we consider the ordered list and the true order as a point $(a_1, a_2, \ldots, a_n)$ and a point $(1, 2, \ldots, n)$ in an n-dimensional Euclidean space, then ED is the Euclidean Distance between these two points, which is $\sqrt{\sum_{i=1}^{n}(a_i - i)^2}$. For simplicity we use the squared value of Euclidean distance as the measure. Then $ED = \sum_{i=1}^{n}(a_i - i)^2$. Clearly, ED is a true-order rank measure.
   For the example in Table 1, It is easy to obtain that $ED = (3-1)^2 + (6-2)^2 + (8-3)^2 + (1-4)^2 + (4-5)^2 + (2-6)^2 + (5-7)^2 + (7-8)^2 = 76$.

2. **Manhattan Distance (MD)**
   This measure MD is similar to ED except that here we sum the absolute values instead of sum squared values. It is also a true-order rank measure. For our order problem $MD = \sum_{i=1}^{n}|a_i - i|$. For the example in Table 1, it is easy to obtain that $MD = |3-1| + |6-2| + |8-3| + |1-4| + |4-5| + |2-6| + |5-7| + |7-8| = 22$.

3. **Sum of Reversed Number (SRN)**
   This is roughly the sum of the reversed pairs in the list. That is, $SRN = \sum_{i=1}^{n} s(i)$. It is clearly a true-order measure.
   For the ith example, its reversed number $s(i)$ is defined as the number of examples whose positions in $\pi(l)$ are greater than i but the actual ranked positions are less than i. For the example in Table 1, we can find that the examples of 1 and 2 are both ranked higher than the first example 3 in $\pi(l)$. Thus $s(1) = 1 + 1 = 2$. Similarly we have $s(2) = 4$, $s(3) = 5$, etc. Therefore the SRN for the ordered list $\pi(l)$ is $SRN = 2 + 4 + 5 + 0 + 1 + 0 + 0 + 0 = 12$.

4. **Area Under Curve (AUC)**

   The Area Under the ROC Curve, or simply AUC, is a single-number measure widely used in evaluating classification algorithms, and it is not a true-order measure for ranking. To calculate AUC for an ordered list, we only need the true classification (positive or negative examples). For a balanced ordered ranked list with n examples (half positive and half negative), we treat any example whose actual ranked position is greater than $\frac{n}{2}$ as a positive example; and the rest as negative. From left to right we assume the ranking positions of positive examples are $r_1, r_2, \ldots, r_{\lceil \frac{n}{2} \rceil}$. Then $AUC = \frac{\sum_{a_{r_i} > n/2}(r_i - i)}{n^2}$ [6].

   In Table 1, 5, 6, 7, and 8 are positive examples positioned at 2, 3, 7, and 8 respectively. Thus, $AUC = \frac{(2-1)+(3-2)+(7-3)+(8-4)}{4 \times 4} = \frac{5}{8}$.

5. **Accuracy (acc)**

   Like AUC, accuracy is also not a true-order rank measure. Similar to AUC, if we classify examples whose rank position above half of the examples as positive, and the rest as negative, we can calculate accuracy easily as $acc = \frac{tp+tn}{n}$, where tp and tn are the number of correctly classified positive and negative examples respectively. In the ordered list $\pi(l)$ in Table 1, 5, 6, 7, and 8 are positive examples, others are negative examples. Thus $tp = 2$, $tn = 2$. $acc = \frac{2+2}{8} = \frac{1}{2}$.

2.4 New Rank Measure for Ordering

We propose a new measure called Ordered Area Under Curve (OAUC), as it is similar to AUC both in meaning and calculation. The only difference is that each term in the formula is weighted by its true order, and the sum is then normalized. Thus, OAUC is a true-order measure. This measure is expected to be better than AUC since it "spreads" its values more widely compared to AUC.

OAUC is defined as follows:

$$OAUC = \frac{\sum a_{r_i}(r_i - i)}{\lfloor \frac{n}{2} \rfloor \sum_{i=1}^{\lceil \frac{n}{2} \rceil}(\lfloor \frac{n}{2} \rfloor + i)}$$

In the ordered list in Table 1, the positive examples are 5, 6, 7, 8 which are positioned at 7, 2, 8 and 3 respectively. Thus $r_1 = 2$, $r_2 = 3$, $r_3 = 7$, $r_4 = 8$, and $a_{r_1} = 6$, $a_{r_2} = 8$, $a_{r_3} = 5$, $a_{r_4} = 7$. $OAUC = \frac{6(2-1)+8(3-2)+5(7-3)+7(8-4)}{4((4+1)+(4+2)+(4+3)+(4+4))} = \frac{31}{52}$.

3 Comparing Rank Measures for Ordering

We first intuitively compare some pairs of measures and analyze whether any two measures satisfy the criteria of consistency and discriminancy. To begin with, we consider ED and MD because these two measures are quite similar in their definitions except that ED sums the squared distance while MD sums the absolute value. We expect that these two measures are consistent in most cases. On the other hand, given a dataset with n examples there are a total of $O(n^3)$ different ED values and $O(n^2)$ different MD values. Thus ED is expected to be more discriminant than MD. Therefore we expect that ED is consistent with and more discriminant than MD.

For AUC and OAUC, since OAUC is an extension of AUC, intuitively we expect that they are consistent. Assuming there are n_1 negative examples and n_0 positive examples, the different values for OAUC is $n_1 \sum_{i=1}^{n_0} (n_1 + i)$, which is greater than the different values of AUC ($n_0 n_1$). We can also expect that OAUC is more discriminant and therefore better than AUC.

However for the rest of the ordering measures we cannot make these intuitive claims because they have totally different definitions or computational methods. Therefore, in order to perform an accurate and detailed comparison and to verify or overturn our intuitions, we will conduct experiments to compare all measures.

3.1 Comparing Rank Measures on Artificial Datasets

To obtain the average degrees of consistency and discriminancy for all possible ranked lists, we use artificial datasets which consist of all possible ordered list of length 8. [1] We assume that the ordered lists are uniformly distributed. We exhaustively compare all pairs of ordered lists and calculate the degree of consistency and degree of discriminancy between two rank measures for ordering.

Table 2 lists the degree of consistency between every pair of six rank measures for ordering. The number in each cell represents the degree of consistency between the measures in the same row and column of the cell. We can find that the degree of consistency between any two measures are greater than 0.5, which indicates that these measures are "similar" in the sense that they are more likely to be consistent than inconsistent.

Table 3 shows the degree of discriminancy among all 6 rank measures. The number in the cell of the ith row and the jth column is the degree of discriminancy for the measure in ith row over the one in jth column.

From these two tables we can draw the following conclusions. First, these results verified our previous intuitive conclusions about the relations between ED and MD, and between AUC and OAUC. The degree of consistency between ED and MD is 0.95, and between AUC and OAUC 0.99, which means that ED and MD, and AUC and OAUC are highly consistent. The degree of discriminancy for ED over MD, and for OAUC over AUC are greater than 1, which means that ED is better than MD, and OAUC is better than AUC.

Table 2. Degree of consistency between pairs of rank measures for ordering

	AUC	SRN	MD	ED	OAUC	acc
AUC	1	0.88	0.89	0.87	0.99	0.98
SRN	0.88	1	0.95	0.98	0.89	0.91
MD	0.89	0.95	1	0.95	0.90	0.95
ED	0.87	0.98	0.95	1	0.88	0.90
OAUC	0.99	0.89	0.90	0.88	1	0.97
acc	0.98	0.91	0.95	0.90	0.97	1

---

[1] There are $n!$ different ordered lists for length n, so it is infeasible to enumerate longer lists.

Table 3. Degree of discriminancy between pairs of rank measures for ordering

	AUC	SRN	MD	ED	OAUC	acc
AUC	1	0.88	1.42	0.21	0.0732	14.0
SRN	1.14	1	1.84	0.242	0.215	9.94
MD	0.704	0.54	1	0.117	0.116	6.8
ED	4.76	4.13	8.55	1	0.87	38.2
OAUC	13.67	4.65	8.64	1.15	1	94.75
acc	0.071	0.10	0.147	0.026	0.011	1

Second, since all values of the degree of consistency among all measures are greater than 0.5, we can decide which measure is better than another only based on the value of degree of discriminancy. Recall (Section 2.1) that a measure f is better than another measure g iff $\mathbf{C_{f,g}} > 0.5$ and $\mathbf{D_{f/g}} > 1$. The best measure should be the one whose degrees of discriminancy over all other measures are greater than 1. From Table 3 we can find that all the numbers in the OAUC row are greater than 1, which means that the measure OAUC's degrees of discriminancy over all other measures are greater than 1. Therefore OAUC is the best measure. In the same way we can find that ED is the second best measure, and SRN is the third best. The next are AUC, MD, and acc is the worst.

Finally we can obtain the following preference order of for all six rank measures for ordering:

$$OAUC \succ ED \succ SRN \succ AUC \succ MD \succ acc$$

From the preference order we can conclude that OAUC, a new measure we design based on AUC, is the best measure. ED is the close, second best. The difference for these two measures are not very large (the degree of discriminancy for OAUC over ED is only 1.15). Therefore we should use OAUC and ED instead of others to evaluate ordering algorithms in most cases. Further, the two none-true-order classification measures AUC and accuracy do not perform well as compared with the true-order measures ED and SRN. This suggests that generally we should avoid using classification measures such as AUC and accuracy to evaluate ordering. Finally, MD is the worst true-order measure, and it is even worse than AUC. It should be avoided.

3.2 Comparing Rank Measures with Ranking Algorithms

In this section, we perform experiments to compare two classification algorithms in terms of the six rank measures. What we hope to conclude is that the better rank measures (such as OAUC and ED) would be more sensitive to the significance test (such as the t-test) than other less discriminant measures (such as MD and accuracy). That is, OAUC and ED are more likely to tell the difference between two algorithms than MD and accuracy can. Note that here we do not care about which rank algorithm predicts better; we only care about the sensitivity of the rank measures that are used to compare the rank algorithms. The better the rank measure (according to our criteria), the more sensitive it would be in the comparison, and the more meaningful the conclusion would be for the comparison.

We choose Artificial Neural Networks (ANN) and Instance-Based Learning algorithm (IBL) as our algorithms as they can both accept and produce continuous target. The ANN that we use has one hidden layer; the number of nodes in the hidden layer is half of the input layer (the number of attributes). We use real-world datasets to evaluate and compare ANN and IBL with the six rank measures. We select three real-world datasets *Wine*, *Auto-Mpg* and *CPU-Performance* from the UCI Machine Learning Repository [7].

In our experiments, we run ANN and IBL with the 10-fold cross validation on the training datasets. For each round of the 10-fold cross validation we train the two algorithms on the same training data and test them on the same testing data. We measure the testing data with six different rank measures (OAUC, ED, SRN, AUC, MD and acc) discussed earlier in the paper. We then perform paired, two-tailed t-tests on the 10 testing datasets for each measure to compare these two algorithms.

Table 4 shows the significance level in the t-test. [2] The smaller the values in the table, the more likely that the two algorithms (ANN and IBL) are significantly different, and the more sensitive the measure is when it is used to compare the two algorithms. Normally a threshold is set up and a binary conclusion (significantly different or not) is obtained. For example, if we set the threshold to be 0.95, then for the artificial dataset, we would conclude that ANN and IBL are statistically significantly different in terms of ED, OAUC and SRN, but not in terms of AUC, MD and acc. However, the actual significance level in Table 4 is more discriminant for the comparison. That is, it is "a better measure" than the simple binary classification of being significantly different or not.

Table 4. The significance level in the paired t-test when comparing ANN and IBL using different rank measures

Measures	Wine	Auto-mpg	CPU
OAUC	0.031	8.64×10^{-4}	1.48×10^{-3}
ED	0.024	1.55×10^{-3}	4.01×10^{-3}
SRN	0.053	8.89×10^{-3}	5.91×10^{-3}
AUC	0.062	5.77×10^{-3}	8.05×10^{-3}
MD	0.053	0.0167	5.97×10^{-3}
acc	0.126	0.0399	0.0269

From Table 4 we can obtain the preference order from the most sensitive measure (the smallest significance level) to the least sensitive measure (the largest significance level) for each dataset is:

- Wine: ED, OAUC, SRN = MD, AUC, acc.
- Auto-mpg: OAUC, ED, AUC, SRN, MD, acc.
- CPU-Performance: OAUC, ED, SRN, MD, AUC, acc.

These preference orders are roughly the same as the preference order of these measures discovered in the last section:

---

[2] The confidence level for the two arrays of data to be statistically different is one minus the values in the table.

$$OAUC \succ ED \succ SRN \succ AUC \succ MD \succ acc$$

The experimental results confirm our analysis in the last section. That is, OAUC and ED are the best rank measures for evaluating orders. In addition, MD and accuracy should be avoided as rank measures. These conclusions will be very useful for comparing and constructing machine learning algorithms for ranking, and for applications such as Internet search engines and data mining for CRM (Customer Relationship Management).

4 Conclusions

In this paper we use the criteria proposed in our previous work to compare five commonly used rank measures for ordering and a new proposed rank measure (OAUC). We conclude that OAUC is actually the best rank measure for ordering, and it is closely followed by the Euclidian distance (ED). Our results indicate that in comparing different algorithms for the order performance, we should use OAUC or ED, and avoid the least sensitive measures such as Manhattan distance (MD) and accuracy.

In our further work, we plan to improve existing rank learning algorithms by optimizing the better measures, such as OAUC and ED, discovered in this paper.

References

1. Provost, F., Domingos, P.: Tree induction for probability-based ranking. Machine Learning **52:3** (2003) 199–215
2. Ferri, C., Flach, P.A., Hernandez-Orallo, J.: Learning decision trees using the area under the ROC curve. In: Proceedings of the Nineteenth International Conference on Machine Learning (ICML 2002). (2002) 139–146
3. Ling, C.X., Zhang, H.: Toward Bayesian classifiers with accurate probabilities. In: Proceedings of the Sixth Pacific-Asia Conference on KDD. Springer (2002) 123–134
4. Bradley, A.P.: The use of the area under the ROC curve in the evaluation of machine learning algorithms. Pattern Recognition **30** (1997) 1145–1159
5. Ling, C.X., Huang, J., Zhang, H.: AUC: a statistically consistent and more discriminating measure than accuracy. In: Proceedings of 18th International Conference on Artificial Intelligence (IJCAI-2003). (2003) 519–526
6. Hand, D.J., Till, R.J.: A simple generalisation of the area under the ROC curve for multiple class classification problems. Machine Learning **45** (2001) 171–186
7. Blake, C., Merz, C.: UCI Repository of machine learning databases [http://www.ics.uci.edu/] [~mlearn/MLRepository.html]. Irvine, CA: University of California, Department of Information and Computer Science (1998)

Dynamic Ensemble Re-Construction for Better Ranking

Jin Huang and Charles X. Ling

Department of Computer Science,
The University of Western Ontario,
London, Ontario, Canada N6A 5B7
{jhuang33, cling}@csd.uwo.ca

Abstract. Ensemble learning has been shown to be very successful in data mining. However most work on ensemble learning concerns the task of classification. Little work has been done to construct ensembles that aim to improve ranking. In this paper, we propose an approach to re-construct new ensembles based on a given ensemble with the purpose to improve the ranking performance, which is crucial in many data mining tasks. The experiments with real-world data sets show that our new approach achieves significant improvements in ranking over the original Bagging and Adaboost ensembles.

1 Introduction

Classification is one of the fundamental tasks in knowledge discovery and data mining. The performance of a classifier is usually evaluated by predictive accuracy. However, most machine learning classifiers can also produce the probability estimation of the class prediction. Unfortunately, this probability information is ignored in the measure of accuracy.

In many real-world data mining applications, however, we often need the probability estimations or ranking. For example, in direct marketing, we often need to promote the most likely customers, or we need to deploy different promotion strategies to customers according to their likelihood of purchasing. To accomplish these tasks we need a ranking of customers according to their likelihood of purchasing. Thus ranking is often more desirable than classification in these data mining tasks.

One natural question is how to evaluate a classifier's ranking performance. In recent years, the area under the ROC (Receiver Operating Characteristics) curve, or simply AUC, is increasingly received attention in the communities of machine learning and data mining. Data mining researchers [1,2] have shown that AUC is a good summary in measuring a classifier's overall ranking performance. Hand and Till [3] present a simple approach to calculating AUC of a classifier for binary classification.

$$\hat{A} = \frac{S_0 - n_0(n_0 + 1)/2}{n_0 n_1},\tag{1}$$

where n_0 and n_1 are the numbers of positive and negative examples respectively, and $S_0 = \sum r_i$, where r_i is the rank of the i_{th} positive example in the ranked list.

Ensemble is a general approach which trains a number of classifiers and then combines their predictions in classification. Many researches [4,5,6] have shown that the ensemble is quite effective in improving the classification accuracy compared with a single classifier. The reason is that the prediction error of an individual classifier can be counteracted by the combination with other classifiers. Bagging [5] and Boosting [7] are two of the most popular ensemble techniques.

Most previous work of ensemble learning is focussed on classification. To our knowledge, there is little work that directly constructs ensembles to improve probability estimations or ranking. [8] compared the probability estimations (ranking) performance of different learning algorithms by using AUC as the comparison measure and demonstrated that Boosted trees and Bagged trees perform better in terms of ranking than Neural Networks and SVMs. [9] used the boosting technique on the general preference learning (ranking) problem and proposed a new ranking boosting algorithm: RankBoost.

In this paper, we propose a novel approach to improve the ranking performance over a given ensemble. The goal of this approach is to select some classifiers from the given ensemble to re-construct new ensembles. It first uses the k-Nearest Neighbor method to find training data subsets which are most similar to the test set, then it uses the measure SAUC (see Section 2.2) as heuristic to dynamically choose the diverse and well performed classifiers. This approach is called DERC (Dynamic Ensemble Re-Construction) algorithm. The new ensembles constructed by this approach are expected to have better ranking performance than the original ensemble.

The paper is organized as follows. In Section 2 we give detailed description for our new algorithm. In Section 3 we perform experiments on real world data sets to show the advantages of the new algorithm.

2 DERC (Dynamic Ensemble Re-Construction) Algorithm

In an ensemble, the combination of the predictions of several classifiers is only useful if they disagree to some degree. Each ensemble classifier may perform diversely during classification. Our DERC algorithm is motivated by this diversity property of ensemble. The diversity implies that each ensemble classifier performs best in probability estimation (ranking) only in a subset of training instances. Thus given a test (sub)set, if we use the k-Nearest Neighbor method to find some training subsets that are most similar to it, the classifiers that perform diversely and accurately on those similar training subsets are also expected to perform well on the test (sub)set. Therefore the new ensembles constructed are expected to have better ranking performance than the original ensemble.

Our DERC algorithm involves two basic steps: finding the most similar training (sub)sets, and selecting the diverse and accurate classifiers.

Now we use Figure 1 to illustrate how DERC algorithm works. Suppose that we are given an ensemble E with multiple classifiers built on a training set S, and we have an unlabeled test set T at hand. Our goal is to select some classifiers from the ensemble E to build one or more new ensembles to perform ranking on test set T.

2.1 Finding the Most Similar Training Subsets

The first step is to stratify the test set to some equal parts and find the most similar training subsets corresponding to test partitions. Since the labels of test instances are unknown, we randomly pick a classifier from ensemble E to classify the test set T to obtain the predicted labels. Assume that we want to construct 3 new ensembles. According to the predicted class labels we stratify (partition with equal class distributions) the test set T into 3 equal sized parts: T_1, T_2, and T_3. We want to select some classifiers from ensemble E to build 3 different new ensembles which are responsible for ranking T_1, T_2 and T_3 respectively.

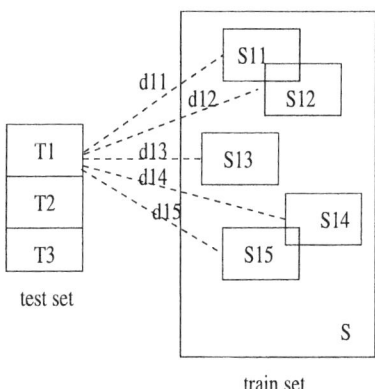

Fig. 1. An example for the similar sets

For each stratified test subset we use the k-Nearest Neighbor method to find k subsets of training set which are most similar to that test set part. For each instance of the test subset, we compute the distances from this instance to all training instances and find the nearest k instances. We use the following method to compute the distance between two instances u and v, which are from the test subset and training dataset, respectively. Suppose that an instance has k_1 nominal attributes A_i and k_2 numerical attributes B_i. We use the simplified VDM measure proposed in [10] to compute the distance of all nominal attributes.

$$VDM(u,v) = \sum_{C} \sum_{i=1}^{k_1} (\frac{N_{A_i=a_u,C=c}}{N_{A_i=a_u}} - \frac{N_{A_i=a_v,C=c}}{N_{A_i=a_v}})^2$$

where $N_{A_i=a_u}$ is the number of instances in test subset holding value a_u on attribute A_i, $N_{A_i=a_u,C=c}$ is the number of instances in test subset which are predicted belonging to class c and hold value a_u on attribute A_i. Here note that since test set is unlabeled, we use the class labels predicted in the first step.

We simply use the Euclidean distance to compute the difference of numerical attributes. $ED(u,v) = \sum_{i=1}^{k_2}(b_{u_i} - b_{v_i})^2$, where b_{u_i} is instance u' value on numerical attribute B_i.

The distance of u and v is

$$d(u,v) = VDM(u,v) + ED(u,v)$$

After the distances are computed, we randomly pick one from the k nearest instances of each test instance and use them to form a training subset. This subset is most similar to the test subset. We can use this method to find a desired number of most similar training subsets. The distance between two similar data sets is simply the average distances of each test subset instance with its corresponding nearest training instance. As shown in Figure 1, assume that S_{11}, S_{12}, S_{13}, S_{14} and S_{15} are T_1's 5 most similar training subsets. Their distances to T_1 are computed as $d_{11}, d_{12}, d_{13}, d_{14}$ and d_{15}, respectively.

2.2 Selecting Diverse and Accurate Classifiers

After the most similar training subsets are found, we use the following strategy to select diverse and accurate classifiers from original ensemble. Instead of directly using AUC as the criterion to choose classifiers, we propose a new measure SAUC (Softened Area Under the ROC Curve) as the heuristic.

For a binary classification task, SAUC is defined as

$$SAUC(\gamma) = \frac{\sum_{i=1}^{m} \sum_{j=1}^{n} U(p_i^+ - p_j^-)^\gamma}{mn} \quad (2)$$

$$U(x) = \begin{cases} x & \text{if } x \geq 0 \\ 0 & \text{if } x < 0 \end{cases}$$

where $\gamma \geq 0$, p_i^+, p_j^- represent the predicted probabilities of being positive for the ith positive and the jth negative examples in all m positive examples and n negative examples, respectively.

We choose a series of measures $SAUC(\gamma_1)$, $SAUC(\gamma_2)$, $\cdots$, $SAUC(\gamma_n)$ as heuristics. We use $SAUCs$ as heuristics for two reasons. First, SAUC with different powers γ may have different sensitivities and robustness to instance ranking variations. Thus using SAUCs with varied power γ as heuristics can more reliably select diverse classifiers in terms of ranking. Second, SAUC is a softened version of AUC and thus it is basically consistent with AUC. From Equation 2 we can see that $SAUC(0) = AUC$. Thus using SAUCs as criteria can select the classifiers with accurate ranking performance.

As shown in Figure 1 we use each classifier C_t of ensemble E to classify $S_{11}, S_{12}, S_{13}, S_{14}$ and S_{15} to obtain the respective $SAUC(\gamma_1)$ as $SA_{11}, SA_{12}, SA_{13}, SA_{14}, SA_{15}$. We then compute a score for C_t, which is the weighted average of the $SAUC(\gamma_1)$ values obtained above. It is $S_t = \sum_{i=1}^{5} \frac{SA_{1i}}{d_{1i}}$. We choose the classifier with the highest score. We repeat the above step n times by using a different $SAUC(\gamma_i)$ each time to select a new classifier.

Finally we use all the classifiers selected to construct a new ensemble. This ensemble is responsible for ranking T_1. The new ensemble combination method is weighted averaging, in which a classifier's weight is its score computed above. Using the same method we can construct two other ensembles which are responsible for ranking T_2 and T_3, respectively. We give the pseudo-code of this algorithm in Table 1.

One natural question about the DERC algorithm is that how many new ensembles should be constructed to give the best ranking performance. Since the number of test set partitions equals to the number of new ensembles, this question is equivalent to

Table 1. The pseudo code for DERC algorithm

DERC(E, S, T, n)
  Input:
  E : An ensemble with classifiers $C_1, \cdots, C_N$
  S : Training data set
  T : Test data set
  n : The number of test set partitions

  choose a classifier from E to classify T
  stratify T into $T_1, T_2, \cdots, T_n$
  for *each partition T_i* **do**
    $E_i^* \leftarrow \phi$
    find the most similar training subsets $S_{i1}, S_{i2}, \cdots, S_{ik}$
    compute the distances $d_{i1}, d_{i2}, \cdots, d_{ik}$ from T_i to $S_{i1}, \cdots, S_{ik}$ respectively
    for *each measure $SAUC(\gamma_u)$* **do**
      for *each classifier C_t* **do**
        run C_t on $S_{i1}, S_{i2}, \cdots, S_{ik}$
        obtain the $SAUC(\gamma_u)$ of C_t as $SA_{t_{i1}}, \cdots, SA_{t_{ik}}$
        compute the ranking score for classifier C_t
        $r_t \leftarrow \sum_{j=1}^{k} \frac{SA_{t_{ij}}}{d_{ij}}$
      endfor
      choose the classifier CC with highest score r_t
      $E_i^* \leftarrow E_i^* \cup CC$
    endfor
  endfor
  return *all E_i^**

how to choose an optimal number of test set partitions. Clearly, a small number of partitions generally means large partitioned test subsets, which corresponds to large similar training subsets. Thus the corresponding new ensemble may not specialize on all instances of the similar training subsets. Therefore our algorithm may not perform best on a small number of partitions. On the contrary for very large number of partitions, the size of similar training subsets will be very small. In this case there is a danger of overfitting. Therefore we can claim that generally too small or too large number of partitions should be avoided. We will perform experiments in the next section to confirm this claim.

3 Experimental Evaluation

To evaluate the performance of our algorithm, we extract 16 representative binary data sets from UCI [11].

We use Bagging and Adaboost as the ensembling methods and Naive Bayes as the base learner. We choose WEKA [12] as the implementations. In order to increase the ensemble diversity, we randomly select half of the training data for each bootstrap in our

Bagging process. This can guarantee that the bagging classifiers are diverse to some degree. We compare the performance of DERC with Bagging and Adaboost respectively.

In our DERC algorithm we use $SAUC(\gamma_i)$ as criteria to select classifiers. We have to determine the suitable number and scores of the powers γ_i by taking into account the tradeoff between the quality of results and computational costs. We test the SAUC with a wide ranges of powers γ by using all the 16 datasets in the our experiments. The analysis of these measures' performance shows that the power range of [0,3] is a good choice for SAUC. We choose 9 different SAUC with the powers of 0, 0.1, 0.4, 0.8, 1.0, 1.5, 2, 2.5, 3 in our experiment.

We follow the procedure below to perform our experiment:

1. We discretize the continuous attributes in all data sets using the entropy-based method described in [13].
2. We perform 5-fold cross validation on each data set. In each fold we train an ensemble with 15 classifiers using Bagging and Adaboost methods, respectively. We then run our DERC algorithm on the ensemble trained. By varying the number of test set partitions, we have a number of different DERC algorithm models.
3. We run the second step 20 times and we compute the average AUC for all the predictions.

We use a common statistic to compare the learning algorithms across all data sets. We performed two tailed paired t-test with 95% confidence level to count in how many datasets one algorithm performs significantly better, same, and worse than another algorithm respectively. We use win/draw/loss to represent this.

The experimental results are listed in Table 2 and Table 3.

Table 2. Comparing the predictive AUC of DERC algorithms with Bagging

Dataset	Bagging	DERC(1)	DERC(2)	DERC(3)	DERC(4)	DERC(6)
breast	98.84 ± 0.56	98.84 ± 0.53	98.83 ± 0.50	98.85 ± 0.59	98.86 ± 0.55	98.81 ± 0.59
cars	93.56 ± 3.0	**94.77 ± 2.2**	**94.9 ± 2.7**	**94.83 ± 2.7**	**94.87 ± 2.9**	**95.02 ± 2.1**
credit	92.89 ± 1.2	**93.43 ± 1.1**	**93.36 ± 1.2**	**93.32 ± 1.2**	**93.3 ± 1.4**	**93.3 ± 1.1**
echocardio	72.34 ± 8.4	72.34 ± 8.4	**74.21 ± 8.3**	**74.11 ± 8.4**	**74.11 ± 8.4**	73.09 ± 8.4
eco	99.28 ± 0.84	99.34 ± 1.1	99.34 ± 1.0	99.32 ± 0.84	99.3 ± 1.0	99.33 ± 0.84
heart	85.89 ± 0.45	86.01 ± 0.5	**86.97 ± 0.5**	**86.81 ± 0.64**	86.06 ± 1.7	85.92 ± 2.6
hepatitis	86.73 ± 2.6	87.06 ± 2.6	87.5 ± 2.9	**89.14 ± 2.6**	**88.59 ± 2.4**	**88.2 ± 1.8**
import	97.75 ± 2.6	97.75 ± 2.6	97.59 ± 2.8	97.72 ± 2.6	97.72 ± 2.6	97.74 ± 2.6
liver	61.77 ± 1.6	61.33 ± 0.45	61.64 ± 0.6	61.4 ± 0.18	61.26 ± 0.3	61.19 ± 3.7
pima	77.27 ± 8.9	**79.33 ± 7.6**	**79.29 ± 7.7**	**79.26 ± 8.0**	**79.14 ± 8.6**	**79.22 ± 8.7**
thyroid	95.12 ± 1.7	95.19 ± 1.6	95.10 ± 1.6	95.16 ± 1.9	**95.24 ± 1.9**	**95.29 ± 1.5**
voting	96.00 ± 0.36	96.08 ± 0.36	96.07 ± 0.36	96.27 ± 0.36	95.99 ± 0.36	96.01 ± 0.36
sick	96.84 ± 1.56	95.20 ± 2.48	•94.27 ± 2.11	•94.27 ± 3.47	•93.99 ± 2.79	•94.08 ± 3.02
ionosphere	94.59 ± 3.21	94.80 ± 3.22	**95.96 ± 3.47**	**95.85 ± 2.63**	**95.84 ± 2.79**	**95.84 ± 3.92**
german	84.26 ± 4.02	**87.58 ± 4.33**	**87.40 ± 4.1**	**87.23 ± 4.21**	**87.44 ± 4.2**	**87.4 ± 4.17**
mushroom	99.89 ± 0.04	99.79 ± 0.04	99.88 ± 0.04	99.90 ± 0.04	99.89 ± 0.04	99.89 ± 0.04
w/d/l		4/12/0	7/8/1	8/7/1	8/7/1	7/8/1

Table 3. Comparing the predictive AUC of DERC algorithms with Adaboost

Dataset	AdaBoost	DERC(1)	DERC(2)	DERC(3)	DERC(4)	DERC(6)
breast	98.99 ± 2.1	98.39± 2.4	98.41 ± 2.1	98.46± 2.1	98.51 ± 2.1	98.53 ± 2.1
cars	91.74 ± 5.0	**93.21± 5.0**	**93.14± 5.0**	**94.72± 5.0**	**93.89± 5.0**	**93.21± 5.0**
credit	92.06 ± 3.7	92.04 ± 3.7	92.06 ± 3.5	92.08± 4.8	92.10 ± 5.3	91.77 ± 4.7
echocardio	72.02 ± 4.8	**73.94 ± 4.8**	**73.94 ± 4.8**	**73.94± 4.8**	**73.94 ± 4.8**	**73.94 ± 4.8**
eco	99.30 ± 1.0	99.13 ± 1.0	99.02 ± 1.0	99.24± 1.0	99.27 ± 1.0	99.62 ± 1.0
heart	88.03 ± 0.28	88.51 ± 0.31	**90.39 ± 0.28**	89.72± 1.22	89.34 ± 1.6	89.58 ± 1.24
hepatitis	85.25 ± 8.6	•83.16 ±5.8	•83.03 ± 5.6	•83.24±8.6	•83.9±8.8	•83.84 ± 5.4
import	98.99 ± 1.7	98.90 ± 0.0	98.98 ± 3.6	98.73± 0.0	98.68 ± 5.2	98.88 ± 0.0
liver	65.45 ± 6.2	66.44 ± 4.1	66.20 ± 5.1	**67.08 ± 5.1**	**67.77 ± 5.1**	66.29 ± 5.1
pima	75.99 ± 8.3	74.92 ± 8.1	74.89 ± 7.2	**77.81± 8.3**	**77.99 ± 6.5**	**78.13± 8.4**
thyroid	95.61 ±0.35	95.55 ± 0.8	95.64 ± 0.27	95.65±0.18	95.58 ± 0.71	95.58 ± 0.35
voting	96.37 ±2.9	96.32 ±2.9	96.39 ± 3.3	96.5 ± 1.4	96.37 ± 1.4	96.37 ± 2.9
sick	97.02 ±1.56	97.08 ± 1.51	97.07 ± 1.43	97.02 ± 1.5	96.99 ± 1.24	97.08 ± 2.54
ionosphere	94.56 ±3.21	94.80 ± 3.47	95.96 ±4.37	95.85±4.26	95.84±3.97	95.84 ± 3.68
german	86.41 ±4.02	**88.24 ± 4.33**	**88.21 ± 4.1**	**88.21±4.21**	**88.19±4.2**	**88.19± 4.17**
mushroom	99.92 ±0.04	99.79 ± 0.04	99.88 ± 0.04	99.90 ± 0.04	99.89 ± 0.04	99.89 ± 0.04
w/d/l		3/12/1	4/11/1	5/10/1	5/10/1	4/11/1

Table 2 shows the AUC values for the Bagging algorithm and the DERC algorithms with different settings on various data sets. We use DERC(i) to denote the corresponding DERC algorithm which generate a number of *i* new ensembles. Each data cell represents the average AUC value of the 20 trials of 5-fold cross validation for the corresponding algorithm and data set. The data in bold shows the corresponding algorithm performs significantly better than Bagging on the corresponding data set. The data with a "•" means it is significantly worse than that of Bagging.

From this table, we can see that DERC outperforms the original Bagging algorithm. The w/d/l statistics shows that all DERCs with different settings have much more wins than losses compared with Bagging algorithm. If we rank them according to the w/d/l number, we can see that the DERC with 3 or 4 partitions performs best, the DERC with 2 or 6 partitions the second best, while the DERC with 1 partition the worst.

We can also see how the partition numbers influences the dynamic re-construction performance. We can observe that generally the dynamic re-constructions with the partition numbers of 3 or 4 perform best. It shows that dynamic re-construction with intermediate number of partitions outperforms that with large or small number of partitions. This result confirms our discussion in the previous section.

We also compare our DERC algorithm with Adaboost and report the results in Table 3. The similar comparisons show that DERC also significantly outperforms Adaboost in terms of AUC. DERC(3) wins in 5 datasets, ties in 10 datasets on loses only in 1 dataset.

4 Conclusions and Future Work

In this paper we propose a novel dynamic re-construction technique which aims to improve the ranking performance of any given ensemble. This is a generic technique which can be applied on any existing ensembles. The advantage is that it is independent of the specific ensemble construction method. The empirical experiments show that this dynamic re-construction technique can achieve significant performance improvement in term of ranking over the original Bagging and Adaboost ensembles, especially with an intermediate number of partitions .

In our current study we use Naive Bayes as the base learner. For our future work, we plan to investigate how other learning algorithms perform with the DERC technique. We also plan to explore whether DERC is also effective when it is applied on other ensemble methods.

References

1. Bradley, A.P.: The use of the area under the ROC curve in the evaluation of machine learning algorithms. Pattern Recognition **30** (1997) 1145–1159
2. Provost, F., Fawcett, T.: Analysis and visualization of classifier performance: comparison under imprecise class and cost distribution. In: Proceedings of the Third International Conference on Knowledge Discovery and Data Mining. AAAI Press (1997) 43–48
3. Hand, D.J., Till, R.J.: A simple generalisation of the area under the ROC curve for multiple class classification problems. Machine Learning **45** (2001) 171–186
4. Bauer, E., Kohavi, R.: An empirical comparison of voting classification algorithms: Bagging, boosting and variants. Machine Learning **36** (1999) 105–139
5. Breiman, L.: Bagging predictors. Machine Learning **24** (1996) 123–140
6. Quinlan, J.R.: Bagging, boosting, and C4.5. In: Proceedings of the Thirteenth National Conference on Artificial Intelligence. (1996) 725 – 730
7. Freund, Y., Schapire, R.E.: Experiments with a new boosting algorithm. In: Proceedings of the Thirteenth International Conference on Machine Learning. (1996) 148 – 156
8. Caruana, R., Niculescu-Mizil, A.: An empirical evaluation of supervised learning for ROC area. In: The First Workshop on ROC Analysis in AI. (2004)
9. Freund, Y., Iyer, R., Schapire, R., Singer, Y.: An efficient boosting algorithm for combining preferences. Journal of Machine Learning Research **4** (2003) 933–969
10. Stanfill, C., Waltz, D.: Toward memory-based reasoning. Communications of the ACM **29** (1986) 1213–1228
11. Blake, C., Merz, C.: UCI repository of machine learning databases. http://www.ics.uci.edu/~mlearn/MLRepository.html (1998) University of California, Irvine, Dept. of Information and Computer Sciences.
12. Witten, I.H., Frank, E.: Data Mining: Practical machine learning tools with Java implementations. Morgan Kaufmann, San Francisco (2000)
13. Fayyad, U., Irani, K.: Multi-interval discretization of continuous-valued attributes for classification learning. In: Proceedings of Thirteenth International Joint Conference on Artificial Intelligence. Morgan Kaufmann (1993) 1022–1027

Frequency-Based Separation of Climate Signals

Alexander Ilin[1] and Harri Valpola[2]

[1] Helsinki University of Technology, Neural Networks Research Centre,
P.O. Box 5400, FI-02015 TKK, Espoo, Finland
Alexander.Ilin@tkk.fi

[2] Helsinki University of Technology, Lab. of Computational Engineering,
P.O. Box 9203, FI-02015 TKK, Espoo, Finland
Harri.Valpola@tkk.fi
http://www.cis.hut.fi/projects/dss/

Abstract. The paper presents an example of exploratory data analysis of climate measurements using a recently developed denoising source separation (DSS) framework. We analysed a combined dataset containing daily measurements of three variables: surface temperature, sea level pressure and precipitation around the globe. Components exhibiting slow temporal behaviour were extracted using DSS with linear denoising. These slow components were further rotated using DSS with nonlinear denoising which implemented a frequency-based separation criterion. The rotated sources give a meaningful representation of the slow climate variability as a combination of trends, interannual oscillations, the annual cycle and slowly changing seasonal variations.

1 Introduction

One of the main goals of statistical analysis of climate data is to extract physically meaningful patterns of climate variability from highly multivariate weather measurements. The classical technique for defining such dominant patterns is principal component analysis (PCA) or empirical orthogonal functions (EOF) as it is called in climatology (see, e.g., [1]). However, many researchers pointed out that the maximum remaining variance criterion used in PCA can lead to such problems as mixing different physical phenomena in one extracted component [2,3]. This makes PCA a useful tool for information compression but limits its ability to isolate individual modes of climate variation.

To overcome this problem, rotation of the principal components proved useful [2]. The different rotation criteria reviewed in [2] are based on the general "simple structure" idea aimed at, for example, spatial or temporal localisation of the rotated components. The rotation of EOFs can be either orthogonal or oblique, which potentially leads to better interpretability of the extracted components.

Independent component analysis (ICA) is a recently developed statistical technique for component extraction which can also be used for rotating principal components. The basic assumption made in ICA is the statistical independence of the extracted components, which may lead to a meaningful data representation in a number of applications (see, e.g., [4] for introduction). ICA is based on higher-order statistics and in this respect bears some similarity to classical rotation techniques such as the Varimax orthogonal rotation [2]. Several attempts to apply ICA in climate research have already been made [5,6].

In this paper, we analyse weather measurements using a novel extension of ICA called denoising source separation [7]. DSS is a general separation framework which does not necessarily exploit the independence assumption but rather looks for hidden components which have "interesting" properties. The interestingness of the properties is controlled by means of a denoising procedure. For example, in [8], the sources with most prominent interannual oscillations were identified using DSS with linear filtering as denoising. The leading components were clearly related to the well-known El Niño–Southern Oscillation (ENSO) phenomenon and several other interesting components were extracted as well.

In the present work, we use DSS with linear denoising as the first, preprocessing step of climate data analysis. A wider frequency band in the denoising filter is used to identify the slow subspace of the climate system. The found slow components are further rotated using an iterative DSS procedure based on nonlinear denoising. The rotation is done such that the extracted components would have distinct power spectra.

The extracted components turned out to represent the subspace of the slow climate phenomena as a linear combination of trends, decadal-interannual oscillations, the annual cycle and other phenomena with distinct spectral contents. Using this approach, the known climate phenomena are identified as certain subspaces of the climate system and some other interesting phenomena hidden in the weather measurements are found.

2 DSS Method

DSS is a general algorithmic framework which can be used for discovering interesting phenomena hidden in multivariate data [7]. Similarly to PCA, ICA or other rotation techniques, DSS is based on the linear mixing model. The basic assumption is that there are some hidden components $\mathbf{s}(t)$ (also called sources or factors) which are reflected in the measurements $\mathbf{x}(t)$ through a linear mapping: $\mathbf{x}(t) = \mathbf{A}\mathbf{s}(t)$. The mapping $\mathbf{A}$ is called the mixing matrix in the ICA terminology or the loading matrix in the context of PCA.

The goal of the analysis is to estimate the unknown components $\mathbf{s}(t)$ and the corresponding loading vectors (the columns of $\mathbf{A}$) from the observed data $\mathbf{x}(t)$. In the climate data analysis, the components usually correspond to the time-varying states of the climate system and the loading vectors are the spatial maps showing the typical weather patterns corresponding to the found components. The components $\mathbf{s}(t)$ are usually normalised to unit variances, and therefore the spatial patterns have a meaningful scale.

The first step of DSS is so-called whitening or sphering (see Fig. 1). The goal of whitening is to uniform the covariance structure of the data in such a way that any linear projection of the data has unit variance. The positive

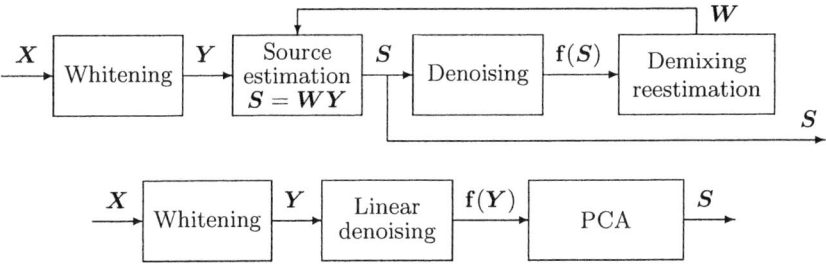

Fig. 1. The steps of the DSS algorithm in the general case (above) and in the case of linear denoising (below)

effect of such transformation is that any orthogonal basis in the whitened space defines uncorrelated sources. Therefore, whitening is used as a preprocessing step in many ICA algorithms, which allows restricting the mixing matrix to be orthogonal afterwards. Whitening is usually implemented by PCA.

The following stage is an orthogonal rotation of the white components Y based on the source separation criterion defined in the form of a denoising procedure. It is in general done using an iterative algorithm with three steps: source estimation, denoising of the source estimates and reestimation of the demixing matrix. Without denoising, this procedure is equivalent to the power method for computing the principal component of Y. Since Y is white, all the eigenvalues are equal and the solution without denoising becomes degenerate. Therefore, even slightest changes made by denoising can determine the DSS rotation. Since the denoising procedure emphasises the desired properties of the sources, DSS can find the rotation where the properties of interest are maximised.

Linear denoising is a simpler case as it does not require the described iterative procedure. DSS based on linear denoising can be performed in three steps: whitening, denoising and PCA on the denoised data (see Fig. 1). The idea behind this approach is that denoising renders the variances of the sphered components different and PCA can identify the directions which maximise the properties of interest. The eigenvalues given by PCA tell the ratio of the variance of the sources after and before filtering which is the objective function in linear denoising. The components are ranked according to the prominence of the desired properties the same way as the principal components in PCA are ranked according to the amount of variance they explain.

More general *nonlinear denoising* can implement more complicated separation criteria (see [7,9] for several examples). The objective function is usually expressed implicitly in the denoising function. Therefore, ranking the components is more difficult in this case and depends on the exact separation criterion used in the denoising procedure.

In the present work, DSS is exploited twice. First, DSS with linear denoising extracts components which exhibit most prominent variability in the slow time scale. Therefore, linear denoising is implemented using a low-pass filter whose frequency response is shown in Fig. 2. A similar approach (but with another

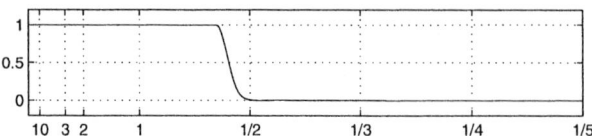

Fig. 2. Frequency response of the filter used in DSS with linear denoising. The abscissa is linear in frequency but is labeled in terms of periods, in years.

type of filter) was introduced in [8] to identify the subspace with most prominent interannual variability.

After that, the slow components are rotated such that they would have distinct power spectra. This is done by DSS with the general iterative procedure in which denoising implements a frequency-based separation criterion. Practically, the denoising procedure is based on whitening smoothed discrete cosine transform (DCT) power spectra of the components and using inverse DCT to calculate the denoised sources $f(S)$. This denoising mechanism is somewhat similar to the whitening-based estimation of the source variance proposed in [9]. The algorithm also tries to order the sources according to their frequencies using topographic ideas somewhat similar to [10].

3 Data and Preprocessing Method

The proposed technique is applied to measurements of three major atmospheric variables: surface temperature, sea level pressure and precipitation. This set of variables is often used for describing global climate phenomena such as ENSO [11]. The datasets are provided by the reanalysis project of the National Centers for Environmental Prediction–National Center for Atmospheric Research (NCEP/NCAR) [12].[1]

The data represent globally gridded measurements over a long period of time. The spatial grid is regularly spaced over the globe with 2.5° × 2.5° resolution. Although the quality of the data is worse for the beginning of the reanalysis period and it considerably varies throughout the globe, we used the whole period of 1948-2004.

The long-term mean was removed from the data and the data points were weighed similarly to [8] to diminish the effect of a denser sampling grid around the poles. Each data point was multiplied by a weight proportional to the square root of the corresponding area of its location. The spatial dimensionality of the data was reduced using the PCA/EOF analysis applied to the weighed data. For each dataset, we retained 100 principal components which explain more than 90% of the total variance. The DSS analysis was then applied to the combined data containing the measurements of the three variables.

---

[1] The authors would like to thank the NOAA-CIRES Climate Diagnostics Center, Boulder, Colorado, USA, for providing NCEP Reanalysis data from their Web site at http://www.cdc.noaa.gov.

4 Results

First, we identified the subspace of slow climate phenomena by applying DSS with low-pass filtering as linear denoising to the daily weather measurements. The time course of the most prominent slow components extracted from highly multidimensional data is shown in the leftmost column of Fig. 3. The annual cycle appears in the two leading components as the clearest slow source of the climate variability. The following components also possess interesting slow behaviour.

However, the sources found at this stage appear to be mixtures of several climate phenomena. For example, the third and the fourth components are mixtures of slow trends and the prominent ENSO oscillations. Similar mixed phenomena can be found in other components as well. This effect is also seen from the power spectra of the components (not shown here). Many components possess very prominent slowest, decadal or close-to-annual frequencies. Except for the two annual cycle sources, none of the components has a clear dominant peak in its power spectrum.

The first sixteen slow components extracted by DSS with linear denoising were further rotated using frequency-based DSS described in Section 2. To discard high-frequency noise, the monthly averages of the slow components were used at this stage. The time course of the rotated sources is presented in Fig. 3 and the spatial patterns corresponding to some of the sources are shown in Fig. 4. The components now have more clear interpretation compared to the original slow components.

The power spectra of the rotated components are more distinct (see the rightmost column of Fig. 3). However, some of the power spectra look quite similar and we can roughly categorise the found sources into three subspaces with different variability time scales: trends (components 1–5), interannual oscillations (components 6–11) and components 12–16 with dominating close-to-annual frequencies in their spectra. The subspaces are identified reliably due to the distinct differences in the corresponding power spectra but the components within the subspaces may remain mixed.

Among the slowest climate trends, the most prominent one is component 3 which has a constantly increasing time course. This component may be related to global warming as the corresponding surface temperature map has mostly positive values all over the globe (see Fig. 4). The highest temperature loadings of this component are mainly concentrated around the North and South Poles, the sea level pressure map has a clear localisation around the South Pole and the precipitation loadings are mostly located in the tropical regions.

Components 6–11 exhibit prominent oscillatory behaviour in the interannual time scale. The most prominent sources here are components 7 and 8 which are closely related to the ENSO oscillations both in the time course and in the corresponding spatial patterns (see Fig. 4). They are very similar to the first two components extracted in [8]: component 7 is similar to the ENSO index and component 8 bears resemblance with the differential ENSO index. Component 6 may be related to the slower behaviour of ENSO. Component 10 has very distinct spatial patterns with a prominent dipole structure over the continents

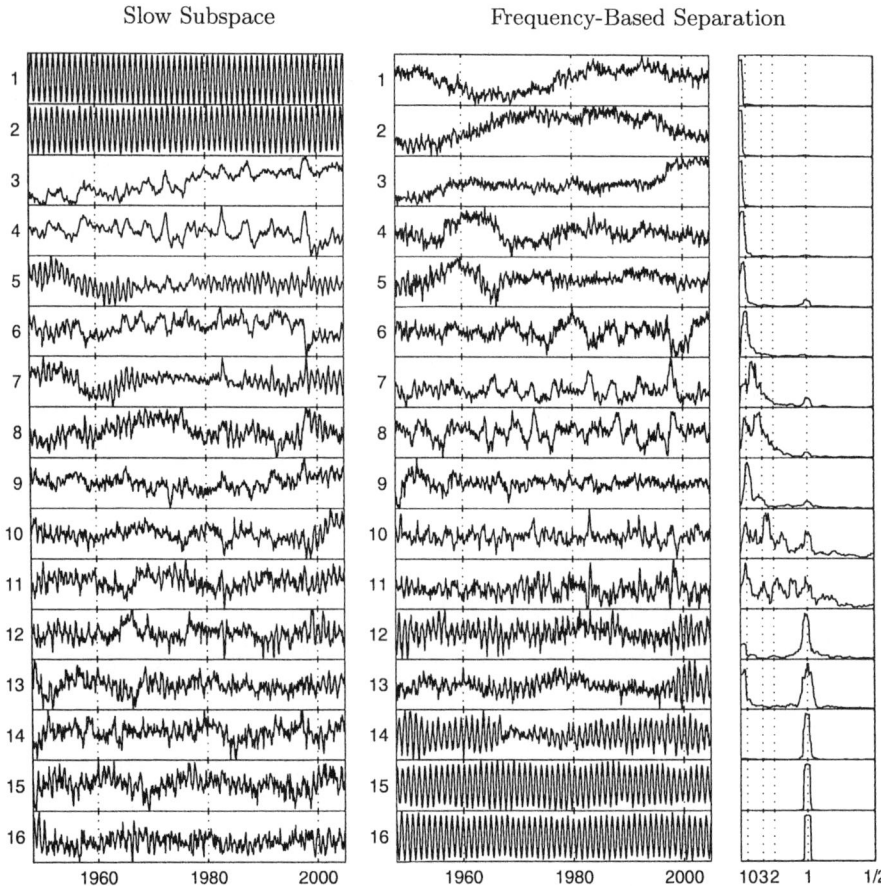

Fig. 3. Left: The monthly averages of the components extracted by DSS with linear denoising. Middle: The rotated slow components estimated by frequency-based DSS. The variances of all the components are normalised to unity. Right: The power spectra of the components found by frequency-based DSS. The abscissa is linear in frequency but is labeled in terms of periods, in years.

in the Northern Hemisphere in the temperature maps. A similar source was also extracted in [8].

Components 12–16 have prominent close-to-annual frequencies in their power spectra. The annual cycle now appears in components 15–16. The rest of the sources resemble the annual oscillations modulated (multipiled) by some slow signals. Thus, this set of components may be related to some phenomena slowly changing the annual cycle. However, as we already pointed out, the found rotation within this subspace may not be most meaningful.

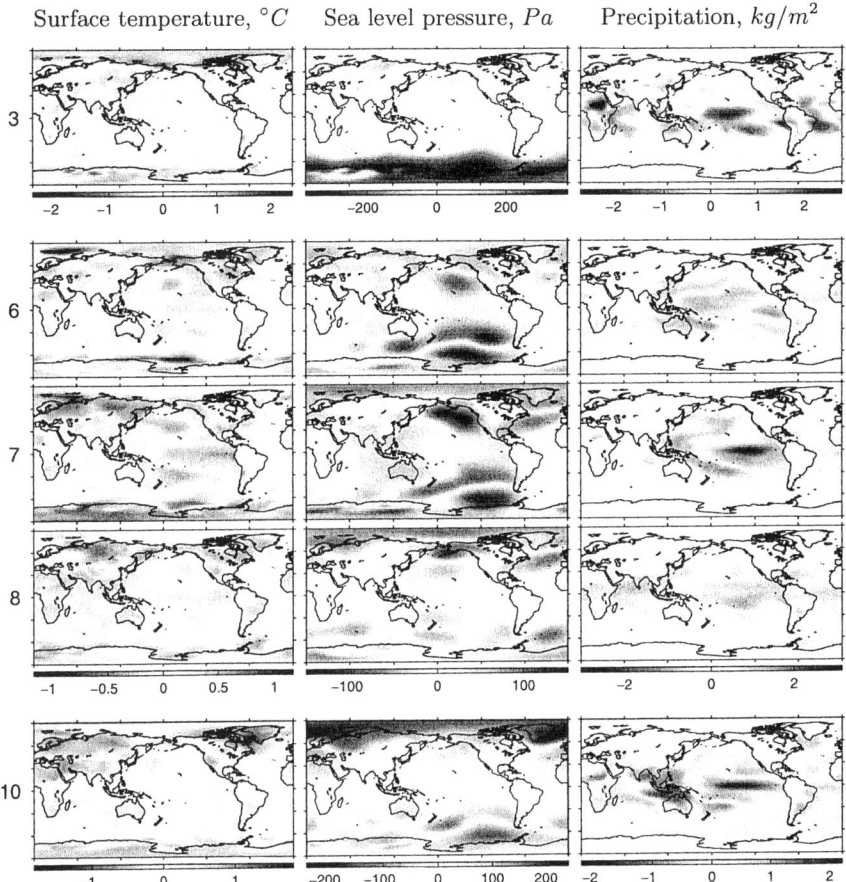

Fig. 4. The spatial patterns of several components found by frequency-based DSS. The label on the left indicates the number of the component in Fig. 3.

5 Discussion

In this paper, we showed how the DSS framework can be applied to exploratory analysis of climate data. We used a frequency-based separation criterion to identify slow varying climate phenomena with distinct temporal behaviour. The presented algorithm can be used for both finding a physically meaningful representation of the data and for an easier interpretation of the complex climate variability. It can also be useful for making predictions of future measurements or for detecting artifacts produced during the data acquisition.

Representing climate variability as a combination of hidden phenomena does not have a unique solution because of the high complexity of the climate system where different phenomena constantly interact with each other. This task always allows some subjectivity where the exact details of the separation procedure depend on the ultimate goal of research. A good example of such subjectivity is choosing the number of extracted components in the proposed DSS procedure. In the presented experiments, we chose this number such that the components would easily be interpretable. According to our experience, increasing the number of components usually results in describing one phenomenon by several components having slightly different frequency contents. This may be useful for better understanding of well-known climate phenomena or for discovering new, not easily observable phenomena, but it may also be counter-productive if the solution becomes overfitted.

Note also that the proposed method may sometimes identify reliably only the subspaces of components having similar power spectra and the rotation within the subspaces may not be most meaningful. Some other separation criteria based on, for instance, dynamical modelling or the interaction with the seasonal variations is an important line of future research.

References

1. H. von Storch and W. Zwiers, Statistical Analysis in Climate Research. Cambridge, U.K.: Cambridge Univ. Press, 1999.
2. M. B. Richman, Rotation of principal components, J. of Climatology, vol. 6, pp. 293-335, 1986.
3. K.-Y. Kim and Q. Wu, A comparison study of EOF techniques: Analysis of non-stationary data with periodic statistics, J. of Climate, vol. 12, pp. 185-199, 1999.
4. A. Hyvärinen, J. Karhunen, and E. Oja, Independent Component Analysis. J. Wiley, 2001.
5. F. Aires, A. Chédin, and J.-P. Nadal, Independent component analysis of multivariate time series: Application to the tropical SST variability, J. of Geophysical Research, vol. 105, pp. 17, 437-17, 455, July 2000.
6. A. Lotsch, M. A. Friedl, and J. Pinzón, Spatio-temporal deconvolution of NDVI image sequences using independent component analysis, IEEE Trans. on Geoscience and Remote Sensing, vol. 41, pp. 2938-2942, December 2003.
7. J. Särelä and H. Valpola, Denoising source separation, Journal of Machine Learning Research, vol. 6, pp. 233-272, 2005.
8. A. Ilin, H. Valpola, and E. Oja, Semiblind source separation of climate data detects El Niño as the component with the highest interannual variability, in Proc. of Int. Joint Conf. on Neural Networks (IJCNN2005), (Montreal, Quebec, Canada), 2005. Accepted.
9. H. Valpola and J. Särelä, Accurate, fast and stable denoising source separation algorithms, in Proc. of Fifth Int. Conf. on Independent Component Analysis and Blind Signal Separation (ICA 2004) (C. G. Puntonet and A. Prieto, eds.), vol. 3195 of Lecture Notes in Computer Science, (Granada, Spain), pp. 65-72, Springer-Verlag, Berlin, 2004.
10. A. Hyvärinen, P. Hoyer, and M. Inki, Topographic independent component analysis, Neural Computation, vol. 13, no. 7, pp. 1525-1558, 2001.
11. K. E. Trenberth and J. M. Caron, The Southern Oscillation revisited: Sea level pressures, surface temperatures, and precipitation, Journal of Climate, vol. 13, pp. 4358-4365, December 2000.
12. E. Kalnay and Coauthors, The NCEP/NCAR 40-year reanalysis project, Bulletin of the American Meteorological Society, vol. 77, pp. 437-471, 1996.

Efficient Processing of Ranked Queries with Sweeping Selection*

Wen Jin[1], Martin Ester[1], and Jiawei Han[2]

[1] School of Computing Science, Simon Fraser University,
{wjin, ester}@cs.sfu.ca
[2] Department of Computer Science, Univ. of Illinois at Urbana-Champaign
hanj@cs.uiuc.edu

Abstract. Existing methods for top-k ranked query employ techniques including sorting, updating thresholds and materializing views. In this paper, we propose two novel index-based techniques for top-k ranked query: (1) indexing the layered skyline, and (2) indexing microclusters of objects into a grid structure. We also develop efficient algorithms for ranked query by locating the answer points during the sweeping of the line/hyperplane of the score function over the indexed objects. Both methods can be easily plugged into typical multi-dimensional database indexes. The comprehensive experiments not only demonstrate that our methods outperform the existing ones, but also illustrate that the application of data mining technique (microclustering) is a useful and effective solution for database query processing.

1 Introduction

Rank-aware query processing is important in database systems. The answer to a top-k query returns k tuples ordered according to a specific score function that combines the values from participating attributes. The combined score function is usually linear and monotone with regard to the individual input. For example, given a hotel database with attributes of x (*distance* to the beach) and y (*price*), and the score function $f(x,y) = 0.3x + 0.7y$, the top-3 hotels are the best three hotels that minimize f.

The straightforward method to answer top-k queries is to first calculate the score of each tuple, and then output the top-k tuples from them. This *fully-ranked* approach is undesirable for querying a relatively small number k of a large number of objects. Several methods towards improving the efficiency of such queries have been developed, but they are either specific to joined relations of two dimensions [16], or incompatible with other database indexing techniques [4,8], or computationally expensive [11]. In this paper, we propose two novel index-based techniques for top-k ranked query. The first method is *indexing the layered skyline based on the skyline operator* [2], whereas the second, motivated by a major data mining technique-microclustering, is *indexing microclusters of the dataset into a grid structure*. We develop efficient algorithms for answering the ranked query by locating the answer points during sweeping the line/hyperplane of the score function over the indexed objects. Both methods can easily be plugged into typical multi-dimensional database indexes. For example, the

---

* The work was supported in part by Canada NSERC and U.S. NSF IIS-02-09199.

layered skyline can be maintained in a multi-dimensional index structure with blocks described by (1) MBR (Minimum Bounded Rectangle) such as R-tree [9] and R*-tree [3], or (2) spherical Microcluster such as CF-tree [19] and SS-tree [18]. The comprehensive experiments not only demonstrate the high performance of our methods over the existing ones, but also illustrate that the microclustering technique is an effective solution to top-k query processing.

The rest of the paper is organized as follows. Section 2 presents the foundations of this paper. Section 3 and Section 4 give KNN-based and Grid-based sweeping algorithms for the ranked queries respectively and illustrate their plug-in adaptations for R-trees and CF-trees. Section 5 reviews related work. We present our experimental results in Section 6 and conclude the paper in Section 7.

2 Foundations

Let a d-dimensional dataset be X, and a linear monotone score function be $f(x) = \sum_{i=1}^{d} a_i \cdot x_i$ where $x \in X$, and a_i is the weight on the attribute value x_i of x such that $\sum_{i=1}^{d} a_i = 1$, $0 \leq a_i \leq 1$. Without loss of generality, we assume the lower the score value, the higher the rank of the object. A top-k ranked query returns a collection $T \subset X$ of k objects ordered by f, such that for all $t \in T$, $x \in X$ and $x \notin T$, $f(t) \leq f(x)$. We say $p = (p_1, \ldots, p_d) \in X$ *dominates* another object $q = (q_1, \ldots, q_d) \in X$, denoted as $p \succ q$, if $p_i \leq q_i$ $(1 \leq i \leq d)$ and at least there is one attribute, say, the jth attribute $(1 \leq j \leq d)$, $p_j < q_j$. Hence, q is a *dominated object*. The skyline operator [2] is defined as objects $\{s_1, s_2, \ldots, s_m\} \subset X$ that are not dominated by any other object in X. A *multilayered skyline* [6], which is regarded as a stratification of the dominating relationship in the dataset, is organized as follows: (1) the first layer skyline L_1 is the skyline of X, and (2) the ith layer $(i > 1)$ skyline L_i is the set of skyline objects in $X - \bigcup_{j=1}^{i-1} L_j$. As for two objects $p \in X, q \in X$, if $p \succ q$ then $f(p_1, \ldots, p_d) < f(q_1, \ldots, q_d)$, so we can derive the following interesting properties.

Lemma 1. *Given X and f, for any $i < j$, if object $q \in L_j$, there exists at least one object $p \in L_i$ s.t. $f(p_1, \ldots, p_d) < f(q_1, \ldots, q_d)$.*

Theorem 1. *Given $K(K \geq k)$ layers of skyline $L_1, \ldots, L_K$, any top-k tuples w.r.t. f must be contained in $\bigcup_{j=1}^{K} L_j$, i.e. in the first K skyline layers.*

Definition 1. *(MicroCluster[19]) The MicroCluster C for n objects is represented as $(n, \overline{CF1(C)}, \overline{CF2(C)}, \overline{CF3(C)}, r)$, where the linear sum and the sum of the squares of the data values are maintained in $\overline{CF1(C)}, \overline{CF2(C)}$ respectively. The centroid of C is $\overline{CF3(C)} = \frac{\overline{CF1(C)}}{n}$, the radius of C is $r = (\frac{\overline{CF2(C)}}{n} - (\frac{\overline{CF1(C)}}{n})^2)^{\frac{1}{2}}$.*

Therefore, maintaining K layers of skylines is enough to answer any top-k ranked query with $k \leq K$. The layered skyline can be organized into rectangle-like and sphere-like *blocks*, hence supported by two types of multi-dimensional database indexes. We choose R-tree and CF-tree as typical representatives of these two types of indexes. The basic storage of a *block* in R-tree is a MBR in a leaf node, and is a

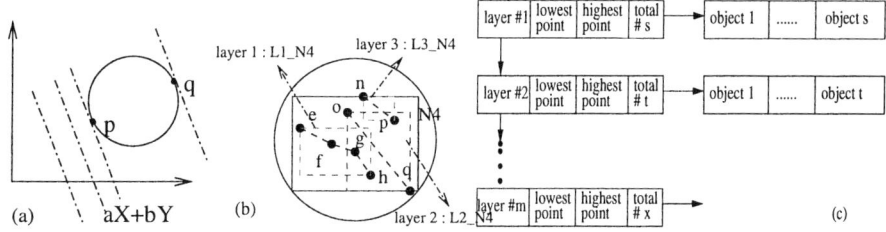

Fig. 1. (a)Contact Points (b) Layered-Skyline in N4 (c) Linked List for Layered Skyline in a Block

microcluster in leaf node in CF-tree [19]. We compute the $K(k \leq K)$ layers of skyline objects by recursively applying existing skyline computing algorithms such as [14], and only store layered skyline as *blocks* in the leaf nodes of index structures.

3 A KNN-Based Sweeping Approach for Top-k Queries

Here we present methods of two levels of a hyperplane sweeping: (1) sweeping over blocks such that the I/O cost of accessing blocks is minimized, and (2) sweeping within a block such that the CPU cost of processing objects within a block is minimized.

(1) Sweeping over blocks. During the sweeping process, the hyperplane always contacts the best point first, the next best point second, and so on. Based on the indexed K skyline layers, we develop an efficient branch-and-bound algorithm for top-k query similar to the optimal KNN search algorithm [15] [16]. Each block has a lowest/highest point corresponding to the lowest/highest score. In the algorithm, a sorted queue Q is used to maintain the processed points and blocks in ascending order of their score. The algorithm starts from the root node and inserts all its blocks to the queue. The score for a data block is the score for its lowest point. The first block in the queue will then be expanded. If the entry is a leaf *block*, we will access its data with some strategy. In the expanding process, we also keep track of how many data points are already present, and if an object or a block is dominated by enough ($> k$) objects (in some blocks) lining in the queue before it, then it can be pruned. The expanding process stops when the queue has k objects in the front.

Algorithm 1 Branch-and-Bound Ranking (BBR) Method.
Input: k, and a multi-dimensional index tree
Output: Top-k answer in Q
Method:

1. $Q :=$ Root Block;
2.   WHILE top-k tuples not found DO
3. $F :=$ the first non-object element from Q;
4. $S :=$ SweepIntoBlock(F); //S is a set of blocks and/or objects
5.   FOR each block/object s in S Do

6.   IF more than k objects in Q having smaller score than s
7.   Discard s;
8.   ELSE Insert s to Q;
9.   Output k objects from Q;

Algorithm 1 have different implementations for **SweepIntoBlock**. It can simply expand the block and access all the belonged objects (noted as **BBR1**). For a MBR in R-tree, the lowest/highest points are the lower/upper right corner points, while for a microcluster in CF-tree, these are two contact points of the sweeping hyperplane to the sphere, shown as p and q in Fig. 1(a). Given a sweeping hyperplane $y = \sum_{i=1}^{d} a_i \cdot x_i$ and a microcluster F with radius R of the origin: $F(x_1,\ldots,x_d) = \sum_{i=1}^{d} x_i^2 - R^2 = 0$ (1). To obtain p and q, we only need solve $\nabla F(x'_1,\ldots,x'_d) = c \cdot \boldsymbol{A}$ (2) together with (1). Here c is a free variable, $\boldsymbol{A} = a_1,\ldots,a_d$, and $\nabla F(x_1,\ldots,x_d)$ which works as the gradient of F at $X(x'_1,\ldots,x'_d)$, is $(\frac{\partial F}{\partial x_1}(x'_1,\ldots,x'_d),\ldots,\frac{\partial F}{\partial x_d}(x'_1,\ldots,x'_d))$.

(2) Sweeping within layered blocks. In order to avoid processing unnecessary objects in each block, we make use of the layered skylines since they give a contour of the data distribution, and develop an efficient sweeping within-layered-blocks method (noted as **BBR2**), as a procedure SweepIntoBlock in Algorithm 1. Suppose the *block* in a leaf node has m layers of skylines, the lowest/highest object as well as the total number of objects for that layer is maintained. As shown the node in Fig. 1(b), objects e, f, g and h are layer-1 skyline objects in node N_4, we put all skyline objects in layer 1 minimally hyperectangle-bounded by a pseudo-node denoted as $L_1\_N_4$. The linked list storage structure for the layered skylines is shown in Fig. 1(c), where the header is the summarization information of the pseudo-node which links to its bounding skyline objects. Now if a leaf *block* is chosen from the queue, we only expand the pseudo-node that has the best lowest point according to the score function.

4 A Grid-Based Sweeping Approach for Top-k Queries

Although the KNN-based approach can efficiently obtain the top-k objects, it may still visit and compare all the objects in a block even when the layering technique is applied. In this section, we present an alternative grid-based method for more efficiently organizing the objects. Since the user-specified weights of a score function will often have a fuzzy rather than a crisp semantic, approximate, query processing seems to be acceptable if

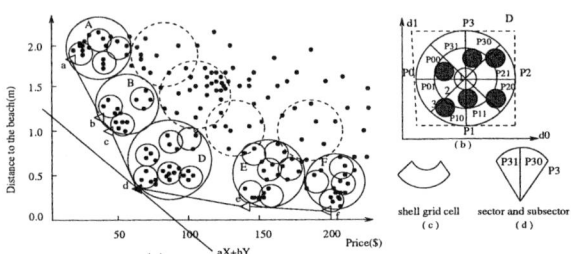

Fig. 2. Shell-Grid Partition of Microclusters

this allows significantly improved response time. The basic idea is to build a grid-like partition of the *blocks* and access the objects within a block along the grid. For CF-tree, a shell-grid partition is made (the R-tree case can be adapted by bounding a

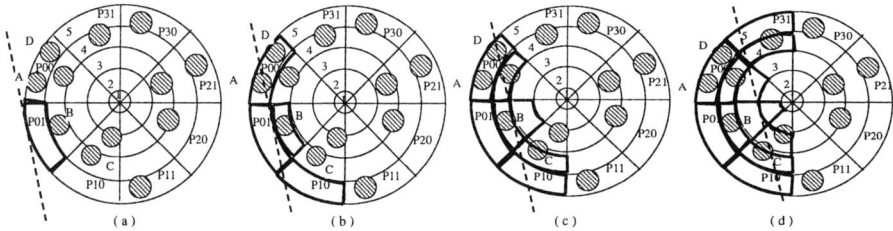

Fig. 3. Sweeping a Shell Grid

MBR over any microcluster). The block entries are then assigned to the grid cells, and the sweeping algorithm is applied. All objects of the current grid cell are accessed before those of other grid cells. This approach reduces the number of comparisons, but it may lead to a non-exact result if k answer objects are found before a further grid cell with better objects is accessed. The overview of CF-tree with the shell grid partition is shown in Fig. 2, where 2(b) depicts the anatomy of an intermediate microcluster node D in 2(a). A single **shell grid cell** (or **cell**) is shown in 2(c). We can enforce the radius of the leaf microcluster to be smaller than ε when building the CF-tree [13]. Motivated by the Pyramid indexing [1], we propose a novel idea of building shell grid partitions for microcluster nodes. The partition made is in partial shells and the center of the sphere is pre-computed. Assume the center is $o = (o_0, \ldots, o_{d-1})$. We split the spherical space into $2^{d-1} \cdot 2d$ fan-shaped partitions, each having the center as the top, and $1/(2^{d-1} \cdot 2d)$ of the $(d-1)$-dimensional spherical surface as the base. The sphere is split into $2d$ **sectors** $P_0, \ldots, P_{2d-1}$ according to the square cube with $2d$ surfaces (dashed in Fig. 2(b) enclosing the sphere is 2-dimension case), as P_3 in Fig. 2(d). Using hyperplane perpendicular to each axis and passing through the sphere center to split the whole space, each sector P_i is divided into 2^{d-1} **subsectors** $P_{i0}, \ldots, P_{i(2^{d-1}-1)}$, as P_{31} and P_{30} in Fig. 2(d). Then the whole space is divided with parallel spherical **shells** starting from the center. We have the following property:

Lemma 2. *For any object x in sector P_i where $i < d$, it satisfies: for any dimension j, $0 \leq j < d$, $i \neq j$, $|o_i - x_i| \leq |o_j - x_j|$; for sector P_i where $i \geq d$, it satisfies: for any dimension j, $0 \leq j < d$, $j \neq (i-d)$, $|o_{i-d} - x_{i-d}| \geq |o_j - x_j|$.*

We further number these subsectors from 0 to $2^{d-1} - 1$ in $(d-1)$ binary format $s_0, \ldots, s_{d-1}$. If $i < d$, the bit s_i does not exist in the binary string, otherwise s_{i-d} is excluded. For each subsector, if $s_j = 1$, the belonged points have $x_j > x_i$, and it is opposite for $s_j = 0$. Each subsector has $2^{d-1} - 1$ direct neighbors in the same sector, and another $(d-1) \cdot 2^{d-2}$ in the neighboring sector. As illustrated in Fig. 3, a useful sweeping property is: *the sweeping process explores first the outmost shell grid cell of the sector which the sweeping hyperplane tangent contacts, then goes to its directed neighboring cells in the same shell. If there is no data in those neighboring cells, sweeping should go to the inner cell in the same sector directly.* Generally we can first calculate the two contacting points and the sector number P_{mn} that the sweeping plane contacts first, and then start the hierarchical sweeping process. A sorted queue

Q is used to store the expanding entities including microclusters and a pseudo-node that has candidate sector number and shell number information.

Algorithm 2 A Shell-Grid Ranking(SGR) Method.
Input: CF-tree with Grid Shell Partitions, k.
Output: Top-k answers in list T.
Method:

1. Calculate standard contacting points and subsector number P_{mn}; $Q = \emptyset$; $T = \emptyset$;
2. Insert into Q the outmost cell of root node of CF;
3. WHILE the first k tuples are not found DO
4.   Remove the first entity E in Q;
5.   IF E is a cell
6.   insert blocks in subsector P_{mn} and its direct neighbors with pseudo-entities;
7.   ELSE IF E is an intermediate node
8.   insert into Q the outmost cell of E;
9.   ELSE IF E is a pseudo-entity
10.   insert into Q blocks in its neighbor subsectors of the same cell with pseudo-entities;
11.   ELSE IF E is a leaf
12.   add E to T;
13. Output k points from T;

To analyze the error bound measured in the score difference of the objects, we observe the sweeping process in Fig. 4, some objects in other microcluster (i.e., MC2) are better than those in the current selected microcluster (i.e., MC1). In the extreme case, $q=(x'_1,\ldots,x'_d)$ is computed as part of the answer instead of w whose score is slightly larger than that of $p = (x_1,\ldots,x_d)$. $f(q) - f(w) < f(q) - f(p) = a_1 \cdot (x'_1 - x_1) + \cdots + a_d \cdot (x'_d - x_d) < 2 \cdot \varepsilon \cdot \sum a_i = 2 \cdot \varepsilon$, so the maximum error is $O(\varepsilon)$.

5 Experiments

In this section, we report the results of our experiments performed on a Pentium III 800MHz machine of 512M RAM running WindowsXP. We implemented our methods

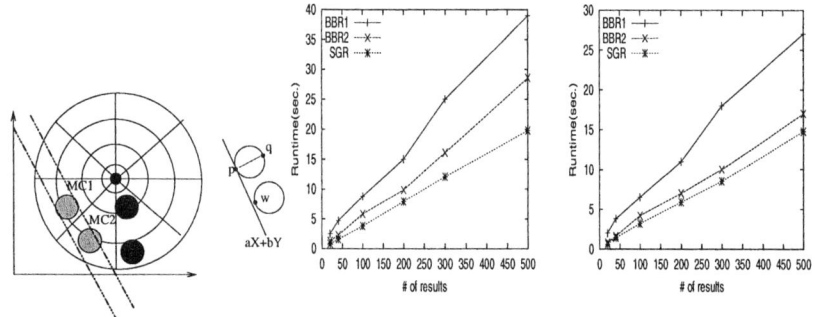

Fig. 4. Error Bound   **Fig. 5.** Query time (1)   **Fig. 6.** Query time (2)

and Onion in C++, and obtain PREFER in www.db.ucsd.edu/prefer. Two types of datasets of 100,000 records with 5 attributes in independent/correlated distribution were generated by the data generator of [2].

(1) Comparison of BBR and SGR. Figs. 5 and 6 show that all the algorithms have better performance for the correlated dataset. BBR2 runs much faster than BBR1 due to its smaller number of visited objects, while SGR ($\varepsilon = 10$) is best and is an order of magnitude faster than BBR2. When ε changes from 5 to 10 and 15 in the independent dataset, runtime decreases due to the decreasing number of visited microclusters, and the error rate as well as coverage rate become relatively higher due to the increasing size of microclusters (Figs. 7, 8 and 9).

(2) Comparison with Onion and PREFER. We compare the time to construct K layers of skylines for BBR, SGR, and K layers of convex hulls for Onion($K = 200$). When the dimension varies from 2 to 5, SGR uses the least time (Fig. 10). BBR2 is much faster than Onion when the dimension increases, as the complexity of computing convex hull increases exponentially with dimensions. When k changes, all perform better on the correlated dataset due to the smaller number of skyline objects. SGR is still the best due to its high efficient sweeping, BBR2 ranks second and PREFER queries faster than Onion (Figs. 11, 12). When k increases, the coverage rate of the answers of SGR is higher than that of PREFER (Figs. 13 and 14). Because correlated dataset has less number of skyline, the size of materialized views for PREFER will be reduced, which leads to the lower coverage rate than the independent dataset.

6 Related Work

The top-k ranked query problem was proposed by Fagin in the context of multimedia database systems [7], and methods can be categorized into three types: **(1) Sorted accessing and ranking** mainly applies some strategies to sequentially search the sorted list of each attributes until the top-k tuples are retrieved. [17] proposed different ways to improve that in [7]. Further, a "threshold algorithm (TA)" [8] is developed to scan all query-relevant index lists in an interleaved manner. **(2) Random accessing and ranking** supports mainly random access over the dataset until the answers have been retrieved. [10] uses foot-rule distance to measure the two rankings and model the

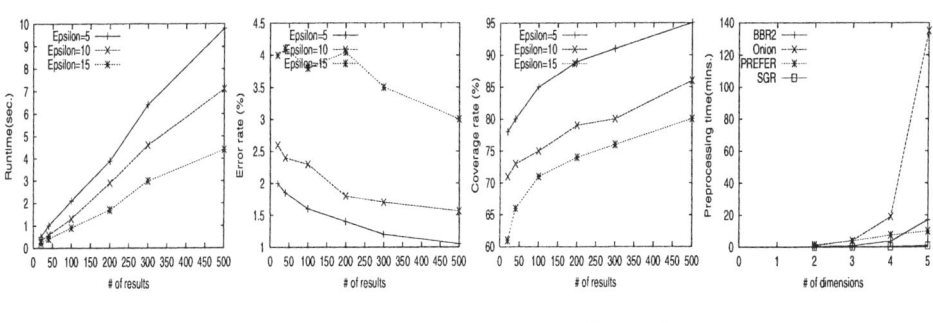

Fig. 7. Effect of Eps(1)   **Fig. 8.** Effect of Eps(2)   **Fig. 9.** Effect of Eps(3)   **Fig. 10.** Preprocessing

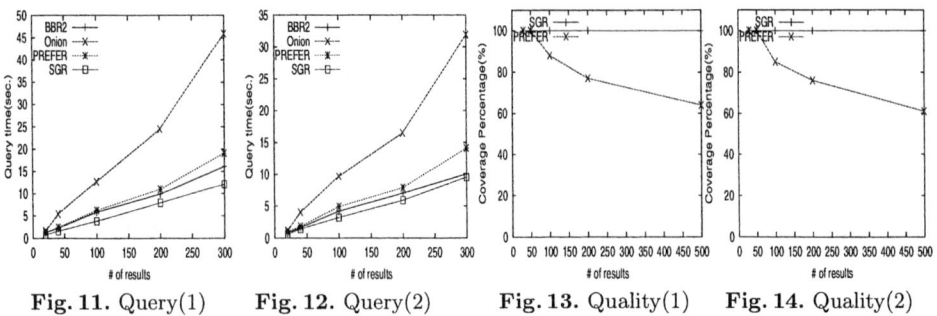

Fig. 11. Query(1) **Fig. 12.** Query(2) **Fig. 13.** Quality(1) **Fig. 14.** Quality(2)

rank problem as the minimum cost perfect matching problem, whereas [5] proposes to translate the top-k query into a range query in database. **(3) Pre-materialization and rank indices** organizes the tuples in a special way, then applies similarity match for the answer of ranked query. In [4], an index is built on layered convex hulls for all points. Such index for large databases is expensive due to the convex hull finding complexity. [11,12] propose to pre-materialize views of different top-k query answers. When the query's score function is close to a view's, a small number of the tuples in that view is necessary for the top tuples, and the query result can be produced in a pipelined fashion. As there is no guarantee how many tuples should each view store for query, it often stores the whole dataset in each view. [16] proposes a ranked index to support top-k query($k \leq K$) but it only applies to two dimensions and can have a huge number of materialized partitions.

7 Conclusions

Rank-aware query processing has recently emerged as an important paradigm in database systems, and only few existing methods exploit materialization and index structures. In this paper, we propose indexing layered skyline and shell-grid microclusters for top-k ranked query, and present methods sweeping the hyperplane of the score function over the indexed objects. Our methods can be easily adapted to existing multi-dimensional index structures. The experimental results demonstrate the strength of our methods and the usefulness of the microclustering technique in top-k query processing.

References

1. S. Berchtold, C. Bhm and H. P. Kriegel. The Pyramid-Technique: Towards Breaking the Curse of Dimensionality. *SIGMOD* 1998.
2. S. Borzsonyi, D. Kossmann, and K. Stocker. The Skyline Operator. *ICDE* 2001.
3. N. Beckmann, H. P. Kriegel, R. Schneider, and B. Seeger. The R*-tree: An Efficient and Robust Access Method for Points and Rectangles *SIGMOD* 1990.
4. Y. C. Chang, L. D. Bergman, V. Castelli, C. S. Li, M. L. Lo and J. R. Smith. Onion Technique:Indexing for Linear Optimization Queries. *SIGMOD* 2000.

5. S. Chaudhuri and L. Gravano. Evaluating Top-k Selection Queries. *VLDB* 1999.
6. T. Cormen, C. E. Leiserson, et al. Introduction to Algorithms, The MIT Press, 2001.
7. R. Fagin. Fuzzy Queries in Multimedia Database Systems. *PODS* 1998.
8. R. Fagin. et al. Optimal Aggregation Algorithms for Middleware. *PODS* 2001.
9. A. Guttman. R-trees: A dynamic index structure for spatial searching. *SIGMOD* 1984.
10. S. Guha, et al. Merging the Results of Approximate Match Operations. *VLDB* 2004.
11. V. Hristidis, N. Koudas, and Y. Papakonstantinou. PREFER: A System for the Efficient Execution of Multi-parametric Ranked Queries. *SIGMOD* 2001.
12. V. Hristidis and Y. Papakonstantinou. Algorithms and applications for answering ranked queries using ranked views. *VLDB J.* 2004.
13. Wen Jin, Jiawei Han, Martin Ester Mining Thick Skylines over Large Databases. *PKDD* 2004.
14. D. Papadias, Y. F. Tao, G. Fu and B. Seeger. An Optimal and Progressive Algorithm for Skyline Queries. *SIGMOD* 2003.
15. N. Roussopoulos, S. Kelley, and F. Vincent. Nearest Neighbor Queries. *SIGMOD* 1995.
16. P. Tsaparas, et al. Ranked Join Indices. *ICDE* 2003.
17. E. L. Wimmers, L. M. Haas, M. T. Roth and C. Braendli. Using Fagin's Algorithm for Merging Ranked Results in Multimedia Middleware. *CoopIS* 1999.
18. D. A. White and R. Jain. Similarity Indexiug with the SS-tree. *ICDE* 1996.
19. T. Zhang, R. Ramakrishnan, and M. Livny BIRCH: an efficient data clustering method for large databases. *SIGMOD* 1996.

Feature Extraction from Mass Spectra for Classification of Pathological States

Alexandros Kalousis, Julien Prados, Elton Rexhepaj, and Melanie Hilario

University of Geneva,
Computer Science Department,
Rue General Dufour, 1211, Geneve, Switzerland
{kalousis, prados, hilario}@cui.unige.ch, rexhepaj@unil.ch

Abstract. Mass spectrometry is becoming an important tool in proteomics. The representation of mass spectra is characterized by very high dimensionality and a high level of redundancy. Here we present a feature extraction method for mass spectra that directly models for domain knowledge, reduces the dimensionality and redundancy of the initial representation and controls for the level of granularity of feature extraction by seeking to optimize classification accuracy. A number of experiments are performed which show that the feature extraction preserves the initial discriminatory content of the learning examples.

1 Proteomics and Mass Spectrometry

Clinical proteomics aims at investigating changes in protein expression in order to discover new disease markers and drug targets. Mass spectrometry is emerging as an important tool for biomarker discovery in proteomics. To discover these biomarker patterns, the data miner must face a number of technical challenges, foremost among which is the extremely high-dimensionality and high redundancy of mass spectra.

In this paper we will present a method of feature extraction from mass spectra that retains as much as possible the initial information content of learning examples, extracts biologically meaningful features, reduces the degree of spatial redundancy, and achieves a significant level of dimensionality reduction.

One of the main problems in preprocessing and feature extraction from mass spectra is the appropriate selection of the preprocessing parameters. This is qualitative relying on systematic experimentation and visual inspection of the resulting spectra. There is a strong need for automatic and objective methods of parameter selection, [1]. The problem is that it is not obvious what should be the measure that a given parameter set should optimize. However when the goal is classification and biomarker discovery one obvious measure to optimize is classification accuracy. We propose a solution that tightly couples the preprocessing and feature extraction steps with the learning task and uses the estimated classification performance of the latter to guide the selection of the appropriate set of parameters. In order to do that we exploit a common practice in machine learning in which cross-validation is used for parameter selection, [2].

The paper is organized as follows: section 2 describes the preprocessing that we apply on a mass spectrometry problem; section 3 explains which steps of preprocessing, and how, are related with feature extraction and how to control the dimensionality in the new representation; in section 4 we investigate the degree of dimensionality reduction brought by feature extraction, we perform a series of classification experiments in order to establish its classification performance, and exhibit how we can automatically select the appropriate parameter values for preprocessing and feature extraction; finally we conclude in section 5.

2 Mass Spectra Preprocessing

A mass spectrum of a biological sample is a one-dimensional signal. The x-axis corresponds to the mass, m/z value, of proteins detected in the biological sample and the y-axis to the intensity of these masses, the latter is strongly related to the concentration of the corresponding proteins in the sample. A mass spectrum is a vector whose dimensionality is equal to the number of m/z values recorded by the spectrometer, the value of each dimension is the intensity of the corresponding m/z value. Intensities of neighboring m/z values are highly correlated, resulting in high spatial redundancy among the features of a mass spectrum.

Mass spectra demand considerable effort for preprocessing, which can be roughly divided to: baseline removal, denoising, smoothing, normalization, peak detection and calibration, [1]. We will describe now how we tackled each of them.

The baseline is an offset of the intensities of masses, which happens mainly at low masses, and varies between different spectra. In order for comparisons between intensities of m/z values to be meaningful it has to be subtracted. To compute the baseline we used a local weighted quadratic fitting on the list of the local minima extracted from the spectrum. On the new fitted values of local minima a new search for local minima was performed, the first fitting smooths out small variations. Using the new local minima the signal is split to piecewise constant parts and the final baseline is simply computed by the reapplication of the initial local weighted quadratic fitting, on the piecewise constant signal.

To denoise and smooth the signal we used wavelet decomposition coupled with a median filter; a detailed description is given in section 3. Signal intensities are normalized, to be less dependent on experimental conditions, via total ion current which is equivalent to normalizing with the L_1 norm of the spectrum.

Peak detection is the detection of local maxima in the mass spectrum. A peak collectively represents all the m/z values that define it, that is: starting from its left closest local minimum and moving to its right closest local minimum. The intensities of all these neighboring m/z values exhibit a high level of redundancy, thus by representing a spectrum only via its peaks we considerably reduce the level of spatial redundancy. Peak calibration establishes which peaks among different spectra correspond to the same peak, i.e. the same protein. We used the approach followed in [3] which is essentially complete linkage hierarchical clustering with some additional domain constraints. The final clusters contain masses from different spectra that correspond to the same peak.

3 Feature Extraction

Feature extraction is the combined effect of all the preprocessing steps. However three of them are central: denoising-smoothing, peak detection and peak calibration. The first step determines how many peaks-features will be preserved in the preprocessed spectrum, it thus, indirectly, determines the dimensionality of the finally constructed feature space (later in this section we will see how this is achieved). The two latter steps are the actual steps of feature extraction. Clusters established in the peak calibration step become the extracted features. The feature value of a learning instance in the new representation for the feature that is associated with a given cluster will be simply the intensity in the preprocessed mass spectrum at the mass value that is associated with that cluster.

Wavelets are very popular in signal processing because they are able to analyze both local and global behavior of functions, a phenomenon dubbed time and frequency localization [4]. Classical tools like the Fourier transform are of global nature and assume stationary signals, which is not the case for the signals found in mass spectra. This is why wavelets have recently received attention as a tool for preprocessing mass spectra [1]. Wavelet decomposition reconstructs a signal as a linear combination of some basis functions. By thresholding the coefficients of the basis functions we get a denoised version of the signal.

We work with the decimated wavelet transform and perform the wavelet decomposition using Mallat's pyramid algorithm, [4]. As wavelet basis we have chosen Daubechies, which has been reported previously to have a good performance on mass spectrometry data, [1]. We have opted for hard thresholding and in order to further smooth the produced signal we apply a moving median window filter (window size equals nine points). Smooth behavior is essential for peak detection, if the signal is not smooth enough a very high number of local maxima would be detected that are rather the result of random fluctuations.

To perform the thresholding we create the distribution-histogram of the wavelet coefficients. By specifying a percentile on that distribution, all the coefficients falling within it will be set to zero and the signal will be reconstructed from the remaining coefficients. The threshold of the wavelet coefficients controls the dimensionality of the finally produced feature space after peak detection. Large values of the threshold result in fewer detected peaks and thus lower dimensionality. The question is how to select the wavelet threshold without relying on a visual and qualitative inspection of the resulting signals. We should strive for a careful balance; as values of the wavelet threshold become higher and higher we do not remove only noise but we also start removing a part of the signal that potentially contains valuable discriminatory information.

4 Experimentation

We worked with three different mass spectrometry datasets, one for ovarian cancer [5], (version 8-07-02), one for prostate cancer [6] and an extended version of the early stroke diagnosis dataset used in [3]. A short description of these datasets is given in table 1. They are all two class problems, diseased vs controls.

The learning algorithms used are: a decision tree algorithm J48, with M=2, C=0.25, a one nearest neighbor algorithm, IBL, and a support vector machine algorithm, SMO, with a simple linear kernel and C=0.5. The implementations of the algorithms were those of the WEKA machine learning environment, [7]. The three learning algorithms were chosen so that they represent a diverse set of learning biases. Performance estimation was done using 10-fold stratified cross-validation and controlling for significant differences using McNemar's test with a level of significance of 0.05.

We explored three issues: the degree of dimensionality reduction achieved by the feature extraction mechanism, the amount of discriminatory information preserved by the different levels of preprocessing, namely denoising and feature extraction, and finally how preprocessing could be optimized.

Dimensionality Reduction. We varied the wavelet threshold from 0.5 to 0.95 with a step of 0.05, and from 0.95 to 0.99 with a step of 0.01. The degree of dimensionality reduction ranges from 60% to 95% of the initial number of features in the complete mass spectrum depending on the threshold and the dataset. Due to lack of space we list some of the results in table 2. The dimensionality reduction is done in such a way that it reflects domain knowledge and reduces the spatial redundancy of the initial representation.

Table 1. Description of mass spectrometry datasets considered

dataset	# controls	#diseased	mass range (Daltons)	# features
ovarian	91	162	0-20k	15154
prostate	253	69	0-20k	15154
stroke	101	107	0-70k	28664

Table 2. Feature reduction for different values of the wavelet threshold. For each dataset and wavelet threshold, θ, we give: the number of features after feature extraction (# features), and the percentage of feature reduction (reduction %).

	prostate		ovarian		stroke	
θ	# features	reduction %	#features	reduction %	#features	reduction %
0.5	3779	75.06	1591	89.50	11983	58.19
0.6	3538	76.65	1371	90.95	11294	60.59
0.7	3223	78.73	991	93.46	9780	65.88
0.8	2616	82.73	865	94.29	6954	75.73
0.9	1668	88.99	775	94.88	3154	88.99
0.99	1009	93.34	668	95.59	1255	95.62

Preprocessing and Discrimination. To see whether the various steps of preprocessing preserve the discriminatory information we evaluated the learning algorithms on three different representations of the classification problems: 1) the initial complete mass spectra where we only performed baseline removal and normalization, *bl-tic*, this is a single dataset; 2) the complete mass spectra where baseline removal, normalization, noise removal with different wavelet thresholds, and smoothing are performed, this is a group of datasets collectively identified as *all*, each dataset corresponds to a specific wavelet threshold; 3) the datasets produced after feature extraction, collectively identified as *peaks*. *bl-tic* will provide us with a performance baseline since it contains all the initially available information. Comparing the performance of learning on *all* and *bl-tic* we can see how denoising and smoothing affects the discriminatory content of the learning examples, while comparisons between *peaks* and *bl-tic* allows us to establish the effect of feature extraction on the discriminatory content. The estimated performances (accuracies) are given in figure 1.

A close examination of figure 1 shows that in general the accuracies of the learning algorithms on the *all* and *peaks* representations are similar to the baseline accuracy on *bl-tic*. There is no clear trend associated with the level of denoising, and no systematic difference that would show a clear advantage or disadvantage of denoising-smoothing and feature extraction.

To establish a precise picture of the effect of denoising-smoothing and feature extraction on discrimination content we computed the significance level of the accuracy differences on the *bl-tic* representation and on each of the *all*, and *peaks* representations, i.e., for each value of the wavelet threshold, the results are summarized in table 3 in terms of significant wins and losses.

Table 3. Significant wins and losses table summarized over the different threshold values. A triplet w/t/l for a pair of representations x vs y gives the number of significant wins (w), ties (t), and significant losses for x.

	bl-tic vs *all*			*bl-tic* vs *peaks*		
	SMO	J48	IBL	SMO	J48	IBL
ovarian	0/14/0	0/14/0	1/13/0	0/14/0	8/6/0	0/14/0
prostate	6/8/0	0/14/0	0/14/0	0/14/0	1/11/2	0/14/0
stroke	2/12/0	0/14/0	0/14/0	1/13/0	0/14/0	1/13/0

Denoising and smoothing in general preserve the discriminating information contained within the learning examples, table 3, column *bl-tic* vs *all*. However there are threshold values for which classification accuracy significantly deteriorates compared to the baseline; a fact that calls for an informed way of selecting the appropriate threshold value. A similar picture arises when we examine the performance of feature extraction for the different values of the wavelet threshold, table 3, column *bl-tic* vs *peaks*. In most of the cases feature extraction retains a discriminatory content similar to that of the *bl-tic* representation, nevertheless here also the wrong choice of the wavelet threshold can lead to a significant drop in classification accuracy compared with the baseline accuracy.

The optimal value of the wavelet threshold depends not only on the dataset but also on the learning algorithm used. A good or bad selection of the wavelet

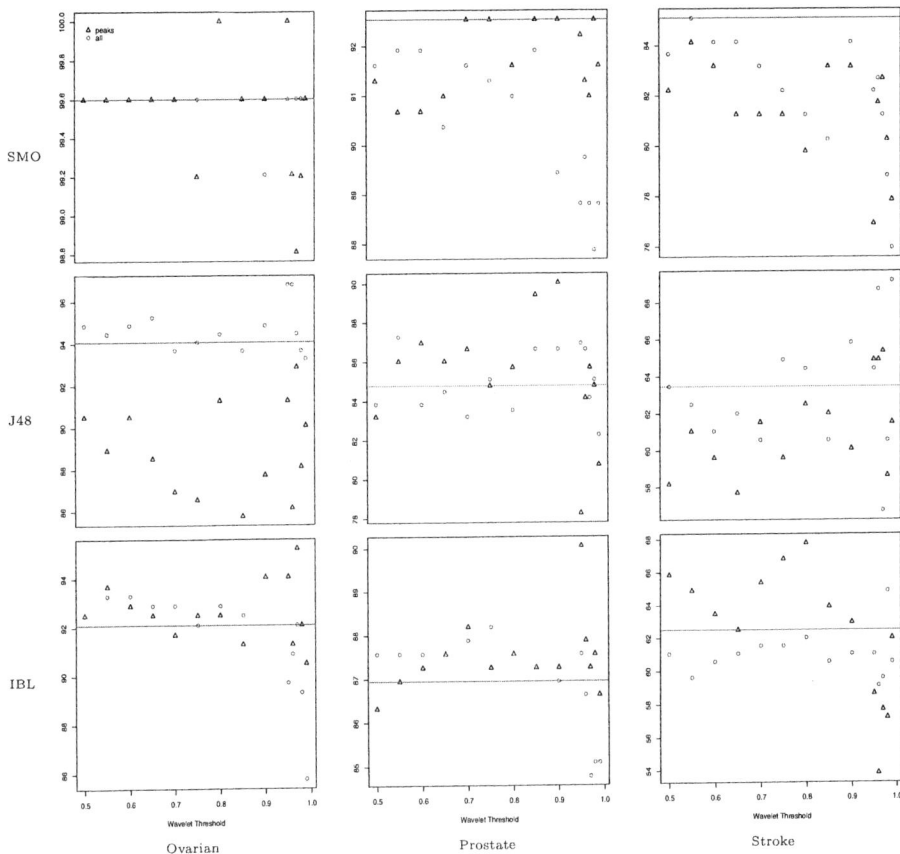

Fig. 1. Accuracy results for each dataset and learning algorithm. Each graph gives the accuracy of a given algorithm for all wavelet thresholds using: the complete spectrum, *all*, only the peaks, *peaks*. The horizontal lines correspond to the accuracy on the *bl-tic* version of the dataset. The y-axis gives the accuracy %, note the differences in scales.

threshold can lead to a performance that is, respectively, significantly better or worse than the baseline performance. Thus all these raise again the issue of the optimal selection of the wavelet threshold in order to optimize feature extraction. Clearly this selection should be guided not only by the dataset but also by the learning algorithm that we are planning to use for classification.

Optimizing Denoising and Feature Extraction. We would like to select that wavelet threshold that would have the highest chances of maximizing classification performance on the testing instances for a given learning algorithm. To achieve that we tightly integrate the whole preprocessing pipeline to the learning process and adopt a commonly used technique for parameter tuning that is based on cross-validation. More precisely the user gives an initial list of interesting parameter values and selects a learning algorithm. Then as a part of the training phase the preprocessing and the learning algorithm are cross validated (ten folds) tightly coupled for each value. The value that gave the highest accuracy is chosen, the training set is preprocessed with the chosen value, the learning algorithm is trained on the preprocessed training set and tested on the test set.

Evaluation was done with ten fold stratified cross validation. The threshold values among which selection is performed are the same as in the previous experiments. For each dataset again two representations were considered: *all* that contains all the features where only denoising and smoothing is performed, and *peaks* where feature extraction was also performed, results are given in table 4.

Denoising with automatic selection of the appropriate wavelet threshold retains the discriminatory information of the learning examples when compared with the *bl-tic* representation. For any learner and any dataset, classification performances on the *all* representation are very similar with those on the *bl-tic* representation, no statistically significant differences are observed (statistical significance data not shown). The same holds when we compare the classification performance of feature extraction with the classification performance on *bl-tic*, performances are similar and no statistical significant difference is observed.

Table 4. Classification accuracy with automatic parameter selection. Automatic parameter selection is only performed for *all* and *peaks*, *bl-tic* is repeated for comparison reasons.

dataset	bl-tic			all			peaks		
	SMO	J48	IBL	SMO	J48	IBL	SMO	J48	IBL
ovarian	99.60	94.07	92.09	99.20	95.25	96.04	99.20	92.49	92.88
prostate	92.54	84.78	86.95	92.23	86.95	86.95	92.54	84.78	90.06
stroke	85.09	63.46	62.50	81.73	65.86	56.73	81.73	62.01	62.01

Overall automatic selection of the wavelet threshold avoids the pitfalls of manual selection. It reduces the chances of selecting a threshold value that would result in a significant deterioration of the classification performance. Equally important it eliminates the need for a visual and qualitative inspection of the results of denoising in order to select the appropriate threshold, thus relieving the analyst from a significant burden, while on the same time it replaces a qualitative approach (visual inspection) with an objective criterion (classification accuracy).

5 Conclusion

Mass spectrometry data are characterized by very high dimensionality and very high levels of redundancy among their features. In this paper we exploited domain knowledge in order to reduce dimensionality and on the same time remove redundancy from the initial representation by feature extraction. One of our central tools was wavelet decomposition. Wavelets have been used before in mass spectrometry [8,9] and in other domains like measuring time series similarities [10]. Nevertheless they all worked with the wavelet coefficients. We return to the initial representation of the signal and extract features from that, thus the extracted features are of direct biological relevance. [1] use wavelets to extract peaks from mass spectra but they did not follow up with classification, moreover they stressed the problem of the appropriate level of denoising and parameter selection. To address these issues we tightly coupled feature extraction with classification and provided an effective and automatic way to control the granularity of feature extraction with a view to maximizing classification accuracy.

We should note here that what we propose is not a feature selection method. Our goal is not to minimize the number of features that can be used to effectively perform classification but to extract a high level, more compact, less redundant and well understood representation of the mass spectra that retains as much as possible the initial discriminatory content of the training examples. The new extracted representation seems, according to the experimental evidence, to retain the discriminatory content of the learning examples. Once the new representation is extracted one may proceed to a typical data analysis scenario where aggressive methods of feature selection could now be used on the new representation.

References

1. Morris, J., Coombes, K., Koomen, J., Baggerly, K., Kobayashi, R.: Feature extraction and quantification for mass spectrometry in biomedical applications using the mean spectrum. Bioinformatics (2005) Advanced publication.
2. Hastie, T., Tibshirani, R., Friedman, J.: The Elements of Statistical Learning. Springer (2001)
3. Prados, J., Kalousis, A., Sanchez, J.C., Allard, L., Carrette, O., Hilario, M.: Mining mass spectra for diagnosis and biomarker discovery of cerebral accidents. Proteomics **4** (2004) 2320–2332
4. Mallat, S.: A wavelet tour of signal processing. Academic Press (1999)
5. Petricoin, E., et al: Use of proteomic patterns in serum to identify ovarian cancer. The Lancet **395** (2002) 572–577
6. Petricoin, E., et al: Serum proteomic patterns for detection of prostate cancer. Journal of the NCI **94** (2002)
7. Witten, I., Frank, E.: Data Mining: Practical Machine Learning Tools and Techniques with Java Implementations. Morgan Kaufmann (1999)
8. Qu, Y., et al: Data reduction using a discrete wavelet transform in discriminant analysis of very high dimensional data. Biometrics **59** (2003) 143–151
9. Lee, K.R., Lin, X., Park, D., Eslava, S.: Megavariate data analysis of mass spectrometric proteomics data using latent variable projection method. Proteomics **3** (2003)
10. Zbigniew R. Struzik, A.S.: The haar wavelet transform in the time series similarity paradigm. In: Principles of Data Mining and Knowledge Discovery, Third European Conference, Springer (1999) 12–22

Numbers in Multi-relational Data Mining

Arno J. Knobbe[1,2] and Eric K.Y. Ho[1]

[1] Kiminkii, Postbus 171, NL-3990 DD, Houten, The Netherlands
{a.knobbe, e.ho}@kiminkii.com
[2] Utrecht University, P.O. box 80 089, NL-3508 TB Utrecht, The Netherlands

Abstract. Numeric data has traditionally received little attention in the field of Multi-Relational Data Mining (MRDM). It is often assumed that numeric data can simply be turned into symbolic data by means of discretisation. However, very few guidelines for successfully applying discretisation in MRDM exist. Furthermore, it is unclear whether the loss of information involved is negligible. In this paper, we consider different alternatives for dealing with numeric data in MRDM. Specifically, we analyse the adequacy of discretisation by performing a number of experiments with different existing discretisation approaches, and comparing the results with a procedure that handles numeric data dynamically. The discretisation procedures considered include an algorithm that is insensitive to the multi-relational structure of the data, and two algorithms that do involve this structure. With the empirical results thus obtained, we shed some light on the applicability of both dynamic and static procedures (discretisation), and give recommendations for when and how they can best be applied.

1 Introduction

Whereas numeric data is at the core of the majority of propositional Data Mining systems, it has been largely overlooked in Multi-Relational Data Mining (MRDM). Most MRDM systems assume that the data is a mixture of symbolic and structural data, and if the source database contains numbers, they will either have to be filtered out or pre-processed into symbolic values. Apart from historical reasons – symbolic representations are popular in the logical roots of MRDM –, the full treatment of numeric data comparable to propositional approaches is mostly ignored for reasons of simplicity and efficiency. MRDM is characterised by large hypothesis spaces, and the inclusion of continuous domains that offer a large range of (very similar) refinements is thought to make MRDM intractable. Most multi-relational systems rely on so-called discretisation procedures to reduce the continuous domains to more manageable symbolic domains of low cardinality, such that the search remains realistic. The resulting loss of precision is assumed to be negligible.

In this paper, we survey a number of existing approaches to dealing with numeric data in MRDM, with the aim of empirically determining the value of each of these approaches. These approaches include a number of pre-processing procedures suggested recently [6, 2], as well as one of the few MRDM algorithms that deal with numbers dynamically, developed by the authors of this paper [2, 4]. The discretisation procedures include a simple algorithm that considers each table in isolation, and discretises each numeric attribute on the basis of the distribution of its values, regardless of any other tables connected to the current table. Two further discretisation procedures do involve the multi-relational structure of the database, and aim at finding good intervals, keeping in mind that the resulting symbolic attributes will be used in the context of the other tables in the database. The algorithm that deals with numbers dynamically does not require any pre-processing of the data. Rather than fixing a number of intervals prior to the analysis, it will consider the numeric data for a hypothesis at hand, and determine thresholds that are optimal for the given context. Especially at deeper levels of the search, where reasonably specific subgroups are considered, relevant thresholds will differ significantly from those determined on the whole dataset.

We test the four approaches experimentally on four well-known multi-relational datasets where numeric attributes play an important role: Mutagenesis (two varieties), Financial and Musk. With these experiments, we aim to shed some light on when and how each approach can best be applied. Furthermore, we hope to get some guidelines for important parameters of the discretisation procedures, such as the coarseness of the discretisation and the choice of representation. The experimental results are compared to those obtained on databases where all numeric information is removed, in order to get a baseline for the procedures that do (to some extent) involve the continuous domains.

2 Foundations

In the class of discrete patterns that we aim at (decision trees, rules, etc.), dealing with numeric data comes down to choosing numeric thresholds that form useful subgroups. Clearly, the distribution of numeric values, and how the target concept depends on this distribution is essential. In propositional data mining, choosing thresholds is fairly straightforward, as there is a one-to-one correspondence between occurring values and individuals. In MRDM however, we are dealing with non-determinate (i.e. one-to-many) relations between tables. In many cases, numeric attributes do not appear in the target table, and multiple values of the attribute are associated with a single structured individual. Whereas in propositional data mining, we can think of the whole database as a 'cloud' of points, in MRDM each individual forms a cloud. The majority of pattern languages in MRDM characterise such individuals by testing for the presence of values that exceed a given threshold. As the following lemma shows, only the largest and smallest values within each individual are relevant to include or exclude an individual on the basis of a single numeric test. Only these values will therefore be candidates for numeric thresholds.

Lemma 1. Let B be a bag of real numbers, and t some real, then
$$\exists v \in B: v \geq t \;\; \textit{iff} \;\; \max(B) \geq t,$$
$$\exists v \in B: v \leq t \;\; \textit{iff} \;\; \min(B) \leq t.$$

Lemma 1 furthermore demonstrates that there is a difference between the set of thresholds appropriate for the $\leq$ and the $\geq$ operator. This means that any procedure that selects thresholds will have to be performed separately for each operator.

Choosing thresholds can roughly be done in two ways: dynamically and statically. A *dynamic* approach (see Section 3) considers the hypothesis at hand, and determines a collection of thresholds on the basis of the information contained in the individuals covered by the hypothesis in question. A *static* approach (see Section 4) on the other hand considers the entire database prior to analysis and determines a collection of thresholds once and for all. Typically these thresholds are then used to pre-process the data, replacing the numeric data with symbolic approximations. We refer to such a pre-processing step as *discretisation*. Clearly, a dynamic approach is preferable from an accuracy standpoint, as optimal thresholds are computed for the situation at hand. On the other hand, dynamic computation of thresholds makes algorithms more complex, and less efficient.

In the context of discretisation, we refer to numeric thresholds as *cut points*. A collection of n-1 cut points splits the continuous domain into n intervals. A group of values falling in a specific interval is referred to as a *bin*.

In MRDM, it makes sense to not just consider the available numeric values in the computation of cut-points, by also the multi-relational structure of the database. In general, a table is connected to other tables by associations, some of which may be non-determinate (a single record in one table corresponds to multiple records in another table). The effect of such associations is thus that records in a table can be divided into *groups*, depending on the relation to records in the associated table. Considering the multi-relational structure in the computation of cut points is hence tantamount to considering the numeric value, as well as the group the value belongs to. In the remainder of this paper, we refer to groups as the sets implied by this multi-relational structure.

3 Dynamic Handling of Numbers

An MRDM algorithm that handles numbers dynamically considers a range of cut points for a given numeric attribute, and determines how each of these tentative cut points influences the quality of a multi-relational hypothesis under consideration. As the optimal cut point depends on the current hypothesis, and many hypotheses are considered by an MRDM algorithm, the set of relevant cut points cannot be determined from the outset. Rather, we will have to consider the subgroup at hand, and query the database for a list of relevant cut points, and associated statistics.

In general, all values for the numeric attribute that occur in the individuals covered by the hypothesis at hand can act as candidate cut points. In theory, this set of values can be quite large, which can make the dynamic generation of cut points very inefficient. The MRDM system Safarii [2, 4] uses an approach that considers only a subset of these values, thus reducing some of the work. It relies on the observation from Lemma 1 that only the extreme values within a bag of numbers are relevant in order to test the presence of values above or below a certain cut point. Safarii uses a database primitive (a predefined query template) called NumericCrossTable [2] that selects the minimum (maximum) value within each individual covered by the current hypothesis, and then groups over these extreme values to produce the desired counts. We thus get a more reasonable number of candidate refinements.

Unfortunately it is still not realistic to continue the search on the basis of each of these refinements. Safarii therefore selects from the reduced set of candidate refinements only the optimal one for further examination. Because the operators $\leq$ and $\geq$ produce two different sets of candidate refinements, we essentially get two refinements per hypothesis and numeric attribute encountered. Note that keeping only the optimal refinements introduces a certain level of greediness into the algorithm.

4 Discretisation

In this section, we briefly outline the three methods for discretising numeric data to be used in our experiments. We refer to [3] for a full description. Conceptually, discretisation entails defining a number of consecutive intervals on the domain of a numeric attribute, and replacing this attribute with a nominal attribute that represents the interval values fall into. The three methods are identical in how numeric attributes are transformed based on the intervals defined. The essential difference between the methods lies in how the cut points between intervals are computed.

The first method presented computes a (user-determined) number of cut points based on the distribution of values of the numeric attribute. It ignores the fact that data in a particular table will generally be considered in the context of that in other tables. The remaining two methods do consider the multi-relational structure of the data, and compute cut points assuming that discretised values will be considered after joining with tables that are directly attached to the table at hand.

Because the numeric data typically appears in tables other than the target table, it is not always straightforward to assign a class (which is related to the target table) to the value. All three methods are therefore class-blind (or *unsupervised*): the methods do not consider a predefined target concept. As a result, the transformed data can be used on a range of class-definitions.

Equal Height Histogram. The first algorithm computes cut points regardless of any multi-relational structure. It simply considers every numeric attribute in every table in turn and replaces it by a nominal attribute that preserves as much of the information in the original attribute as possible. A collection of cut points is computed that produces bins of (approximately) equal size. Such a procedure is known as *equal interval frequency*, or *equal height histogram*, which is the term we will adopt.

Equal Weight Histogram. The second discretisation procedure involves an idea proposed by Van Laer et al. [6]. The algorithm considers not only the distribution of numeric values present, but also the groups they appear in. It is observed that larger groups have a larger impact on the choice of cut points because they have more contributing numeric values. In order to compensate for this, numeric values are weighted with the inverse of the size of the group they belong to. Rather than producing bins of equal size, we now compute cut points to obtain bins of equal weight.

Aggregated Equal Height Histogram. Like the EqualWeight algorithm, the AggregatedEqualHeight algorithm proposed in [2] takes the multi-relational structure of the database into account in the computation of the cut points. The algorithm is centred around the idea that not all values within a group are relevant when inquiring about the presence of numeric values above or below some threshold. As was outlined in Section 2, it suffices to consider the minimum and maximum value within a group. The idea of the AggregatedEqualHeight algorithm is hence to take the minimum value per group and compute an equal height histogram on these values, in order to discretise all values. The process is then repeated for the maximum per group. We thus get two new attributes per numeric attribute.

Representation. In our discussion of the different discretisation procedures, we have assumed that the outcome is a collection of nominal attributes, where each value represents one of the computed intervals. In fact when we produce n nominal values, we do not only lose some amount of precision (which we assume to be minimal), but also the inherent order between intervals. Although the inability to handle ordered domains (numeric or ordinal) is part of our motivation for applying discretisation, we can choose a representation that preserves the order information without having to accommodate for it explicitly. This representation involves n-1 binary attributes per original numeric attribute, one for each cut point. Rather than representing each individual interval, the binary attributes represent overlapping intervals of increasing size. By adding such attributes as conjuncts to the hypothesis through repeated refinements, a range of intervals can be considered. A further advantage of this representation is that the accuracy is less sensitive to the number of intervals as the size of the intervals does not decrease with the number of intervals. An important disadvantage of this representation is the space it requires. Especially with larger numbers of intervals, having n-1 new binary attributes per original attribute can become prohibitive.

In our experiments, we will consider both the nominal and the binary representation, and compare the results to determine the optimal choice. We will refer to the latter representation as *cumulative binary*.

5 Experiments

Although we have multiple approaches to dealing with numeric data to test, we have chosen to apply a single mining algorithm. This allows us to sensibly compare results. The algorithm of choice is the Rule Discovery algorithm contained in the Safarii MRDM package produced by the authors [2, 4]. This algorithm produces a set of independent multi-relational rules. The algorithm includes the dynamic strategy for dealing with numbers described in Section 3. In order to test the discretisation procedures, we have pre-processed the different databases by generating the desired discretised attributes, and removing the original numeric attributes. The different discretisation procedures were implemented in the pre-processing companion to Safarii, known as ProSafarii.

Although a range of evaluation measures and search strategies is available in Safarii, we have opted for rules of high *novelty*, discovered by means of *beam search* (beam width 100, maximum depth 6). A time limit of 30 minutes per experiment was selected. The algorithm offers filtering of rules by means of a computed convex hull in ROC space [2]. The area under the ROC curve gives a good measure of the quality of the discovered rule set, as it is insensitive to copies or redundant combinations of rules. We will use this measure (values between 0.5 and 1) to compare results.

We will test the different algorithms on the following three well-known multi-relational databases:

- **Mutagenesis [5].** A database containing structural descriptions of molecules. We use two varieties, called B2 and B3. B2 contains symbolic and structural information as well as a single numeric attribute describing the charge of each atom. B3 contains two additional attributes on the molecule-level.
- **Financial [7, 2].** A database containing seven tables, describing various activities of customers of a Czech bank.
- **Musk [1].** A database describing 166 continuous features of different conformations molecules may appear in.

In [3] we present a detailed overview of the results obtained. We summarize the main conclusions in the paragraphs below.

Discretisation Procedures. Let us begin by considering how well the discretisation procedures perform. The table below summarises how often each procedure is involved in a win or a tie (no other procedure is superior). Procedures are compared per setting of the number of bins, in order to get comparable results. It turns out that AggregatedEqualHeight is clearly the best choice for Financial and Musk. Surprisingly, the propositional procedure EqualHeight performs quite well on Mutagenesis B2. The results for EqualHeight and EqualWeight on Mutagenesis B3 are virtually identical, which should come as no surprise, as this database contains two powerful attributes in the target table. The multi-relational data is mostly ignored.

In every case, the use of discretised attributes is better than not using the numeric information altogether, although in a few cases the advantage was minimal.

	EqualHeight	EqualWeight	AggregatedEqualHeight
Mutagenesis B2	62.5%	50.0%	37.5%
Mutagenesis B3	75.0%	87.5%	75.0%
Financial	0%	12.5%	87.5%
Musk	0%	25%	75.0%

Discretisation vs. Dynamic Handling. So can the discretisation procedures compete with the dynamic approach to numeric data, or is it always best to use the latter? In the table below, we compare the performance of the collection of discretisation procedures to dynamic handling of numbers. Each row shows in how many of the 3×4×2=24 runs discretisation outperforms the dynamic approach. In the majority of cases, the dynamic approach outperforms the discretisation procedures, as was expected. However, for every database, there are a number of choices of algorithm, representation and number of bins, for which discretisation can compete, or even give slightly better results (see [3] for details).

If the set of cut points considered by the dynamic approach in theory is a superset of that considered by any discretisation procedure, how can we explain the moderate performance of the dynamic algorithm in such cases? The main reason is that the dynamic algorithm is more greedy than the discretisation procedures, because of the way numeric attributes are treated. Of the many refinements made possible by the numeric attribute, only the optimal pattern is kept for future refinements. Therefore, good rules involving two or more numeric conditions may be overlooked. On the other hand, the nominal attributes resulting from discretisation produce a candidate for each occurring value, rather than only the optimal one. Because beam search allows several candidates to be considered, it may occur that sub-optimal initial choices may lead to optimal results in more complex rules.

	discretisation	dynamic
Mutagenesis B2	5	19
Mutagenesis B3	9	15
Financial	0	24
Musk	1	23

Choice of representation. The comparison between the two proposed representations is clear-cut: the cumulative binary representation generally gives the best results (see table below). The few cases where the nominal representation was (slightly) superior can be largely attributed to lower efficiency caused by the larger hypothesis space of the cumulative binary approach.

Although the cumulative binary representation is very desirable from an accuracy point of view, in terms of computing resources and disk space, the cumulative binary approach can become quite impractical, especially with many bins. Particularly in the Musk database, which contains 166 numeric attributes, several limits of the database technology used were encountered.

	nominal	cumulative binary	ties
Mutagenesis B2	3	5	4
Mutagenesis B3	0	9	3
Financial	2	6	4
Musk	5	4	7

Effect of Number of Bins. As has become clear, the number of bins is an important parameter of the discretisation procedures considered. Can we say something about the optimal value for this parameter? It turns out that the answer to this question depends on the choice of representation. Let us consider the cumulative binary representation. The performance roughly increases as more cut points are added (see the diagrams on the next page). This is because extra cut points just add extra opportunities for refinement and thus extra precision. The only exception to this rule is when severe time constraints are present. Because of the larger search space, there may be no time to reach the optimal result. For the nominal representation, there appears to be an optimal number of cut points that depends on specifics of the database in question. Having fewer cut points has a negative effect on the precision, whereas too many cut points results in rules of low support, because each nominal value only represents a small interval. For the Mutagenesis and Musk database, the optimal value is relatively low: between 2 and 4. The optimal value for Financial is less clear.

6 Conclusion

In general, we can say that the dynamic approach to dealing with numbers outperforms discretisation. This should come as no surprise, as the dynamic approach is more precise in choosing the optimal numeric cut points. It is surprising however to observe that in some cases, it is possible to choose parameters and set up the discretisation process such that it is superior. Unfortunately, it is not immediately clear when faced with a new database what choice of algorithm, representation and

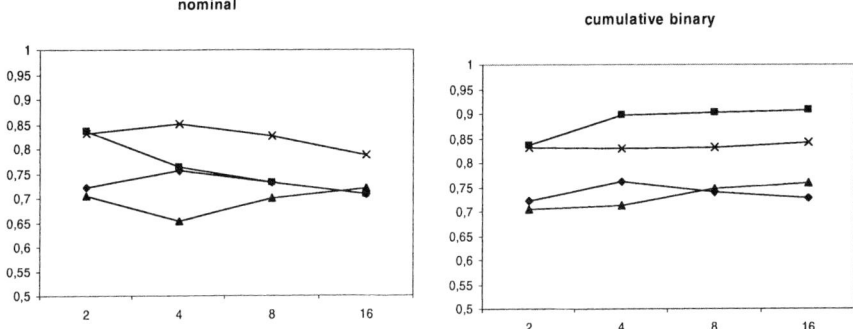

coarseness produces the desired result. Essentially, it is a matter of some experimentation to come up with the right settings. Even then, there is no guarantee that the extra effort of pre-processing the data provides a substantial improvement over the dynamic approach. Of course, when working with a purely symbolic MRDM system, discretisation is mandatory.

For discretisation, we recommend that the AggregatedEqualHeight procedure be tried first, as it has proven to give good results. It is worth the effort to consider EqualHeight as an alternative. The added value of the EqualWeight procedure over EqualHeight is negligible, and can therefore be ignored.

Our experimentation shows that in general, the simple nominal representation commonly used in MRDM projects is sub-optimal. Moreover, this representation is rather sensitive to the selected number of bins. In most cases the cumulative binary representation is preferable. This representation should be applied with as many bins as is realistic, given space and time limitations. Only when time restrictions can be expected to have a detrimental effect on the search depth, should lower numbers be considered.

References

1. Dietterich, T., Lathrop, R., Lozano-Pérez, T. *Solving the multiple-instance problem with axis-parallel rectangles*, Artificial Intelligence, 89(1-2):31-71, 1997
2. Knobbe, A.J. *Multi-Relational Data Mining*, Ph.D. dissertation, 2004, http://www.kiminkii.com/thesis.pdf
3. Knobbe, A.J., Ho, E.K.Y. *Numbers in Multi-Relational Data Mining*, 2005, http://www.kiminkii.com/publications/pkdd2005long.pdf
4. *Safarii, the Multi-Relational Data Mining engine*, Kiminkii, 2005, http://www.kiminkii.com/safarii.html
5. Srinivasan, A., Muggleton, S.H., Sternberg, M.J.E., King, R.D., *Theories for mutagenicity: A study in first-order and feature-based induction*, Artificial Intelligence, 85(1,2), 1996
6. Van Laer, W., De Raedt, L., Džeroski, S., *On multi-class problems and discretization in inductive logic programming*, In Proceedings ISMIS '97, LNAI 1325, Springer-Verlag, 1997
7. Workshop notes on Discovery Challenge PKDD '99, 1999

Testing Theories in Particle Physics Using Maximum Likelihood and Adaptive Bin Allocation

Bruce Knuteson[1] and Ricardo Vilalta[2]

[1] Laboratory for Nuclear Science, Massachusetts Institute of Technology,
77 Massachusetts Ave. Cambridge, MA 02139-4307, USA
knuteson@mit.edu
[2] Department of Computer Science, University of Houston,
4800 Calhoun Rd., Houston TX 77204-3010, USA
vilalta@cs.uh.edu

Abstract. We describe a methodology to assist scientists in quantifying the degree of evidence in favor of a new proposed theory compared to a standard baseline theory. The figure of merit is the log-likelihood ratio of the data given each theory. The novelty of the proposed mechanism lies in the likelihood estimations; the central idea is to adaptively allocate histogram bins that emphasize regions in the variable space where there is a clear difference in the predictions made by the two theories. We describe a software system that computes this figure of merit in the context of particle physics, and describe two examples conducted at the Tevatron Ring at the Fermi National Accelerator Laboratory. Results show how two proposed theories compare to the Standard Model and how the likelihood ratio varies as a function of a physical parameter (e.g., by varying the particle mass).

1 Introduction

Common to many scientific fields is the problem of comparing two or more competing theories based on a set of actual observations. In particle physics, for example, the behavior of Nature at small distance scales is currently well described by the Standard Model. But compelling arguments suggest the presence of new phenomena at distance scales now being experimentally probed, and there exists a long array of proposed extensions to the Standard Model.

The problem of assessing theories against observations can be solved in various ways. Some previous work bearing an artificial intelligence flavor has attempted to use observations to explain processes in both particle physics and astrophysics [4]. From a statistical view, a common solution is to use a maximum-likelihood approach [1,2], that selects the theory T maximizing $P(\mathcal{D}|T)$ (i.e., the conditional probability of a set of actual observations $\mathcal{D}$ assuming T is true). Implicit to this methodology is the–often false–assumption that the form of the distributions characterizing the set of competing theories is known. In practice, a scientist suggests a new theory in the form of new equations or new parameters (e.g., new suggested mass for an elementary particle). In particle physics, a software is then used to simulate the response of the particle detector if the new proposed theory T were true, resulting in a data file made of Monte Carlo events from which one can estimate the true distribution characterizing T. At that point one can compare how close T matches the actual observations (stored in $\mathcal{D}$) obtained from real particle colliders.

To estimate the true distribution of a theory T, we take the Monte Carlo data and follow a histogram approach [5]. We create a series of bins $\{b_k\}$ over the variable space and attempt to predict the number of events expected in every bin b_k if theory T were true. The novelty of our approach lies in the adaptive mechanism behind this bin allocation. Bins are selected to emphasize regions where the number of events predicted by T is significantly different from those predictions generated by competing theories, in a sense discovering regions in the variable space where a discrepancy among theories is evident.

This paper is organized as follows. Section 2 provides background information and notation. Section 3 provides a general description of the mechanism to compute likelihood ratios. Section 4 describes a solution to the problem of adaptive bin allocation. Section 5 reports on the experimental analysis. Lastly, Section 6 gives a summary and discusses future work.

2 Background Information and Notation

In modern particle accelerators, collisions of particles travelling at nearly the speed of light produce debris that is captured by signals from roughly one million channels of readout electronics. We call each collision an *event*. Substantial processing of the recorded signals leads to an identification of the different objects (e.g., electrons ($e^{\pm}$), muons ($\mu^{\pm}$), taus ($\tau^{\pm}$), photons (γ), jets (j), b-jets (b), neutrinos (ν), etc.) that have produced any particular cluster of energy in the detector. Each object is characterized by roughly three variables, corresponding to the three components of the particle's momentum. An event is represented as the composition of many objects, one for each object detected out of the collision. These kinematic variables can be usefully thought of as forming a *variable space*.

We store events recorded from real particle accelerators in a dataset $\mathcal{D} = \{\mathbf{e}_i\}$, where each event $\mathbf{e} = (a_1, a_2, \cdots, a_n) \in A_1 \times A_2 \times \cdots \times A_n$ is a variable vector characterizing the objects identified on a particular collision. We assume numeric variables only (i.e., $a_i \in \Re$) and that $\mathcal{D}$ consists of independently and identically distributed (i.i.d.) events obtained according to a fixed but unknown joint probability distribution in the variable space.

We assume two additional datasets, $\tilde{D}_n$ and $\tilde{D}_s$, made of discrete Monte Carlo events generated by a detector simulator designed to imitate the behavior of a real particle collider. The first dataset assumes the realization of a new proposed theory T_N; the second dataset is generated under the assumption that the Standard Model T_S is true. Events follow the same representation on all three datasets.

3 Overview of Main Algorithm

In this section we provide a general description of our technique. To begin, assume a physicist puts forth an extension to the Standard Model through a new theory T_N. We define our metric of interest as follows:

$$\mathcal{L}(T_N) = \log_{10} \frac{P(\mathcal{D}|T_N)}{P(\mathcal{D}|T_S)} \quad (1)$$

where $\mathcal{D}$ is the set of actual observations obtained from real particle colliders. Metric $\mathcal{L}$ can be conveniently thought of as units of evidence for or against theory T_N. The main challenge behind the computation of $\mathcal{L}$ lies in estimating the likelihoods $P(\mathcal{D}|\cdot)$. We explain each step next.

3.1 Partitioning Events into Final States

Each event (i.e., each particle collision) may result in the production of different objects, and thus it is appropriate to represent events differently. As an example, one class of events may result in the production of an electron; other events may result in the production of a muon. The first step consists of partitioning the set of events into subsets, where each subset comprises events that produced the same type of objects. This partitioning is orthogonal; each event is placed in one and only one subset, also called *final state*. Let m be the number of final states; the partitioning is done on all three datasets: $\mathcal{D} = \{\mathcal{D}_i\}_{i=1}^m$, $\tilde{D}_n = \{\tilde{D}_{ni}\}_{i=1}^m$, and $\tilde{D}_s = \{\tilde{D}_{si}\}_{i=1}^m$. Each particular set of subsets $\{\mathcal{D}_i, \tilde{D}_{ni}, \tilde{D}_{si}\}$ is represented using the same set of variables. Estimations obtained from each set of subsets are later combined into a single figure (Section 3.3).

3.2 Computation of Binned Likelihoods

The second step consists of estimating the likelihoods $P(\mathcal{D}|\cdot)$ adaptively by discovering regions in the variable space where there is a clear difference in the number of Monte Carlo event predictions made by T_N and T_S. Since we treat each subset of events (i.e., each final state) independently (Section 3.1), in this section we assume all calculations refer to a single final state (i.e. a single set of subsets of events $\{\mathcal{D}_i, \tilde{D}_{ni}, \tilde{D}_{si}\}$).

We begin by putting aside for a moment the real-collision dataset $\mathcal{D}_i$. The discrete Monte Carlo events predicted by T_N and T_S in datasets $\tilde{D}_{ni}$ and $\tilde{D}_{si}$ are used to construct smooth probability density estimates $P_i(e|T_N)$ and $P_i(e|T_S)$. Each density estimate assumes a mixture of Gaussian models:

$$P_i(e|T) = P_i^T(e) = \sum_{l=1}^r \alpha_l \phi(e; \mu_l, \Sigma_l) \quad (2)$$

where r is the number of Gaussian models used to characterize the theory T under consideration. The mixing proportions α_l are such that $\sum_l \alpha_l = 1$, and $\phi(\cdot)$ is a multivariate normal density function:

$$\phi(e; \mu, \Sigma) = \frac{1}{(2\pi)^{d/2}|\Sigma|^{1/2}} \exp\left[-\frac{1}{2}(e-\mu)^t \Sigma^{-1}(\mathbf{x}-\mu)\right] \quad (3)$$

where e and μ are d-component vectors, and $|\Sigma|$ and Σ^{-1} are the determinant and inverse of the covariance matrix.

At this point we could follow the traditional approach to Maximum Likelihood estimation by using the real-collision dataset $\mathcal{D}_i$ and the above probability density estimates:

$$P(\mathcal{D}_i|T) = \prod_j P_i(\mathbf{e}_j|T) = \prod_j P_i^T(\mathbf{e}_j) \quad (4)$$

where T takes on T_N or T_S and the index j goes along the events in $\mathcal{D}_i$.

The densities $P_i(\mathbf{e}|T)$ can in principle be used to compute an unbinned likelihood ratio. But in practice, this ratio can suffer from systematic dependence on the details of the smoothing procedure. Over-smoothed densities cause a bias in favor of distributions with narrow Gaussians, while the use of under-smoothed densities cause undesired dependence on small data irregularities. The calculation of a binned likelihood ratio in the resulting discriminant reduces the dependence on the smoothing procedure, and has the additional advantage that it can be used directly to highlight regions in the variable space where predictions from the two competing theories T_N and T_S differ significantly. We thus propose to follow a histogram technique [5] as follows.

Constructing a Binned Histogram

We begin by defining the following discriminant function:

$$D(\mathbf{e}) = \frac{P_i(\mathbf{e}|T_N)}{P_i(\mathbf{e}|T_N) + P_i(\mathbf{e}|T_S)} \quad (5)$$

The discriminant function D takes on values between zero and unity, approaching zero in regions in which the number of events predicted by the Standard Model T_S greatly exceeds the number of events predicted by the new proposed theory T_N, and approaching unity in regions in which the number of events predicted by T_N greatly exceeds the number of events predicted by T_S. We employ function D for efficiency reasons: it captures how the predictions of T_N and T_S vary in a single dimension.

We use D to adaptively construct a binned histogram. We compute the value of the discriminant D at the position of each Monte Carlo event predicted by T_N (i.e., every event contained in $\tilde{D}_n$) and T_S (i.e. every event contained in $\tilde{D}_s$). The resulting distributions in D are then divided into a set of bins that maximize an optimization function. This is where our adaptive bin allocation strategy technique is invoked (explained in detail in Section 4). The result is a set of bins that best differentiate the predictions made by T_N and T_S. The output of the Adaptive-Bin-Allocation algorithm is an estimation of the conditional probability $P(\mathcal{D}_i|T)$.

As an illustration, Figure 1 (left) shows the resulting binned histogram in D for a real scenario with a final state e^+e^- (i.e., electron and positron). The Adaptive-Bin-Allocation algorithm chooses to consider only two bins, placing a bin edge at D = 0.4. Note events from T_S (line L2) tend to lie at values for which D(e) is small, and events from T_N (line L3) tend to lie at values for which D(e) is large.

Figure 1 (right) shows how the two bins in the discriminant map back onto the original variable space defined on $m_{e^+e^-}$ (the invariant mass of the electron positron pair), and positron pseudorapidity. The dark region corresponds to points e in the variable space for which D(e) < 0.4; similarly the light region corresponds to points e

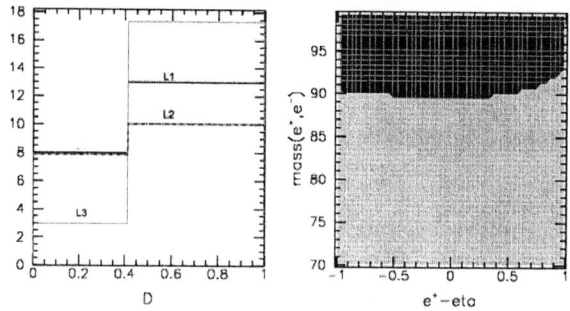

Fig. 1. (Left) The optimally-binned histogram of the discriminant D for the predictions of T_S (line L2), T_N (line L3), and real data $\mathcal{D}$ (line L1). (Right) The mapping of the bins in D back into regions in the original variable space. The dark region corresponds to points e in the variable space for which $D(\mathbf{e}) < \theta$; the light region corresponds to points e in the variable space for which $D(\mathbf{e}) > \theta$ ($\theta = 0.4$).

for which $D(\mathbf{e}) > 0.4$. Each region is assigned a binned probability (Section 4); all probabilities are then combined into a final state probability $P(\mathcal{D}_i|T)$.

3.3 Combining Likelihoods and Incorporating Systematic Errors

Once we come up with an estimation of $P(\mathcal{D}_i|T)$, the next step consists of combining all probabilities from individual final states into a single probability for the entire experiment through the product $P(\mathcal{D}|T) = \prod_i P(\mathcal{D}_i|T)$, where T takes on T_N or T_S and the index i goes along all final states. As a side note, a single particle accelerator has normally several experiments running that can also be combined through such products.

Finally, systematic uncertainties are introduced into the analysis to reflect possible imperfections in the modelling of the response of the physical detector. There are usually roughly one dozen sources of systematic error, ranging from possible systematic bias in the measurements of particle energies to an uncertainty in the total amount of data collected.

4 Adaptive Bin Allocation

We now explain in detail our approach to estimate the likelihood $P(\mathcal{D}_i|T)$ for a particular final state. To begin, assume we have already computed the value of the discriminant D at the position of each Monte Carlo event predicted by T_N and T_S (Section 3.2), and decided on a particular form of binning that partitions D into a set of bins $\{b_k\}$. Let $\mu_{k|T}$ be the number of events expected in bin k if theory T is true[1]. Often in the physical sciences the distribution of counts in each bin is Poisson; this is assumed in what follows. The probability of observing λ_k events in a particular bin k is defined as:

---

[1] Recall T is either the new theory T_N or the Standard Model T_S.

$$P(\lambda_k|T) = \frac{e^{-\mu_k|T}\mu_k|T^{\lambda_k}}{\lambda_k!} \quad (6)$$

Now, the probability of observing the real data D_i assuming the correctness of T and neglecting correlated uncertainties among the predictions of T in each bin, is simply:

$$P(D_i|T) = \prod_k P(\lambda_k|T) \quad (7)$$

where the index k runs along the bins and λ_k is the number of events observed in the real data D_i within bin k.

The question we now pose is how should the bins be chosen? Many finely spaced bins allow finer sampling of differences between T_N and T_S, but introduce a larger uncertainty in the prediction within each bin (i.e., the difference in the events predicted by T_N and T_S under finely spaced bin comes with low confidence levels). On the other hand, a few coarsely spaced bins allow only coarse sampling of the distributions predicted by T_N and T_S, but the predictions within each bin are more robust. The question at hand is not only how many bins to use, but also where to place their edges along the discriminant D [3].

4.1 Searching the Space of Binnings

In selecting an optimal binning we focus our analysis on the two theories T_N and T_S exclusively (choosing a set of optimal bins is independent of the real data used for theory validation). Our goal is to produce a set of bins $\{b_k\}$ that maximize the difference in predictions between the two theories. We start by defining an optimization function over the space of binnings. We merit partitions that enhance the expected evidence in favor of T_N, $\mathcal{E}(T_N)$, if T_N is correct, plus the expected evidence in favor of T_S, $\mathcal{E}(T_S)$, if T_S is correct. Given a particular set of bins, $\{b_k\}_{k=1}^v$, the proposed optimization function is defined as follows:

$$\mathcal{O}(\{b_k\}) = \mathcal{E}(T_N, \{b_k\}) + \mathcal{E}(T_S, \{b_k\}) \quad (8)$$

The evidence for each theory is as follows:

$$\mathcal{E}(T_N, \{b_k\}) = \sum_{\lambda_1}\sum_{\lambda_2}\cdots\sum_{\lambda_v} \left(\prod_k P(\lambda_k|T_N)\right) \times \log_{10}\left(\frac{\prod_k P(\lambda_k|T_N)}{\prod_k P(\lambda_k|T_S)}\right) \quad (9)$$

and similarly for $\mathcal{E}(T_S, \{b_k\})$. Each summation on the left varies over the range $[0, \infty]$. The evidence for each theory has a straightforward interpretation. Recall that $\prod_k P(\lambda_k|T) = P(D_i|T)$ and therefore each evidence $\mathcal{E}$ is the relative entropy of the data likelihoods (if $\log_{10}$ is replaced with $\log_2$), averaged over all possible outcomes on the number of real events observed on each bin. The two components in equation 8 are necessary because relative entropy is not symmetric. The representation for $\mathcal{O}$ can be simplified as follows:

Algorithm 1: Adaptive-Bin-Allocation
Input: D, $\tilde{D}_{ni}, \tilde{D}_{si}$
Output: Set of bins $\{b_k\}$
ALLOCATE-BINS(D,$\tilde{D}_{ni},\tilde{D}_{si}$)
(1)   Evaluate D at each discrete Monte Carlo event in $\tilde{D}_{ni}$ and $\tilde{D}_{si}$.
(2)   Estimate probability densities $f(\mu_{k|T})$ for $T = T_N$ and $T = T_S$.
(3)   Initialize set of bins $\{b_0\}$, where b_0 covers the entire domain of D.
(4)   **repeat**
(5)   Search for a cut point c over D that maximizes function $\mathcal{O}$.
(6)   Replace the bin b_k where c falls with the two corresponding new bins.
(7)   **until** The value o^* maximizing $\mathcal{O}(\cdot)$ is such that $o^* < \epsilon$
(8)   **end**
(9)   **return** $\{b_k\}$

Fig. 2. Steps to generate a set of bins that maximize the distance between the events predicted by theory T_N and theory T_S

$$\mathcal{O}(\{b_k\}) = \sum_k \sum_{\lambda_k} (P(\lambda_k|T_N) - P(\lambda_k|T_S)) \times (\log_{10} P(\lambda_k|T_N) - \log_{10} P(\lambda_k|T_S)), \tag{10}$$

In practice one cannot evaluate $\mathcal{O}$ by trying all possible combinations in the number of real events observed on each bin. Instead we estimate the expected number of events in bin k if theory T is true, $\mu_{k|T}$, and consider $\pm s$ standard deviations (s is user-defined) around that expectation, which can be quickly evaluated with arbitrary accuracy by explicitly computing the sum for those bins with expectation $\mu_{k|T} \leq 25$ and using a gaussian approximation for those bins with expectation $\mu_{k|T} > 25$.

Although in principle maximizing $\mathcal{O}$ requires optimizing the positions of all bin edges simultaneously, in practice it is convenient to choose the bin edges sequentially. Starting with a single bin encompassing all points, this bin is split into two bins at a location chosen to maximize $\mathcal{O}$. At the next iteration, a new split is made that improves $\mathcal{O}$. The algorithm continues iteratively until further division results in negligible or negative change in $\mathcal{O}$. Figure 2 (Algo. 1) illustrates the mechanism behind the binning technique. The complexity of the algorithm is linear in the size of the input space (i.e., in the size of the two datasets $\tilde{D}_{ni}$ and $\tilde{D}_{si}$).

4.2 Example with Gaussians of Varying Width

To illustrate the mechanism behind the bin-allocation mechanism, assume a scenario with two Gaussian distributions of different widths over a variable x. Figure 3(left) shows the true (but unknown) distributions $f_1(x)$ and $f_2(x)$, where $f_i(x) = \frac{n}{\sqrt{2\pi}\sigma_i} e^{(-(x-\mu)^2/2\sigma_i^2)}$ with $i = \{1, 2\}$ and parameter values $n = 100$, $\mu = 25$, $\sigma_1 = 5$, and $\sigma_2 = 8$. The units on the vertical axis are the number of events expected in the data per unit x. We used one thousand points randomly drawn from $f_1(x)$ and from $f_2(x)$. These points are shown in the histogram in Fig. 3(right), in bins of unit width in x. The algorithm proceeds to find edges sequentially before halting, achieving a final

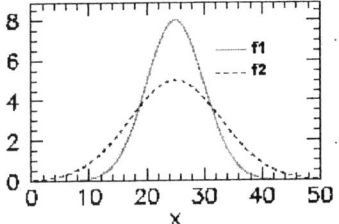
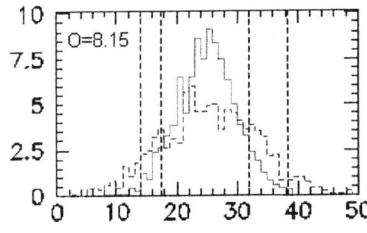

Fig. 3. (Left) Two Gaussian distributions, f_1 and f_2, with same mean but different variance. (Right) The bin-allocation mechanism identifies those regions where f_1 and f_2 cross.

figure of merit. The resulting bins are concentrated in the regions $x \approx 20$ and $x \approx 30$, where $f_1(x)$ and $f_2(x)$ cross.

5 Experiments

We describe two examples conducted at the Tevatron ring at the Fermi National Accelerator Laboratory in Chicago, Illinois. The accelerator collides protons and anti-protons at center of mass energies of 1960 GeV (i.e., giga electron volts). A typical real-collision dataset of this collider is made of about 100 thousand events.

We divide each of the Monte Carlo data sets $\tilde{D}_n$, and $\tilde{D}_s$ into three equal-size subsets. The first subset is used to compute the probability densities $P_i(e|T_N), P_i(e|T_S)$ (Section 3.2); the second subset is used to run the adaptive bin-allocation mechanism (Section 4); the last subset is used to estimate the figure of merit $\mathcal{L}(T_N) = \log_{10} \frac{P(\mathcal{D}|T_N)}{P(\mathcal{D}|T_S)}$ (Section 3). Each experiment produces several hundreds of final states. The running time for each experiment was approximately one hour on a Linux machine with a Pentium 3 processor and 1 GB of memory.

Searching for Leptoquark Pair Production. The first experiment is motivated by a search for leptoquark pair production as a function of assumed leptoquark mass. We show how a theory that advocates leptoquarks with small masses –that if true would result in an abundance of these particles compared to their heavier counterparts– is actually disfavored by real data. Figure 4 (left) shows the log likelihood ratio $\mathcal{L}(T_N)$ (equation 1) for different leptoquark masses. Units on the horizontal axis are GeV. The new proposed theory is disfavored by the data for small mass values, but becomes identical to the Standard Model for large mass values. Figure 4 (second left) shows the posterior distribution $p(M_{LQ}|\mathcal{D})$ obtained from a flat prior and the likelihood on the left.

Searching for a Heavy Z' Particle. The second experiment is similar in spirit to the previous one. Figure 4(third from left) shows a search for a heavy Z' as a function of assumed Z' mass. Z's with small masses, which would be more copiously produced in the Tevatron than their heavier counterparts, are disfavored by the data. The posterior probability $p(m_{Z'}|\mathcal{D})$ flattens out beyond $m_{Z'} \approx 250$ GeV (Figure 4, right), indicating that the data is insufficiently sensitive to provide evidence for or against Z's at this mass.

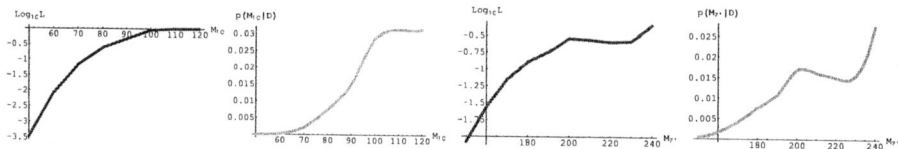

Fig. 4. (left) The log likelihood ratio $\mathcal{L}(T_N)$ (equation 1) for different leptoquark masses. (second left) The posterior distribution $p(M_{LQ}|\mathcal{D})$ obtained from a flat prior and the likelihood on the left. (third left) The log likelihood ratio for different Z' masses. (right) The posterior probability $p(m_{Z'}|\mathcal{D})$ flattens out beyond $m_{Z'} \approx 250$ GeV. Units on the horizontal axis are GeV.

6 Conclusions and Future Work

This paper describes an approach to quantify the degree of evidence in favor of a new proposed theory compared to a standard baseline theory. The mechanism adaptively allocates histogram bins that emphasize regions in the variable space where there is a clear difference in the predictions made by the two theories. The proposed mechanism carries two important benefits: 1) it simplifies substantially the current time needed to assess the value of new theories, and 2) it can be used to assess a family of theories by varying a particular parameter of interest (e.g., particle mass).

We expect the procedure outlined here to have widespread application. The calculation of likelihood ratios is common practice in the physical and social sciences; the main algorithm can be easily adapted to problems stemming from other scientific fields. One barrier lies in generating Monte Carlo data to model a theory distribution. Particle physicists have invested huge amounts of effort in producing a detector simulator designed to imitate the behavior of real particle colliders.

Acknowledgments. This material is based upon work supported by the National Science Foundation under Grants no. IIS-431130 and IIS-448542.

References

1. Duda R. O., Hart P. E., Stork D. G.: Pattern Classification. John Wiley Ed. 2nd Edition (2001).
2. Hastie T., Tibshirani R., Friedman J.: The Elements of Statistical Learning. Springer-Verlag Ed. (2001).
3. Knuteson, Bruce: Systematic Analysis of HEP collider data. Conference on Statistical Problems in Particle Physics, Astrophysics, and Cosmology. Stanford CA.
4. Kocabas S., Langley P.: An Integrated Framework for Extended Discovery in Particle Physics. Proceedings of the 4th International Conference on Discovery Science, pp. 182-195. Springer Verlag. (2001).
5. Scott D.W.: Multivariate Density Estimation: Theory, Practice, and Visualization. Wiley Series in Probability and Statistics Ed. (1992).

Improved Naive Bayes for Extremely Skewed Misclassification Costs

Aleksander Kołcz and Abdur Chowdhury

AOL, Inc., 44900 Prentice Drive, Dulles VA 20166, USA
{arkolcz, cabdur}@aol.com

Abstract. Naive Bayes has been an effective and important classifier in the text categorization domain despite violations of its underlying assumptions. Although quite accurate, it tends to provide poor estimates of the posterior class probabilities, which hampers its application in the cost-sensitive context. The apparent high confidence with which certain errors are made is particularly problematic when misclassification costs are highly skewed, since conservative setting of the decision threshold may greatly decrease the classifier utility. We propose an extension of the Naive Bayes algorithm aiming to discount the confidence with which errors are made. The approach is based on measuring the amount of change to feature distribution necessary to reverse the initial classifier decision and can be implemented efficiently without over-complicating the process of Naive Bayes induction. In experiments with three benchmark document collections, the decision-reversal Naive Bayes is demonstrated to substantially improve over the popular multinomial version of the Naive Bayes algorithm, in some cases performing more than 40% better.

1 Introduction

In certain binary classification problems one is interested in very high precision or very high recall with respect to the target class, especially if the cost of false-positive or false negative misclassifications is disproportionally high. Even though probabilistic cost-sensitive classification frameworks have been proposed, the complicating factor of their successful deployment is uncertainty of precise misclassification costs and the fact that estimation of posterior class probabilities is often inaccurate, especially when dealing with problems involving large numbers of attributes, such as text. As a result, the region within which a classifier can actually benefit the target application may be quite narrow.

In this work, we focus on the problem of extending the utility of the Naive Bayes classifier for problems involving extremely asymmetric misclassification costs. Concentrating on text applications we discuss why certain misclassification errors may be committed with an apparent high confidence and propose an effective method of adjusting the output of Naive Bayes at classification time so as to decrease its overconfidence.

2 Classification with Extremely Asymmetric Misclassification Costs

Let us assume a two-class problem $\{(x, y) : y \in \{0, 1\} \text{ and } x \in \mathcal{X}\}$, where $y = 1$ designates that x belongs to class C (target) and $y = 0$ designates that $x \in \overline{C}$. Assuming no costs associated with making the correct decision, the expected misclassification cost of a classifier F over input domain $\mathcal{X}$ is defined as

$$cost(F) = c_{01} P(F = 0 \wedge x \in C) + c_{10} P(F = 1 \wedge x \in \overline{C})$$

where c_{01} is the cost of misclassifying the target as non-target and c_{10} is the cost of making the opposite mistake. If accurate estimates of $P(C|x)$ are available, the optimum class assignment for input x results from minimizing the expected loss. In problems with highly asymmetric misclassification costs, assigning x to the more expensive class may be preferable even if its posterior probability is quite low. If $c_{10} \gg c_{01}$, the application dictates very low tolerance for *false positives* and an acceptable classifier needs to be close to 100% correct when assigning objects to class C. Conversely, if $c_{01} \gg c_{10}$ then *false-negatives* are highly penalized and an acceptable classifier needs to be characterized by nearly perfect recall in detecting objects belonging to C. Perfect precision in detecting C is equivalent to perfect recall in detecting $\overline{C}$, but while perfect recall is always possible, perfect precision may not be, especially if the target class is also the one with least examples. In this work we will focus on the problem on achieving near-perfect recall with respect to the target class.

3 Sources of Overconfidence in Naive Bayes Classification

3.1 The Multinomial Model

Naive Bayes (NB) is one of the most widely used classifiers, especially in the text domain where it tends to perform quite well, despite the fact that many of its model assumptions are often violated. Several variants of the classifier have been proposed in the literature [1] but in applications involving text, the multinomial model has been found to perform particularly well [2].

Naive Bayesian classifiers impose the assumption of class conditional feature independence which, although rarely valid, has proved to be of surprisingly little significance from the standpoint of classification accuracy [3]. Given input x, NB computes the posterior probability of class C using the Bayes formula $P(C|x) = P(C) \frac{P(x|C)}{P(x)}$. Input x is assigned to the class with the highest expected misclassification cost which, assuming feature independence and when only two classes are present, is determined by the log-odds score:

$$score(x) = const + \sum_i \log \frac{P(x_i|C)}{P(x_i|\overline{C})} \qquad (1)$$

3.2 Overconfidence in Decision Making

It has been recognized that Naive Bayes, while being often surprisingly accurate as a classifier (in terms of the 0/1 loss), tends to be poor when it comes to assessing the confidence of its decisions [4][3]. In particular, the class-probability estimates of NB tend to be clustered near the extreme values of 0 and 1. As shown in [5], this is particularly true in the text domain. When classifying documents with many features, their correlations may compound each other, thus leading to exponential growth in the odds. This effect can intensify in areas only sparsely populated by the training data. Since the log odds in (1) depend on the ratio of class-conditional probabilities, they can be quite high even if the values of the probabilities themselves are very low. But probability estimates for features that were seen relatively rarely in the training data are likely to be more "noisy" than the ones obtained for features with substantial presence. This may result in NB outcomes that appear quite confident even if the neighborhood the test input was only weakly represented in the training set.

Figure 1 illustrates the scatter of NB scores for erroneously classified documents vs. the maximum document frequency (DF) for features contained by these documents. The maximum DF of features in x provides a rough measure of how well the region of containing x was represented by the training data.

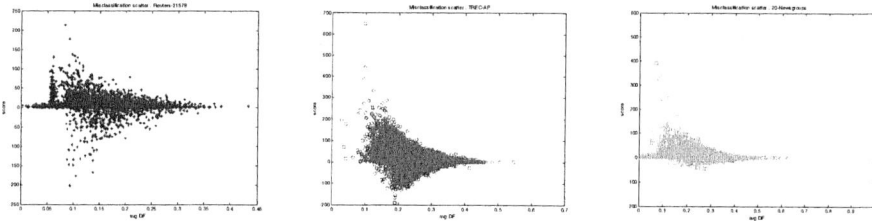

Fig. 1. Scores of Naive Bayes misclassifiications (at default decision threshold) vs. maximum training-set document frequency for features belonging to the misclassified documents (for the collections of: Reuters-21578, 20-Newsgroups and TREC-AP). Misclassifications of documents falling into sparsely populated regions are likely to be made with higher confidence (signified by high absolute score values) than those made for documents for which the training data contained much of related content.

Thus scarcity by itself appears to be a good indicator of overconfidence, although in practice it may be interacting with other factors, such as local class imbalance and document length (e.g., a large number of "noisy" features).

In [6] it was argued that the trust put in the posterior probability estimates of a classifier should decrease with a suitably defined distance between the test input and the training data. In [7] it was suggested that for learners capable of fast incremental learning, the reliability of their posterior estimates can be improved within the framework of transductive learning. Given that a classifier assigns x to class C, it is assumed that a confident decision is one that is little affected by adding x to the training pool of C. With augmented training data, an updated estimate of $P(C|x)$ is obtained, where its difference to the original is used to gauge the sensitivity of the classifier. The techniques of [6] and [7] both rely on modulating the posterior probability estimate of the base classifier with a normalized reliability indicator, which is interpreted as probability, i.e.,

$$\widehat{P}(C|x) = P(C|x) \cdot R(C|x) \qquad (2)$$

where $R(C|x)$ monotonically approaches 1 as the reliability increases.

4 Changing Naive Bayes' Mind: A New Reliability Measure

The log-odds score of NB has the natural geometric interpretation of the projection of the input onto the weight vector normal to the decision hyperplane. On the other hand, the reliability metrics of [6] and [7], while providing a measure of classifier uncertainty, do not offer a similar interpretation of a margin within which a particular classification is made. We propose a novel reliability metric for Naive Bayes, based on the concept of gauging the difficulty of *reversing* the classification outcome of NB for a given input. Our motivation comes from applying Naive Bayes to on-line learning. Unlike discriminative models such as decision trees, generative learners such as NB can be expected to be stable under small adjustments of the training data. Thus, in order for NB to correct itself, a more extensive change to the distribution of the training data may be needed.

To provide a concrete example, let us consider applying a NB classifier to the problem of spam detection, where a user is given a way to correct classifier mistakes by adding a particular email message to the appropriate pool of training data. Take a scenario where arrival of a spam message finds prompts a corrective action. If an "identical" spam appears again the user responds with another training event, and so on until the classifier correctly identifies the message as spam. In this scenario, the confidence of NB in its initial (mistaken) decision can be linked to the number of training events necessary to correct its outcome, which in turn translates to the amount of change to the training distribution needed for decision reversal. Intuitively, decisions that are confident will require more extensive adjustment of the distribution than less confident ones.

Thus, given that the classifier declares that x belongs to class C, we want to ask how much training with x would it take to reverse its opinion. Since the classifier outcome (1) is determined by its score and assuming the decision threshold of 0 and that the perturbation of the training data does not alter class priors, in order to achieve a decision reversal, one needs to satisfy

$$\log P(x|C) - \log P(x|\overline{C}) = \log \widetilde{P}(x|\overline{C}) - \log \widetilde{P}(x|C) - score \qquad (3)$$

where *score* is the original output score, while $\widetilde{P}(x|C)$ and $\widetilde{P}(x|\overline{C})$ denote estimates over the altered training data.

A question arises as how best to measure the effected change to the training distribution. Here we consider the Kullback-Leibler (KL) divergence, i.e.,

$$rdist(x) = KL\left(P\left(x|\overline{C}\right), \widetilde{P}\left(x|\overline{C}\right)\right) = \sum_{x_i} P\left(x_i|\overline{C}\right) \log \frac{P\left(x_i|\overline{C}\right)}{\widetilde{P}\left(x_i|\overline{C}\right)} \quad (4)$$

Once the KL divergence (4) is computed, a straightforward combination method is to scale (see eq. (2)) the original posterior estimate (for the predicted class) with a suitably defined function of the KL divergence, similarly to the approaches taken in [6] and [7]. Here the difficulty lies in an appropriate choice of the normalization function $R(C|x) : rdist(x) \rightarrow [0,1]$, but an additional problem with such an approach in the context of Naive Bayes is that the original posterior estimates produced by the NB are already very close to 1 or very close to 0. Thus the modulation of (4) essentially boils down to substituting $R(C|x)$ for $P(C|x)$[1]. Given that in the case of extreme misclassification costs one is primarily interested in the narrow region where posterior probabilities are close to 1 or 0, the substitution effect may be undesirable since one loses the original degree-of-confidence information. Therefore, we consider directly modulating the raw log-odds score returned by NB, which typically have a much larger dynamic range:

$$\widehat{score}(x) = score(x) \cdot rdist(x) \quad (5)$$

Other score transformations could be considered. In this work we will also use a function KL distance in the form of:

$$\widehat{score}(x) = score(x) \cdot \exp\left(-\gamma \cdot rdist(x)\right) \quad (6)$$

as an alternative to (5).

5 Experimental Setup

In the experiments described below we compare classifiers at the point where they achieve 100% test-set recall for the target class. At this operating setting, a classifier's utility is measured by its *specificity* (true-negative rate), i.e., the fraction of non-target documents that are classified correctly. Arguably, this measure is very sensitive to class noise and in practice one would have to account for such a possibility, e.g., via interactive or automatic data cleansing procedures.

We compared the proposed *decision-reversal* extension to Naive Bayes (labeled as NB-KL) with the following:

- NB: Unmodified multinomial Naive Bayes (baseline).
- NB-Trans: Kukar's transductive reliability estimator [7] (this is the method closest in spirit to the one proposed here).

---

[1] In fact [7] does it directly by substituting the posterior estimate of $P(C|x)$ with $prec \cdot R(C|x)$, where $prec$ refers to the overall precision of the classifier.

Table 1. Steps involved in the decision-reversal Naive Bayes. The most computationally expensive part is step 2, in which one needs to estimate how many corrective events need to take place before the initial decision of the classifier is changed. A naive implementation would keep on generating such events and updating the model, but since in some cases the number of events may be on the order of hundreds or more, this would add significantly to the evaluation time. Instead, we treat the score as a function of the corrective event count a and identify the zero-crossing of score(alpha). In our implementation of the Newton method, usually only 1–7 iterations are needed.

Algorithm

1. Classify input x using a trained NB model.

2. Estimate the multiplicity α with which x needs to be added to the opposite class to achieve decision reversal.

3. Measure the KL divergence (eq.(4)) between the original and the perturbed distribution of features for the class opposite to the one originally predicted.

4. Modulate the original score (eq.(5) or (6)).

Multi-class problems were treated as a series of two-class tasks, with one class serving as the target and the remaining categories ones as the anti-target, i.e., one-against-the-rest. The results obtained by each classifier and for each dataset are reported by macro-averaging the specificity obtained in the constituent two-class tasks.

5.1 Data Sets

We chose three document collections that have often been extensively used in text categorization literature. In each case the collection was split (in the standard way for these collections) into a training set and a test set, which were defined as follows:

- Reuters-21578 (101 categories, 10,724 documents): We used the standard mod_apte split of the data.
- 20 Newsgroups (20 categories, 19,997 documents): A random sample of 2/3 of the dataset was chosen for training with the remaining documents used for testing.
- TREC-AP (20 categories, 209,783 documents): The training/test split described in [?] was used.

Features were extracted by removing markup and punctuation, breaking the documents on whitespace, and converting all characters to lowercase. No stopword removal or stemming was performed. In a modification of the standard bag of words representation, in-document frequencies of terms were ignored.

Table 2. Macro-averaged classification performance (non-target specificity) captured at the point of perfect target recall. The decision-reversal variant of Naive Bayes consistently outperformed the baseline, while the transductive method consistently underperfomed in all three cases.

Dataset	NB	NB-Trans	NB-KL	$\frac{\Delta(\text{NB-KL}-\text{NB})}{\text{NB}}$ [%]
Reuters-21578	0.4743	0.3070	**0.6693**	41
20 Newsgroups	0.4033	0.3297	**0.5379**	33
TREC-AP	0.5004	0.1871	**0.5954**	19

In all two-class experiments, the feature set was reduced to the top 5,000 attributes with the highest values of Mutual Information (MI) between the feature variable and the class variable estimated over the training set.

6 Results

Table 2 shows the results. For all three datasets, NB-KL provided a substantial improvement over the baseline NB. The transductive method [7] generally underperformed the baseline NB. With hindsight, this is perhaps not too surprising. To achieve high specificity at 100% target class recall, one needs to discount errors for the target class where classification is made with an apparently high confidence. In such cases, the probability of a test document belonging to the target class is estimated by NB to be almost one. The transductive step will increase the probability even further, but this is likely to produce only a very small difference between the original and the final class-probability distributions. Thus the original decision made by NB proves in such cases to be quite stable. It appears therefore that the utility of the transductive method may be highest in cases where the apparent confidence of NB decisions is low.

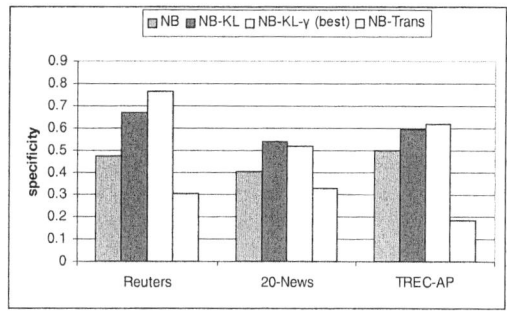

Fig. 2. at the point of perfect target recall. The results for best parameter settings in (6) are compared to the baseline NB, NB-Trans and the default settings of NB-KL. In the case of Reuters-21578 and TREC-AP exponential discounting results in substantial increase in specificity. For 20-Newsgroups, however, the original formulation of NB-KL works better.

To examine the effect of an alternative form of score transformation, we evaluated the performance of NB-KL using the exponential formula (6) with the choice if γ in $[0.001, 50]$. The best optimization results obtained for NB-KL parametrized according to (6) are compared in Figure 2 with the baseline NB, NB-Trans and the default results for NB-KL. In some cases parameter optimization can substantially improve the performance of NB-KL. The parametric formula (6) was unable however to outperform the regular NB-KL in the case of the 20-Newsgroups dataset. The optimum way of incorporating the decision-reversal information may thus need to be investigated further.

7 Conclusions

The decision-reversal NB proved to be effective in increasing Naive Bayes specificity and countering its native overconfidence. Although the original form of the algorithm performed quite well, further improvements were achieved (in 2 out of 3 datasets) by considering an alternative exponential form of discounting the perturbation distance. The dependence of the effectiveness of incorporating the decision reversal information on the form of the discounting function will be the subject of future work. We are also intending to investigate the effects of combining the proposed method of curbing the overconfidence with techniques motivated by explicit reduction of feature interdependence (e.g., as realized by feature selection).

References

1. Lewis, D.D.: Naive (Bayes) at forty: the independence assumption in information retrieval. In: Proceedings of the 10th European Conference on Machine Learning. (1998) 4–15
2. McCallum, A., Nigam, K.: A comparison of event models for Naive Bayes text classification. In: Proceedings of the AAAI-98 Workshop on Learning for Text Categorization. (1998)
3. Domingos, P., Pazzani, M.: On the optimality of the simple bayesian classifier under zero-one loss. Machine Learning **29** (1997) 103–130
4. Webb, G., Pazzani, M.: Adjusted probability naive bayesian induction. In: Proceedings of the 11th Australian Joint Conference on Artificial Intelligence. (1998)
5. Bennett, P.N.: Assessing the calibration of Naive Bayes posterior estimates. Technical Report CMU-CS-00-155, Computer Science Department, School of Computer Science, Carnegie Mellon University (2000)
6. Wu, Y.L., Goh, K.S., Li, B., You, H., Chang, E.Y.: The anatomy of a multimodal information filter. In: Proceedings of the Ninth ACM SIGKDD International Conference on Knowledge Discovery and Data Mining (KDD-2003). (2003) 462–471
7. Kukar, M.: Transductive reliability estimation for medical diagnosis. Artificial Intelligene in Medicine **29** (2003) 81–106
8. Lewis, D.D., Schapire, R.E., Callan, J.P., Papka, R.: Training algorithms for linear text classifiers. In: Proceedings of SIGIR-96, 19th ACM International Conference on Research and Development in Information Retrieval. (1996) 298–306

Clustering and Prediction of Mobile User Routes from Cellular Data

Kari Laasonen

Basic Research Unit, Helsinki Institute for Information Technology,
Department of Computer Science, University of Helsinki
Kari.Laasonen@cs.Helsinki.FI

Abstract. Location-awareness and prediction of future locations is an important problem in pervasive and mobile computing. In cellular systems (e.g., GSM) the serving cell is easily available as an indication of the user location, without any additional hardware or network services. With this location data and other context variables we can determine places that are important to the user, such as work and home. We devise online algorithms that learn routes between important locations and predict the next location when the user is moving. We incrementally build clusters of cell sequences to represent physical routes. Predictions are based on destination probabilities derived from these clusters. Other context variables such as the current time can be integrated into the model. We evaluate the model with real location data, and show that it achieves good prediction accuracy with relatively little memory, making the algorithms suitable for online use in mobile environments.

1 Introduction

Location awareness has a large role in ubiquitous computing. Several applications have been proposed that rely on knowing or predicting the location of the user. In this paper we present an algorithm for predicting user movement with respect to cell-based location data. Such location data consists of a sequence of cells, with no regard to physical locations or topology. With this data the task is to learn, on user's personal mobile device, places that are personally important to that user, and to make predictions about the place the user is moving to. Such predictions are useful in, e.g., a presence service, which makes the whereabouts of the user available to other people. Many other proactive applications, such as early-reminder systems [1,2] and traffic planning [3] become possible if we can anticipate the future location of the user.

This paper works with the conceptual model presented in Laasonen *et al.* [4]. The contribution of the present paper is a novel algorithm for predicting routes. The algorithm analyzes whole paths using clustering techniques, instead of relying on the short path fragments of the earlier paper. This both conserves memory and offers better prediction accuracy. The presented approach also respects users' privacy by doing all processing on the mobile phone.

Most previous work on determining user locations and routes uses GPS coordinate data [1,2,3]. However, GPS can be problematic in urban areas due to signal shadowing. GPS receivers are also nowhere as ubiquitous as mobile phones. Ashbrook and Starner [1] cluster coordinate data to infer locations, but movement times can be used as well [5]. Alternative methods of prediction of future locations include first or second-order Markov models [1], and Bayes classifiers [2,6].

Our data takes the form of a sequence of cell identifiers. An interesting approach to clustering sequences is with probabilistic suffix trees [7]. Such methods unfortunately require too much memory and processing capacity to be useful in mobile phones.

2 Problem Setting

A GSM phone communicates over the air with a base station. In any given location there may be several base stations whose radio signal reaches the phone. The phone chooses one of them, and switches transparently over to a new base station as needed. A *cell* is the area covered by a single base station; when we say the phone is in some cell, we mean that the phone is in the area of the corresponding base station.

In this paper we work with GSM cell data, for a number of reasons. Mobile phones are ubiquitous and cellular networks are present almost everywhere. Since no operators or external service infrastructure are involved, data gathering is easy and inexpensive. On the other hand, cells may overlap, they vary widely in size, and signal shadowing can make cells appear non-contiguous. Finally, a certain physical location does not have one-to-one correspondence to cells because of radio interference, phone network load and various other issues.

The data consists of *cell transitions*. At the lowest level each cell is represented by an opaque numeric identifier (e.g., "*Sonera.3286.15754*"). Our location data is a time-stamped sequence of such identifiers. We can visualize the data by a graph where the vertices are the observed cells, and there is an edge (c_i, c_j) if (and only if) a transition occurred from cell c_i to c_j. A fragment of such a graph is shown in Fig. 1. This graph shows both the author's daily commute from home ("Vuosaari") to work and trips from home to downtown Helsinki. It does not include transitions in the opposite direction. (For illustrative purposes, some of the cells have been named.)

From our earlier work we will be building on the concepts of cell clusters and bases. If overlapping cells have approximately equal signal strength, the phone may hop between cells even when the user is not moving. This oscillation is handled by *clustering* cells with our earlier method [4]. Intuitively, a cell cluster is a group of nearby cells where most transitions happen within the cluster.

A *location* is either a cell cluster or a single cell. Locations are identifiable in the sense that we can reliably detect the user entering and leaving them. Finally, locations that are important to the user are called *bases*. A location is considered to be a base when the time spent there as a portion of the total time the software

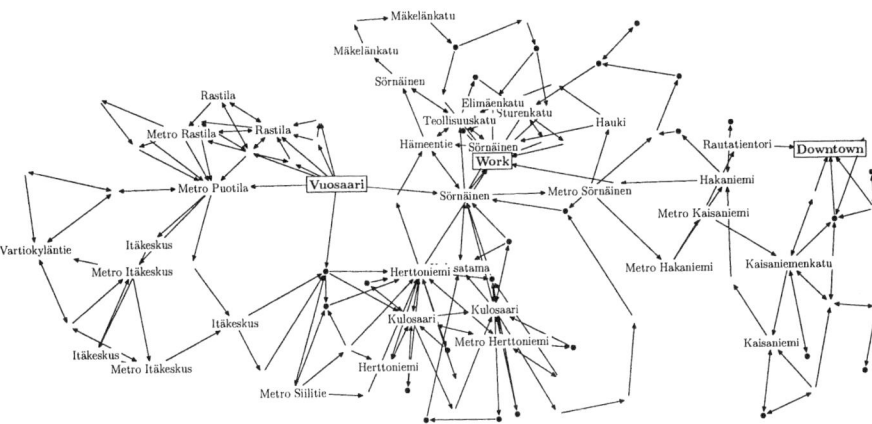

Fig. 1. A partial cell transition graph for routes from "Vuosaari" to either "Work" or "Downtown." Unlabeled dots represent cells that have not been named. The data comes from 69 separate trips.

has been run goes above a certain threshold. The set of bases can change over time, as new places become important or old ones are visited less often. The problem of determining bases is covered in [4]; in this paper we work with a set of known bases. In an actual implementation, learning bases and routes occurs in parallel.

We can now define the problem of *route prediction* as follows: when the user is not in a base, what is the most probable next base? A secondary task is to give some useful characterization of the direction of movement. Furthermore, because the prediction software is run on a mobile phone, there will be tight constraints on the amount of memory and processing power that is available.

Perhaps the most important consequence of using cell-based location data is that we lack the physical topology of the cell network. This includes both the correspondence between cells and physical locations, and all indications of direction. The approach of the present paper is to look at entire routes between two bases, and attempt to learn all different physical routes as strings of cell identifiers. Whenever the user completes a route r between bases a and b, we determine if an existing route between a and b is similar to r. If such a route is found, the two routes are clustered together. Figure 2 shows the effect of applying such route clustering to the data of Fig. 1. There are five different physical routes; the two most frequently traveled are shown in the figure. The graph is obviously much simpler, and furthermore corresponds quite closely to the routes actually traveled in the real world.

Using entire paths makes it also possible to detect *fork points*, which are places where overlapping paths diverge, such as "Sörnäinen" in Fig. 2. When there are several good similarity matches, we can offer a fork prediction as an insurance against the actual base prediction going amiss. From the point of a presence service, a high-confidence prediction of the fork is probably more useful than several low-confidence base predictions.

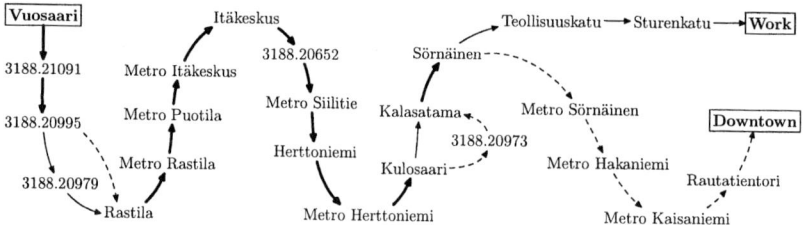

Fig. 2. The most frequent composite routes from "Vuosaari" to "Work" (thin line) or to "Downtown" (dashed line). Edges appearing on both routes are shown with a heavy line. Unnamed cells have numeric identifiers only.

3 Prediction Algorithm

The goal is to predict the next actual base b^*, given that the user's last base was a and since then we have seen a cell sequence $c_1, \ldots, c_k$. When the user is not in any base, at each cell transition we make a prediction, which is a set of pairs (b, p), where b is a possible future base and p the probability of the user going there. When the user arrives at base b^*, the entire route $a, c_1, \ldots, c_n, b^*$ is used to make better subsequent predictions.

3.1 Route Clustering

A *route* is simply a string of cell identifiers. For each pair (a, b) of bases we maintain a set of routes R_{ab}. Instead of storing in R_{ab} all cell paths between a and b, we aim to keep only "typical" paths. Not only does this decrease the memory requirements substantially, but it also proves crucial in estimating the relevance of a given route.

A new route $p = a, c_1, \ldots, c_n, b$ is added to the database when the user arrives at base b, using incremental clustering (Algorithm 1). First p is processed so that only unique cells remain: nearby duplicate cells are collapsed into one. Next (line 3) we determine the similarity of the new path against the existing routes in R_{ab}. If p is similar enough with some route r^*, it is merged with it; otherwise we add p as a new distinct route between a and b.

The similarity function $\text{sim}(r, p)$ tries to approximate the scheme described by Mannila and Moen [8], who use edit distance coupled with item-level similarity. Our version is a heuristic that resembles the Jaccard measure $|r \cap p| / |r \cup p|$, but enforces ordering for the items. That is, strings r and p are considered equivalent if every element in p appears in r in the same order. Elements in r but not in p are ignored. This asymmetry derives from the fact that a route cluster typically contains more cells than there are in any actual instance of that route. (Algorithm for computing $\text{sim}(r, p)$ is omitted due to space constraints.)

The purpose of merging two paths is to produce a composite path that retains the features of both (similar) participants. We first find the optimal alignment of the two path strings (line 4). Computing the alignment of two strings inserts

ADD-ROUTE(p)
Input: Cell sequence $p = a, c_1, \ldots, c_n, b$, routes R_{ab} between a and b
1   Collapse nearby duplicate cells in p
2   $r^* = \operatorname{argmax}\{\operatorname{sim}(r,p) \mid r \in R_{ab}\}$
3   **if** $\operatorname{sim}(r^*, p) > \sigma$
4   **then** $r_1, p_1 \leftarrow \operatorname{align}(r^*, p)$ ▷ Merge p with r^* (see text)
5         $X \leftarrow$ set of letters in $r_1 \cup p_1$
6   **for each** $x \in X$ **do** $v(x) \leftarrow$ average position of x in r_1 and p_1
7   Replace r^* with an ordering of all $x_i \in X$ such that $v(x_i) \leq v(x_{i+1})$
8   **else** $R_{ab} \leftarrow R_{ab} \cup \{p\}$ ▷ Add a new distinct route

Algorithm 1. Clustering routes

empty elements ("spaces") into both strings so that identical elements will appear, as much as possible, in the same position [9]. For example, the alignment of "timers" and "tries" yields "t⎵imers" and "tri⎵e⎵s". Finally, the merging is completed by ordering all cell identifiers in ascending order by average position in the aligned strings (lines 5–7).

3.2 Making Predictions

Predictions are computed by Algorithm 2, using the previous base a and a history h of m most recently encountered cells. We start by finding S, a set of candidate bases. If $b \in S$, a trip $a \to b$ has been observed. For each b, line 3 computes the similarity of the history h against all possible routes leading to b. A simple prediction system would stop here, and predict that the next base b is the one that maximizes s_b. However, several routes can have nearly equal similarities and still lead to different destinations.

PREDICT-BASE(h, a, A, C, R)
Input: Recent history h, previous base a, context A, context model C, routes R
1   $S = \{b \mid R_{ab} \neq \varnothing\}$ ▷ Set of candidate bases
2   **for each** $b \in S$
3   **do** $s_b = \max\{\operatorname{sim}(r, h) \mid r \in R_{ab}\}$
4   Given a and b, find past context data $C_{ab} \in C$
5   Compute $p_b = s_b P(b \mid a, A, C_{ab})$ ▷ See text
6   $b = \operatorname{argmax}_{b \in S} p_b$
7   **return** $(b, p_b / \sum_{k \in S} p_k)$ ▷ Return the prediction and its probability

Algorithm 2. Prediction of the next base b

We can choose between destinations by conditioning on additional context variables, such as time of day, weekday and route frequency. We maintain a context database C that stores information from past instances of trips between pairs of bases. In the most straightforward model we set $C_{ab} = \langle n, T_d(a), T_w(a) \rangle$; this means that for each base pair (a, b) we store n, the number of trips, followed by $T_d(a)$ and $T_w(a)$, distributions of time of day and weekday when the trip started (user left base a). In this case the current context A in algorithm 2 is simply the current time $t = (t_d, t_w)$. We have

$$P(b \mid a, t, C_{ab}) \propto P(b, t \mid a, C_{ab}) = P(t \mid a, b, C_{ab}) P(b \mid a, C_{ab})$$
$$\propto P(t \mid a, b, C_{ab}) \cdot n,$$

by the definition of conditional probability and the chain rule.

The remaining task is to find the probability of being on the given route at time t. A simple assumption is that the time of day t_d of any given route follows a normal distribution, so we need to store in $T_d(a)$ the sum and the square sum of the previous event times. This allows for later reconstruction of the distribution. For the weekday t_w the normality assumption works less well, so the frequency is used instead. Since t_d and t_w are not really independent, we maintain a separate normal distribution for each weekday, and in the end compute the joint probability as $P(t_w, t_d) = P(t_w) P(t_d | t_w)$.

4 Evaluation

The algorithms were evaluated on the real dataset presented in [4]. The data was collected during six months in 2003 with an early version of the ContextPhone software [10] running on a Nokia 7650 phone. The movements of three volunteers were tracked both at work and at leisure. The movement patterns range from very simple (daily commute, weekend and holiday trips) to moderately complex.

The baseline algorithm is the fragment-based method [4], which was tested with several window sizes k. Since the algorithms are intended for small devices, their memory consumption is also investigated. To simplify the evaluation, both algorithms were tested with offline-given bases, i.e., we did not try to learn both bases and routes at the same time. The algorithms received cell transition events one at a time, supplying a ranked set of predictions for the next base. The top-ranked prediction was then compared to the actual base. Following [4, sect. 4.4], we exclude cases when the user is apparently not moving (stationary).

Figure 3 shows how the different methods compare. Each graph shows how the various prediction algorithms performed. The F_2 and F_4 are the fragment method with a window size of 2 and 4, respectively. The symbol C denotes the route prediction algorithm described in Section 3.2. The model C' additionally includes all intermediate cells and their time distributions. Finally, the bar C_3 shows what happens when the algorithm is allowed to learn each route for the first two times it was seen: prediction results for these instances are not included in the score.

A prediction is *correct* if it matches the actual next base and the probability of the given prediction is larger than $u = 0.3$. A *low correct* prediction is one that is correct, but probability is less than u, or the second-best prediction is

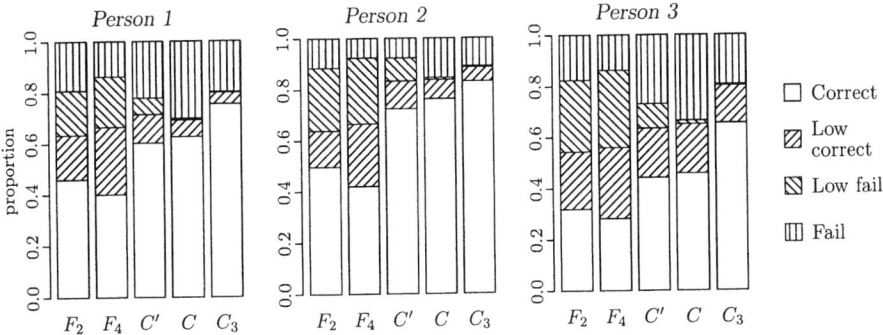

Fig. 3. Route recognition accuracy. Methods F_2 and F_4 are fragment-based. Method C' augments C with intermediate cells; C_3 ignores the first two trips between bases.

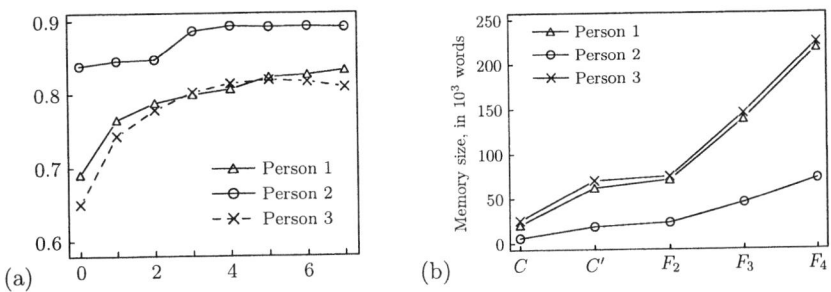

Fig. 4. (a) Prediction accuracy as function of number of learning trips. (b) Comparison of the memory consumption of the algorithms.

correct with nearly equal probability (e.g., $p_1 = 0.55$ and $p_2 = 0.44$), or the fork point was predicted correctly. A *low fail* prediction was wrong, but with a low probability value. Finally, a *fail*-type prediction was a high-confidence prediction that went wrong, or no prediction at all.

For all persons, the route-based method yielded more predictions that succeeded. Taking into account also the low-confidence correct predictions, however, we see more moderate improvements. However, the results so far do not allow for *learning*. As the last (C_3) case of Fig. 3 shows, the prediction quality improves if the algorithm is given time to learn each route. The score is tallied when the pair (a, b) has been seen at least $q = 3$ times. As Fig. 4(a) shows, accuracy improves rapidly with increasing q. The conclusion is that the route-based method is a clear improvement on the fragment method when it comes to prediction accuracy.

Although models C and C' are very similar in their prediction accuracy, the former uses much less memory, as shown in Fig. 4(b). But even model C' consumes less memory than any fragment-based method. For the latter, memory use consists of the fragments themselves and the associated storage for context predictors. The route-based method needs less predictor memory, preferring compact route descriptions.

Because the proposed algorithm is a combination of two separate predictors, it is fairly oblivious to parameter changes. The tests were run with $\sigma = 0.7$ and $m = 12$, which provide a good compromise between quality and efficient use of memory.

5 Conclusion

We have presented a method for predicting user movement from cellular data gathered with user's own mobile phone. The algorithm tackles the problem by attempting to recognize physical routes traveled by the user. The idea is that distinct physical routes correspond to clusters of cell sequences. Later predictions are based on matching the current cell history against known routes. The current time provides additional context to aid prediction. Evaluation of the method with real dataset shows that the method is able to learn and predict routes with good accuracy, while still consuming little memory.

References

1. Ashbrook, D., Starner, T.: Using GPS to learn significant locations and predict movement across multiple users. Personal and Ubiquitous Computing **7** (2003) 275–286
2. Marmasse, N., Schmandt, C.: A user-centered location model. Personal and Ubiquitous Computing **6** (2002) 318–321
3. Harrington, A., Cahill, V.: Route profiling: putting context to work. In: Proceedings of the 2004 ACM symposium on Applied computing (SAC'04), New York, NY, USA, ACM Press (2004) 1567–1573
4. Laasonen, K., Raento, M., Toivonen, H.: Adaptive on-device location recognition. In: Pervasive Computing: Second International Conference. Volume 3001 of LNCS., Springer Verlag (2004) 287–304
5. Kang, J.H., Welbourne, W., Stewart, B., Borriello, G.: Extracting places from traces of locations. In: WMASH'04: Proceedings of the 2nd ACM international workshop on Wireless mobile applications and services on WLAN hotspots, New York, NY, USA, ACM Press (2004) 110–118
6. Patterson, D.J., Liao, L., Fox, D., Kautz, H.: Inferring high-level behavior from low-level sensors. In: UbiComp 2003. Volume 2864 of LNCS., Springer Verlag (2003) 73–89
7. Yang, J., Wang, W.: CLUSEQ: efficient and effective sequence clustering. In: Proceedings of the 19th International Conference on Data Engineering, IEEE Computer Society (2003) 101–112
8. Mannila, H., Moen, P.: Similarity between event types in sequences. In: Data Warehousing and Knowledge Discovery: First International Conference. Volume 1676 of LNCS., Springer Verlag (1999) 271–280
9. Gusfield, D.: Algorithms on strings, trees, and sequences. Cambridge University Press (1997)
10. Raento, M., Oulasvirta, A., Petit, R., Toivonen, H.: ContextPhone: a prototyping platform for context-aware mobile applications. IEEE Pervasive Computing **4** (2005) 51–59

Elastic Partial Matching of Time Series

L.J. Latecki[1], V. Megalooikonomou[1], Q. Wang[1], R. Lakaemper[1],
C.A. Ratanamahatana[2], and E. Keogh[2]

[1] Computer and Information Sciences Dept.,
Temple University, Philadelphia, PA 19122
{latecki, vasilis, qwang, lakaemper}@temple.edu
[2] Computer Science and Engineering Dept.,
University of California, Riverside, CA 92521
{ratana, eamonn}@cs.ucr.edu

Abstract. We consider a problem of elastic matching of time series. We propose an algorithm that automatically determines a subsequence b' of a target time series b that best matches a query series a. In the proposed algorithm we map the problem of the best matching subsequence to the problem of a cheapest path in a DAG (directed acyclic graph). Our experimental results demonstrate that the proposed algorithm outperforms the commonly used Dynamic Time Warping in retrieval accuracy.

1 Motivation

For many datasets we can easily and accurately extract the beginning and ending of patterns of interest. However in some domains it is non-trivial to define the exact beginning and ending of a pattern within a longer sequence. This is a problem because if the endpoints are incorrectly specified they can swamp the distance calculation in otherwise similar objects. For concreteness we will consider an example of just such a domain and show that Minimal Variance Matching (MVM), proposed in this paper, can be expected to outperform Dynamic Time Warping (DTW) and Euclidean distance. There is increasing interest in indexing sports data, both from sports fans who may wish to find particular types of shots or moves, and from coaches who are interested in analyzing their athletes performance over time. Let us consider the high jump. We can automatically collect the athletes center of mass information from video and convert to time series. In Fig. 1, we see 3 time series automatically extracted from 2 athletes.

Both sequence **A** and **B** are from one individual, a tall male, and **C** is from a (relatively) short female with a radically different style. The difference in their technique is obvious even to a non-expert, however **A** and **C** where automatically segmented in such a way that the bounce from the mat is visible, whereas in **B** this bounce was truncated. In Fig. 1(*middle*) we can see that DTW is forced to map this bounce section to the end of sequence **B**, even though that sequence clearly does not have a truly corresponding section. In contrast MVM is free to ignore the sections that do not have a natural correspondence. It is this difference that enables MVM to produce the more natural clustering shown in

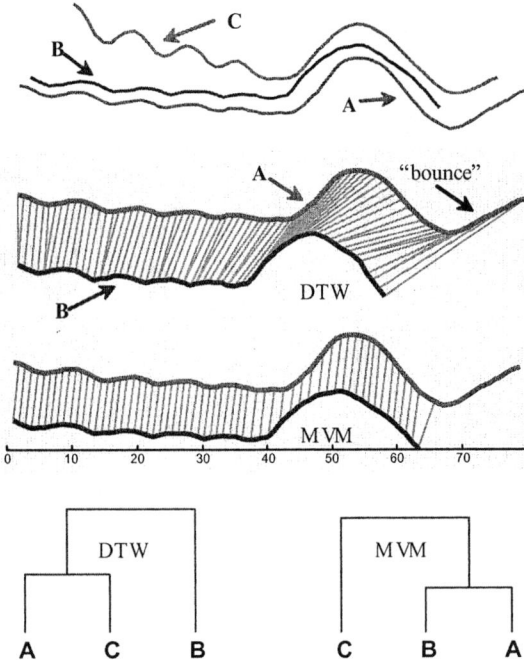

Fig. 1. (*top*) Three examples of athletes trajectories as they attempt a high jump. The sequence shows the height of their center of mass (with possible parallax effects). Reading left to right we can see their bounding run followed by the takeoff and landing. (*middle*) The alignment achieved by DTW and MVM on two of the sequences. (*bottom*) The clustering achieved by DTW and MVM.

Fig. 1(*bottom*). While this is a somewhat contrived example on a specialized domain, similar remarks apply to many commercially important domains including medical data mining and oil exploration.

2 Related Work

Because time series are a ubiquitous and increasingly prevalent type of data, there has been much research effort devoted to time series data mining in recent years. Many data mining algorithms have similarity measurement at their core. Examples include motif discovery [1], anomaly detection [2], rule discovery [3], classification [4] and clustering [5]. In this paper we deal with computation of time series distances based on elastic time series matching.

As many researchers have mentioned in their work [3,4,6], the Euclidean distance is not always the optimal distance measure for similarity searches. For example, in some time series, different parts have different levels of significance in their meaning. Also, the Euclidean distance does not allow shifting in time axis, which is not unusual in real life applications.

To solve the problem of time scaling in time series, Dynamic Time Warping (DTW) [7,8] aligns the time axis prior to the calculation of the distance. DTW distance between time series is the sum of distances of their corresponding elements. Dynamic programming is used to find corresponding elements so that this distance is minimal. The DTW distance has been shown to be superior to the Euclidean in many cases [5,9,10,11]. See [12] for a detailed discussion of DTW. As illustrated in Section 1, DTW requires the matched time series to be well aligned, and it is particularly sensitive to outliers, since it is not able to skip any elements of the target series. DTW always matches the query time series to the whole target time series.

The Longest Common Subsequence (LCSS) measure has been used in time series [13,14] to deal with the alignment and outliers problems. Given a query and a target series, LCSS determines their longest common subsequence, i.e., LCSS finds subsequences of the query and target (of the same length) that best correspond to each other. The distance is based on the ratio between the length of longest common subsequence and the length of the whole sequence. The subsequence does not need to consist of consecutive points, the order of points is not rearranged, and some points can remain unmatched. When LCSS is applied to time series of numeric values, one needs to set a threshold that determines when values of corresponding points are treated as equal [14]. The performance of LCSS heavily depends on correct setting of this threshold, which may be a particularly difficult problem for some applications.

The proposed MVM (Minimal Variance Matching) computes the distance value between two time series directly based on the distances of corresponding elements, just as DTW does, and it allows the query sequence to match to only subsequence of the target sequence, just as LCSS does. The main difference between LCSS and MVM is that LCSS optimizes over the length of the longest common subsequence (which requires the distance threshold), while MVM directly optimizes the sum of distances of corresponding elements (without any distance threshold). The main difference between DTW and MVM is that MVM can skip some elements of the target series when computing the correspondence.

While DTW requires that each point of the query sequence is matched to each element of the target sequence, MVM allows skipping elements of the target sequence. LCSS allows skipping elements of both query and target sequence. Therefore, MVM should be used when one is interested in finding the best matching part of the target sequence for a given query sequence, since it guarantees that the whole query sequence will be matched. This is, for example, the case, when the query is a model sequence, one wants to find in a given data set. However, when the query sequence contains outliers and skipping them is allowed, then LCSS should be used.

3 Minimal Variance Matching

We now present an algorithm for elastic matching of two time series of different lengths m and n, which we will call **Minimal Variance Matching (MVM)**.

More specifically, for two finite sequences of real positive numbers $a = (a_1, ..., a_m)$ and $b = (b_1, ..., b_n)$ with $m < n$, the goal is to find a subsequence b' of b of length m such that a best matches b'. Thus, we want to find the best possible correspondence of sequence a to a subsequence b' of b. Formally we define a **correspondence** as a monotonic injection $f : \{1, ..., m\} \to \{1, ..., n\}$, (i.e., a function f such that $f(i) < f(i+1)$) such that a_i is mapped to $b_{f(i)}$ for all $i \in \{1, ..., m\}$. The set of indices $\{f(1), ..., f(m)\}$ defines the subsequence b' of b. Recall that in the case of DTW, the correspondence is a relation on the set of indices $\{1, ..., m\} \times \{1, ..., n\}$, i.e., a one-to-many and many-to-one mapping.

Once the correspondence is known, it is easy to compute the distance between the two sequences. We do not have any restrictions on distance functions, i.e., any distance function is possible. To allow for comparison to the existing time series matching techniques, we use the Euclidean distance in this paper:

$$d(a, b, f) = \sqrt{\sum_{i=1}^{m} (b_{f(i)} - a_i)^2}. \qquad (1)$$

Our goal is to find a correspondence f so that $d(a, b, f)$ is minimal. More precisely, an optimal correspondence $\hat{f}$ of numbers in series a to numbers in series b is defined as the one that yields the global minimum of $d(a, b, f)$ over all possible correspondences f:

$$\hat{f} = \mathrm{argmin}\{d(a, b, f) : f \text{ is a correspondence}\}. \qquad (2)$$

Finally, the optimal distance is obtained as $d(a, b) = d(a, b, \hat{f})$, i.e., $d(a, b)$ is the global minimum over all possible correspondences.

We can also state the correspondence problem in a statistical framework. Let us assume that there is a subsequence b' of b that is a noisy version of a such that $a \sim b' - \mathcal{N}(0, v)$, where $\mathcal{N}(0, v)$ denotes a zero-mean Gaussian noise variable with variance v, i.e., $b' = (b_{f(i)})_i$ for $i \in \{1, ..., m\}$. Since the mean of the differences $(b_{f(i)} - a_i)_i$ is zero, i.e., $b' - a \sim \mathcal{N}(0, v)$, the variance σ^2 of difference sequence $(b_{f(i)} - a_i)_i$ is given by

$$\sigma^2(a, b, f) = \frac{1}{m} \sum_{i=1}^{m} (b_{f(i)} - a_i)^2. \qquad (3)$$

Clearly, $\sigma^2(a, b, f) = v$ (the variance of the Gaussian noise). Observe that in this case the variance corresponds to the Euclidean distance (1). Thus, the variance of the difference sequence is minimal when mapping f establishes a correct correspondence of elements of both sequences.

Now we describe the method used to minimize (3). We first form the difference matrix

$$r = (r_{ij}) = (b_j - a_i).$$

It is a matrix with m rows and n columns with $m < n$. For example, the difference matrix for two time series $t_1 = (1, 2, 8, 6, 8)$ and $t_2 = (1, 2, 9, 3, 3, 5, 9)$ is shown

$$r = \begin{bmatrix} \boxed{0} & 1 & 8 & 2 & 2 & 4 & 8 \\ -1 & \boxed{0} & 7 & 1 & 1 & 3 & 7 \\ -7 & -6 & \boxed{1} & -5 & -5 & -3 & 1 \\ -5 & -4 & 3 & -3 & -3 & \boxed{-1} & 3 \\ -7 & -6 & 1 & -5 & -5 & -3 & \boxed{1} \end{bmatrix}$$

Fig. 2. In order to compute $\hat{f}$ for t_1=(1, 2, 8, 6, 8) and t_2=(1, 2, 9, 3, 3, 5, 9), we first form the difference matrix with rows corresponding to elements of t_1 and columns to elements of t_2

in Fig. 2. Observe that t_1 and t_2 are similar if we ignore the two elements in t_2 with value 3.

Clearly, (r_{ij}) can be viewed as a surface over a rectangle of size m by n, where the height at point (i, j) is the value r_{ij}. We obtain the correspondence with minimal variance by solving the least-value path problem on the difference matrix. To obtain the solution, we treat (r_{ij}) as a directed graph with the links: r_{ij} is directly linked to r_{kl} if and only if **(1)** $k - i = 1$ and **(2)** $j < l$. When traversing the obtained directed graph, the meaning of both conditions is as follows. For any two consecutive points r_{ij}, r_{kl} in each path (1) means that we always go to the next row, while (2) means that we can skip some columns, but cannot go backwards.

Our goal is to have a least-value path with respect to the following cost function for each directed link: $linkcost(r_{ij}, r_{kl}) = (r_{kl})^2$. Each path can start in first row, between columns 1 and $n - m$, i.e., at r_{1j} for $j = 1, ..., n - m$ and the path can end at r_{mj} for $j = n - m, ..., n$. The conditions (1) and (2) imply that we can obtain a DAG (directed acyclic graph) G whose nodes are the elements of $(r_{ij})_{ij}$ and weights are defined by the function $linkcost$. It is well known that we can solve the least-value path problem using the shortest path algorithm on G. The obtained least-value path defines exactly correspondence $\hat{f}$, which minimizes (3) in accordance with (2).

The shortest path for the example matrix in Fig. 2 is marked with boxes. Following the boxes, the optimal correspondence $\hat{f}$ is given by

$$\hat{f}(1) = 1, \ \hat{f}(2) = 2, \ \hat{f}(3) = 3, \ \hat{f}(4) = 6, \ \hat{f}(5) = 7.$$

Finally, from (1) we obtain the distance $d(t_1, t_2) = \sqrt{3} \approx 1.732$.

The obtained optimal correspondence $\hat{f}$ automatically determines a subsequence $b' = \hat{f}(a)$ of a target time series b that best matches a query series a. In particular, two cases are possible: **Whole Sequence Matching:** Subsequence b' is dense in b, which indicates a similarity of a to b. **Subsequence Matching:** Subsequence b' is not dense in b but is dense in part of b which indicates a similarity of a to part of b.

4 Experimental Results

We compare the results of MVM to the DTW results on three data sets Face, Leaf, and Gun from [12]. A detailed description of these data sets is given in [12]. We briefly mention that Face dataset is composed of 112 sequences representing head profiles of 4 different individuals. The length of each sequence ranges from 107 to 240 points. Leaf dataset is composed of 442 sequences representing contours of six different leaf species. The length of each sequence ranges from 22 to 475 points. Gun dataset is composed of 200 sequences representing gun drawing events by two different actors. The length of each sequence is 150 points.

Following [12], we measure the classification accuracy of 1-NN (Nearest Neighbor) classifier applied to the distance matrices obtained by the evaluated methods. The obtained results are shown in Table 1. As can be seen MVM systematically outperforms DTW. The DTW results are cited from [12], where all possible sizes of warping windows for DTW were examined and the optimal warping window size was determined for each data set. We did not use any warping or correspondence window bound for MVM.

Table 1. 1-NN classification accuracy. The DTW results are cited from [12].

	Face	Gun	Leaf
MVM	98.21	100	97.29
DTW	96.43	99.00	96.38

Although the proposed method does not require any length normalization, we used length normalized time series in order to allow for a comparison to the results in [12]. When calculating the distance between a pair of time series with MVM, we resampled the query series so that its length is approximately 75% of the length of the target series. This means that the total elasticity amount for MVM is about 25% of the length of the second time series. We obtained nearly identical results with elasticity varying from 25% to 50%. The values of each time series were zero-mean normalized in the standard way. That is, each time series X is normalized as: $X = \frac{(X-\mu(X))}{\sigma(X)}$, where $\mu(X)$ is the mean value of X and $\sigma(X)$ is its standard deviation.

The superior performance of MVM reported in Table 1 is due to MVM ability to correctly align matched sequences in that *bad matching* elements of the target sequence are excluded from the correspondence. One example of this fact is given in Section 1. Here we use the Face dataset in order to directly link this fact to the superior classification accuracy of MVM. The face dataset is a particularly good dataset on which to demonstrate this fact. It consists of head profiles converted to time series representing the curvature at sample points. Because the face is intrinsically elastic, as the subject smiles or grimaces, Euclidean distance in unsuitable here, and we therefore would consider a more elastic distance measure such as DTW or MVM.

As one might imagine, the parts of the signal that correspond to the face contain the most useful information, and the parts of the signal that correspond to the back of the head contain much less information. In fact the parts of the signal that correspond to the back of the head may actually contain misleading information, since the texture of hair causes problems for the time series generation algorithm. The problem with DTW is that it is forced to align everything, thus may be forced to align this poor quality data from the back of the head in one signal to poor quality data in another. A small amount of such poor quality data can rapidly swamp the distance calculation. In contrast, MVM has the ability to simply ignore the poor quality data, as shown in Fig. 3. Note that in this simple contrived example we might be able to achieve better results simply by truncating the back of the head section of the signals. However this begs a nontrivial question of finding a good segmentation algorithm. In any case in data mining we generally assume that we do not have such a priori knowledge about the domain in question.

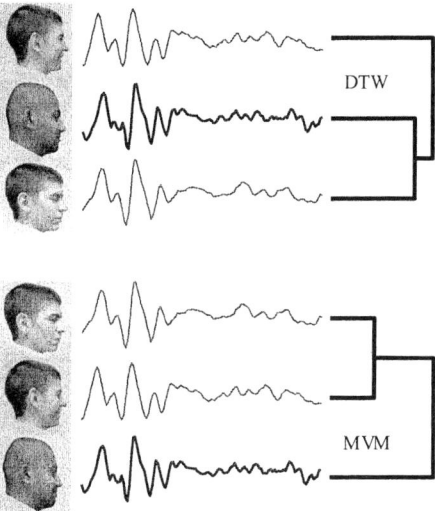

Fig. 3. Three time series from Face dataset, derived from profiles, compared using DTW and MVM. Two time series of the same person are correctly identified by MVM which is not the case for DTW.

5 Conclusions

The proposed new method for time series matching, called MVM, performs the following tasks simultaneously (**1**) automatically determines whether the query sequence best matches the whole target sequence or only part of the target sequence, (**2**) automatically skips outliers that are present in the target sequence, (**3**) minimizes the statistical variance of dissimilarities of corresponding elements. The reported experiments show that this method outperforms DTW.

By mapping the problem of elastic matching of sequences to finding a cheapest path in a DAG, we provide an efficient algorithm to compute MVM.

Acknowledgments

This work was supported in part by NSF under Grant No. IIS-0237921, and by NIH under Grant No. R01MH68066-01A1. We would like to thank the two models that appear in Figure 3, Sandeep Gupta and Paul Dilorenzo. Finally we acknowledge the helpful comments of the reviewers.

References

1. Chiu, B., Keogh, E., Lonardi, S.: Probabilistic discovery of time series motifs. In: Proc. ACM SIGKDD Int. Conf. on Knowledge Discovery and Data Mining, Washington (2003)
2. Keogh, E., Lonardi, S., Ratanamahatana, C.: Towards parameter-free data mining. In: Proc. ACM SIGKDD Int. Conf. on Knowledge Discovery and Data Mining, Seattle (2004)
3. Höppner, F.: Discovery of temporal patterns. learning rules about the qualitative behavior of time series. In: Proc. European Conf. on Principles and Practice of Knowledge Discovery in Databases, Freiburg (2001) 192–203
4. Rafiei, D.: On similarity-based queries for time series data. In: Proc. Int. Conf. on Data Engineering, Sydney (1999) 410–417
5. Aach, J., Church, G.: Aligning gene expression time series with time warping algorithms. Bioinformatics **17** (2001) 495–508
6. Megalooikonomou, V., Wang, Q., Li, G., Faloutsos, C.: A multiresolution symbolic representation of time series. In: Proc. IEEE Int. Conf. on Data Engineering (ICDE05), Tokyo (2005) 668–679
7. Sakoe, H., Chiba, S.: Dynamic programming algorithm optimization for spoken word recognition. IEEE Trans. on Acoustics, Speech, and Signal Processing **26** (1978) 43–49
8. Berndt, D., Clifford, J.: Using dynamic time warping to find patterns in time series. In: Proc. AAAI-94 W. on Knowledge Discovery and Databases. (1994) 229–248
9. Kollios, G., Vlachos, M., Gunopoulos, D.: Discovering similar multidimensional trajectories. In: Proc. Int. Conf. on Data Engineering, San Jose (2002) 673–684
10. Chu, S., Keogh, E., Hart, D., Pazzani, M.: Iterative deepening dynamic time warping for time series. In: Proc. SIAM Int. Conf. on Data Mining. (2002)
11. Yi, B., Jagadish, K., Faloutsos, C.: Efficient retrieval of similar time sequences under time warping. In: Proc. Int. Conf. on Data Engineering. (1998) 23–27
12. Ratanamahatana, C.A., Keogh, E.: Everything you know about dynamic time warping is wrong. In: W. on Mining Temporal and Sequential Data, Seattle (2004)
13. Das, G., Gunopoulos, D., Mannila, H.: Finding similar timie series. In: Proc. 1st PKDD Symposium. (1997) 88–100
14. Vlachos, M., Hadjieleftheriou, M., Gunopoulos, D., Keogh, E.: Indexing multi-dimensional time-series with support for multiple distance measures. In: Proc. of ACM SIGKDD, Washington (2003) 216–225

An Entropy-Based Approach for Generating Multi-dimensional Sequential Patterns

Chang-Hwan Lee

Department of Information and Communications,
DongGuk University,
Seoul, Korea 100-715
chlee@dgu.ac.kr

Abstract. This paper proposes a new method for generating multi-dimensional sequential patterns. While the current sequential pattern methods are generating patterns within a single attribute, the proposed method is able to detect them among different attributes. We employ an information theoretic method for generating multi-dimensional sequential patterns with the use of Hellinger entropy measure. A number of theorems are proposed to reduce the computational complexity of the sequential pattern systems. The proposed method is tested on some synthesized transaction databases.

1 Introduction

Among many techniques in data mining, sequential pattern is a technique which can discover more meaningful information by considering time attribute, together with other traditional attributes. Sequential patterns can be widely used in many different applications, such as mining banking patterns from bank accounts, and predicting certain kind of disease from history of symptoms.

Almost all of the current methods for mining sequential patterns are based on the Apriori algorithm [1], SPIRIT [6], FreeSpan [4], PrefixSpan [5], SPADE [9], CloSpan [8], and TSP [7]. After that, a series of Apriori-like algorithms have been proposed. However, one of the limitations of the current sequential pattern algorithms is that they mine only one dimension. They only consider one attribute, and thus can not detect sequential patterns hiding across different attributes.

On the other hand, multi-dimensional sequential pattern mining attempts to find sequential patterns across several dimensions of attribute. By incorporating the additional attributes, the sequential patterns found are richer and more informative to the user. The aim of this multi-dimensional sequential pattern mining is to get more interesting sequential patterns with different dimensional attributes.

However, there is very little study on mining sequential patterns in multi-dimensional circumstances(e.g. UniSeq [3]).

In this paper, we propose a new paradigm for generating multi-dimensional sequential patterns. We use an entropy function, called Hellinger measure, as an underlying tool for developing multi-dimensional sequential patterns. Therefore this method could provide more theoretic background in sequential pattern generation. Also, we replaced the traditional measures of sequential patterns(like support and confidence) by more sophisticated, information-theoretic measure. The proposed method could calculate the significance of each sequential pattern(called H measure) as a numeric value, and those sequential patterns are given in a sorted order. The H measure can be interpreted as the importance or significance of sequential patterns.

2 Problem Description

The format of sequential patterns generated in this paper is as follows:

$$A = a \wedge B = b \wedge \cdots \rightarrow T = t \quad \text{with} \quad \alpha, \ \beta, \text{ and } H$$

where A, B and T are attributes with a, b and t being values in their respective discrete alphabets. We restrict the right-hand expression to being a single value assignment expression while the left-hand side may be a conjunction of such expressions. The semantics of above format is that if a person does an action(e.g., purchase) based on the condition(left-hand side) of above pattern at a given time, then he/she will later do an action described in right-hand side with high possibility H. Each sequential pattern comes with three numeric values such as α, β, and H. The α, β, and H represent the *information content*, the *generality*, and the *significance* of sequential pattern, respectively. The interpretation of these numeric terms will be explained in the following section. The final sequential patterns generated from the database are sorted based on the H value.

Since our sequential pattern method handles multi-dimensional databases, the format of database is different from the format used by traditional sequential pattern methods. Each transaction of database is associated with different attributes for multi-dimensional sequential patterns mining.

The transaction database is in its first normal form(each attribute, including the items, contains only one value). The database consists of a set of tuples

$$< cid, \ tid, \ a_1, \ a_2, \ \ldots, \ a_n, \ c >$$

where cid is an identification of the customer and tid the time. Let $a_1, a_2, \ldots, a_n$ denote the multi-dimensional attributes with respect to the customer, product, or transaction, and c means the item bought by the customer cid. In case multiple items are purchased together, each of them is represented in different tuples with the same cid and tid. In addition, the entire transaction database is sorted based primarily on customer-id(cid) and secondly on transaction-time(tid).

Information Contents of Sequential Patterns

The basic idea of sequential pattern generation in this paper starts with the assumption that the value assignments in the left hand side of each sequential pattern affects the probability distribution of the right-hand side(target attribute).

Intuitively speaking, if a certain value assignment has significantly changed the probability distribution of the target, it is clear that the given value assignment plays an important role in determining the class values of the target attribute. Therefore, it is a natural definition, in this paper, that the significance of a sequential pattern is interpreted as the degree of dissimilarity between a priori probability distribution and a posteriori probability distribution of the target attribute.

In this paper, this dissimilarity is defined as instantaneous information, which is the information content of the sequential pattern given that the left-hand side happens. The critical part now is how to define or select a proper measure which can correctly measure the instantaneous information.

We employ an entropy function, called Hellinger measure, as a tool for defining the information content of sequential pattern rules. The Hellinger measure was originally introduced by Beran [2], and is defined as

$$\sqrt{\sum_i \left(\sqrt{p(t_i)} - \sqrt{p(t_i|a)}\right)^2} \qquad (1)$$

where t_i denotes the value of attribute T. It becomes zero if and only if both a priori and a posteriori distributions are identical, and ranges from 0 to 1. Unlike other information measures, this measure is applicable to every possible case of probability distributions. It can be interpreted as a distance measure where distance corresponds to the amount of divergence between a priori and a posteriori distribution. Therefore, we employ Hellinger measure as a measure of divergence, which will be used as the information amount of sequential patterns.

3 Contents of *H* Measure

In terms of the probabilistic sequential pattern rules, let us interpret the event $A = a$ as the target concept to be learned and the event(possibly conjunctive) $B = b$ as the hypothesis describing this concept. The *information content* of the sequential pattern rule is defined as

$$\left[\sqrt{P(a|b)} - \sqrt{P(a)}\right]^2 + \left[\sqrt{1 - P(a|b)} - \sqrt{1 - P(a)}\right]^2 \qquad (2)$$

where $P(a|b)$ means the conditional probability of $A = a$ under the condition $B = b$ has happened beforehand. Notice that Equation (2) has a different form of definition from that of Equation (1). In sequential pattern generation, one particular value of class attribute appears in the right hand side of the pattern, and thus the probabilities for all other values are included in $1 - P(a)$. In addition, we squared the original form of Hellinger measure because, by squaring the original form of Hellinger measure, we could derive a boundary of the Hellinger measure, which allows us to reduce drastically the search space of possible sequential pattern rules.

Another criteria we have to consider is the *generality* of the sequential patterns. The generality is similar to the *support* in Apriori-like methods. The basic idea behind generality is that the more often left-hand side occurs for a sequential pattern, the more useful the pattern becomes. In this paper, we use $\sqrt{P(b)}$ to represent the probability that the sequential pattern will occur and, as such, can be interpreted as the measure of sequential pattern generality. The reason for using the square root form of the original probability is that, by using the square root form, we could derive some boundaries of H measure.

As a result, by multiplying the generality with the information content of the sequential pattern rules, we have the following term

$$\sqrt{P(b)}[\left(\sqrt{P(a|b)} - \sqrt{P(a)}\right)^2 + \left(\sqrt{1-P(a|b)} - \sqrt{1-P(a)}\right)^2]$$

which possesses a direct interpretation as a multiplicative measure of the generality and information content of a given sequential pattern rule. In this paper, we call above multiplicative term as H *measure* of sequential patterns.

4 Sequential Pattern Generation

We will now define the algorithm and discuss its basic ideas. The algorithm takes time-related database in the form of discrete attribute vectors and generates a set of K sequential patterns, where K is a user-defined parameter. The set of generated sequential patterns are the K most informative(significant) sequential patterns from the database as defined by the H measure.

The algorithm employs branch-and-bound with depth-first search over possible left-hand sides. The algorithm first generates all possible cases of first-order sequential patterns. The first-order sequential patterns are sequential patterns that have single value assignment in left-hand side, described as follows,

$$B_i = b_{ij} \rightarrow A = a_k$$

where B_i, B_{ij}, and A_k represent i-th attribute, the value of i-th attribute, and the target value, respectively.

The algorithm proceeds then calculating the H measures of each first-order sequential patterns, finding K most informative sequential patterns in terms of H measure, and then placing these K sequential patterns in an ordered list, called BEST. The smallest H measure, that of the Kth element of BEST, is then defined as the running minimum H_*. The critical part of the algorithm is the specialization criterion since it determines how much of the exponentially large hypothesis space actually needs to be explored by the algorithm.

From that point onwards, new patterns which are candidates for inclusion in the sequential pattern set have their H measure compared with H_*. If greater than H_*, they are inserted in the list and the Kth sequential pattern is deleted. And H_* is updated with the value of the H measure of whatever sequential pattern is now Kth on the list. The algorithm systematically tries to specialize all first-order sequential patterns and terminates when it has determined that no more sequential patterns exist which can be specialized to achieve a higher H measure than H_*.

Figure 1 describes the pseudo code of decision whether to continue specializing or to back-up on the depth-first search. The H measure of each sequential pattern can be considered as the weight of the sequential pattern.

> **if** success rate of $H_g \neq 1$
> **then**
>   calculate the value of H_s using Theorem 1
>   **if** $H_s \leq H_*$ **then** cease to specialize; /* Theorem 1 */
> **else**
>   cease to specialize; /* Theorem 2 */

Fig. 1. Algorithm for specialization

5 Characteristics of H Measure

The characteristic of the specialization behavior is critical to the performance of the algorithm. Therefore, it is important to derive some quantitative bounds on the nature of specialization, which can be used to improve computational performance.

Specialization is the process by which we try to increase a sequential pattern's information content by adding an extra condition to the pattern's left-hand side. The consequent necessary decrease in generality of the sequential pattern should be less than an increase in the information content to the extent that the overall H measure is increased. We will examine specialization, using the H measure as the definition of sequential pattern goodness, with $\sqrt{p(a)}$ corresponding to generality and Equation (2) corresponding to information content. If we define H_s and H_g as the H measures of the specialized and general sequential patterns, respectively, is it possible to find a bound of H_s in terms of H_g ?

Suppose we have a sequential pattern

$$B = b \rightarrow A = a \ . \tag{3}$$

We would like to specialize this sequential pattern by adding a condition $C = c$ so that we have a specialized sequential pattern

$$B = b \wedge C = c \rightarrow A = a \ . \tag{4}$$

For the sake of illustration, sequential patterns in formula (3) and (4) are denoted as R_g and R_s, respectively. In this paper, we deal with a sequential pattern which contains only one condition and try to specialize it. More general cases which have more than one condition in the left hand side can be easily understood. Suppose H_g and H_s are the H measures of the sequential patterns R_g and R_s, respectively. Our goal is to answer the question "Can we describe the bound of H_s in terms of H_g ?" In other words, is it possible to estimate the maximum value of H_s without knowing any information about attribute C ? The motivation for bounding H_s in this manner is two-folds. Firstly, it produces some theoretical insight into specialization, while secondly, the bound can be used by the sequential pattern algorithm to search the search space(hypothesis space) efficiently.

Consider that we are given a general sequential pattern whose H measure, H_g, is defined as

$$H_g = \sqrt{P(b)} \left[2 - 2\sqrt{P(a|b)P(a)} - 2\sqrt{(1-P(a|b))(1-P(a))}\right]$$

We try to calculate the bound of

$$H_s = \sqrt{P(c|b)}\sqrt{P(b)}[2 - 2\sqrt{P(a|bc)P(a)} - 2\sqrt{(1-P(a|bc))(1-P(a))}]$$

Given no information about C, we can state the following results.

Theorem 1. *If the H measure of a specialized pattern satisfies the following boundary:*

$$H_s \leq \max\{\sqrt{P(a|b)}\sqrt{P(b)}\left[2\sqrt{m} - 2\sqrt{P(a)}\right],$$
$$2\sqrt{P(b)} - \sqrt{1-P(a|b)}\sqrt{P(b)}\left[2\sqrt{P(a)} + 2\sqrt{1-P(a)}\right]\}$$

where m represents the number of class in the target attribute, the general pattern discontinues specializing.

Proof is omitted due to space limit. As a special case of Theorem 1, if the success rate, conditional probability($P(a|b)$), of general pattern becomes 1, the H measure of the specialized pattern is always less than or equal to that of general pattern.

Theorem 2. *If the conditional probability($P(a|b)$) of general pattern is 1, H measure of specialized pattern cannot be greater than that of general pattern. Therefore, the general pattern discontinues specializing.*

Proof is omitted due to space limit. As a consequence of these theorems we note that since the bound of specialized sequential pattern is achievable without further information about C, we can decide in advance that the specialized sequential pattern cannot be improved with respect to H Measure. The logical consequence of this statement is that it precludes using the bound to discontinue specializing based on the value of H_g alone. In particular, if the bound is less than the information content of the worst sequential pattern, then specialization cannot possibly find any better sequential pattern. This principle will be the basis for restricting the search space of the system.

6 Experimental Results

In order to test the functionality of the algorithm proposed in this paper, we assumed an artificial transaction database with 14 attributes, and synthesized two sets of artificial databases. The proposed algorithm was tested on two synthetic databases. Each database contains 20,000 records, and data values are generated

Table 1. Sequential patterns using database I

Sequential Patterns	Conf.	H
Price=20-29 → Item=P07	0.13137	0.00023
Gender=male & Item=P06 → Item=P02	0.11015	0.00021
Qty=1 → Item=P09	0.11800	0.00021
Item=P09 → Item=P01	0.11067	0.00019
SaleorNot=sale → Item=P00	0.11207	0.00017
Age=20-29 & Qty=over 5 → Item=P03	0.10592	0.00015
Price=30-39 & Qty=1 → Item=P02	0.10559	0.00015

using random numbers. For each data set, the entire data set is read and then 100 most informative sequential patterns were generated.

The topmost 6 sequential patterns from the first database is shown in Table 1. For each pattern in Table 1, its corresponding values for confidence(Conf.) and H measure are shown, and the resulting patterns are sorted based on their H measure values. The confidence means the number of transactions satisfying both left-hand side and right-hand side of the pattern divided by the number of transactions satisfying left-hand side only.

The topmost pattern in Table 1 means that customers who purchased items (whatever the items are) of which price are between 20-29 later purchase item P07. This type of patterns can not be acquired from traditional sequential pattern methods. The 4th pattern shows a sequential pattern equivalent to the one generated from Apriori-like method. It illustrates that the functionality of our method includes that of traditional sequential pattern methods.

The second database also contains 20,000 records, and data values are generated using random numbers. However, in the second database, we assumed that there are a number of sequential patterns hiding in the real world, and the database is generated based on those sequential patterns. The sequential patterns we have assumed are as follows.

- Color=white & Qty=1 → Item=P05
- Region=city & Item=10-19 → Item=P08

The goal of this experiment is to verify whether the proposed algorithm is able to detect these sequential patterns hidden in the database. For the second

Table 2. Sequential patterns using database II

Sequential Patterns	Conf.	H
Item=1 → Item=P07	0.17834	0.000181
Color=white & Qty=1 → Item=P05	0.15481	0.000103
Price=10-19 → Item=P03	0.13250	0.000065
Price=30-39 → Item=P09	0.14624	0.000051
Region=city & Item=10-19 → Item=P08	0.11951	0.000040
Color=white & Qty=1 → Item=P08	0.11440	0.000040

experiment, the entire data set is read and then 100 most informative sequential patterns were generated. The topmost 6 sequential patterns from the second database is shown in Table 2. The sequential patterns we have assumed are generated from the system and shown in Table 2 as the 2nd and 5th pattern, respectively. We could also see many other multi-dimensional sequential patterns in Table 2. This experiment illustrates that our proposed algorithm is able to effectively detect the sequential patterns hidden within the database.

7 Conclusion

In this paper we have introduced a new method for generating multi-dimensional sequential patterns from transaction databases. We developed an information theoretic measure, called H measure, which becomes the criteria for selecting and sorting inductive sequential patterns generated. The boundary of the H measure is analyzed and two heuristics are developed to reduce the computational complexity of the system. In addition, missing values can be handled by considering them as separate categories. The algorithm is applied to some synthetic transaction databases. The resulting sequential patterns generated from the data sets show how the system detects the hidden multi-dimensional sequential patterns of data sets effectively.

References

1. R. Agrawal and R. Srikant, *Mining sequential patterns*, Int. Conf. on Data Engineering, 1995, pp. 3–14.
2. R. J. Beran, *Minimum hellinger distances for parametric models*, Ann. Statistics **5** (1977), 445–463.
3. J. Pei-K. Wang Q. Chen H. Pinto, J. Han and U. Dayal, *Multi-dimensional sequential pattern mining*, Int. Conf. on Information and Knowledge Management, 2001.
4. B. Mortazavi-Asl Q. Chen U. Dayal J. Han, J. Pei and M-C. Hsu, *Freespan: Frequent pattern-projected sequential pattern mining*, Int. Conf. Knowledge Discovery and Data Mining (KDD00), 2000.
5. B. Mortazavi-Asl H. Pinto Q. Chen U. Dayal J. Pei, J. Han and M.-C. Hsu, *Prefixspan: Mining sequential patterns efficiently by prefix-projected pattern growth*, Int. Conf. on Data Engineering, 2001.
6. R. Rastogi M. Garofalaskis and K. Shim, *Spirit:sequential pattern mining with regular expression constraints*, Int. Conf. on Very Large Databases, 1999.
7. Jiawei Han Petre Tzvetkov, Xifeng Yan, *Tsp: Mining top-k closed sequential patterns*, Int. Conf. on Data Mining, 2003.
8. Ramin Afshar Xifeng Yan, Jiawei Han, *Clospan: Mining closed sequential patterns in large databases*, Int. Conf. on Data Mining, 2003.
9. M. J. Zaki, *Spade: An efficient algorithm for mining frequent sequences*, Machine Learning **42** (2001), 31–60.

Visual Terrain Analysis of High-Dimensional Datasets

Wenyuan Li[1], Kok-Leong Ong[2], and Wee-Keong Ng[1]

[1] Nanyang Technological University, Centre for Advanced Information Systems,
Nanyang Avenue, N4-B3C-14, Singapore 639798
liwy@pmail.ntu.edu.sg, awkng@ntu.edu.sg

[2] School of Information Technology, Deakin University,
Waurn Ponds, Victoria 3217, Australia
leong@deakin.edu.au

Abstract. Most real-world datasets are, to a certain degree, skewed. When considered that they are also large, they become the pinnacle challenge in data analysis. More importantly, we cannot ignore such datasets as they arise frequently in a wide variety of applications. Regardless of the analytic, it is often that the effectiveness of analysis can be improved if the characteristic of the dataset is known in advance. In this paper, we propose a novel technique to preprocess such datasets to obtain this insight. Our work is inspired by the resonance phenomenon, where similar objects resonate to a given response function. The key analytic result of our work is the *data terrain*, which shows properties of the dataset to enable effective and efficient analysis. We demonstrated our work in the context of various real-world problems. In doing so, we establish it as the tool for preprocessing data before applying computationally expensive algorithms.

1 Introduction

The subfield of data analysis is essentially a collection of algorithms that focused on analyzing large datasets of high-dimensionality. Often than not, the cornerstone of these algorithms is to address the dimensionality curse when trying to provide effective and efficient results for a given user query. Towards this, there have been many research done; including cluster analysis to find clusters embedded in subspaces (also known as *subspace clustering* or *biclustering*), and dimensionality reduction.

In cluster analysis, most models are based on distance or similarity measures, or correlation measures of feature subsets or objects. While they unveil the details of subspace clusters, most are of no interest to the user. For example, more than $10,000$ clusters were obtained through OP-clustering [1] on a drug activity dataset with a dimension of $10,000 \times 30$. Clearly, this is overwhelming to the user trying to find insights about the data in question, e.g., the relationship among patterns rather than a list of patterns. Usually, closer inspection would suggest close relationships among clusters. And if high level insights is what the user is after, then this level of pattern redundancy would be inappropriate. Yet, a combinatorial explosion of patterns (satisfying the query) occur as the size and dimensionality of the dataset increases. Dimension reduction is one alternative to 'curb' the combinatorial explosion of patterns by passing a reduced space to the analytical algorithms. The drawback, however, is the loss of patterns embedded

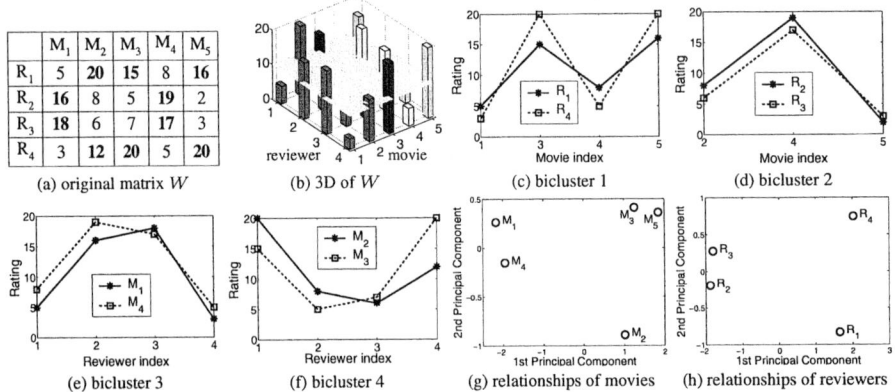

Fig. 1. The rating matrix for 4 reviewers and 5 movies, biclusters and PCA 2-dimensional space

in the subspace of the original space. This happens because most reduction techniques made use of distance or similarity measures over the full dimension, and therefore lack the mechanism to find the embedded patterns that are subtle but important.

In this paper, we introduce the concept of *data terrain* to visualize high-dimensional datasets while overcoming the limitations of subspace clustering and dimension reduction. Our proposal effectively reveals the relationship among subspace clusters, and allows the user to explore the data at different levels of details. We show, by means of real-world applications (e.g., biclusters, outliers, and frequent itemsets), how the data terrain can help discover generic patterns that can be utilized to effectively analyze the patterns embedded in the original space. Unfortunately, to find this data terrain under varying conditions proved to be NP-hard. Thus, our contribution in this paper includes the proposal of efficient techniques to find the data terrain. We next show a motivating example to illustrate the relevance of data terrains in analysis. We then introduce the resonance model in Section 3 and summarize our work in Section 4.

2 Motivating Example

We begin by introducing the concept of data terrain and show by means of an example, how it facilitates better data analysis; and why it is better than other techniques like biclustering and dimension reduction. Our example is based on the survey of popular movies. Fig. 1(a) shows the rating matrix W of 4 reviewers (R_i) on 5 movies (M_j), where each movie is rated on a scale of 1 to 20.

We first use biclustering to analyze the relationship between the reviewers and the movies. If requiring biclusters with at least 2 rows and columns, more than 10 biclusters can be discovered. Fig. 1(c) – (f) are the distinct biclusters found in this case. While these biclusters precisely characterized the reviewers' 'rating style' on movies, there is too much redundancy in the solution for such a small dataset. In real-world situations where the dataset is much larger, it will take much longer before the an-

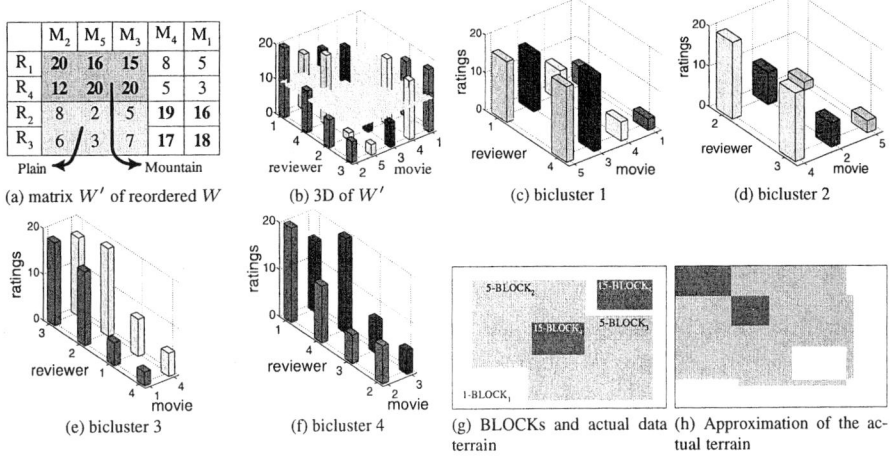

Fig. 2. (a) – (f): Reordered rating matrix given by Fig. 1; (g) & (h): Illustrations of σ-BLOCKs

alyst is able to come to a conclusion. Worse, the analyst is likely to become confused on the real relationship between the reviewers and the movies if they were to work at this level of detail. The other approach would be to use dimension reduction techniques. We consider using a popular technique known as Principal Component Analysis (PCA). The PCA's output is shown in Fig. 1(g) and (h). Again, the relationship between the reviewers and movies is revealed. However, as PCA runs across the full dimension of the data, we loose many subtle and local insights about reviewers and movies. For example, movies M_2, M_4 and M_5 are perceived as different in Fig. 1(g). Yet, if we check back on our analysis using biclustering, we can see from Fig. 1(d) that they are actually quite similar if we consider the ratings from reviewer R_2 and R_3. Thus, if only PCA is performed, we will not be able to arrive at this conclusion.

Interestingly, if we view W in 3D space, we can capture the relationships that both biclustering and PCA revealed. As Fig. 1(b) shows, a direct 3D 'plot' of W does not seem to reveal any interesting insights – but if we were to reorder W into W' as shown in Fig. 2(a) (and also Fig. 2(c) – (f); where every bicluster can be shown in this manner [1]), we have a 3D terrain of W' as depicted in Fig. 2(b). Notably, this terrain provides the insights that earlier requires both biclustering and PCA analysis.

To illustrate this, notice that any bicluster from Fig. 2(c) – (f) can be obtained by selecting some points from the 'mountains' and 'plains' in this terrain. At the same time, we can also make conclusions that would otherwise be obtained through PCA: (i) there are primarily two groups of movies and reviewers; (ii) M_3 and M_5 have higher similarity than M_2 despite being in the same group; and (iii) the 'rating style' of R_2 and R_3 is opposite that of R_1 and R_4. Thus, the terrain captures both local and global relationships about the data in an intuitive and effective manner. Of course, real-world datasets are much more complex that result in more complicated terrains. Consequently, trying to discover such a terrain proved to be a NP-hard problem.

3 Discovering Data Terrains

Conceptually, moving from W to W' is simply the reordering of the matrix to form the 'mountains' and 'plains'. Yet, this ordering on both dimensions can be difficult to achieve efficiently on massive datasets. To prove the hardness of this problem, we first give the following definitions. Let $\mathcal{O}$ be a set of objects, where $o \in \mathcal{O}$ is defined by a set of attributes $\mathcal{A}$. Further, let w_{ij} be the magnitude of o_i over $a_j \in \mathcal{A}$. Then we can represent the relationship of all objects and their attributes in a matrix $W = (w_{ij})_{|\mathcal{O}| \times |\mathcal{A}|}$ for the weighted bipartite graph $G = (\mathcal{O}, \mathcal{A}, E, W)$, where E is the set of edges. Thus, discovering the 'mountains' transforms into the problem of evaluating subgraphs where the magnitude of all its edges are above some 'altitude', i.e., $w_{ij} \geqslant \sigma$. Formally, the concept of a 'mountain' in this data terrain is called a BLOCK.

Definition 1. *Given a weighted bipartite graph G, a σ-BLOCK (or simply σ-B) is a subgraph $G' = (\mathcal{O}', \mathcal{A}', E', W')$ of G satisfying $w_{ij} \geqslant \sigma$ for any $i \in \mathcal{O}'$ and $j \in \mathcal{A}'$.*

From Definition 1, σ-B can be intuitively viewed as a plane (or a transverse section) with a specified altitude σ that 'cuts' across W. In the case of Fig. 1(b), we set the plane at σ=10 to obtain two 10-Bs as shown in Fig. 2(b): $\{R_1, R_4, R_2\} \times \{M_2, M_5\}$ and $\{R_2, R_3\} \times \{M_4, M_1\}$. Therefore, a series of σ-Bs can be generated when considering planes with different σ values. Once this set of BLOCKs relevant to G is found, we can order them to find the data terrain.

Definition 2. *Given a bipartite graph $G = (\mathcal{O}, \mathcal{A}, E, W)$ and a set of BLOCKs $\{B_1, B_2, \ldots, B_k\}$ found from G, the terrain of W is two ordered sequences of $\mathcal{O}$ and $\mathcal{A}$, such that these BLOCKs are placed consecutively in the reordered W.*

It is interesting to note that sorting both dimensions, i.e., $\mathcal{O}$ and $\mathcal{A}$, is an extension of sorting a single dimensional array to determine its distribution. However, sorting both dimensions simultaneously to get the 2-dimensional distribution is practically infeasible, i.e., finding the σ-BLOCKs by iteratively decreasing σ from the maximum value of W is NP-hard. In fact, finding a single σ-B is NP-hard.

Theorem 1. *Finding the largest σ-BLOCK ($|\mathcal{O}'| \times |\mathcal{A}'|$) is NP-hard.*

Proof. Our problem can be reduced from the *maximum edge biclique* [2], which is NP-complete. Details of this proof can be referred to [3].

Given the difficulty of finding σ-Bs, we seek alternative methods to discover the data terrain. Since our objective is to find the 'mountains' and 'plains' but *not* where they are on the terrain, then some approximation to the actual terrain (that is computationally efficient) should suffice. The insignificance of the specific locations of the 'mountains' and 'plains' can be demonstrated from Fig. 2(g) and (h), where the same set of insights are obtained from both figures. As this terrain is approximated, we called it the macro-view[1]. To obtain the macro-view of a terrain for a dataset, we used a novel

---

[1] The complete work of this paper includes a *micro*-view of the data terrain. Together, they provide a complete solution for analysis of high-dimensional datasets. Due to space constraints, the reader is referred to [3] for the details.

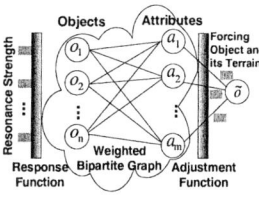

	M_3	M_2	M_4	M_5	M_1
R_1	15	20	8	16	5
R_4	20	12	5	20	3
R_3	7	6	17	3	18
R_2	5	8	19	2	16

(a) the abstract resonance model (b) reordered matrix from Fig. 1(a) given by the linear model R($\mathcal{O}, \mathcal{A}, W, \mathtt{I}, \mathtt{I}$)

Fig. 3. The model and the effectiveness of the linear instance

model inspired by the physics of resonance. This resonance model is very efficient even on very large and high-dimensional datasets. Instead of checking every σ-B, we can simulate a resonance experiment by injecting a response function to elicit objects of interest to the analyst. Proofs of all theorems in this section are omitted and refer to [3].

3.1 The Model

To simulate a resonance phenomenon, we require a forcing object $\tilde{o}$, such that when an appropriate response function $\mathbf{r}$ is applied, $\tilde{o}$ will resonate to elicit those objects $\{o_i, \ldots\} \subset \mathcal{O}$ in G, whose 'natural frequency' is similar to $\tilde{o}$. This 'natural frequency' represents the characteristics of both $\tilde{o}$ and the objects $\{o_i, \ldots\}$ who resonated with $\tilde{o}$ when $\mathbf{r}$ was applied. For the weighted bipartite graph $G = (\mathcal{O}, \mathcal{A}, E, W)$ and $W = (w_{ij})_{|\mathcal{O}| \times |\mathcal{A}|}$, this 'natural frequency' of $o_i \in \mathcal{O}$ is $\mathbf{o_i} = (w_{i1}, w_{i2}, \ldots, w_{i|\mathcal{A}|})$. Since a one-dimensional array (or vector) can be sorted to obtain its own terrain, we also refer $\mathbf{o_i}$ as the terrain of the object o_i. Likewise, the terrain of the forcing object $\tilde{o}$ is defined as $\mathbf{\tilde{o}_i} = (\tilde{w}_1, \tilde{w}_2, \ldots, \tilde{w}_{|\mathcal{A}|})$.

Put simply, if two objects of the same 'natural frequency' will resonate and therefore, should have a similar terrain. The evaluation of resonance strength between objects o_i and o_j is given by the response function $\mathbf{r}(\mathbf{o_i}, \mathbf{o_j}) : \mathbb{R}^n \times \mathbb{R}^n \to \mathbb{R}$. We defined this function abstractly to support different measures of resonance strength. For example, one existing measure to compare two terrains is the well-known *rearrangement inequality theorem*, where $\mathtt{I}(\mathbf{x}, \mathbf{y}) = \sum_{i=1}^n x_i y_i$ is maximized when the two positive sequences $\mathbf{x} = (x_1, \ldots, x_n)$ and $\mathbf{y} = (y_1, \ldots, y_n)$ are ordered in the same way (i.e. $x_1 \geqslant x_2 \geqslant \cdots \geqslant x_n$ and $y_1 \geqslant y_2 \geqslant \cdots \geqslant y_n$) and is minimized when they are ordered in the opposite way (i.e. $x_1 \geqslant x_2 \geqslant \cdots \geqslant x_n$ and $y_1 \leqslant y_2 \leqslant \cdots \leqslant y_n$).

Notice if two vectors maximizing $\mathbf{I}(\mathbf{x}, \mathbf{y})$ are put together to form $M = [\mathbf{x}; \mathbf{y}]$ (in MATLAB format), we obtain the terrain. More importantly, all σ-Bs are immediately obtained from this terrain with the need to search every σ-B! This is why the model is efficient – it only needs to consider the resonance strength among objects once the appropriate response function is selected. For example, the response function $\mathtt{I}$ is a suitable candidate to characterize the similarity of terrains of two objects. Likewise, $\mathtt{E}(\mathbf{x}, \mathbf{y}) = \exp(\sum_{i=1}^n x_i y_i)$ is also an effective response function.

To find the 'mountains' and 'plains', the forcing object $\tilde{o}$ evaluates the resonance strength of every objects o_i against itself to locate a 'best fit' based on the contour of its terrain. By running this iteratively, those objects that resonated with $\tilde{o}$ are discovered and placed together to form the 'mountains' within the 2-dimensional matrix W. In the same fashion, the 'plains' are discovered by combining those objects that resonated weakly with $\tilde{o}$. This iterative learning process between $\tilde{o}$ and G is outlined below.

Initialization. Set up $\tilde{o}$ with a uniform distribution: $\tilde{o} = (1, 1, \ldots, 1)$; normalize it as $\tilde{o} = \texttt{norm}(\tilde{o})^2$; then let $k = 0$; and record this as $\tilde{o}^{(0)} = \tilde{o}$.

Apply Response Function. For each object $o_i \in \mathcal{O}$, compute the resonance strength $\texttt{r}(\tilde{o}, o_i)$; store the results in a vector $\mathbf{r} = \big(\texttt{r}(\tilde{o}, o_1), \texttt{r}(\tilde{o}, o_2), \ldots, \texttt{r}(\tilde{o}, o_{|\mathcal{O}|})\big)$; and then normalize it, i.e., $\mathbf{r} = \texttt{norm}(\mathbf{r})$.

Adjust Forcing Object. Using $\mathbf{r}$ from the previous step, adjust the terrain of $\tilde{o}$ for all $o_i \in \mathcal{O}$. To do this, we define the adjustment function $\texttt{c}(\mathbf{r}, \mathbf{a_j}) : \mathbb{R}^{|\mathcal{O}|} \times \mathbb{R}^{|\mathcal{O}|} \to \mathbb{R}$, where the weights of the j-th attribute is given in $\mathbf{a_j} = (w_{1j}, w_{2j}, \ldots, w_{|\mathcal{O}|j})$. For each attribute a_j, $\tilde{w}_j = \texttt{c}(\mathbf{r}, \mathbf{a_j})$ integrates the weights from $\mathbf{a_j}$ into $\tilde{o}$ by evaluating the resonance strength recorded in $\mathbf{r}$. Again, c is abstract, and can be materialized using the inner product $\texttt{c}(\mathbf{r}, \mathbf{a_j}) = \mathbf{r} \bullet \mathbf{a_j} = \sum_i w_{ij} \cdot \texttt{r}(\tilde{o}, o_i)$. Finally, we compute $\tilde{o} = \texttt{norm}(\tilde{o})$ and record it as $\tilde{o}^{(k+1)} = \tilde{o}$. We denote the resonance model as $R(\mathcal{O}, \mathcal{A}, W, \texttt{r}, \texttt{c})$, where the instances of functions r and c can be either I or E.

Test Convergence. Compare $\tilde{o}^{(k+1)}$ against $\tilde{o}^{(k)}$. If the result converges, go to the next step; else apply r on $\mathcal{O}$ again (i.e., forcing resonance), and then adjust $\tilde{o}$.

Macro-View of Terrain. Sort the objects $o_i \in \mathcal{O}$ by the coordinates of $\mathbf{r}$ in descending order; and sort the attributes $a_i \in \mathcal{A}$ by the coordinates of $\tilde{o}$ in descending order.

3.2 Properties of the Model

The abstract view of the general model is given in Fig. 3(a). Depending on the response and adjustment function, the abstract model instantiates into different implementations. In practice, we have the linear model $R(\mathcal{O}, \mathcal{A}, W, \texttt{I}, \texttt{I})$, and the non-linear model $R(\mathcal{O}, \mathcal{A}, W, \texttt{E}, \texttt{E})$. We shall discuss some important properties of our model in this section. In particular, we show that the model gives a good approximation to the actual terrain, and that its iterative process converges quickly.

Approximation to Actual Terrain. Using the synthetic data from Fig. 2(g), we can see how well both implementations approximate the actual terrain. The linear and non-linear model converges to a precision of $\epsilon = 0.001$, i.e., once $\|\tilde{o}^{k+1} - \tilde{o}^k\| \leqslant \epsilon$, terminates. The reordered matrices are the same as Fig. 2(h). We then performed the same test on the movie-rating example. Result of linear model is shown in Fig. 3(b) and the non-linear model in Fig. 2(a). Obviously, $R(\mathcal{O}, \mathcal{A}, W, \texttt{E}, \texttt{E})$ gives a better approximation, where the 'mountains' and 'plains' are easily distinguishable. Thus, we can conclude that the different instances of R may give an approximate of the actual terrain. These conclusions are also empirically proven in [3].

Convergence. Since the resonance model is iterative, it is essential that it converges quickly to be efficient. Essentially, the model can be seen as a type of *discrete dynamical system* [4]. The convergence of linear and non-linear models is proven below.

Theorem 2. $R(\mathcal{O}, \mathcal{A}, W, \texttt{r}, \texttt{c})$, *where* r, c *are* I *or* E, *converges in limited iterations.*

---

[2] $\texttt{norm}(\mathbf{x}) = \mathbf{x}/\|\mathbf{x}\|_2$, where $\|\mathbf{x}\|_2 = (\sum_{i=1}^n x_i^2)^{1/2}$ is 2-norm of vector $\mathbf{x} = (x_1, \ldots, x_n)$.

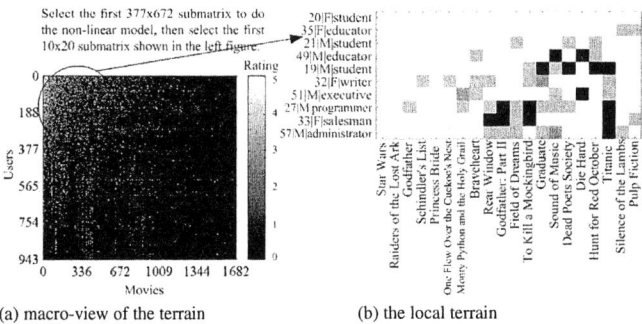

Fig. 4. A case of data analysis by macro-view of the terrain: visualization of MovieLens data (with 943 users, 1,682 movies and 100,000 ratings on the scale of 1 to 5. It is available at http://www.grouplens.org), and analysis of its local terrain. We obtain the macro-view using R($\mathcal{O}, \mathcal{A}, W, \mathtt{I}, \mathtt{I}$) to get the terrain in (a). It is possible that the user is not satisfied with an overview of the dataset and might be interested in further analysis. A case of 'zoom in' on the 'crowded' but a local terrain of the macro-view is shown in (b).

In practice, the model is very efficient because we are only interested in the convergence of orders of coordinates in $\tilde{\mathbf{o}}^k$ and $\mathbf{r}^k$. With k iterations, the complexity is $O(k \times |\mathcal{O}| \times |\mathcal{A}|)$. In our experiments, our model converges within 50 iterations even on the non-linear configurations giving a time complexity of $O(|\mathcal{O}| \times |\mathcal{A}|)$. In all cases, the complexity is sufficiently low to efficiency handle large datasets.

Average Inter-resonance Strength $\frac{1}{\binom{k}{2}} \sum_{\substack{i \neq j \in \mathcal{O}' \\ |\mathcal{O}'|=k}} \mathbf{r}(\mathbf{o}_i, \mathbf{o}_j)$ **among Objects.** Theorem 3 is in fact an optimization process to find the best k objects, whose average inter-resonance strength is the largest among any subset of k objects. Next, we exploit this property to unveil the relationship between the macro-view of the data terrain and the biclusters.

Theorem 3. *Given the macro-view terrain W', the average inter-resonance strength $\frac{1}{\binom{k}{2}} \sum_{1 \leqslant i \neq j \leqslant k} \mathbf{r}(\mathbf{o}_i, \mathbf{o}_j)$ of the first k objects, w.r.t. the resonance strength with $\tilde{\mathbf{o}}$, is largest for any subset with k objects.*

Approximation to Maximum Edge Biclique (MEB). The non-linear configuration of our model, i.e., R($\mathcal{O}, \mathcal{A}, W, \mathtt{E}, \mathtt{E}$) has such capability. Details refer to [3].

3.3 Real World Examples

A demonstration of how a macro-view of the data terrain can help the user in analysis is shown by a real-world case in Fig. 4. Next we show how it can have applications in data mining for finding frequent itemsets and biclustering in theory. All empirical evidences refer to [3].

Finding Frequent Itemsets. A transaction dataset can be constructed as a matrix, where each transaction is an object, and each item is an attribute whose value w_{ij} in $W_{|\mathcal{O}| \times |\mathcal{A}|}$ is 1 if the j-th item occurs in the i-th record, and 0 otherwise. We therefore have the following that relates frequent itemsets and BLOCKs.

Theorem 4 (Frequent Itemsets and BLOCKs). *A frequent itemset is the attribute set of a 1-BLOCK, and its support is the number of objects in the BLOCK.*

Discovering Biclusters. A popular measure for biclusters [5] is defined as Eqn. (1). The residue $\text{H}(W)$ of given a matrix W is a δ-bicluster if $\text{H}(W) \leqslant \delta$.

$$\text{H}(W) = \frac{1}{mn} \sum_{\substack{1 \leqslant i \leqslant m \\ 1 \leqslant j \leqslant n}} (w_{ij} - w_{iJ} - w_{Ij} + |W|)^2 \qquad (1)$$

where $w_{iJ} = \frac{1}{n}\sum_{j=1}^{n} w_{ij}$, $w_{Ij} = \frac{1}{m}\sum_{i=1}^{m} w_{ij}$, and $|W| = \frac{1}{mn}\sum_{\substack{1 \leqslant i \leqslant m \\ 1 \leqslant j \leqslant n}} w_{ij}$.

Theorem 5 (Bicluster and Average Resonance Strength of Macro-View Terrains). *Given a matrix $W = (w_{ij})_{m \times n}$, where $\mathcal{O}$ are the rows and $\mathcal{A}$ are columns, we have the inverse relation of the average inter-resonance strength and $\text{H}(W)$ as follows*

$$\text{H}(W) = \|W\|^2 + |W|^2 - \frac{1}{n}\bar{\mathbf{r}}(W) - \frac{1}{m}\bar{\mathbf{r}}(W^T) \qquad (2)$$

where $\|W\| = \sqrt{\frac{1}{mn}\sum_{\substack{1 \leqslant i \leqslant m \\ 1 \leqslant j \leqslant n}} w_{ij}^2}$, $\bar{\mathbf{r}}(W) = \frac{1}{\binom{m}{2}}\sum_{\substack{1 \leqslant i,j \leqslant m \\ i \neq j}} \text{I}(\mathbf{w}_{i\_}, \mathbf{w}_{j\_})$ *is the average inter-resonance strength among* $\mathbf{w}_{i\_}$, *and* $\bar{\mathbf{r}}(W^T) = \frac{1}{\binom{n}{2}}\sum_{\substack{1 \leqslant i,j \leqslant n \\ i \neq j}} \text{I}(\mathbf{w}_{\_i}, \mathbf{w}_{\_j})$ *is the average inter-resonance strength among* $\mathbf{w}_{\_j}$, *and* $\mathbf{w}_{i\_}$ *is the i-row vector of W with* $\mathbf{w}_{\_j}$ *the j-column vector of W.*

It can be interpreted as follows. Since $\|W\|$ and $|W|$ are sum of W in different forms, we can consider them as fixed constant. If the average inter-resonance strength of W and W^T, i.e., $\bar{\mathbf{r}}(W)$ and $\bar{\mathbf{r}}(W^T)$, is higher, then $\text{H}(W)$ is lower and thus, W behaves like a bicluster. For $\text{R}(\mathcal{O}, \mathcal{A}, W, \text{I}, \text{I})$ and $\text{R}(\mathcal{A}, \mathcal{O}, W^T, \text{I}, \text{I})$, we conclude that if we select the first k rows and columns of W with large resonance strength $\mathbf{r}(\mathbf{o}_i, \tilde{\mathbf{o}})$ to form W', it is straightforward that we will have a smaller $\text{H}(W')$ and thus, W a bicluster.

4 Summary

In this paper, we proposed the data terrain as a means to visualize and analyze high-dimensional datasets. With this terrain, patterns in subspaces can be visualized and analyzed. We provided a novel solution to obtain the the macro-view of a terrain efficiently, and demonstrated its real-world application.

References

1. Liu, J., Wang, W.: Op-cluster: Clustering by tendency in high dimensional space. In: Proceedings of ICDM, Melbourne, Florida (2003) 187
2. Peeters, R.: The maximum edge biclique problem is NP-complete. Disc. App. Math. **131** (2003) 651–654
3. Li, W., Ong, K.L., Ng, W.K.: Visual terrain analysis of high dimensional datasets. Technical Report (www.deakin.edu.au/ leong/tr0406) (TRC04/06), Deakin University (2005)
4. Sandefur, J.T.: Discrete Dynamical Systems. Oxford: Clarendon Press (1990)
5. Cheng, Y., Church, G.M.: Biclustering of expression data. In: Proceedings of the 8th International Conference on Intelligent System for Molecular Biology. (2000)

An Auto-stopped Hierarchical Clustering Algorithm for Analyzing 3D Model Database

Tian-yang Lv[1,2], Yu-hui Xing[2], Shao-bing Huang[1], Zheng-xuan Wang[2], and Wan-li Zuo[2]

[1] College of Computer Science and Technology, Harbin Engineering University, Harbin, China
[2] College of Computer Science and Technology, Jilin University, Changchun, China
raynor1979@163.com, wanli@mail.jlu.edu.cn

Abstract. In the research of shape-based 3D model retrieval, the analysis and classification of 3D model database is an important topic for improving the retrieval performance. However, it encounters difficulties due to lack of valuable prior knowledge and the semantic gaps exist in 3D model retrieval. The paper proposes a new auto-stopped hierarchical clustering algorithm overcome these problems, which combines outlier detection with clustering. The Princeton Shape Benchmark along with 2 data sets from UCI is employed to evaluate the performance of the algorithm. And the new algorithm outperforms other auto-stopped algorithms and obtains better classification of 3D model database.

Keywords: shape-based 3D model retrieval; clustering; outlier detection.

1 Introduction

With the proliferation of 3D models and their wide spread through internet, 3D model retrieval, especially shape-based 3D model retrieval, becomes a new emerging research field [1]. However, as an important subtopic, the analysis and organization of the 3D model databases encounters difficulties due to lack of the valuable domain knowledge. For instance, little is known about the number of models' classes. Moreover, the two-level semantic gaps exist in 3D model retrieval: one is the gap between the shape of model and its feature, which means models with similar shape have great different feature; the other is the gap between the shape of model and its meaning in real-life, which results in the mistakes in manually classifying the 3D model database.

The paper explores the application of the clustering techniques in analyzing 3D model database. And the clustering result is treated as the classification of 3D model database, since models of the same cluster have similar feature.

The topic has not been thought much in the previous works. For example, it is very difficult to pre-decide an appropriate number of final clusters k for 3D model database, while k is required by many traditional clustering algorithms, such as the hierarchical clustering algorithms CURE [2] and the partitioning algorithm K-means.

Thus, the paper proposes an auto-stopped hierarchical clustering algorithm, which integrates a new outlier mining method in clustering and cancels the parameter k. It is based on the following observation: the distances among data or clusters not only show their similarity degree, but also demonstrate the dissimilarity. With the progressing of clustering, the dissimilarity $D(C_{NN-A}, C_{NN-B})$ between the two most similar clusters C_{NN-A} and C_{NN-B} at present is increasing. And the clustering should stop at the moment when C_{NN-A} and C_{NN-B} are so diverse from each other. The outlier-mining process can provide that suitable "diverse degree" since outliers are detected according to their "great difference" from the others.

The rest of paper is organized as follows: after introducing related works in section 2, section 3 proposes the new clustering algorithm; section 4 gives the experimental result; finally, section 5 summarizes the paper.

Table 1. Important Notations

Notation	Description
N	Total number of Data
M	Dimensionality of Data
C_i	The i th Cluster
$D(C_i, C_j)$	Distance between C_i and C_j

2 Related Works

CURE algorithm [2] employs the novel concept of *representative* to represent a cluster and *r representatives* are shrunk towards the cluster's centroid by a fraction α before computing clusters' distance to avoid noise. However, CURE needs the parameter k and does not consider clusters' density in merging decisions.

Some researches try to make clustering algorithm optimally estimate k. [3] proposes a method based on dissimilarity increment. But it is short at handling outlier and detecting clusters with complex shape, like the linearly inseparable datasets.

To achieve the property of rotation invariance, [1] states a method using spherical harmonic transformation on voxel descriptors of 3D model. Its overview is: first, the 3D model is projected into a $2R \times 2R \times 2R$ voxel grid and set the corresponding value of a voxel 1, if it contains point of polygonal surface, otherwise set the value of 0; then, normalize the model with translation and scale; thus, for each sphere with the radius r, the spherical function of a 3D model can be defined as:

$$f_r(\theta, \varphi) = Voxel(r\sin(\theta)\cos(\varphi) + R, r\cos(\theta) + R, r\sin(\theta)\sin(\varphi) + R) \quad (1)$$

where $\theta \in [0, \pi], \varphi \in [0, 2\pi]$ and $r \in [0, R]$. And for each spherical harmonic function f_r can be decomposed as the sum of different frequencies, like:

$$f_r(\theta, \varphi) = \sum_{L=0}^{B-1} f_r^l(\theta, \varphi), \quad f_r^l(\theta, \varphi) = \sum_{m=-l}^{l} a_{l,m} Y_l^m(\theta, \varphi) \quad (2)$$

Where $Y_l^m(\theta, \varphi)$ is the harmonic homogeneous polynomial of l. Combining the signature $\{\|f_r^0\|, \|f_r^1\|, ...\}$ for f_r with different r, the shape descriptor for the 3D model is obtained, whose dimensionality depends on B and R with R usually equals 32.

3 An Auto-stopped Hierarchical Clustering Algorithm

Based on the traditional hierarchical clustering process, the Auto-Stopped Clustering Algorithm using Representatives ASCAR shows its uniqueness in three aspects: (1) adopts a new distance-based outlier detection method before clustering to detect outliers and exclude their disturbance for the clustering process; (2) employs the representatives and considers clusters' density in computing clusters' distance; (3) stops clustering automatically according to the dissimilarity reflected by the outliers.

3.1 The Outlier Detection Method Based on Even-Distribution Pattern

The basic idea of distance-based outlier detection method is: if the distances between data a and most other data are larger than the threshold D_{out}, a is an outlier [4]. It is critical but usually difficult to decide an appropriate D_{out}. This method also ignores the local distribution feature of one data.

The new method decides D_{out} according to the even distribution pattern of data. It is a very useful reference, since clusters and outliers exist only if the real-life data distribute unevenly. In that case, the distances $\overline{D}_{NN}$ between each data and its nearest neighbor are the same. $\overline{D}_{NN}$ is approximately decided according to equation 3, where $a_{max}^{(i)}$ and $a_{min}^{(i)}$ is the maximum and the minimum of all data's i th-dimension. And $D_{out} = \overline{D}_{NN}/\beta$, where β is a parameter to describe the diversity of the realistic distribution situation from the even pattern.

$$\overline{D}_{NN} = \sqrt{\sum_{i=1}^{M}((a_{max}^{(i)} - a_{min}^{(i)})/\sqrt[M]{N})^2} \qquad (3)$$

Factor ξ is adopted to evaluate the local distribution feature of a data. For data a, $\xi(a) = D_{NN}(a)/D_{NN}(b)$, where $D_{NN}(a)$ is the distance between a and its nearest-neighbor and so is $D_{NN}(b)$. The value of $\xi(a)$ shows the isolation degree of a from its neighbors. Special method is used for very similar or duplicate data. And, the equation of $\xi(a)$ is:

$$\xi(a) = \begin{cases} D_{NN}(a)/D_{NN}(b) & if(D_{NN}(b) > 10^{-4}) \\ 1 & else \end{cases} \qquad (4)$$

Therefore, the outlier evaluation criterion is stated as follows:

Data a is an outlier, if $D_{NN}(a) * \xi(a) > (\dfrac{\sqrt{\sum_{i=1}^{M}((a_{max}^{(i)} - a_{min}^{(i)})/\sqrt[M]{N})^2}}{\beta})$

Since outliers are extremely far away from the others while the normal are relatively near to each other, a method is proposed to decide the appropriate β:

(1) name β_{Step} as the *step length* and $\beta_{Step} = D_{NN}(a_{farest}) \times \xi(a_{farest})/\overline{D}_{NN}$, where a_{farest} satisfies $D_{NN}(a_{farest}) \times \xi(a_{farest}) \geq D_{NN}(b) \times \xi(b)$ for any b; (2) observe the increasing speed V of the detected outlier number n_{out} under different value of β,

viz. $V = \nabla n_{out} / \nabla \beta = \nabla n_{out} / \beta_{Step}$, where $\beta = l \times \beta_{Step}$ and $l=\{1, 2 ...\}$, call l the *step Num.*; (3) if V reaches its first peak when $l_i \times \beta_{Step}$, $\beta = (l_i - 1) \times \beta_{Step}$.

3.2 The Computation of the Clusters' Distance

The algorithm ASCAR adopts *representative* from CURE algorithm to improve clustering performance. But ASCAR excludes the influence of "noise" by adopting a professional outlier mining method without using the parameter α. And ASCAR also considers the cluster's density in deciding whether clusters should be merged.

The algorithm decides the distance $D(C_i, C_j)$ between C_i and C_j according to two factors: first, the distance $D_{min}(C_i, C_j)$ of the nearest *representatives* coming from C_i and C_j respectively; second, the factor δ measuring the change of cluster's density *Den*. The density of C_i or C_j approximately equals the average distances among its *representatives*. For the new-borne cluster C_{new} created by merging C_i and C_j, $Den(C_{new})=D_{min}(C_i,C_j)$. Then, $\delta(C_i)$ is defined as follows and so is $\delta(C_j)$:

$$\delta(C_i) = \begin{cases} Den(C_i)/Den(C_{New}) & if\ (Den(C_i) > Den(C_{New})) \\ Den(C_{New})/Den(C_i) & otherwise \end{cases} \quad (5)$$

Since it is impossible to compute the density of the cluster with only one data, define $D(C_i,C_j)=D_{min}(C_i,C_j)$ in that case. And the way to compute $D(C_i,C_j)$ is:

$$D(C_i,C_j) = \begin{cases} D_{Min}(C_i,C_j) \times (\delta(C_i)+\delta(C_j))/2 & if\ (n_i > 1)\ \&\&(n_j > 1) \\ D_{Min}(C_i,C_j) & otherwise \end{cases} \quad (6)$$

Factor δ reflects the influence of cluster's density on merging decision. That is, the bigger the difference between the density of C_i or C_j with that of C_{new}, the less possibility for C_i and C_j to be merged.

3.3 Automatic Stop

Without user-specified condition to stop clustering, it is necessary to extract this information from the processed data. As stated in former parts, it is a suitable opportunity to stop clustering if the clusters to be merged are too dissimilar. We propose D_{out} as the dissimilarity threshold to decide this opportunity for two reasons: (1) D_{out} is used to detect outliers, while the major characteristic of outliers is their dissimilarity from the others; (2) D_{out} is decided according to the even distribution pattern, which is also a useful reference for cluster analysis since the existence of clusters shows the diversity of the realistic distribution situation from the even pattern.

And the stop criterion of clustering is :

Suppose C_{NN-A} and C_{NN-B} are the most similar clusters at present, stop clustering if $D(C_{NN-A},C_{NN-B})>D_{out}$.

3.4 Complexity Analysis and Overview of ASCAR

The complexity of traditional hierarchical algorithm is $O(N^2)$. Since ASCAR is constructed on the traditional method, it is only necessary to analyze the complexity of each change. The complexity increases by $O(N)$ to perform one more scan to detect outliers. The complexity increases by $O((r*n_i+r^2)*(N-k))$ at most in computing clusters' distance. Thus, the complexity increases by $O((r*n_i+r^2)*(N-k)+N)$ in total. Since $r^2<N$ in most cases, the complexity of ASCAR equals $O(N^2)$.

The overview of the proposed algorithm ASCAR is listed in Figure 1.

```
Algorithm ASCAR( r, β )
1.{ Read all data and decide vector a_max and a_min;
2.  Treat each data as a separate cluster;
3.  Compute each cluster's nearest-neighbor;
4.  Determine the value of β_Step;
5.  D_out = outlier(a_max, a_min, β_Step);
6.  Name the nearest clusters at present as C_NN-A, C_NN-B;
7.  while (D(C_NN-A, C_NN-B) <= D_out)
8.  {  Merge clusters C_NN-A and C_NN-B;
9.     Update C_NN-A and C_NN-B; }
10. }   //End of ASCAR
```

Fig. 1. The Auto-stopped Hierarchical Clustering Algorithm ASCAR

4 Experiment and Analysis

The evaluation of the new algorithm is undertook in two aspects: first, the real-life datasets *Iris* and *Wine* of UCI Machine Learning Repository [5] are adopted; then ASCAR is applied in analyzing the Princeton Shape Benchmark[8].

4.1 Data Sets of UCI

The criterions *Entropy* and *Purity* of [6] are adopted to measure the clustering results' quality for Iris and Wine datasets. And the better the clustering result, the smaller is

Table 2. Overview of the clustering results of ASCAR and other algorithms

	Dataset	Parameters	k	*Entropy*	*Purity*	n_{out}
ASCAR	Iris	β=2.2, r=5	5	0.3542	0.8121	5
	Wine	β=1.8, r=4	14	0.1837	0.8864	3
Frozen	Iris	A=4.0	2	0.4206	0.6667	--
	Wine	A=0.5	13	0.4998	0.7247	--
DBScan	Iris	ε=0.7, *MPts*=3	2	0.4077	0.6867	--
	Wine	ε=35, *MPts*=3	6	0.5866	0.6798	--

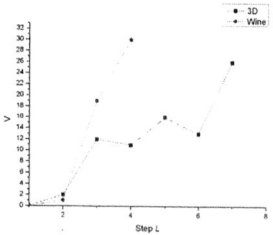 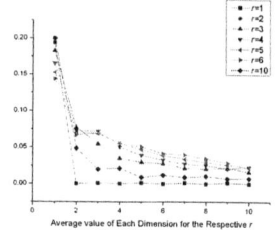

Fig. 2. Changes of V with the Increase of l **Fig. 3.** Average of each dimension f

the *Entropy* and the bigger is the *Purity*. The clustering performance of ASCAR is listed in Table 2 along with the detected number of outlier. To be more persuasive, Table 2 also gives the best clustering results of DBScan[7] and Frozen. Figure 2 shows the change of V with the increasing of *step num. l*.

4.2 Princeton Shape Benchmark

The feature extraction method of section 2 with $R=32$ and $B=10$ is applied to obtain the shape feature from 3D models. Obviously, the model feature with the dimensionality 320 will greatly reduce the clustering performance. Figure 3 shows that the first element of the transformation result for each $f_r(\theta,\varphi)$ plays the most important role in distinguishing model. Therefore, we select that element and obtain the shape feature with $M=32$. In experiment, the Euclidean distance is adopted.

Table 3. Cluster's detail of C_{90} and C_{160}

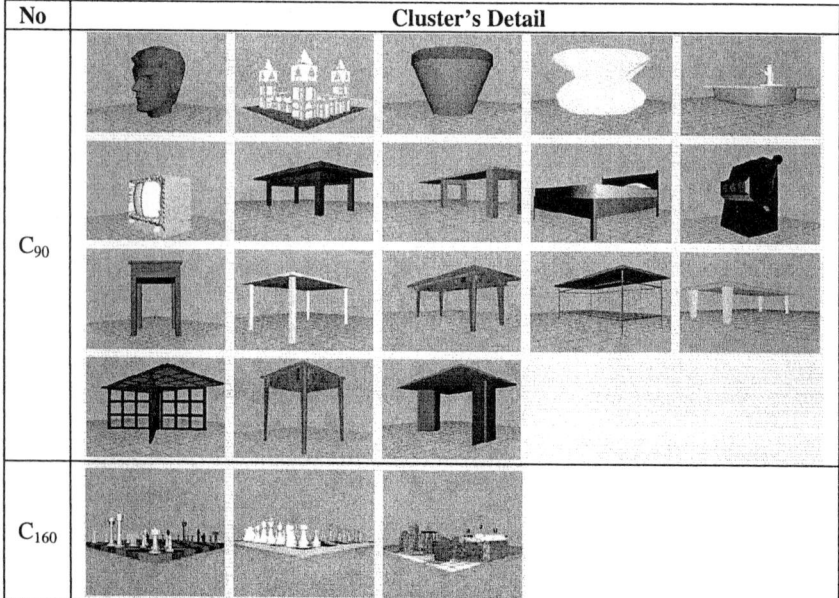

Table 4. Part of the detected outliers and the respective value of *step num l*

M741 ($L=1$)	M737($L=2$)	M416($L=2$)	M1401($L=3$)	M286($L=3$)

Since there is no valuable knowledge of the classification of the models in PSB, we have to list the details of the result cluster. When $r=4$ and $\beta=0.8*5$, ASCAR obtains 160 clusters with the smallest size of 2. Due to space limit, Table 3 just lists the details of C_{90} and C_{160}. Comparing to the manual classification result of PSB, ASCAR achieves very similar classes' number. But, ASCAR clusters the models with similar shape together no matter what real-life meaning they represent, especially if the feature extraction method satisfies the request that models with similar shape have similar feature. However, this cannot is not always satisfied and clustering mistakes can be observed in Table 3. Table 4 lists part of the detected outliers along with *step num l*, under which they are pruned.

We also applied the auto-stopped algorithms DBScan and Frozen in analyzing PSB. Under all possible value of parameters, Frozen algorithm obtains over 1200 clusters, while DBScan tends to obtain a little huge clusters. For instance, when $\varepsilon=0.2$ and *MPts*=2, DBScan gets 66 clusters with $n_0=715$, $n_1=1028$, $n_2, n_3...n_7=2$, and $n_8, n_9...n_{65}=1$. Obviously, these results are not acceptable as a classification of the database.

5 Conclusion

To analyze the 3D model database, the paper proposes a new strategy that integrates outlier detection with clustering and introduces an auto-stopped hierarchical clustering algorithm ASCAR. Experimental results show ASCAR's good performance in clustering the Princeton Shape Benchmark and 2 datasets from UCI. The future works will concentrate on the study of using the representations of the clustering result to establish the index of 3D model database.

Acknowledgements

This work is sponsored by the Natural Science Foundation of China under grant number 60373099 and the Natural Science Research Foundation of Harbin Engineering University under the grant number HEUFT05007.

References

1. T.Funkhouser, et al. A Search Engine for 3D Models. ACM Transactions on Graphics.22 (1), (2003) 85-105.
2. S. Guha, R. Rastogi, K. Shim: CURE: an Efficient Clustering Algorithm for Large Database. In: Laura M. Haas and Ashutosh Tiwary, eds. Proceedings of the ACM SIGMOD Conference on Management of Data. Seattle, Washington: ACM Press (1998) 73-84.

3. Ana L.N. Fred, José M.N. Leitão: A new Cluster Isolation criterion Based on Dissimilarity Increments. IEEE Transactions on Pattern Analysis and Machine Intelligence, VOL. 25, No. 8 (August 2003) 944-958
4. Edwin M. Knorr, Raymond T. Ng: Finding Intensional Knowledge of Distance-Based outliers. In: Proceedings of the 25th Very Large Data Bases conference. Edinburgh, Scotland (1999) 211 - 222
5. Hettich, S. & Blake, C.L. & Merz, C.J. (1998). UCI Repository of machine learning databases [http://www.ics.uci.edu/~mlearn/MLRepository.html]. Irvine, CA: University of California, Department of Information and Computer Science.
6. Ying Zhao, George Karypis: Criterion Functions for Document Clustering: Experiment and Analysis. Technical Report #01-40, University of Minnesota (2001) 1 – 40
7. M. Ester, H.-P. Kriegel, J. Sander, and X. Xu: A Density-Based Algorithm for Discovering Clusters in Large Spatial Databases with Noise. In: Proceedings of 2nd International Conference on Knowledge Discovery and Data Mining, Portland, OR (1996) 226-231
8. Philip Shilane, Patrick Min, Michael Kazhdan, and Thomas Funkhouser: The Princeton Shape Benchmark. Shape Modeling International, Genova, Italy, (June 2004)

A Comparison Between Block CEM and Two-Way CEM Algorithms to Cluster a Contingency Table

Mohamed Nadif[1] and Gérard Govaert[2]

[1] LITA EA3097, Université de Metz, Ile du Saulcy, 57045 Metz, France
mohamed.nadif@iut.univ-metz.fr
[2] HEUDIASYC, UMR CNRS 6599, Université de Technologie de Compiègne,
BP 20529, 60205 Compiègne Cedex, France
gerard.govaert@utc.fr

Abstract. When the data consists of a set of objects described by a set of variables, we have recently proposed a new mixture model which takes into account the block clustering problem on the both sets and have developed the *block CEM* algorithm. In this paper, we embed the block clustering problem of contingency table in the mixture approach. In using a Poisson model and adopting the classification maximum likelihood principle we perform an adapted version of block CEM. We evaluate its performance and compare it to a simple use of CEM applied on the both sets separately. We present detailed experimental results on simulated data and we show the interest of this new algorithm.

1 Introduction

Cluster analysis is an important tool in a variety of scientific areas such as pattern recognition, information retrieval, micro-array, data mining, and so forth. Although many clustering procedures such as hierarchical clustering, k-means or self-organizing maps, aim to construct an optimal partition of objects or, sometimes, of variables, there are other methods, called block clustering methods, which consider simultaneously the two sets and organize the data into homogeneous blocks.

A wide variety of procedures have been proposed for finding patterns in data matrices. These procedures differ in the pattern they seek, the types of data to which they apply, and the assumption on which they rely. Let us mention the works of Hartigan (1975), Bock (1979), Garcia and Proth (1986), Marchotorchino (1987), Govaert (1983, 1995), Arabie and Hubert (1990), Duffy and Quiroz (1991) and Ritschard et al. (2001) who have proposed some algorithms dedicated to different kinds of matrices.

These last years, block clustering (also called biclustering) has become an important challenge in data mining context. In the text mining field, Dhillon (2001) has proposed a spectral block clustering method by exploiting the duality between rows (documents) and columns (words). In the analysis of micro-array data where data are often presented as matrices of expression levels of genes under different conditions, block clustering of genes and conditions has permitted to overcome the problem of the choice of similarity on the both sets found in conventional clustering methods (Cheng and Church, 2000). Also, these kinds of methods have practical importance in a wide of variety of applications such as text and market basket data analysis. Typically, the data that arises in these applications is arranged as a two-way contingency or co-occurrence table.

In this paper, we will focus on these kinds of data. The data which we consider is noted $\mathbf{x}$; it is a $r \times s$ data matrix defined by $\mathbf{x} = \{(x_{ij}); i \in I, j \in J\}$, where I is a categorical variable with r categories and J a categorical variable with s categories. In exploiting the duality between I and J, we will study the block clustering problem in embedding it in the mixture approach. We will propose a *block mixture model* which takes into account the block clustering situation and perform an innovative co-clustering algorithm. This one is based on the alternated application of Classification EM (Celeux and Govaert, 1992) on intermediate data matrices. To propose this algorithm, we set this problem in the classification maximum likelihood (CML) approach (Symons, 1981). This paper deals to compare block CEM and two-way CEM, i.e. CEM applied separately on I and J. Results on simulated data are given, confirming that block CEM gives much better performance than two-way CEM.

The paper is organized as follows. In Section 2, we give the necessary background CML approach and we describe the CEM algorithm and its steps when the data is arranged as a two-way contingency. In Section 3, we start by recalling our block mixture model and we describe the block CEM algorithm. In order to compare two-way CEM and block CEM, in Section 4, we perform numerical Monte Carlo simulations. A final section summarizes and indicates the recommended algorithm.

2 Mixture Model and Clustering

For convenience, we represent a partition of I into g clusters by $\mathbf{z} = (\mathbf{z}_1, \ldots, \mathbf{z}_r)$ where $\mathbf{z}_i$, which indicates the component of the row i, is represented by $\mathbf{z}_i = (z_{i1}, \ldots, z_{ig})$ with $z_{ik} = 1$ if row i is in cluster k and 0 otherwise. Then, the kth cluster corresponds to the set of rows i such that $z_{ik} = 1$. We will use similar notation for a partition $\mathbf{w}$ into m clusters of the set J. In the following, to simplify the notation, the sums and the products relating to rows, columns or clusters will be subscripted respectively by letters i, j or k without indicating the limits of variation, which will be thus implicit. Thus, for example, the sum $\sum_i$ stands for $\sum_{i=1}^{r}$ or $\sum_{i,j,k,\ell}$ stands for $\sum_{i=1}^{r} \sum_{j=1}^{s} \sum_{k=1}^{g} \sum_{\ell=1}^{m}$.

2.1 CML Approach and the CEM Algorithm

In the model-based clustering (see for instance (McLachlan and Peel, 2000), it is assumed that the data are generated by a mixture of underlying probability distributions, where each component k of the mixture represents a cluster. Thus, the density of the observed data x is expressed as

$$f(\mathbf{x};\boldsymbol{\theta}) = \prod_i \sum_k \pi_k \varphi_k(\boldsymbol{x}_i; \alpha_k) \qquad (1)$$

where $\boldsymbol{\theta} = (\pi_1,...,\pi_g,\alpha_1,...,\alpha_g)$, $(\pi_1,...,\pi_g)$ are the mixing proportions and $(\alpha_1,...,\alpha_g)$ are the parameters of the density components φ_k.

The clustering problem can be studied under mixture model using two different approaches: the maximum likelihood (ML) approach and the classification maximum likelihood (CML) approach (Symons, 1981). In this paper we focus on the second approach.

The ML approach estimates the parameters of the mixture and the partition is derived from these parameters using the maximum a posteriori principle (MAP). In the CML, the partition is added to the parameters to be estimated. The CML approach consists in estimating the parameters of the mixture and the partition. The maximum likelihood estimation of these new parameters leads to optimize in $\boldsymbol{\theta}$ and z the complete data log-likelihood

$$L_C(\mathbf{z},\boldsymbol{\theta}) = L(\boldsymbol{\theta};\mathbf{x},\mathbf{z}) = \log f(\mathbf{x},\mathbf{z};\boldsymbol{\theta}) = \sum_{i,k} z_{ik} \log\left(p_k \varphi_k(\boldsymbol{x}_i;\alpha_k)\right).$$

This optimization can be done by the Classification EM (CEM) algorithm (Celeux and Govaert, 1992), a variant of EM (Dempster, Laird and Rubin, 1977), which converts the posterior probabilities t_{ik}'s to a discrete classification in a C-step before performing the M-step.

2.2 Application to Contingency Table

In this situation, the contingency table x is a $r \times s$ data matrix defined by $\mathbf{x} = \{(x_{ij}); i \in I, j \in J\}$, where I and J are categorical variables with r and s categories. The sum of each row i will be denoted x_i. Thus, if we note $\boldsymbol{\theta} = (\pi_1,...,\pi_g,\alpha_{11},...,\alpha_{gs})$ the parameter of the model and φ is the multinomial distribution of the k-th component, the log-likelihood (up to a constant) can be written as $L(\boldsymbol{\theta};\mathbf{x}) = \sum_i \log \sum_k \pi_k \alpha_{k1}^{x_{i1}} \ldots \alpha_{ks}^{x_{is}}$, and the complete data log-likelihood as $L(\boldsymbol{\theta};\mathbf{x},\mathbf{z}) = \sum_{i,k} z_{ik} \left(\ln \pi_k + \sum_j x_{ij} \log \alpha_{kj} \right)$.

In clustering context, the use of the mixture model deals to find the component from which each row arises. The CEM algorithm allows us to achieve this goal and the different steps of CEM in this situation are

- E-step: compute the posterior probabilities $t_{ik}^{(c)} \propto \pi_k \alpha_{k1}^{x_{i1}} \ldots \alpha_{ks}^{x_{is}}$;
- C-step: the kth cluster of $\mathbf{z}^{(c+1)}$ is defined with $z_{ik}^{(c+1)} = 1$ if $k = \operatorname{argmax}_{k=1,...,g} t_{ik}^{(c)}$ and $z_{ik}^{(c+1)} = 0$ otherwise;
- M-step: by standard calculations, one arrives at the following re-estimations parameters $\pi_k^{(c+1)} = \frac{n_k^{(c+1)}}{r}$ and $\alpha_{kj}^{(c+1)} = \frac{x_{kj}}{x_k}$ where $n_k^{(c+1)}$ is the cardinality of the kth cluster of $\mathbf{z}^{(c+1)}$, $x_{kj} = \sum_i z_{ik}^{(c+1)} x_{ij}$ and $x_k = \sum_j x_{kj}$.

Having found the estimate of the parameters and noting $f_{kj} = \frac{x_{kj}}{x_{..}}$ where $x_{..} = \sum_{i,j} x_{ij}$, we can show that, when the proportions are fixed, the maximization of $L(\theta; \mathbf{x}, \mathbf{z})$ is equivalent to the maximization of the mutual information $I(\mathbf{z}, J) = \sum_{k,j} f_{kj} \log \frac{f_{kj}}{f_{k.}f_{.j}}$ and approximately equivalent to the maximization of the chi-square criterion $\chi^2(\mathbf{z}, J) = x_{..} \sum_{k,j} \frac{(f_{kj}-f_{k.}f_{.j})^2}{f_{k.}f_{.j}}$. Hence the use of the both criteria $\chi^2(\mathbf{z}, J)$ and $I(\mathbf{z}, J)$ supposes implicitly that the data arise from a mixture of multinomial distributions. To tackle the block clustering problem, we can obviously use the CEM on I and J separately (noted 2CEM) but unfortunately it is unaware of the correspondence between I and J. It will be seen later that this process is ineffective to detect homogeneous blocs.

3 Block Mixture Model for Contingency Table

To study the block clustering problem, we have extended (Govaert and Nadif, 2003) the mixture model to propose a block mixture model defined by the following probability density function

$$f(\mathbf{x}; \boldsymbol{\theta}) = \sum_{(\mathbf{z},\mathbf{w}) \in \mathcal{Z} \times \mathcal{W}} \prod_i \pi_{z_i} \prod_j \rho_{w_j} \prod_{i,j} \varphi(x_{ij}; \boldsymbol{\alpha}_{z_i w_j})$$

where $\boldsymbol{\theta} = (\boldsymbol{\pi}, \boldsymbol{\rho}, \boldsymbol{\alpha}_{11}, \ldots, \boldsymbol{\alpha}_{gm})$, $\boldsymbol{\pi} = (\pi_1, \ldots, \pi_g)$ and $\boldsymbol{\rho} = (\rho_1, \ldots, \rho_m)$ are the mixing proportions and $\varphi(x, \boldsymbol{\alpha}_{k\ell})$ is a probability density function defined on the real set $\mathbb{R}$.

Counts in the $r \times s$ cells of a contingency table are typically modelled as random variables. In our situation, we assume that for each block $k\ell$ the values x_{ij} are distributed according the Poisson distribution $\mathcal{P}(\alpha_i \beta_j \delta_{k\ell})$ for which the probability mass function is

$$\frac{e^{-\alpha_i \beta_j \delta_{k\ell}} (\alpha_i \beta_j \delta_{k\ell})^{x_{ij}}}{x_{ij}!}.$$

The Poisson parameter is split into α_i and β_j the effects of the row i and the column j and $\delta_{k\ell}$ the effect of the block $k\ell$. Because the aim is to maximize the complete data log-likelihood not only depending on $\boldsymbol{\theta}$ but on $\mathbf{z}, \mathbf{w}$, an adapted re-parametrization of the Poisson distribution becomes necessary. To this end, we impose some constraints and we assume that $\sum_\ell \beta_\ell \delta_{k\ell} = 1$ and $\sum_k \alpha_k \delta_{k\ell} = 1$ with $\alpha_k = \sum_{i,k} z_{ik} \alpha_i$ $\beta_\ell = \sum_{j,\ell} w_{j\ell} \beta_j$.

To tackle the simultaneous partitioning problem, we will use the CML approach, which aims to maximize the classification log-likelihood called complete data log-likelihood associated to the block mixture model. With our model, the complete data are $(\mathbf{z}, \mathbf{w}, \mathbf{x})$ and the classification log-likelihood is given by

$$L_c(\mathbf{z}, \mathbf{w}, \boldsymbol{\theta}) = L(\boldsymbol{\theta}; \mathbf{x}, \mathbf{z}, \mathbf{w}) = \log(p(\mathbf{z}; \boldsymbol{\theta}) p(\mathbf{w}; \boldsymbol{\theta}) f(\mathbf{x}|\mathbf{z}, \mathbf{w}; \boldsymbol{\theta})).$$

To maximize $L_c(\mathbf{z}, \mathbf{w}, \boldsymbol{\theta})$, like in Govaert and Nadif (2003) we propose to maximize alternatively the classification log-likelihood with $\mathbf{w}$ and $\boldsymbol{\rho}$ fixed and then with $\mathbf{z}$ and $\boldsymbol{\pi}$ fixed. By noting $x_{i\ell} = \sum_j w_{j\ell} x_{ij}$, the classification log-likelihood can be written as

$$L_c(\mathbf{z},\mathbf{w},\boldsymbol{\theta}) = \sum_{i,k} z_{ik} \log \pi_k + \sum_{j,\ell} w_{j\ell} \log \rho_\ell + \sum_{i,k} z_{ik} \sum_\ell x_{i\ell} \log \delta_{k\ell}.$$

If we note $\mathbf{u}_i = (x_{i1}, \ldots, x_{i\ell}, \ldots, x_{im})$ and $\gamma_{k\ell} = x_{.\ell} \delta_{k\ell}$, the classification log-likelihood can be decomposed into two terms

$$L_c(\mathbf{z},\mathbf{w},\boldsymbol{\theta}) = L_c(\mathbf{z},\boldsymbol{\theta}/\mathbf{w}) + g(\mathbf{x},\mathbf{w},\boldsymbol{\rho})$$

where the first one, can be written as

$$L_c(\mathbf{z},\boldsymbol{\theta}/\mathbf{w}) = \sum_{i,k} z_{ik} \log(\pi_k \Phi(\mathbf{u}_i, \boldsymbol{\gamma}_k))$$

where $\Phi(\mathbf{u}_i, \boldsymbol{\gamma}_k)$ is the multinomial distribution for $x_{i1}, \ldots, x_{im}$ with the probabilities $\gamma_{k1}, \ldots, \gamma_{km}$ and the second one can be written as

$$g(\mathbf{x},\mathbf{w},\boldsymbol{\rho}) = \sum_{j,\ell} w_{j\ell} \log \rho_\ell - \sum_\ell x_{.\ell} \log x_{.\ell}.$$

Hence, $L_c(\mathbf{z},\boldsymbol{\theta}/\mathbf{w})$, called in the followings conditional classification log-likelihood, corresponds to the complete log-likelihood associated to a classical mixture model defined on the samples $\mathbf{u}_1, \ldots, \mathbf{u}_r$. As $g(\mathbf{x},\mathbf{w},\boldsymbol{\rho})$ does not depend on $\mathbf{z}$, maximizing $L_c(\mathbf{z},\mathbf{w},\boldsymbol{\theta})$ for $\mathbf{w}$ fixed is equivalent to maximize the conditional classification log-likelihood $L_c(\mathbf{z},\boldsymbol{\theta}/\mathbf{w})$, which can be done by the CEM algorithm applied to the multinomial mixture model. The different steps of CEM are

- E-step: compute the posterior probabilities $t_{ik}^{(c)}$;
- C-step: the kth cluster of $\mathbf{z}^{(c+1)}$ is defined with $z_{ik}^{(c+1)} = 1$ if $k = \operatorname{argmax}_{k=1,\ldots,g} t_{ik}^{(c)}$ and $z_{ik}^{(c+1)} = 0$ otherwise.
- M-step: by standard calculations, one arrives at the following re-estimations parameters

$$\pi_k^{(c+1)} = \frac{\# z_k^{(c+1)}}{r} \quad \text{and} \quad \delta_{k\ell}^{(c+1)} = \frac{x_{k\ell}}{x_{k.} x_{.\ell}}$$

where $\#$ denotes the cardinality and

$$x_{k\ell} = \sum_i z_{ik}^{(c+1)} x_{i\ell} = \sum_{ij} z_{ik}^{(c+1)} w_{j\ell} x_{ij}.$$

In the same way, we can show that

$$L_c(\mathbf{z},\mathbf{w},\boldsymbol{\theta}) = L_c(\mathbf{w},\boldsymbol{\theta}/\mathbf{z}) + g(\mathbf{x},\mathbf{z},\boldsymbol{\pi})$$

where

$$g(\mathbf{x},\mathbf{z},\boldsymbol{\pi}) = \sum_{i,k} z_{ik} \log \pi_k - \sum_k x_{k.} \log x_{k.}.$$

does not depend on $\mathbf{w}$ and $L_c(\mathbf{w}, \boldsymbol{\theta}/\mathbf{z})$ corresponds to the complete log-likelihood associated to a classical mixture model defined on the samples $\mathbf{v}_1, \ldots, \mathbf{v}_s$ where $\mathbf{v}_j = (x_{1j}, \ldots, x_{kj}, \ldots, x_{gj})$ with $x_{kj} = \sum_i z_{ik} x_{ij}$ and therefore develop the different steps of the CEM algorithm applied on $\mathbf{v}_1, \ldots, v_s$ to maximize $L_c(\mathbf{z}, \mathbf{w}, \boldsymbol{\theta})$ for $\mathbf{z}$ fixed.

Finally, we can describe easily the different steps of the algorithm called block CEM and noted BCEM:

1. Start from an initial position $(\mathbf{z}^{(0)}, \mathbf{w}^{(0)}, \boldsymbol{\theta}^{(0)})$.
2. Computation of $(\mathbf{z}^{(c+1)}, \mathbf{w}^{(c+1)}, \boldsymbol{\theta}^{(c+1)})$ starting from $(\mathbf{z}^{(c)}, \mathbf{w}^{(c)}, \boldsymbol{\theta}^{(c)})$:
   (a) Computation of $\mathbf{z}^{(c+1)}, \boldsymbol{\pi}^{(c+1)}, \delta^{(c+\frac{1}{2})}$ using the CEM algorithm on the data $(\mathbf{u}_1, \ldots, \mathbf{u}_r)$ starting from $\mathbf{z}^{(c)}, \boldsymbol{\pi}^{(c)}, \delta^{(c)}$.
   (b) Computation of $\mathbf{w}^{(c+1)}, \boldsymbol{\rho}^{(c+1)}, \delta^{(c+1)}$ using the CEM algorithm on the data $(\mathbf{v}_1, \ldots, \mathbf{v}_s)$ starting from $\mathbf{w}^{(c)}, \boldsymbol{\rho}^{(c)}, \delta^{(c+\frac{1}{2})}$.
3. Iterate the steps 2 until the convergence.

4 Numerical Experiments

To illustrate the behavior of our algorithms BCEM and 2CEM, we studied their performances on simulated data. We selected twenty five kinds of data arising from 3×2-component Poisson block mixture in considering firstly the situation where the proportions are equal proportions ($\pi_1 = \pi_2 = \pi_3$ and $\rho_1 = \rho_2$). These data are obtained by varying the following parameters: the degree of overlapping which depends on the parameters $\boldsymbol{\theta} = (\boldsymbol{\pi}, \boldsymbol{\rho}, \delta)$, and the sizes r and s. This degree of overlapping can be measured by the Bayes error corresponding to our model. Its computation being theoretically difficult, we used Monte Carlo simulations and evaluated this error by comparing the simulated partitions and those we obtained by applying a C-step. Five degrees of overlapping have been considered and are approximatively equal to $6\%, 11\%, 16\%, 18\%, 20\%$. Concerning the size, we took $r \times s = (30 \times 100), (50 \times 100), (100 \times 100), (500 \times 100)$ and (1000×100).

For each of these 25 data structures, we generated 30 samples and for each sample, we ran BCEM and CEM 100 times starting from random situations and selected the best solution for each method. In order to summarize the behavior of these algorithms, we used the proportion of misclassified points "error rate" occurring for each sample.

The results obtained are displayed in Table 1. For each data set and each algorithm, we summarize the 30 trials with the means and standard deviations of error rates obtained by comparing the partitions obtained by the both methods and the simulated partitions. In Table 2, we report the means and standard deviations of running times.

From these experiments, the main point arising are the following.

- The version 2CEM working on the two sets separately is suitably effective only when the clusters are well separated. This shows the risk of the use of such methods when the clusters are ill-separated.

Table 1. Comparison of BCEM and 2CEM for 30 kinds of data : means and standard deviations of error rates

Size		Overlap				
		1	2	3	4	5
30	BCEM	0.177 (0.084)	0.321 (0.186)	0.560 (0.164)	0.665 (0.106)	0.657 (0.135)
	2CEM	0.309 (0.066)	0.427 (0.134)	0.625 (0.124)	0.663 (0.092)	0.678 (0.101)
50	BCEM	0.105 (0.055)	0.239 (0.076)	0.488 (0.126)	0.707 (0.116)	0.682 (0.146)
	2CEM	0.262 (0.066)	0.350 (0.090)	0.581 (0.103)	0.701 (0.086)	0.710 (0.102)
100	BCEM	0.063 (0.024)	0.155 (0.015)	0.335 (0.062)	0.449 (0.160)	0.623 (0.155)
	2CEM	0.183 (0.056)	0.281 (0.049)	0.477 (0.101)	0.570 (0.086)	0.658 (0.124)
500	BCEM	0.061 (0.011)	0.123 (0.011)	0.166 (0.019)	0.198 (0.022)	0.255 (0.040)
	2CEM	0.098 (0.019)	0.195 (0.024)	0.277 (0.043)	0.375 (0.070)	0.446 (0.080)
1000	BCEM	0.065 (0.005)	0.118 (0.007)	0.162 (0.012)	0.187 (0.016)	0.212 (0.029)
	2CEM	0.083 (0.012)	0.190 (0.022)	0.247 (0.025)	0.300 (0.052)	0.376 (0.037)

Table 2. Comparison of BCEM and 2CEM for 30 kinds of data : means and standard deviations of running times

Size		Overlap				
		1	2	3	4	5
30	BCEM	2.102 (0.126)	2.162 (0.187)	1.934 (0.176)	1.870 (0.131)	1.871 (0.094)
	2CEM	1.422 (0.058)	1.490 (0.048)	1.565 (0.198)	1.476 (0.039)	1.461 (0.037)
50	BCEM	2.314 (0.274)	2.901 (0.173)	2.823 (0.553)	2.394 (0.098)	2.444 (0.084)
	2CEM	2.440 (0.150)	2.418 (0.141)	2.689 (0.693)	2.437 (0.149)	2.349 (0.157)
100	BCEM	2.282 (0.147)	3.386 (0.192)	3.785 (0.335)	3.230 (0.225)	2.827 (0.169)
	2CEM	4.599 (0.071)	4.607 (0.070)	4.685 (0.061)	4.653 (0.104)	4.560 (0.053)
500	BCEM	6.346 (0.435)	7.387 (0.758)	8.784 (0.933)	7.800 (0.833)	6.868 (0.729)
	2CEM	26.760 (0.250)	26.430 (0.227)	26.719 (0.364)	26.540 (0.436)	26.407 (0.183)
1000	BCEM	9.460 (1.130)	10.521 (0.981)	10.189 (0.874)	8.382 (0.609)	7.916 (0.626)
	2CEM	54.566 (0.280)	54.453 (0.318)	54.387 (0.348)	54.796 (0.443)	54.277 (0.186)

- Incontestably BCEM outperforms 2CEM. The results are very encouraging and its performance increases with the size of data.
- It appears clearly that BCEM is undoubtedly faster as soon as the size is large enough.

We carried out other simulations on large data sets with proportions dramatically different, not included in this text, which confirms these remarks.

5 Conclusion

Setting the problem of block clustering under the CML approach, we have compared block CEM and two-way CEM. The first one gives encouraging results on simulated data and real data and is therefore strongly recommended : it is faster and better than two-way CEM. Currently, we are evaluating block CEM on other large real data sets. In this paper, we have considered the block clustering for contingency tables under the CML approach and, as in Govaert and Nadif (2005a, 2005b) for binary data, it would be interesting to study the block clustering of contingency table under the ML and fuzzy approaches.

References

Arabie, P., J., H.L.: The bond energy algorithm revisited. IEEE Transactions on Systems, Man, and Cybernetics **20** (1990) 268–274

Bock, H.: Simultaneous clustering of objects and variables. In Diday, E., ed.: Analyse des Données et Informatique, INRIA (1979) 187–203

Celeux, G., Govaert, G.: A classification em algorithm for clustering and two stochastic versions. Computational Statistics and Data Analysis **14** (1992) 315–332

Cheng, Y., Church, G.: Biclustering of expression data. In: Proceedings of the Eighth International Conference on Intelligent Systems for Molecular Biology (ISMB). (2000) 93–103

Dempster, A.P., Laird, N.M., Rubin, D.B.: Maximum likelihood from incomplete data via the em algorithm (with discussion). Journal of the Royal Statistical Society **B 39** (1977) 1–38

Dhillon, I.: Co-clustering documents and words using bipartite spectral graph partioning. In: ACM SIGKDD Conference, San Francisco, USA. (2001) 269–274

Duffy, D.E., Quiroz, A.J.: A permutation-based algorithm for block clustering. Journal of Classification **8** (1991) 65–91

Garcia, H., Proth, J.M.: A new cross-decomposition algorithm: The GPM comparison with the bond energy method. Control and Cybernetics **15** (1986) 155–165

Govaert, G.: Classification croisée. Thèse d'état, Université Paris 6, France (1983)

Govaert, G.: Simultaneous clustering of rows and columns. Control and Cybernetics **24** (1995) 437–458

Govaert, G., Nadif, M.: Clustering with block mixture models. Pattern Recognition **36** (2003) 463–473

Govaert, G., Nadif, M.: An EM algorithm for the block mixture model. IEEE Transactions on Pattern Analysis and Machine Intelligence **27** (2005) 643–647

Govaert, G., Nadif, M.: Fuzzy clustering to estimate the parameters of block mixture models. Soft Computing (in press, 2005)

Hartigan, J.A.: Clustering Algorithms. Wiley, New York (1975)

Marchotorchino, F.: Block seriation problems: A unified approach. Applied Stochastic Models and Data Analysis **3** (1987) 73–91

McLachlan, G.J., Peel, D.: Finite Mixture Models. Wiley, New York (2000)

Ritschard, G. Zighed, D., Nicoloyannis, N., Maximisation de l'association par regroupement de lignes ou de colonnes d'un tableau croisé. Revue de Mathématiques & Sciences Humaines **39** (2001) 81–97

Symons, M.J.: Clustering criteria and multivariate normal mixture. Biometrics **37** (1981) 35–43

An Imbalanced Data Rule Learner

Canh Hao Nguyen and Tu Bao Ho

School of Knowledge Science,
Japan Advanced Institute of Science and Technology,
1-1 Asahidai, Nomi, Ishikawa, 923-1292, Japan
{canhhao, bao}@jaist.ac.jp

Abstract. Imbalanced data learning has recently begun to receive much attention from research and industrial communities as traditional machine learners no longer give satisfactory results. Solutions to the problem generally attempt to adapt standard learners to the imbalanced data setting. Basically, higher weights are assigned to small class examples to avoid their being overshadowed by the large class ones. The difficulty determining a reasonable weight for each example remains. In this work, we propose a scheme to weight examples of the small class based solely on local data distributions. The approach is for categorical data, and a rule learning algorithm is constructed taking the weighting scheme into account. Empirical evaluations prove the advantages of this approach.

1 Introduction

It a practical sense, applying standard machine learning methods to real world tasks when their class distribution is imbalanced is problematic. This is the case of a data set, in which the number of examples of one class is substantially smaller than the others. For instance,if one class accounts for only 2% of the total number of examples in the data set, a classifier can get a high accuracy of 98% just by assigning all its examples to the large class. However, in such a case, the classifier completely fails to learn the small class, which is usually of interest. In practice, researchers have encountered this problem in many domains, including the detection of fraudulent transactions [1], network intrusion detections [2] and oil spills in satellite radar images [3].

The reason that standard classifiers can no longer give a satisfactory performance on such data sets is because they make the fundamental assumption that frequencies of classes are equally distributed. Adaptations to imbalanced data sets are usually made by giving small class examples higher weights. One simple way is resampling, which duplicates small class examples or selects only a subset of large class ones. Such approaches do not have high performance, as reported in [4], because various examples are affected differently by the class imbalance problem. SMOTE [5] combines synthetic example generation with downsampling, but the resampling degree is not specified. Resampling to reflect relative weights between examples or classes still remains an art.

It is believed to be more promising to weight examples differently, and various approaches have been proposed. Kubat et al. [3] insist that large class examples in a mixed region should be weighted zero as long as that increases performance measure. Learning on a cluster basis is used [6] to weight examples accordingly. Learning decision trees (DT) [7] is made independent of class frequencies by using the Area Under the ROC Curve (AUC) as a splitting criterion, which is equivalent to weighting examples according to their distribution in the set of examples covered by the splitting nodes. A general way to optimally weight examples (in Bayes risk sense) is using MetaCost [8], by bagging and then probability estimation. However, in highly imbalanced data sets, examples of small class are rarely learned, making their optimal costs extremely high. Again, it is still a challenge to weight examples optimally for the imbalanced data problem.

We propose a method to estimate the optimal weight of each small class example basing solely on local data distributions. The intuition is that by looking more closely into local data distribution, we have more chance to reveal useful information about the effect of class imbalance. To this end, we first define the concept of *vicinity*, which characterizes local data distribution and then determines examples' weights with the aiming of maximizing AUC in the vicinity. The weight is integrated into a rule induction algorithm at the rule pruning step.

The paper is organized as follows. Section 2 is the foundation and formulation of our locally adaptive weighting scheme. Integration of the weighting scheme into our rule learning algorithm is described in Section 3. In Section 4, we show experimental evaluations of the scheme to other imbalanced data classifiers. Conclusions are presented and future work is discussed in section 5.

2 Locally Adaptive Weighting Scheme

We approach the imbalanced data problem by giving a weight adaptively for each small class example, while keeping the weights of large class examples at a default value (i.e. 1). The key idea is to weight each small class based on its local neighborhood (hence, it is locally adaptive), which is defined as the vicinity of the rule covering it. This section will define the concept of vicinity and derive the formulation of example weighting based on vicinity using AUC as the criterion.

Vicinity: The idea behind vicinity is as follows. Consider two rules $R_i, i = 1, 2$ with the same coverage for every class (R_i covers n_i, p_i examples from large and small classes respectively, $n_1 = n_2, p_1 = p_2$). Conventionally, the two rules are evaluated as the same goodness (e.g., precision for small class $\frac{p_i}{p_i+n_i}$). Assume that we have some way to define the surround of a rule, called a neighborhood. If R_1 is likely to be pruned to a better one, then in its neighborhood there must be some examples of the same class as the predicting class of R_1. On the other hand, R_2 is surrounded by examples from other classes, hence it cannot be pruned to a better one. Our idea is to evaluate the two rules differently, R_1 to be higher than R_2, reflecting their ability to be pruned. This different evaluation is based on the fact that there is a set of examples in each rule's neighborhood, which creates the difference in pruning ability. By vicinity, then, we mean this set of examples. We define vicinity based on the concept of k-vicinity.

Definition: *The distance from a rule to an example is the minimum number of attribute value pairs in the body of the rule that must be removed in order to make the rule cover the example.*

Definition: *The k-vicinity of a rule R for a training data set D is the set of examples in D that are less than or equal to a distance of k to the rule.*

$$k\text{-}vicinity(R) = \{x \mid x \in D, Distance(R, x) \leq k\} \quad (1)$$

K-vicinity is a subset the training data set, which is potentially covered by the rule after k steps of generalization. The smaller k is, the higher the influence the examples in k-vicinity may have on the generalization (pruning) ability of the rule. For example, 0-vicinity is the set of examples covered by the rule, m-vicinity is the whole data set if m is the number of attribute-value pairs in the rule body. The set of all k-vicinities is a nested chain of subsets of the data set, meaning: $0\text{-}vicinity \subseteq 1\text{-}vicinity \subseteq ... \subseteq m\text{-}vicinity$. We define vicinity using this chain with weights. Formally, vicinity is a function f over k-vicinities.

$$vicinity = f\{1\text{-}vicinity, 2\text{-}vicinity, ..., m\text{-}vicinity\} \quad (2)$$

Estimating vicinity is a difficult task. However, the way around the problem is to let the vicinity of a rule remain a virtual concept. We only need to calculate the *ratio of class distribution* in the vicinity as in the weighting scheme discussed in the next section.

Example Weighting and Rule Evaluation: As a vicinity is expected to contain examples that influence the pruning ability of a rule, we use this assumption to define the *best* rule as the one giving optimal classification within its vicinity. Our idea is to weight examples in the vicinity such that the optimal classification coincides with the lowest misclassification cost. Defining optimal classification on a vicinity results in a locally adaptive weighting scheme.

We define optimal classification as the one that gives the largest AUC. AUC is a popular metric for comparing classifiers' performance [9, 10] when the misclassification costs are unknown. When a classifier is a set of rules, as in Figure 1 (a), the ROC curve contains a set of line segments. Here, the classifier is assumed to have four rules, sorted in decreasing order of their precision for a class. For simplicity, we assume that there is only one rule for small class in a vicinity. Then, the ROC curve of a classifier (by R or $R2$) in its vicinity would look like Figure 1 (b). The classifier here consists of a rule (say R) and the default rule predicting the large class. Suppose R covers p small and n large class examples, and the vicinity contains P small and N large class examples. The rule evaluation metric, defined to be AUC above, is calculated [7] as:

$$AUC(R) = \frac{p}{2P} - \frac{n}{2N} + \frac{1}{2} \quad (3)$$

The above formula implies that the weight of a small class example in this vicinity is $\frac{N}{P}$ when the weight of a large class example is the default value 1.

The rule evaluation metric is used to compare different rules for search bias. However, it is not natural to compare AUC in different contexts (vicinities).

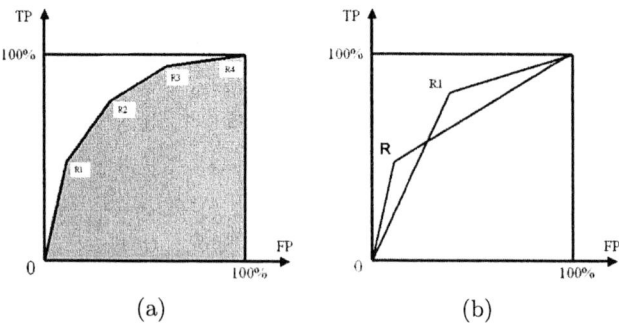

Fig. 1. The ROC space: (a) plot of a rule learner. (b) for rule comparison in a vicinity.

Hence, we propose a comparison strategy that rule R_1 is considered better than rule R if and only if it gives a higher AUC in the vicinity of R. Equivalently, R_1 is considered better than R if their AUC difference (in formula 4) is positive.

$$AUC(R_1) - AUC(R) = \tfrac{1}{2}(\tfrac{p_1-p}{P} - \tfrac{n_1-n}{N}) = \tfrac{1}{2}[(p_1-p)\tfrac{N}{P} - (n_1-n)]\tfrac{1}{N} \quad (4)$$

For the purpose of comparing rules for search bias, it is sufficient to know $\tfrac{N}{P}$. This is the reason we allow vicinity to remain an abstract concept, while we heuristically estimate $\tfrac{N}{P}$ directly. From equation 2, we propose to estimate class distribution ratio $\tfrac{N}{P}$ in the vicinity to be:

$$\frac{N}{P} = \sum_{k=1}^{m} w_k * \frac{N_k}{P_k} \quad (5)$$

where m is the number of vicinities. In this formula, $\tfrac{N_k}{P_k}$ is class distribution ratio in k-vicinity and w_k is its associated weight, with $\sum w_k = 1$. This just smooths out the class distribution ratios of different k-vicinities to estimate that of the vicinity, in similar fashion to a shrinkage estimator, to make it robust. The definition of vicinity is tunable by its weighting scheme, namely the set $\{w_k\}$. If w_k is large for small k-s, vicinity reflects more local information. If w_k is large for large k-s, vicinity is more global. If we want to define vicinity to be the whole data set, then set $w_m = 1$. Having tunable set of $\{w_k\}$ is a generalization of a simple cost sensitive classification. In this algorithm, we fix the default as:

$$w_k = \frac{1}{m}, k = \overline{1,m} \quad (6)$$

Such weights can also be set adaptively by users.

Discussion: The key point which makes this example weighting scheme suitable for imbalanced data is the use of a local neighborhood. Having a myopic view around a rule gives us a better picture of how much the imbalance may hinder classification rules. Examples far away from the boundaries of classes may not participate in any vicinity, also not affecting class discrimination (this is similar to the idea behind SVMs).

IDL
1. Generate a candidate rule set
2. Prune rules from high coverage to low

GenerateRuleSet
1. Generate a decision tree
2. Stop when leaf nodes contain only example from one class
3. Extract the set of leaf nodes that contain only examples of small class
4. Convert those nodes into rules and return

PruneRules
1. Sort rules according to coverage
2. From high to low coverage rule do
3. Remove *best* attribute value pair
4. Until no more AUC is gained
5. Return pruned rules

Fig. 2. IDL algorithm

3 IDL: Imbalanced Data Learner

IDL is a rule induction algorithm, which uses an one-sided selection strategy, taking example weights into account to learn rules for a small class. IDL consists of two steps. First, it generates a set of candidate rules for the small class by growing a decision tree, which are meant to be complete and of high precision. Then it prunes these rules by greedily removing attribute value pairs to make them robust. Example weighting is used in rule pruning step using Formula 3 as the rule evaluation metric. The overall strategy is depicted in Figure 2.

In the candidate rule set generation step, IDL grows a decision tree and only stops when the leaf nodes contain examples from one class. As recommended in [11], IDL takes the impurity $(2\sqrt{p(1-p)})$ gain as the splitting criterion. After the decision tree is fully grown, the set of leaf nodes that contain only small class examples are collected and turned into a set of rules. In the second step, the collected rules for the small class are sorted in decreasing order of coverage. Starting from the highest coverage one, each rule is pruned by removing the *best* attribute value pair (the one having the highest AUC difference), according to formula 4. It stops when removing does not improve either AUC or when the precision of the rule (calculated without taking weights into account) falls under a certain threshold. The examples covered by a rule are marked so that if they are covered again, only half of their weights are retained. This makes the rules overlap, and also greatly improves the recall of the classifier. The threshold represents the minimum precision a rule should achieve, reflecting the amount of noise in the data. This threshold is generally set by the users, and is estimated in IDL as follows. First, set it to 80%, then do a 10-fold stratified cross-validation on the data set to estimate its difficulty to learn. Taking the F-measure on the small class, say f (percent), then the threshold takes the value $max(50, f - 10)$.

In the first step, IDL constructs an unpruned decision tree, which is of $O(ea)$ time complexity, where e is the number of examples and a is the number of attributes. In the second step, suppose it generates k rules, each has maximum n_k attribute value pairs. As each pruning operation requires a pass of the data set to calculate the class distribution ratio in the vicinity, the time complexity of this step is at most $O(ekn_k)$.

4 Experimental Evaluation

We evaluated IDL on its ability to learn a small class, and compared it to other approaches. The first of these was SMOTE-NC (over C4.5) [5], the nominal categorical version of what is arguably best method (SMOTE) for learning imbalanced data. Since SMOTE is sensitive to its degree of sampling parameters, we ran it on three degrees of small class upsampling, namely N=100%, N= 300% and N=700%. We also compared IDL to a general classifier of C4.5, with and without cost sensitive setting. In cost sensitive setting (C.S.), the relative cost is just the ratio of class distribution of the data set. Boosting is also capable of enhancing imbalanced data learners [12], so AdaBoost over C4.5 was also compared. We used these classifiers from WEKA[1]. All algorithms ran with their default parameters. We used F-measure on the small class as the performance criterion, instead of the AUC measure (since we learn only small class rules).

$$F - measure = \frac{2 * precision * recall}{precision + recall} \quad (7)$$

We evaluated those algorithms on selected fifteen UCI data sets [2], where smallest class was chosen to be small class, and the other classes were merged to become the large class. As the algorithm is for categorical data, all data sets were discretized. We split data sets with a ratio of 75-25 randomly in a stratified manner, the large parts were used for training and the small parts for testing. Table 1 shows the result of testing on the small part of the data. The columns are names, percentage of small class proceeded with class index and then classifiers (SMOTE is tested with three parameters). All numbers are in percentage. The last line shows average performance of on all data sets.

The table shows that our approach outperforms general classifiers by a large margin, and is competitive to all three parameters for SMOTE. IDL shows an improvement of 11.74% in terms of F measure on the small class compared to a standard classifier of C4.5. For the cost sensitive setting of C4.5 (C.S), it also improves by 3.81%. Compared to AdaBoost, IDL's accuracy is 2.85% higher. This means that IDL is more suitable for imbalanced data than general classifiers. Comparing IDL to a imbalanced data learner of SMOTE (SMOTE-NC version), IDL is also competitive to the three parameter settings; the average

---
[1] www.cs.waikato.ac.nz/ml/weka/
[2] http://www.ics.uci.edu/ mlearn/MLRepository.html

Table 1. Comparison of Classifiers on UCI data

Name	%	C4.5	C.S.	SMOTE				A.Boost	IDL
				100	300	700	average		
annealing1	11.0	73.9	62.3	77.4	71.0	66.7	71.7	70.6	96.2
car3	3.7	66.7	84.2	77.4	80.0	80.0	79.1	80.0	76.9
flare4	8.0	0.0	36.9	30.4	36.1	38.6	35.0	32.7	29.0
glass3	13.5	93.3	73.7	93.3	82.4	82.4	86.0	80.0	85.7
hypo0	5.0	85.7	81.9	85.7	83.1	83.1	84.0	84.6	84.6
inf0	6.3	100.0	100.0	100.0	100.0	100.0	100.0	100.0	100.0
krkopt16	0.9	85.7	82.8	88.4	88.4	90.1	89.0	84.1	87.1
krkopt4	0.7	58.0	69.1	66.7	71.8	66.7	68.4	61.3	75.9
led7	8.4	59.8	59.9	63.9	61.6	50.5	58.7	50.7	62.4
letter0	3.9	91.0	90.0	92.0	91.8	89.7	91.2	96.0	92.0
satimage3	9.7	51.5	52.8	57.9	51.8	51.3	53.7	57.9	50.3
segmentation5	14.1	100.0	100.0	100.0	100.0	100.0	100.0	100.0	100.0
sick1	6.5	82.8	69.6	82.8	81.8	73.8	79.5	82.8	80.9
vowel5	9.1	0.0	75.2	67.8	69.4	65.5	68.2	80.0	60.0
yeast4	3.4	0.0	30.8	33.3	44.4	40.0	39.2	21.1	43.5
Average	6.94	63.23	71.16	74.59	74.24	71.89	73.57	72.12	74.97

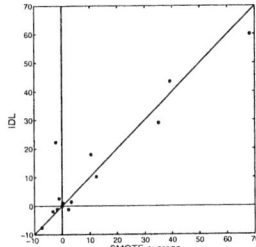

Fig. 3. Improvement of IDL versus SMOTE

performance of SMOTE is 1.40% lower than that of IDL. It is noteworthy that there is no systematical way to determine the resampling degree for SMOTE.

It is interesting to look at the improvement of IDL over C4.5 compared to the average improvement of that of SMOTE over C4.5 in Figure 3. X axis is the performance improvement of SMOTE (averaging all parameters), while y axis is for IDL. The set of points shows a near linear relation. This means that improvement of IDL is proportionate with that of SMOTE, meaning that IDL is consistently similar to SMOTE.

5 Conclusion

We have proposed a method to weight examples for a small class based on their local neighborhood. Neighborhood is defined as the virtual concept of vicinity, while computation is based on k-vicinities. The algorithm is clearly more accurate than general classifiers, including Adaboost and MetaCost, and is competitive to SMOTE while having the advantage of not requiring resampling parameters. From this, we can conclude that the information of neighborhood of an example is useful for weighting it, in order to compensate imbalanced data.

The clear limitation of this method is how to define the weighting scheme for a vicinity. For the moment, computational complexity is its main problem, which should be reduced for large data sets. Applying the weighting scheme to other classifiers for imbalanced data is a natural extension. Whether local data distribution can be used to improve classifiers in general is an open question.

References

[1] Fawcett, T., Provost, F.: Combining data mining and machine learning for effective user profiling. In Simoudis, Han, Fayyad, eds.: The Second International Conference on Knowledge Discovery and Data Mining, AAAI Press (1996) 8–13

[2] Lazarevic, A., Ertoz, L., Ozgur, A., Srivastava, J., Kumar, V.: "evaluation of outlier detection schemes for detecting network intrusions". In: Third SIAM International Conference on Data Mining. (2003)

[3] Kubat, M., Holte, R.C., Matwin, S.: Machine learning for the detection of oil spills in satellite radar images. Machine Learning **30** (1998) 195–215

[4] Japkowicz, N.: The class imbalance problems: Significance and strategies. In: Proceedings of the 2000 International Conference on Artificial Intelligence (IC-AI'2000). Volume 1. (2000) 111–117

[5] Chawla, N.V., Bowyer, K.W., Hall, L.O., Kegelmeyer, W.P.: Smote: Synthetic minority over-sapling technique. Journal of Artificial Intelligence Research **16** (2002) 321–357

[6] Nickerson, A., Japkowicz, N., Milios, E.: Using unsupervised learning to guide re-sampling in imbalanced data sets. In: Eighth International Workshop on AI and Statitsics. (2001) 261–265

[7] Ferri, C., Flach, P., Hernandez-Orallo, J.: Learning decision trees using the area under the roc curve. In Sammut, C., ed.: Nineteenth International Conference on Machine Learning ICML'02, Morgan Kaufmann (2002)

[8] Domingos, P.: Metacost: A general method for making classifiers cost-sensitive. In: Knowledge Discovery and Data Mining. (1999) 155–164

[9] Provost, F., Fawcett, T.: Robust classification for imprecise environments. Machine Learning **42** (2001) 203–231

[10] Furnkranz, J., Flash, P.: An analysis of rule evaluation metrics. In: The Twentieth International Conference on Machine Learning (ICML'03), AAAI Press (2003) 202–209

[11] Elkan, C.: The foundations of cost-sensitive learning. In: Seventeenth International Joint Conference on Artificial Intelligence (IJCAI'01). (2001) 973–978

[12] Joshi, M.V., Agarwal, R.C., Kumar, V.: Predicting rare classes: can boosting make any weak learner strong? In: Proceedings of the eighth ACM international conference on Knowledge discovery and data mining, ACM Press (2002) 297–306

Improvements in the Data Partitioning Approach for Frequent Itemsets Mining

Son N. Nguyen and Maria E. Orlowska

School of Information Technology and Electrical Engineering,
The University of Queensland, QLD 4072, Australia
{nnson, maria}@itee.uq.edu.au

Abstract. Frequent Itemsets mining is well explored for various data types, and its computational complexity is well understood. There are methods to deal effectively with computational problems. This paper shows another approach to further performance enhancements of frequent items sets computation.

We have made a series of observations that led us to inventing data pre-processing methods such that the final step of the Partition algorithm, where a combination of all local candidate sets must be processed, is executed on substantially smaller input data. The paper shows results from several experiments that confirmed our general and formally presented observations.

Keywords: Association rules, Frequent itemset, Partition, Performance.

1 Introduction

Since the association rules mining introduction by Argawal et al. [5], many algorithms and their subsequent improvements have been proposed to solve association rules mining, especially frequent itemsets mining problems.

In this paper, we review the state of the art in association rules mining with a focus on frequent itemsets mining. There are many well-accepted approaches such as "Apriori" by Argawal et al. [1], ECLAT by Zaki [7], and more recently "FP-growth" by Han et al. [8]. Another interesting class of solutions is based on the data partitioning approach. This fundamental concept was originally proposed as a Partition algorithm by Savaserse et al. [2], and it was improved later in AS-CPA by Lin et al. [4] and ARMOR by Pudi et al. [11]. A common feature of these results is their target, namely the limitation of I/O operations by considering data subsets dictated by the main memory size.

An intriguing question is whether we could improve the overall performance of mining large data sets by a smarter but not too 'expensive' design of the data fragments - rather than determine them by a sequential transaction allocation based on the fragment size only.

The main goal of this paper is to demonstrate our observations, generalize, and specify corresponding data pre-processing for the Partitioning approach in order to improve the performance. Our study is supported by a series of experiments which indicate a dramatic improvement in the performance of the Partitioning approach with our fragmentation method, in contrast to the traditional one [2].

The remainder of the paper is organised as follows. Section 2 introduces the basic concepts related to frequent itemsets mining. Section 3 reviews the current state of art in the field, especially for frequent itemsets mining and the Partitioning approach. Section 4 presents our observations and open issues. We propose the pre-processing data fragmentation solution in section 5. Section 6 shows the result from our experiment, and finally, we present our concluding remarks in section 7.

2 Preliminary Concepts

For the completeness of this presentation and to establish our notation, this section gives a formal description of the problem of mining frequent itemsets. It can be stated as follows [1]:

Let $I = \{i_1, i_2, ..., i_m\}$ be a set of m distinct literals called items. Transaction database D is a set of variable length transactions over I.

Each transaction contains a set of items $\{i_j, i_k, ..., i_h\} \subseteq I$, $i_j < i_k < ... < i_h$. Each transaction has an associated unique identifier called TID.

For an itemset $X \subseteq I$, the support is denoted $\sup_D(X)$, equals to the fraction of transactions in D containing X.

The problem of mining frequent itemsets is to generate all frequent itemsets X that have $\sup_D(X)$ no less than user specified minimum support threshold.

3 Review Frequent Itemsets Mining

Throughout the last decade, there have been many attempts and well-known algorithms that target an efficient solution of the frequent itemsets mining problem. However, the performance of these algorithms depends on many, often very specific input data features and additionally, implementation environments. As a result, several claims made in earlier papers were later debated by other authors.

3.1 Partitioning Approach for Frequent Itemsets Mining

Savaserse et al. [2] proposed the Partition algorithm based on the following principle. A fragment $P \subseteq D$ of the database is defined as any subset of the transactions contained in the database D. Further, any two different fragments are non-overlapping. *Local support* for an itemset is the fraction of transactions containing that itemset in a fragment. *Local candidate itemset* is being tested for minimum support within a given fragment. A *Local frequent itemset* is an itemset whose local support in the fragment is no less than the minimum support. *Global support, Global candidate itemset, Global frequent itemset* are defined as above except they are in the context of the entire database. The goal is to find all *Global frequent itemsets*.

The following Lemma 1 supports the main principle of the Partition algorithm.

Lemma 1: If X is a frequent itemset in database D, which is partitioned into n fragments $P_1, P_2, ..., P_n$, then X must be a frequent itemset in at least one of the n fragments.

Proof: Due to the limit space, the proof can be seen in [10]

The Partition algorithm divides D into n fragments. The algorithm first scans fragment P_i in the main memory at a time, for i = 1,...,n, to find the set of all *Local frequent itemsets* in P_i, denoted as LP_i. Then, by taking the union of LP_i, a set of candidate itemsets over D is constructed, denoted as C^G. Based on *Lemma 1*, C^G is a superset of the set of all *Global frequent itemsets* in D. Finally, the algorithm scans each fragment for the second time to calculate the support of each itemset in C^G and to find the *Global frequent itemsets*.

3.2 Related Work in Partitioning Approach

One of the Partition algorithm derivatives is AS-CPA (*Anti-Skew Counting Partition Algorithm*) by Lin et al. [4]. Recently, there has been another development based on the partitioning approach in the ARMOR algorithm by Pudi et al. [11].

All the above algorithms mainly attempt to reduce the number of false candidates as early as possible. However, they do not consider any features and characteristics of data sets in order to partition the original data set more suitably for further processing.

Further in this paper, we demonstrate that looking more closely into the data itself may deliver good gains in overall performance. As a result, the *Local frequent itemsets* can be dramatically reduced. Furthermore, in many cases that leads to a larger number of common Global candidates among fragments. Finally, as a consequence, this approach reduces substantially the *Global candidates* (C^G set).

4 Observations in the Partitioning Approach

We begin by considering the first and very obvious measurable data-partitioning attribute – the size of fragments and their impact on the efficiency of the frequent items search process. Further on we examine more closely the composition of fragments at the design time to ensure that selection of transactions satisfy some desired properties.

4.1 Reasoning About Size of Fragment

It is not hard to observe that the size of the fragments is inverse-proportional to the size of the output of Local computation. Hence, the question is: *What is a 'good' fragment size?* We consider several heuristic methods to identify the suitable size of fragments.

We note the following observation: the smaller fragment generates a more negative effect on the number of *Local frequent itemsets*. Clearly, the best partitioning of data set D into n fragments is defined as a method that generates the smallest number of Global candidates. We denote this smallest number as G_n. Note that the perfect solution would have to exhibit the following property; *every fragment of the data generates identical Local frequent itemsets*.

We generalise these observations as follows;

Lemma 2: If database D is partitioned into (n+1) fragments $P_1, P_2, ... , P_{n+1}$ then the number of Global candidates, denoted $|C^G_{n+1}|$, is always greater than or equal to G_n; $|C^G_{n+1}| \geq G_n$

Proof: Due to the limit space, the proof can be seen in [10]

As a consequence, the size of a fragment should maintain proper balance in order to control the number of *Local frequent itemsets*.

4.2 Some Characteristics of Fragment Data

Data skew has a negative impact on the Partitioning approach. Basically, data skew causes the *Local frequent itemsets* generated from different fragments to have very few common elements. In such situations, the number of *Global candidates* (being the union of all LP_i) is rather large.

Obviously, fragments that have many dissimilar transactions (transactions with small or empty intersections) generate a small number of *Local frequent itemsets*. In this paper we call them ***dissimilar fragments***.

These observations confirm our initial hypothesis that there are some relationships between the composition of fragments and the amount of computation required at the end. We illustrate the fact that a larger number of fragments increase the size of the computation space. In addition, for given number of fragments n, a different partition also impacts on the number of *Global candidates*. Furthermore, the gap in performance is increased dramatically when the support threshold is decreased and the number of fragments is increased.

5 Data Set Pre-processing

We present the following algorithms for original data pre-processing.

5.1 Naive Algorithm

One of the simplest techniques to be considered is the skipping technique. Before formalising this concept we show a simple example to illustrate its main principle.

Consider data set D represented by a straight line on figures below. We partition D into 2 fragments as illustrated on the Figure 1. When a Skipping technique is used then D is initially divided into 4 small sequential parts. Each fragment is created by taking the union of 2 small skipping parts as it is shown on Figure 2.

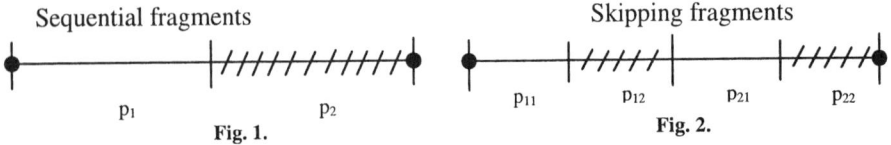

Sequential fragments p_1 p_2

Fig. 1.

Skipping fragments p_{11} p_{12} p_{21} p_{22}

Fig. 2.

One can easily generalise such partition process for any higher number of expected fragments.

5.2 An incremental Clustering Algorithm

The incremental clustering algorithm is our idea for pre-processing. The data set will be scanned only once and all clusters (fragments) containing mostly dissimilar transactions are generated at the end of that scan. We introduce some basic definitions.

Definition 5.1: Cluster Centroid is a set of all Items in the cluster, we denote it $C_i = \{I_1, I_2,.., I_n\}$. Additionally, each item in C_i has its associated weight which is its number of occurrences in the cluster; $\{w_1, w_2, ..., w_n\}$

Definition 5.2: Similarity function between two item sets, in particular a transaction and a cluster centroid, is denoted Sim (T_i, C_j) and defined as follows;
Sim $(T_i, C_j) \rightarrow R^+$; Calculation of this function:
   1. Let S be the intersection between the arguments of Sim function, $S = T_i \cap C_j$
   2. If $S = \emptyset$ then Sim $(T_i, C_j) = 0$. Otherwise, $S = \{ I_1, I_2, ..., I_m \}$ with the corresponding weights $\{ w_1, w_2, ..., w_m \}$ in cluster C_j, respectively, therefore Sim $(T_i, C_j) = w_1 + w_2 + ... + w_m$

Cluster Construction:

Informally, each transaction is evaluated in terms of the following criteria;
   a) We sssign a new transaction T_i to cluster C_j which has the minimum Sim(T_i, C_j) value among open clusters (*a cluster is open if has not exceeded its expected size in terms of number of transactions*).
   b) Each new allocation to a cluster C_j, updates the cluster centroid C_j. All already existing common items' weight is increased by 1, and the other new items are added to C_j with the weight of value 1.

Reasoning about the size of clusters: based on the observation in section 4.1, the cluster sizes should be well balanced.

The pseudo incremental clustering algorithm is described as the following

Input: Transaction database: D; k – number of output clusters
Output: k clusters based on the above criteria for Partition approach.
   Begin
   1. Assign the first k transactions to all k clusters, and initialize the all Cluster Centroids: $\{C_1, C_2, ..., C_k\}$
   2. Consider the next k transactions: $\{T_1, T_2, ..., T_k\}$. These k transactions are assigned to k different clusters. These operations are done based on the following criteria: (i) the minimum similarity between the new transaction and the suitable clusters; (ii) the sizes of these clusters are controlled to keep the balance. The following are more detail about this processing.
      Let $C^{run} = \{C_1, C_2, ..., C_k\}$ is a set of all k clusters; $T^{run} = \{T_1, T_2, ..., T_k\}$
      For each transaction T_i in T^{run} : T_1 to T_k
      Begin
      a) Calculate the similar functions between T_i and all the clusters in C^{run}; determine the minimum similar function value, denoted Sim(T_i, C_j)
      b) Assign T_i to cluster C_j which has the minimum Sim(T_i, C_j) value. Update the cluster centroid C_j
      c) Remove C_j from the set of all the suitable clusters in order to keep the same size constraint. $C^{run} = C^{run} - \{ C_j\}$;
      End
   3. Repeat step 2 till all transactions in D are clustered
   End

The time complexity of this incremental clustering algorithm is about $O(|D| * k * m)$ where $|D|$ is the number of all transactions, k is the given number of clusters, and m is the number of all items in D.

6 Experiments

In this section, we conducted experiments on: one synthetic data set [1], and 3 real data sets [13]. These data sets are converted to format as the above definitions.

Table 1. The characteristics of data sets

Data sets	Transactions	Items	DB Size (~MB)
T10I4D100K	100K	870	4
WebView-1	26K	492	0.7
WebView-2	52K	3335	2
BMS-POS	435K	1657	10

Our goal is to compare the cardinality of the outputs from two phases of the Partitioning algorithm; at the Local level and the Global level, before and after application of our pre-processing. Firstly, data set is partitioned into fragments; secondly the Apriori algorithm (by Zhu T. [12]) is applied to find *Local frequent itemsets* (LP_i) for each fragment. Subsequently, union of these LPi generates the *Global candidates*.

Resulting figures for each data set are represented in following template table 2. The 2nd, 3rd and 4th columns' names indicate three techniques for data preparation: *Sequent* fragments correspond to loading clusters with original data, *Skipping* fragments are constructed as described in section 5.1; and the *Clustering* fragments are the pre-processed data as presented by our clustering method described in section 5.2.

The data sets used are indicated on the top of each table segment. We present three different scenarios; each data set is partitioned into 1, 2 and 5 fragments. The *Sequent* column represents the numbers of the Local level (LP_1, LP_2, ..., LP_n), the number of *Global candidates*. Note that this figure is presented by showing its two components; for example, **16 + (378)** indicates that there are **16** candidates to be checked and **378** common candidates don't need additional check.

Using the same convention, the *Skipping* and *Clustering* columns represent the figures for the *Skipping* technique and the *Clustering* pre-processing, respectively.

As can be seen from Table 2 and 3, there are big gains from the careful data preprocessing. Further, to discuss the impact of threshold level, let us denote the cardinality of checked Global candidate set as $|C_n^G|$, where n is the number of fragments. $|C_n^G|$ is reduced for all data sets for all support thresholds. For example, if T10I4D100k is partitioned into 2 fragments, $|C_2^G|$ decreases from **16** for *Sequent* to **3** for *Clustering* with the support threshold **0.01**. This reduction is also present when considering other real data sets that are partitioned into 2 fragments. Its value reduces from **1,820** to **348** with the threshold **0.005** for very large data set BMS-POS. Moreover, if data sets are partitioned into 5 fragments, this gap among 3 techniques is even greater. For example, if T10I4D100k is partitioned into 5 fragments, $|C_5^G|$ decreases from **48** for *Sequent* to **24** for *Clustering* with the threshold **0.01**, and **698** to **373** with

Table 2. The figures with a threshold 0.01

	Sequent	Skipping	Clustering
	T10I4D100K		
1-fragment: **385** Frequent Itemsets			
2 fragments			
LP1	385	387	385
LP2	387	386	386
C_2^G	*16+ (378)*	*17 + (378)*	*3 + (384)*
5 fragments			
LP1	392	386	387
LP2	381	388	387
LP3	393	388	384
LP4	386	387	388
LP5	390	391	388
C_5^G	*48+ (366)*	*57 + (362)*	*24+ (375)*
	WebView-1		
1-fragment: **208** Frequent Itemsets			
LP1	227	241	210
LP2	229	201	213
C_2^G	*152+152)*	*116 + (163)*	*17+ (203)*
5 fragments			
LP1	284	250	226
LP2	197	230	221
LP3	241	254	213
LP4	255	242	207
LP5	266	205	205
C_5^G	*425+ (92)*	*228 + (141)*	*74+ (181)*
	WebView-2		
1-fragment: **186** Frequent Itemsets			
LP1	279	156	192
LP2	221	236	179
C_2^G	*292+(104)*	*120 + (136)*	*19+ (176)*
5 fragments			
LP1	133	197	188
LP2	558	182	209
LP3	384	184	193
LP4	244	180	169
LP5	227	247	195
C_5^G	*756+ (55)*	*157 + (135)*	*64+ (160)*
	BMS-POS		
1-fragment: **1,503** Frequent Itemsets			
LP1	1,400	1,353	1,512
LP2	1,662	1,680	1,498
C_2^G	*390+ (1,336)*	*341+ (1,346)*	*60+ (1,475)*
5 fragments			
LP1	1,996	1,719	1,150
LP2	1,334	1,146	1,471
LP3	744	1,639	1,864
LP4	1,348	1,810	1,822
LP5	2,885	1,377	1,364
C_5^G	*2,263+ (689)*	*950+ (1,067)*	*894+ (1,121)*

Table 3. The figures with threshold 0.005

	Sequent	Skipping	Clustering
	T10I4D100K		
1-fragment: **1,073** Frequent Itemsets			
2 fragments			
LP1	1,079	1,101	1,068
LP2	1,101	1,077	1,092
C_2^G	*158 + (1,011)*	*148 +(1,015)*	*70 + (1,045)*
5 fragments			
LP1	1,150	1,181	1,089
LP2	1,141	1,074	1,110
LP3	1,248	1,091	1,059
LP4	1,110	1,122	1,135
LP5	1,120	1,135	1,098
C_5^G	*698 + (893)*	*578 + (889)*	*373 + (941)*
	WebView-1		
1-fragment: **633** Frequent Itemsets			
LP1	644	774	659
LP2	755	612	641
C_2^G	*503 + (448)*	*416 + (485)*	*94 + (603)*
5 fragments			
LP1	1,107	771	779
LP2	489	842	733
LP3	839	941	676
LP4	894	769	663
LP5	977	517	597
C_5^G	*1,806 + (271)*	*1,069 + (374)*	*497 + (493)*
	WebView-2		
1-fragment: **996** Frequent Itemsets			
LP1	1,980	738	1,064
LP2	1,058	1,422	941
C_2^G	*2,150 + (444)*	*808 + (676)*	*191 + (907)*
5 fragments			
LP1	682	1,130	1,067
LP2	8,546	997	1,355
LP3	2,899	911	986
LP4	1,271	957	791
LP5	1,257	1,412	1,069
C_5^G	*10,007+(230)*	*1,114 + (625)*	*751 + (723)*
	BMS-POS		
1-fragment: **6,017** Frequent Itemsets			
LP1	5,419	5,311	6,024
LP2	6,709	6,729	5,972
C_2^G	*1,820+ (5,154)*	*1,468+ (5,286)*	*348+ (5,824)*
5 fragments			
LP1	8,480	7,014	4,339
LP2	4,975	4,290	5,932
LP3	2,541	6,619	7,530
LP4	5,177	7,315	7,443
LP5	12,755	5,287	5,289
C_5^G	*10,718+ (2,346)*	*4,353+ (3,956)*	*4,075+ (4,191)*

the threshold **0.005**, respectively. Exceptional performance for WebView-2 data set with the threshold **0.005** the reduction is from **10,007** to only **751** when data set is partitioned into 5 fragments.

Hence naturally, another interesting and encouraging trend can be found in the growth of the number of common candidates between LP_i for fragmented data sets. For example, if data sets are partitioned into 5 fragments, this common number increases from **689** to **1,121** for BMS-POS with the threshold **0.01** as well as from **230** to **723** for WebView-2 with the threshold **0.005**.

In summary, the figures from 2 above tables show that the *Clustering* pre-processing technique can significantly improve the Partitioning approach. It is delivered in form of two strongly related benefits; reduction of the number of *Global candidates* requiring the final check and increase of the common candidates numbers that don't require any additional checks.

7 Conclusion

This paper considers a new approach for further performance improvements in frequent itemsets computation. Based on the original Partition algorithm, we show that the composition of fragments and the number of fragments generated, impact on the size of the data used by this algorithm.

We propose a pre-processing method (an incremental clustering algorithm), mainly to demonstrate that there is potential in the direction of performance improvement. Figures from the experiments show that this pre-processing offers good benefits already. The main question which still deserves consideration is related to the identification of methods that will deliver an even better partition for the original data sets.

Acknowledgment. We wish to thank the Data Mining group at ITEE School - The University of Queensland and the anonymous reviewers for suggestions.

References

[1] Agrawal R., Srikant R.: *Fast algorithms for mining association rules*. Proc. 20th Int. Conf. Very Large Data Bases, Morgan Kaufmann, 1994 (487 - 499)

[2] Savasere A., Omiecinski E., Navathe S.: *An efficient algorithms for mining association rules in large database*. Proc. 21th Int. Conf. Very Large Data Bases, Swizerland, 1995

[3] Goethals B.: *Survey on frequent pattern mining*. University of Helsinki, 2002

[4] Lin J.L., Dunham M.H.: *Mining association rules: Anti-skew algorithms*. Proc. 14th IEEE Int. Conf. on Data Engineering, Florida, 1998

[5] Agrawal R., Imielinski T., Swami A.N.: *Mining association rules between sets of items in lagre database*. Proc. 1993 ACM SIGMOD Int. Conf. on Management of Data, 1993

[6] Brin S., Motwani R., Ullman D.J., Tsur S.: *Dynamic Itemset Counting and implication rules for masket basket data*. Proc. ACM SIGMOD 1997 Int. Conf. on Management of Data, 1997 (255 - 264)

[7] Zaki M.J.: *Scalable algorithms for association mining*. IEEE Transactions on Knowledge and Data Engineering, 12(3): 372-390, 2000

[8] Han J., Pei J., Yin Y., Mao R.: *Mining frequent patterns without candidate generation: A frequent-pattern tree approach*. Data Mining and Knowledge Discovery, Kluwer Academic Publishers, (8): 53-87, 2004
[9] Mueller A.: *Fast sequential and parallel algorithm for association rules mining: A comparison*. Technical Report CS-TR-3515, University of Maryland, 1995
[10] Son N. Nguyen: *Data partitioning approach into selected data mining problems*. PhD Confirmation report, The University of Queensland, Australia, 2005
[11] Pudi V., Haritsa J.: *ARMOR: Association rule mining based on Oracle*. Workshop on Frequent Itemset Mining Implementations (FIMI'03 in conjunction with ICDM'03), 2003
[12] Zhu T.: *The Apriori algorithm implementation,* http://www.cs.ualberta.ca/~tszhu/
[13] Ron Kohavi, Carla Brodley, Brian Frasca, Llew Mason, and Zijian Zheng. *KDD-Cup 2000 organizers' report: Peeling the onion.* SIGKDD Explorations, 2(2):86-98, 2000

On-Line Adaptive Filtering of Web Pages

Richard Nock[1] and Babak Esfandiari[2]

[1] GRIMAAG, Université Antilles-Guyane, Schoelcher, France
rnock@martinique.univ-ag.fr
[2] Dept of Systems and Computer Engineering,
Carleton University, Ottawa, Canada
babak@sce.carleton.ca

Abstract. We present a browser extension to dynamically learn to filter unwanted Uniform Resource Locators (such as advertisements or flashy images) based on minimal user feedback. Our extension builds upon one of the top ten of Mozilla firefox plug-ins which filters URLs *without* learning capabilities. We apply a weighted majority-type learning algorithm working on regular expressions. Experimental results confirm that the accuracy of the predictions converges quickly to very high levels, with other key parameters: recall, specificity and precision.

1 Introduction

Many attempts have been made to make Web browsing more pleasant by allowing the user to remove big pictures and unwanted animations that interfere with reading. Some browsers such as Netscape or Mozilla allow the user to collapse such pictures or even create blacklists of internet domains that supply them.

But the most sophisticated approach so far has been proposed by the developers of AdBlock. AdBlock [1] is, according to "Mozdev update" data [6], in the top ten of the most popular extension to the Mozilla Firefox web browser [5], with about 100000 downloads. To use AdBlock, the user has to come up with a collection of regular expressions that describe the URL patterns of images that they want to see filtered. As a result, whenever the browser is pointed to an item whose URL is matched by a regular expression, it is simply ignored, which not only "cleans up" the web page, but also makes page downloading faster.

```
/[^a-z\d=+%](\w*\d+x\d)?\d*(show)?(\w{3,}%20|alligator|avs|barter|blog|box|central|d?html|i?frame|front|fuse|get|house|inline|
instant|live|main|net|partner|primary|provider|rotated?|secure|side|smart|sponsor|story|text|view)?_?ads?(v?(bot|brite|broker|
bureau|butler|center|click|client|creative|content|coun(cil|t)|data|engage|er(tis\w+|t(pro)?|ve?r?)|farm|force|frame|gif|group
id|head|id|ima?ge?|info|js|juggler|legend|link|log|man(ager)?|max|mentor|meta\.com|net|optimi[sz]er|pic|popup|proof|q\.nextag|
quest\.nl|redire?c?t?|remote|revolver|rotator|sale|sdk|sfac|solution|sonar|source|space|srv|stat.*\.asp|sys|track|trix|view|ty
pe|zone))?\d*(s|status)?\d*[\W_](?!\w+\.edu|aware|adurl=|block|login|nl/|.*(&sbc|\.(wmv|rm)))/
```

Fig. 1. Example of a long regular expression found on AdBlock's forum

However, as often discussed in the AdBlock online forum, coming up with regular expressions is a difficult task, especially for the non-computer savvy. Writing and mastering them accurately requires extensive readings [2], and those published on-line can be especially hard to read and understand. Figure 1 presents the example of a regular expression posted on AdBlock's forum. Most of the regular expressions posted are smaller than this one, but some of them appear to be much more complicated to understand. Thus, the user faces the risk of obtaining unwanted browsing/blocking behaviors, sometimes without really knowing how to correct them. The problem cannot be solved from a global standpoint, as it would be impossible to come up with a general set of filters that would satisfy every user. Finally, as the advertisement suppliers and browsing habits change, so should the set of regular expressions that are needed. The behavior of AdBlock is too static to be suited to these dynamic interactions, but moving to a dynamic interaction between the user and the filter is everything but trivial. The comfort the extension brings to the user has to be greater than its eventual drawbacks, and the complexity of the algorithm is clearly such a potential drawback.

To address this problem, we propose a fast machine learning approach that would create filters based on minimal interaction with the user. The user is not required to know how to create regular expressions; all that is required is for the user to click on URLs (*e.g.* images) that he/she wants to see blocked. Conversely, from time to time the user will need to unblock URLs that shouldn't have been blocked by the adaptive filter. Based on this simple feedback, our proposed method, an adaptation of the well known Weighted Majority algorithm [4], builds a set of "experts" (simple regular expressions on URLs) that vote on whether a given URL should be blocked or not.

In the following Section, we review some works related to our topic. Then, the next Section is devoted to a formal presentation of the algorithm. After a Section presenting the browser extension, a Section presents and discusses experimental results. A last Section concludes and presents relevant issues on the topic.

2 Related Work

Our approach is inspired by the concept of Interface Agents [3]. An interface agent is a piece of software that assists a user of a complex system by observing his/her behavior and detecting patterns that it could reproduce in order to automate tedious tasks. Typically such programs use some kind of incremental machine learning algorithm to build the knowledge base. [3] devised interface agents that used k-nearest neighbor to classify mail and even share the filters with other agents. In a previous work, we used an adaptation of the Version Spaces algorithm to automate simple network management tasks [7].

But the closest work is perhaps the use of Bayesian filtering for detecting email spam [8], which is now a standard feature in mainstream email programs. Bayesian methods for filtering emails have the advantage of being conceptually simple, and a great body of previous work has made them tailored to common text classification tasks.

In our case, however, the setting makes them *a priori* not the best classification tool suited for web browsing. Classification is indeed made *on-line*. This is a crucial remark because the frequency of browsing through URLs is much higher than that of email receipt for the average user. This makes it necessary to have an ultra-fast classification tool with easy updates on the classifier, to filter the URLs as they come. In the case of email spam detection, it is already necessary to have efficient feature selection algorithms to reduce the vector space to a small set prior to using Bayesian methods [8]. Making the additional heavy-weight online updates for URL filtering, such as the computation of the probability table for *each* feature, would rapidly slow down the browser and make its use very uncomfortable. Furthermore, Bayesian methods rely on independence assumptions (at best partially relaxed) on the features to make the classification sound [8]. This is clearly not a desirable assumption for URL classification, since it partially omits token positions and contexts in an URL.

3 Theoretical Setting

Very informally, the algorithm can be reduced to the following infinite loop: get an *example*, update a *set of experts* and update the *weights* of each expert. Any time during the algorithm, a prediction is possible on an *observation* by using a *weighted majority* over the current set of experts.

More formally, each observation belongs to a set X, which contains all possible observations. Each observation is an URL (Uniform Resource Locator). From the user's standpoint, X can be partitioned into two subsets. The first one contains the URLs he would like to block, *i.e.* refrain from loading. The other one contains all the other URLs, *i.e.* those he wishes to leave unblocked. To each URL can thus be associated a status which we call a *class* or *label* (block/unblock), and our objective is to predict the class of each URL as accurately as possible with respect to the user, given that any two different users may probably correspond to different partitions of X. Our algorithm builds therefore a decision function (or classifier) from X onto $\{-1, +1\}$, with "+1" denoting the class of the URLs to be blocked (also called the *positive* class), and "-1" the class of the URLs to leave unblocked (the *negative* class).

We denote a couple (observation, class) obtained from the user as an *example*. We let $(x_1, y_1), (x_2, y_2), ...$ denote the stream of examples observed from the user, and (x_t, y_t) is thus the tth example of the stream. We build a set of experts **E** which is growing with time; to keep notations clear, we do not use the time subscript on **E**: it should be clear from context which set of experts we use. Each expert of **E** is a couple (hypothesis, weight). An hypothesis is a function $h : X \to \{-1, 0, +1\}$ which is allowed to *abstain* (this is the output "0"). More precisely, each hypothesis' output is either $\{-1, 0\}$ or $\{0, +1\}$, which means that the corresponding expert is authorized to say "I don't know", thus delegating the decision on the class of an observation to the other experts. The weight associated to hypothesis h is denoted $w_t(h) \in \mathbb{R}^+$. It is a function of t since it is updated each time an example is received. At the very beginning of the algorithm, prior to seeing the first example, we initialize the following set of parameters:

- $\beta \in (0, 1)$ is a learning constant chosen by the user,
- $\mathbf{E} \leftarrow \emptyset$ is the initial set of experts,
- $t \leftarrow 1$ is the "time stamp" labeling the examples received.

Algorithm 1 below displays more formally what happens when example (x_t, y_t) is received.

Algorithm 1: Receive_New_Example$((x_t, y_t))$

   Input: example (x_t, y_t)
   $\mathbf{N} \leftarrow$ Create_Hypotheses$((x_t, y_t))$;
   Update_Experts $(\mathbf{N})$;
   foreach $(h, w_t(h)) \in \mathbf{E}$ **do**
      $\lfloor\ w_{t+1}(h) \leftarrow w_t(h) \times u(\beta, h, t)$;
   $t \leftarrow t + 1$;

There are two possible choices for function $u(\beta, h, t)$:

$$u(\beta, h, t) = \frac{1 + y_t h(x_t)}{2\beta} + \frac{(1 - y_t h(x_t))\beta}{2} \ . \qquad (1)$$

There are two procedures in Algorithm 1. Create_Hypotheses(.) takes an example as input, and outputs a set of hypotheses (*i.e.* regular expressions). Since the theory underlying the algorithm does not depend on this procedure, we postpone the details and its implementation to the experimental section.

Update_Experts (.) takes as input a set of hypotheses, and creates a set of experts which is used to grow $\mathbf{E}$. In other words, it initializes the weights of the hypotheses. Details are given in Algorithm 2 (here, "0" denotes the function which is zero everywhere in $\mathbb{R}$).

Algorithm 2: Update_Experts$(\mathbf{N})$

   Input: hypothesis set $\mathbf{N}$
   foreach $h \in \mathbf{N}$ **do**
      $w_t(h) \leftarrow (u(\beta, 0, t))^{t-1}$;
      $\mathbf{E} \leftarrow \mathbf{E} \cup \{(h, w_t(h))\}$;

Weight initialization for new experts makes it possible to consider from the theoretical standpoint that each of them was created at the beginning of the algorithm, as everything is like if it were abstaining until "really" put into $\mathbf{E}$. There remains to give the way $\mathbf{E}$ is used to classify an observation $x \in X$. Just prior to receiving example $t + 1$, the decision made out of $\mathbf{E}$, $H_{\mathbf{E},t}$, relies on an ordinary majority vote: $\forall x \in X, H_{\mathbf{E},t}(x) = \text{sign}(\sum_{(h,w_t(h)) \in \mathbf{E}} w_t(h) \times h(x))$.

4 Design of the Browser Extension

The Mozilla Firefox Web browser [5] is an open source product with an architecture specifically designed for allowing 3rd party extensions. This makes it possible to easily modify the browser behavior by overriding or augmenting the existing UI components, intercepting and reacting to browser events, and accessing environment variables. Our filtering algorithm and the test drivers were both implemented as such extensions in Javascript.

4.1 User Interface Elements

As a principle, a learning interface agent must remain as unobtrusive as possible, and therefore the user interface additions were kept to a minimum. We have only provided two extra menu items in the browser's context menu:

- one called "Block Me" which appears only when the user right-clicks on an URL (*e.g.* an image) that he/she wishes to block;
- the other called "Unblock" which is always available should the user want to unblock an URL that appears to be blocked by mistake. Selecting this item brings up the list of blocked items for the page, and the user can then choose which URL needs to be unblocked.

The "Block Me" button is the way the user provides the positive examples to the algorithm, while the "Unblock" button provides the negative ones. One could envisage that the non-blocked items that were *correctly* classified as such should also be fed to the algorithm (once the user has left the given page, thus confirming that they were correctly left unblocked) for weight reinforcement purposes, but we have decided against it, as we thought that if the user is the sole trigger for example provision, he/she will have a better feel for what is happening behind the scenes. This remark also holds for the blocked items that were *not* unblocked by the user. Finally, this way the user has control over the creation and potential proliferation of experts, which otherwise could slow down the browser without much benefit. Notice that updates of the expert weights occur only when receiving misclassified examples: false positives decrease the weight of "positive" experts (voting for the "Block" class), while false negatives decrease the weight of negative experts (voting for the "Unblock" class).

4.2 Implementation of Create_Hypotheses(.)

To generate the new set of experts **N** in Algorithm 1, we tokenize the example URLs using the character "/" as delimiter. The tokens obtained represent items such as domain names, folders, but exclude file names. In that last case indeed, filenames are often generated automatically for the URLs to block (*e.g.* by advertisement sites), and the resulting filenames generally have little significance. Furthermore, this helps to keep the list simple to manipulate manually. This very simple choice of tokenization seems to be chosen by a significant proportion of users sharing their regular expressions on AdBlock's forums. Notice that "http" is also a resulting token. The user may view its weight as the balance between the rate of false positives and the rate of false negatives achieved through learning, or, similarly, as an indication of the ratio between precision and recall.

The obtained tokens are then compared with the corresponding existing set of experts. By "corresponding" we mean that tokens obtained from positive (resp. negative) examples are compared to the "positive" (resp. negative) set of experts. If no match is found, the new token is added to the corresponding list of experts, and its weight is initialized using Algorithm 2. More tokens could obviously be generated. For instance, we could also use the full URL itself as an expert. Also, the character "." could be used as a delimiter, to help identify the parts of a domain name that are key to its classification (e.g. host name or domain extension). Finally, one could create experts that capture the importance of the *order* in which significant tokens appear in a URL. The factor to consider however is to avoid the proliferation of experts.

5 Experimental Results

In our experiments, we have fixed $\beta = 1/\sqrt{e} \approx 0.61$ in update rule (1). In order to obtain results that are independent from any particular browsing habit, we needed to provide a test setting that could be used seamlessly by any kind of user. To do so, in addition to providing the standalone extension described in the previous section, we embedded our algorithm inside the AdBlock extension code.

The AdBlock user is asked to set up filters as usual in the form of regular expressions, creating as a result an oracle for the embedded learner. The AdBlock filters override the learner's classification in order to remain transparent to the end user. This means that to the user, the extension is behaving no differently than the regular AdBlock. However, all learner misclassifications (*i.e.* false positives and false negatives) are fed back as such to the algorithm, leading to the expert creation and weight adjustments described above.

At each step consisting of k observations (*e.g.* visited image URLs), we freeze a copy of the learner's knowledge base up to that point. While the *unfrozen* version keeps evolving and accepting feedback from the oracle, the *frozen* copy is used to evaluate the learning accuracy of the accumulated knowledge so far by populating a confusion matrix based on its predictions on the *incoming* examples. After n such steps, and for a total of $n \times k$ observations, the user is notified that the testing is finished, and the logs are collected. We can therefore compare the learner at each step and observe the evolution of its ability to classify the upcoming observations. However, as we get close the final steps of each test, the number of observations available to the more recent learners decreases, and the statistical confidence in the more recent results decreases as well. To reduce this phenomenon, we allow some more observations to be collected after the last step.

5.1 AdBlocking on a Single Commercial Website

Our first set of tests were designed to see whether our algorithm was able to correctly predict which URLs to block on a single "busy" (*i.e.* littered with annoying images) web page, and if so, after how many visits. We used a common

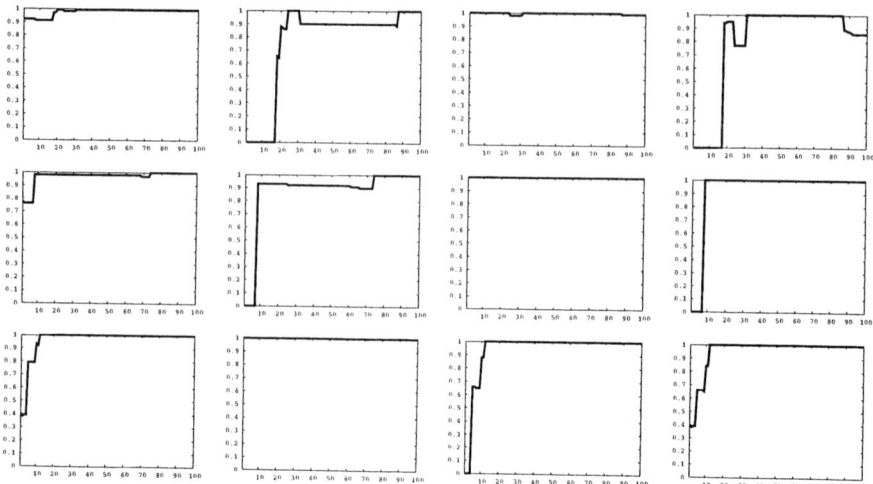

Fig. 2. From left to right: accuracy, recall, specificity and precision. From up to bottom: websites of CNN, Fox News and MSN (x-axis=step number, see text for details).

set of AdBlock regular expressions, as collaboratively devised on AdBlock discussion forums, as oracle. On three popular and large commercial websites, we have run AdBlockLearner with $k = 1$ and $n = 100$. The total 100 observations were usually reached very quickly. Figure 2 plots the evolution of four key parameters throughout learning. If we denote by TP the number of true positives, TN the number of true negatives, FP the number of false positives and FN the number of false negatives, then the *specificity* is $TN/(TN + FP)$, the *recall* is $TP/(TP + FN)$, and the *precision* is $TP/(TP + FP)$. As can be seen, the algorithm converges quickly to very good prediction, in terms of *all* four parameters. This is good given that commercial web sites use dynamic loading of advertisements using cookies, and as a result hitting reload usually brings up a different set of images and URLs. However it is important to point out that to obtain similar results in a non-test setting, the misclassifications that were detected by the oracle would have to correspond to as many direct feedbacks by the user. In practice, in the absence of so many interactions, the four parameters can be suboptimal, but it is definitely acceptable.

5.2 AdBlocking While Surfing to Different Websites

The next set of tests measures robustness to overfitting. How does the learned knowledge "travel" over to other web sites, are the rules learned so far useful to new websites, and how much more learning is left to do? Our intuition was that the amount of misclassifications would decrease over time, as usually the providers of invasive advertisements are the same in many different commercial sites. We asked the users to simply follow their usual browsing habits, and we

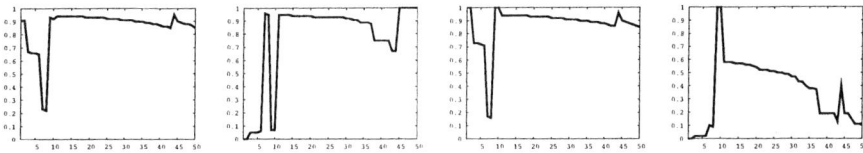

Fig. 3. Evolution of the four parameters during a typical browsing session (conventions follow figure 2, see text for details)

set $k = 10$ (to absorb some of the variability) and $n = 50$. To make the task harder, we requested that each chosen web site had to be visited *only once*. The results are charted on figure 3. It is quite remarkable that the accuracy, the specificity *and* the recall all generally remained at very high values after a short period, given the tough experimental setting. However, the fact that the precision decreases tends to indicate that there is a significant increase in FP (to make the precision decrease) and in TN (to make the specificity remains at high values). This may display the fact that the number of negative examples tends to increase, but the experts might be too simple to fit to the growing amount of information, to discriminate all the examples that come from various websites. In case false positives are deemed unacceptable by the user, i.e. the user does not want to have to manually unblock erroneously blocked URLs, it is possible to alter the utility of the weighted majority vote by giving more weight to negative votes, as also indicated by [8]. The trade-off would be a drop in overall accuracy and an increase in the rate of false negatives.

6 Conclusion and Future Work

In this paper, we have experimentally demonstrated the efficiency of a careful adaptation of weighted majority. Compared to usual weighted majority, our setting makes use of the fact that experts may abstain instead of always predicting a class. This raises an important theoretical issue, as the efficiency of weighted majority is usually measured with respect to its number of *mistakes* [4]. In our setting, we would certainly appreciate this quantity to be as small as possible, *but* we would *also* appreciate the number of abstention to be small. Since mistake bounds do not take into account the number of abstentions, this raises both the problem of finding accurate quantities to minimize, and relevant bounds that our adaptation of weighted majority satisfies.

Acknowledgments and Code Availability

We would like to thank "Rue", chief developer of AdBlock, for his timely help and enthusiasm, and for welcoming us to his team. R. Nock would like to warmly thank Ottawa University and Stan Matwin for an invitation grant, during which part of this work was achieved. Both the standalone extension

(AdBlockLearner) and the test driver (AdBlockLearnerTest) are available at http://adblocklearner.mozdev.org , including source code and documentation. They are compatible with most versions of Mozilla Firefox.

References

1. AdBlock, 2005. http://adblock.mozdev.org.
2. Jeffrey E. F. Friedl. *Mastering Regular Expressions*. O' Reilly, 1997.
3. Y. Lashkari, M. Metral, and P. Maes. Collaborative interface agents. In *Proc. of AAAI-94*, pages 444–449, 1994.
4. N. Littlestone and M. Warmuth. The weighted majority algorithm. *Information and Computation*, pages 212–261, 1994.
5. Mozilla Firefox, 2005. http://mozilla.org/products/firefox.
6. Mozilla Firefox Extensions, 2005. http://update.mozilla.org/extensions/.
7. R. Nock and B. Esfandiari. Oracles and assistants : machine learning applied to network supervision. In *Canadian Artificial Intelligence Conference*, number 1418 in Lecture Notes in Computer Science, pages 86–98. Springer-Verlag, 1998.
8. M. Sahami, S. Dumais, D. Heckerman, and E. Horvitz. A bayesian approach to filtering junk email. In *AAAI Workshop on Learning for Text Categorization*. AAAI Press, 1998.

A Bi-clustering Framework for Categorical Data

Ruggero G. Pensa[1], Céline Robardet[2], and Jean-François Boulicaut[1]

[1] INSA Lyon, LIRIS CNRS UMR 5205,
F-69621 Villeurbanne cedex, France
[2] INSA Lyon, PRISMa EA INSA-UCBL 2058,
F-69621 Villeurbanne cedex, France
{Ruggero.Pensa, Celine.Robardet, Jean-Francois.Boulicaut}@insa-lyon.fr

Abstract. Bi-clustering is a promising conceptual clustering approach. Within categorical data, it provides a collection of (possibly overlapping) bi-clusters, i.e., linked clusters for both objects and attribute-value pairs. We propose a generic framework for bi-clustering which enables to compute a bi-partition from collections of local patterns which capture locally strong associations between objects and properties. To validate this framework, we have studied in details the instance CDK-MEANS. It is a K-MEANS-like clustering on collections of formal concepts, i.e., connected closed sets on both dimensions. It enables to build bi-partitions with a user control on overlapping between bi-clusters. We provide an experimental validation on many benchmark datasets and discuss the interestingness of the computed bi-partitions.

1 Introduction

Many data mining techniques have been designed to support knowledge discovery from categorical data which can be represented as Boolean matrices: the rows denote objects and the columns denote Boolean attributes that enable to record object properties as attribute-value pairs. For instance, given **r** in Table 1, we say that object t_2 satisfies properties g_2 and g_5. Clustering is one of the major data mining tasks and it has been studied extensively, including for the special case of categorical or Boolean data. Its main goal is to identify a partition of objects and/or properties such that an objective function which specifies its quality is optimized [1]. Thanks to local search optimizations, many efficient algorithms can provide good partitions but suffer from the lack of explicit cluster characterization. It has motivated the research on conceptual clustering [2] and bi-clustering [3,4,5] whose goal is to compute bi-clusters, i.e., associations of (possibly overlapping) sets of objects with sets of properties. An example of an interesting bi-partition in **r** (Table 1) is $\{\{\{t_1, t_3, t_4\}, \{g_1, g_3, g_4\}\}, \{\{t_2, t_5, t_6, t_7\}, \{g_2, g_5\}\}\}$. The first bi-cluster indicates that $\{t_1, t_3, t_4\}$ almost always share properties $\{g_1, g_3, g_4\}$. A major problem is that most of the bi-clustering algorithms compute non overlapping bi-partitions, while in many application domains, it makes sense to have objects and properties belonging to more than one bi-cluster. It motivates more research on the computation of relevant collections of possibly

Table 1. A Boolean context **r**

	g_1	g_2	g_3	g_4	g_5
t_1	1	0	1	1	0
t_2	0	1	0	0	1
t_3	1	0	1	1	0
t_4	0	0	1	1	0
t_5	1	1	0	0	1
t_6	0	1	0	0	1
t_7	0	0	0	0	1

overlapping bi-clusters. Formal concept analysis [6] might be a solution. Informally, a formal concept is a bi-set (T, G) where the set of objects T and the set of properties G form a maximal (combinatorial) rectangle of true values, e.g., $(\{t_1, t_3\}, \{g_1, g_3, g_4\})$ in **r**. Unfortunately, we generally get huge collections of formal concepts which are difficult to interpret by end-users. In our example, $\{\{t_1, t_3, t_4\}, \{g_1, g_3, g_4\}\}$ is not a formal concept $((t_4, g_1) \notin \mathbf{r})$ but can be built from $\{\{t_1, t_3\}, \{g_1, g_3, g_4\}\}$ and $\{\{t_1, t_3, t_4\}, \{g_3, g_4\}\}$ which are "similar enough" formal concepts. It provides the intuition of our approach.

The contribution of this paper is twofold. First we propose a new bi-clustering framework which enables to compute bi-partitions by grouping local patterns which capture locally strong associations between objects and properties, i.e., bi-sets which satisfy some user-defined constraints. Various local patterns are candidates for such a process, e.g., frequent sets of properties associated to their supporting set of objects, formal concepts, etc. Secondly, we study one instance of this framework, the CDK-MEANS algorithm, which builds simultaneously linked partitions on objects and properties. More precisely, we apply a K-MEANS-like algorithm to a collection of bi-sets (formal concepts in our experiments). As a result, objects and properties are intrinsically associated to clusters, depending on their weights in the finally computed centroids. Our experimental validation confirms the added-value of CDK-MEANS w.r.t. other (bi-)clustering algorithms.

In Section 2, we set up our clustering framework and we survey related work. Section 3 discusses our experimental validation methodology and it contains many experimental results on various benchmark datasets. A comparison between CDK-MEANS, two bi-clustering algorithms (COCLUSTER [4] and BI-CLUST [3]), and two classical clustering algorithms (K-MEANS and EM [1]) is given. Scalability issues are discussed and Section 4 concludes.

2 Clustering Model

Assume a set of objects $\mathcal{O} = \{t_1, \ldots, t_m\}$ and a set of Boolean properties $\mathcal{P} = \{g_1, \ldots, g_n\}$. The Boolean context to be mined is $\mathbf{r} \subseteq \mathcal{O} \times \mathcal{P}$, where $r_{ij} = 1$ if property g_j is satisfied by object t_i. We define the bi-clustering task as follows: we want to compute a partition of K clusters of objects (say $\{C_1^o \ldots C_K^o\}$) and a partition of K clusters of properties (say $\{C_1^p \ldots C_K^p\}$) with a mapping between both partitions such that each cluster of objects is characterized by a cluster of properties. Our idea is that bi-partitions can be computed from bi-sets and it will be instantiated later on formal concepts. Formally, a bi-set is an element $b_j = (T_j, G_j)$ $(T_j \subseteq \mathcal{O}, G_j \subseteq \mathcal{P})$ and we assume that a collection of a priori interesting bi-sets denoted $\mathcal{B}$ has been extracted from $\mathbf{r}$ beforehand. Let us now describe b_j by the Boolean vector $<\mathbf{t}_j>, <\mathbf{g}_j> = <t_{j1}, \ldots, t_{jm}>, <g_{j1}, \ldots, g_{jn}>$ where $t_{jk} = 1$ if $t_k \in T_j$ (0 otherwise) and $g_{jk} = 1$ if $g_k \in G_j$ (0 otherwise). We are looking for K clusters of bi-sets $\{C_1, \ldots, C_K\}$ $(C_i \subseteq \mathcal{B})$. Let us define the centroid of a cluster of bi-sets C_i as $\mu_i = <\boldsymbol{\tau}_i>, <\boldsymbol{\gamma}_i> = <\tau_{i1}, \ldots, \tau_{im}>, <\gamma_{i1}, \ldots, \gamma_{in}>$ where τ and γ are the usual centroid components:

$$\tau_{ik} = \frac{1}{|C_i|} \sum_{b_j \in C_i} t_{jk}, \quad \gamma_{ik} = \frac{1}{|C_i|} \sum_{b_j \in C_i} g_{jk}$$

We now define our distance between a bi-set and a centroid:

$$d(b_j, \mu_i) = \frac{1}{2} \left(\frac{|\mathbf{t}_j \cup \boldsymbol{\tau}_i| - |\mathbf{t}_j \cap \boldsymbol{\tau}_i|}{|\mathbf{t}_j \cup \boldsymbol{\tau}_i|} + \frac{|\mathbf{g}_j \cup \boldsymbol{\gamma}_i| - |\mathbf{g}_j \cap \boldsymbol{\gamma}_i|}{|\mathbf{g}_j \cup \boldsymbol{\gamma}_i|} \right)$$

It is the mean of the weighted symmetrical differences of the set components. We assume $|\mathbf{t}_j \cap \boldsymbol{\tau}_i| = \sum_{k=1}^{m} a_k \frac{t_{jk} + \tau_{ik}}{2}$ and $|\mathbf{t}_j \cup \boldsymbol{\tau}_i| = \sum_{k=1}^{m} \frac{t_{jk} + \tau_{ik}}{2}$ where $a_k = 1$ if $t_{jk} \cdot \tau_{ik} \neq 0$, 0 otherwise. Intuitively, the intersection is equal to the mean between the number of common objects and the sum of their centroid weights. The union is the mean between the number of objects and the sum of their centroid weights. These measures are defined similarly on properties.

Objects t_j (resp. properties g_j) are assigned to one of the K clusters (denoted i) for which τ_{ij} (resp. γ_{ij}) is maximum. We can enable that a number of objects and/or properties belong to more than one cluster by controlling the size of the overlapping part of each cluster. Thanks to our definition of cluster membership determined by the values of τ_i and γ_i, we just need to adapt the cluster assignment step. For this purpose, let us introduce parameters δ_o and δ_p in [0,1] to quantify the membership of each element to a cluster. We say that an object t_j belongs to a cluster C_i^o if $\tau_{ij} \geq (1 - \delta_o) \cdot max_i(\tau_{ij})$. Analogously, a property g_j belongs to a cluster C_i^p if $\gamma_{ij} \geq (1 - \delta_p) \cdot max_i(\gamma_{ij})$. Obviously the number of overlapping objects (resp., properties) depends on the distribution of the values of τ_i (resp. γ_i). Notice that if overlapping is allowed, $\delta = 0$ does not imply that each object or property is assigned to a single cluster. The choice of a relevant value for δ is clearly application-dependent. When a bi-clustering structure holds in the data, little values of δ are not enough to provide relevant overlapping. On another hand, in noisy contexts, even little values of δ can give rise to significant overlapping zones.

We can now provide details about the studied instance of this framework: a bi-clustering based on formal concepts. Many efficient algorithms have been developed that can extract complete collections of formal concepts under constraints. We use D-MINER [7].

Our instance CDK-MEANS is presented in Table 2. It computes a bi-partition of a dataset $\mathbf{r}$ given a collection of bi-sets $\mathcal{B}$ extracted from $\mathbf{r}$ beforehand (e.g., formal concepts), the desired number of clusters K, the threshold values for δ_o

Table 2. CDK-MEANS pseudo-code

CDK-MEANS ($\mathbf{r}$ is a Boolean context, $\mathcal{B}$ is a collection of bi-sets in $\mathbf{r}$, K is the number of clusters, MI is the maximal iteration number, δ_o and δ_p are thresholds values for controlling overlapping)

1. Let $\mu_1 \ldots \mu_K$ be the initial cluster centroids. $k := 0$.
2. Repeat
   (a) For each bi-set $c \in \mathcal{B}$, assign it to cluster C s.t. $d(c, \mu_i)$ is minimal.
   (b) For each cluster C_i, compute τ_i and γ_i.
   (c) $k := k + 1$.
3. Until centroids are unchanged or $k = MI$.
4. If overlap is allowed, for each $t_j \in \mathcal{O}$ (resp. $g_j \in \mathcal{P}$), assign it to each cluster C_i^o (resp. C_i^p) s.t. $\tau_{ij} \geq (1 - \delta_o) \cdot max_i(\tau_{ij})$ (resp. $\gamma_{ij} \geq (1 - \delta_p) \cdot max_i(\gamma_{ij})$).
5. Else, for each $t_j \in \mathcal{O}$ (resp. $g_j \in \mathcal{P}$), assign it to the first cluster C_i^o (resp. C_i^p) s.t. τ_{ij} (resp. γ_{ij}) is max.
6. Return $\{C_1^o \ldots C_K^o\}$ and $\{C_1^p \ldots C_K^p\}$

---

and δ_p, and a maximum number of iterations MI. On our example $\mathbf{r}$, CDK-MEANS provides the bi-partition given in Section 1. The complexity is linear in $\mathcal{B}$ and scalability issues are discussed in Section 3.

Related work. [3] and [4] bi-clustering methods alternatively refine a partition when the other one is fixed, optimizing respectively the Goodman-Kruskal's τ coefficient and the loss in mutual information. The first interesting difference is instead of considering objects and properties as separated entities during the bi-clustering task (even if objective functions are computed on both sets), CDK-MEANS considers their associations as the elements to process. The second one is that CDK-MEANS can easily compute partitions with overlapping clusters.

3 Experimental Validation

Different techniques can be used to evaluate the quality of a partition. An external criterion consists in comparing the computed partition with a "correct" one. It means that data instances are already associated to some correct labels and that one quantifies the agreement between computed labels and correct ones. A popular measure is the Jaccard coefficient [1].

To evaluate the quality of our bi-clustering using an internal criterion we use Goodman and Kruskal's τ coefficient [8]. It is evaluated in a co-occurrence table $\mathbf{p}$ and it discriminates well bi-partitions w.r.t. the intensity of the functional link between both partitions [3]. p_{ij} is the frequency of relations between an object of a cluster C_i^o and a property of a cluster C_j^p. $p_{i.} = \sum_j p_{ij}$ and $p_{.j} = \sum_i p_{ij}$. The $\tau_\mathcal{O}$ coefficient evaluates the proportional reduction in error given by the knowledge of C^o on the prediction of C^p ($\tau_\mathcal{O}$ will denote the measure when exchanging the partitions):

$$\tau_Q = \frac{\sum_i \sum_j \frac{p_{ij}^2}{p_{i.}} - \sum_j p_{.j}^2}{1 - \sum_j p_{.j}^2}$$

We report on experiments using eight well-known datasets taken from the UCI ML Repository[1] and from the JSE Data Archive[2]. All the experiments have been performed on a PC with 1 Gb RAM and a 3.0 GHz P4 processor. First, without considering the class variable, we have processed each dataset with D-MINER [7]. Minimal set size constraints have been used for mushroom and credit-a (minimal itemset and objectset sizes $(13, 15)$ and $(6, 15)$) to obtain complete collections of formal concepts before using CDK-MEANS.

Table 3. Goodman-Kruskal's coefficient values for different bi-clustering algorithms (MR-2 and MR-5 refer to mushroom with 2 and 5 clusters)

Dataset	Dim.	BI-CLUST Max	COCLUSTER Max	COCLUSTER Mean	CDK-MEANS Max	CDK-MEANS Mean
voting	435×48	**0.320**	0.320	0.315±0.002	0.311	0.311±0.000
titanic	2201×8	**0.332**	0.321	0.226±0.076	0.314	0.160±0.109
iris-2	150×8	0.543	0.543	0.357±0.195	0.543	0.474±0.056
iris-3	150×8	**0.544**	0.390	0.379±0.045	0.523	0.329±0.080
zoo-2	101×16	0.191	0.186	0.157±0.034	**0.192**	0.165±0.020
zoo-7	101×16	-	0.080	0.065±0.009	**0.083**	0.049±0.015
breast-w	699×18	**0.507**	0.507	0.474±0.121	0.498	0.498±0.000
credit-3	690×52	0.104	0.014	0.003±0.003	**0.110**	0.091±0.015
credit-2	690×52	-	0.012	0.006±0.004	**0.096**	0.055±0.011
mr-2	8124×126	-	0.198	0.158±0.026	0.176	0.157±0.017
mr-5	8124×126	**0.187**	0.119	0.097±0.009	0.116	0.112±0.004
ads	3279×1555	-	0.006	0.003±0.001	**0.538**	0.137±0.109

We compared CDK-MEANS bi-partitions with those obtained by COCLUSTER [4], and BI-CLUST [3]. As the initialization of these algorithms is randomized, we executed them 100 times on each dataset and we selected the result which returned the best Goodman-Kruskal's coefficient. The number of desired clusters for each experiment has been set to the number of class variable values, except for BI-CLUST which automatically determines the number of clusters. BI-CLUST is available within WEKA[3] and we were not able to process internet-ads (more than 1500 properties). We summarize these results in Table 3. We provide only the τ_Q coefficients. The corresponding τ_O coefficients are equal or not significantly different. Notice that, when CDK-MEANS has the worst results, the Goodman-Kruskal's coefficient is not significantly dissimilar from other algorithm coefficients. On the other hand, for internet-ads, the coefficient obtained with CDK-MEANS is considerably higher than the one obtained with COCLUSTER. This is due to the high dimension of the dataset which is not well handled by the other algorithms. Also the average behavior is similar to the one of CO-CLUSTER. The average values of the two algorithms are often similar, as well

---
[1] http://www.ics.uci.edu/~mlearn/MLRepository.html
[2] http://www.amstat.org/publications/ jse/jse_data_archive.html
[3] http://www.cs.waikato.ac.nz/ml/weka/

Table 4. Jaccard coefficient values w.r.t. class variable for different algorithms

Dataset	Bi-Clust	Cocluster	K-Means	EM	CDK-means
voting	0.6473	0.6473	0.6027	0.6459	**0.6737**
titanic	0.4281	0.4651	0.3697	0.3697	**0.4745**
iris-2	0.4992	0.4992	**0.5117**	0.4992	0.4992
iris-3	0.4932	0.5240	**0.5394**	**0.5394**	0.5144
zoo-2	0.5141	**0.5630**	0.5027	0.5179	0.5141
zoo-7	-	0.1647	0.1843	**0.2325**	0.2212
breast-w	0.8246	0.8287	0.7777	**0.8328**	0.7666
credit-3	0.4233	0.3869	0.3765	0.3405	**0.4452**
credit-2	-	0.4360	0.4698	0.4442	**0.4915**
mr-2	-	0.6819	0.3496	**0.6976**	0.6356
mr-5	0.5068	0.3450	0.3192	0.3364	0.3375
ads	-	0.4317	-	-	**0.8019**

as the standard deviation values. Notice that for voting-records and breast-w, CDK-Means has always produced the same bi-partition.

CDK-Means generally needs for more execution time than the other algorithms because it processes possibly large collections of formal concepts. In these benchmarks, the extraction of formal concepts by itself is not that expensive (from 1 to 20 seconds). Using minimal size constraints during the formal concept extraction phase enables to reduce the collection size and it will be discussed later. For titanic, iris, and zoo, CDK-Means performs in less than one second, while for breast-w, credit-a and internet-ads, the average execution time is less than one minute. For mushroom, the average execution time is about seven minutes since more than 50 000 formal concepts have to be processed.

We also used the Jaccard index to compare the agreement of the object partitions with those determined by the class variables. Here again, we have selected the bi-partition with the highest τ_Q coefficient[4]. We provide the comparisons in Table 4. Again, our algorithm is competitive w.r.t. the other bi-clustering methods. With the exception of breast-w, our algorithm always performs as or better than Bi-Clust, and most of the times better than Cocluster.

Finally, we have compared our results, w.r.t. two classical clustering algorithms, the WEKA implementations of K-Means and EM (see Table 4). Except for breast-w, our algorithm is competitive w.r.t the other ones. For most datasets, CDK-Means performs better than K-Means and EM. Once again, when our algorithm obtains the best result, the difference with the score of the others is significant (except on breast-w). These results show that our clustering of formal concepts is a relevant approach for both partitioning and bi-partitioning tasks.

Scalability Issues. Collections of formal concepts are usually huge, especially in intrinsically noisy data. Since CDK-Means has a linear complexity in the number of bi-sets, it can be time-consuming. An obvious solution is to select some formal concepts, for instance the ones which involve enough objects and/or properties. Interestingly, such minimal size constraints can be pushed into formal concept mining algorithms [7]. Not only it enables the extraction in hard

---

[4] Clearly, it does not lead to the highest Jaccard's index.

Table 5. Clustering results on ads-internet with different minimal size constraints

(σ_p,σ_o)	$|\mathcal{B}|$	time(s)	τ(mean)	τ(max)	J-class	J-ref
(0,0)	7682	33	0.137 ± 0.109	0.538	0.8019	1
(4,4)	2926	8	0.194 ± 0.137	0.565	0.6763	0.6737
(5,5)	2075	5	0.254 ± 0.148	0.565	0.6862	0.7490
(5,10)	1166	2.5	0.223 ± 0.119	0.511	0.6745	0.7405
(7,10)	873	2	0.204 ± 0.095	0.549	0.6172	0.6658
(10,10)	586	1.5	0.227 ± 0.125	0.543	0.6080	0.7167

contexts, but also, intuitively, it removes formal concepts which might be due to noise. We therefore guess that this can increase the quality of the clustering result. Let σ_o be the minimal size of the object set and σ_p be the minimal size of the property set. Properties (resp. objects) that are in relation with less than σ_o objects (resp. σ_p properties) will not be included in any formal concept. As our bi-partitioning method is based only on a post-processing of these patterns, these objects and/or properties can not be included in the final bi-partition. This is not necessarily a problem if we prefer a better robustness to noise. However, one can be interested in finding a bi-partition that includes all objects and properties. An obvious solution is to add the top and bottom formal concepts $(\mathcal{O},\emptyset)$ and $(\emptyset,\mathcal{P})$. This has been done in some experiments (mushroom, credit-a) and we noticed that the decrease of the Jaccard and Goodman-Kruskal's coefficients were not significant. We made further experiments to understand the impact of using minimal size constraints on both the execution time and the quality of the computed bi-partition. We have considered internet-ads as the most suitable for these experiments (high cardinality for both object and property sets). We extracted formal concepts by setting some combinations of constraints ($0 \leq \sigma_p < 10$ and $0 \leq \sigma_o < 10$) and by adding $(\mathcal{O},\emptyset)$ and $(\emptyset,\mathcal{P})$. The results are summarized in Table 5. It shows that, increasing the minimal size threshold considerably reduces the number of extracted formal concepts and thus the average execution time. Also the extraction time decreases from 4 seconds (for $\sigma_p = \sigma_o = 0$) to one second (for $\sigma_p = \sigma_o = 10$). Moreover, the maximum Goodman-Kruskal's coefficient does not change significantly. In some cases it is greater than the coefficient computed when no size constraint is used. Also the average values of the Goodman-Kruskal's measures are better in general (while standard deviation values are similar). We then computed the Jaccard index of the different partitions w.r.t. the class variable (J-class column) and the partition obtained without setting any constraint (J-ref column). The slight variability of the Jaccard indexes and the high values of the τ measures show that they are still consistent w.r.t. the class one. Finally, results are always better than those obtained by using COCLUSTER (see Fig. 3 and Fig. 4) whose average execution time is about 4.2 seconds. In other terms, increasing σ_p and σ_o can eliminate the impact of noise due to sparse sub-matrices. In particular, grouping larger formal concepts can improve the relevancy of bi-partitions. Notice that if we do not add $(\mathcal{O},\emptyset)$ and $(\emptyset,\mathcal{P})$, we get better results involving a subset of the original matrix: constraints can be triggered to trade-off between the coverage of the bi-partition and the quality of the result.

4 Conclusion and Future Work

We have introduced a new bi-clustering framework which exploits local patterns in the data when computing a collection of (possibly overlapping) bi-clusters. The instance CDK-MEANS builds simultaneously a partition on objects and a partition on properties by applying a K-MEANS-like algorithm to a collection of extracted formal concepts. Our experimental validation has confirmed the added-value of CDK-MEANS w.r.t. other (bi-)clustering algorithms. We demonstrated that such a "from local patterns to a relevant global pattern" approach can work. Due to the lack of space, we omitted the experimental results on real data and the study of overlapping clusters [9]. Many other instances of the framework might be studied. For instance, given extracted local patterns, alternative clustering techniques can be considered. Also, other kinds of local patterns (i.e., relevant bi-sets which are not formal concepts) could be considered. Finally, an exciting challenge concerns constraint-based clustering. Our framework gives rise to opportunities for pushing constraints at two different levels, i.e., during local pattern mining but also when building bi-partitions from them.

Acknowledgements. The authors want to thank Luigi Mantellini for his technical support. This research is partially funded by CNRS (ACI MD 46 Bingo).

References

1. Jain, A., Dubes, R.: Algorithms for clustering data. Prentice Hall, Englewood cliffs, New Jersey (1988)
2. Fisher, D.H.: Knowledge acquisition via incremental conceptual clustering. Machine Learning **2** (1987) 139–172
3. Robardet, C., Feschet, F.: Efficient local search in conceptual clustering. In: Proceedings DS'01. Number 2226 in LNCS, Springer-Verlag (2001) 323–335
4. Dhillon, I.S., Mallela, S., Modha, D.S.: Information-theoretic co-clustering. In: Proceedings ACM SIGKDD 2003, Washington, USA, ACM Press (2003) 89–98
5. Madeira, S.C., Oliveira, A.L.: Biclustering algorithms for biological data analysis: A survey. IEEE/ACM Trans. Comput. Biol. Bioinf. **1** (2004) 24–45
6. Wille, R.: Restructuring lattice theory: an approach based on hierarchies of concepts. In Rival, I., ed.: Ordered sets. Reidel (1982) 445–470
7. Besson, J., Robardet, C., Boulicaut, J.F., Rome, S.: Constraint-based concept mining and its application to microarray data analysis. Intelligent Data Analysis **9(1)** (2005) 59–82
8. Goodman, L.A., Kruskal, W.H.: Measures of association for cross classification. Journal of the American Statistical Association **49** (1954) 732–764
9. Pensa, R.G., Robardet, C., Boulicaut, J.F.: Using locally relevant bi-sets for categorical data conceptual clustering. Research report, LIRIS CNRS UMR 5205 - INSA Lyon, Villeurbanne, France (2005) Submitted to a journal.

Privacy-Preserving Collaborative Filtering on Vertically Partitioned Data*

Huseyin Polat and Wenliang Du

Department of Electrical Engineering and Computer Science,
Syracuse University, CST 3-114, Syracuse, NY 13244-1240, USA
{hpolat, wedu}@ecs.syr.edu

Abstract. Collaborative filtering (CF) systems are widely used by E-commerce sites to provide predictions using existing databases comprised of ratings recorded from groups of people evaluating various items, sometimes, however, such systems' ratings are split among different parties. To provide better filtering services, such parties may wish to share their data. However, due to privacy concerns, data owners do not want to disclose data. This paper presents a privacy-preserving protocol for CF grounded on vertically partitioned data. We conducted various experiments to evaluate the overall performance of our scheme.

1 Introduction

Collaborative filtering (CF) is a recent technique that helps users cope with information overload using other users' preferences. It is widely used by E-commerce, direct recommendations, and search engines [1,2]. The goal is to predict how well a user (*an active user*) will like an item that he/she did not buy before based on other users' preferences [4].

Data collected for CF purposes might be vertically partitioned between different parties where the parties hold disjoint sets of items' ratings collected from the same users. An individual's preferences for products might be split among different E-commerce companies such as Amazon.com and MovieFinder.com. Online vendors can produce better referrals if they share information about their customers with other vendors. Joint data is beneficial for E-commerce sites because customers prefer returning to stores with better referrals and they search for more products to purchase. Shared information will also benefit customers by making it more likely to receive more accurate and reliable recommendations. Combining vertically partitioned data (VPD) is helpful when CF systems have limited rated items. To find more reliable matchings and provide more accurate referrals, the overlap between users should be large enough; this might be achieved by integrating VPD. However, due to privacy concerns, data owners do not want to collaborate and disclose their data to each other.

---

* This work was supported by Grants ISS-0219560 and ISS-0312366 from the United States National Science Foundation.

VPD-based CF is essential and can be achieved if privacy measures are introduced to data owners. We study the privacy-preserving collaborative filtering (PPCF) on VPD problem: *To maximize mutual advantages, two online vendors (A and B) holding disjoint sets of items' ratings gathered from the same users, want to provide CF services to their future customers using the joint data. How can they provide such services using the integrated data while preserving their privacy?*

We propose a protocol to achieve PPCF on VPD. Privacy, accuracy, and efficiency are conflicting goals. Therefore, the proposed protocol should achieve a good balance between them. Our scheme consists of off-line and online computation components. We conduct some computations off-line to achieve data exchange between parties with privacy. During the online computation, a new customer (an active user a) communicates with both parties. They then perform data exchange through a with privacy. The company that does not hold ratings of the target item (the item that a is looking for a prediction) finds prediction and tells a. Since data exchanges are required whenever a new customer wants a prediction and either party can act as an active user in multiple scenarios to learn about other party's data, the proposed protocol should be secure against such attacks coming from both parties.

Canny proposes two schemes for PPCF [1,2]. A community of users can compute personalized recommendations without exposing individual data using such schemes. Polat and Du use randomized perturbation techniques for PPCF [6,7]. Vaidya and Clifton [8,9,10] present privacy-preserving methods for association rule mining, naïve Bayes classifier, and K-means clustering based on VPD. We used the CF algorithm proposed by [4]. If v_{ij} is user i's vote on item j, and $\overline{v_i}$ and σ_i are the mean vote and the standard deviation of the user i's ratings, respectively, then the z-scores (z_{ij}) can be defined as $z_{ij} = (v_{ij} - \overline{v_i})/\sigma_i$. Herlocker et. al find predictions as follows where n is the number of users:

$$p_{aq} = \overline{v_a} + \sigma_a \cdot \frac{\sum_{i=1}^{n} w_{ai} \cdot z_{iq}}{\sum_{i=1}^{n} w_{ai}} \qquad w_{ai} = \sum_{k} z_{ak} \cdot z_{ik} \qquad (1)$$

where k is the item set both a and the user i have rated and q is the target item. σ_a and σ_i are standard deviations of a's ratings and i's ratings, respectively. p_{aq} is the prediction for a on q and w_{ai} is similarity between a and i. We used homomorphic property for our proposed protocol: $E_k(x) * E_k(y) = E_k(x+y)$. Many such systems exist, and an example is the system by Paillier [5]. A useful property of homomorphic encryption schemes is that an addition operation can be conducted based on the encrypted data without decrypting them.

2 PPCF on VPD

Without privacy as a concern, data owners exchange their data to provide CF services. However, with privacy as a concern, the companies should not be able to learn each other's data. We can write Eq. (1) as $p_{aq} = \overline{v_a} + \sigma_a \cdot P$ where P can be defined as follows:

$$P = \frac{\sum_{i=1}^{n}\left[\sum_{k} z_{ak} z_{ik}\right] z_{iq}}{\sum_{i=1}^{n}\sum_{k} z_{ak} z_{ik}} = \frac{\sum_{k} z_{ak}\left[\sum_{i=1}^{n} z_{ik} z_{iq}\right]}{\sum_{k} z_{ak}\left[\sum_{i=1}^{n} z_{ik}\right]} \quad (2)$$

CF systems can tell whether a will like q or not, rather than telling how much he/she will like it. To do this, p_{aq} is compared with a threshold (τ). If $p_{aq} \geq \tau$, q is recommended as like, otherwise it is recommended as dislike. If the ratings vary from 1 to 5, τ is set to 3.5 while it is set to 2 if they range from -10 to 10. Since A's and B's data is used to calculate P, Eq. (2) can be written as:

$$P = \frac{\sum_{k_A} z_{ak_A}\left[\sum_{i=1}^{n} z_{ik_A} z_{iq}\right] + \sum_{k_B} z_{ak_B}\left[\sum_{i=1}^{n} z_{ik_B} z_{iq}\right]}{\sum_{k_A} z_{ak_A}\left[\sum_{i=1}^{n} z_{ik_A}\right] + \sum_{k_B} z_{ak_B}\left[\sum_{i=1}^{n} z_{ik_B}\right]} = \frac{A_N + B_N}{A_D + B_D} \quad (3)$$

where $k = k_A + k_B$, and k_A and k_B represent the item sets both a and i have rated among the items held by A and B, respectively.

2.1 Off-Line Computation

The denominator part in Eq. (3) can be easily computed because A and B can find A_D and B_D using their own data. However, the party who does not own q needs to have z_{iq} values for all $i = 1, \ldots, n$ to compute $\sum_{i=1}^{n} z_{ik_j} z_{iq}$ necessary for the numerator. Since A and B follow the same steps, we only explain the procedure for A. A horizontally divides its $n \times m_A$ data matrix into c_A sub-matrices where each sub-matrix consists of n/c_A users and their ratings for items held by A, where m_A is the number of items A owns. It then disguises data in each sub-matrix independently. For $i = 1, \ldots, c_A$, A performs the followings:

Step 1. Permutes m_A column vectors using a permutation function Π_{Ai}.
Step 2. For $j = 1, \ldots, m_A$, divides the permuted column vector $\Pi_{Ai}(I_{ij})$ into d_{ij} random vectors where $\Pi_{Ai}(I_{ij}) = \sum_{z=1}^{d_{ij}} X_{ijz}$ and d_{ij} is an integer chosen with uniform random distribution over the range $[1, \beta_A]$.
Step 3. Permutes $X_{i11}, X_{i12}, \ldots, X_{i1d_{i1}}, X_{i21}, X_{i22}, \ldots, X_{i2d_{i2}}, \ldots, X_{im_A1}$, $X_{im_A2}, \ldots, X_{im_A d_{im_A}}$ random vectors found in step 2 using π_{Ai}.
Step 4. A sends D_{Ai} permuted random vectors to B where $D_{Ai} = d_{i1} + d_{i2} + \ldots + d_{im_A}$. B computes the scalar products between these permuted random vectors and its m_B column vectors using the corresponding parts of them and finds $D_{Ai} m_B$ scalar product results.
Step 5. B encrypts the scalar product results using a homomorphic encryption scheme and its public key e_b and sends $D_{Ai} m_B$ encrypted values to A.

Step 6. Since A knows Π_{Ai} and π_{Ai} and homomorphic encryption is used, it finds the scalar product results of its m_A and B's m_B column vectors in encrypted forms using homomorphic encryption property. After conducting these steps for all $i = 1, \ldots, c_A$, A gets encrypted scalar product results for its all c_A sub-matrices. Since A's data is horizontally divided, it again uses homomorphic encryption property to find the final scalar product results in encrypted forms.

A creates a matrix Σ_A consisting of $e_b(\Sigma_{ij})$ for $i = 1, \ldots, m_A$ and $j = 1, \ldots, m_B$ where $e_b(\Sigma_{ij})$ represents the encrypted scalar product between i^{th} column vector of B and the jth column vector of A. It generates large enough v_{ij} random numbers for $i = 1, \ldots, m_B$ and $j = 1, \ldots, m_A$, encrypts them using e_b, and adds them to the $e_b(\Sigma_{ij})$ values using homomorphic encryption property. It finds matrix Σ'_A consisting of $e_b(\Sigma'_{ij})$ where $\Sigma'_{ij} = \Sigma_{ij} + v_{ij}$ and stores v_{ij} values in a matrix V_A. It sends Σ'_A to B that decrypts the encrypted values and finds the matrix Σ''_A consisting of Σ'_{ij} values and stores it. By following the same procedure, B finds matrices Σ_B, Σ'_B, and V_B. B stores V_B and A finds Σ''_B, and then stores it. A and B compute the item mean votes and store them in $m_A \times 1$ and $m_B \times 1$ matrices, respectively.

2.2 Online Computation

Since either party can act as an active user in multiple scenarios, online component should be secure against such attacks. The steps are as follows:

Step 1. a sends his/her data and a query to the company that owns q. Assume that B owns q. B computes B_N and B_D. However, since A can act as an active user in multiple scenarios to learn them, B uses private B_N & B_D computation protocol, which is explained in the following, to compute them.

Step 2. B can compute A'_N value using the data from the qth row of the matrix Σ''_A and a's corresponding data where $A'_N = A_N + R_q$. The data from the qth row of the matrix Σ''_A represents $\sum_{i=1}^{n} z_{ik_A} z_{iq}$ values disguised by v_{qk_A} random numbers for all k_A. A can compute $R_q = \sum_{k_A} z_{ak_A} v_{qk_A}$ where k_A represents the items rated by a among the items held by A.

Step 3. B computes $A_N + R_q + B'_N$ and B'_D and sends them together with a's new mean vote, standard deviation, and the z-scores for those items rated by a among items held by A to A through a. A computes R_q, finds $A_N + B'_N = A'_N + B'_N - R_q$ and A_D, and estimates P' using Eq. 3 based on the query.

A computes p'_{aq}, tells a whether he/she will like q or not by comparing p'_{aq} with τ. Since B can act as an active user, A uses a random threshold to prevent B from learning A_D and A_N. It generates a uniform random number $(r_{A\tau})$ from a range $[-\alpha_A, \alpha_A]$, finds $\tau + r_{A\tau}$, and uses it as a random threshold.

Our scheme can be extended to multi-party. Each vendor exchanges data off-line with others and stores it as in two-party scheme. During online phase, a sends his/her data to the party that owns q. That party computes the required data like it does in two-party scheme and sends results to a. Then one party acts as a master site. Other parties computes the values required for numerator and denominator parts. Each company creates a large enough uniform random number from a range $[-\gamma, \gamma]$, adds it to the values for numerator and denominator parts, and sends them through a to master site, which estimates the prediction.

Private B_N & B_D Computation Protocol. We explain the protocol only for B. After B gets a' data, it finds the number of rated items (C_B) that a rated among the items it holds. If C_B is less than $\lfloor m_B/2 \rfloor$, then B finds the items that a did not rate among the items B owns. B generates a uniform random integer S_{Ba} from the range $[1, m_B - C_B]$, randomly selects S_{Ba} unrated items among the items it owns, and fills their cells in the a's ratings vector with their mean votes. If C_B is bigger than $\lfloor m_B/2 \rfloor$, B finds the items that a rated among the items B owns and creates a uniform random integer S_{Br} from the range $[1, C_B]$. It randomly selects S_{Br} rated items and removes their ratings from a's ratings vector. B computes B'_N and B'_D using the new ratings vector of a and finds a ratings' new mean and standard deviation and computes the z-scores.

3 Privacy and Overhead Costs Analysis

In this section we first investigated privacy.

Claim 1. *After B gets the required data from A, the probability of guessing the A's data for B is 1 out of $\left(m_A! D_A! (\beta_A)^{m_A}\right)^{c_A}$*. Since A uses Π_{Ai}s for all $i = 1, \ldots, c_A$ to permute its m_A column vectors in each sub-matrix, for B, the probability of guessing the correct positions of them is 1 out of $m_A!$. A divides each of its permuted column vectors into random vectors where it decides how many random vectors a permuted vector be divided into based on a uniform random integer from the range $[1, \beta_A]$. Therefore, the probability of guessing the number of random vectors that each vector is divided into is 1 out of $(\beta_A)^{m_A}$. A uses π_{Ai}s for all $i = 1, \ldots, c_A$ to permute random vectors. Therefore, guessing their correct positions is 1 out of $D_A!$ with the assumption that all D_{Ai} values are same and equal to D_A. Since A horizontally divides its data into c_A parts, the probability of guessing the A's data for B is 1 out of $\left(m_A! D_A! (\beta_A)^{m_A}\right)^{c_A}$.

Claim 2. *A is not able to learn B_N and B_D due to the private B_N & B_D computation protocol.* Since B uses random integer S_{Ba} and randomly selects S_{Ba} unrated items among the items it owns, the probability of guessing the correct S_{Ba} and which S_{Ba} unrated items are selected is 1 out of $((m_B - C_B)(m_B - C_B)!)/(S_{Ba}!(m_B - C_B - S_{Ba})!)$. B also fills unrated items' cells in a's ratings vector with their mean votes, which are only known by it.

Claim 3. *B is not able to learn A_N and A_D when it acts as an active user in numerous scenarios.* Since A tells a that he/she will like or dislike q and produces referrals using a random threshold, B will not learn A_N and A_D.

Claim 4. *A will not learn scalar product results conducted by B due to encryption and B is not able to derive Σ_{ij}s from Σ'''_{ij}s due to random numbers.*

Unlike off-line communication cost, online communication cost is vital and the number of online communications is only 4 for our scheme. The additional storage costs due to privacy issues are $O(m_A m_B + m_A)$ and $O(m_A m_B + m_B)$ for

A and B, respectively. Although off-line computation cost is not critical, online computation cost is essential.

Claim 5. *Additional computation costs due to privacy concerns during the online phase are insignificant.*

Claim 6. *The costs due to the encryptions and the decryptions done off-line are $O(c_B D_B + m_A m_B)$ and $O(c_A D_A + m_A m_B)$ for A and B, respectively where D_A and D_B represent the average number of random vectors for each sub-matrix.*

Claim 7. *The costs due to the multiplications done off-line are $O(nm_A D_B)$ and $O(nm_B D_A)$ for A and B, respectively.*

4 Experiments

We used Jester and MovieLens (ML) data sets. Jester [3] has 100 jokes and records of 17,988 users. The ratings range from -10 to +10. ML (www.cs.umn.edu/research/Grouplens) consists of ratings for 3,592 movies made by 7,463 users. Ratings are made on a 5-star scale. We measured the accuracy of our approach using classification accuracy (CA) and F-measure (FM), which is a weighted combination of precision and recall.

4.1 Methodology

We randomly selected 2,000 users for training from Jester and ML. Since we conducted different sets of experiments with varying number of rated items (M), we found those users who rated certain number of items and randomly selected 400 test users among them for each experiment. For each test user, we randomly selected 5 rated items, withheld a single rated item for each test user, and tried to predict its value given all other ratings. We did this for all 5 test items. We replaced the test item's entry as null. We ran the selection of the subset of rated or unrated items protocol 10 times for each test item. We created r_τ uniform random numbers while varying the range to evaluate random threshold. For each test item, we created 10 uniform random numbers for those experiments testing accuracy with random threshold. We converted the withheld items' ratings into binary ratings. We then compared the recommendations on our scheme with the withheld items' converted ratings and found CA and FM values.

4.2 Experimental Results

Number of Rated Items (M). We hypothesize that since prediction quality improves with increasing M, when two parties conduct CF on the joint data, accuracy improves. To show the effects of different M values, we conducted experiments using ML data while varying M. Table 1 shows CAs and FMs with varying M. Based on the settings of each experiment, we selected those users for testing who rated M number of items. Overall performance increases with increasing M. If there is limited number of rated items, with increasing M values, we gain significant improvement. However, the improvement becomes

Table 1. Prediction Quality vs. M

M	$M < 50$	$40 < M < 100$	$100 \leq M < 200$	$200 \leq M < 400$
CA	0.6645	0.7010	0.7100	0.7140
FM	0.7313	0.7686	0.7713	0.7743

stable when enough ratings are available. CA is 0.6645 when M is less than 50 while it increases to 0.7010 when $40 < M < 100$. Besides CA, FM also increases from 0.7313 when $M < 50$ to 0.7686 when $40 < M < 100$.

Number of Removed Ratings (S_r). We conducted experiments while varying S_r using Jester and ML, and showed CAs in Fig. 1. 400 test users were randomly selected among those users who rated more than 200 and 80 items for ML and Jester, respectively. As seen from Fig. 1, accuracy slightly becomes worse with increasing S_r because the available ratings are decreasing. When we increased S_r from 0 to 100, we lost 0.0065 accuracy for ML. This means that if there are significantly large number of ratings available, removing some of them does not affect accuracy too much.

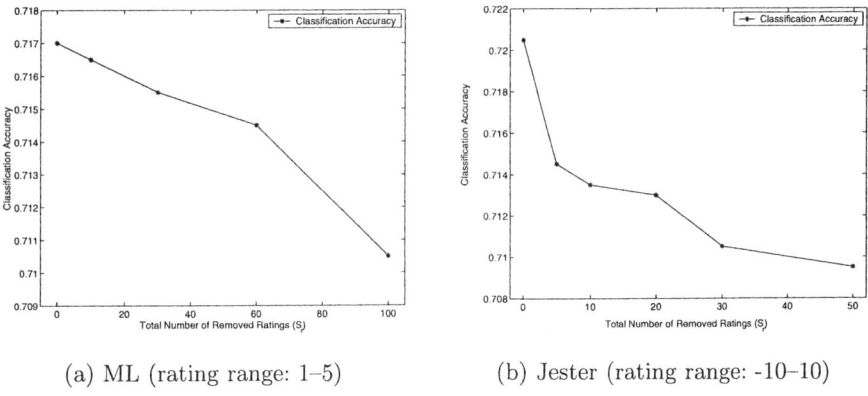

(a) ML (rating range: 1–5) (b) Jester (rating range: -10–10)

Fig. 1. Prediction Quality vs. S_r

Number of Appended Ratings (S_a). Since accuracy improves with increasing available ratings, appending more ratings may improve accuracy. However, since empty cells are filled with the item mean votes, which can be considered default votes for a and may not match with his/her true ratings for those items, that might make accuracy worse. We performed experiments using ML with varying S_a. 400 test users were randomly selected among those users who rated more than 40 and less than 100 items. CA improves with increasing S_a up to 60 appended ratings. When S_a is 100, CA becomes worse and it is 0.7035; but it is still better than the CA, which is 0.7010, when S_a is 0.

Range of Uniform Random Values (α). To show how different α values affect our results, we ran experiments while using uniformly created r_τ values from the range $[-\alpha, \alpha]$ with varying α values using ML. 400 test users were randomly selected among those users who rated more than 40 and less than 100 items. The results slightly become worse with increasing α because with increasing range, r_τ values become larger, the random threshold ($\tau + r_\tau$) fluctuates more and causes a loss in the performance. When we increased α from 0 to 0.1, CA degrades by 0.0015 while FM decreases by 0.0032.

5 Conclusions and Future Work

We have presented a solution to the PPCF based on VPD problem. Our solution makes it possible for two parties to conduct filtering services using their joint data without disclosing their data to each other. Our experiment results have shown that our solution produces accurate referrals compared with the true ratings. We will study multi-party scheme in detail and show how accuracy and privacy change compared to two-party scheme and with varying number of parties.

References

1. J. Canny. Collaborative filtering with privacy. In *Proceedings of the IEEE Symposium on Security and Privacy*, pages 45–57, Oakland, CA, USA, May 2002.
2. J. Canny. Collaborative filtering with privacy via factor analysis. In *Proceedings of the 25th Annual International ACM SIGIR Conference on Research and Development in Information Retrieval*, pages 238–245, Tampere, Finland, 2002.
3. D. Gupta, M. Digiovanni, H. Narita, and K. Goldberg. Jester 2.0: A new linear-time collaborative filtering algorithm applied to jokes. In *Workshop on Recommender Systems Algorithms and Evaluation, 22nd International Conference on Research and Development in Information Retrieval*, Berkeley, CA, USA, August 1999.
4. J. L. Herlocker, J. A. Konstan, A. Borchers, and J. T. Riedl. An algorithmic framework for performing collaborative filtering. In *Proceedings of the 1999 Conference on Research and Development in Information Retrieval*, August 1999.
5. P. Paillier. Public-key cryptosystems based on composite degree residue classes. In *Advances in Cryptology – EUROCRYPT'99*, pages 223–238, 1999.
6. H. Polat and W. Du. Privacy-preserving collaborative filtering using randomized perturbation techniques. In *Proceedings of the 3rd IEEE International Conference on Data Mining (ICDM'03)*, Melbourne, FL, USA, November 19–22 2003.
7. H. Polat and W. Du. SVD-based collaborative filtering with privacy. In *Proceedings of the 20th ACM Symposium on Applied Computing Special Track on E-commerce Technologies*, Santa Fe, NM, USA, March 13–17 2005.
8. J. Vaidya and C. Clifton. Privacy preserving association rule mining in vertically partitioned data. In *Proceedings of the 8th ACM SIGKDD*, pages 639–644, 2002.
9. J. Vaidya and C. Clifton. Privacy preserving k-means clustering over vertically partitioned data. In *Proceedings of the 2003 ACM SIGKDD*, Washington, DC, USA, Agustos 24–27 2003.
10. J. Vaidya and C. Clifton. Privacy preserving naïve bayes classifier for vertically partitioned data. In *Proceedings of the 2004 SIAM Conference on Data Mining*, Orlando, FL, USA, May 2004.

Indexed Bit Map (IBM) for Mining Frequent Sequences

Lionel Savary and Karine Zeitouni

PRiSM Laboratory, 45 Avenue des Etats-Unis,
78035 Versailles, France
{Lionel.Savary, Karine.Zeitouni}@prism.uvsq.fr

Abstract. Sequential pattern mining has been an emerging problem in data mining. In this paper, we propose a new algorithm for mining frequent sequences. It processes only one scan of the database thanks to an indexed structure associated to a bit map representation. Thus, it allows a fast data access and a compact storage in main memory. The experimental results show the efficiency of our method compared to existing algorithms. It has been tested on synthetic data and on real data containing sequences of activities of a urban population time-use survey.

1 Introduction

The problem of mining sequential patterns was first introduced in the context of customer transactions analysis [2]. It aims to retrieve frequent patterns in the sequences of products purchased by customers through time ordered transactions. Several algorithms have been proposed in order to improve the performances and to reduce required space in memory [5], [9], [6]. Other works have concerned mining frequent sequences in DNA [8] or Web Usage Mining [3]. Finally, notice the use of bit map structure in providing a compact representation and good performances [5].

The target application in this paper is related to population time-use analysis and more precisely their daily displacements [4]. Our data are related to daily activities carried out by each surveyed person at the scale of a whole urban area. Thus, for each person of a surveyed household, it captures the activity program [7], the transport mode used between two activities, the departure time, and the duration of the trip. For example, during a day, an individual can leave home, take children to school, go to work, pick children up from school, and come back home. Activity programs of most individuals may be the same or be similar. Each activity program could be seen as a sequence of single values, making it possible to discover frequent activity sequences that characterise groups of the surveyed individuals. This allows analyzing the mobility of this urban population. Likewise, when considering transport mode, schedules or duration sequences, it would be possible to determine a typology of used transport modes, schedules, and so on.

Existing algorithms are either inappropriate or not enough efficient to our specific case. Most works [1], [2], [6] make multiple scan of the database, which can be considered as the main bottleneck of algorithms of frequent sequence mining. Furthermore, unlike the analysis of sequential transactions where each transaction is an item set, our context only focuses on the analysis of sequences of items.

Although existing works [9], [10], [12] can be applied in this context, we propose here a new algorithm more appropriate to this particular case. This algorithm only makes one scan of the database. The indexed bit map structure needs few spaces in the main memory and allows a fast access to the data. The experimental results, using real or synthetic data, show that our algorithm outperforms existing ones.

The paper is organised as follows: section 2 presents related works, then, section 3 describes the proposed algorithm, section 4 proposes an optimisation, section 5 relates the experimentation and performance study, and finally, a general conclusion summarizes our contribution and traces some perspectives.

2 Related Works

Most works related to mining frequent sequences are in the field of customer transaction analysis. Early work on frequent patterns -*Apriori* algorithm- only considered transactions, not sequence of transactions [1]. This algorithm is costly because it carries out multiple scans of the database to determine frequent subsets of items. Three algorithms dealing with sequence of transactions are presented and compared in [2]: *AprioriAll*, *AprioriSome* and *DynamicSome*. *AprioriAll* algorithm is an adaptation of *Apriori* to sequences where candidate generation and support are computed differently. *AprioriAll*, and *AprioriSome* only compute maximal frequent sequences. Their principle is to jump to candidates of size k+next(k) in the next scan, where next(k)>1. Maximum frequent sequences of lower size that have not been calculated are given in the backward phase. The value of next(k) increases with $P_k = |L_k|/|C_k|$, where L_k stands for frequent sequences of size k, and C_k the whole generated candidates of size k. *DynamicSome* algorithm is based on *AprioriSome* but uses a jump by a multiple of user defined *step*.

SPAM algorithm [5] uses a bitmap representation of transaction sequences once the entire database has been loaded in a lexicographic tree. But this algorithm considers that the entire database and all used data structures should completely fit into main memory, and then do not adapt for large datasets.

The *GSP* algorithm [6] exploits the property that all contiguous subsequences of a frequent sequence also have to be frequent. As *Apriori*, it generates frequent sequences, then candidate sequences by adding one or more items.

PrefixSpan [10] first finds the frequent items after scanning the database once. The sequence database is then projected, according to the frequent items, into several smaller databases. Finally, all sequential patterns are found by recursively growing subsequence fragments in each projected database. Employing a divide-and-conquer strategy with the *PatternGrowth* methodology, *PrefixSpan* efficiently mines the complete set of patterns.

3 IBM Algorithm

We are now going to focus on the specific case where the considered sequences are basic since they are composed of single items, not of a set of items. This is the case in DNA [8], Web usage data [3] or activity program sequences [7]. Our algorithm will be compared to PrefixSpan, one of the most efficient among the above mentioned methods.

A sequence is said frequent if it is included in a number of sequences greater than a support given by the user. The inclusion between two sequences $s1 = (a_1, .., a_n)$ and $s2 = (b_1, ..., b_n)$: $s1 \subset s2$ is defined by : $\exists\ b_{i1} = a_1,..., b_{in} = a_n$ such that $i1 < i2 < ... < in$.

3.1 Principle of the Algorithm

The proposed approach is two phases. The first stage is the data encoding into a memory resident data structures. The second one is the frequent generation that in turn is composed of candidate generation, and candidate support checking.

The data structure is based on four components: (i) a Bit Map (IBM) is a binary matrix representing the distinct sequences of the database, (ii) an SV vector encodes all the ordered combinations of sequences, (iii) an index (INDEX) on the Bit Map allows a direct access to sequences according to their size, (iv) an NB table associated to the Bit Map which informs about the frequency of each distinct sequence (figure 1).

This algorithm only makes one scan of the database during which the total number of distinct sequences, the frequency of these sequences and the number of sequence by size are computed. This allows computing the support of each generated sequence. These sequences are classified by decreasing size in the IBM and only distinct sequences are stored in the Bit Map. An index by size allows a direct access to sequences according to their size. This structure provides an optimisation since a generated sequence s of size t will be directly compared with the sequences of the same or greater size stored in the IBM (figure 1).

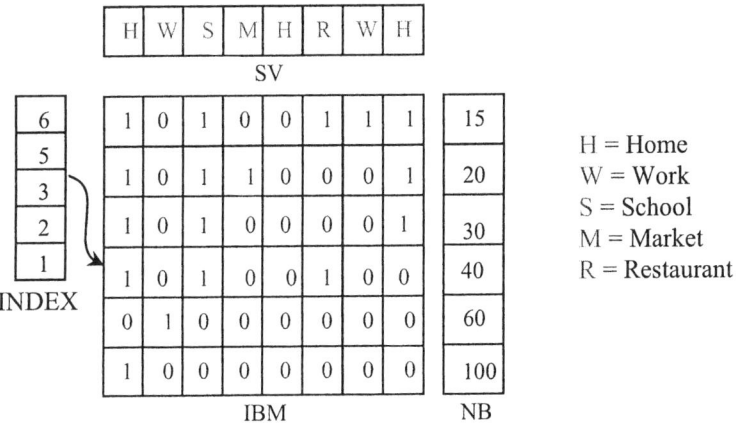

Fig. 1. The data structure

In order to simplify the notations, we represent each activity by a specific character, e.g. HSWSH standing for (Home, School, Work, School, Home). In the figure 1, the sequence vector (SV) is made of 5 ordered activities (H,W,S,M,H). In this example one supposes that the database is composed of six distinct sequences of size 1 to 5 encoded in the IBM. The bit *1* indicates the items present in the sequence according to the SV and bit *0*, those that are not. Here, there are 6 distinct sequences: (H), (W), (HRS), (HSH), (HSMH) and (HSRWH). In the above example (figure 1), each cell of the INDEX indicates the first line where the corresponding size of sequence is stored. For example, the cell number 5 (with value 6) corresponds to the line number 6 of the first sequence of size 5 encoded in the IBM. The table NB associates to the IBM stores the frequency of each distinct sequence. Thus the sequence (HSMH) of size 4 occurs 20 times in the database. In this algorithm, INDEX, SV, NB and IBM are built on the fly during one pass. At each insertion of a sequence, the IBM may become larger, and a set of shifting operations are applied to the bit values stored in this table.

```
IBM (sequence database DB, threshold t)
00 For each sequence s in DB
01   Gen-sequence-vector(s)
02   Encode and Insert s in the IBM
03   Update NB
04   Update INDEX
05 End For
06 Integer k := 1;
07 While exists frequent sequence of size k
08   k := k+1;
09   Generate Ck
10   Get-frequent-sequences (t)
11 End While
```

Fig. 2. IBM algorithm

Figure 2 shows the general IBM algorithm that takes as parameters: the database of sequences *DB* and a threshold *t*. This value (*t*) stands for the minimum frequency of the sequences which will be taken into account for the generation of the candidates. Then for each sequence *s* reads from the database during the scan, the SV (line 01) is generated using a merging process (see section 3.2). If the sequence already exists in SV, only the NB table is updated (line 03): the line corresponding to this sequence in NB (and encoded in the IBM) is incremented. So, the frequency corresponding to this value is incremented. Else, if the sequence is not presented in SV, it is generated by the Gen-sequence-vector(s) function (section 3.2). The height of the IBM is increased to one line (line 02), the length is increased to the SV length, and the INDEX (line 04) is updated. Then, a set of shifting operations is applied to the IBM in order to preserve the initial values of existing sequences while encoding the new one.

Once all the data have been encoded in this structure (SV, IBM, NB, INDEX), new candidates (line 09) are generated (see section 3.3) and compared to the data stored in the IBM (line 10) with a fast access thanks to the index (INDEX).

3.2 Generation of the Sequence Vector

The sequence vector is generated during the unique scan of the database according to the algorithm of figure 3. Here, *s* stands for a sequence of the database read during the scan, and position(x) stands for the cell number of value x in the SV. If an item *a* of *s* already exists in SV, then there is nothing to do, otherwise, there are two possibilities: if there exists an item *b* such that the cell number of *b* is greater than the cell number of *a* and *b* is in SV (line 04 and 05), then *a* is inserted before the value *b* in SV; otherwise, *a* is inserted at the end of SV (line 06). Thus all the distinct sequences of the database are represented in the SV using a merging process.

```
Gen-sequence-vector(s):
00 var SV := ϕ; {SV empty at the beginning};
01 Integer current_position := 0; {position in SV};
02 For each item a of s
03    If a ∉ SV
04       If ∃ b ∈ s such that (b ∈ SV and position(b) >
       position(a) in s and position(b) > current_position)
05          Insert a before b
06       Else insert a at the end of SV
07       current_position := position(a) in SV;
08 End For
```

Fig. 3. Sequence Vector generation

3.3 Candidate Generation

During the scan, the frequencies of all items are computed. Those whose support is underneath the one specified by the user are deleted. Then, candidates are generated from these frequent items, using the fusion process as in GSP algorithm [6].

3.4 Candidate Support Counting

For a given candidate C of size S, the algorithm first accesses the first sequence of size S encoded in IBM, which corresponds to the line l=INDEX(S). For each line starting from the line l to the last line of IBM table, the algorithm determines using the SV vector if C is contained in each line of IBM. If so, the corresponding frequency of this sequence stored in the NB table, is added to the frequency of the candidate. After the comparison with each line until the last one, the support of C is computed.

4 Implementation and Optimization

The IBM algorithm has been implemented in Java. It takes few spaces in the main memory. But whereas the bit variable is not provided in programming languages like Java or C++, some shifting operations are required to access the target value stored in the bit map and corresponding to the value stored in SV. In order to avoid these superfluous computations, we have proposed a variant with IBM2 algorithm, where the bit map is replaced by a Boolean matrix, i.e. where cells are declared of Boolean type, which takes 8 bits for each cell. Although this solution requires more space in memory, the access to the target value stored in the Boolean matrix is done directly without shifting computations. The result of their respective performances is detailed in the next section and compared with PrefixSpan.

5 Experimental Results

The experiments were performed on a 2.5 GHz Pentium IV with 1.5 GB of memory running Microsoft Windows XP Professional. Our implementation of IBM and IBM2 has been compared with PrefixSpan, based on the package PrefixSpan-0.4.tar.gz[1]. This test has concerned the scalability of the algorithm, by measurements of runtime and memory occupancy while varying the dataset size, and the support threshold. Moreover, we have tested the impact of the number of distinct items. Four synthetic datasets have been generated for the experimentations, with different sizes: 100000, 300000, 600000 and 1000000 rows. The size of sequences is randomly generated from 2 to 60, and the number of distinct items is about 10 for figures 4 to 7. This number has been pushed to 35 distinct items in order to test its impact.

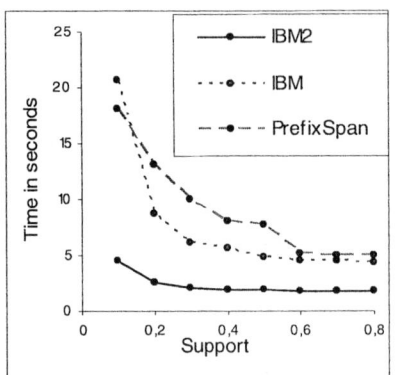
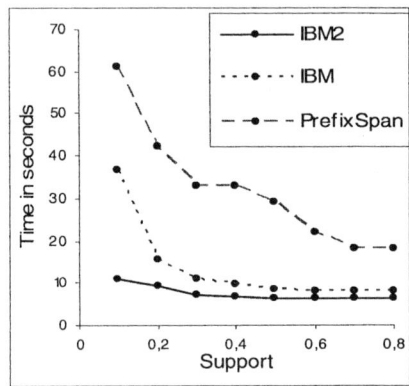

Fig. 4. Performances with 100,000 rows

Fig. 5. Performances with 300,000 rows

Although IBM and IBM2 have been implemented in Java, and PrefixSpan in C++ - *a priori* more optimal than Java -, IBM and IBM2 outperform PrefixSpan. The experimentations show that the larger is the database size, the more IBM and IBM2 win PrefixSpan (Figures 4 to 7). This is because IBM and IBM2 make only one scan of the database and the Indexed Bit Map structure allows a faster access to the sequences than the structure used in PrefixSpan. Moreover, as the support threshold decreases, the gap between IBM and PrefixSpan increases. Concerning the resource consumption, the size of the bit map depends on the size of SV, which may increase with the number of distinct sequences. Notice that SV size only increases when the encountered sequence can not be encoded using the current SV. Moreover, not all the items of the inserted sequence are added in SV, but only those that are not present in the

---

[1] http://chasen.org/~taku/software/prefixspan/

same order. Finally, since the probability to find common ordered items between SV and the current sequence becomes high as the building process advances, SV size becomes stable regardless of the size of the database. For instance, with a database composed of 600,000 rows, SV contains about 265 values for 90,000 distinct rows. The size of the Boolean Map is then equal to: 265*90,000 = 23.85 Mega Bytes. As IBM is 8 times more compact, the size of the Bit Map is less than 3 MB. With 1,000,000 rows (figure 7), SV contains 370 elements for 160,000 distinct rows. Then, the size of the Boolean Map reaches 59.2 MB, whereas the size of the Bit Map fits in 7.5 MB. Concerning the impact of distinct item number, for 100,000 rows until 20 distinct items, IBM and IBM2 perform better than PrefixSpan. Between 20 and 35 distinct items, IBM2 performs better than PrefixSpan, which becomes faster than IBM. But above 35 distinct items, PrefixSpan is faster than IBM and IBM2.

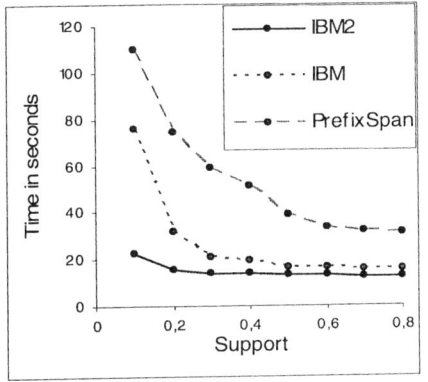 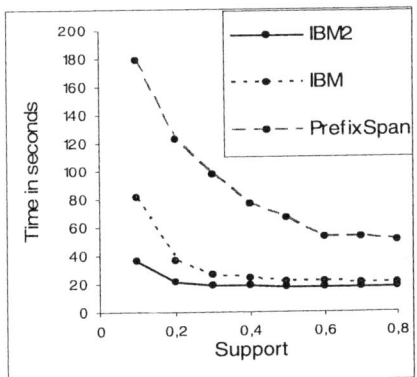

Fig. 6. Performances with 600,000 rows   **Fig. 7.** Performances with 1,000,000 rows

These results also show that IBM is more appropriate than IBM2 for very large databases, due to data compression. However, IBM2 runs faster than IBM. This is due to the costs of shifting operations necessary to access target values, while IBM2 directly accesses the target sequences.

6 Conclusion and Perspectives

This paper has presented a new algorithm IBM and its variant IBM2. The aim of this algorithm is to find all frequent sequences in item sequences. It has been applied to discover all frequent activity sequences in the time use mobility database within an urban environment. IBM only makes one scan of the database and provides an efficient data structure saving runtime and memory space. The use of the specified index provides another optimization of comparisons during candidate counting. Experimental results show that in most cases, IBM2 outperforms IBM, which in turn outperforms PrefixSpan for large and very large databases, with limited distinct items. Extensive experiments have been conducted that attest for the effectiveness and the efficiency of the proposed method, and are detailed in [11]. In perspective, IBM will be extended to multidimensional sequences (e.g. with attributes) and spatial sequences (such as trajectories). Other application fields will be explored, like pattern mining from DNA, Web Usage Mining or extension to customer transaction analysis. Finally, the proposed data structure adapts to similarity analysis of sequences and may be a good basis for efficient sequence clustering.

References

1. Agrawal, R., Srikant, R.: Fast Algorithms for Mining Association Rules. In Proc. of the 20th Int. Conf. Very Large Data Bases (VLDB), Santiago, Chile, September (1994)
2. Agrawal, R., Srikant, R.: Mining sequential patterns. In Proc. of the 11th Int'l Conference on Data Engineering, Taipei, Taiwan, March (1995)
3. Spiliopoulou, M., Faulstich Lukas, C., Winkler, K.: A data miner analyzing the navigational behaviour of web users. In Proc. Of the Workshop on Machine Learning in User Modelling of the ACAI'99 Int. Conf., Creta, Greece, July (1999)
4. Ministère de l'Equipement, des Transports et du Logement. L'enquête ménages déplacements « méthode standard ». Collections du Certu. Octobre (1998), ISSN 1263-3313
5. Jay, A., Johannes, G., Tomi, Y., Jason F.: Sequential Pattern Mining using A Bitmap Representation. SIGMOD pp 429-435, July (2002), Edmonton, Alberta, Canada
6. Srikant, R., Agrawal, R.: Mining Sequential Patterns : Generalizations and Performance Improvements. Proc. 5th EDBT, Mars 25-29, (1996). Avignon, France. pp 3-17
7. Wang, D., Tao, C.: A spatio-temporal data model for activity-based transport demand modeling.. International Journal of Geographical Information Science, (2001), 15(6), pp 561-585
8. Han, J., Jamil, H. M., Lu, Y., Chen, L., Liao, Y., Pei, J.: DNA Miner: A system prototype for mining DNA sequences. In the proc. Of the ACM SIGMOD International Conference on the management of data, Day 21-24,(2001), Santa Barbara, CA, USA
9. Zaki, M. J.: Efficient Enumeration of Frequent Sequences. Int. Conference on Information and Knowledge Management, November(1998), Washington DC
10. 10. Pei, J., Han, J., Mortazavi-Asl, B., and H., Pinto.: Prefixspan: Mining sequential patterns efficiency by prefix-projected pat tern growth. In Proc. of the International Conference on Data Engineering (ICDE), pp 215–224, (2001)
11. Savary, L., Zeitouni, K.: Indexed Bit Map (IBM) for Mining Frequent Sequences. PRiSM Laboratory Technical Report N° 2005/82, August (2005), Versailles University, France. http://www.prism.uvsq.fr/rapports/bin/bibliography.php?id=300

STochFS: A Framework for Combining Feature Selection Outcomes Through a Stochastic Process

Jerffeson Teixeira de Souza[1], Nathalie Japkowicz[2], and Stan Matwin[2]

[1] Computer Science Department, Federal University of Ceará,
Fortaleza, 60455-760, Brazil
jeff@lia.ufc.br
[2] School of Information Technology and Engineering, University of Ottawa,
Ottawa, K1N 6N5, Canada
{nat,stan}@site.uottawa.ca

Abstract. The *Feature Selection* problem involves discovering a subset of features such that a classifier built only with this subset would have better predictive accuracy than a classifier built from the entire set of features. Ensemble methods, such as Bagging and Boosting, have been shown to increase the performance of classifiers to remarkable levels but surprisingly have not been tried in other parts of the classification process. In this paper, we apply the ensemble approach to feature selection by proposing a systematic way of combining various outcomes of a feature selection algorithm. The proposed framework, named STochFS, have been shown empirically to improve the performance of well-known feature selection algorithms.

1 Introduction

The *Feature Selection* problem involves discovering a subset of features such that a classifier built only with this subset would have better predictive accuracy than a classifier built from the entire set of features.

Ensemble methods aim at improving the predictive performance of a given learning algorithm. Its general principle is to construct a combination of some learning models through a systematic process, instead of using a single model. Several combination methods, such as bagging [4] and boosting [6], have been shown to improve the performance of classifiers to remarkable levels. Given the benefits of classifier combinations of the type described above, it is surprising that other parts of the classification process (and in particular, Feature Selection) have not been tried in a systematic combined manner. In this paper, we will apply the ensemble approach to feature selection by proposing a systematic way of combining various outcomes of a feature selection algorithm.

2 Feature Selection

Feature subset selection algorithms can be classified into three broad categories based on whether or not feature selection is performed independently of the learning algorithm used to construct the classifier. If feature selection is performed independently of the learning algorithm, the technique is said to follow a *filter* approach. Otherwise, it is said to follow a *wrapper* approach. While the filter approach is generally computationally more efficient than the wrapper approach, its major drawback is that an optimal selection of features may not be independent of the inductive and representational biases of the learning algorithm that is used to construct the classifier. The wrapper approach on the other hand, involves the computational overhead of evaluating candidate feature subsets by executing a selected learning algorithm on the dataset represented using each feature subset under consideration.

A combination of these two approaches, that is, the use of two evaluation methods (a filter-type evaluation function and a classifier) creates a *hybrid* solution. Hybrid solutions attempt to combine the good characteristics of both filters and wrappers[1]. The combinations of approaches performed by hybrid feature selection algorithm, however, are heuristic in nature and cannot be systematically applied or analyzed.

In this paper, we propose the systematic combination of the outcomes of feature selection algorithms using the Bagging technique via a stochastic process.

3 The Framework

The STochFS framework combines the results of a feature selection algorithm in a stochastic manner by summarizing these outcomes in a single structure and using it as a seed in the generation of new feature selection subsets which are evaluated with a learning algorithm.

Initially, the $NumOuts$ best subsets returned by a single run of a feature selection system fs (or the single results of $NumOuts$ different runs, if such an algorithm returns only one best subset per execution) are stored into a two-dimensional array, see Figure 1. This array will then be condensed into a new array, called $Adam$, that will simply store the number of times each feature appeared in the $NumOuts$ best subsets. Next, STochFS will iteratively ($NumIter$ times) generate new subsets of features in a stochastically guided fashion using $Adam$ as a seed and evaluate them with a learning system over the dataset D. The generation of a new subset is such that features with high value in $Adam$ have a better chance of being selected than those with a low one at each iteration. At the end, the subset with best accuracy will be returned. If subsets tie it terms of accuracy, the one with the lowest cardinality is returned.

Each of the procedures used in this framework are described next.

GenerateOutcomes($fs, D, NumOuts$) executes the feature selection algorithm, fs, $NumOuts$ times and stores its outcomes in O. This procedure works differently, as described in the next two sections, depending on whether the feature selection algorithm to be used in probabilistic or deterministic.

---

[1] A description of recently proposed hybrid feature selections algorithms can be found in [13], [14], [2],[5] and [12].

```
STochFS(fs, D, NumIter, NumOuts)

O = GenerateOutcomes(fs, D, NumOuts)
Adam = CalculateAdam(O)
for j = 1 to NumIter
    S = GenerateSubset(Adam)
    if Error(S, D) < Error(S_best, D) then
        S_best = S
    else
        if Error(S, D) = Error(S_best, D) and
           Card(S) < Card(S_best) then
            S_best = S
return S_best
```

Fig. 1. The STochFS Framework

CalculateAdam(O) uses the following equation:

$$Adam = \{a_i, 1 \leq i \leq n\}$$

where: $a_i = \sum o_{ji}$, with $1 \leq j \leq k$ and $1 \leq i \leq n$.
to create the *Adam* vector. *Adam* stores the number of occurrences of each feature in O, which represent the number of times each feature was selected by fs as a relevant feature.

GenerateSubset(Adam) generates a new subset of features S in a stochastically guided fashion using *Adam* as a seed. The generation process works as described below. Let i denote a particular feature in *Adam*. Let S be a vector of n elements where n is the total number of features in O. Element S_i (of S) = 1 if feature i is included in the subset of features represented by S. $S_i = 0$, otherwise. Vector S is computed as follows:

$S_i = 1$, if $a_i > random(k)$ and $S_i = 0$ otherwise,
where $random(k)$ returns a random number between 0 and k.

This procedure is such that features with high frequency have a better chance of being selected than those with a low one at each iteration.

Error(S, D) makes use of a learning algorithm, inputting the subset S to generate a prediction model and receiving the error rate calculated for this model over dataset D.

In order to deal with the problem generated by deterministic feature selection algorithms, we have adapted STochFS to combine outcomes from both probabilistic and deterministic algorithms. Section 3.1 describes its use with probabilistic algorithms while section 3.2 details its use with deterministic algorithms.

3.1 Combining Outcomes of Probabilistic Feature Selection Algorithms

In the case the feature selection algorithms used in STochFS, fs, is a probabilistic algorithm, GenerateOutcomes($fs, D, NumOuts$) proceeds as follows:

GenerateOutcomes($fs, D, NumOuts$)

for $i = 1$ to $NumOuts$
    $O[i]$ = FeatureSelection(fs, D)

Fig. 2. GenerateOutcomes() for Probabilistic Algorithms

where

FeatureSelection(fs, D) runs a feature selection system fs over D and stores its outcome in $O[i]$.

Since probabilistic algorithms present random components, we apply such algorithms directly in our framework.

3.2 Combining Outcomes of Deterministic Feature Selection Algorithms

On the other hand, if fs is deterministic, the outcomes are generates as follows:

GenerateOutcomes($fs, D, NumOuts$)

for $i = 1$ to $NumOuts$
    $D[i]$ = Resample(D)
    $O[i]$ = FeatureSelection($D[i]$)

Fig. 3. GenerateOutcomes() for Deterministic Algorithms

where

Resample(D) creates samples of the original dataset D by *bootstrap aggregation*. [4]. Each bootstrap replicate, stores in $D[i]$, contains on the average 63.2% of the instances in D[2].

FeatureSelection($D[i]$) runs a feature selection system fs over $D[i]$ and stores its outcome in $O[i]$.

This procedure would add some randomness to the selection process and allow for a more open space to be considered by the stochastic search based on *Adam* performed in STochFS.

---

[2] This is the same sampling technique used in Bagging.

4 STochFS Evaluation

In order to evaluate STochFS we have selected four selection systems that vary according to how stochastic they are. First, the LVF algorithm [11] uses a Las Vegas approach to generate new subsets and can in fact be considered the most random selection algorithm of all. The Relief algorithm [9] uses randomness to select an instance which will be used to update the relevance weights for all features. Therefore, it deals with chance but clearly less directly as in LVF. We have also considered two deterministic algorithms, Focus [1] and RelieveD [8], where Focus finds the smallest subset that perfectly represents the original dataset and RelieveD is the deterministic version of Relief that uses all instances in the dataset to update the feature weights. For these two approaches, we have added randomness by using bootstrap aggregation, as described in section 3.2.

For each feature selection algorithm we performed a series of experiments using three different classifiers (C4.5, Naive Bayes and k-Nearest Neighbor) and 13 datasets from the UCI Repository [3]: Credit (15 features, 690 instances), Labor (16, 57), Vote (16, 435), Primary Tumor (17, 339), Lymph (18, 148),

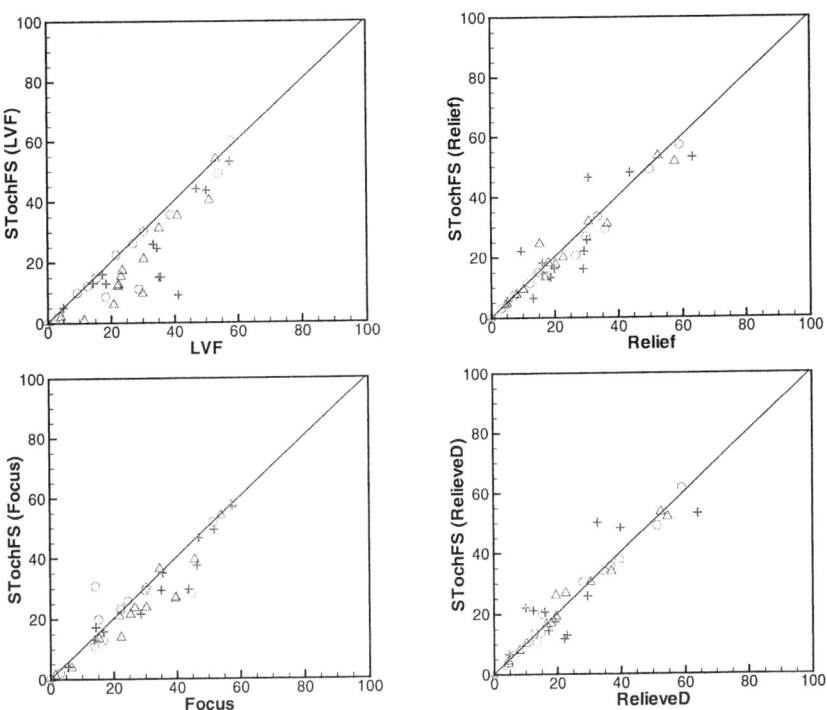

Fig. 4. Summary of the experimental results (error rates). Points under the line indicate that STochFS performed better than its underlying algorithm. Red circles indicate results for C4.5, blue triangles indicate results for Naive Bayes and green plus signs results for kNN.

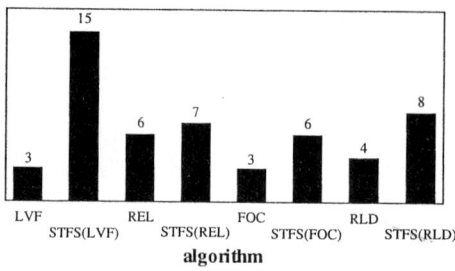

Fig. 5. Number of experiments (out of 39) which each algorithm performed the best or tied with the best. LVF = LVF, STFS(LVF) = STochFS using LVF, REL = Relief, STFS(REL) = STochFS using Relief, FOC = Focus, STFS(FOC) = STochFS using Focus, RLD = RelieveD and STFS(RLD) = STochFS using RelieveD.

Mushroom (22, 8124), Colic (23, 368), Autos (25, 205), Ionosphere (34, 351), Soybean (35, 683), Splice (60, 3190), Sonar (60, 208), Audiology (69, 226).

The following configurations were used in the experiments: for LVF the inconsistency threshold is the initial inconsistency of the dataset and the number of iterations is $77 \cdot N^5$ (both as suggested in [11]), where N is the number of features in the original dataset. The number of iterations in Relief is half of the instances in the dataset, the number of $NearHits$ and $NearMisses$ considered was set to 10 ([10]) and the selection threshold to 0.01 ([7]). For RelieveD, the Relief algorithms was executed using all instances. Finally, for STochFS the number of iterations was set to $10 \cdot N$ and $NumOuts$ was set to 10^3.

The results of our experiments are summarized in Figures 4 and 5 and Table 1. Figure 4 shows that combining the subsets that result from LVF, Relief and Focus executions can indeed generate more accurate outcomes in most cases. In addition, compared to the RelieveD algorithm, STochFS does not seem to result in improved subsets. These observations are confirming when analyzing Table 1, which shows the number of experiments each algorithm performed better within each significance level calculated with the student's t-test. From this table we can verify the drastic selection improvement obtained by STochFS when combining LVF outcomes. Also, as suggested by Figure 4, the gains achieved by the application of STochFS over Relief and Focus are also significant. Finally, the experimental tests could not demonstrate any relevant improvement when considering RelieveD outcomes.

By considering the influence of randomness in each underlying algorithm, we can try to explain some of the experimental results. First, as discussed earlier, random approaches will result in open search spaces, which could allow for better predictive improvements. This suggestion was clearly confirmed in the experiments. For the extremely random LVF algorithm, for instance, STochFS

---

[3] To get to this number, we have tried different values for several datasets of small and medium sizes (up to 69 features) and the results showed that the STochFS performance is hurt, in several cases, if we use less than ten outcomes. Furthermore, using more than ten does not improve its performance in most situations.

Table 1. Score of the number of experiments (out of 39) each algorithm performed better within each significance level (calculated with the student's t-test). A score "A x B" for a certain algorithm f and significance level s means that STochFS performed better than f within s A times. Similarly, it also means that algorithm f outperformed STochFS B times within s.

	<0.001	<0.005	<0.01
STochFS vs LVF	19 x 0	4 x 0	4 x 1
STochFS vs Relief	12 x 3	3 x 0	4 x 2
STochFS vs Focus	7 x 3	6 x 1	3 x 0
STochFS vs RelieveD	6 x 6	2 x 1	2 x 2

generated the most significant improvements. Yet for Relief, a less stochastic solution, the improvements were less visible. In addition, STochFS was less aggressive when dealing with the two deterministic algorithms Focus and RelieveD.

In order to measure the overall effectiveness of STochFS in selecting relevant features and not only its power in boosting the performance of other feature selection algorithm, we have identified for each pair dataset-classifier the best subset generated by all selection algorithms, including LVF, Relief, Focus and RelieveD and the STochFS variants employing these four algorithms as underlying solutions. The results are summarized in Figure 5, which shows the number of times each algorithm performed the best or tied with the best. This table brings an interesting observation. Out of the 39 experiments (combination of 3 classifiers and 13 datasets), the application of STochFS in one algorithm or another generated the best subset of all (or tied with the best) in 36 cases. This impressive result can be explained by the use of the learning algorithm as evaluation function and the effective search heuristic employed by STochFS.

5 Conclusion

In this paper, we have applied the ensemble approach to feature selection by proposing a systematic way of combining various outcomes of a feature selection algorithm. The proposed framework, named STochFS, is similar to the process of Bagging in that randomness is used to generate various outcomes of a feature selection algorithm on the same problem, but it is different in the way these various outcomes are combined: instead of voting we use a stochastic process seeded in a summary of these outcomes.

Experimental results have shown that the performance of well-known feature selection algorithms can be significantly improved, especially when randomness plays a significant role in such underlying solutions. In addition, STochFS achieved an overall superior performance when compared to the considered base selection algorithms.

The research results presented and discussed in this paper can give a considerable but partial understanding of the proposed framework. For future research directions, we expect to consider not only the selection power of STochFS but also its time efficiency. In addition, in this paper we only consider the combination of outcomes generated from the same feature selection algorithm. However, the STochFS framework can be used to compose results from different selection systems. This study could identify complementary solutions that together may work better than far apart.

References

1. H. Almuallim and T.G. Dietterich. Learning with many irrelevant features. In *Proceedings of the Ninth National Conference on Artificial Intelligence (AAAI'91)*, volume 2, pages 547–552, Anaheim, CA, 1991. AAAI Press.
2. J. Bala, K. DeJong, J. Huang, H. Vafaie, and H. Wechsler. Using learning to facilitate the evolution of features for recognizing visual concepts. *Evolutionary Computation*, 4(3):297–311, 1996.
3. C.L. Blake and C.J. Merz. UCI repository of machine learning databases, 1998. http://www.ics.uci.edu/~mlearn/MLRepository.html.
4. L. Breiman. Bagging predictors. *Machine Learning*, 24(2):123–140, 1996.
5. S. Das. Filters, wrappers and a boosting-based hybrid for feature selection. In *Proceedings of the Eighteenth International Conference on Machine Learning*, 2001.
6. Y. Freund and R.E. Schapire. Experiments with a new boosting algorithm. In *Proceedings of the Thirteenth International Conference on Machine Learning (ICML'96)*, pages 148–156, 1996.
7. M.A. Hall. Correlation-based feature selection for discrete and numeric class machine learning. In *Proceedings of the Seventeenth International Conference on Machine Learning*. Stanford University, CA, Morgan Kaufmann Publishers, 2000.
8. G.H. John, R. Kohavi, and K. Pfleger. Irrelevant features and the subset selection problem. In *Proceedings of the Eleventh International Conference on Machine Learning (ICML'94)*, pages 121–129, 1994.
9. K. Kira and L.A. Rendell. A practical approach to feature selection. In *Proceedings of the Ninth International Workshop on Machine Learning*, pages 249–256, Aberdeen, Scotland, 1992. Morgan-Kaufmann.
10. I. Kononenko, M. Robnik-Sikonia, and U. Pompe. *ReliefF for estimation and discretization of attributes in classification.*, pages 31–40. Artificial Intelligence: Methodology, Systems, Applications. IOS Press, 1996.
11. H. Liu and R. Setiono. A probabilistic approach to feature selection - a filter solution. In *Proceedings of the Thirteenth International Conference on Machine Learning (ICML'96)*, pages 319–327, 1996.
12. M. Richeldi and P. Lanzi. ADHOC: A tool for performing effective feature selection. In *Proceedings of the International Conference on Tools with Artificial Intelligence*, pages 102–105, 1996.
13. M. Sebban and R. Nock. A hybrid filter/wrapper approach of feature selection using information theory. *Pattern Recognition*, (35):835 846, 2002.
14. E.P. Xing, M.I. Jordan, and R.M. Karp. Feature selection for high-dimensional genomic microarray data. In *18th International Conference on Machine Learning*, pages 601–608, San Francisco, CA, 2001. Morgan Kaufmann.

Speeding Up Logistic Model Tree Induction

Marc Sumner[1,2], Eibe Frank[2], and Mark Hall[2]

[1] Institute for Computer Science,
University of Freiburg,
Freiburg, Germany
sumner@informatik.uni-freiburg.de
[2] Department of Computer Science,
University of Waikato,
Hamilton, New Zealand
{eibe, mhall}@cs.waikato.ac.nz

Abstract. Logistic Model Trees have been shown to be very accurate and compact classifiers [8]. Their greatest disadvantage is the computational complexity of inducing the logistic regression models in the tree. We address this issue by using the AIC criterion [1] instead of cross-validation to prevent overfitting these models. In addition, a weight trimming heuristic is used which produces a significant speedup. We compare the training time and accuracy of the new induction process with the original one on various datasets and show that the training time often decreases while the classification accuracy diminishes only slightly.

1 Introduction

Logistic Model Trees (LMTs) are born out of the idea of combining two complementary classification schemes: linear logistic regression and tree induction. It has been shown that LMTs perform competitively with other state-of-the-art classifiers such as boosted decision trees while being easier to interpret [8]. However, the main drawback of LMTs is the time needed to build them. This is due mostly to the cost of building the logistic regression models at the nodes. The LogitBoost algorithm [6] is repeatedly called for a fixed number of iterations, determined by a five fold cross-validation. In this paper we investigate whether cross-validation can be replaced by the AIC criterion without loss of accuracy. We also investigate a weight trimming heuristic and show that it improves training time as well.

The rest of this paper is organized as follows. In Section 2 we give a brief overview of the original LMT induction algorithm. Section 3 describes the modifications made to various parts of the algorithm. In Section 4, we evaluate the modified algorithm and discuss the results, and in Section 5 we draw some conclusions.

2 Logistic Model Tree Induction

The original LMT induction algorithm can be found in [8]. We give a brief overview of the process here, focusing on the aspects where we have made im-

---

LogitBoost (J classes)

1. Start with weights $w_{ij} = 1/n$, $i = 1,\ldots,n$, $j = 1,\ldots,J$, $F_j(x) = 0$ and $p_j(x) = 1/J$ $\forall j$
2. Repeat for $m = 1,\ldots,M$:
   (a) Repeat for $j = 1,\ldots,J$:
       i. Compute working responses and weights in the jth class
       $$z_{ij} = \frac{y_{ij}^* - p_j(x_i)}{p_j(x_i)(1 - p_j(x_i))}$$
       $$w_{ij} = p_j(x_i)(1 - p_j(x_i))$$
       ii. Fit the function $f_{mj}(x)$ by a weighted least-squares regression of z_{ij} to x_i with weights w_{ij}
   (b) Set $f_{mj}(x) \leftarrow \frac{J-1}{J}(f_{mj}(x) - \frac{1}{J}\sum_{k=1}^{J} f_{mk}(x))$, $F_j(x) \leftarrow F_j(x) + f_{mj}(x)$
   (c) Update $p_j(x) = \frac{e^{F_j(x)}}{\sum_{k=1}^{J} e^{F_k(x)}}$
3. Output the classifier $\mathrm{argmax}_j \; F_j(x)$

Fig. 1. LogitBoost algorithm

provements. We begin this section with an overview of the underlying foundations and conclude it with a synopsis of the original LMT induction algorithm.

2.1 Logistic Regression

Linear logistic regression models the posterior class probabilities $Pr(G = j|X = x)$ for the J classes via functions linear in x and ensures that they sum to one and remain in $[0, 1]$. The model is of the form

$$Pr(G = j|X = x) = \frac{e^{F_j(x)}}{\sum_{k=1}^{J} e^{F_k(x)}}, \tag{1}$$

where $F_j(x) = \beta_j^T \cdot x$. Numeric optimization algorithms that approach the maximum likelihood solution iteratively are used to find the estimates for β_j.

One such iterative method is the LogitBoost algorithm [6], shown in Figure 1. In each iteration, it fits a least-squares regressor to a weighted version of the input data with a transformed target variable. Here, y_{ij}^* are the binary pseudo-response variables which indicate group membership of an observation like this

$$y_{ij}^* = \begin{cases} 1 & \text{if } y_i = j, \\ 0 & \text{if } y_i \neq j \end{cases}, \tag{2}$$

where y_i is the observed class for instance x_i.

If we constrain f_{mj} to be linear in x, then we achieve linear logistic regression if the algorithm is run until convergence. If we further constrain f_{mj} to be a linear function of only the attribute that results in the lowest squared error, then we arrive at an algorithm that performs automatic attribute selection. By using cross-validation to determine the best number of LogitBoost iterations M, only those attributes are included that improve the performance on unseen instances. This method is called "SimpleLogistic" [8].

2.2 Logistic Model Trees

For the details of the LMT induction algorithm, the reader should consult [8]. Here is a brief summary of the algorithm:

- First, LogitBoost is run on all the data to build a logistic regression model for the root node. The number of iterations to use is determined by a five fold cross-validation. In each fold, LogitBoost is run on the training set up to a maximum number of iterations (200). The number of iterations that produces the lowest sum of errors on the test set over all five folds is used in LogitBoost on all the data to produce the model for the root node and is also used to build logistic regression models at all nodes in the tree.
- The data is split using the C4.5 splitting criterion [10]. Logistic regression models are then built at the child nodes on the corresponding subsets of the data using LogitBoost. However, the algorithm starts with the committee $F_j(x)$, weights w_{ij} and probability estimates p_{ij} inherited from the parent.
- As long as at least 15 instances are present at a node and a useful split is found (as defined in the C4.5 splitting scheme), then splitting and model building is continued in the same fashion.
- The CART cross-validation-based pruning algorithm is applied to the tree [3].

3 Our Modifications

In the following we discuss the additions and modifications we made to the algorithms that make up LMT induction.

Weight Trimming. The idea of weight trimming in association with Logit-Boost is mentioned in [6]. It is a very simple, yet effective method for reducing computation of boosted models. In our case, only training instances carrying $100 \cdot (1 - \beta)\%$ of the total weight mass are used for building the simple linear regression model, where $\beta \in [0, 1]$. Typically $\beta \in [0.01, 0.1]$. We used $\beta = 0.1$. In later iterations more of the training instances become correctly classified with a higher confidence; hence, more of them receive a lower weight and the number of instances carrying $100 \cdot (1 - \beta)\%$ of the weight becomes smaller.

The computation needed to build the simple linear regression model thus decreases as the iterations proceed. Of course, the computational complexity is still $O(n \cdot a)$; however, n is reduced by a potentially large constant factor and in practice, a reduction in computation time is achieved without sacrificing predictive accuracy (see Section 4).

Automatic Iteration Termination. In the original induction algorithm for LMTs the number of LogitBoost iterations for all nodes is determined by a five fold cross-validation at the root and used for all nodes. This, of course, is a time consuming process when the number of attributes and/or instances in the training data is large. The aim is to terminate the LogitBoost iterations when the model performs well on the training data, yet does not overfit it.

A common alternative to cross-validation for model selection is the use of an in-sample estimate of the generalization error, such as Akaike's Information Criterion (AIC) [1]. We investigated its usefulness in selecting the optimum number of LogitBoost iterations and found it to be a viable alternative to cross-validation in terms of classification accuracy and far superior in training time.

AIC provides an estimate of the generalization error when a negative log-likelihood loss function is used. Let this function be denoted as *loglik* and let N be the number of training instances. Then AIC is defined as

$$AIC = -\frac{2}{N}loglik + 2\frac{d}{N}, \tag{3}$$

where d is the number of inputs or basis functions. If the basis functions are chosen adaptively, then d must be adjusted upwards [7]. In this case, d denotes the effective number of parameters, or degrees of freedom, of the model. It is not clear what value to use for d in SimpleLogistic. In [4] the effective number of parameters is computed for boosting methods using the "boosting operator" which is a linear combination of the hat matrices of the basis functions. Our implementation of this method led to a large computational overhead which actually made it slower than cross-validation.

Intuitively, the logistic regression model built via SimpleLogistic should be penalized (i.e. d should increase) each time a new attribute is introduced to the model. We could just use the number of attributes used in a committee function F_j for any class j. However, when an iteration introduces no new attributes, the model is not penalized, although it has become more complex. At the other end of the spectrum, we could penalize each iteration equally which leads to another estimate, $d = i$, where i is the iteration number. As i increases, the first term in Equation 3 decreases (because LogitBoost performs quasi-Newton steps approaching the maximum log-likelihood [6]) and the second term (the penalty term) always increases. Empirically, this was found to be a good estimate (see Section 4).

The optimal number of iterations is the i which minimizes AIC. So, in order to determine the optimal number i^*, LogitBoost must be run up to a maximum number of iterations (in LMT induction, this is 200), and then run again for i^* iterations. Just as with the cross-validation process in the original LMT induction algorithm, we performed the AIC procedure at the root node of the logistic model tree and used i^* throughout the tree. Also, if no minimum was found for 50 iterations, then no more iterations were performed and the iteration that produces the minimum AIC was used as i^*. This is analogous to the heuristic employed for the cross-validation method in [8]. However, we found that we can modify this process to achieve a speed-up without loss of accuracy.

We tested the AIC-based model selection method on 13 UCI datasets [2] (see Section 4) and observed that, for every dataset, AIC only had one global minimum over all iterations. Hence we can attempt to speed up the aforementioned model selection method. We stop searching as soon as AIC no longer decreases. Determining the optimal number of iterations in this fashion will be called the *First AIC Minimum* (FAM) method in the remainder of this paper.

FAM allows us to efficiently compute (an approximation to) the optimal number of iterations at *each* node in the tree. This is advantageous in two ways:

- Instead of iterating up to a maximum number of iterations five times (in the cross-validation case) or once (in the AIC case) and then building the model using the optimal number of iterations i^*, LogitBoost can be stopped immediately when i^* is found.
- As we move down the tree, the sets of training instances become smaller (they are subsets of the instances observed at the parent node). AIC is inversely related to the number of instances N, so as N becomes smaller, not only will simpler models be selected, but training time decreases as fewer iterations are needed. It is also much more intuitive that a different number of iterations is appropriate for different datasets occuring in a tree.

In addition, we no longer need to set a maximum number of iterations to be performed. As mentioned in [8] the limit of 200 was appropriate for the observed datasets; however it is not clear whether this is enough for all datasets.

4 Experiments

This section evaluates the modified LMT induction algorithm and its base learner SimpleLogistic by comparing them against the original implementation. For the evaluation we measure training times and classification accuracy. All experiments are ten runs of a ten-fold stratified cross-validation. The mean and standard deviation over the 100 results are shown in all tables presented here. In all experimental results, a corrected resampled t-test was used [9] instead of the standard t-test to test the difference in training times and accuracy, at a 5% significance level. This corrects for the dependencies in the estimates of the different data points, and is thus less prone to false-positives.

We used 13 datasets with a nominal class variable available from the UCI repository [2]. We used only datasets that have approximately 1000 training instances or more (vowel was the exception with 990 instances). Both numeric and nominal attributes appear in all of the UCI datasets.

All experiments were run using version 3.4.4 of the Weka machine learning workbench [11]. Almost all experiments were run on a pool of identical machines with an Intel Pentium 4 processor with 2.8GHz and 512MB ram[1], Linux kernel 2.4.28 and Java 1.5.0-b64.

---

[1] All LMT algorithms required more memory for the adult dataset and were thus run on an Intel Pentium 4 processor with 3.0GHz and 1GB ram.

Table 1. Training time and accuracy for SimpleLogistic and SimpleLogistic using weight trimming

Dataset	Training Time			Accuracy	
	SimpleLog.	SimpleLog. (WT)		SimpleLog.	SimpleLog. (WT)
vowel	77.94±23.59	39.67±12.72	•	81.98±4.10	82.07±3.82
german-credit	7.97±1.94	6.79±1.55		75.37±3.53	75.35±3.48
segment	50.55±14.82	20.02±5.61	•	95.10±1.46	86.71±25.67
splice	253.96±38.83	79.02±9.55	•	95.86±1.17	95.87±1.09
kr-vs-kp	57.28±15.09	25.98±8.35	•	97.06±0.98	97.07±0.92
hypothyroid	104.76±27.17	47.88±10.72	•	96.61±0.71	96.55±0.72
sick	25.40±6.10	12.09±3.39	•	96.68±0.71	96.63±0.70
spambase	119.28±18.73	43.19±4.38	•	92.75±1.12	92.40±1.24
waveform	65.53±9.31	25.42±3.77	•	86.96±1.58	86.90±1.55
optdigits	659.33±123.68	111.32±21.35	•	97.12±0.67	97.17±0.67
pendigits	489.51±148.34	257.86±84.23	•	95.44±0.62	95.51±0.61
nursery	266.51±25.56	119.19±11.36	•	92.61±0.68	92.60±0.77
adult	2953.77±849.82	1866.15±344.05	•	85.61±0.38	85.56±0.38

• statistically significant improvement

4.1 SimpleLogistic

We first evaluated the SimpleLogistic learner for logistic regression, measuring the effects of weight trimming and investigating the use of FAM for determining the optimum number of LogitBoost iterations.

Weight Trimming in SimpleLogistic. From Table 1 it can be seen that weight trimming consistently reduces the training time of SimpleLogistic on all datasets (with the exception of german-credit) while not affecting classification accuracy. The greatest effect of weight trimming was seen on the optdigits dataset. Here, a speedup of almost 6 was recorded. Overall, weight trimming is a safe heuristic (i.e. it does not affect accuracy) that can result in significant speedups.

FAM in SimpleLogistic. This section deals with the evaluation of SimpleLogistic implemented with FAM, introduced in Section 3. SimpleLogistic with FAM is compared with the original cross-validation-based approach.

Table 2 shows the training time and classification accuracy for both algorithms on the 13 UCI datasets. FAM consistently produced a significant speedup on all datasets. This ranged from 3.3 on the splice dataset to 14.8 on the segment dataset. Looking at the classification accuracy, we can see that FAM performs significantly worse on two datasets (kr-vs-kp and hypothyroid), but the degradation is within reasonable bounds.

4.2 Logistic Model Trees

We can now observe the impact of our modifications on the LMT induction algorithm. Table 3 compares the original LMT version with the modified LMT induction algorithm (using FAM and weight trimming). As expected, our modifications result in the algorithm being much faster on all of the datasets. The

Table 2. Training time and accuracy for SimpleLogistic using cross-validation and FAM

Dataset	Training Time		Accuracy	
	SimpleLog. (CV)	SimpleLog. (FAM)	SimpleLog. (CV)	SimpleLog. (FAM)
vowel	77.94±23.59	6.87±0.31 •	81.98±4.10	80.85±3.69
german-credit	7.97±1.94	0.59±0.05 •	75.37±3.53	75.34±3.70
segment	50.55±14.82	3.42±0.45 •	95.10±1.46	94.67±1.66
splice	253.96±38.83	77.48±3.69 •	95.86±1.17	95.87±1.06
kr-vs-kp	57.28±15.09	6.69±0.37 •	97.06±0.98	96.38±1.14 ○
hypothyroid	104.76±27.17	8.89±1.16 •	96.61±0.71	95.89±0.65 ○
sick	25.40±6.10	1.57±0.14 •	96.68±0.71	96.50±0.76
spambase	119.28±18.73	15.74±1.28 •	92.75±1.12	92.69±1.19
waveform	65.53±9.31	7.75±0.39 •	86.96±1.58	86.84±1.59
optdigits	659.33±123.68	135.61±26.47 •	97.12±0.67	97.12±0.66
pendigits	489.51±148.34	59.43±1.58 •	95.44±0.62	95.45±0.62
nursery	266.51±25.56	49.36±1.42 •	92.61±0.68	92.58±0.68
adult	2953.77±849.82	381.92±10.13 •	85.61±0.38	85.59±0.38

•, ○ statistically significant improvement or degradation

Table 3. Training time and accuracy for LMT and LMT using FAM and weight trimming

Dataset	Training Time		Accuracy	
	LMT	LMT (FAM+WT)	LMT	LMT (FAM+WT)
vowel	408.11±80.95	15.86±0.84 •	94.06±2.40	93.56±2.94
german-credit	32.74±10.87	3.25±0.16 •	75.50±3.65	71.83±3.40 ○
segment	143.75±52.64	10.58±1.77 •	97.06±1.31	97.06±1.25
splice	785.51±202.14	71.55±1.42 •	95.89±1.14	95.19±1.19 ○
kr-vs-kp	250.79±64.58	12.17±0.36 •	99.64±0.33	99.57±0.37
hypothyroid	405.73±94.04	7.39±0.64 •	99.54±0.36	99.61±0.30
sick	139.31±50.79	6.83±0.73 •	98.95±0.58	98.93±0.62
spambase	746.71±123.57	54.93±1.65 •	93.56±1.14	93.58±1.13
waveform	175.53±63.26	43.67±0.80 •	86.86±1.60	86.49±1.52
optdigits	3162.37±781.49	133.15±7.08 •	97.38±0.57	97.36±0.64
pendigits	3535.06±765.34	185.15±4.96 •	98.58±0.33	98.73±0.33
nursery	634.96±85.82	72.44±7.08 •	98.95±0.34	98.64±0.32 ○
adult	26935.85±9112.20	1429.93±54.76 •	85.58±0.42	85.43±0.37

•, ○ statistically significant improvement or degradation

greatest speedup recorded was 55 on the hypothyroid dataset. Most common was a speedup between 10 and 25. Only on the german-credit, waveform, and nursery datasets was the speedup around 10 or less (10.1, 4.0, and 8.8, respectively).

On german-credit, splice and nursery the modified version's classification performance was significantly worse than that of the original version, although only on german-credit was the performance worse by more than one percent. Otherwise, the modified version performed competitively with the original version.

As a closing note to our experiments, we would like to compare the modified version of logistic model trees to boosted C4.5 decision trees. For the comparison we chose AdaBoost [5] using 100 iterations and the LMT version using FAM and weight trimming. The results can be seen in Table 4. 100 iterations are too many for a few of the datasets, but moving from 10 to 100 iterations results in an improvement in accuracy in many cases [8].

The training time of the two algorithms is fairly equal, with a slight advantage for the modified LMT algorithm. It was faster on 9 of the 13 datasets, with a

Table 4. Training time and accuracy for AdaBoost using C4.5 with 100 iterations and LMT using FAM and weight trimming

Dataset	Training Time			Accuracy	
	AdaBoost	LMT (FAM+WT)		AdaBoost	LMT (FAM+WT)
vowel	29.32±0.38	15.86±0.84	•	96.74±1.89	93.56±2.94 o
german-credit	7.42±0.17	3.25±0.16	•	74.40±3.23	71.83±3.40
segment	45.53±0.67	10.58±1.77	•	98.58±0.76	97.06±1.25 o
splice	11.89±5.46	71.55±1.42	o	94.94±1.24	95.19±1.19
kr-vs-kp	21.14±6.44	12.17±0.36	•	99.60±0.31	99.57±0.37
hypothyroid	19.07±11.46	7.39±0.64	•	99.70±0.31	99.61±0.30
sick	49.40±2.32	6.83±0.73	•	99.06±0.45	98.93±0.62
spambase	70.22±63.21	54.93±1.65		95.34±0.87	93.58±1.13 o
waveform	463.38±4.18	43.67±0.80	•	85.01±1.77	86.49±1.52 •
optdigits	402.52±3.06	133.15±7.08	•	98.55±0.50	97.36±0.64 o
pendigits	274.59±2.72	185.15±4.96	•	99.41±0.26	98.73±0.33 o
nursery	24.90±0.48	72.44±7.08	o	99.79±0.14	98.64±0.32 o
adult	796.01±64.89	1429.93±54.76	o	82.18±0.46	85.43±0.37 •

•, o statistically significant improvement or degradation

speedup of usually between 1 and 2. The exceptions are the sick dataset (7.2) and the waveform dataset (10.6). AdaBoost was faster on three datasets, the greatest improvement being on the splice dataset (6.0). In terms of classification accuracy, the new LMT version exhibits results similar to those reported in [8] for the original LMT algorithm. On six datasets AdaBoost performed significantly better, while on two datasets LMT was the better classifier.

5 Conclusions

We have proposed two modifications to the SimpleLogistic algorithm employed by LMT that are designed to improve training time. The use of AIC instead of cross-validation to determine an appropriate number of LogitBoost iterations resulted in a dramatic speedup. It resulted in a small but significant decrease in accuracy in only two cases when performing stand-alone logistic regression. The simple heuristic of weight trimming consistently improved the training time while not affecting accuracy at all.

The use of AIC and weight trimming in LMT have resulted in training times up to 55 times faster than the original LMT algorithm while, in most cases, not significantly affecting classification accuracy. These results were measured on datasets of relatively low size and dimensionality. We would expect the speedup to be even greater on larger and high-dimensional datasets.

References

1. H. Akaike. Information theory and an extension of the maximum likelihood principle. In *Second Int Symposium on Information Theory*, pages 267–281, 1973.
2. C.L. Blake and C.J. Merz. UCI repository of machine learning databases, 1998. [www.ics.uci.edu/~mlearn/MLRepository.html].
3. L. Breiman, H. Friedman, J. A. Olshen, and C. J. Stone. *Classification and Regression Trees*. Wadsworth, 1984.

4. Peter Bühlmann and Bin Yu. Boosting, model selection, lasso and nonnegative garrote. Technical Report 2005-127, Seminar for Statistics, ETH Zürich, 2005.
5. Yoav Freund and Robert E. Schapire. Experiments with a new boosting algorithm. In *Proc In. Conf on Machine Learning*, pages 148–156. Morgan Kaufmann, 1996.
6. Jerome Friedman, Trevor Hastie, and Robert Tibshirani. Additive logistic regression: a statistical view of boosting. *The Annals of Statistics*, 38(2):337–374, 2000.
7. Trevor Hastie, Robert Tibshirani, and Jerome Friedman. *The Elements of Statistical Learning: Data Mining, Inference, and Prediction*. Springer-Verlag, 2001.
8. Niels Landwehr, Mark Hall, and Eibe Frank. Logistic model trees. *Machine Learning*, 59(1/2):161–205, 2005.
9. C. Nadeau and Yoshua Bengio. Inference for the generalization error. In *Advances in Neural Information Processing Systems 12*, pages 307–313. MIT Press, 1999.
10. R. Quinlan. *C4.5: Programs for Machine Learning*. Morgan Kaufmann, 1993.
11. I. H. Witten and E. Frank. *Data Mining: Practical Machine Learning Tools and Techniques with Java Implemenations*. Morgan Kaufmann, San Francisco, 2000.

A Random Method for Quantifying Changing Distributions in Data Streams

Haixun Wang[1] and Jian Pei[2]

[1] IBM T. J. Watson Research Center, USA
haixun@us.ibm.com
[2] Simon Fraser University, Canada
jpei@cs.sfu.ca

Abstract. In applications such as fraud and intrusion detection, it is of great interest to measure the evolving trends in the data. We consider the problem of quantifying changes between two datasets with class labels. Traditionally, changes are often measured by first estimating the probability distributions of the given data, and then computing the distance, for instance, the K-L divergence, between the estimated distributions. However, this approach is computationally infeasible for large, high dimensional datasets. The problem becomes more challenging in the streaming data environment, as the high speed makes it difficult for the learning process to keep up with the concept drifts in the data. To tackle this problem, we propose a method to quantify concept drifts using a universal model that incurs minimal learning cost. In addition, our model also provides the ability of performing classification.

1 Introduction

In this paper, we study *the distance between two data distributions* instead of two vectors or two sequences. Assume tuples in a training set D are drawn from an unknown distribution $F(\mathbf{x}, t)$. Each tuple is of the form $(\mathbf{x}, t)$, where $\mathbf{x}$ is a vector and t is the class label of $\mathbf{x}$. The task of supervised learning or classification is to learn the unknown relationship between $\mathbf{x}$ and t, that is, to find a model $f^*(\mathbf{x})$, such that the averaged difference between $f^*(\mathbf{x})$ and t is minimum.

We assume there are concept drifts in the unknown data distribution $F(\mathbf{x}, t)$. How do we quantify the concept drift by defining and computing the distance between the original dataset D and a new data set D', which is drawn from the changed unknown distribution? Furthermore, how quantified changes can be used to tune the model $f^*(\mathbf{x})$ we learned before so that it maintains high accuracy on the changed data?

In the field of information theory, relative entropy, or the Kullback Leibler (K-L) divergence, has been suggested as an appropriate measure for comparing data distributions [5]. However, such methods are not computationally feasible for large, high dimensional datasets, or data coming from continuous streams. In the field of data mining, several works have studied how to *detect* changes of data distributions over streams and sequences [1,10]. However, more often than not, change detection only serves to trigger a costly learning process, and the change itself is not used to mend the current prediction model directly. Recently, several works [8,13] have studied how to update the current model $f^*(\mathbf{x})$ in response to the concept drifts in data streams, for instance, by assimilating new instances in D' and forgetting old instances in D. These can be very costly undertakings since they do not handle changes directly on the probability distribution level, but rely on a lot of learning and re-learning.

We aim at devising an efficient method to measure distribution changes in high-dimensional, labeled datasets. We assign a *signature* to each dataset, and compare distribution changes by comparing the signatures. Furthermore, the signature should also enable us to make predictions.

2 A Model-Based Naive Approach

In this section, we introduce a naive but computationally feasible method for measuring distances between two datasets. We analyze the prediction error of this naive approach through bias/variance decomposition, and we study its impact on the distance measure. In the next section, we introduce a general approach based on the lessons learned here.

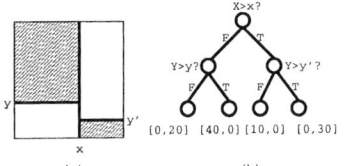
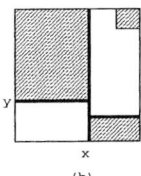

Fig. 1. Model-based description

Fig. 2. Distribution Changes

2.1 Measuring Distribution Changes Using a Classification Model

Assume we are given a dataset D which consists of a set of tuples $(\mathbf{x}, t)$, where $\mathbf{x}$ is a vector and t is the class label of $\mathbf{x}$. We learn a decision tree classifier T_D from D. The decision tree classifier T_D can be regarded as a summarization of the class distribution of dataset D. More specifically, let $n_1, n_2, ..., n_k$ be the leaf nodes of T_D. Each leaf node n_i is associated with a class distribution (number of objects belonging to each class). Together, $(n_1, n_2, ..., n_k)$ forms a special histogram of frequency counts.

For instance, in Figure 1(a) we show a two dimensional dataset where the shaded areas in the top-left and bottom-right corner are populated with objects of one class, and the rest of the area is populated with objects of the other class. In the rest of the paper, we assume the number of objects in an area is proportional to the size of the area.

From the dataset, we learn a decision tree classifier, which partitions the two dimensional space into 4 areas, each represented by a leaf node as shown in Figure 1(b). Each leaf node is associated with the number of objects of each class in that area. For instance, the second leaf node to the left represents the top-left area, where we assume [40,0] are the number of objects of the two classes in that area. All together, we can use the class distribution of the objects in the leaf nodes to describe the dataset. We call it the *signature* of the data:

$$([0, 20], [40, 0], [10, 0], [0, 30]) \qquad (1)$$

Assume now there is some distribution change in the underlying dataset. In one case, the boundary of the shaded area moved from x to x^* horizontally and from y to y^* vertically, as shown in Figure 2(a).

We want to quantify the change using the model we learned from the original dataset. Here, we use the decision tree to classify the changed data set, and use the classification error to quantify the change. To a certain extent, the classification error represents the magnitude of the change, but certainly not the change itself. Because, for instance, datasets in Figure 2(a) and 2(b) will have the same classification error (compared with the original data set in Figure 1(a), they have the same amount of shaded area "out of the place"), but they have very different data distributions. Apparently, the error-based distance measure cannot be used to replace or tune the predictions made by the original decision tree for the changed data.

To ensure that the measure can represent, to a certain extent, the distribution of the change so that it can be used to help make predictions without learning a new model from the changed dataset, we simply 'throw' the objects in Figure 2(a) into the decision tree learned from the original dataset. The class distribution in the leaf nodes is now the signature of the changed dataset:

$$([0, 20], [38, 2], [10, 0], [2, 28]) \quad (2)$$

Now, the dataset in Figure 2(b) results in a different signature: $([0, 20], [40, 0], [10, 0], [4, 26])$, which means signatures are better than prediction errors in representing distributions.

Although we didn't learn a decision tree from the new datasets, the signature, which combines the original decision tree structure and the new class distributions in the leaf nodes, give us some ability to make predictions. Take the dataset in Figure 2(a) and its signature Eq (2) for example. If a test object is classified into the 2nd leaf node to the left, the prediction that the object belong to the positive class will be the probability output $\frac{n_1}{n_1+n_2} = \frac{38}{38+2}$, where n_1 and n_2 are the number of positive and negative nodes in that leaf node respectively.

The signatures also enable us to quantify the differences between the two datasets. If we treat the signature as a vector, we can use any L_p metric to compute their distance. For example, the distance function Eq (3) between two signatures a and b is based on the Manhattan distance:

$$Dist_s(a,b) = \frac{1}{2} \sum_{j=1}^{n} \sum_{k=1}^{c} |\frac{n_{a,j,k}}{N_a} - \frac{n_{b,j,k}}{N_b}| \quad (3)$$

where n is the number of leaf nodes, c is the number of different classes, $n_{a,j,k}$ is the number of nodes in the j-th leaf node that are of class label k, and N_a is the total number of objects in dataset a. For any two signatures a and b, we have $0 \leq Dist_s(a,b) \leq 1$.

This naive approach gives us the following benefits. First, it is computationally efficient to compare the differences of two data distributions. Second, the data descriptors can be used to make predictions. However, this naive method is also flawed.

2.2 Error Analysis

In the naive method, the model used to describe other datasets is partially learned from a dataset which may have a very different data distribution. This can result in significant prediction error and create problems for the distance measure. In this section, we first reveal such problems, then we use bias-variance decomposition to study their cause.

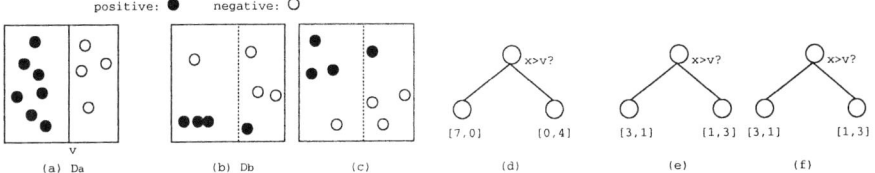

Fig. 3. A Greedy Learner

From D_a in Figure 3(a), we learn a decision tree, and we show the tree hierarchy in Figure 3(d). We then populate the leaf nodes of the decision tree with objects in other datasets. Figure 3(b) and 3(c) represent two very different data distributions. However, because of the tree structure learned from D_a, a same signature, $([3,1],[1,3])$, will be assigned to both datasets. Thus, the distance between the two very different distributions is 0. The signature is thus inaccurate because of the possible large variance introduced by training datasets such as D_a.

For similar reasons, using signatures assigned by the naive method for prediction is also flawed. We populate the leaf nodes of the decision tree learned from D_a in Figure 3(a) with objects in dataset D_b in Figure 3(b). This results in a signature of $([3,1],[1,3])$. Such a signature apparently has large prediction error – even when it is applied on D_b itself, the error can be as large as 25% under zero-one loss.

Clearly, this is due to the fact that D_a's data distribution is very different from D_b's. Decision trees are built in a divide-and-conquer, greedy manner, and in this case, there is no need to make a split on the Y axis for training set D_a, although such a split will result in the largest information gain as far as training set D_b is concerned. The difference of the two data distributions, combined with the greedy nature of the decision tree construction process, results in a large prediction error.

We observe samples $(\mathbf{x},t)$ drawn independently from some unknown distribution. We want to learn the unknown relationship between $\mathbf{x}$ and t. That is, we want to find a function, $f^*(\mathbf{x})$, that minimizes a certain loss function $L(t, f^*(x))$, where L can be zero-one loss, square loss, absolute loss, etc.

We use the notation $f^*(\mathbf{x}|D)$ to indicate that the prediction model we learn depends on the training dataset D. We decompose the expected prediction error (EPE) into three terms: noise (σ^2), bias, and variance:

$$EPE(\mathbf{x}) = \sigma^2 + Bias(f^*(\mathbf{x}|D))^2 + Var(f^*(\mathbf{x}|D))$$

Let $E_D(f^*(\mathbf{x}|D))$ be the predicted value for sample $\mathbf{x}$ averaged over all the training datasets. The variance can be expressed by:

$$E_D(E_D f^*(\mathbf{x}|D) - f^*(\mathbf{x}|D))^2$$

The variance term measures how sensitive the predicted value at **x** is to random fluctuations in the training dataset. Traditionally, a model is learned from a training dataset D drawn from the data distribution we try to learn. In our case, we have two training sets, D_a and D_b. From D_a we learn the structure of the histogram (or equivalently the hierarchy of a decision tree), and from D_b we learn the data distributions within the structure or within the hierarchy. The variance can thus be expressed by:

$$E_{D_a,D_b}(E_{D_a,D_b}f^*(\mathbf{x}|D_a,D_b) - f^*(\mathbf{x}|D_a,D_b))^2$$

Since D_a might be drawn from a data distribution different from the distribution of D_b, which is the distribution we want to learn, by including both D_a and D_b in the condition, the variance is increased because of the added fluctuations.

3 A Universal Model

As discussed in the previous section, the majority of variance and bias is introduced due to training set D_a, from which we learn the structure of a histogram, or a hierarchy of a decision tree. Furthermore, it constitutes the major part of the learning cost. When the change of data distributions between D_a and D_b is non-trivial, the benefits of learning the tree structure from D_a becomes insignificant, since there is no guarantee that such a tree structure will fit the training dataset of D_b well. In this case, it becomes obvious that using an arbitrary tree structure not only serves the same purpose but at the same time eliminates the cost of learning such a structure.

Our goal is to find such an 'arbitrary' structure. It must be general and universal so that it can fit 'any' dataset D_b well, thus we can avoid the bias and variance component in the prediction error such as those introduced by one particular dataset D_a.

3.1 Distance by Random Signatures

A decision tree assigns a signature to a dataset. A signature can be regarded as a special histogram. Each bin, which corresponds to a leaf node in the decision tree, is 'cut out' or defined by the splitting conditions on the path from the root node to that leaf node. The learning procedure determines those conditions as well as their applying order through the computation of information gain.

Take the training set D_a in Figure 4 as an example. It is a two dimensional dataset with two class labels. From D_a, we learn a decision tree, which partitions the two dimensional space into a set of 'bins', each of which is in fact a leaf node in the decision tree. The signature is created by an *entropy-based partition*, since a decision tree is often constructed through the computation of information gain. Note that this learning procedure has super-linear complexity.

We propose to create signatures by randomly partitioning the multi-dimensional space into a set of bins. Figure 5 is such an example. The positions and the order of the splits are totally random, and instead of creating one histogram, we create multiple histograms, each of which is independently and randomly partitioned. In the following, we study two different ways of random partitioning.

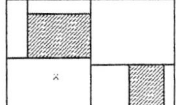 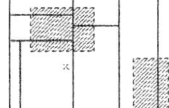 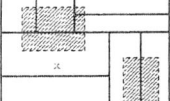 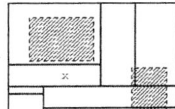

Fig. 4. A Decision Tree Histogram

Fig. 5. Random Forest Histograms

Random Forest. We use the following procedure to create a random decision tree for a training dataset D.

1. $partition(D)$: randomly pick an unused attribute to partition D into $D_1, \cdots, D_n$;
2. for each partition D_i ($1 \leq i \leq n$), recursively invoke $partition(D_i)$ till the k-th recursive level.

We repeat this process N times to create a forest of N random trees [3]. Each tree defines a signature, and the random forest consists of N signatures for the dataset.

Random Histograms. We use the following procedure to create a random histogram for a training dataset D.

1. Randomly pick k attributes, $a_1, \cdots, a_k$, as well as one value for each attribute, such that $\{a_1 = v_1, \cdots, a_k = v_k\}$ defines a bin in the histogram.
2. Repeat the above step M times so that we have a histogram of M bins.

We repeat the above process N times to create N random histograms.

Each of the above methods creates N random structures. Given a dataset D_x, we populate the random structures with objects in D_x, which results in N signatures $S_{x,1}, \cdots, S_{x,N}$ for D_x. We use the same random structures for all datasets. Clearly, for any two datasets, D_a and D_b, signatures $S_{a,i}$ and $S_{b,i}$ have the same number of bins and each bin defines the same subspace in the multi-dimensional data space. We then define the distance between two datasets D_a and D_b as: $Dist(D_a, D_b) = \frac{1}{N} \sum_{i=1}^{N} Dist_s(S_{a,i}, S_{b,i})$, where $Dist_s$ is the distance between two signatures defined in Eq (3), and we have $0 \leq Dist(D_a, D_b) \leq 1$.

The difference between this method and the naive method is that in this method, i) the structure of a signature does not rely on one dataset (which is known as D_a in the naive method), and ii) instead of having one signature, it uses multiple signatures. As will be discussed in detail in the following sections, the multiple random signatures is capable of 'fitting' any dataset, which means the distance metric and the prediction model will have high accuracy.

3.2 Classification by Random Signatures

A signature is composed of a set of histograms, each of which can be expressed by a vector $[n_1, \cdots, n_c]$, where n_i is the number of objects that belong to class i.

The signature is used for prediction: an testing object that falls into a bin with class histogram $[n_1, \cdots, n_c]$ is classified to be of class i if $i = \arg\max_i \frac{n_i}{\sum n_i}$. However, a random signature is often a "weak" classifier.

The weakness of a single random signature can often be averted as our random methods create N signatures for a training dataset. The final prediction is a voted combination of all signatures. In other words, each signature is a classifier, and the N signatures form a classifier ensemble.

Combining an ensemble of classifiers is an established research area [2,6,12]. Particularly, for random forests, the prediction accuracy is shown to be no less than that of normal decision trees. Although each random signature is possibly a very "weak" classifier, it has been shown that if each classifier in the ensemble is independent in the production of its error, the expected error of the ensemble can be reduced to zero as the number of the classifiers goes to infinity [7].

3.3 Signatures' Structural Diversity

Whether the signature-based distance metric and prediction model are meaningful depends on whether the random signatures can "fit" any dataset. The strength of an ensemble comes from its diversity [9]. In this section, we discuss how to guarantee signatures' structural diversity.

In an ensemble, a classifier is valuable if it disagrees on some inputs with the other classifiers. Building a diverse ensemble in which each hypothesis is as different as possible is important to an ensemble method. Normally, diversity is measured by prediction disagreements among ensemble classifiers. In our case, random structures are created without a training dataset, which means we can only measure diversity by directly studying the differences of their internal structures. In a signature, each bin corresponds to a set of attribute values. We use the number of different attribute combinations as a measure of diversity. Let A be the number of attributes of the datasets. For simplicity, in our discussion we assume each attribute has v unique values.

- In a *random forest*, each tree of height k has v^k leaf nodes. The path from the root node to any leaf node has $k-1$ edges. Thus, the diversity of attribute combinations in one random tree is at most $\min(v^{k-1}, \binom{k-1}{A})$. In the worst case, all leaf nodes (bins) share one attribute combination.
  Furthermore, attribute combinations may be correlated.
- For *random histograms*, each bin is defined independently by k attribute values. To compare with the above methods, we create v^k bins. The diversity can be as high as $\min(v^k, \binom{k}{A})$. In the worst case, all bins share one attribute combination. This occurs when all attributes are used ($k = A$), or each random selection returns the same set of attributes.

In summary, random histograms provide the most diverse set of attribute combinations with low correlation.

Our second question is how many bins should we keep in each random structure? We answer this question for random histograms. For random histograms, the number of attribute combinations is at most $\min(v^k, \binom{k}{A})$. Note that $\binom{k}{A}$ reaches maximum when $k = A/2$. Thus, when $v^{A/2} > \binom{A/2}{A}$, we shall use $k = A/2$ attributes for random histograms; otherwise, we shall use k attributes where k satisfies $v^k >= \binom{k}{A}$ and $v^{k-1} < \binom{k-1}{A}$.

4 Conclusion

The ability to quantify the similarity between two datasets is important to many applications, especially data stream applications that deal with time-changing data distributions. Statistical methods, such as K-L divergence and Kriging, are usually not computationally feasible for large, high speed datasets. In this paper, we propose a new approach based on the theory of random forests and classifier ensemble. To measure the difference between two data distributions, our approach measures the difference between the models derived from the datasets. To do this, we must use models that can truthfully represent the dataset, and models that can be trained efficiently. The models we propose for this purpose is the random histograms. The random histograms assign datasets signatures, which serve for two purposes: i) to measure distance between datasets by directly comparing signatures; and ii) to perform classification.

References

1. Charu C. Aggarwal. A framework for diagnosing changes in evolving data streams. In *SIGMOD*, 2003.
2. Eric Bauer and Ron Kohavi. An empirical comparison of voting classification algorithms: Bagging, boosting, and variants. *Machine Learning*, 36(1-2):105–139, 1999.
3. L. Breiman. Random forests. *Machine Learning*, 45(1):5–32, 2001.
4. Sergey Brin. Near neighbor search in large metric spaces. In *VLDB*, Switzerland, 1995.
5. Thomas M. Cover. *Elements of Information Theory*. Wiley-Interscience, 1991.
6. Yoav Freund and Robert E. Schapire. Experiments with a new boosting algorithm. In *ICML*, pages 148–156, 1996.
7. L. Hansen and P. Salamon. Neural network ensembles. *IEEE Transactions on Pattern Analysis and Machine Intelligence*, 12:993–1001, 1990.
8. G. Hulten, L. Spencer, and P. Domingos. Mining time-changing data streams. In *SIGKDD*, pages 97–106, San Francisco, CA, 2001. ACM Press.
9. A. Krogh and J. Vedelsby. Neural network ensembles, cross validation, and active learning. In *Advances in Neural Information Processing Systems*, volume 7, pages 231–238. MIT Press, 1995.
10. Junshui Ma and Simon Perkins. Online novelty detection on temporal sequences. In *SIGKDD*, 2003.
11. M. A. Oliver and R. Webster. Kriging: a method of interpolation for geographical information systems. *International Journal Geographic Information Systems*, 4(3), 1990.
12. Kagan Tumer and Joydeep Ghosh. Error correlation and error reduction in ensemble classifiers. *Connection Science*, 8(3-4):385–403, 1996.
13. Haixun Wang, Wei Fan, Philip S. Yu, and Jiawei Han. Mining concept-drifting data streams using ensemble classifiers. In *SIGKDD*, 2003.

Deriving Class Association Rules Based on Levelwise Subspace Clustering

Takashi Washio, Koutarou Nakanishi, and Hiroshi Motoda

I.S.I.R., Osaka University, 8-1,
Mihogaoka, Ibaraki City,
Osaka, 567-0047 Japan

Abstract. Most approaches of Class Association Rule (CAR) based classification have not intensively addressed the classification of instances including numeric attributes. In this paper, a levelwise subspace clustering method deriving hyper-rectangular clusters is proposed to efficiently provide quantitative, interpretative and accurate CARs.

1 Introduction

"*Class Association Rules* (CARs)" for classification has been proposed in recent studies [1,2,3]. These rules have the form "$\{<p_1:q_1>,...,<p_m:q_m>\} \Rightarrow cl$" where $<p:q>$ is an item and cl a class. p represents an attribute and q its value. An example is "$\{<Age:[30,39]>,<Married:Yes>,<NumCars:[2,2]>\} \Rightarrow Houseowner$" stating "A person who is in his/her thirties, married, and owns two cars belongs to the class *Houseowner*." Here, a "*numeric item*" has a numeric interval value whereas a "*categorical item*" has a categorical value. A numeric item $<p:q>$ in the CAR is supported by a numeric item $<p_t:q_t>$ in an instance t if $p_t = p$ and $q_t \subseteq q$ where $\subseteq$ states that the range of q_t is within the range of q. Hence, "$t_1 = \{<Age:[35,37]>,<Married:Yes>,<NumCars:[2,2]>,<Child:[3,3]>\}$" supports the aforementioned rule body, whereas "$t_2 = \{<Age:[29,31]>,<Married:Yes>,<NumCars:[2,2]>,<Child:[3,3]>\}$" does not, because $<Age:[29,31]>$ is not within $<Age:[30,39]>$. Given a training data set D which is a table (or a set) of class labeled instances (transactions), let D_{cl} be a set of all instances having a class cl in D. The body of a CAR including numeric items is a "*quantitative frequent itemset*," QFI in short, which is supported by D_{cl} more frequently than a "*minimum support (minsup)*" threshold. The numeric part of a QFI corresponds to an axis-parallel and hyper-rectangular region in an attribute subspace of D.

CBA, CMAR and CAEP are the representatives of the CAR based classification [1,2,3]. Especially CAEP, using the strength of all CARs, widely shows the best performance among many rule-based classifiers including C4.5. However, it discretizes each numeric attribute by an entropy measure without considering the dependency of the distributions among multiple attributes, and thus a cluster of instances having the same class can often be fragmented. An effective solution for this issue is the introduction of the clustering in every numeric attribute subspace to derive strong rule bodies. CLIQUE, DOC and SUBCLU seek dense clusters in every subspace [4,5,6]. However, they do not provide QFIs covering both numeric and categorical attributes and corresponding to axis-parallel and hyper-rectangular clusters in an efficient manner. The approaches to mine quantitative association rules have addressed this issue [7,8]. However, they discretize each attribute without considering the dependency among attributes, and thus can result in the aforementioned fragmentation.

In this study, a novel and efficient approach to the exhaustive, axis-parallel and hyper-rectangular subspace clustering is proposed. Moreover, the combined use of this clustering and CAEP is evaluated. The proposed clustering algorithm has a levelwise structure. While this is similar to SUBCLU, our approach can derive clusters on both numeric and categorical items, and the numeric items having interval values can be processed.

2 CAEP

CAEP is briefly explained at first [3]. The training phase of CAEP consists of two processes. The first is to derive all rule bodies. Let the support of an itemset a by D_{cl} be $support_{D_{cl}}(a) = |\{t \in D_{cl} | a \in t\}|/|D_{cl}|$. For every cl, a set of QFIs, $LQFI(cl)$, in which every itemset a satisfies $support_{D_{cl}}(a) \geq minsup$, is derived from D_{cl}. Subsequently, for every $a \in LQFI(cl)$, the following "*growth rate*" for a class cl is calculated. Let $\bar{D}_{cl} = D - D_{cl}$ be the opponent instances of cl.

Growth rate:
If $support_{\bar{D}_{cl}}(a) \neq 0$, $growth\_rate_{\bar{D}_{cl} \to D_{cl}}(a) = \frac{support_{D_{cl}}(a)}{support_{\bar{D}_{cl}}(a)}$,
If $support_{\bar{D}_{cl}}(a) = 0$ and $support_{D_{cl}}(a) \neq 0$, $growth\_rate_{\bar{D}_{cl} \to D_{cl}}(a) = \infty$,
Otherwise $growth\_rate_{\bar{D}_{cl} \to D_{cl}}(a) = 0$.
When the growth rate of a is more than a "*growth rate threshold*" $\rho(>1)$, i.e., $growth\_rate_{\bar{D}_{cl} \to D_{cl}}(a) \geq \rho$, a is called an "*emerging pattern* (EP)" and selected as a rule body where its head has the class cl, i.e., $a \Rightarrow cl$. Let $LEP(cl)$ be a set of all EPs selected from $LQFI(cl)$ under this measure. The underlying principle here is to select the rule bodies having the strength to differentiate the class cl from the others. Even if the rule confidence is high in D_{cl}, the rule can match many instances in $\bar{D}_{cl}$. Such rules are weak for classification.

The second process is to derive a "*base score*." First, the strength of an EP a based on the relative difference between $support_{D_{cl}}(a)$ and $support_{\bar{D}_{cl}}(a)$ is introduced as $support_{D_{cl}}(a)/(support_{D_{cl}}(a) + support_{\bar{D}_{cl}}(a)) = growth\_rate_{\bar{D}_{cl} \to D_{cl}}(a) /(growth\_rate_{\bar{D}_{cl} \to D_{cl}}(a) + 1)$. The following "*aggregate score*" represents the possibility of t to be classified into cl by EPs in $LEP(cl)$.

Aggregate score:
$$score(t, cl) = \sum_{a \subseteq t, a \in LEP(cl)} \frac{growth\_rate(a)}{growth\_rate(a) + 1} * support_{D_{cl}}(a). \quad (1)$$

Because the number of EPs for each cl may not be balanced, instances may get higher scores for some classes. A base score is introduced to eliminate this bias.

Base score:

$base\_score(cl)$ is the median of all aggregate scores in $\{score(t, cl)|t \in D_{cl}\}$. The testing phase uses $base\_score(cl)$, $growth\_rate(a)$ and $support_{D_{cl}}(a)$ obtained in the training phase. Given a test instance t, its aggregate score for cl, $score(t, cl)$, is computed from these results and Eq.(1). Then, it is normalized by $base\_score(cl)$ to eliminate the aforementioned bias as follows.

Normalized score:

$$norm\_score(t, cl) = \frac{score(t,cl)}{base\_score(cl)}.$$

cl having the maximum normalized score is assigned to the class of t. Except the derivation of $LQFI(cl)$ for all cl, the computational complexity of CAEP is $O(N)$ where $N = |D|$, since it scans the training data only twice.

3 Mining Rule Bodies of CARs

3.1 Levelwise Subspace Clustering

First we focus on the clustering of instances consisting of numeric items only. In our approach, the density of instances in a subspace is defined on their projected distribution to every attribute axis while ensuring the exhaustive finding of the clusters. The computational complexity is around $O(N \log N)$ as discussed later.

Let t and t' be instances sharing a numeric attribute p with interval values q and q' respectively. The "Δ_p-neighborhood" $N_{\Delta_p}(t)$ on p is defined by $N_{\Delta_p}(t) = \{t' \in D_{cl} | Dist_p(q, q') \leq \Delta_p\}$ where "permissible range" Δ_p is a real positive number on p. If intervals q and q' overlap, then $Dist_p(q, q') = 0$, otherwise $Dist_p(q, q')$ is the distance between their boundaries facing each other. An instance $t \in D_{cl}$ is called a "core instance" on p if $N_{\Delta_p}(t)$ contains at least $MinPts$ instances, i.e., $|N_{\Delta_p}(t)| \geq MinPts$. When a core instance t is contained in $N_{\Delta_p}(t')$ of another core instance t', t and t' have a "connection."

Definition 1 (Density-Connected Set). *A non-empty subset $C \subseteq D_{cl}$ is a "density-connected set" on p if C is the union of the Δ_p-neighborhoods of core instances where all the core instances are in a chain of connections on p.*

Definition 2 (Dense Cluster). *A "dense cluster" $C^S \subseteq D_{cl}$ in a subspace formed by a set of numeric attributes S is defined as a maximal set of instances which is a density-connected set on every $p \in S$ in D_{cl}.*

Definition 3 (Quantitative Frequent Itemset). *Let $C^S \subseteq D_{cl}$ be a dense cluster in a subspace S and $a(C^S) = \{<p : q> \mid p \in S, q = [\min_p(C^S), \max_p(C^S)]\}$ an itemset where $\min_p(C^S)$ and $\max_p(C^S)$ are the minimum and the maximum interval boundaries of instances in C^S on p. If $|C^S| \geq minsup$, i.e., $support_{D_{cl}}(a(C^S)) \geq minsup$, then $a(C^S)$ is a "quantitative frequent itemset (QFI)." When the dimension of S is k, it is called a k-QFI.*

A QFI is a dense, axis-parallel and monotone hyper-rectangular region having a maximal volume in the subspace. Similarly to the dense clusters of SUBCLU, the following (anti-)monotonicity property of QFIs holds.

Table 1. An example of transaction data set of $cl = Houseowner$; $D_{Houseowner}$

$t_1 = (\{< Age : [20, 23] >, < Child : [2, 3] >, < NumCars : [2, 2] >\}, Houseowner)$
$t_2 = (\{< Age : [30, 30] >, < Child : [4, 5] >, < NumCars : [1, 1] >,$
$\quad < Savings : [10K, 10K] >\}, Houseowner)$
$t_3 = (\{< Age : [30, 30] >, < Child : [2, 2] >, < NumCars : [5, 5] >,$
$\quad < Savings : [11K, 11K] >\}, Houseowner)$
$t_4 = (\{< Age : [30, 35] >, < Child : [5, 5] >, < NumCars : [1, 1] >\}, Houseowner)$
$t_5 = (\{< Age : [35, 37] >, < Child : [2, 2] >, < NumCars : [2, 2] >,$
$\quad < Savings : [5K, 5K] >\}, Houseowner)$
$t_6 = (\{< Age : [36, 39] >, < Child : [2, 2] >, < NumCars : [2, 3] >\}, Houseowner)$

Table 2. Process of levelwise subspace clustering of $D_{Houseowner}$

1-QFIs
$(\{< Age : [30, 39] >\}, \{t_2, t_3, t_4, t_5, t_6\}), (\{< Child : [2, 5] >\}, \{t_1, t_2, t_3, t_4, t_5, t_6\})$
$(\{< NumCars : [1, 3] >\}, \{t_1, t_2, t_4, t_5, t_6\}), (\{< Savings : [10K, 11K] >\}, \{t_2, t_3\})$
2-QFIs
$(\{< Age : [30, 39] >, < Child : [2, 2] >\}, \{t_3, t_5, t_6\})$
$(\{< Age : [30, 35] >, < Child : [4, 5] >\}, \{t_2, t_4\})$
$(\{< Age : [30, 39] >, < NumCars : [1, 3] >\}, \{t_2, t_4, t_5, t_6\})$
$(\{< Age : [30, 30] >, < Savings : [10K, 11K] >\}, \{t_2, t_3\})$
$(\{< Child : [2, 5] >, < NumCars : [1, 3] >\}, \{t_1, t_2, t_4, t_5, t_6\})$
3-QFIs
$(\{< Age : [35, 39] >, < Child : [2, 2] >, < NumCars : [2, 3] >\}, \{t_5, t_6\})$
$(\{< Age : [30, 35] >, < Child : [4, 5] >, < NumCars : [1, 1] >\}, \{t_2, t_4\})$

Lemma 1 (Monotonicity). $\forall T \subseteq S$, if $a(C^S)$ is a QFI in S, then a QFI $a(C^T)$ supported by $a(C^S)$, i.e., $a(C^S) \subseteq a(C^T)$, exists in T.

Proof. Because C^S is a density-connected set on $\forall p \in S$, it is a density-connected set on $\forall p \in T$, and hence $C^S \subseteq C^T$. Therefore, $\forall p \in T$, $[\min_p(C^S), \max_p(C^S)] \subseteq [\min_p(C^T), \max_p(C^T)]$, and $a(C^T)$ is supported by $a(C^S)$. ∎

Accordingly, a levelwise bottom up approach is applicable to search all QFIs. We exemplify its operation by using the dataset in Table 1. Each instance (transaction) t_i consists of a numeric itemset and a class $cl = Houseowner$. We perform the clustering under parameters of $\Delta_{Age} = 5$, $\Delta_{Child} = 1$, $\Delta_{NumCars} = 1$, $\Delta_{Savings} = 1K$, $MinPts = 1$ and $minsup = 2$. First, the items in each t_i are lexicographically ordered by the attribute names. This has been already done in this table. Subsequently, 1-QFIs are searched, where the instances are maximally density-connected on an attribute. For Age, a 1-QFI, $\{< Age : [30, 39] >\}$, exists since the items densely range from 30 to 39 under $\Delta_{Age} = 5$, and its support 5 is more than $minsup$. This 1-QFI with its *"transaction id list* (TID-List)*"* is indicated in Table 2. Each attribute has an 1-QFI in this example.

In the next step, the levelwise search for k-QFIs $(k > 1)$ starts. Index lists named $TID-List$ are used to point instances in D_{cl} similarly to AprioriTid algorithm [9]. Assuming that all $(k-1)$-QFIs are known, the following *"Candidate-Generation"* operation derives all candidate k-QFIs.

Definition 4 (Candidate-Generation).
Join Phase: *For two $(k-1)$-QFIs sharing $k-2$ attributes,*
$((k-1) - QFI = \{< p_1 : q_1 >, < p_2 : q_2 >, ..., < p_{k-2} : q_{k-2} >, < p_{k-1} : q_{k-1} >\}, TID-List)$,
$((k-1) - QFI' = \{< p_1 : q'_1 >, < p_2 : q'_2 >, ..., < p_{k-2} : q'_{k-2} >, < p_k : q'_k >\}, TID-List')$,
their join is derived as follows:
$(candidate - k - QFI = \{< p_1 : q^c_1 >, ..., < p_{k-1} : q^c_{k-1} >, < p_k : q^c_k >\}, TID-List^c)$.

```
QFI-Count(candidate − k − QFI, TID − List^c);
(1) k − QFIS = φ, TIDLS = φ;
(2) If |TID − List^c| < minsup return k−QFIS;
(3) S = {p| < p : q >
        ∈ candidate − k − QFI, p is numeric.};
(4) TIDLS.temp = {TID − List^c};
(5) while TIDLS ≠ TIDLS.temp do begin
(6)   TIDLS = TIDLS.temp;
(7)   forall p ∈ S do begin
(8)     TIDLS.temp =
            MDCS(TIDLS.temp, p);
(9)   end
(10) end
(11) forall TID − List ∈ TIDLS do begin
(12)   k − QFIS = k − QFIS+
            (QFI(S, TID − List), TID − List);
(13) end
(14) return k − QFIS;
```

Fig. 1. Algorithm of QFI-Count

```
(1) For each numeric attribute, create an index
    list sorted with the ascending order of D.
    Sort items in each t ∈ D lexicographically.
(2) L_1 = {(1 − QFI, TID − List)};
(3) for (k=2; L_{k−1} ≠ φ; k++) do begin
(4)   C_k =
        {(candidate − k − QFI, TID − List^c)} =
        Extended − Candidate−
        Generation(L_{k−1});
(5)   forall (candidate − k − QFI,
              TID − List^c) ∈ C_k do begin
(6)     L_k = L_k ∪
              QFI − Count(candidate − k − QFI,
              TID − List^c)
(7)   end
(8) end
(9) Answer L = ⋃_k L_k;
```

Fig. 2. Entire algorithm

where q_i^c is the intersection of the two intervals $q_i \cap q_i'$ for $i = 1, ..., k − 2$, $q_{k−1}^c = q_{k−1}$, $q_k^c = q_k'$ and $TID − List^c = TID − List \cap TID − List'$. If some $q_i^c = \phi$, the two $(k − 1)$-QFIs are not joined.

Prune Phase: For all $(k−1)$-subsets s of this candidate-k-QFI, if the following $(k − 1)$-QFI exists:

$$\forall < p_i : q_i^c > \in s, \exists < p_i : q_i > \in (k − 1) − QFI, q_i^c \cap q_i \neq \phi, \qquad (2)$$

the candidate-k-QFI is retained, and $TID − List^c$ is a candidate dense cluster $\hat{C}^S$ where $|S| = k$. Otherwise the candidate-k-QFI is pruned.

This prune phase is based on Lemma 1. As far as q_i^c intersects with q_i in Eq.(2), the possibility that s and $(k − 1) − QFI$ shares more than $minsup$ transactions is not negligible. Thus the candidate k-QFI is retained under this condition. In Table 2, a candidate-2-QFI, $\{< Age : [30, 39] >, < Child : [2, 5] >\}$ with $TID − List^c = \{t_2, t_3, t_4, t_5, t_6\}$ is derived from two 1-QFIs, $\{< Age : [30, 39] >\}$ and $\{< Child : [2, 5] >\}$. This passes the prune phase.

"QFI-Count" shown in Fig.1 derives dense clusters $C^S = TID − List$ and their corresponding k-QFIs, if they exist, by assessing the density of instances in $\hat{C}^S$ based on Definition 2 and 3. In the inside loop from (7) to (9), a maximal density-connected set C is searched on p within $\hat{C}^S$ at first in a function $MDCS$ along with Definition 1 under given Δ_p and $MinPts$. Multiple C can be found when multiple dense clusters are included in $\hat{C}^S$. $MDCS$ repeats to update C on p from every C derived and kept in $TIDLS.temp$ at the previous loop iteration. C having a size less than $minsup$ is discarded in $MDCS$. This update continues in the outer loop from (5) to (10), until each C converges to C^S where each C^S is independent of the convergence process due to the (anti-)monotonicity property. In the loop from (11) to (13), each QFI corresponding to $C^S = TID − List$ is computed by Definition 3 in a function QFI and returned as the output. In the example, the candidate-2-QFI, $\{< Age : [30, 39] >, < Child : [2, 5] >\}$

with $TID - List^c = \{t_2, t_3, t_4, t_5, t_6\}$ is given to this QFI-Count. In the inside loop, $MDCS$ derives $TIDLS.temp = \{\{t_2, t_3, t_4, t_5, t_6\}\}$ on Age. Next, it derives $TIDLS.temp = \{\{t_3, t_5, t_6\}, \{t_2, t_4\}\}$ on $Child$. Further applications of $MDCS$ do not change $TIDLS.temp$. Since the sizes of candidates are more or equal to $minsup = 2$, two 2-QFIs, $(\{< Age : [30, 39] >, < Child : [2, 2] >\}, \{t_3, t_5, t_6\})$ and $(\{< Age : [30, 35] >, < Child : [4, 5] >\}, \{t_2, t_4\})$, are derived.

3.2 Deriving QFIs of Numeric and Categorical Items

Candidate-Generation is extended to derive QFIs consisting of numeric and categorical items. The categorical items in the joined itemset are given in the same way as in the AprioriTid algorithm. In the join phase of Definition 4, if $q_i^c = \phi$ for some numeric item or $q_i \neq q_i'$ for some categorical item, the two given $(k-1)$-QFIs are not joined. Otherwise they are joined as $q_i^c = q_i \cap q_i'$ for each numeric item and $q_i^c = q_i = q_i'$ for each categorical item. In the prune phase, the condition $q_i^c = q_i$ for a categorical item is applied in addition to $q_i^c \cap q_i \neq \phi$ for a numeric item in Eq.(2). The algorithm QFI-Count of Fig.1 is also altered. When the candidate-k-QFI consists of categorical items only, the loop from (5) to (10) is skipped, and $TIDLS = TIDLS.temp$ is applied. The function QFI at step (12) is also altered. For a categorical attribute p_i, its value is set to be $q_i^c = q_i = q_i'$.

The entire algorithm to derive QFIs from D is indicated in Fig.2. Required parameters are Δ_p for all numeric attributes, $MinPts$ and $minsup$. First, some index lists are created for the efficient processing in Extended-Candidate-Generation and QFI-Count. Subsequently, all QFIs are computed in L by the adaptation of the AprioriTid Algorithm. In the implementation, the inversed indexing $(t_i, \{candidate - k - QFI\})$ from each t_i to its containing candidate-k-QFIs is used instead of $(candidate - k - QFI, TID - List^c)$ similarly to the standard AprioriTid. This approach is applied to D_{cl} of every class cl to derive $LQFI(cl)$ required by CAEP described in the previous section.

4 Experimental Evaluation

4.1 Computational Efficiency

The most expensive task is the derivation of QFIs for CAR's bodies. Thus, its computational efficiency is evaluated by using artificial data sets. First, a set of seed items, SSI, is randomly generated where $r_n\%$ of them are numeric and the rest categorical. Second, a set of seed QFIs, $SQFI$, is generated by randomly selecting seed items from SSI. The size of each QFI is determined by uniform random distribution having its average at $\overline{|QFI|}$. Third, a set of instances (transactions) D is generated where each instance t is made by randomly selecting a QFI from $SQFI$ and further randomly adding extra $2\overline{|QFI|}$ seed items taken from SSI in the average. Finally, the values of numeric items in each t are distorted by introducing Gaussian noise having 5% amplitude. Our algorithm is tested on a Pentium 4 2.7 GHz PC with 2GB RAM. The default parameters for the test are $|SSI| = 1000$, $r_n = 50\%$, $|SQFI| = 10$, $\overline{|t|} = 12$, $N = |D| = 40000$,

Table 3. Complexity of clustering

Para-meter	Range of Assessment	Dependency of Comp.Time	Dependency of Mem.Cons.
$|SSI|$	[20, 20000]	constant	constant
r_n	[0%, 100%]	constant	constant
$|SQFI|$	[1, 50]	$O(|SQFI|)$	$O(|SQFI|)$
$|t|$	[8, 100]	exp. inc.	exp. inc.
$minsup$	[0.2%, 10%]	exp. dec.	exp. dec.
Δ_p	[0.1%, 100%]	inc. const.	inc. const.
$MinPts$	[1, 8000]	dec. const.	dec. const.
N	[200, 10^6]	$O(N \log N)$	$O(N)$

exp. inc./dec. : exponential increase/decrease.
inc./dec. const. : increase/decrease and saturation.

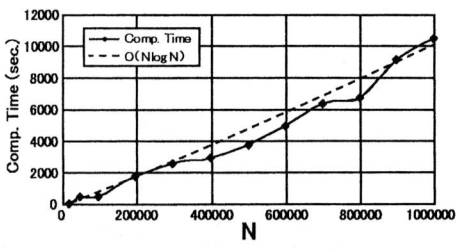

Fig. 3. N vs. Comp. Time

$minsup = 5\%$, $MinPts = 1$ and $\Delta_p = 20\%$ (in relative width of the maximum and minimum values of the instances on each numeric attribute).

The qualitative dependencies of computation time and memory consumption on the parameters summarized in Table 3 are similar to the AprioriTid. Their rapid increases are observed under the small values of Δ_p and $MinPts$ where the values are comparable with the average gap among two different instances values. The dependency of the computation time on N shown in Fig.3 is almost $O(N \log N)$ up to 1 million instances, while SUBCLU requiring the range query of each instance shows $O(N^2)$ where a dense cluster is such that for each instance in the dense cluster the neighborhood of a given radius ϵ has to contain at least a minimum number of $MinPts$ instances [6]. The memory consumption is $O(N)$, because the size of the inversed TID-List is proportional to N.

4.2 Classification Performance and Interpretability

CAEP combined with QFI derivation is called LSC-CAEP in this paper. Its 10CV accuracy has been compared with C4.5, CBA, CMAR and the original CAEP by using the UCI repository data as indicated in Table 4. These data sets were selected so that the accuracy by various approaches has been reported [1,2,3]. The parameters of LSC-CAEP was set as $minsup = 12\%$, $MinPts = 1$, $\Delta_p = 16\%$ and $\rho = 1.1$, where their optimality has been confirmed through empirical surveys. The bold faces in the table are the best. The standard deviations of the accuracies over 10CV by C4.5, CBA and LSC-CAEP are 2.6%, 3.4% and 4.0% respectively, while the average discrepancies of the accuracies of LSC-CAEP are 5.6% from C4.5 and 4.5% from CBA in Table 4. LSC-CAEP performs moderately better than C4.5 and better or equal at least to CBA. As shown in the last column of Table 4, the computation times of LSC-CAEP ranges from 0.1 to 87sec, while C4.5 and CBA are from 0.3 to 2.2sec. The speed of LSC-CAEP usually remains practical due to its aforementioned good scalability, though it is affected by the dependency among attributes similarly to the Apriori algorithm.

The interpretability of the rules is important but a quite subjective matter. The following two QFIs having large support values are found by LSC-CAEP in Labor Relations Database in UCI repository.

Table 4. Comparison of accuracies

dataset	num. of records	num. of attributes(numeric)	num. of classes	C4.5	CBA	CMAR	CAEP	LSC-CAEP [comp. time (sec)]
Cleve	303	13(5)	2	.782	**.828**	.822	.833	.789 [38]
Ecoli	336	8(7)	8	.824	-	-	-	**.831** [22]
Heart	270	13(6)	2	.808	.819	.822	.837	**.845** [87]
Hepatitis	155	19(6)	2	.806	.818	.805	.830	**.852** [26]
Iris	150	4(4)	3	.953	.947	.940	.947	**.967** [0.1]
Glass	214	9(9)	7	.687	**.739**	.701	-	.681 [19]
Labor	57	16(8)	2	.793	.863	.897	-	**.943** [0.1]
Wine	178	13(13)	3	.927	.950	.950	.971	**.972** [52]
Zoo	101	16(0)	7	.922	.968	**.971**	-	.911 [19]

support=19: {class:good, duration-years:[2,2], working-hours:[33,40], wage-inc.-2nd-year(%):[4.0,5.8]}.

support=16: {class:good, duration-years:[3,3], working-hours:[35,40], wage-inc.-2nd-year(%):[3.5,5.0]}.

These QFIs suggest an assumption that the increase of job stability from 2 years to 3 years balances with admitting slightly longer working hours and 0.5% ~ 0.8% less wage increase. The following two CARs having high growth rates are found in Iris data. The insights on the species of iris can be learned.

growth rate=4.5: petal width:[1.4-2.5]) → class:virginica

growth rate=1.9: sepal length:[4.9-7.0], sepal width:[2.0-3.4] → class:setosa.

The fine granularity of the interval boundaries helps the interpretation.

5 Discussion and Conclusion

To check the applicability of LSC-CAEP to large data sets, LSC-CAEP, C4.5 and CBA were applied to Census-Income data containing 199523 instances and 40 attributes (numeric: 7 categorical: 33) in UCI KDD Archive, and confirmed that the accuracy achieved by LSC-CAEP is 92.4%, which is comparable with 94.3% and 94.0% of C4.5 and CBA respectively. Further improvement of the performance of LSC-CAEP will be addressed in future study.

The most expensive tasks in LSC-CAEP are the sort which is $O(N \log N)$ and QFI-Count of Fig.1. The maximal density-connected sets on every numeric attribute p are easily derived in one scan of $TIDLS.temp$ in $MDCS$ by using the sort index list built at the first step in Fig.2. Hence it is $O(N)$ at maximum. The iteration of the outer loop from (5) to (10) in QFI-Count varies extensively. In the worst case, an instance is removed in each loop path from the edge of a region where instances are ranged in a periodic manner, and the loop becomes $O(N^2)$. However, in the most likely case which is an exponential density distribution, a portion $0 < r < 1$ of the instances in the average are retained in each loop path. The loop finishes by the time $r^m N$ becomes less than $minsup$ where m is the number of loop paths. Thus $minsup \leq r^m N$, and m is around $O(\log N)$. Accordingly, the entire algorithm is expected to be $O(N \log N)$.

Our proposal enabled efficient subspace clustering on the mixtures of numeric and categorical data in a levelwise algorithm. Further new approaches of clustering and classification for large data sets can be developed along this line.

Acknowledgement. The authors wish to thank Dr. Alexandre Termier in ISIR, Osaka Univ. for his extensive support to write this paper. This research was partially supported by the Japan Society for the Promotion of Science (JSPS) Grant-in-Aid for Scientific Research (B), 16300045, 2005.

References

1. Liu, B., Hsu, W., Ma, Y.: Integrating classification and association rule mining. Proc. of Fourth International Conference on Knowledge Discovery and Data Mining (1998)
2. Li, W., Han, J., Pei, J.: Cmar: Accurate and efficient classification based on multiple class-association rules. Proc. of First IEEE International Conference on Data Mining (2001) 369–376
3. Dong, G., Zhang, X., Wong, L., Li, J.: Caep: Classification by aggregating emerging patterns. Proc. of Second International Conference on Discovery Science, Lecture Notes in Computer Science **1721** (1999) 30–42
4. Agrawal, R., Gehrke, J., Gunopulos, D., Raghavan, P.: Automatic subspace clustering of high dimensional data for data mining applications. Proc. of the 1998 ACM SIGMOD international conference on Management of data (1998) 94–105
5. Procopiuc, C.M., Jones, M., Agarwal, P.K., Murali, T.M.: A monte carlo algorithm for fast projective clustering. Proceedings of the 2002 ACM SIGMOD international conference on Management of data (2002) 418–427
6. Kailing, K., Kriegel, H.P., Kroger, P.: Density-connected subspace clustering for high-dimensional data. Proc. Fourth SIAM International Conference on Data Mining (SDM'04) (2004) 246–257
7. Srikant, R., Agrawal, R.: Mining quantitative association rules in large relational tables. Proc. of 1996 ACM SIGMOD Int. Conf. on Management of Data (1996) 1–12
8. Wang, K., Hock, S., Tay, W., Liu, B.: Interestingness-based interval merger for numeric association rules. Proc. of 4th Int. Conf. on Knowledge Discovery and Data Mining (KDD) (1998) 121–128
9. Agrawal, R., Srikant, R.: Fast algorithms for mining association rules. Proc. of 20th Int. Conf. on Very Large Data Bases (VLDB) (1994) 487–499

An Incremental Algorithm for Mining Generators Representation

Lijun Xu and Kanglin Xie

Department of Computer Science and Engineering, Shanghai JiaoTong University, 282#,
No.1954 HuaShan Road, Shanghai, China, 200030
lijunxu@sjtu.edu.cn, xie-kl@cs.sjtu.edu.cn

Abstract. This paper presents an efficient algorithm for maintaining the generator representation in dynamic datasets. The generators representation is a kind of lossless, concise representation of the set of frequent itemsets. Furthermore, the algorithm utilizes a novel optimization based on generators borders for the first time in the literature. Generators borders are the borderline between frequent generators and other itemsets. New frequent generators can be generated through monitoring them. Experiments show that our algorithm is more efficient than previous solutions.

1 Introduction

Frequent itemsets mining [1] is an important subject in many data mining applications, such as the discovery of association rules, correlations, sequential rules and episodes. A lot of algorithms have been proposed for this domain. But most algorithms assume that all transactions are available prior to the execution of the algorithm. However, in most cases this assumption does not hold. Many datasets are updated with blocks of data at regular time intervals. Recognizing the importance of the problem, many researchers [2-6, 11, 12] have proposed their solutions and efficient algorithms. The first incremental frequent itemsets mining algorithm, FUP, was proposed by Cheung et al. [3]. FUP2 [4] algorithm, adapted from FUP, can simultaneously handle deletions and additions. Two algorithms both adopt a level-wise search strategy like Apriori algorithm [1] and use the previous result for guiding the update. Feldman et al. [6] and Thomas et al. [11] proposed two similar algorithms respectively. The main idea of two algorithms is to keep track of frequent itemsets and the negative border that contains the itemsets form the borderline between frequent itemsets and infrequent itemsets. New frequent itemsets can be found by monitoring the negative border. Ayan et al. [2] presented UWEP algorithm, which follows the approach of FUP2. UWEP prunes the itemsets that will become infrequent by a look-ahead pruning strategy. ZIGZAG algorithm [12] is enlightened by GenMax [7] algorithm, an algorithm for discovering maximal frequent itemsets. It incrementally computes maximal frequent itemsets combining previous knowledge. But it may scan the dataset again to compute support values of some frequent itemsets that are not maximal frequent itemsets. Chi et al. [5] proposed Moment algorithm, which uses an in-memory data structure to monitor frequent closed itemsets and the itemsets that form the boundary between the frequent closed itemsets and the rest of the itemsets. Moment handles new transactions or deleted transactions one by one, which may cause frequent changes of the boundary and affect the performance of the algorithm.

In this paper we present an efficient algorithm, called GBorder2, to maintain the generators representation in dynamic datasets. The generators representation is a kind of lossless, concise representation of the set of frequent itemsets. The usage of the generators representation can significantly reduce the times of data scans and the number of candidates in that the generators representation can be orders of magnitude smaller than the set of all frequent itemsets. Moreover, to the best of our knowledge, the algorithm introduces a novel optimization utilizing generators borders for the first time. Generators borders are the borderline between frequent generators and other itemsets. This optimization provides significantly computational or I/O savings as new frequent generators can be generated through monitoring generators borders.

2 Problem Definition

Let I be a set of items. A subset $X \subseteq I$ is called an itemset. An itemset with k items is called k-itemset. Let D be a transactional database, where each transaction is a subset of I. The number of transactions in D is denoted by $|D|$. During each update, obsolete transactions are removed and new transactions are added. Let d^+ be the set of newly added transaction, d^- be the set of deleted transactions and N be the updated dataset, i.e. $N=(D-d^-) \cup d^+$.

The support value of an itemset X, Sup(X), is the number of the transactions that contain X. An itemset is frequent if it satisfies the minimum support threshold (θ). Let F be the set of frequent itemsets, i.e. $F=\{X|Sup(X) \geq \theta|D|\}$.

An itemset is a generator if none of its proper subsets has the same support as it has. We denote the set of generators by G and the set of frequent generators by FG, i.e. $FG = F \cap G$. Negative generators border, GB^-, is defined as the set of infrequent generators whose proper subsets are frequent generators. Positive generators border, GB^+, is defined as the set of frequent non-generators whose proper subsets are generators. The generators representation consists of two components: (a) FG enriched by the support value for each itemset $X \in FG$; (b) GB^-. The following lists two important conclusions. Please refer to [8, 9] for more details.

Theorem 1. $X \in G \rightarrow \forall S \subset X, S \in G; X \notin G \rightarrow \forall S \supset X, S \notin G$.

Theorem 2. Let $X \subseteq I$. If $\exists Z \in GB^-$ and $Z \subseteq X$, then $X \notin F$. Otherwise, $X \in F$ and $Sup(X) = \min(\{Sup(S)|S \in FG \land S \subseteq X\})$.

3 GBorder2 Algorithm

GBorder2 algorithm is enlightened by the idea of the negative border [6, 10, 11]. GBorder2 maintains two kinds of generators borders: GB^- and GB^+. GB^- defines the borderline between frequent generators and infrequent generators, and GB^+ defines the borderline between frequent generators and frequent non-generators. Most itemsets do not change their status (from frequent to infrequent, from infrequent to frequent, from generator to non-generator or from non-generator to generator) when a small number of new transactions are added or a small portion of the dataset is removed. If the itemset does not change its status, nothing needs to be done except for updating its support value. Otherwise, as we shall present, the changes must come through generators borders.

Theorem 3. Let ChangedGB be a set of itemsets that belong to FG in N and belong to GB^+ or GB^- in D. If X is a frequent generator in N and is not a frequent non-generator in D, then there exists a subset $Y \subseteq X$, $Y \in$ ChangedGB.

Proof: There are two possible cases for X:

1. X is a frequent non-generator in D. Let Y be the smallest subset of X that is a frequent generator in N but a frequent non-generator in D. As Y has minimal size, all its proper subsets are frequent generators in D. Thus Y belongs to GB^+ in D and Y belongs to ChangedGB.
2. X is an infrequent itemset in D. Let Y be the smallest subset of X that is a frequent generator in N but an infrequent generator in D. As Y has minimal size, all its proper subsets are frequent generators in D. Thus Y belongs to GB^- in D and Y belongs to ChangedGB.

3.1 Algorithm Description

The pseudo-code for GBorder2 algorithm is given in Fig. 1. We assume that each itemset X that belongs to frequent generators or generators borders (OldFG, OldGB$^-$ or OldGB$^+$) and its support value in D, sup(X, D), are already known.

The approach starts by scanning d^+, d^- and computing the support values of all itemsets of OldFG, OldGB$^-$ and OldGB$^+$ in d^+ and d^- respectively (Lines 1-3). Since the addition of new transactions and the deletion of obsolete transactions, some itemsets of OldFG, OldGB$^-$ or OldGB$^+$ may change their status. Thus the frequent generators and the generators borders are determined again (Lines 4-6). ChangedGB contains the new frequent generators that originally belong to the generators borders in D (Line 7). It is used to generate candidates in the later steps.

Next, the candidates are generated and tested level by level like the classical Apriori algorithm [1] (Lines 8-26). (i+1)-candidates (C_{i+1}), is generated based on i-itemsets of ChangedGB (ChangedGB$_i$), new i-generators calculated in the last while-loop steps (G_i), i-generators (NewFG$_i$) (Line 12). For each candidate X, the algorithm first determines Sup(X,d^+) and Sup(X,d^-) by scanning d^+ and d^- (Line 14). Then there are two possible cases when Sup(X,D) is calculated. If X is infrequent in D, the algorithm has to scan D and determines its support value (Line 15-16). Otherwise, its support value can be directly retrieved from OldFG according to Theorem 2 (Lines 17-18). Finally the qualified candidates are added into NewFG (Line 23), NewGB$^-$ (Line 21) or NewGB$^+$ (Line 25) respectively after updating their support values.

The while-loop steps (Lines 10-26) are performed only if ChangedGB is not empty. Thus unnecessary computing and I/O requirements are avoided if there is no new generator generated. Furthermore, the number of candidates can be considerably reduced even though these steps are performed.

```
Input: OldFG, OldGB⁻, OldGB⁺, N (N=(D-d⁻)∪d⁺) and θ
Output: NewFG, NewGB⁻ and NewGB⁺
1)  for X∈OldFG∪OldGB⁻∪OldGB⁺
2)    Scan d⁺, d⁻ and calculate Sup(X, d⁺), Sup(X, d⁻)
3)    Sup(X, N)=Sup(X,D)+Sup(X, d⁺)-Sup(X, d⁻)
4)  NewFG={X|X∈OldFG∪OldGB⁻∪OldGB⁺∧Sup(X,N)≥θ|N|∧∀S⊂X,
Sup(X,N) <Sup(S,N)}
5)  NewGB⁻={X|X∈OldFG∪OldGB⁻∪OldGB⁺∧Sup(X,N)<θ|N|∧∀S⊂X,
S∈NewFG∧∀S⊂X, Sup(X,N)<Sup(S,N)}
6)  NewGB⁺={X|X∈OldFG∪OldGB⁻∪OldGB⁺∧Sup(X,N)≥θ|N|∧∀S⊂X,
S∈NewFG∧∃S⊂X, Sup(X,N)=Sup(S,N)}
7)  ChangedGB={X|X∈OldGB⁻∪OldGB⁺∧X∈NewFG}
8)  n=max({i|ChangedGBᵢ≠∅}),
9)  G₀=∅, i=0
10) while (Gᵢ≠∅∨i≤n)
11)   Gᵢ₊₁=∅
12)   Cᵢ₊₁={X||X|=i+1∧∃ i-subset S⊂X, S∈ChangedGBᵢ∪Gᵢ∧∀
i-subset S⊂X, S∈NewFGᵢ∪ChangedGBᵢ}
13)   for X∈Cᵢ₊₁
14)     Scan d⁺, d⁻ and calculate Sup(X, d⁺), Sup(X, d⁻)
15)     if ∃S⊂X∧S∈OldGB⁻ then
16)       Scan D and calculate Sup(X, D)
17)     else
18)       Sup(X,D)=min{Sup(S,D)|S⊂X∧S∈OldFG}
19)     Sup(X,N)= Sup(X,D)+Sup(X, d⁺)-Sup(X, d⁻)
20)     if Sup(X,N)<θ|N| then
21)       Add X into NewGB⁻
22)     else if ∀S⊂X, Sup(X,N)<Sup(S,N) then
23)       Add X into Gᵢ₊₁
24)     else
25)       Add X into NewGB⁺
26)   NewFG=NewFG∪Gᵢ₊₁, i=i+1
```

Fig. 1. GBorder2 Algorithm

3.2 Discussions

GBorder2 handles the general case for transaction insertions as well as deletions. For the add-only case ($d^+\neq\emptyset$ and $d^-=\emptyset$) or the delete-only case ($d^+=\emptyset$ and $d^-\neq\emptyset$), there exists some improvements on the implementation of the algorithm.

For the add-only case, as we shall present in Theorem 4, a generator in D is still a generator in N. Then we can optimize GBorder2 by modifying Line 4-6 in Fig.1. The changes are shown in Fig. 2.

Theorem 4. Let X be a generator in D. If $d^-=\emptyset$, i.e. $N=D\cup d^+$, then X is still a generator in N.

4) NewFG={X|X∈OldFG∪OldGB⁻∧Sup(X,N)≥θ|N|}∪{X|X∈OldGB⁺∧Sup(X,N)≥θ|N|∧∀S⊂X, Sup(X,N)<Sup(S,N)}
5) NewGB⁻={X|X∈OldFG∪OldGB⁻∧Sup(X,N)<θ|N|∧∀S⊂X, S∈NewFG}∪{X|X∈OldGB⁺∧Sup(X,N)<θ|N|∧∀S⊂X, S∈NewFG∧ ∀S⊂X, Sup(X,N)<Sup(S,N)}
6) NewGB⁺={X|X∈OldGB⁺∧Sup(X,N)≥θ|N|∧∀S⊂X, S∈NewFG∧∃S⊂X, Sup(X,N)=Sup(S,N)}

Fig. 2. Optimizations for add-only case

Proof. Let S be an arbitrary subset of X. According the definition of generators, Sup(X,D)<Sup(S,D). AS S is a subset of X, Sup(X,d⁺)≤Sup(S,d⁺). Then Sup(X,N)=Sup(X,D)+Sup(X,d⁺)<Sup(S,D)+Sup(S,d⁺) =Sup(S,N).

So X is a generator in N.

For the delete-only case, a non-generator in D is still a non-generator in N (See Theorem 5). So any new generator must be infrequent generator in D. We have two improvements over the pseudo-code of GBorder2. The first one is presented in Fig. 3. The second one is that Lines 15-18 are replaced with Line 16 as none of the candidates are frequent in D.

4) NewFG={X|X∈OldFG∪OldGB⁻∧Sup(X,N)≥θ|N|∧∀S⊂X, Sup(X,N)<Sup(S,N)}
5) NewGB⁻={X|X∈OldFG∪OldGB⁻∧Sup(X,N)<θ|N|∧ ∀S⊂X, S∈NewFG∧ ∀S⊂X, Sup(X,N)<Sup(S,N)}
6) NewGB⁺={X|X∈OldFG∪OldGB⁻∧Sup(X,N)≥θ|N|∧∀S⊂X, S∈NewFG∧∃S⊂X, Sup(X,N)=Sup(S,N)}∪ {X|X∈OldGB⁺∧Sup(X,N)≥θ|N|∧∀S⊂X, S∈NewFG}
7) ChangedGB={X|X∈OldGB⁻∧X∈NewFG}

Fig. 3. Optimizations for delete-only case

Theorem 5. Let X be a non-generator in D. If $d^+=\emptyset$, i.e. N=D−d⁻, then X is still a non-generator in N.

Proof. Let S be an proper subset of X and Sup(X,D)=Sup(S,D). Obviously, any transaction in D that contains S also contain X. d⁻ is a portion of D and thus Sup(X,d⁻)=Sup(S,d⁻). Then Sup(X,N)= Sup(X,D)−Sup(X,d⁻)= Sup(S,D)−Sup(S,d⁻)= Sup(S,N). So X is a non-generator in N.

4 Experimental Results

We performed extensive experiments to evaluate GBorder2 algorithm. We compared it with FUP2 algorithm. We implemented two algorithms using Microsoft Visual C++ 6.0. We used the same data structures and subroutines in order to minimize any performance differences caused by minor differences in implementation. The two algorithms are not fully optimized due to the time limitation. They were performed on a Pentium 1.2G processor with 1G MB, running Windows 2000.

We choose four datasets for the performance tests, which are publicly available from IBM Almaden Research Center (www.almaden.ibm.com/cs/quest/demos.html). The T10I4D100K dataset and the T40I10D100K dataset are synthetic datasets, while the connect dataset and the gazelle dataset are real-world datasets. Their characteristics are shown in Table 1.

Table 1. Characteristics of four datasets

Dataset	#Items	#Trans.	Avg. Trans. Len.	Max. Trans. Len.
T10I4D100K	1,000	100,000	3.7	31
T40I10D100K	1,000	100,000	8.5	77
gazelle	498	59,601	2.5	267
connect	130	67,557	43	43

We first conducted several experiments to evaluate the speed up of GBorder2 over FUP2. Without loss of generality, let $|D|$=100K and $|d^+|=|d^-|$=10K. We duplicated and randomized each original dataset to obtain 110K transactions. Fig. 4 shows the results over different datasets. There are two interesting trends we observe:

1) For synthetic datasets, GBorder2 shows better performance for high support thresholds than low support thresholds. The reason is that the probability of generators borders expanding is higher at low support thresholds and as a result GBorder2 may have to san the whole dataset.

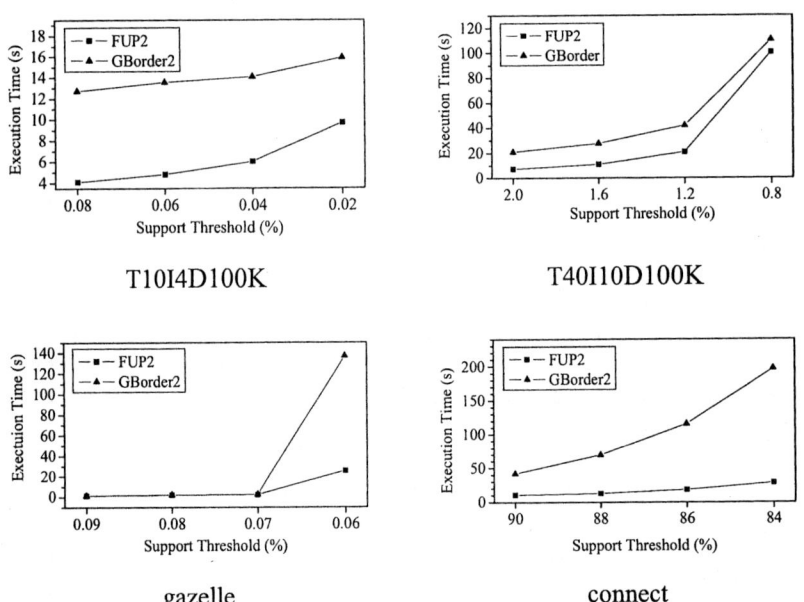

Fig. 4. Performance experiments

2) For real-world datasets, GBorder2 outperforms FUP2 throughout the entire range. Moreover, the performance gain of GBorder2 is larger for higher support thresholds. The phenomenon should be caused by the characteristics of real-world datasets. Real-world datasets are always strongly correlated datasets and a large number of frequent itemsets are non-generators for them. On the contrary, most frequent itemsets are generator for synthetic datasets.

Next, we conducted some experiments to find out if GBorder2 is able to deal with large datasets. Let $|D|=x$ and $|d^+|=|d^-|=x/10$, where x is varied in the experiments. We used a support threshold of 0.02% for the T10I4D100K dataset and 0.06% for the gazelle dataset. The results are plotted in Fig. 5. Obviously, the execution time of GBorder2 increases linearly as x increase, which implies that GBorder2 can handle large datasets well.

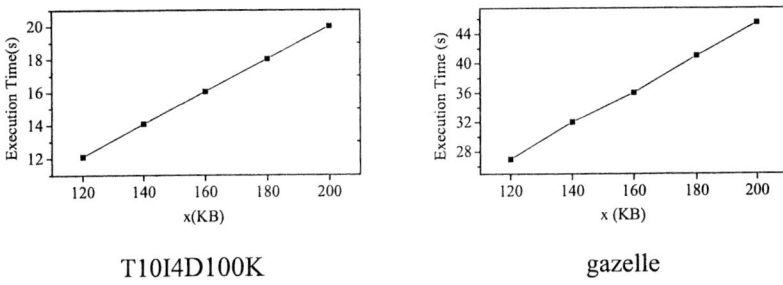

Fig. 5. Scale-up experiments

5 Conclusion

The paper focuses on the problem of frequent itemsets mining in dynamic datasets. Unlike existing incremental approaches, we propose an efficient algorithm to discover the generators representation using generators borders. The generators representation is a lossless, concise representation of frequent itemsets. New frequent generators can be computed by monitoring generators borders alone. To the best of our knowledge, it is the first incremental approach that combines the border technique and the generators representation. The usage of two techniques provides significant computational and I/O savings. Extensive experimental results show the efficiency of our approach.

A number of lossless concise representations have been proposed [8]. All these representations, except for frequent closed itemsets, consist of two components: one main component and several borders. All border representations, except for the generators representation, are about two orders of magnitude more concise than frequent closed itemsets in practice. Due to the common characteristics of all border representations, our algorithm can be extended to update other border representations in an incremental manner.

Reference

1. R. Agrawal and R. Srikant. Fast algorithms for mining association rules. In Proc. of VLDB'94 (1994) 487-499.
2. N. Ayan, A. Tansel and E. Arkun. An efficient algorithm to update large itemsets with early pruning. In Proc. of the 5th ACM-SIGKDD (1999) 287-291.
3. D. Cheung, J. Han, V. Ng and C. Y. Wong. Maintenance of discovered association rules in large databases: An incremental updating technique. In Proc. of the 12th Data Engineering (1996) 106-114.
4. D. Cheung, S. Lee, and B. Kao. A general incremental technique for maintaining discovered association rules. In Proc. of the 5th Database Systems for Advanced Applications (1997) 1-4.
5. Y. Chi, H. Wang, P. S. Yu and R. R. Muntz. Moment: maintaining closed frequent itemsets over a stream sliding window. In Proc. of ICDM'04 (2004) 59-66.
6. R. Feldman, Y. Aumann, A. Amir, and H. Mannila. Efficient algorithms for discovering frequent sets in incremental databases. In Proc. of ACM SIGMOD Workshop on Research Issues on Data Mining and Knowledge Discovery (1997) 59-66.
7. K. Gouda and M. Zaki. Efficiently mining maximal frequent itemsets. In Proc. of ICDM'01 (2001) 163-170.
8. M. Kryszkiewicz. Concise representations of association rules. In Proc. of the ESF Exploratory Workshop on Pattern Detection and Discovery (2002) 92-109.
9. N. Pasquier, Y. Bastide, R. Taouil and L. Lakhal. Pruning closed itemset lattices for association rules. In Proc. of BDA'98 (1998) 177-196.
10. H. Toivonen. Sampling large databases for association rules. In Proc. of VLDB'96 (1996) 134-145.
11. S. Thomas, S. Bodagala, K. Alsabti, and S. Ranka. An efficient algorithm for the incremental updation of association rules. In Proc. of KDD'97 (1997) 263-266.
12. A. Veloso, W. Meira Jr., M. B. de Carvalho, B. Pôssas, S. Parthasarathy, and M. Zaki. Mining frequent itemsets in evolving databases. In Proc. of SIAM'02 (2002).

Hybrid Technique for Artificial Neural Network Architecture and Weight Optimization

Cleber Zanchettin and Teresa Bernarda Ludermir

Center for Informatics, Federal University of Pernambuco (UFPE),
P.O. Box 7851, 50.732-970, Recife, PE, Brazil
{cz, tbl}@cin.ufpe.br

Abstract. This work presents a technique that integrates the heuristics tabu search, simulated annealing, genetic algorithms and backpropagation. This approach obtained promising results in the simultaneous optimization of the artificial neural network architecture and weights.

1 Introduction

Optimization is the process of finding the best solution for a problem from a group of possible solutions. An optimization problem has an objective function and a group of restrictions, both related to the decision variables of the problem. Genetic Algorithms (AG) [3], Simulated Annealing (SA) [1] and Tabu Search (TS) [2] are iterative algorithms used to solve different combinatorial optimization problems. These three algorithms are the most popular from a class of optimization algorithms known as general iterative algorithms. All three optimization heuristics have similarities [4]: (1) They are approximation (heuristic) algorithms, i.e., they do not assure the finding of an optimal solution; (2) They are blind in that they do not know when they have reached an optimal solution, and therefore, must be told when to stop; (3) They have a "hill climbing" property, i.e., they occasionally accept uphill (bad) moves; (4) They are general, i.e., they can easily be engineered to implement any combinatorial optimization problem; all that is required is to have a suitable solution representation, a cost function, and a mechanism to traverse the search space; and (5) Under certain conditions, they asymptotically converge to an optimal solution.

This paper presents a new technique that integrates the main potentialities of these three heuristics. This technique is evaluated in the simultaneous optimization of the number of connections and weight connection values among processing units of the Multi-Layer Perceptron neural network (MLP) [5].

The MLP trained by the backpropagation algorithm (BP) is one of the most used connectionist models in the literature. To obtain successful use, the network topology plays a very important role. A lack of connections can render the network incapable of solving the investigated problem as a result of the inadequacy of adjustable parameters, whereas an excess of connections can cause overfitting in the training data and fail to have an adequate generalization capacity. In general, the training of the MLP neural networks is accomplished through successive attempts with different network topologies until reaching satisfactory results for the problem. Besides consuming time, this process can establish network architectures with unnecessary connections and nodes. Moreover, the larger the topology, the more complex the value adjustment of these connections becomes. Thus, the simultaneous optimization of architectures and weights of artificial neural networks is an interesting approach to the generation of efficient networks with small topologies.

2 Search Heuristics Description

The genetic algorithm is characterized by a parallel search of the state space as against a point-by-point search through conventional optimization techniques. The parallel search is accomplished by keeping a set of possible solutions for the optimization problem, called population. An individual in the population is a string of symbols and is an abstract representation of the solution. The symbols are called genes and each string of genes is termed a chromosome. The individuals in the population are evaluated through a fitness measure. The population of chromosomes evolves from one generation to the next through the use of two types of genetic operators: (1) unary operators, such as mutation and inversion, which alter the genetic structure of a single chromosome; and (2) higher-order operator, referred to as crossover, which consists of obtaining a new individual by combining genetic material from two selected parent chromosomes. The parent chromosomes are chosen by way of selection techniques [3].

In the experiments performed, each chromosome is represented as described in Section 3. The initial population was defined with a size of 10 chromosomes. The chromosomes are classified by Rank Based Fitness Scaling [8]. The parents chosen for the next generation is accomplished in a probabilistic manner, using Universal Stochastic Sampling [8]. Elitism was also used, with a probability of 10%. For the combination of the parent chromosomes, the crossover operator Uniform Crossover [9] was used, with a probability of 80%. The mutation operator used was the Gaussian Mutation [6], with a probability of 10%. The stop criteria were: (1) the GL_5 criterion, this criterion provides an idea of the generalization loss during training and it is sufficiently useful to avoid overfitting. It is defined as the increase in the validation error in relation to the minimum validation error; and (2) a maximum number of 500 generations.

The simulated annealing method is different of the others search methods in that uphill moves are occasionally accepted to escape of local minima. The search process consists of a sequence of iterations. Each iteration consists of randomly changing the current solution to create a new solution in its neighborhood. Once a new solution is created, the corresponding change in the cost function is computed to decide if the new solution can be accepted. If the new solution cost is lower than the current solution cost, is accepted. Otherwise, the Metropolis's criterion is verified [10], based on the Boltzmann probability. A random number d in $[0,1]$ interval is generated from a uniform distribution. If $\delta \leq e^{\frac{\Delta E}{T}}$, where ΔE is the change in the cost function and T is a parameter called temperature, then the new solution is accepted as the current solution. If not, the current solution is unchanged and the process continues from the current solution.

The algorithm was originally derived from thermodynamic simulations. Thus, the parameter T is referenced as temperature and the temperature reduction process is called the cooling process. The chosen cooling strategy was *geometric cooling rule*. According to this rule, the new temperature is equal to the current temperature multiplied by a temperature factor (smaller than one, but close to one) [11]. The initial temperature is set to 1, and the temperature factor is set to 0.9. The temperature is decreased at each 10 iterations, with a maximum number of 1.000 iterations. The stop criterion GL_5 also was used.

Tabu search is an iterative search algorithm characterized by the use of a flexible memory. In this method, each iteration consists of the evaluation of a certain amount of new solutions (neighborhood moves). The best of these solutions (in terms of cost function) is accepted. However, the best candidate solution may not improve the current solution. Thus, the algorithm chooses the new solution that produces the largest improvement or the smallest deterioration in the cost function. This strategy allows the method to escape from local minima. A tabu list is used to store a certain amount of recently visited solutions. The solutions in tabu list are marked as forbidden to subsequent iterations. The tabu list registers T last visited solutions. When the list is full, a new movement is registered in substitution to the older solution kept on the list.

In the present work, a neighborhood with 20 solutions is used, and the algorithm chooses the best non-tabu solution. The proximity criterion [6] was used to compare solutions. A new solution is considered identical to the tabu solution one if: (1) each connectivity bit in the new solution is identical to the corresponding connectivity bit in the tabu solution; and (2) each connection weight in the new solution is within $\pm N$ of the corresponding connection weight in the tabu solution. The parameter N is a real number with a value of 0.001. A maximum number of 100 iterations is allowed. The stop criterion employed was also GL_5.

3 Integration of Simulated Annealing, Tabu Search and Genetic Algorithms

The simulated annealing method has the ability to escape from local minima through the choice between accepting or discarding a new solution that increases cost (uphill moves). The tabu search method, in contrast, evaluates one group of new solutions at each iteration (instead of only one solution as in simulated annealing). This makes a tabu search faster, as it generally needs less iterations to converge. The genetic algorithm evolution, in turn, involves a sequence of iterations, where a group of solutions evolves through selection processes and reproduction. This process, which is more elaborate than the other algorithms, can result in solutions with a better quality.

These observations motivated the proposal of an optimization technique (GaTSa) that combines the main potentialities of genetic algorithms, simulated annealing and tabu search in an effort to avoid their limitations. In general terms: at each iteration, a group of new solutions is generated, starting from the micro-evolution of the current population, as in genetic algorithms. The cost of each solution is evaluated, and the best solution is chosen, as in tabu search. However, differently from a tabu search, this solution is not always accepted. The acceptance criterion is the same used in the simulated annealing algorithm - if the chosen solution has a smaller cost than the current solution, it is accepted; otherwise, it can either be accepted or not, depending on a probability calculation. This probability is given by the same expression used in the simulated annealing method. The visited solutions are marked as tabu, as in a tabu search. During the optimization process, only the best solution found is stored, that is, the final solution comes back through the method.

Algorithm 1. Proposed algorithm Pseudo-code
---
1. $P_0 \leftarrow$ initial population with K solutions s_k
2. $T_0 \leftarrow$ initial temperature
3. $I_T \leftarrow$ iterations number
4. Update S_{BSF} with s_k of the P_0 (best solution found so far)
5. For $i = 0$ to $I_{max} - 1$
6.   If $i + 1$ is not a multiple of I_T
7. $T_{i+1} \leftarrow T_i$
8.   Else
9. $T_{i+1} \leftarrow$ new temperature
10.   If validation based stopping criteria are not satisfied
11.   Stop global search execution
12.   For $j = 0$ to g_n
13.   Generate a new population P' from P_i
14. $P_i \leftarrow P'$
15.   Choose the best solution s_k from P_i
16.   If $f(s') < f(s_k)$
17. $s_{k+1} \leftarrow s'$
18.   Else
19. $s_{k+1} \leftarrow s'$ with probability $e^{\frac{f(s')-f(s_k)}{T_{i+1}}}$
20.   If $f(s_{k+1}) < f(S_{BSF})$
21.   Update S_{BSF}
22. End For
23. Keep the topology contained in S_{BSF} constant and use the weights as initial ones for training with the backpropagation algorithm
---

The proposed method pseudo-code is presented in Algorithm 1. Let S be a group of solutions and f a real cost function, the proposed algorithm searches the global minimum s, such that $f(s) \leq f(s')$, $\forall\ s' \in S$. The process finishes after I_{max} iterations or if the stop criterion based on the validation error is satisfied. The best found solution S_{BSF} (*best so far*) is returned. The cooling process updates the temperature T_i of the iteration i at each I_T algorithm iterations. At each iteration, a new population with k solutions is generated. A genetic micro-evolution of g_n generations is used to generate this population from the current population. Moreover, at the end of the global search (GaTSa), a hybrid training is used, combining the proposed method with a local search technique. The local search technique can be implemented, for instance, by the well-known backpropagation algorithm.

Each solution is codified in a vector. This vector represents the connections among the processing units of the MLP artificial neural network. Each of these connections is specified by two parameters: (a) the connectivity bit, a boolean value that simbolyzes the existence or absence of a connection; and (b) the connection weight, which is a real number. If the connectivity bit is equal to zero, its associated weight is not considered, for the connection does not exist in the network. All possible connections among adjacent layers are considered.

Different from the constructive algorithms that only generate one solution at the end of the process, iterative algorithms originate possible (candidate) solutions at each iteration. The cost function is used to evaluate the performance among consecutive iterations and select the solution that minimizes (or maximizes) an objective function.

The cost function for the investigated problem is the arithmetic average between: (1) the classification error of the training set (percentage of incorrectly classified training patterns); and (2) the percentage of connections used by the artificial neural network. Therefore, the algorithms try to minimize both network performance and processing complexity. Only valid networks (i.e., networks with at least one unit in the hidden layer) were considered.

The operator for the generation of neighbors is used to derive new solutions from the current solution. The method used in simulations is defined as follows: (1) the connectivity bits for the current solution are changed according to a given probability, which in the present work is set to 20%. This operation deletes some network connections and creates new ones. Next, a random number taken from a uniform distribution in [-1; +1] is added to each connection weight. These two steps can change both topology and connection weights to produce a new neighbor solution.

4 Experiments and Results

Real data is used in the experiments. The problem aims to classify odor patterns obtained through an artificial nose. The odorant compositions analyzed are from three different vintages (years 1995, 1996 and 1997) of the same commercial red wine (Almadm, Brazil) produced with merlot-type grapes. The artificial nose used is composed of six distinct conducting polymer sensors constructed with an electrochemical deposition of polypyrrole using different types of dopants. Three data acquisitions were performed. In each acquisition for each wine vintage, the resistance value of each sensor was recorded for five seconds. A set of six values from the six sensors at the same time was considered a pattern. Thus, each acquisition contains 1.800 patterns (600 from each vintage). There were three acquisitions and 5.400 patterns of data.

In previous works with this data base, the best performance obtained by the MLP was achieved by an architecture with 6 processing units in the input layer, 4 processing units in the hidden layer and 3 processing units in the output layer [7]. This topology was keep constant as the maximum architecture in the optimization experiments performed. In all investigated algorithms, the parameter configurations were maintained at the standard configuration or adjusted based on previous experiments. The values used may not be the best values for the problem, but the objective of the present paper is to demonstrate the potentialities of the techniques and not the ideal algorithms configuration.

Table 1 presents the average performance of each investigated optimization technique. These results were obtained for each technique in the optimization of the number of connections and weight connection values of an MLP artificial neural network. The parameters evaluated were: (1) Squared Error Percentage (SEP) and the classification error (Class) of training, validation and test sets; (2) algorithm iteration number; (3) artificial neural network connection number; and (4) the temperature value. The following table displays the average results of 10 simulations. Each simulation contains 30 different runs of the algorithm.

Table 1. Optimization techniques performance

Technicque	Training SEP Class	Validation SEP Class	Test SEP Class	Iterations	Connections	Temperature
TS	18,74 5,44	18,86 5,88	18,75 5,3805	51	11,42	-
SA	19,65 6,91	19,76 7,47	19,65 6,9331	715	11,77	0,0085
GA	21,66 15,88	21,73 16,52	21,66 15,9240	315	16,64	-
GaTSa	18,69 3,58	18,76 3,81	18,69 3,5664	46	8,33	0,7098
GaTSa + BP	4,78 -	2,41 -	2,14 2,8684	86	8,33	0,7098
BP	6,25 -	3,15 -	2,84 6,7854	90	36	-

The technique that combines the heuristics of tabu search, simulated annealing and genetic algorithms obtained the best result performance. This technique was better even without using the local search heuristic to optimize the artificial neural network connection values. The average classification error obtained was 2.87%, with an average of 8 connections from 36 possible connections in a fully connected neural network. Using a full connected network, the local optimization technique backpropagation obtained an average error of 6.78%.

The genetic algorithms, tabu search and simulated annealing methods incorporate domain specific knowledge in their own search heuristics. They also tolerate some elements of non-determinism, which helps the search escape from local minima. They rely on the use of a suitable cost function that provides feedback to the algorithm as the search progresses. The main difference among them is how and where domain-specific knowledge is used. For example, in simulated annealing such knowledge is mainly included in the cost function. Solutions involved in a perturbation are selected randomly, and perturbations are accepted or rejected according to a probability.

In the case of genetic algorithms, domain specific knowledge is exploited in all phases. The fitness of individual solutions, the reproduction selection, genetic operators, as well as the generation of the new population, incorporate domain-specific knowledge. Tabu search is different from the above heuristics in that it has an explicit memory component. At each iteration, the neighborhood of the current solution is partially explored, and a move is made to the best non-tabu solution in that neighborhood. The neighborhood function, together with the size and content of the tabu list, is problem specific. The direction of the search is also influenced by memory structures.

The proposed integration uses a larger amount of information on the problem domain and uses this information in practically all search phases. This is possible through the integration of the main potentialities of the three investigated search

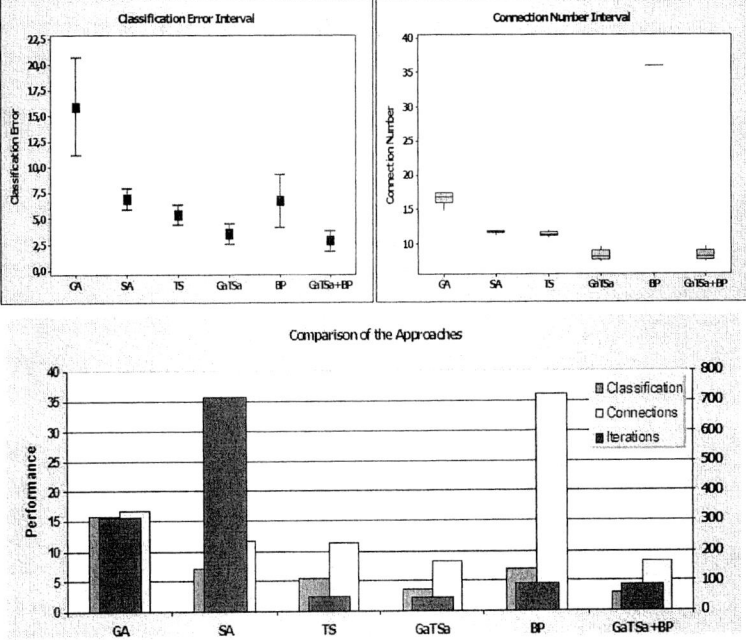

Fig. 1. Result analysis

heuristics. Moreover, the proposed technique has two well-defined stages: a global search phase, where it makes use of the capacity for generating new solutions for the genetic algorithms, the cooling process and cost function of the simulated annealing as well as the memory characteristics of the tabu search technique; and a local search phase, where it makes use of characteristics such as gradient descending for a more precise solution adjustment. These characteristics can obtain better solutions for the investigated problems, with a short search time, low computational cost and minimal investigated search space.

Figure 1 presents graphs comparing the performance of the investigated techniques. The proposed technique obtained the best results regarding the classification error, final network connection number and the number of iterations needed for architecture optimization.

5 Final Remarks

This work presented a technique that integrates the heuristics of tabu search, simulated annealing, genetic algorithms and backpropagation. In the simultaneous optimization of the connection number and connection values of the Multi-Layer Perceptron neural network, this technique obtained promising results in comparison with the isolated techniques. The proposed technique combines strategies of global and local searches, presenting promising results regarding the investigated solution space, computacional cost and search time. The investigated problem involves a critical subject, the stability versus plasticity relation in the training of artificial neural networks.

Without a deeper investigation, it is not possible to say if these results can be extended to other problem classes. An interesting theoretical study proved a number of theorems stating that the average performance of any pair of iterative (deterministic or non-deterministic) algorithms across all problems is identical. Thus, if an algorithm performs well on a certain class of investigated problems, it necessarily pays for that with degraded performance on the remaining set of problems [12]. Future investigations should consider this presupposition and verify the performance of this optimization technique on other problems.

Acknowledgments

The authors would like to thank CNPq, CAPES and FINEP (Brazilian research agencies) for their financial support.

References

1. Kirkpatrick, S., Gellat Jr, C. D., Vecchi, M. P.: Optimization by simulated annealing. Science **220** (1983) 671–680
2. Glover, F.: Future paths for integer programming and links to artificial intelligence. Computers and Operation Research **13** (1986) 533–549
3. Goldberg, D. E.: Genetic Algorithms in Search, Optimization and Machine Learning. Addison-Wesley Longman Publishing Co., Inc. (1989) 372 pages
4. Sait, S. M., Youssef, H.: Iterative Computer Algorithms with Applications in Engineering: Solving Combinatorial Optimization Problems. IEEE Computer Society Press (1999) 387 pages
5. Rumelhart, D. E., Hilton, G. E., Williams, R. J.: Learning Representations by Backpropagation Errors. Nature **323** (1986) 533–536
6. Sexton, R J., Alidaee, B., Dorsey, R. E., Johnson, J. D.: Global Optimization for Artificial Neural Networks: A Tabu Search Application. European Journal of Operational Research **106:2-3** (1998) 570–584
7. Yamazaki, A., de Souto, M.C.P., Ludermir, T.B.: Optimization of Neural Network Weights and Architectures for Odor Recognition using Simulated Annealing. In Proceedings International Joint Conference on Neural Networks (2002) 547–552
8. Baker, J. E.: Reducing bias and inefficiency in the selection algorithm. In Proceedings of the Second International Conference on Genetic Algorithms and their application. Lawrence Erlbaum Associates (1987) 14–21
9. Sywerda, G.: Uniform crossover in genetic algorithms. In Proceedings of international conf. on Genetic algorithms. Morgan Kaufmann Publishers Inc. (1989) 2–9
10. Metropolis, N., Rosenbluth, A. W., Rosenbluth, M. N., Teller, A. H., Teller, E.: Equation of state calculations by fast computing machines. The Journal of Chemical Physics **21:6** 1087–1092
11. Pham, D. T., Karaboga, D.: Intelligent Optimisation Techniques. Springer-Verlag New York (1998) 312 pages
12. Wolpert, D. H., Macready, W. G.: No free lunch theorems for optimization. IEEE Transactions on Evolutionary Computation **1:1** (1997) 67–82

Author Index

Abu-Mostafa, Yaser S. 157
Albertí, Pere 462
Appice, Annalisa 169
Atzori, Maurizio 10
Avesani, Paolo 343
Azevedo, Paulo J. 96

Bahamonde, Antonio 462
Barajas, Jorge Mario 429
Beerenwinkel, Niko 285
Berthold, Michael R. 1
Bistarelli, Stefano 22
Bodon, Ferenc 437
Bonchi, Francesco 10, 22
Borges, José 34
Boulicaut, Jean-François 651
Bringmann, Björn 46

Cai, Deng 445
Cardie, Claire 2
Carvalho, Deborah R. 453
Castelli, Vittorio 355
Ceccherini-Silberstein, Francesca 285
Ceci, Michelangelo 169
Chakrabarti, Deepayan 133
Chakraborti, Sutanu 380
Chen, Shyh-Kwei 368
Choong, Yeow Wei 205
Chowdhury, Abdur 561
Cunningham, Pádraig 486

Däumer, Martin 285
Davidson, Ian 59
De Raedt, Luc 3
de Souza, Jerffeson Teixeira 667
del Coz, Juan José 462
Degenhard, Andreas 331
Demichelis, Francesca 343
Díez, Jorge 462
Ding, Chris 71
Domingos, Pedro 297
Dong, Lin 84
Du, Wenliang 643

Ebecken, Nelson 453
Esfandiari, Babak 634
Ester, Martin 527

Faloutsos, Christos 133
Fätkenheuer, Gert 285
Feldman, Ronen 217
Ferreira, Pedro Gabriel 96
Fischer, Ingrid 392
Frank, Eibe 84, 240, 675
Freitas, Alex A. 453
Fresko, Moshe 217

Gábor, Bálint 470
Geurts, Pierre 478
Giannotti, Fosca 10
Govaert, Gérard 609
Greene, Derek 486
Greiner, Russell 121
Guo, Jun 264
Gyenes, Viktor 470

Hall, Mark 675
Han, Jiawei 404, 445, 527
He, Xiaofei 445
He, Xiaofeng 71
Hilario, Melanie 536
Ho, Eric K.Y. 544
Ho, Tu Bao 321, 617
Hoffmann, Daniel 285
Holmes, Geoffrey 495
Huang, Jin 503, 511
Huang, Shao-bing 601

Ilin, Alexander 519

Japkowicz, Nathalie 667
Jin, Wen 527

Kaiser, Rolf 285
Kalousis, Alexandros 536
Keogh, Eamonn 6, 577
Kirkby, Richard 495
Kleinberg, Jon 133
Knobbe, Arno J. 544

Knuteson, Bruce 552
Kohavi, Ron 7
Kołcz, Aleksander 561
Korn, Klaus 285
Koychev, Ivan 380
Kramer, Stefan 84
Kriegel, Hans-Peter 417

Laasonen, Kari 569
Lakaemper, Rolf 577
Latecki, Longin Jan 577
Laurent, Anne 205
Laurent, Dominique 205
Law, Yan-Nei 108
Lee, Chang-Hwan 585
Lee, Chi-Hoon 121
Lengauer, Thomas 285
Leskovec, Jurij 133
Levene, Mark 34
Li, Haiquan 146
Li, Jinyan 146
Li, Ling 157
Li, Qunxia 264
Li, Wenyuan 593
Li, Xue 429
Lin, Hsuan-Tien 157
Ling, Charles X. 274, 503, 511
Liu, Gang 264
Lőrincz, András 470
Lothian, Rob 380
Ludermir, Teresa Bernarda 709
Lv, Tian-yang 601

Malerba, Donato 169
Matias, Yossi 8
Matwin, Stan 667
Mavroeidis, Dimitrios 181
McGinty, Lorraine 228
Megalooikonomou, Vasilis 577
Meinl, Thorsten 392
Motoda, Hiroshi 692

Nadif, Mohamed 609
Nakanishi, Koutarou 692
Nattkemper, Tim 331
Ng, Wee-Keong 593
Nguyen, Canh Hao 617
Nguyen, Son N. 625
Nock, Richard 634

Oette, Mark 285
Olivetti, Emanuele 343
Ong, Kok-Leong 593
Orlowska, Maria E. 625

Paşca, Marius 193
Pedreschi, Dino 10
Pei, Jian 684
Pensa, Ruggero G. 651
Perno, Carlo-Federico 285
Pfahringer, Bernhard 495
Philippsen, Michael 392
Plantevit, Marc 205
Polat, Huseyin 643
Prados, Julien 536
Pratap, Amrit 157

Ratanamahatana, Chotirat Ann 577
Ravi, S.S. 59
Reilly, James 228
Rexhepaj, Elton 536
Robardet, Céline 651
Rockstroh, Jürgen K. 285
Rosenfeld, Benjamin 217

Salamó, Maria 228
Sañudo, Carlos 462
Satou, Kenji 321
Savary, Lionel 659
Schmidberger, Gabi 240
Schmidt, Mark 121
Schmidt-Thieme, Lars 437
Schneider, Karl-Michael 252
Shao, Zheng 445
Shen, Haifeng 264
Sheng, Shengli 274
Sing, Tobias 285
Singla, Parag 297
Smyth, Barry 228
Soh, Donny 146
Sumner, Marc 675
Svicher, Valentina 285

Talia, Domenico 309
Teisseire, Maguelonne 205
Theobald, Martin 181
Tran, Tuan Nam 321
Tresp, Volker 417
Trunfio, Paolo 309
Tsatsaronis, George 181

Vagena, Zografoula 355
Valpola, Harri 519
Varini, Claudio 331
Vazirgiannis, Michalis 181
Veeramachaneni, Sriharsha 343
Verta, Oreste 309
Vilalta, Ricardo 552
Vlachos, Michail 355, 368

Walter, Hauke 285
Wang, Haixun 684
Wang, Qiang 577
Wang, Zheng-xuan 601
Washio, Takashi 692
Wehenkel, Louis 478
Weikum, Gerhard 181
Wiratunga, Nirmalie 380
Wong, Limsoon 146

Wörlein, Marc 392
Wu, Kun-Lung 368

Xie, Kanglin 701
Xing, Yu-hui 601
Xu, Lijun 701

Yan, Xifeng 445
Yin, Xiaoxin 404
Yu, Kai 417
Yu, Philip S. 355, 368
Yu, Shipeng 417

Zanchettin, Cleber 709
Zaniolo, Carlo 108
Zeitouni, Karine 659
Zimmermann, Albrecht 46
Zuo, Wan-li 601

Lecture Notes in Artificial Intelligence (LNAI)

Vol. 3734: S. Jain, H.U. Simon, E. Tomita (Eds.), Algorithmic Learning Theory. XII, 490 pages. 2005.

Vol. 3721: A. Jorge, L. Torgo, P. Brazdil, R. Camacho, J. Gama (Eds.), Knowledge Discovery in Databases: PKDD 2005. XXIII, 719 pages. 2005.

Vol. 3720: J. Gama, R. Camacho, P. Brazdil, A. Jorge, L. Torgo (Eds.), Machine Learning: ECML 2005. XXIII, 769 pages. 2005.

Vol. 3717: B. Gramlich (Ed.), Frontiers of Combining Systems. X, 321 pages. 2005.

Vol. 3702: B. Beckert (Ed.), Automated Reasoning with Analytic Tableaux and Related Methods. XIII, 343 pages. 2005.

Vol. 3698: U. Furbach (Ed.), KI 2005: Advances in Artificial Intelligence. XIII, 409 pages. 2005.

Vol. 3690: M. Pěchouček, P. Petta, L.Z. Varga (Eds.), Multi-Agent Systems and Applications IV. XVII, 667 pages. 2005.

Vol. 3684: R. Khosla, R.J. Howlett, L.C. Jain (Eds.), Knowledge-Based Intelligent Information and Engineering Systems, Part IV. LXXIX, 933 pages. 2005.

Vol. 3683: R. Khosla, R.J. Howlett, L.C. Jain (Eds.), Knowledge-Based Intelligent Information and Engineering Systems, Part III. LXXX, 1397 pages. 2005.

Vol. 3682: R. Khosla, R.J. Howlett, L.C. Jain (Eds.), Knowledge-Based Intelligent Information and Engineering Systems, Part II. LXXIX, 1371 pages. 2005.

Vol. 3681: R. Khosla, R.J. Howlett, L.C. Jain (Eds.), Knowledge-Based Intelligent Information and Engineering Systems, Part I. LXXX, 1319 pages. 2005.

Vol. 3673: S. Bandini, S. Manzoni (Eds.), AI*IA 2005: Advances in Artificial Intelligence. XIV, 614 pages. 2005.

Vol. 3662: C. Baral, G. Greco, N. Leone, G. Terracina (Eds.), Logic Programming and Nonmonotonic Reasoning. XIII, 454 pages. 2005.

Vol. 3661: T. Panayiotopoulos, J. Gratch, R. Aylett, D. Ballin, P. Olivier, T. Rist (Eds.), Intelligent Virtual Agents. XIII, 506 pages. 2005.

Vol. 3658: V. Matoušek, P. Mautner, T. Pavelka (Eds.), Text, Speech and Dialogue. XV, 460 pages. 2005.

Vol. 3651: R. Dale, K.-F. Wong, J. Su, O.Y. Kwong (Eds.), Natural Language Processing – IJCNLP 2005. XXI, 1031 pages. 2005.

Vol. 3642: D. Ślezak, J. Yao, J.F. Peters, W. Ziarko, X. Hu (Eds.), Rough Sets, Fuzzy Sets, Data Mining, and Granular Computing, Part II. XXIII, 738 pages. 2005.

Vol. 3641: D. Ślezak, G. Wang, M. Szczuka, I. Düntsch, Y. Yao (Eds.), Rough Sets, Fuzzy Sets, Data Mining, and Granular Computing, Part I. XXIV, 742 pages. 2005.

Vol. 3632: R. Nieuwenhuis (Ed.), Automated Deduction – CADE-20. XIII, 459 pages. 2005.

Vol. 3630: M.S. Capcarrere, A.A. Freitas, P.J. Bentley, C.G. Johnson, J. Timmis (Eds.), Advances in Artificial Life. XIX, 949 pages. 2005.

Vol. 3626: B. Ganter, G. Stumme, R. Wille (Eds.), Formal Concept Analysis. X, 349 pages. 2005.

Vol. 3625: S. Kramer, B. Pfahringer (Eds.), Inductive Logic Programming. XIII, 427 pages. 2005.

Vol. 3620: H. Muñoz-Avila, F. Ricci (Eds.), Case-Based Reasoning Research and Development. XV, 654 pages. 2005.

Vol. 3614: L. Wang, Y. Jin (Eds.), Fuzzy Systems and Knowledge Discovery, Part II. XLI, 1314 pages. 2005.

Vol. 3613: L. Wang, Y. Jin (Eds.), Fuzzy Systems and Knowledge Discovery, Part I. XLI, 1334 pages. 2005.

Vol. 3607: J.-D. Zucker, L. Saitta (Eds.), Abstraction, Reformulation and Approximation. XII, 376 pages. 2005.

Vol. 3596: F. Dau, M.-L. Mugnier, G. Stumme (Eds.), Conceptual Structures: Common Semantics for Sharing Knowledge. XI, 467 pages. 2005.

Vol. 3593: V. Mařík, R. W. Brennan, M. Pěchouček (Eds.), Holonic and Multi-Agent Systems for Manufacturing. XI, 269 pages. 2005.

Vol. 3587: P. Perner, A. Imiya (Eds.), Machine Learning and Data Mining in Pattern Recognition. XVII, 695 pages. 2005.

Vol. 3584: X. Li, S. Wang, Z.Y. Dong (Eds.), Advanced Data Mining and Applications. XIX, 835 pages. 2005.

Vol. 3581: S. Miksch, J. Hunter, E. Keravnou (Eds.), Artificial Intelligence in Medicine. XVII, 547 pages. 2005.

Vol. 3577: R. Falcone, S. Barber, J. Sabater-Mir, M.P. Singh (Eds.), Trusting Agents for Trusting Electronic Societies. VIII, 235 pages. 2005.

Vol. 3575: S. Wermter, G. Palm, M. Elshaw (Eds.), Biomimetic Neural Learning for Intelligent Robots. IX, 383 pages. 2005.

Vol. 3571: L. Godo (Ed.), Symbolic and Quantitative Approaches to Reasoning with Uncertainty. XVI, 1028 pages. 2005.

Vol. 3559: P. Auer, R. Meir (Eds.), Learning Theory. XI, 692 pages. 2005.

Vol. 3558: V. Torra, Y. Narukawa, S. Miyamoto (Eds.), Modeling Decisions for Artificial Intelligence. XII, 470 pages. 2005.

Vol. 3554: A. Dey, B. Kokinov, D. Leake, R. Turner (Eds.), Modeling and Using Context. XIV, 572 pages. 2005.

Vol. 3550: T. Eymann, F. Klügl, W. Lamersdorf, M. Klusch, M.N. Huhns (Eds.), Multiagent System Technologies. XI, 246 pages. 2005.

Vol. 3539: K. Morik, J.-F. Boulicaut, A. Siebes (Eds.), Local Pattern Detection. XI, 233 pages. 2005.

Vol. 3538: L. Ardissono, P. Brna, A. Mitrovic (Eds.), User Modeling 2005. XVI, 533 pages. 2005.

Vol. 3533: M. Ali, F. Esposito (Eds.), Innovations in Applied Artificial Intelligence. XX, 858 pages. 2005.

Vol. 3528: P.S. Szczepaniak, J. Kacprzyk, A. Niewiadomski (Eds.), Advances in Web Intelligence. XVII, 513 pages. 2005.

Vol. 3518: T.B. Ho, D. Cheung, H. Liu (Eds.), Advances in Knowledge Discovery and Data Mining. XXI, 864 pages. 2005.

Vol. 3508: P. Bresciani, P. Giorgini, B. Henderson-Sellers, G. Low, M. Winikoff (Eds.), Agent-Oriented Information Systems II. X, 227 pages. 2005.

Vol. 3505: V. Gorodetsky, J. Liu, V. Skormin (Eds.), Autonomous Intelligent Systems: Agents and Data Mining. XIII, 303 pages. 2005.

Vol. 3501: B. Kégl, G. Lapalme (Eds.), Advances in Artificial Intelligence. XV, 458 pages. 2005.

Vol. 3492: P. Blache, E. Stabler, J. Busquets, R. Moot (Eds.), Logical Aspects of Computational Linguistics. X, 363 pages. 2005.

Vol. 3490: L. Bolc, Z. Michalewicz, T. Nishida (Eds.), Intelligent Media Technology for Communicative Intelligence. X, 259 pages. 2005.

Vol. 3488: M.-S. Hacid, N.V. Murray, Z.W. Raś, S. Tsumoto (Eds.), Foundations of Intelligent Systems. XIII, 700 pages. 2005.

Vol. 3487: J. Leite, P. Torroni (Eds.), Computational Logic in Multi-Agent Systems. XII, 281 pages. 2005.

Vol. 3476: J. Leite, A. Omicini, P. Torroni, P. Yolum (Eds.), Declarative Agent Languages and Technologies II. XII, 289 pages. 2005.

Vol. 3464: S.A. Brueckner, G.D.M. Serugendo, A. Karageorgos, R. Nagpal (Eds.), Engineering Self-Organising Systems. XIII, 299 pages. 2005.

Vol. 3452: F. Baader, A. Voronkov (Eds.), Logic for Programming, Artificial Intelligence, and Reasoning. XI, 562 pages. 2005.

Vol. 3451: M.-P. Gleizes, A. Omicini, F. Zambonelli (Eds.), Engineering Societies in the Agents World V. XIII, 349 pages. 2005.

Vol. 3446: T. Ishida, L. Gasser, H. Nakashima (Eds.), Massively Multi-Agent Systems I. XI, 349 pages. 2005.

Vol. 3445: G. Chollet, A. Esposito, M. Faundez-Zanuy, M. Marinaro (Eds.), Nonlinear Speech Modeling and Applications. XIII, 433 pages. 2005.

Vol. 3438: H. Christiansen, P.R. Skadhauge, J. Villadsen (Eds.), Constraint Solving and Language Processing. VIII, 205 pages. 2005.

Vol. 3430: S. Tsumoto, T. Yamaguchi, M. Numao, H. Motoda (Eds.), Active Mining. XII, 349 pages. 2005.

Vol. 3419: B. Faltings, A. Petcu, F. Fages, F. Rossi (Eds.), Recent Advances in Constraints. X, 217 pages. 2005.

Vol. 3416: M. Böhlen, J. Gamper, W. Polasek, M.A. Wimmer (Eds.), E-Government: Towards Electronic Democracy. XIII, 311 pages. 2005.

Vol. 3415: P. Davidsson, B. Logan, K. Takadama (Eds.), Multi-Agent and Multi-Agent-Based Simulation. X, 265 pages. 2005.

Vol. 3403: B. Ganter, R. Godin (Eds.), Formal Concept Analysis. XI, 419 pages. 2005.

Vol. 3398: D.-K. Baik (Ed.), Systems Modeling and Simulation: Theory and Applications. XIV, 733 pages. 2005.

Vol. 3397: T.G. Kim (Ed.), Artificial Intelligence and Simulation. XV, 711 pages. 2005.

Vol. 3396: R.M. van Eijk, M.-P. Huget, F. Dignum (Eds.), Agent Communication. X, 261 pages. 2005.

Vol. 3394: D. Kudenko, D. Kazakov, E. Alonso (Eds.), Adaptive Agents and Multi-Agent Systems II. VIII, 313 pages. 2005.

Vol. 3392: D. Seipel, M. Hanus, U. Geske, O. Bartenstein (Eds.), Applications of Declarative Programming and Knowledge Management. X, 309 pages. 2005.

Vol. 3374: D. Weyns, H. V.D. Parunak, F. Michel (Eds.), Environments for Multi-Agent Systems. X, 279 pages. 2005.

Vol. 3371: M.W. Barley, N. Kasabov (Eds.), Intelligent Agents and Multi-Agent Systems. X, 329 pages. 2005.

Vol. 3369: V. R. Benjamins, P. Casanovas, J. Breuker, A. Gangemi (Eds.), Law and the Semantic Web. XII, 249 pages. 2005.

Vol. 3366: I. Rahwan, P. Moraitis, C. Reed (Eds.), Argumentation in Multi-Agent Systems. XII, 263 pages. 2005.

Vol. 3359: G. Grieser, Y. Tanaka (Eds.), Intuitive Human Interfaces for Organizing and Accessing Intellectual Assets. XIV, 257 pages. 2005.

Vol. 3346: R.H. Bordini, M. Dastani, J. Dix, A.E.F. Seghrouchni (Eds.), Programming Multi-Agent Systems. XIV, 249 pages. 2005.

Vol. 3345: Y. Cai (Ed.), Ambient Intelligence for Scientific Discovery. XII, 311 pages. 2005.

Vol. 3343: C. Freksa, M. Knauff, B. Krieg-Brückner, B. Nebel, T. Barkowsky (Eds.), Spatial Cognition IV. XIII, 519 pages. 2005.

Vol. 3339: G.I. Webb, X. Yu (Eds.), AI 2004: Advances in Artificial Intelligence. XXII, 1272 pages. 2004.

Vol. 3336: D. Karagiannis, U. Reimer (Eds.), Practical Aspects of Knowledge Management. X, 523 pages. 2004.

Vol. 3327: Y. Shi, W. Xu, Z. Chen (Eds.), Data Mining and Knowledge Management. XIII, 263 pages. 2005.

Vol. 3315: C. Lemaître, C.A. Reyes, J.A. González (Eds.), Advances in Artificial Intelligence – IBERAMIA 2004. XX, 987 pages. 2004.

Vol. 3303: J.A. López, E. Benfenati, W. Dubitzky (Eds.), Knowledge Exploration in Life Science Informatics. X, 249 pages. 2004.

Vol. 3301: G. Kern-Isberner, W. Rödder, F. Kulmann (Eds.), Conditionals, Information, and Inference. XII, 219 pages. 2005.

Vol. 3276: D. Nardi, M. Riedmiller, C. Sammut, J. Santos-Victor (Eds.), RoboCup 2004: Robot Soccer World Cup VIII. XVIII, 678 pages. 2005.